Science of Synthesis Reference Library

The **Science of Synthesis Reference Library** comprises volumes covering special topics of organic chemistry in a modular fashion, with six main classifications: (1) Classical, (2) Advances, (3) Transformations, (4) Applications, (5) Structures, and (6) Techniques. Volumes in the **Science of Synthesis Reference Library** focus on subjects of particular current interest with content that is evaluated by experts in their field. **Science of Synthesis**, including the **Knowledge Updates** and the **Reference Library**, is the complete information source for the modern synthetic chemist.

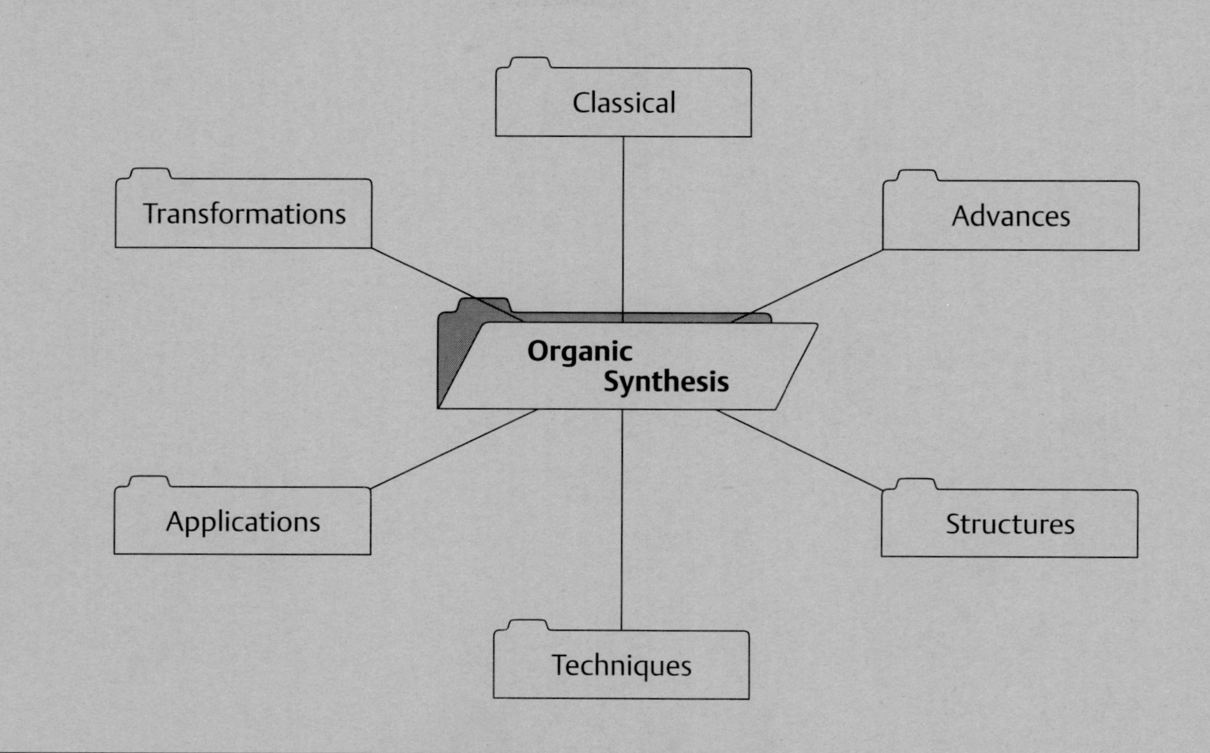

Science of Synthesis

Science of Synthesis is the authoritative and comprehensive reference work for the entire field of organic and organometallic synthesis.

Science of Synthesis presents the important synthetic methods for all classes of compounds and includes:
- Methods critically evaluated by leading scientists
- Background information and detailed experimental procedures
- Schemes and tables which illustrate the reaction scope

Science of Synthesis

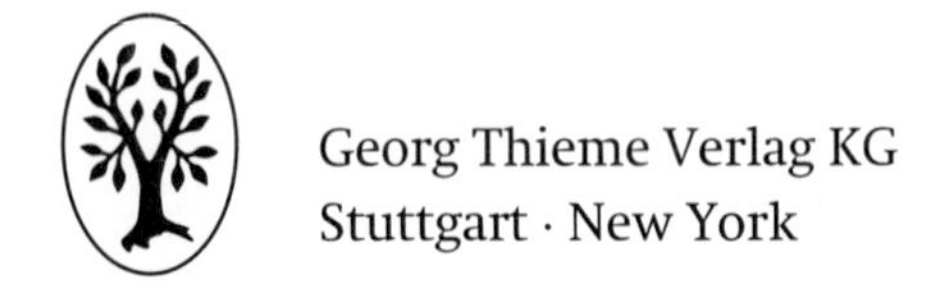

Georg Thieme Verlag KG
Stuttgart · New York

Science of Synthesis

Metal-Catalyzed Cyclization Reactions 1

Reference Library 2016/4a

Volume Editors

S. Ma

S. Gao

Authors

J. M. Alderson
E. M. Beccalli
A. Bonetti
S. Gao
P. J. Guiry
S. Jammi
A. Mazza
C. Nottingham
A. M. Phelps
J. M. Schomaker
M. Shi
X. Tang
D. Wang
Y. Yamamoto
S.-L. You
L. Zhang
X. Zhang

2016
Georg Thieme Verlag KG
Stuttgart · New York

© 2016 Georg Thieme Verlag KG
Rüdigerstrasse 14
D-70469 Stuttgart

Printed in Germany

Typesetting: Ziegler + Müller, Kirchentellinsfurt
Printing and Binding: AZ Druck und Datentechnik GmbH, Kempten

Bibliographic Information published by Die Deutsche Bibliothek

Die Deutsche Bibliothek lists this publication in the Deutsche Nationalbibliografie; detailed bibliographic data is available on the internet at <http://dnb.ddb.de>

Library of Congress Card No.: applied for

British Library Cataloguing in Publication Data

A catalogue record for this book is available from the British Library

ISBN 978-3-13-199861-3

Date of publication: June 15, 2016

Preface

As the pace and breadth of research intensifies, organic synthesis is playing an increasingly central role in the discovery process within all imaginable areas of science: from pharmaceuticals, agrochemicals, and materials science to areas of biology and physics, the most impactful investigations are becoming more and more molecular. As an enabling science, synthetic organic chemistry is uniquely poised to provide access to compounds with exciting and valuable new properties. Organic molecules of extreme complexity can, given expert knowledge, be prepared with exquisite efficiency and selectivity, allowing virtually any phenomenon to be probed at levels never before imagined. With ready access to materials of remarkable structural diversity, critical studies can be conducted that reveal the intimate workings of chemical, biological, or physical processes with stunning detail.

The sheer variety of chemical structural space required for these investigations and the design elements necessary to assemble molecular targets of increasing intricacy place extraordinary demands on the individual synthetic methods used. They must be robust and provide reliably high yields on both small and large scales, have broad applicability, and exhibit high selectivity. Increasingly, synthetic approaches to organic molecules must take into account environmental sustainability. Thus, atom economy and the overall environmental impact of the transformations are taking on increased importance.

The need to provide a dependable source of information on evaluated synthetic methods in organic chemistry embracing these characteristics was first acknowledged over 100 years ago, when the highly regarded reference source **Houben–Weyl Methoden der Organischen Chemie** was first introduced. Recognizing the necessity to provide a modernized, comprehensive, and critical assessment of synthetic organic chemistry, in 2000 Thieme launched **Science of Synthesis, Houben–Weyl Methods of Molecular Transformations**. This effort, assembled by almost 1000 leading experts from both industry and academia, provides a balanced and critical analysis of the entire literature from the early 1800s until the year of publication. The accompanying online version of **Science of Synthesis** provides text, structure, substructure, and reaction searching capabilities by a powerful, yet easy-to-use, intuitive interface.

From 2010 onward, **Science of Synthesis** is being updated quarterly with high-quality content via **Science of Synthesis Knowledge Updates**. The goal of the **Science of Synthesis Knowledge Updates** is to provide a continuous review of the field of synthetic organic chemistry, with an eye toward evaluating and analyzing significant new developments in synthetic methods. A list of stringent criteria for inclusion of each synthetic transformation ensures that only the best and most reliable synthetic methods are incorporated. These efforts guarantee that **Science of Synthesis** will continue to be the most up-to-date electronic database available for the documentation of validated synthetic methods.

Also from 2010, **Science of Synthesis** includes the **Science of Synthesis Reference Library**, comprising volumes covering special topics of organic chemistry in a modular fashion, with six main classifications: (1) Classical, (2) Advances, (3) Transformations, (4) Applications, (5) Structures, and (6) Techniques. Titles will include *Stereoselective Synthesis*, *Water in Organic Synthesis*, and *Asymmetric Organocatalysis*, among others. With expert-evaluated content focusing on subjects of particular current interest, the **Science of Synthesis Reference Library** complements the **Science of Synthesis Knowledge Updates**, to make **Science of Synthesis** the complete information source for the modern synthetic chemist.

The overarching goal of the **Science of Synthesis** Editorial Board is to make the suite of **Science of Synthesis** resources the first and foremost focal point for critically evaluated information on chemical transformations for those individuals involved in the design and construction of organic molecules.

Throughout the years, the chemical community has benefited tremendously from the outstanding contribution of hundreds of highly dedicated expert authors who have devoted their energies and intellectual capital to these projects. We thank all of these individuals for the heroic efforts they have made throughout the entire publication process to make **Science of Synthesis** a reference work of the highest integrity and quality.

The Editorial Board July 2010

Science of Synthesis Reference Library

Metal-Catalyzed Cyclization Reactions (2 Vols.)
Applications of Domino Transformations in Organic Synthesis (2 Vols.)
Catalytic Transformations via C—H Activation (2 Vols.)
Biocatalysis in Organic Synthesis (3 Vols.)
C-1 Building Blocks in Organic Synthesis (2 Vols.)
Multicomponent Reactions (2 Vols.)
Cross Coupling and Heck-Type Reactions (3 Vols.)
Water in Organic Synthesis
Asymmetric Organocatalysis (2 Vols.)
Stereoselective Synthesis (3 Vols.)

Volume Editors' Preface

Metal-catalyzed reactions, especially cyclizations, remain the most useful methods for the efficient construction of cyclic compounds and thus have continuously attracted attention. These methodologies have also been comprehensively applied as the key steps in various novel strategies for the synthesis of natural products and drug molecules. Many new discoveries and advances have been reported, which provide new tools for synthetic organic chemists, medicinal chemists, and even materials chemists. A timely summary is now required of the well-established advances that will shape the future development of this field and its application. On the basis of these considerations, the Editorial Board of *Science of Synthesis* planned two volumes in the Reference Library that focus on metal-catalyzed cyclization reactions. After a careful selection made by the Volume Editors, some of the most significant and practical metal-catalyzed reactions for modern organic synthesis are presented. The organization of these two volumes is based on the types of reaction, which mainly include metal-catalyzed C—C, C—O, and C—N bond formations, as well as epoxidation, aziridination, cyclopropanation, Pauson–Khand reactions, cycloadditions, radical reactions, and metathesis. We hope that these volumes can serve as a reference work for chemists in related areas to inspire future research and the development of new applications.

We would like to take this opportunity to express our sincere thanks for the support and contributions from all of the outstanding authors; without their dedication and professionalism, this project would not have been possible. We are also grateful to Alex Russell and Joe P. Richmond, scientific editors at *Science of Synthesis*, who solved a lot of problems to enable the project to proceed smoothly over the past two and a half years. It has been a very pleasant journey to work with the editorial team at Thieme, especially Michaela Frey and Guido F. Herrmann. We appreciate the professional spirit and passion of all members of this team, which have brought the manuscripts together into a two-volume set.

Volume Editors

Shengming Ma and Shuanhu Gao — Shanghai, May 2016

Abstracts

p 1

1.1 Metal-Catalyzed Intramolecular Coupling Reactions

S. Gao and D. Wang

This chapter presents metal-catalyzed or -promoted intramolecular cross-coupling reactions for C—C bond formation in the preparation of cyclic compounds. Examples of synthetic applications in natural products syntheses are discussed to illustrate the potential of these methodologies.

Keywords: cross coupling • reductive coupling • intramolecular coupling • C—C bond formation • palladium • chromium • copper • nickel • Suzuki–Miyaura coupling • Stille coupling • Negishi coupling • Sonogashira coupling • Hiyama coupling • dehydrogenative coupling • Nozaki–Hiyama–Kishi coupling • ene–yne coupling

p 55

1.2 Intramolecular Heck Reactions

S. Jammi, C. Nottingham, and P. J. Guiry

This chapter presents the best methods for non-enantioselective and enantioselective intramolecular Heck reactions to form cyclic molecules.

Keywords: intramolecular Heck cyclization • non-enantioselective • enantioselective • chiral • cross coupling • palladium(0) • halide • pseudohalide • alkene

p 95

1.3 Metal-Catalyzed Intramolecular Allylic Substitution Reactions

X. Zhang and S.-L. You

Metal-catalyzed intramolecular allylic substitution reactions provide diverse carbocycles and heterocycles. Various types of nucleophiles such as carbon, nitrogen, and oxygen can be employed.

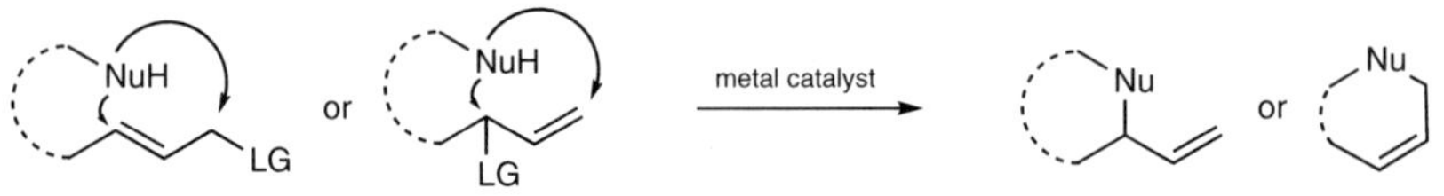

LG = leaving group; Nu = C, N, O

Keywords: allylic substitution · amines · gold · indoles · intramolecular reactions · iridium · palladium · phenols · ruthenium

p 151

1.4 Metal-Catalyzed Intramolecular Cyclizations Involving Cyclopropane and Cyclopropene Ring Opening

X. Tang and M. Shi

Due to the ring strain of cyclopropane rings, transition-metal catalysts can easily undergo an oxidative addition with the cyclopropane moiety of methylenecyclopropanes (MCPs) to give trimethylenemethane (TMM) intermediates. Subsequent intramolecular cyclization with unsaturated systems takes place to give cyclized products. Moreover, in the presence of a chiral ligand, high enantioselectivity can be achieved. The ring-fused products are versatile building blocks in organic synthesis.

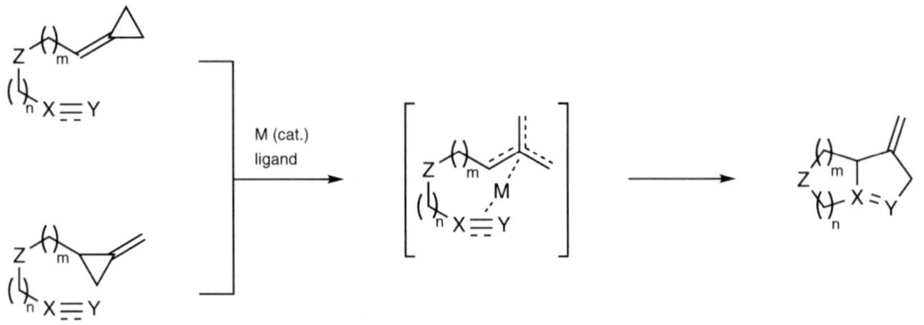

Z = O, CR^1R^2, NR^1; n = m = 0, 1, 2; M = Pd, Ni, Rh, Ru

Keywords: cyclopropanes · methylenecyclopropanes · ring-opening reactions · cyclization · alkenes · alkynes · transition-metal catalysts · trimethylenemethanes · palladium · nickel · rhodium · ruthenium · chiral synthesis

p 259

1.5 Cyclization Reactions of Alkenes and Alkynes

L. Zhang

Discussed in this chapter are two classes of metal-catalyzed cyclization reactions of alkenes and alkynes, namely one where unactivated carbon–carbon double or triple bonds act as nucleophiles to attack tethered electrophiles, and the other where the π-system is activated by a metal-based π-acid and is subsequently attacked by carbonucleophiles. In the former scenario, the in situ generation of electrophiles is typically promoted by hard Lewis acid catalysts, which initiate Prins, aza-Prins, or carbonyl-ene reactions. In the latter scenario, the coordination of a carbon–carbon double or triple bond to a soft Lewis acidic metal catalyst lowers the energy of the π* orbital and thereby enables attack by nucleophiles. A large array of cyclic structural motifs are accessible, many in a stereoselective manner, via such metal catalysis. These motifs, including tetrahydrofurans, tetrahydropyrans, cycloalkenes, dihydronaphthalenes, carbazoles, coumarins, quinolinones, benzopyrans, dihydroquinolines, and phenanthrenes, are essential components of various bioactive compounds and natural products. Exemplary applications of these methods in the syntheses of natural products and relevant structures are also discussed.

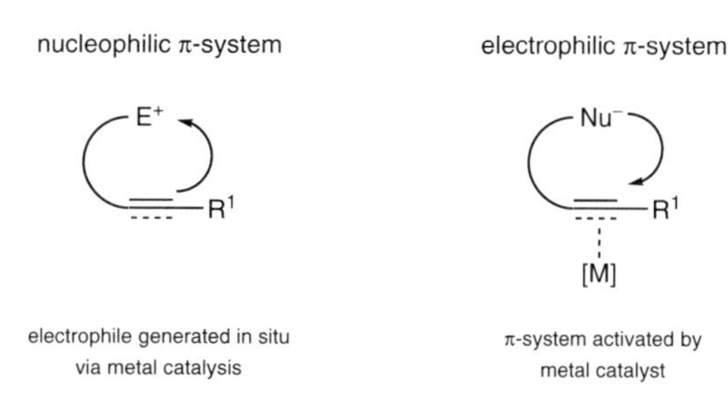

Keywords: cyclization • alkenes • alkynes • catalysis • Prins reaction • carbonucleophiles • gold • iron • mercury • palladium • natural products • tetrahydropyrans • dihydronaphthalenes • carbazoles • coumarins • quinolinones • benzopyrans • dihydroquinolines • phenanthrenes

p 317

1.6 Metal-Catalyzed Cyclization Reactions of Allenes

A. M. Phelps, J. M. Alderson, and J. M. Schomaker

Allenes represent a unique scaffold for powerful cyclization reactions due to the ability to incorporate additional functionality into the newly formed ring. Metal-catalyzed cyclization reactions of allenes represent a powerful strategy for the synthesis of highly functionalized heterocycles and carbocycles. A variety of metals can be used to facilitate cyclization; the nature of the metal influences which allene carbon is attacked by an internal nucleophile, leading to convenient access to multiple ring systems from a simple precursor. The unique axial chirality of allenes can dictate the stereochemistry featured in the ring through axial-to-point chirality transfer.

$AuCl_3$ (2 mol%), CH_2Cl_2, rt

Keywords: allenes • chirality transfer • carbocycles • heterocycles • pyrrolidines • benzocycles • cyclopropanes

p 349

1.7 Cycloisomerizations of Substrates with Multiple Unsaturated Bonds

Y. Yamamoto

Transition-metal complexes catalyze cycloisomerizations of substrates bearing multiple unsaturated carbon–carbon bonds. Carbo- and heterocycles are obtained via selective carbon–carbon bond formations with concomitant hydrogen-atom transfer under mild conditions.

$Pd(OAc)_2$•$(Ph_3P)_2$ (5 mol%)
benzene, 60 °C, 1 h
73%

MeO_2C CO_2Me → MeO_2C H CO_2Me

Keywords: transition-metal catalysts • cycloisomerization • alkynes • alkenes • allenes • 1,3-dienes • carbocycles • heterocycles • regioselectivity • enantioselectivity

p 385

1.8 Metal-Catalyzed Intramolecular C—N and C—O Bond Formation

E. M. Beccalli, A. Bonetti, and A. Mazza

Transition-metal-catalyzed intramolecular C—N and C—O bond formation using unsaturated alkene and alkyne systems containing tethered nitrogen nucleophiles (such as amines, amides, sulfonamides, amidines, azides, carbamates, guanidines, hydrazones, imines, and ureas) or oxygen nucleophiles (such as alcohols, ketones, phenols, hydroxylamines, and carboxylic acids) represents an efficient method for the preparation of heterocycles. Various reaction types may be involved, including amination, hydroamination, oxidative amination, carbamoylation, carboamination, alkoxylation, hydroalkoxylation, oxidative alkoxylation, hydroacyloxylation, carboalkoxylation, and alkoxycarbonylation. Depending on the type of reaction, the choice of transition-metal complex to be used plays a fundamental role in obtaining a successful reaction.

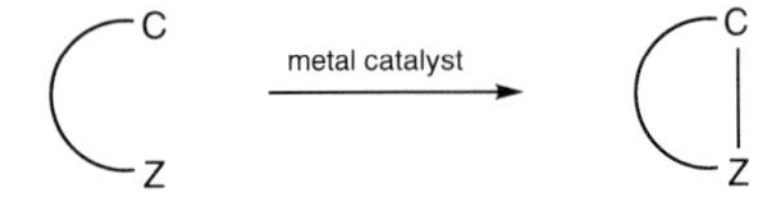

Z = NR^1, O

Keywords: alkoxycarbonylation • alkoxylation • amination • carbamoylation • carboalkoxylation • carboamination • hydroacyloxylation • hydroalkoxylation • hydroamination • intramolecular C—N bond formation • intramolecular C—O bond formation • oxidative amination • oxidative alkoxylation • transition-metal catalysis

Metal-Catalyzed Cyclization Reactions 1

Table of Contents

1.1 **Metal-Catalyzed Intramolecular Coupling Reactions**
S. Gao and D. Wang

1.1	**Metal-Catalyzed Intramolecular Coupling Reactions**	1
1.1.1	Palladium-Catalyzed Intramolecular Coupling Reactions	1
1.1.1.1	Palladium-Catalyzed Intramolecular Suzuki–Miyaura Coupling Reaction and Applications	1
1.1.1.1.1	Palladium-Catalyzed Intramolecular Suzuki–Miyaura Reaction with sp^3 Organoboron Reagents	2
1.1.1.1.2	Palladium-Catalyzed Intramolecular Suzuki–Miyaura Reaction with sp^2 Organoboron Reagents	7
1.1.1.2	Palladium-Catalyzed Intramolecular Stille Coupling and Applications	12
1.1.1.3	Palladium-Catalyzed Intramolecular Negishi Coupling and Applications	16
1.1.1.4	Palladium-Catalyzed Intramolecular Sonogashira Coupling and Applications	18
1.1.1.5	Palladium-Catalyzed Intramolecular Hiyama Reactions and Applications	22
1.1.1.6	Palladium-Catalyzed Intramolecular Arene–Alkene Coupling Through Dehydrogenative Cross Coupling	24
1.1.2	Intramolecular Nozaki–Hiyama–Kishi Coupling Reactions	31
1.1.2.1	Nickel-Catalyzed Intramolecular Alkenylchromium Cyclization and Applications	32
1.1.2.2	Nickel-Catalyzed Intramolecular Alkynylchromium Cyclization and Applications	35
1.1.2.3	Intramolecular Allylchromium Cyclization and Applications	36
1.1.3	Copper-Catalyzed Intramolecular Coupling Reactions	37
1.1.3.1	Copper-Catalyzed Intramolecular Ene–Yne Coupling Reactions and Applications	37
1.1.3.2	Copper-Catalyzed Stille-Type Cross Coupling and Applications	41
1.1.4	Nickel-Catalyzed Intramolecular Reductive Coupling between Organo Halides	44
1.1.5	Conclusions and Future Perspectives	50

1.2 **Intramolecular Heck Reactions**
S. Jammi, C. Nottingham, and P. J. Guiry

1.2	**Intramolecular Heck Reactions**	55
1.2.1	Non-Enantioselective Reactions	56
1.2.1.1	Formation of Small-Sized Rings (3 and 4)	57
1.2.1.1.1	3- and 4-*exo-trig* Processes	57
1.2.1.1.1.1	3-*exo-trig* Cyclization	57

1.2.1.1.1.2 4-*exo-trig* Cyclization 58

1.2.1.2 Formation of Common-Sized Rings (5 and 6) 59

1.2.1.2.1 5- and 6-*exo-trig* Processes 59

1.2.1.2.1.1 5-*exo-trig* Cyclization 59

1.2.1.2.1.2 6-*exo-trig* Cyclization 65

1.2.1.2.2 5- and 6-*endo-trig* Processes 69

1.2.1.2.2.1 5-*endo-trig* Cyclization 69

1.2.1.2.2.2 6-*endo-trig* Cyclization 72

1.2.1.3 Formation of Medium-Sized Rings (7 to 11) 73

1.2.1.3.1 7- and 8-*exo-trig* Processes 74

1.2.1.3.1.1 7-*exo-trig* Cyclization 74

1.2.1.3.1.2 8-*exo-trig* Cyclization 74

1.2.1.3.2 7- and 8-*endo-trig* Processes 75

1.2.1.3.2.1 7-*endo-trig* Cyclization 75

1.2.1.3.2.2 8-*endo-trig* Cyclization 76

1.2.1.4 Formation of Macrocycles (>11) 76

1.2.2 Enantioselective Reactions 77

1.2.2.1 Formation of Tertiary Carbon Centers 78

1.2.2.1.1 Tertiary Centers from Acyclic Alkenes 78

1.2.2.1.1.1 Reactions from Aryl Halides and Pseudohalides 78

1.2.2.1.2 Tertiary Centers from Cyclic Alkenes 81

1.2.2.1.2.1 Reactions from Aryl Halides and Pseudohalides 81

1.2.2.1.2.2 Reactions from Vinyl Halides and Pseudohalides 81

1.2.2.2 Formation of Quaternary Carbon Centers 85

1.2.2.2.1 Quaternary Centers from Acyclic Alkenes 85

1.2.2.2.1.1 Reactions from Aryl Halides and Pseudohalides 85

1.2.2.2.1.2 Reactions from Vinyl Halides and Pseudohalides 87

1.2.2.2.2 Quaternary Centers from Cyclic Alkenes 88

1.2.2.2.2.1 Reactions from Aryl Halides and Pseudohalides 88

1.2.3 Conclusions 91

1.3 **Metal-Catalyzed Intramolecular Allylic Substitution Reactions**

X. Zhang and S.-L. You

1.3 **Metal-Catalyzed Intramolecular Allylic Substitution Reactions** 95
1.3.1 Reactions with Carbon Nucleophiles 96
1.3.1.1 Resonance-Stabilized Enolates 96
1.3.1.1.1 Malonic Esters 96
1.3.1.1.2 β-Keto Esters 99
1.3.1.1.3 Acetamides 101
1.3.1.2 Aliphatic Aldehydes 102
1.3.1.3 Ketones 104
1.3.1.4 Aliphatic Nitro Compounds 105
1.3.1.5 Indoles and Pyrroles 106
1.3.1.5.1 Friedel–Crafts-Type Reactions 106
1.3.1.5.1.1 Indoles with an Allylic Linkage at the C2-Position 106
1.3.1.5.1.2 Indoles with an Allylic Linkage at the C4-Position 110
1.3.1.5.2 Dearomatization Reactions 112
1.3.1.5.2.1 Allylic Dearomatization Reactions 112
1.3.1.5.2.2 Allylic Dearomatization/Migration Reactions 115
1.3.1.6 Phenols 118
1.3.1.6.1 Friedel–Crafts-Type Reactions 118
1.3.1.6.2 Dearomatization Reactions 119
1.3.1.7 Simple Arenes 121
1.3.2 Reactions with Nitrogen Nucleophiles 126
1.3.2.1 Protected Amines 126
1.3.2.1.1 Palladium Catalysis 126
1.3.2.1.2 Iridium Catalysis 128
1.3.2.1.3 Ruthenium Catalysis 131
1.3.2.1.4 Gold Catalysis 132
1.3.2.1.5 Catalysis by Other Metals 134
1.3.2.2 Pyridines and Pyrazines 136
1.3.3 Reactions with Oxygen Nucleophiles 137
1.3.3.1 Phenols 137
1.3.3.2 Aliphatic Alcohols 139
1.3.3.3 Esters 141
1.3.3.4 Acetimidates 143

1.3.4 Applications in the Syntheses of Natural Products and Drug Molecules 144
1.3.4.1 Allylic Alkylation 144
1.3.4.2 Allylic Amination 146
1.3.4.3 Allylic Etherification 147
1.3.5 Conclusions and Future Perspectives 148

1.4 Metal-Catalyzed Intramolecular Cyclizations Involving Cyclopropane and Cyclopropene Ring Opening

X. Tang and M. Shi

1.4 Metal-Catalyzed Intramolecular Cyclizations Involving Cyclopropane and Cyclopropene Ring Opening 151
1.4.1 Intramolecular Cyclizations of Cyclopropanes 151
1.4.1.1 Methylenecyclopropanes 151
1.4.1.1.1 Palladium-Catalyzed Cycloisomerization of Methylenecyclopropanes 151
1.4.1.1.2 Platinum-Catalyzed Cycloisomerization of Methylenecyclopropanes 159
1.4.1.1.3 Rhodium-Catalyzed Cycloisomerization of Methylenecyclopropanes 160
1.4.1.1.4 Nickel-Catalyzed Cycloisomerization of Methylenecyclopropanes 165
1.4.1.1.5 Copper-Catalyzed Cycloisomerization of Methylenecyclopropanes 166
1.4.1.1.6 Magnesium Chloride Catalyzed Cycloisomerization of Methylenecyclopropanes 167
1.4.1.1.7 Gold- or Silver-Catalyzed Cycloisomerization of Methylenecyclopropanes 168
1.4.1.2 Donor–Acceptor Cyclopropanes 176
1.4.1.2.1 Metal Trifluoromethanesulfonate Catalyzed Cycloisomerization of Donor–Acceptor Cyclopropanes 176
1.4.1.2.2 Gold- or Nickel-Catalyzed Cycloisomerization of Alkynylcyclopropyl Ketones 194
1.4.1.2.3 Copper-Catalyzed Cycloisomerization of Donor–Acceptor Cyclopropanes 196
1.4.1.2.4 Pentacarbonyltungsten-Catalyzed Cycloisomerization of Alkynylcyclopropyl Ketones or Oximes 197
1.4.1.3 Vinylidenecyclopropanes 199
1.4.1.3.1 Tin-Catalyzed Cycloisomerization of Vinylidenecyclopropanes 199
1.4.1.3.2 Gold-Catalyzed Cycloisomerization of Vinylidenecyclopropanes 200
1.4.1.3.3 Titanium-Catalyzed Cycloisomerization of Vinylidenecyclopropanes 206
1.4.1.3.4 Rhodium-Catalyzed Cycloisomerization of Vinylidenecyclopropanes 208
1.4.1.3.5 Copper-Catalyzed Cycloisomerization of Vinylidenecyclopropanes 209
1.4.1.4 Other Cyclopropanes 210
1.4.1.4.1 Gold-Catalyzed Cycloisomerization of Functionalized Cyclopropanes 210
1.4.1.4.2 Ruthenium-Catalyzed Cycloisomerization of Functionalized Cyclopropanes 217
1.4.1.4.3 Palladium-Catalyzed Cycloisomerization of Cyclopropane Aminals 221

1.4.1.4.4 Iron-Catalyzed Cycloisomerization of 1-(Alkynylcyclopropyl)alkanols 222
1.4.1.4.5 Silver-Catalyzed Cycloisomerization of Cyclopropanes 223
1.4.2 Intramolecular Cyclizations of Cyclopropenes 227
1.4.2.1 Rhodium-Catalyzed Cycloisomerization of Cyclopropenes 227
1.4.2.2 Ruthenium-Catalyzed Cycloisomerization of Cyclopropenes 234
1.4.2.3 Zinc-Catalyzed Cycloisomerization of Cyclopropanol-Substituted Cyclopropenes 235
1.4.2.4 Copper-Catalyzed Cycloisomerization of Cyclopropenes 237
1.4.2.5 Gold-Catalyzed Cycloisomerization of Cyclopropenes 241
1.4.3 Applications in the Syntheses of Natural Products and Drug Molecules 250

1.5 **Cyclization Reactions of Alkenes and Alkynes**
L. Zhang

1.5 **Cyclization Reactions of Alkenes and Alkynes** 259
1.5.1 Unactivated Alkenes and Alkynes as Nucleophiles 259
1.5.1.1 Oxocarbenium-Based Electrophiles 259
1.5.1.1.1 Prins Cyclization between Homoallylic Alcohols and Aldehydes 260
1.5.1.1.1.1 Highly Stereoselective Iron(III)-Catalyzed Synthesis of *cis*-2-Alkyl-4-halotetrahydropyrans 261
1.5.1.1.1.2 Highly Stereoselective Rhenium(VII)-Catalyzed Synthesis of Tetrasubstituted Tetrahydro-2*H*-pyran-4-ols 261
1.5.1.1.1.3 Indium(III) Chloride Catalyzed Synthesis of Spirotetrahydropyrans 263
1.5.1.1.1.4 Indium(III) Bromide Catalyzed Synthesis of 3-Oxaterpenoids 264
1.5.1.1.1.5 Highly Stereoselective Iron(III) Chloride Catalyzed Synthesis of Tetrahydro-2*H*-pyran-4-ols with Complete OH Selectivity 266
1.5.1.1.2 Prins Cyclization Involving Nucleophilic Alkynes 268
1.5.1.1.2.1 Iron(III)-Catalyzed Cyclization of Alkynyl Aldehyde Acetals 268
1.5.1.1.3 Carbonyl-Ene Reactions 270
1.5.1.1.3.1 Chiral Copper(II)-Catalyzed Enantioselective Intramolecular Carbonyl-Ene Reactions of Unsaturated α-Oxo Esters 270
1.5.1.1.3.2 Chromium(III)-Catalyzed Enantioselective Catalytic Carbonyl-Ene Cyclization Reactions 271
1.5.1.1.3.3 Chromium(III)-Catalyzed Enantioselective Transannular Ketone-Ene Reactions 273
1.5.1.2 Iminium-Based Electrophiles in Aza-Prins Reactions 274
1.5.1.2.1 Iron(III)-Catalyzed Aza-Prins Cyclizations 274
1.5.1.2.2 Scandium(III) Trifluoromethanesulfonate Catalyzed Aza-Prins Cyclizations for the Synthesis of Heterobicycles 276

1.5.2 Metal-Activated Alkenes and Alkynes as Electrophiles 277
1.5.2.1 1,3-Dicarbonyl-Based Carbon Nucleophiles 278
1.5.2.1.1 Cyclizations onto Alkenes 278
1.5.2.1.1.1 Palladium-Catalyzed 6-*endo-trig* Cyclization of Alkenyl 1,3-Dicarbonyls 279
1.5.2.1.1.2 Gold-Catalyzed 5-*exo-trig* Cyclization of Alkenyl β-Oxo Amides 280
1.5.2.1.1.3 Gold-Catalyzed *exo-trig* Cyclization of Alkenyl Ketones 281
1.5.2.1.2 Cyclizations onto Alkynes 283
1.5.2.1.2.1 Cationic Gold(I) Catalyzed Conia-Ene Reactions 283
1.5.2.1.2.2 Zinc-Catalyzed Conia-Ene Reactions 286
1.5.2.1.2.3 Iron(III)-Catalyzed Stannyl Conia-Ene Cyclization 287
1.5.2.2 Arene-Based Carbon Nucleophiles (Friedel–Crafts-Type) 288
1.5.2.2.1 Cyclizations onto Alkenes 288
1.5.2.2.1.1 Ruthenium(III)-Catalyzed Intramolecular Hydroarylation of Unactivated Alkenes 288
1.5.2.2.1.2 Metal-Catalyzed Cyclization of Aryl Allylic Alcohols 290
1.5.2.2.2 Cyclizations onto Alkynes 292
1.5.2.2.2.1 Formation of Dihydronaphthalenes 293
1.5.2.2.2.2 Formation of Phenanthrenes and Other Related Arenes 294
1.5.2.2.2.3 Dehydrative Formation of Carbazoles 299
1.5.2.2.2.4 Formation of 2*H*-1-Benzopyran-2-ones and Quinolin-2(1*H*)-ones 300
1.5.2.2.2.5 Formation of 2*H*-1-Benzopyrans and Dihydroquinolines 302
1.5.2.2.2.6 Formation of Seven-Membered-Ring Annulated Indoles 305
1.5.2.2.2.6.1 Annulation with Nitroenynes 305
1.5.2.2.2.6.2 Annulation with Enynones 307
1.5.3 Applications in the Synthesis of Natural Products and Relevant Structures 309
1.5.3.1 Based on the Prins Cyclization 309
1.5.3.2 Based on Nucleophilic Addition to Metal-Activated Alkenes and Alkynes 311
1.5.4 Conclusions and Future Perspectives 313

1.6 **Metal-Catalyzed Cyclization Reactions of Allenes**
A. M. Phelps, J. M. Alderson, and J. M. Schomaker

1.6 **Metal-Catalyzed Cyclization Reactions of Allenes** 317
1.6.1 Intramolecular C—O Bond Formation 321
1.6.1.1 Palladium-Catalyzed Synthesis of 2,3-Dihydrofurans from Allenic β-Oxo Esters 321
1.6.1.2 Palladium-Catalyzed Synthesis of 2,5-Dihydrofurans from α-Hydroxyallenes 321
1.6.1.3 Gold-Catalyzed Cyclization of Allenoates To Form Functionalized Butenolides 322
1.6.1.4 Gold-Catalyzed Cyclization of α-Hydroxyallenes to 2,5-Dihydrofurans 323

1.6.2 Intramolecular C—N Bond Formation 324

1.6.2.1 Rhodium-Catalyzed Synthesis of Nitrogen-Containing Stereotriads from Allenes 324

1.6.2.2 Gold-Catalyzed Enantioselective Intramolecular Hydroamination of Allenes with Ureas 326

1.6.2.3 Gold-Catalyzed Enantioselective Intramolecular Hydroamination of Allenes with Sulfonamides 327

1.6.2.4 Silver-Catalyzed Synthesis of 4-Vinyloxazolidin-2-ones from Allenyl Carbamates 328

1.6.2.5 Silver-Catalyzed Dynamic Kinetic Enantioselective Intramolecular Hydroamination of Allenes 329

1.6.2.6 Gold-Catalyzed Cycloisomerization of α-Aminoallenes to 2,5-Dihydropyrroles 330

1.6.3 C—C Bond Formation 331

1.6.3.1 Palladium-Catalyzed Carbocyclization of Allenes 331

1.6.3.2 Rhodium- or Copper-Catalyzed Divergent Carbene Reactivity with Allenic Diazo Esters 333

1.6.3.3 Gold-Catalyzed Intramolecular Hydroarylation of 2-Allenic Indoles 334

1.6.3.4 Gold-Catalyzed Hydroarylation of Allenic Anilines and Phenols 336

1.6.3.5 Gold-Catalyzed Intramolecular Hydroarylation of Allenes 337

1.6.3.6 Palladium-Catalyzed Synthesis of Cyclopropanes via Coupling/Cyclization of Allenic Malonates with Organic Halides 338

1.6.3.7 Palladium-Catalyzed Synthesis of Eight- to Ten-Membered Lactones from Allenic 3-Oxoalkanoates and Organic Halides 340

1.6.3.8 Palladium-Catalyzed Synthesis of Nine- to Twelve-Membered Cyclic Compounds from Allenes Containing a Nucleophilic Functionality and Organic Halides 341

1.6.4 C—S Bond Formation 342

1.6.4.1 Cycloisomerization of Sulfanylallenes to 2,5-Dihydrothiophenes 342

1.6.5 Total Synthesis Using Allene Cyclization 343

1.6.5.1 Gold-Catalyzed Cycloisomerization of Dihydroxyallenes to 2,5-Dihydrofurans 343

1.6.5.2 Silver-Catalyzed Cyclization of Allenic Hydroxylamines 344

1.6.5.3 Synthesis of Flinderoles B and C through Gold-Catalyzed Allene Hydroarylation 344

1.6.5.4 Palladium-Catalyzed Domino Cyclization of Allenes Bearing Amino and Bromoindolyl Groups 345

1.6.6 Conclusions 346

1.7 **Cycloisomerizations of Substrates with Multiple Unsaturated Bonds**
Y. Yamamoto

1.7 **Cycloisomerizations of Substrates with Multiple Unsaturated Bonds** 349

1.7.1 Cycloisomerizations of α,ω-Enynes 349

1.7.1.1 Palladium-Catalyzed Cycloisomerization of α,ω-Enynes 351

1.7.1.2 Titanium-Catalyzed Cycloisomerization of α,ω-Enynes 352

1.7.1.3 Ruthenium-Catalyzed Cycloisomerization of α,ω-Enynes 354

1.7.1.4 Iron-Catalyzed Cycloisomerization of α,ω-Enynes 355

1.7.1.5 Enantioselective Cycloisomerization of α,ω-Enynes 357

1.7.2 Cycloisomerizations of α,ω-Dienes 360

1.7.2.1 Palladium-Catalyzed Cycloisomerization of α,ω-Dienes 361

1.7.2.2 Titanium-Catalyzed Cycloisomerization of α,ω-Dienes 363

1.7.2.3 Ruthenium-Catalyzed Cycloisomerization of α,ω-Dienes 365

1.7.2.4 Enantioselective Cycloisomerization of α,ω-Dienes 366

1.7.3 Cycloisomerizations Involving 1,2- and 1,3-Dienes 368

1.7.3.1 Cycloisomerizations of Allenynes 368

1.7.3.2 Cycloisomerizations of Allenenes and Bisallenes 371

1.7.3.3 Cycloisomerizations of (1,3-Diene)enes 374

1.7.3.4 Cycloisomerizations of Bis(1,3-dienes) 379

1.7.4 Applications of Cycloisomerizations to Natural Product Synthesis 380

1.7.5 Conclusions and Future Perspectives 382

1.8 **Metal-Catalyzed Intramolecular C—N and C—O Bond Formation**
E. M. Beccalli, A. Bonetti, and A. Mazza

1.8 **Metal-Catalyzed Intramolecular C—N and C—O Bond Formation** 385

1.8.1 Reaction Mechanisms of Transition-Metal-Catalyzed C—N and C—O Bond Formation 385

1.8.2 C—N Bond Formation 388

1.8.2.1 Hydroamination of Alkynes 389

1.8.2.1.1 Copper Catalysis 389

1.8.2.1.2 Gold Catalysis 391

1.8.2.1.3 Palladium Catalysis 393

1.8.2.1.4 Rhodium Catalysis 395

1.8.2.1.5 Ruthenium Catalysis 396

1.8.2.1.6 Silver Catalysis 397

1.8.2.1.7 Silver–Gold Heterobimetallic Catalysis 397

1.8.2.1.8 Zinc Catalysis 398
1.8.2.2 Hydroamination of Alkenes 401
1.8.2.2.1 Copper Catalysis 401
1.8.2.2.2 Gold Catalysis 402
1.8.2.2.3 Iridium Catalysis 403
1.8.2.2.4 Iron Catalysis 404
1.8.2.2.5 Platinum Catalysis 405
1.8.2.2.6 Rhodium Catalysis 406
1.8.2.3 Oxidative Amination of Alkenes 408
1.8.2.3.1 Palladium Catalysis 409
1.8.2.3.2 Ruthenium Catalysis 413
1.8.2.4 Oxidative Amination of (Het)Arenes 414
1.8.2.4.1 Copper Catalysis 415
1.8.2.4.2 Palladium Catalysis 417
1.8.2.5 Amination of Aryl Halides 427
1.8.2.5.1 Copper Catalysis (Intramolecular Ullmann Condensation) 427
1.8.2.5.2 Palladium Catalysis (Buchwald–Hartwig Reaction) 431
1.8.2.6 Amination of C(sp^3) Centers 433
1.8.2.6.1 Palladium Catalysis 433
1.8.2.6.2 Rhodium Catalysis 436
1.8.2.7 Amination and Hydroamination Involving Azides as Nucleophiles 438
1.8.2.7.1 Rhodium Catalysis (Amination of Alkenes and Arenes) 439
1.8.2.7.2 Ruthenium Catalysis 442
1.8.2.7.3 Zinc Catalysis 443
1.8.2.7.4 Platinum Catalysis (Hydroamination of Alkynes) 443
1.8.2.7.5 Silver–Gold Heterobimetallic Catalysis 444
1.8.2.8 Carbamoylation of Arenes 445
1.8.2.8.1 Palladium Catalysis 445
1.8.2.8.2 Ruthenium Catalysis 446
1.8.2.9 Carboamination of Alkynes 447
1.8.2.9.1 Palladium Catalysis 448
1.8.2.10 Carboamination of Alkenes 450
1.8.2.10.1 Palladium Catalysis 451
1.8.2.11 Applications in the Syntheses of Natural Products and Drug Molecules 454
1.8.3 C—O Bond Formation 461
1.8.3.1 Hydroalkoxylation of Alkynes 461
1.8.3.1.1 Copper Catalysis 461

1.8.3.1.2 Gold Catalysis 462
1.8.3.1.3 Iridium Catalysis 465
1.8.3.1.4 Palladium Catalysis 466
1.8.3.1.5 Platinum Catalysis 467
1.8.3.1.6 Rhodium Catalysis 470
1.8.3.1.7 Ruthenium Catalysis 471
1.8.3.1.8 Zinc Catalysis 474
1.8.3.2 Hydroacyloxylation of Alkynes 474
1.8.3.2.1 Gold Catalysis 475
1.8.3.3 Hydroalkoxylation of Alkenes 475
1.8.3.3.1 Iron Catalysis 476
1.8.3.3.2 Platinum Catalysis 476
1.8.3.3.3 Ruthenium Catalysis 478
1.8.3.3.4 Silver Catalysis 478
1.8.3.4 Hydroacyloxylation of Alkenes 480
1.8.3.4.1 Silver Catalysis 480
1.8.3.5 Oxidative Alkoxylation of Alkenes 481
1.8.3.5.1 Palladium Catalysis 481
1.8.3.6 Oxidative Alkoxylation of Arenes 484
1.8.3.6.1 Copper Catalysis 484
1.8.3.7 Alkoxylation of Aryl or Vinyl Halides 485
1.8.3.7.1 Copper Catalysis 486
1.8.3.7.2 Palladium Catalysis (Buchwald-Type Reaction) 489
1.8.3.8 Alkoxycarbonylation of Alkynes 490
1.8.3.8.1 Palladium Catalysis 490
1.8.3.9 Carboalkoxylation of Alkynes 491
1.8.3.9.1 Palladium Catalysis 491
1.8.3.10 Carboalkoxylation of Alkenes 494
1.8.3.10.1 Palladium Catalysis 494
1.8.3.11 Applications in the Syntheses of Natural Products and Drug Molecules 495
1.8.4 Conclusions 498

Keyword Index 503

Author Index 523

Abbreviations 535

1.1 Metal-Catalyzed Intramolecular Coupling Reactions

S. Gao and D. Wang

General Introduction

Cyclic compounds, including carbocycles and heterocycles, are ubiquitous in natural products and drug molecules. Therefore, methodologies for efficient ring construction are essential in synthetic organic chemistry. The most appropriate methodology for a given synthetic plan depends mainly on the structural features, functionalities, and ring size of the target. Installing electrophilic and nucleophilic coupling components in a single molecule facilitates atom-economic ring construction, and various metal-catalyzed intramolecular coupling reactions to accomplish this have been developed over the past two decades. In this chapter the reported methodologies for metal-catalyzed or -promoted intramolecular cross-coupling reactions for C—C bond formation in the preparation of cyclic compounds are summarized. Reactions are presented according to the metal catalyst involved, together with selected synthetic applications, particularly for natural product synthesis.

1.1.1 Palladium-Catalyzed Intramolecular Coupling Reactions

1.1.1.1 Palladium-Catalyzed Intramolecular Suzuki–Miyaura Coupling Reaction and Applications

The palladium-catalyzed Suzuki–Miyaura cross-coupling reaction has become one of the most useful methodologies for C—C bond formation.[1,2] Advantages of this reaction include: (1) it proceeds under mild conditions; (2) the diverse boronic acids needed are commercially available and environmentally safe; and (3) the boron-containing byproducts are easy to handle and remove, especially in large-scale syntheses.

The intramolecular Suzuki–Miyaura coupling reactions can be divided into two types according to the newly formed bonds: (1) C(sp^2)—C(sp^2) bond formation between a vinyl- or arylborane and an aryl or vinyl halide or trifluoromethanesulfonate, or enol phosphate, to give products **1**; and (2) C(sp^2)—C(sp^3) bond formation between an aryl or vinyl halide or trifluoromethanesulfonate, or enol phosphate, and an alkylborane (*B*-alkyl Suzuki–Miyaura coupling) to give products of type **2** (Scheme 1). This valuable protocol has been used to couple complex molecular fragments during the synthesis of natural products and drug molecules.

Scheme 1 The Intramolecular Suzuki–Miyaura Coupling Reaction

BR^1_2 X — Pd catalyst → **1**

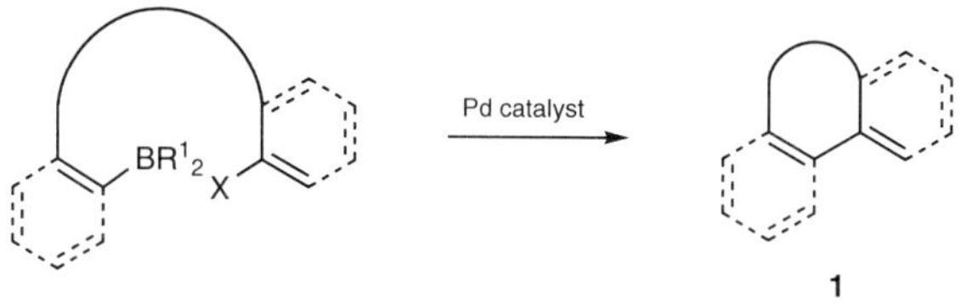

for references see p 51

Pd catalyst

BR^1_2 X

2

X = Cl, Br, I, OTf, $OP(O)(OR^2)_2$

The palladium-catalyzed Suzuki–Miyaura reaction involves a sequence of oxidative addition, transmetalation, and reductive elimination (Scheme 2). The catalytic cycle starts with the oxidative addition of R^1—X to a palladium(0) complex to form an R^1—Pd(II)—X intermediate, which is always the rate-limiting step. Electron-withdrawing groups facilitate oxidative addition more than substrates with electron-donating groups. The order of reactivity of electrophilic partners has been established to be I >> Br > OTf > Cl. The base plays a key role in transmetalation. In pathway A, the base is proposed to enhance the rate of transmetalation by forming an electronically activated anionic boron complex, which can interact with the palladium center to generate an R^1—Pd(II)—R^2 species.[3] In pathway B, the base is proposed to replace the halide in the coordination sphere of the palladium complex, where it facilitates intramolecular transmetalation.[4] Then, the R^1—Pd(II)—R^2 complex undergoes reductive elimination to afford the corresponding coupling product and regenerate the palladium(0) catalyst.

Scheme 2 The General Mechanism of Suzuki–Miyaura Coupling

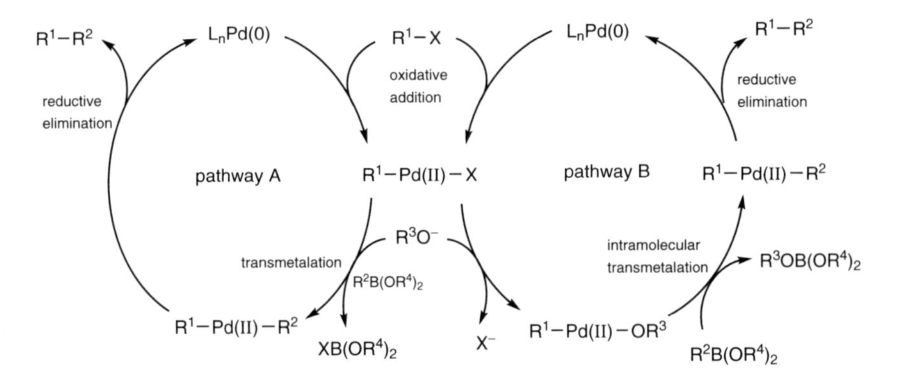

1.1.1.1.1 Palladium-Catalyzed Intramolecular Suzuki–Miyaura Reaction with sp^3 Organoboron Reagents

Palladium-catalyzed intramolecular coupling of alkylboranes with electrophilic coupling components allows reliable construction of C(sp^3)—C(sp^2) bonds.[5–8] The alkylboranes are normally prepared in situ from a terminal alkene through hydroboration with dialkylboranes such as 9-borabicyclo[3.3.1]nonane (9-BBNH), disiamylborane, or dicyclohexylborane. This reaction follows *cis* anti-Markovnikov rules, and tolerates various functional groups. Substrates containing aryl or vinyl halide and terminal alkene functional groups linked through a two- or three-atom tether can be cyclized to form six-membered rings (Table 1, entries 1–4) and five-membered rings (entries 5–10).[7,8] [1,1′-Bis(diphenylphosphino)ferrocene]dichloropalladium(II) [$PdCl_2$(dppf)] is the catalyst used most often in this cross-coupling reaction. The bidentate 1,1′-bis(diphenylphosphino)ferrocene ligand on the square-planar palladium(II) complex **3** is thought to facilitate reductive elimination by enforcing a *cis* geometry between the vinyl (aryl) and alkyl groups.

Table 1 Palladium-Catalyzed Intramolecular Suzuki–Miyaura Reaction To Form Five- and Six-Membered Rings[7,8]

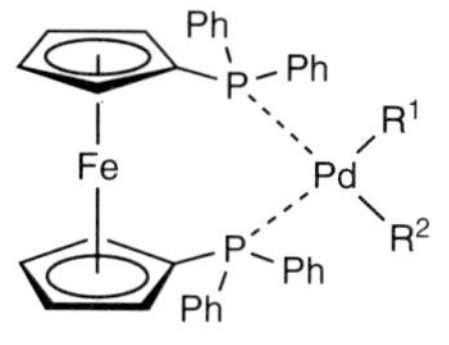
Ph
Ph
Fe
Pd
R1
R2
Ph
Ph

3

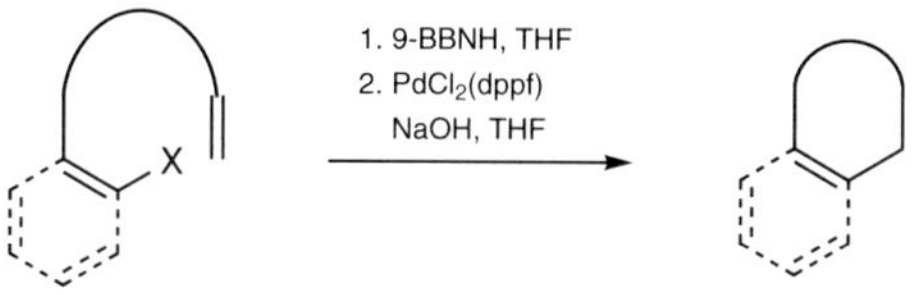
1. 9-BBNH, THF
2. PdCl2(dppf)
NaOH, THF
X

Entry	Substrate	Product	Yield (%)	Ref
1	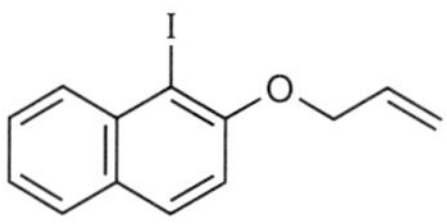	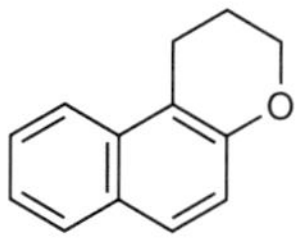	70	[7]
2	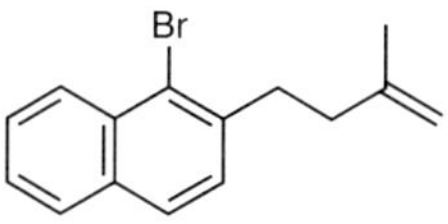	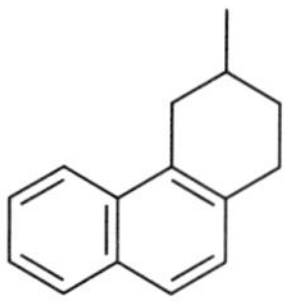	77	[7]
3	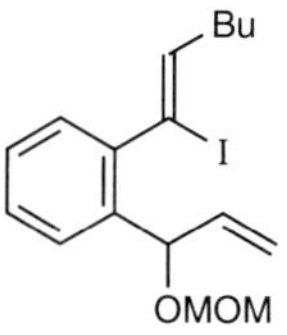	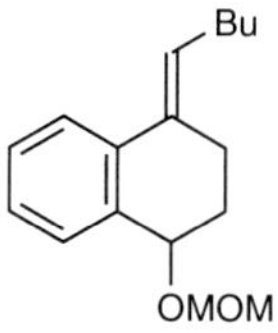	51	[8]
4	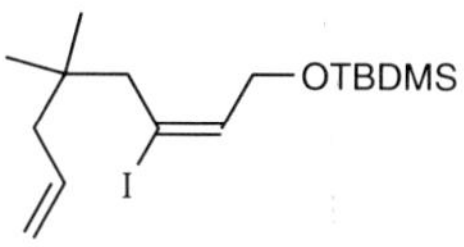	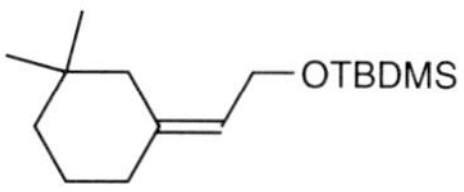	60	[8]
5	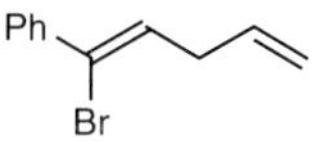	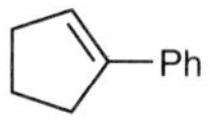	67	[7]
6	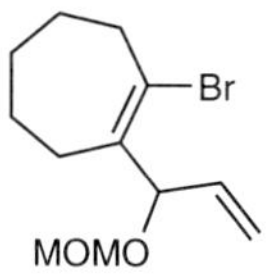	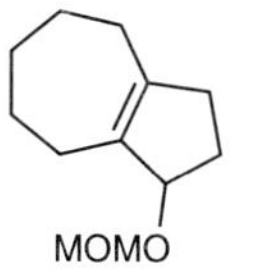	68	[7]

for references see p 51

Table 1 (cont.)

Entry	Substrate	Product	Yield (%)	Ref
7	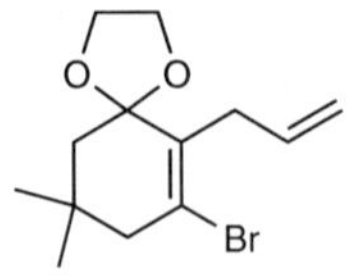	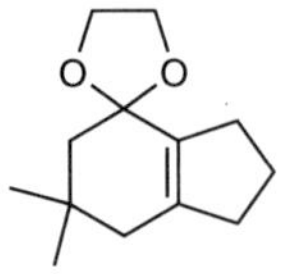	84	[7]
8	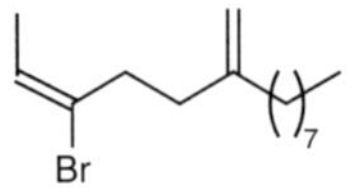	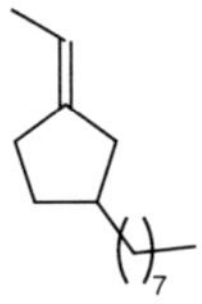	71	[8]
9	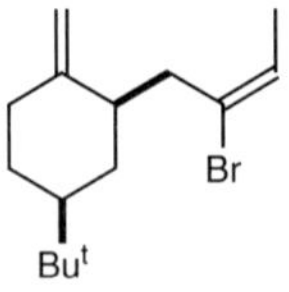	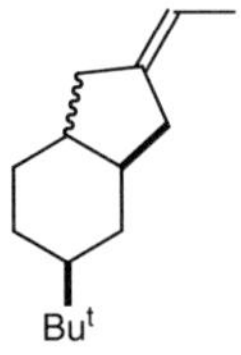	68	[8]
10	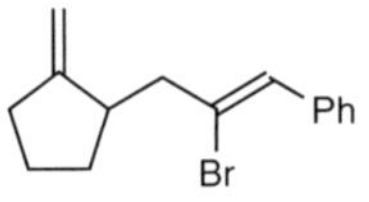	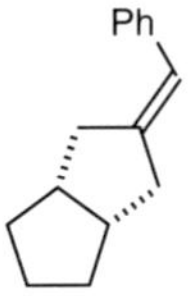	83	[8]

Cyclization reactions aimed at constructing rings with more than six members compete with intermolecular cross-coupling reactions leading to oligomerization. Therefore, these cyclizations are always performed under highly dilute conditions, with slow addition of the alkylborane intermediate to the catalyst solution.[9–11] Palladium-catalyzed macrocyclization affords medium rings (Table 2, entries 1 and 2) and large rings (entry 3) in moderate yield. Transannular macrocyclization of substrates with *Z*- or *E*-alkenyl iodide moieties furnishes compounds useful for preparing natural products (entries 4–6). Danishefsky and co-workers have found the combination [1,1′-bis(diphenylphosphino)ferrocene]dichloropalladium(II) with triphenylarsine to catalyze this coupling effectively.[11] Reaction yield is higher with the base thallium(I) ethoxide than with other bases. Roush and co-workers have systematically studied the effect of thallium(I) ethoxide in the Suzuki–Miyaura coupling reaction, comparing it with thallium(I) hydroxide.[12] Although the role of the thallium base remains unclear, it is thought that a soluble intermediate generated from the reaction of the vinyl- or arylboronic acid and thallium(I) ethoxide is crucial to the coupling reaction and accelerates the desired macrocyclization reaction at the expense of competing oligomerization processes.[13,14]

Table 2 Palladium-Catalyzed Intramolecular Suzuki–Miyaura Reaction To Form Medium and Large Rings[11]

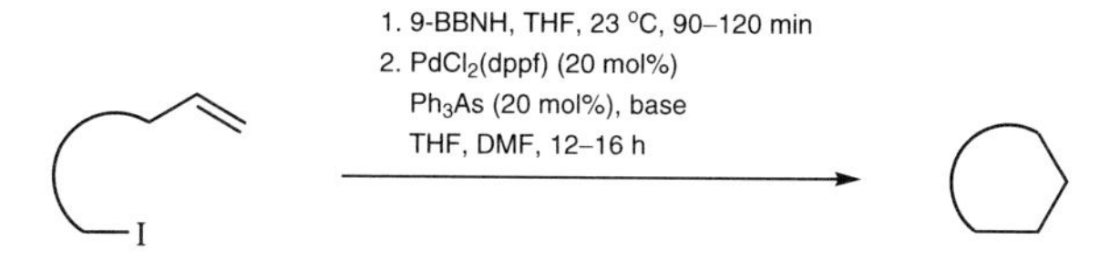

Entry	Substrate	Base	Product	Yield[a] (%)	Ref
1		–		23	[11]
2		TlOEt		22	[11]
3		–		41	[11]
4		–		40	[11]
5		TlOEt		60	[11]
6		TlOEt		46	[11]

[a] Isolated yield after column chromatography.

for references see p 51

In 2009, Williams and co-workers reported the total synthesis of 4-hydroxydictyolactone using an intramolecular *B*-alkyl Suzuki–Miyaura cyclization.[15] This approach was used to produce the nine-membered ring in 66% yield by coupling the terminal alkene and alkenyl iodide moieties in intermediate **4** (Scheme 3).

Scheme 3 Total Synthesis of 4-Hydroxydictyolactone Using an Intramolecular *B*-Alkyl Suzuki–Miyaura Cyclization[15]

1. 9-BBNH, THF, rt, 12 h
2. $Pd(PPh_3)_4$, NaOH
MeCN/H_2O (15:1)
85 °C, 18 h
3. aq AcOH, THF, 85 °C

66%; dr 80:20 at C19

4

4-hydroxydictyolactone

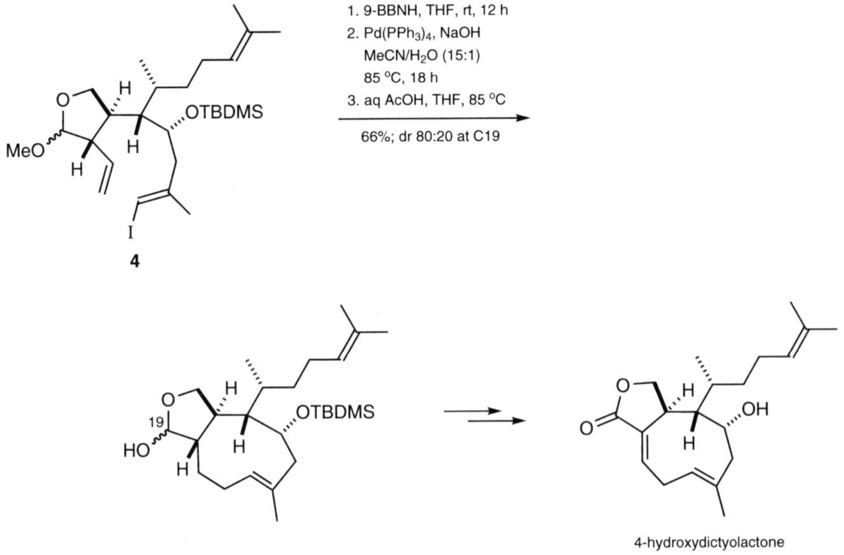

2,3-Dihydro-1*H*-benzo[*f*]benzopyran (Table 1, Entry 1):[7]
A dry 25-mL flask equipped with a magnetic stirrer bar, a septum inlet, an oil bubbler, and a reflux condenser was flushed with N_2. To the flask were added 2-allyloxy-1-iodonaphthalene (1 mmol) and dry THF (1 mL), and then a 0.5 M soln of 9-BBNH (1.05 mmol) in THF at 0 °C. The mixture was slowly warmed to rt and stirred for 3 h. To the borane soln thus obtained were added $PdCl_2$(dppf) (0.015 mmol), 3 M aq NaOH (1 mL), and additional THF (3 mL) at rt. The mixture was heated at 60 °C overnight (ca. 14–16 h). The mixture was diluted with benzene (20 mL) (**CAUTION:** *carcinogen*), and the residual borane was oxidized for 1 h at rt by addition of 3 M NaOH (1 mL) and 30% H_2O_2 (0.5 mL). The product was isolated by chromatography (silica gel, hexane); yield: 70%.

3,4,5,6,7,8-Hexahydro-1*H*-benzo[c]oxecin-1-one (Table 2, Entry 1); Typical Procedure without Thallium(I) Ethoxide:[11]
Hex-5-enyl 2-iodobenzoate (0.083 g, 0.261 mmol) in THF (1.3 mL) at 23 °C was treated with solid 9-BBNH dimer (0.073 g, 0.595 mmol wrt monomer, 2.3 equiv) and the resulting mixture was stirred for 2 h. This soln was added dropwise over 2.5 h to a soln of $PdCl_2$(dppf) (0.042 g, 0.052 mmol, 0.2 equiv) and Ph_3As (0.016 g, 0.052 mmol, 0.2 equiv) in THF (34 mL) and DMF (17 mL) under argon. After 16 h, the mixture was diluted with H_2O (20 mL) and extracted with Et_2O/hexane (1:1). The combined organic layers were washed with H_2O (10 mL), dried ($MgSO_4$), filtered, and concentrated under reduced pressure. Chromatography (silica gel, Et_2O/hexane 1:49 to 5:45 gradient) afforded the title lactone; yield: 0.012 g (23%).

Tricyclic Alkene (Table 2, Entry 5); Typical Procedure Using Thallium(I) Ethoxide:[11]

CAUTION: *Thallium salts are toxic when inhaled, ingested, or absorbed through the skin.*

A soln of 2-allyl-5-(*tert*-butyldimethylsiloxy)-5-[(*E*)-4-iodobut-3-enyl]-2-methoxy-4a-methyl-1,2,3,4,4a,5,6,7-octahydronaphthalene (0.050 g, 0.094 mmol) in THF (0.5 mL) was treated with 0.72 M 9-BBNH in THF (0.2 mL, 0.143 mmol, 1.5 equiv) at 23 °C. After being stirred for 1.5 h, the reaction was quenched with H_2O (0.010 mL, 0.56 mmol, 6 equiv) and then treated with TlOEt (0.020 mL, 0.282 mmol, 3 equiv). This mixture was diluted with THF (4 mL) and the resulting soln was added dropwise over 3 h using a syringe pump to a soln of $PdCl_2$(dppf) (0.015 g, 0.019 mmol, 0.2 equiv) and Ph_3As (0.006 g, 0.019 mmol, 0.2 equiv) in THF (28 mL) and DMF (3 mL) under argon. This mixture was stirred for an additional 12 h, and then was diluted with pH 7.2 phosphate buffer (6 mL) and treated with 30% H_2O_2 (2 mL). After being stirred for 0.5 h, the mixture was extracted with Et_2O/hexane (1:1; 2 × 50 mL) and the combined organic layer was washed with H_2O (10 mL) and dried ($MgSO_4$), filtered, and concentrated under reduced pressure. Chromatography of the crude oil (silica gel, Et_2O/hexane 1:19 to 3:17 gradient) afforded a colorless oil; yield: 23 mg (60%).

1.1.1.1.2 Palladium-Catalyzed Intramolecular Suzuki–Miyaura Reaction with sp^2 Organoboron Reagents

Palladium-catalyzed intramolecular macrocyclizations of a linear dihalide or aryl halide and arylboronate are used to synthesize biaryl-containing macrocycles.[16,17] When cyclization involves a linear dihalide, sequential Miyaura borylation of one of the aryl iodides followed by intramolecular Suzuki–Miyaura reaction generates the aryl–aryl bond. The combination of catalytic [1,1′-bis(diphenylphosphino)ferrocene]dichloropalladium(II) and potassium acetate in dimethyl sulfoxide effectively converts linear diiodide **5** into the 15-membered *m*,*m*-cyclophane **7** in the presence of bis(pinacolato)diboron (**6**) (Scheme 4). Potassium acetate plays a key role in this coupling reaction, perhaps by facilitating formation of an acetoxypalladium(II) intermediate **8**, which undergoes facile transmetalation with the organoboron compound.[16] Substrates containing both aryl iodide and arylboronate groups are used most often as precursors for intramolecular aryl–aryl bond formation. Treating a 0.02 M solution of substrate **9** in toluene/water (30:1) at 90 °C in the presence of a preformed palladium complex [$PdCl_2(SPhos)_2$; SPhos = 2-(dicyclohexylphosphino)-2′,6′-dimethoxybiphenyl] and sodium hydrogen carbonate affords macrocycle **10** in 54% yield.[17] The yield of this coupling reaction is higher with the ligand SPhos than with other ligands.

for references see p 51

Scheme 4 Intramolecular Suzuki–Miyaura Macrocyclization To Construct Biaryl Macrocycles[16,17]

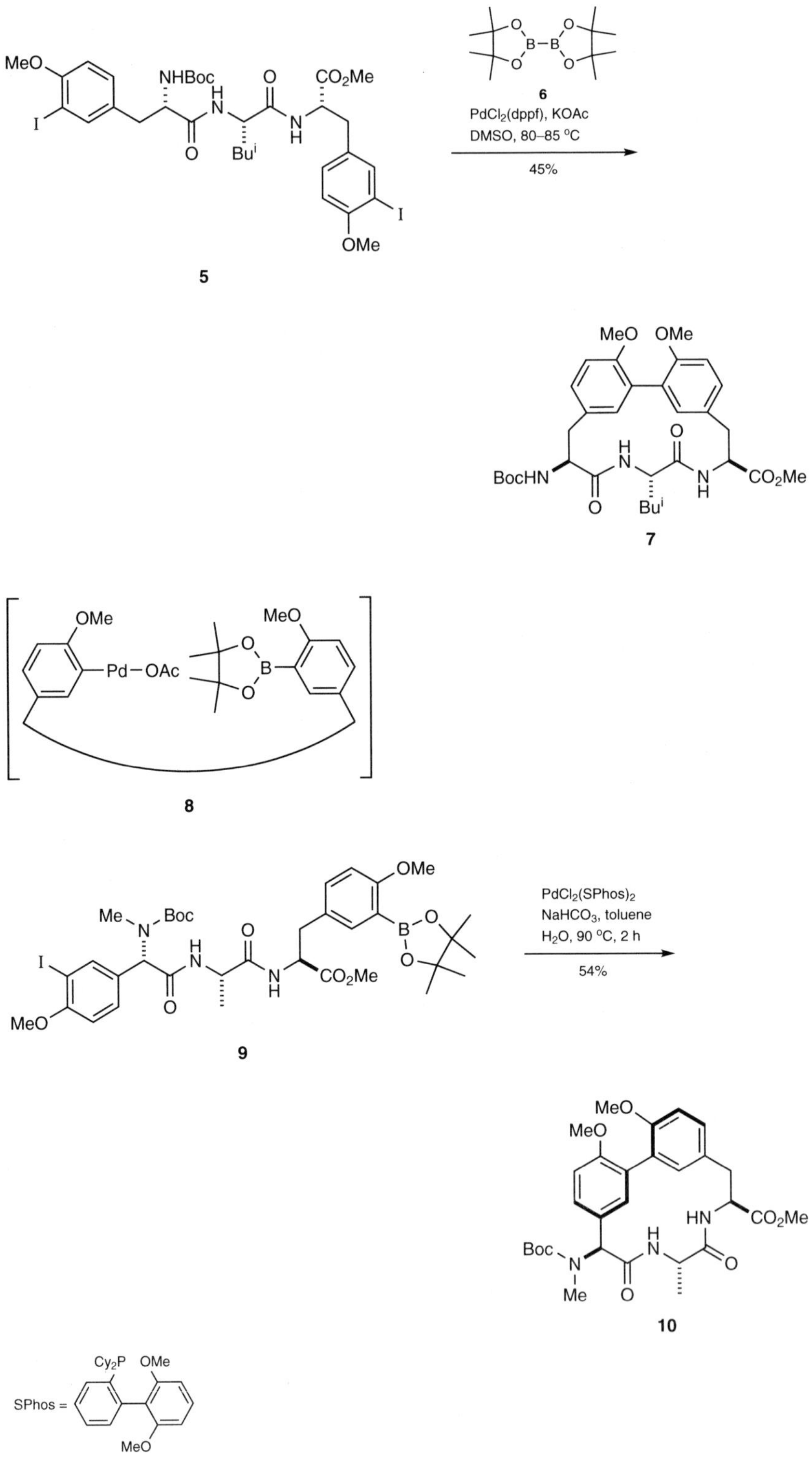

6
$PdCl_2(dppf)$, KOAc
DMSO, 80–85 °C
45%

5

7

8

$PdCl_2(SPhos)_2$
$NaHCO_3$, toluene
H_2O, 90 °C, 2 h
54%

9

10

SPhos =

The total synthesis of complestatin has been accomplished using the intramolecular Suzuki–Miyaura macrocyclization as a key transformation (Scheme 5).[18,19]

Scheme 5 Total Synthesis of Complestatin Using an Intramolecular Suzuki–Miyaura Macrocyclization[18,19]

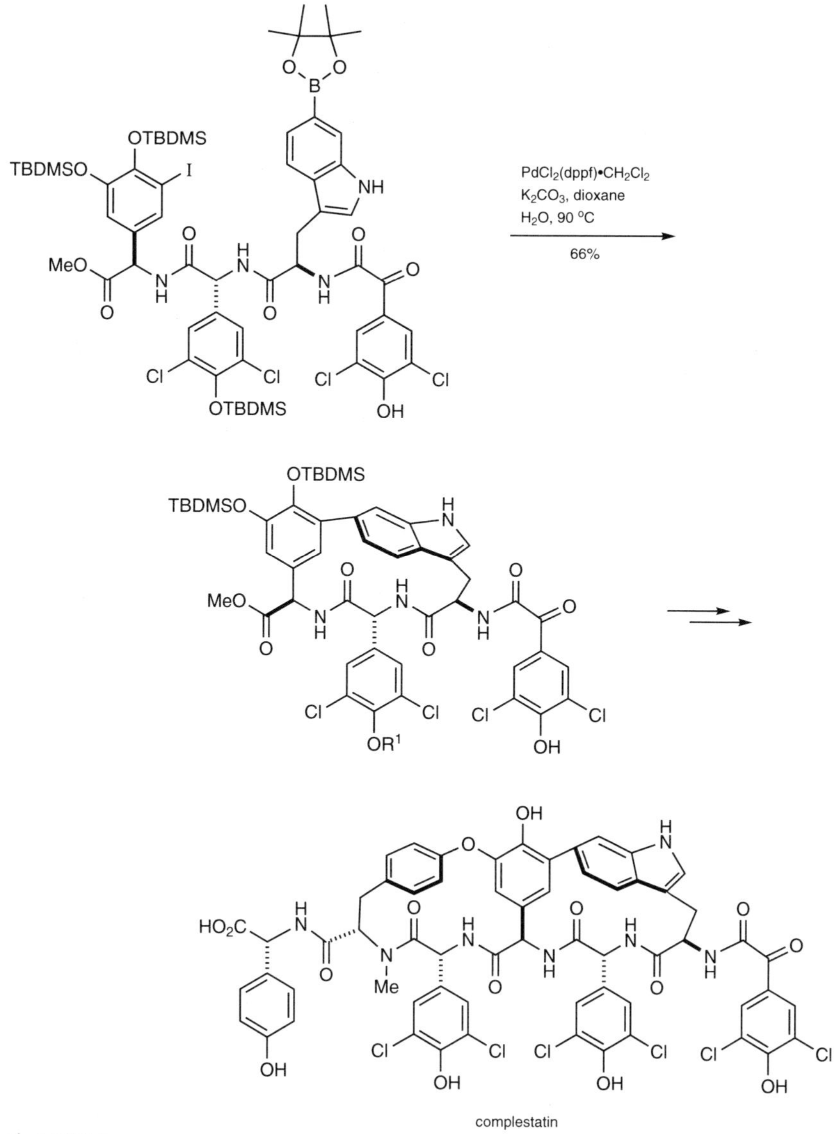

$PdCl_2(dppf)\cdot CH_2Cl_2$
K_2CO_3, dioxane
H_2O, 90 °C
66%

complestatin

R^1 = TBDMS, H

Intramolecular Suzuki–Miyaura macrocyclizations between vinyl iodide and vinylboronate moieties have been used to synthesize apoptolidinones and superstolide A.[20–23] Treating substrate **11** with tetrakis(triphenylphosphine)palladium(0) in the presence of thallium(I) ethoxide in a solution of tetrahydrofuran/water generates macrocycle **12** (60% yield), which was used in the synthesis of the polyketide natural product apoptolidinone A (Scheme 6).[20,21] Under similar conditions, substrate **13** is converted into macrocycle **14** via intramolecular Suzuki–Miyaura reaction followed by a transannular Diels–Alder reaction.[22,23] The use of thallium(I) ethoxide or thallium(I) hydroxide as base in these reaction systems dramatically accelerates the Suzuki–Miyaura reaction.[12–14]

for references see p 51

Scheme 6 Total Synthesis of Apoptolidinone A and Superstolide A Using Intramolecular Suzuki–Miyaura Macrocyclizations[20–23]

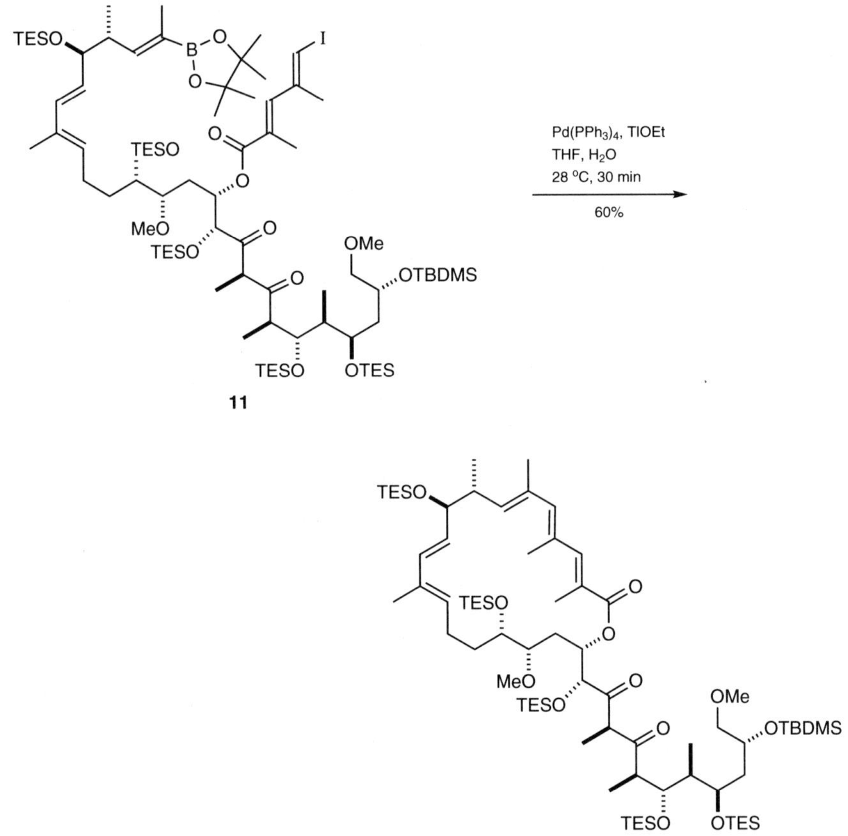

TESO, TESO, MeO, TESO, OMe, OTBDMS, TESO, OTES, I

11

$Pd(PPh_3)_4$, TlOEt
THF, H_2O
28 °C, 30 min
60%

TESO, TESO, MeO, TESO, OMe, OTBDMS, TESO, OTES

12

1. $Pd(PPh_3)_4$, TlOEt, THF, H_2O, 23 °C, 2 h
2. $CDCl_3$, 5 d or toluene, 80 °C, 2 h

30–35%

13

14

apoptolidinone A

superstolide A

Macrocycle 10:[17]

Under argon, $NaHCO_3$ (43.6 mg, 518.7 μmol, 7 equiv) and $PdCl_2(SPhos)_2$ (3.7 mg, 3.7 μmol, 5 mol%) were added successively to a degassed soln of substrate **9** (60.0 mg, 74.1 μmol) in toluene (3.60 mL) and H_2O (0.12 mL). The mixture was heated to 90 °C and stirred for 2 h, and then cooled, diluted with sat. aq NH_4Cl, and extracted with EtOAc. The combined organic layers were washed with brine, dried (Na_2SO_4), and concentrated under reduced pressure. The crude residue was filtered by short flash-column chromatography (alumina, EtOAc/heptane 2:1). Final purification by preparative TLC (silica gel, CH_2Cl_2/MeOH 20:1) afforded a white solid; yield: 22.2 mg (54%).

Macrocycle 12:[20]

CAUTION: *Thallium salts are toxic when inhaled, ingested, or absorbed through the skin.*

To a soln of vinylboronate **11** (4 mg, 2.45 μmol) in THF/H_2O (3:1; 3.6 mL) was added $Pd(PPh_3)_4$ (0.5 mg, 0.44 μmol). The resulting yellow soln was stirred for 5 min before TlOEt (0.3 μL, 3.7 μmol) was introduced dropwise. The soln was stirred for 30 min and turned from yellow to grey. The reaction was quenched with sat. $NaHCO_3$ soln (2 mL)

for references see p 51

and the aqueous layer was extracted with EtOAc (3 × 5 mL). The combined organic layers were dried ($MgSO_4$), filtered, and concentrated under reduced pressure. The residue was purified by flash chromatography (hexanes/EtOAc 60:1) to afford a colorless oil; yield: 2.0 mg (60%).

1.1.1.2 Palladium-Catalyzed Intramolecular Stille Coupling and Applications

Palladium-catalyzed Stille cross coupling between an organotin reagent (alkyl, vinyl, alkynyl, and aryl tin reagents) and an organic electrophile has proven to be a powerful method to construct C(sp^2)—C(sp^2, sp^3, or sp) bonds.[24–27] As in the Suzuki–Miyaura reaction, Stille coupling is also generally thought to involve oxidative addition, transmetalation, and reductive elimination steps (Scheme 7, pathway A). The rate-determining step in this catalytic cycle is transmetalation between the newly formed R^1—Pd(II)—X and the organostannane, in which the R^2 group transfers from the organostannane to the palladium(II) intermediate to generate an R^1—Pd(II)—R^2 intermediate. The rate of transfer from the organostannane to the palladium(II) intermediate depends on the properties of the group attached to the tin. The relative rate of migration for different groups is alkynyl > vinyl > aryl > allyl ~ benzyl >> alkyl. As a result, the most frequently used organotin reagents are mixed organostannanes containing three methyl or butyl spectator groups. After reductive elimination, the coupling product forms and the palladium(0) catalyst is regenerated. Corey and co-workers have reported that addition of copper(I) chloride and lithium chloride facilitate the transmetalation step of Stille coupling.[28] Copper(I) chloride and lithium chloride convert the organostannanes into the more reactive organocopper(I) species (R^2CuLiCl), which undergo faster transmetalation to form the R^1—Pd(II)—R^2 intermediate (Scheme 7, pathway B).

Scheme 7 The General Mechanism of Stille Coupling

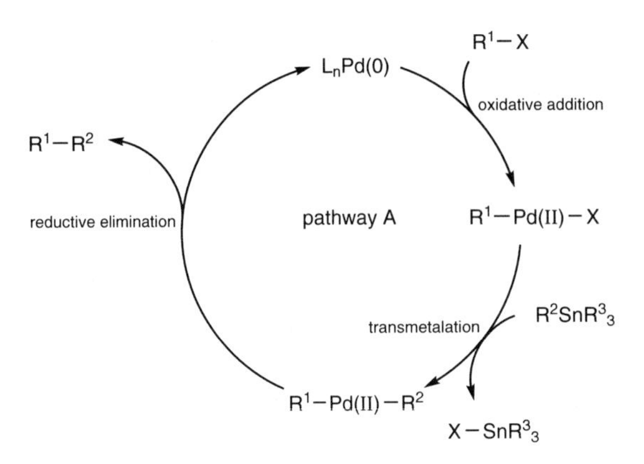

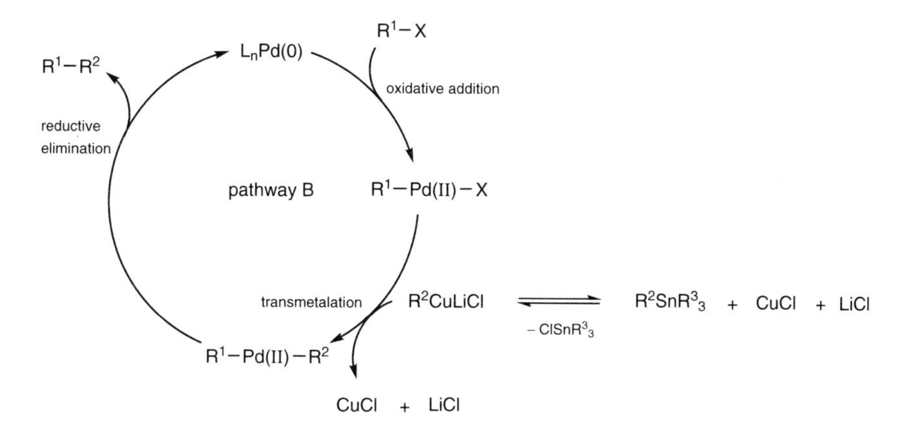

The advantages of Stille coupling include mild reaction conditions, easily available coupling components, and a very broad scope of substrates and functional groups, including ones with highly complex stereochemistry. Additionally, organostannanes are not sensitive to moisture and oxygen, in contrast to other organometallic reagents. Intramolecular Stille coupling between vinyl halides (or trifluoromethanesulfonates) and vinylstannanes can be used to construct a range of cyclic compounds containing 1,3-dienes, highly strained four-membered rings (Table 3, entry 1), fused rings (entries 2 and 3), and macrolactones (entries 4–7).[29–32]

Table 3 Intramolecular Stille Coupling Reactions[30–32]

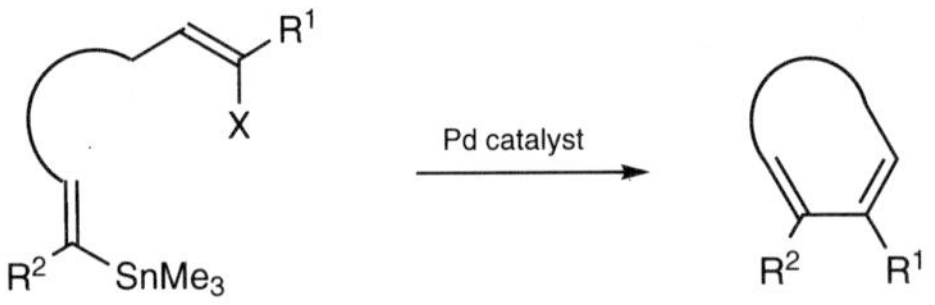

Entry	Substrate	Conditions	Product	Yield (%)	Ref
1	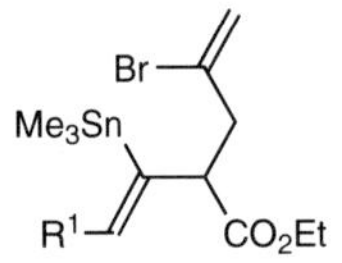 R^1 = Me, $(CH_2)_2OCH_2OMe$, $(CH_2)_3OTBDMS$, H	$Pd(PPh_3)_4$ (5 mol%), DMF, 80 °C, 1 h	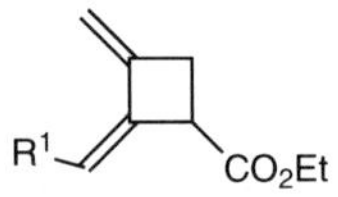	70–95	[30]
2	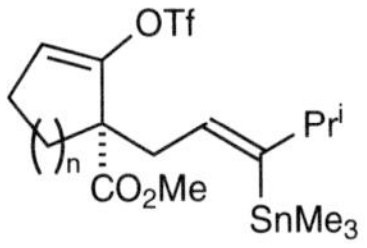 n = 1–3	$Pd(PPh_3)_4$ (5 mol%), MeCN, reflux	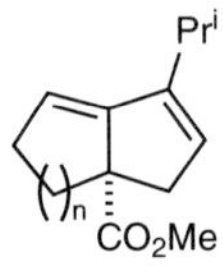	81–84	[31]
3	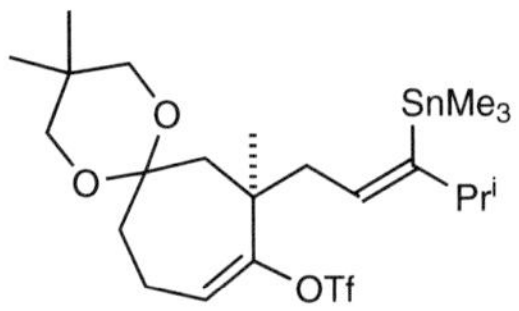	$Pd(PPh_3)_4$ (cat.), rt to 30 °C, 5 min	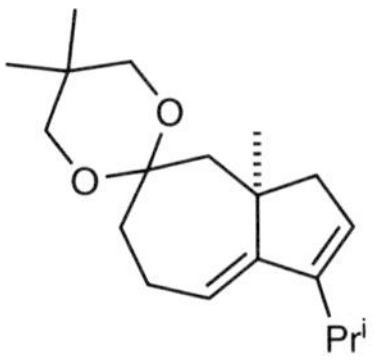	81	[31]
4	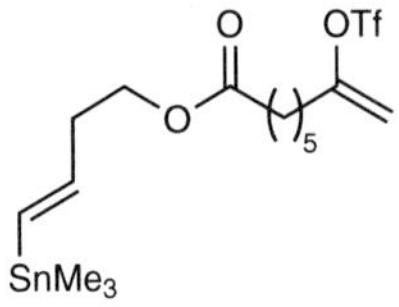	$Pd(PPh_3)_4$ (2 mol%), LiCl (3.0 equiv), THF, reflux	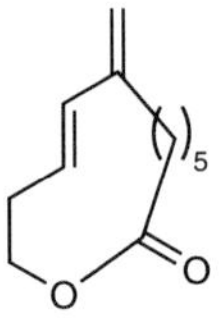	57	[32]

Table 3 (cont.)

Entry	Substrate	Conditions	Product	Yield (%)	Ref
5	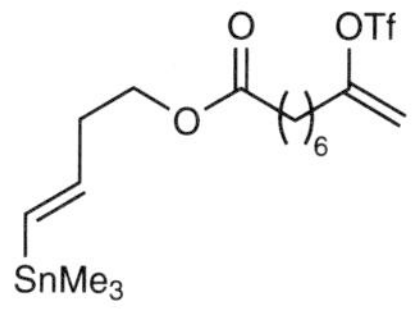	$Pd(PPh_3)_4$ (2 mol%), LiCl (3.0 equiv), THF, reflux	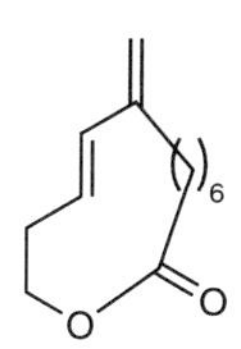	56	[32]
6	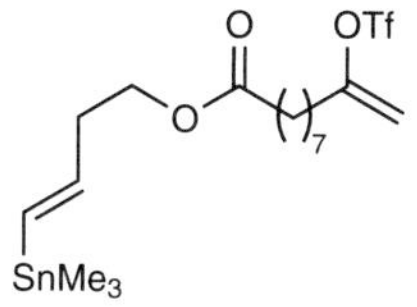	$Pd(PPh_3)_4$ (2 mol%), LiCl (3.0 equiv), THF, reflux	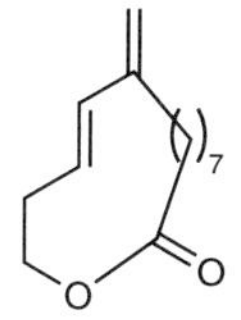	57	[32]
7	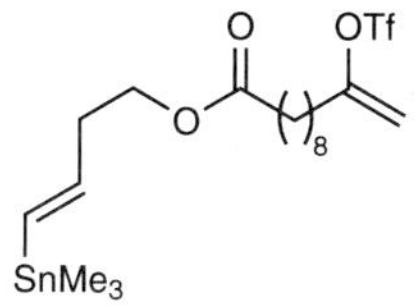	$Pd(PPh_3)_4$ (2 mol%), LiCl (3.0 equiv), THF, reflux	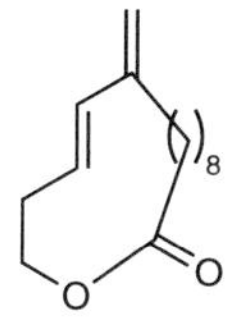	56	[32]

The intramolecular Stille macrocyclization between a vinyl iodide and a vinylstannane, e.g. in substrate **15**, has also been used in the formation of macrocycles, e.g. **16**, as part of the syntheses of sanglifehrin A[33] and sarain A[34,35] (Scheme 8).

Scheme 8 Total Syntheses of Sanglifehrin A and Sarain A Using Intramolecular Stille Macrocyclizations[33–35]

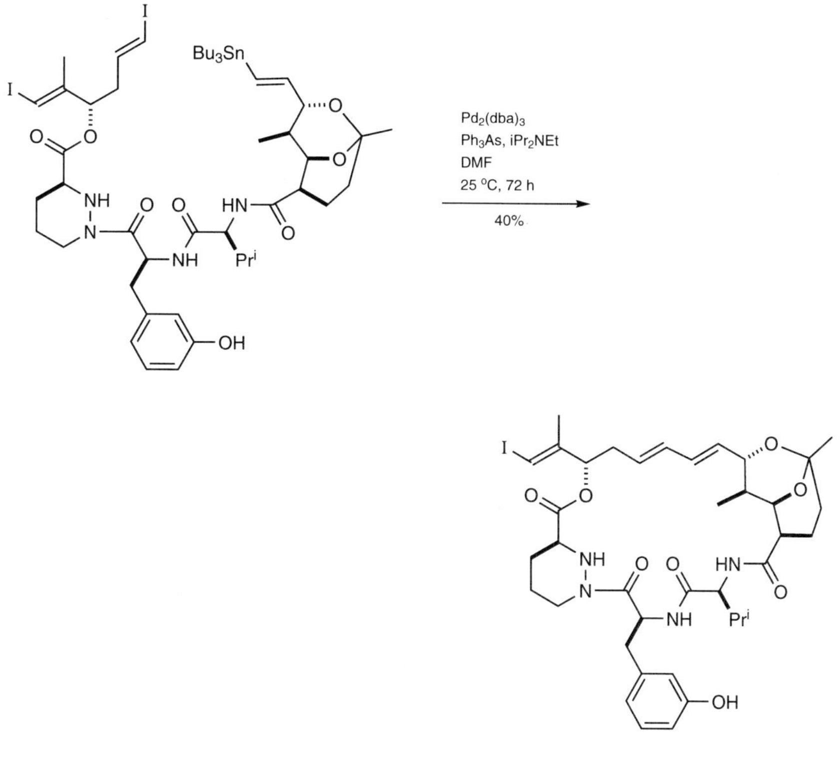

$Pd_2(dba)_3$, Ph_3As, iPr_2NEt, DMF, 25 °C, 72 h, 40%

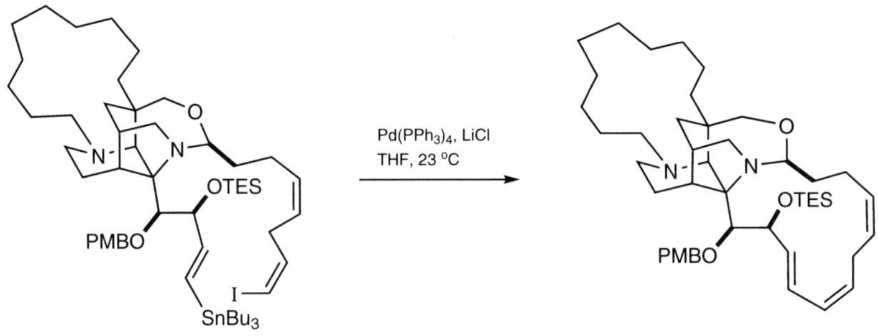

Pd(PPh3)4, LiCl
THF, 23 °C

15

16

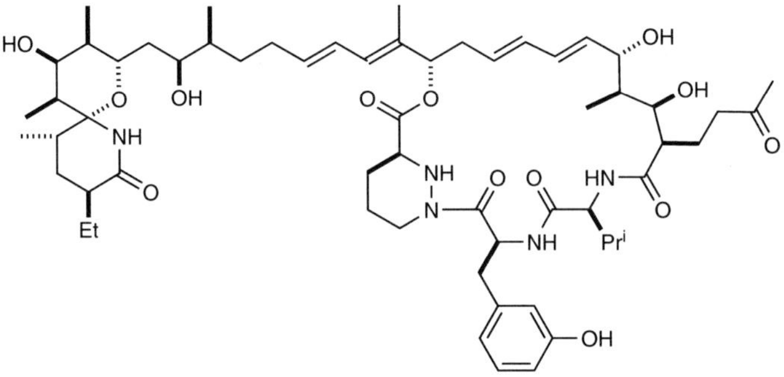

sanglifehrin A

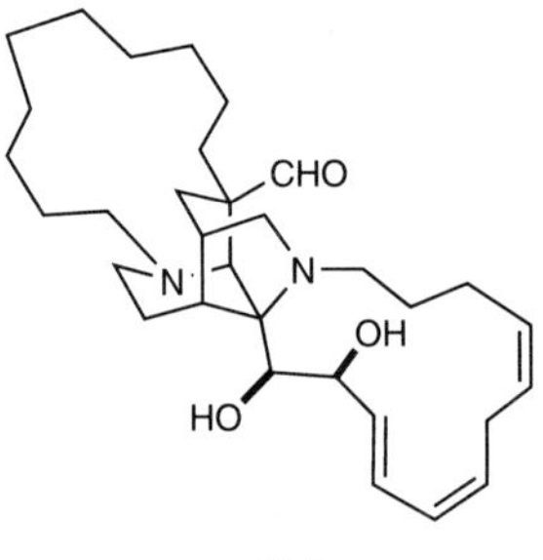

sarain A

Macrocycle 16:[35]
In a glovebox, a soln of $Pd(PPh_3)_4$ (3.0 mg, 0.00258 mmol) in THF (200 μL) was added to vinyl iodide **15** (20.0 mg, 0.0172 mmol) and LiCl (10.9 mg, 0.258 mmol) in THF (11.5 mL) at rt. After 7 d, the reaction vessel was removed from the glovebox and the solvent was evaporated under reduced pressure. The residue was passed over a plug of silica gel (hexanes/EtOAc 4:1 containing 2% Et_3N), and the solvent was evaporated. Purification by flash chromatography (hexanes/EtOAc 40:1 then 19:1 then 9:1) afforded the coupled product **16** [11.2 mg (ca. 87% yield)], which was contaminated with a byproduct believed to be the des-iodo derivative of substrate **15**.

1.1.1.3 Palladium-Catalyzed Intramolecular Negishi Coupling and Applications

The palladium-catalyzed Negishi cross-coupling reaction was developed by Negishi and co-workers in the 1970s.[36–42] This methodology provides an efficient cross coupling of organozinc reagents with aryl, alkenyl, or alkynyl halides. The advantage of organozinc reagents lies in their greater functional-group tolerance compared with organolithiums and Grignard reagents. As shown in Scheme 9, the catalytic cycle of the Negishi coupling reaction is a similar process to the Suzuki–Miyaura and Stille cross couplings, including oxidative addition, transmetalation, and reductive elimination. Notably, organozinc reagents are among the most reactive of nucleophilic species in palladium-catalyzed cross-coupling reactions. Organozinc reagents undergo rapid transmetalation with transition-metal salts compared with organoboron and organotin compounds. The most frequently used organozinc reagents in Negishi couplings include diorganozinc species ($R^1{}_2Zn$) and organozinc halides (R^1ZnX), which can be prepared by either direct reaction of the organic halide with activated zinc metal or by transmetalation of the corresponding organolithium or Grignard reagent with a zinc halide (ZnX_2).

Scheme 9 The General Mechanism of Negishi Coupling

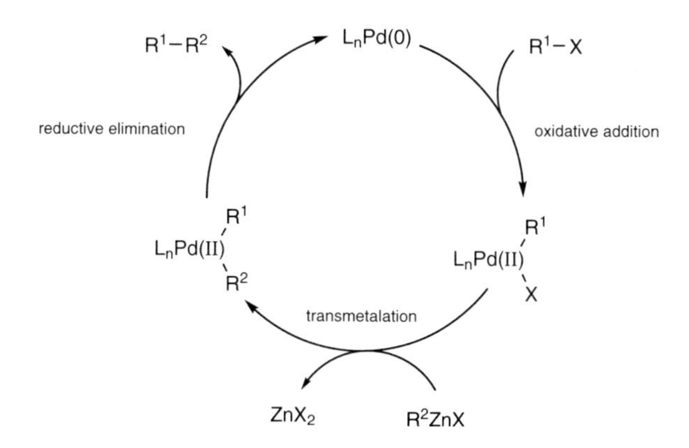

Because of the reactivity and stability of organozinc species, Negishi coupling is normally used in intermolecular cross coupling for the formation of C(sp^2)—C(sp^2) as well as C(sp^2)—C(sp) and C(sp^2)—C(sp^3) bonds.[43] The intramolecular Negishi reaction would appear to have been relatively underutilized. One example was reported by Jackson and co-workers in the total synthesis of K-13.[44] As shown in Scheme 10, the ring closure is accomplished using the intramolecular Negishi coupling between an aromatic iodide and the organozinc reagent. Treatment of substrate **17** with zinc (activated by treatment with iodine), and then addition of this solution of the presumed organozinc reagent to a dilute solution of a catalyst derived from tris(dibenzylideneacetone)dipalladium(0) and tri-2-tolylphosphine gives the desired macrocyclic peptide **18** in 35% yield. It is believed that the insertion of zinc into the aromatic C—I bond is normally slower than insertion of zinc into an aliphatic C—I bond.

Scheme 10 Total Synthesis of K-13 Using Intramolecular Negishi Coupling[44]

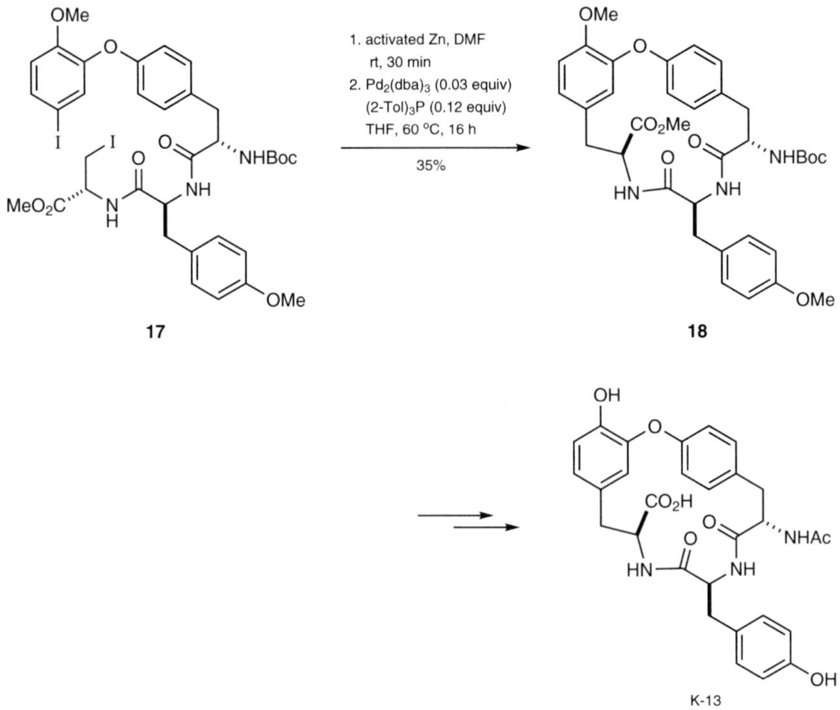

1. activated Zn, DMF, rt, 30 min
2. $Pd_2(dba)_3$ (0.03 equiv), $(2\text{-Tol})_3P$ (0.12 equiv), THF, 60 °C, 16 h

35%

17 → 18 → K-13

Macrocyclic Peptide 18:[44]

I_2 (4 mg, 0.016 mmol) was added to a rapidly stirred suspension of Zn (24 mg, 0.38 mmol) in DMF (0.5 mL) at rt under N_2. After 30 min, a soln of the diiodide **17** (300 mg, 0.33 mmol) in DMF (0.5 mL) was added, and the mixture was stirred for a further 30 min. The soln of the organozinc iodide was then added dropwise to a stirred soln of $Pd_2(dba)_3$ (9.2 mg, 0.01 mmol, 3 mol%) and $(2\text{-Tol})_3P$ (12.2 mg, 0.04 mmol, 12 mol%) in THF (140 mL), and the mixture was heated at 60 °C and stirred for 16 h. The solvent was removed under reduced pressure to give a crude oil, which was dissolved in EtOAc (25 mL), washed with H_2O (2 × 10 mL) and sat. brine (10 mL), dried (Na_2SO_4), and filtered. The solvent was removed under reduced pressure to give a solid, which was purified by column chromatography; yield: 75 mg (35%).

1.1.1.4 Palladium-Catalyzed Intramolecular Sonogashira Coupling and Applications

The palladium and copper cocatalyzed Sonogashira coupling is regarded as a general method for the construction of C(sp)—C(sp^2) bonds between terminal alkynes and aryl or alkenyl halides or trifluoromethanesulfonates.[45–50] The mechanism of this cross coupling is assumed to proceed through the combination of palladium-catalyzed and copper-catalyzed cycles (Scheme 11). In the palladium-catalyzed cycle, in situ reduction of palladium(II) complexes may give rise to a catalytically active species $Pd(0)L_n$, which undergoes the oxidative addition with R^1X. Next, in the copper-catalyzed cycle, the base is proposed to abstract the acetylenic proton from a π-alkyne–copper complex, formed by the interaction of the terminal alkyne with copper(I) salt, to generate a copper acetylide intermediate. This copper acetylide species undergoes transmetalation, which is the rate-determining step in the overall coupling, to generate a $R^1Pd(C{\equiv}CR^2)L_2$ species. This then undergoes reductive elimination to yield the final coupled alkyne and regenerate the $Pd(0)L_n$ catalyst.

Scheme 11 The Proposed Mechanism of Sonogashira Coupling

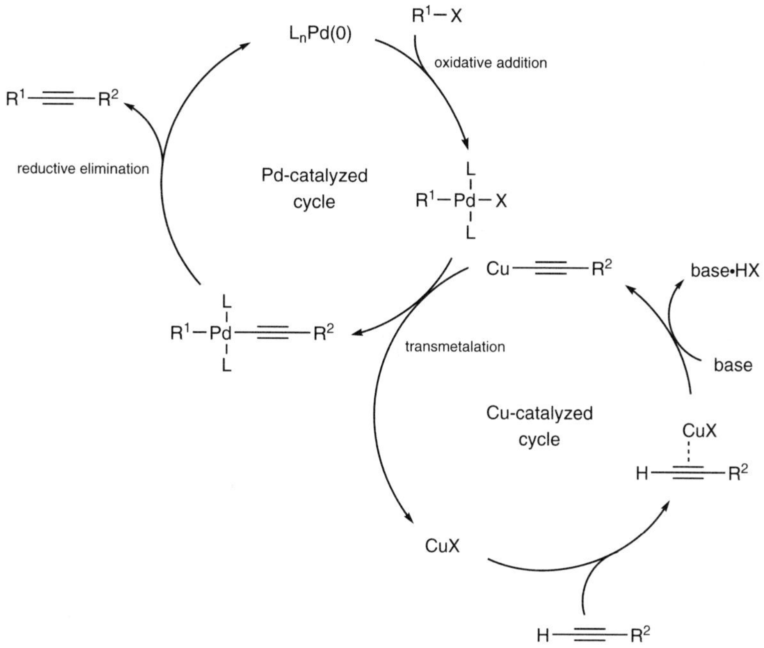

Intramolecular Sonogashira coupling has been reported less often in organic synthesis, because it is effective only for constructing macrocycles of ring size greater than 15 due to the conjugated enyne ring strain. Iqbal and co-workers have reported the preparation of conformationally constrained cyclic peptides using an intramolecular Sonogashira coupling (Scheme 12).[51] Under palladium/copper-cocatalyzed conditions, the oxidative dimerization of the alkynes **19** was also observed. However, under copper-free Sonogashira coupling conditions, with a bulky, electron-rich phosphine ligand [$(2\text{-Tol})_3$P] in acetonitrile, the desired cyclic peptides **20** are formed in moderate or poor isolated yield.

Scheme 12 Preparation of Cyclic Peptides Using Intramolecular Sonogashira Coupling[51]

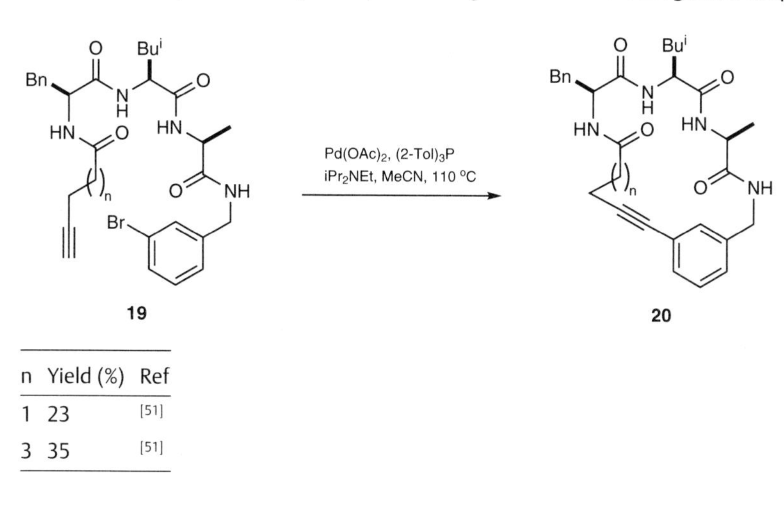

n	Yield (%)	Ref
1	23	[51]
3	35	[51]

for references see p 51

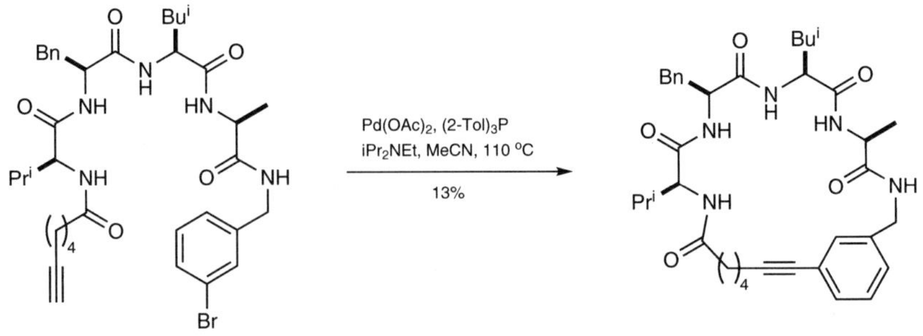

Pd(OAc)$_2$, (2-Tol)$_3$P
iPr$_2$NEt, MeCN, 110 °C
13%

An intramolecular Sonogashira reaction of alkyne **21** catalyzed by tetrakis(triphenylphosphine)palladium(0) and copper(I) iodide in diethylamine solvent furnishes product **22** containing a 30-membered macrocyclic ring in 35% yield, which was used in the synthesis of penarolide sulfate A$_1$ (Scheme 13).[52] The intramolecular coupling of substrate **23** in the presence of tris(dibenzylideneacetone)dipalladium(0) (0.5 equiv), copper(I) iodide (2 equiv), and diisopropylethylamine (30 equiv) in degassed dimethylformamide (2 mM) at room temperature gives the desired coupling product **24**, which contains the core skeleton of kedarcidin chromophore, as a single atropisomer in 90% yield (Scheme 13).[53]

Scheme 13 Synthesis of Penarolide Sulfate A$_1$ and Core Skeleton of Kedarcidin Chromophore Using Intramolecular Sonogashira Coupling[52,53]

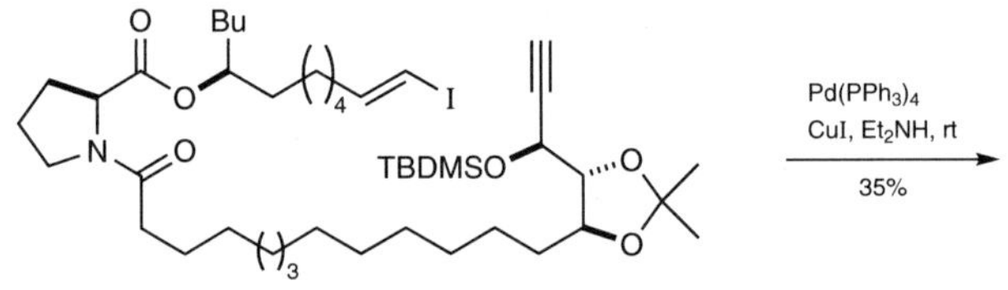

Pd(PPh$_3$)$_4$
CuI, Et$_2$NH, rt
35%

21

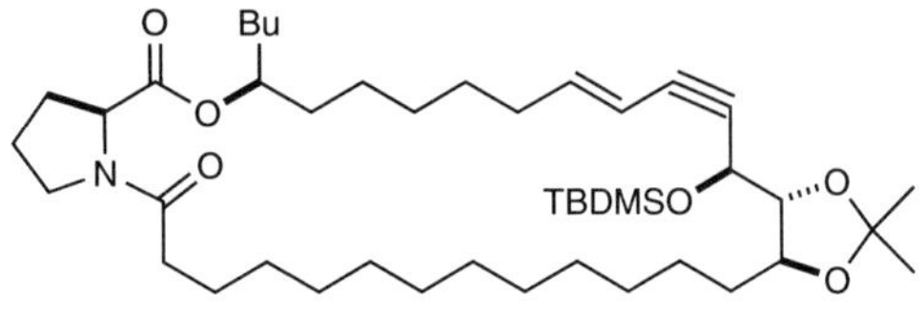

22

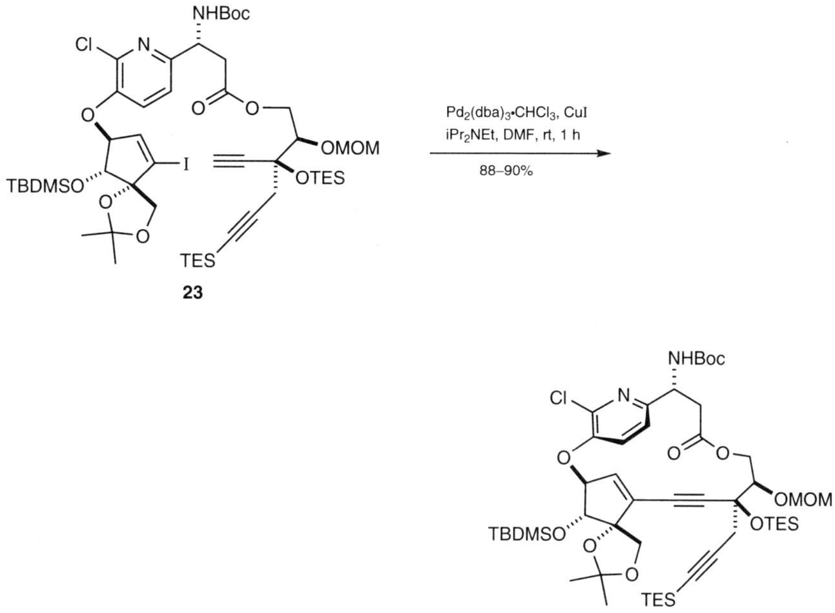

Pd$_2$(dba)$_3$•CHCl$_3$, CuI
iPr$_2$NEt, DMF, rt, 1 h
88–90%

23 → **24**

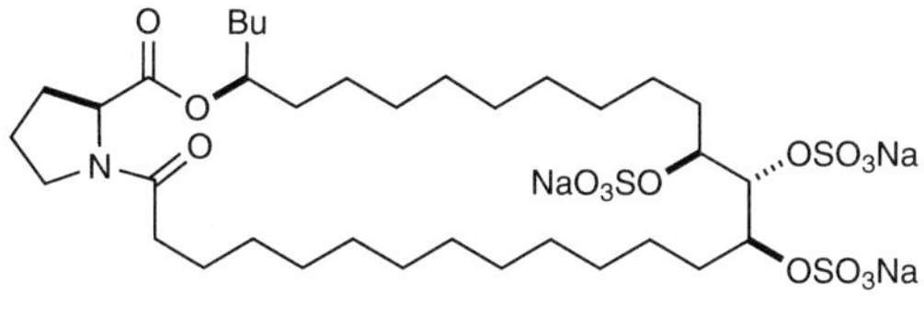

penarolide sulfate A$_1$

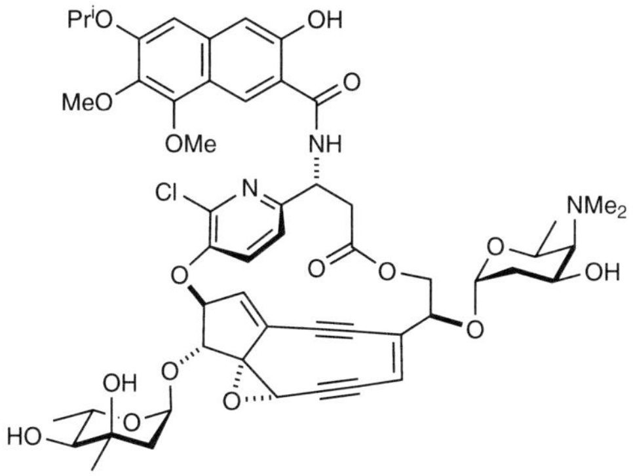

kedarcidin chromophore

Cyclic Peptide 20 **(n = 3):**[51]
Pd(OAc)$_2$ (72 mg, 0.32 mmol) and (2-Tol)$_3$P (195 mg, 0.64 mmol) were added to warm HPLC-grade MeCN (1.2 L) and refluxed at 110 °C for 30 min. Then, acyclic peptide **19** (n = 3; 500 mg, 0.80 mmol) was added in a single portion, and the resulting mixture was stirred for 15 min at the same temperature. Finally, iPr$_2$NEt (0.7 mL, 4 mmol) was added. After 15 h, the mixture was filtered through a pad of Celite, washing with hot MeCN (100 mL). The filtrate was concentrated, and the residue was purified by flash chromatography (silica gel, CH$_2$Cl$_2$/MeOH 49:1) to afford a white solid; yield: 243 mg (35%, as reported).

for references see p 51

Macrocycle 24:[53]
A soln of substrate **23** (51.7 mg, 0.05 mmol) and iPr_2NEt (238.3 μL, 1.37 mmol) in DMF (22.8 mL) was degassed by freeze–pump–thaw cycles (3 ×). After addition of CuI (17.4 mg, 0.091 mmol), the mixture was vigorously stirred at rt in the dark for 0.5 h, producing a colorless soln. $Pd_2(dba)_3$•$CHCl_3$ (26.0 mg, 0.025 mmol) was added to the resulting soln. The mixture was stirred at rt for 1 h in the dark. Then, it was diluted with Et_2O and quenched with sat. aq NH_4Cl followed by H_2O at 0 °C. The mixture was vigorously stirred at rt until the aqueous phase turned dark blue (ca. 1 h). After being diluted with Et_2O, the layers were separated, and the organic phase was washed with H_2O. The aqueous layer was extracted with Et_2O. The combined organic extracts were washed with brine, dried ($MgSO_4$), and concentrated. The residue was purified by flash column chromatography (silica gel).

1.1.1.5 Palladium-Catalyzed Intramolecular Hiyama Reactions and Applications

Organosilicon reagents are generally stable and less reactive toward electrophiles than other organometallic nucleophiles because of their less-polarized C—Si bonds. Active pentacoordinate silicates are reactive species in the palladium-catalyzed Hiyama coupling reaction, and are normally prepared in situ by the nucleophilic attack of a fluoride ion on silicon,[54–60] as shown in Scheme 14. This active hypervalent intermediate interacts with an R^1—Pd(II)—X species, formed by oxidative addition of the organic halide to palladium(0), forming a four-membered cyclic transition state **25**. The key transmetalation then occurs to generate R^1—Pd(II)—R^2, involving the migration of a reacting organic group (R^2) from silicon to the palladium(II) center. The most commonly used activators for the formation of hypervalent silicates include tris(dimethylamino)sulfonium difluorotrimethylsilane (TASF) and tetrabutylammonium fluoride (TBAF).

Scheme 14 The Proposed Mechanism of Silicon-Assisted Cross-Coupling Reactions

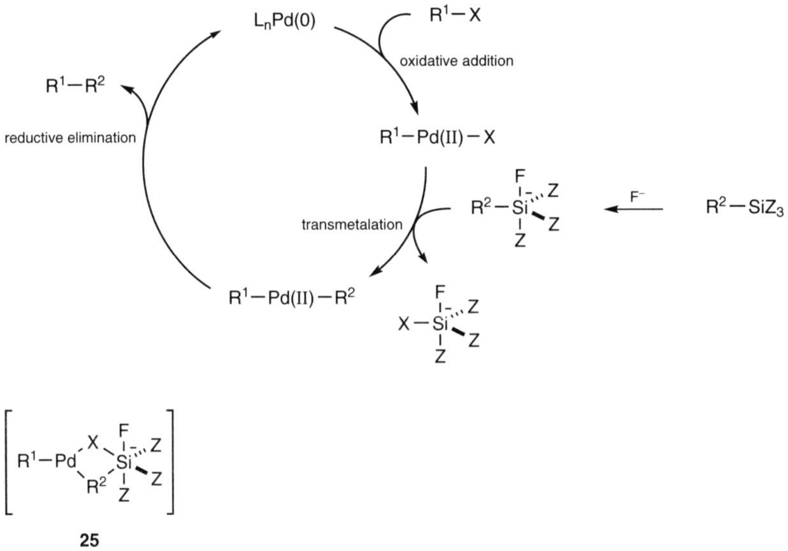

Denmark and co-workers have studied the palladium-catalyzed intramolecular Hiyama coupling reactions between vinyl iodides and cycloalkenylsiloxanes.[61,62] It was found that the siloxanes are highly effective coupling partners with various aryl and alkenyl halides; in the presence of palladium(0), medium-sized rings with an internal Z,Z-1,3-diene unit can be obtained in high yield and with good functional-group compatibility. Low concentration (0.01 M) and slow addition are crucial for the reliability of the yields of these couplings. Under the optimized conditions, the intramolecular coupling of the substrates leads to the formation of a series of cycloalkadienes **26** (Table 4).

Table 4 Palladium-Catalyzed Intramolecular Hiyama Coupling Reactions[61,62]

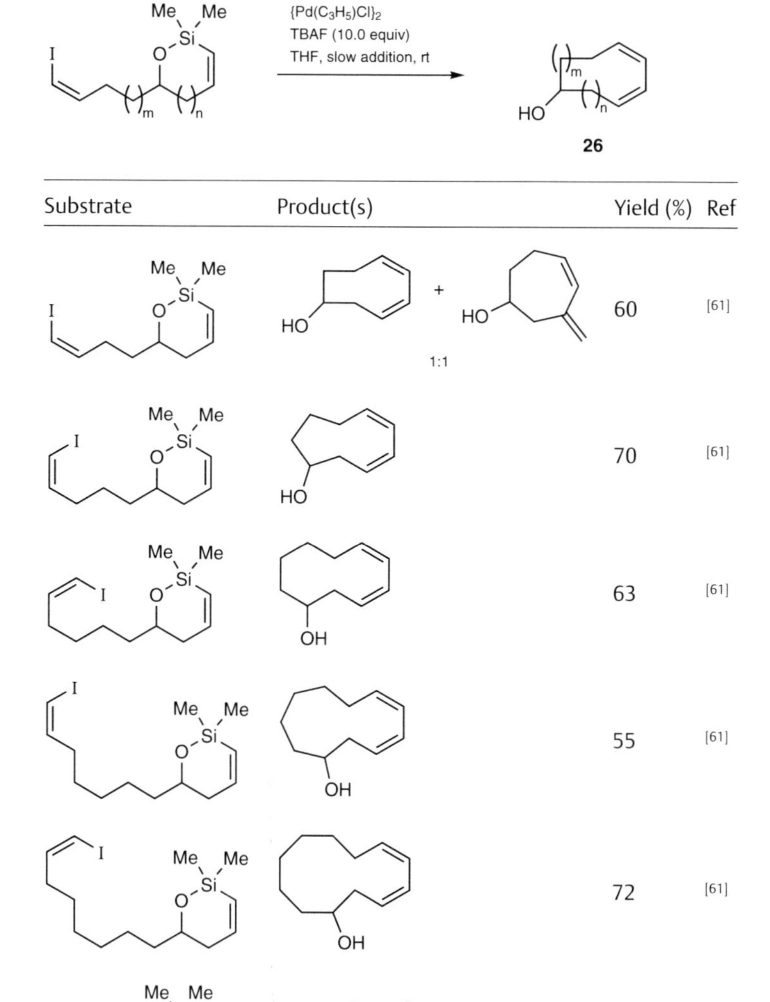

Substrate	Product(s)	Yield (%)	Ref
		60	[61]
		70	[61]
		63	[61]
		55	[61]
		72	[61]
		71	[61]

Denmark and co-workers subsequently reported the total synthesis of (+)-brasilenyne using this methodology.[63,64] As shown in Scheme 15, a nine-membered cyclic ether skeleton containing a *Z*,*Z*-1,3-diene unit was efficiently constructed under the optimal conditions.

Scheme 15 Total Synthesis of (+)-Brasilenyne Using a Hiyama Coupling Reaction[63,64]

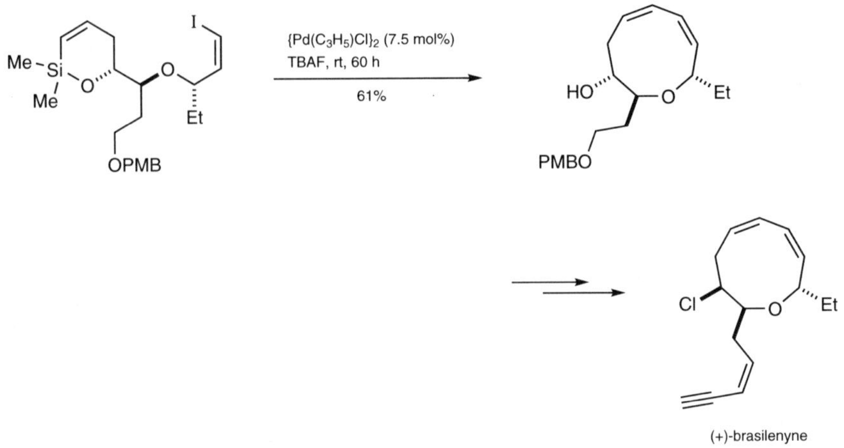

$\{Pd(C_3H_5)Cl\}_2$ (7.5 mol%)
TBAF, rt, 60 h
61%

(+)-brasilenyne

Cycloalkadienols 26; General Procedure:[61,62]

$\{Pd(C_3H_5)Cl\}_2$ (0.075–0.10 equiv) was dissolved in a soln of TBAF (10.0 equiv) in THF under a N_2 atmosphere at ambient temperature. To the mixture was added slowly a 0.1 M soln of the cycloalkenylsiloxane substrate in THF by syringe pump. After complete addition, the deep brown soln was stirred for an additional 2 h. The solvent was removed by rotary evaporation and EtOAc/hexane (5 mL) was then added. The mixture was filtered through a short column of silica gel, which was then eluted with EtOAc/hexane (300–400 mL). The eluate was concentrated by rotary evaporation to give a crude product, which was purified by chromatography (silica gel) followed by Kugelrohr distillation.

1.1.1.6 Palladium-Catalyzed Intramolecular Arene–Alkene Coupling Through Dehydrogenative Cross Coupling

Palladium-mediated cross coupling between arenes and alkenes always undergoes an oxidative Heck-type process.[65] As shown in Scheme 16, the palladium catalysts used in this catalytic cycle normally pass through the valence states of Pd(II) → Pd(0) → Pd(II). An arylpalladium intermediate is firstly generated through a palladium-catalyzed selective C—H bond metalation. Subsequent carbopalladation of the alkene leads to an alkylpalladium complex, which undergoes *syn*-β-H elimination to yield the alkenylation product and palladium(0). The palladium(II) is finally regenerated by the oxidation of palladium(0) using copper salts, silver salts, benzo-1,4-quinone, oxygen, or peroxides, etc. The palladium-catalyzed inter- and intramolecular cross couplings between arenes and alkenes using catalytic dehydrogenative cross coupling have been extensively studied and optimized to facilitate the generality needed in organic synthesis.[66–69]

Scheme 16 The Proposed Mechanism of Catalytic Dehydrogenative Cross Coupling

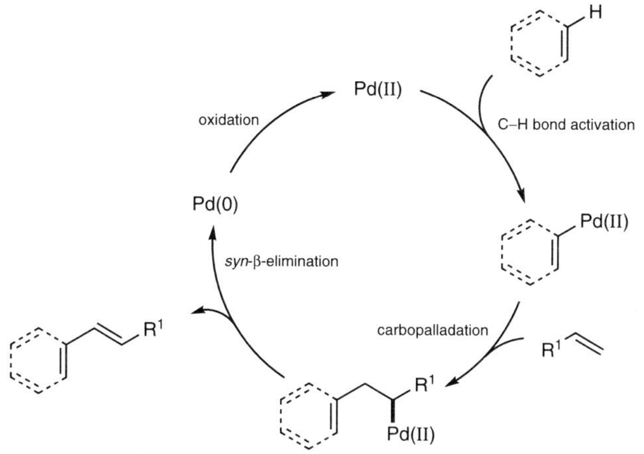

Stoltz and co-workers have studied the palladium-catalyzed oxidative indole annulation using a palladium/pyridine system, which uses dioxygen as an oxidant (Table 5).[70,71] It was found that the ligand and solvent play key roles in this intramolecular coupling. More-electron-rich pyridine ligands shut down the reaction, while switching to pyridines substituted with electron-withdrawing groups increases the reactivity; using ethyl nicotinate (ethyl pyridine-3-carboxylate) as ligand gives the best yield. Better yields are obtained when more-polar solvents are used compared with nonpolar aromatic solvents. This suggests the potential stabilization of charged intermediates in the catalytic cycle. Addition of acetic acid as cosolvent solved the oxidative decomposition of indoles under the aerobic conditions. Cyclization of indole substrate **27** under the optimized conditions gives the desired product **28** containing a quaternary center in 82% yield (Table 5, entry 1). It was proposed that the reaction proceeds through initial palladation at C2, which is followed by alkene insertion and β-hydride elimination. Substituted indoles participate in this oxidative cyclization reaction, forming five- and six-membered rings (Table 5).

for references see p 51

Table 5 Palladium-Catalyzed Oxidative Indole Annulation[70,71]

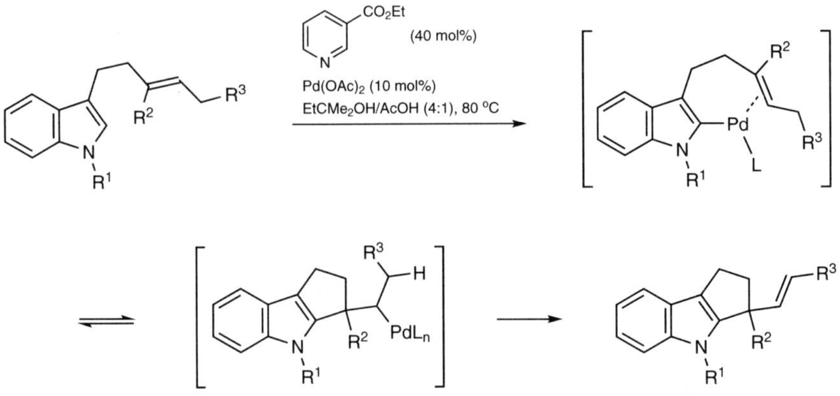

ethyl nicotinate (CO_2Et) (40 mol%)
$Pd(OAc)_2$ (10 mol%)
$EtCMe_2OH$/AcOH (4:1), 80 °C

Entry	Substrate	Product	Time (h)	Yield (%)	Ref
1	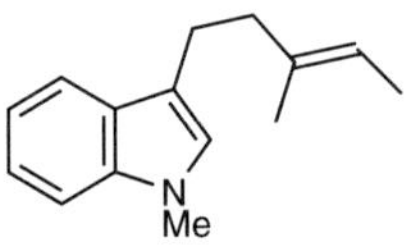 **27**	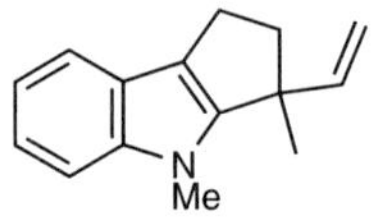 **28**	24	82	[70]
2	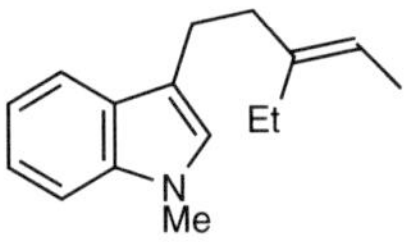	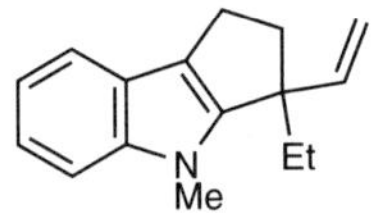	18	74	[70]
3	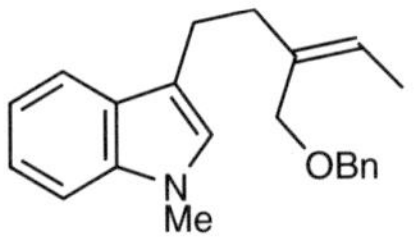	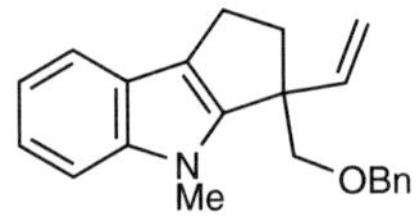	24	60	[70]
4	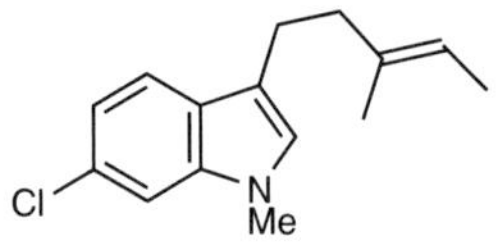	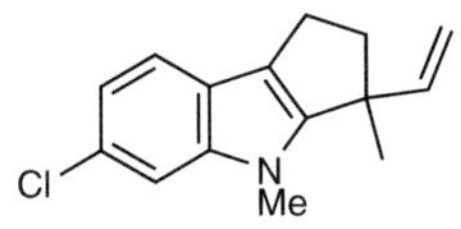	32	62	[70]
5	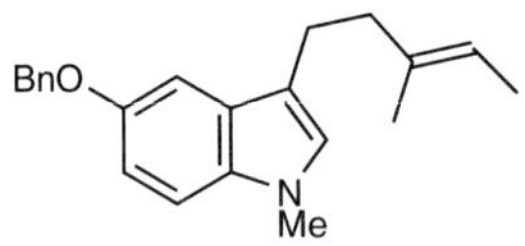	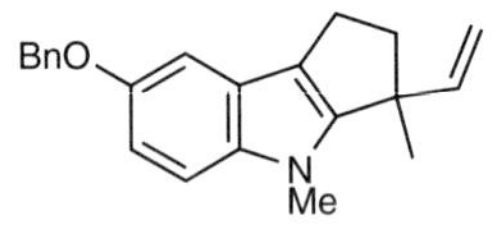	20	73	[70]
6	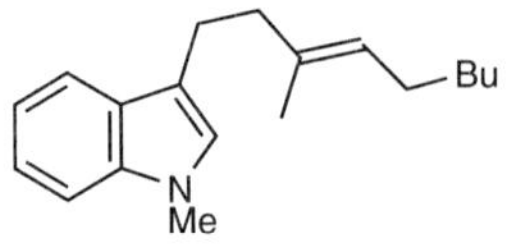	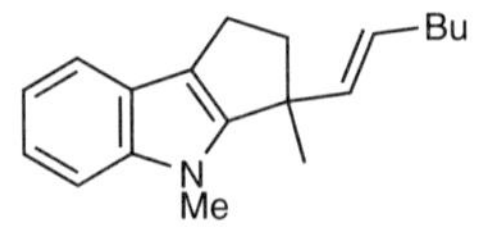	30	79	[70]
7	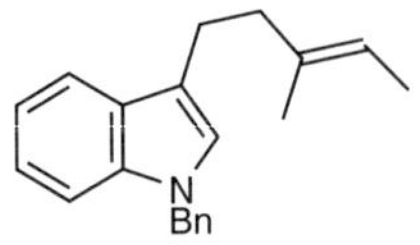	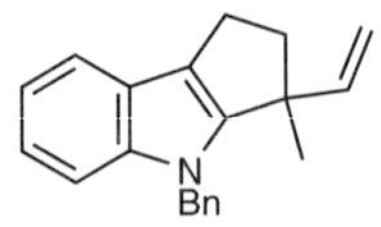	48	69	[70]

Table 5 (cont.)

Entry	Substrate	Product	Time (h)	Yield (%)	Ref
8	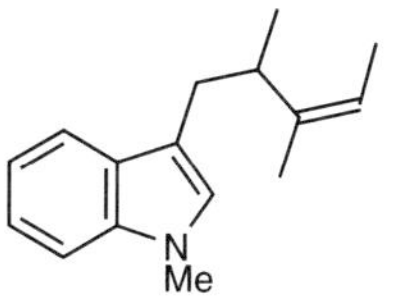	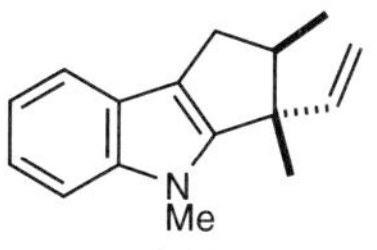 dr 6:1	18	76	[70]
9	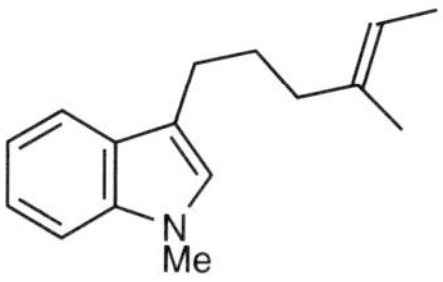	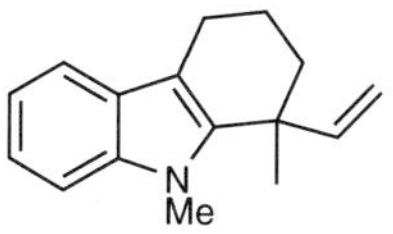	39	66	[70]
10	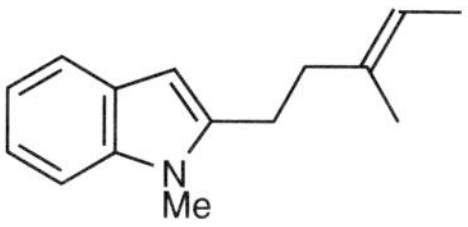	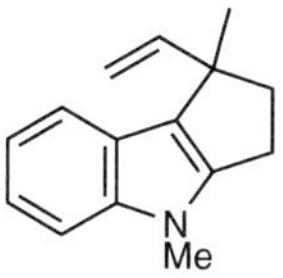	6	73	[70]
11	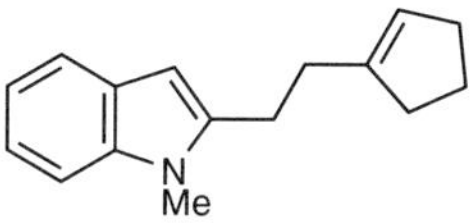	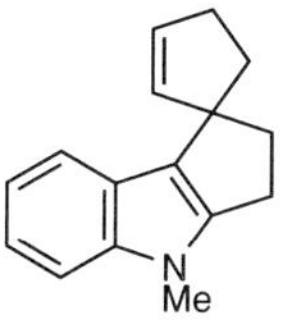	5	68	[70]
12	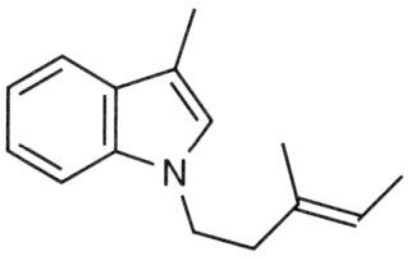	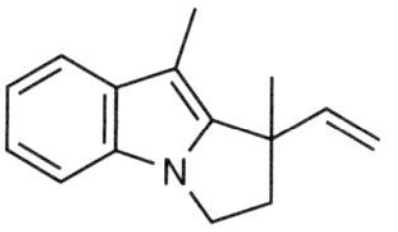	18	74	[70]

Stoltz and co-workers further explored the preparation of electron-rich, highly substituted benzo[*b*]furan and dihydrobenzo[*b*]furan derivatives based on an intramolecular oxidative C—C bond formation that relies on palladium(II) catalysis (Scheme 17 and Table 6).[71,72] The optimal conditions for the cyclization involve palladium(II) acetate, ethyl nicotinate, benzo-1,4-quinone, and sodium acetate. Benzo-1,4-quinone is a better oxidant compared to aerobic conditions. This methodology is suitable for the efficient synthesis of dihydrobenzo[*b*]furan derivatives with quaternary centers (Table 6).

Scheme 17 Palladium(II)-Catalyzed Oxidative Benzo[*b*]furan Synthesis[71,72]

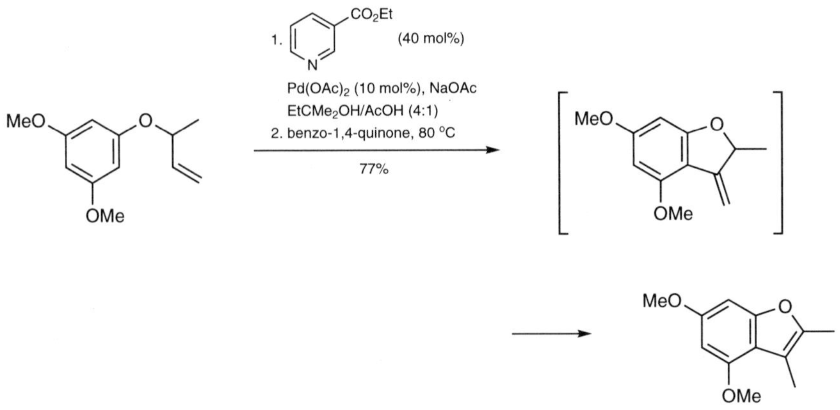

1. ethyl nicotinate (40 mol%)
$Pd(OAc)_2$ (10 mol%), NaOAc
$EtCMe_2OH/AcOH$ (4:1)
2. benzo-1,4-quinone, 80 °C
77%

Table 6 Palladium(II)-Catalyzed Oxidative Dihydrobenzo[*b*]furan Synthesis[71]

1. ethyl nicotinate
$Pd(OAc)_2$, NaOAc
$EtCMe_2OH/AcOH$ (4:1)
2. benzo-1,4-quinone, 100 °C

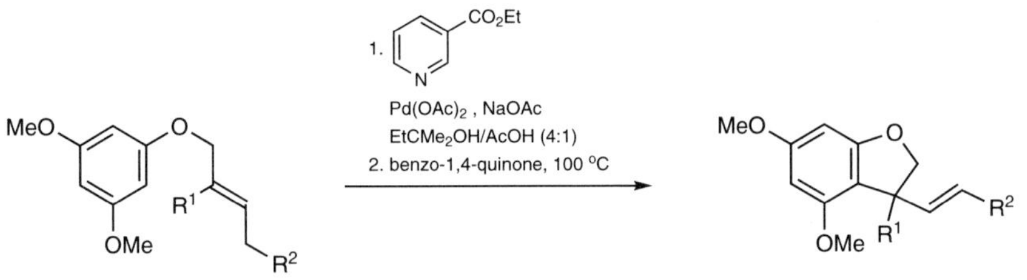

Entry	Substrate	Product	Time (h)	Yield (%)	Ref
1	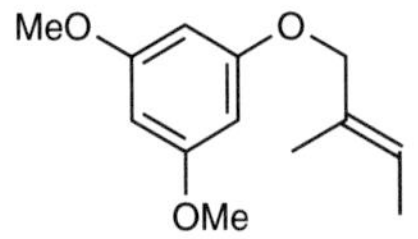	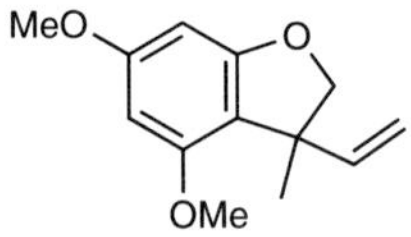	16	74	[71]
2	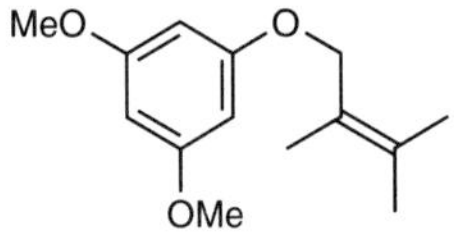	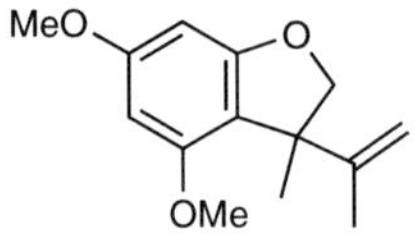	12	71	[71]
3	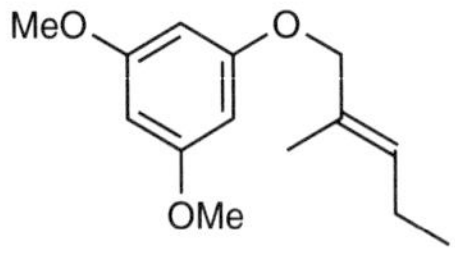	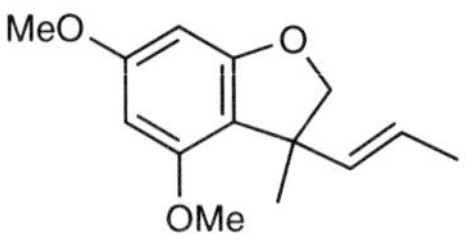	30	58	[71]
4	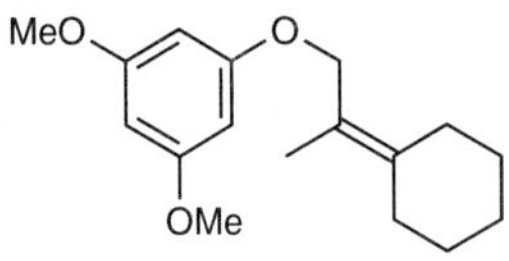	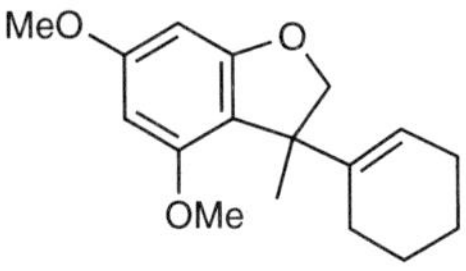	28	55	[71]
5	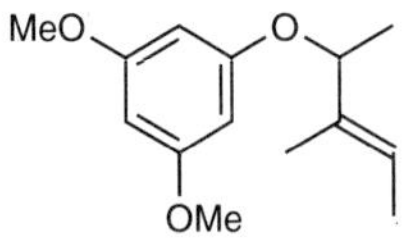	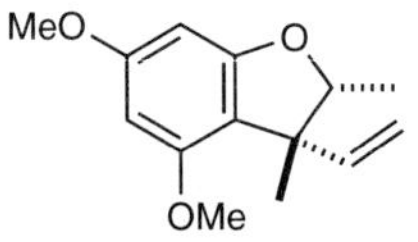	15	74	[71]

Table 6 (cont.)

Entry	Substrate	Product	Time (h)	Yield (%)	Ref
6	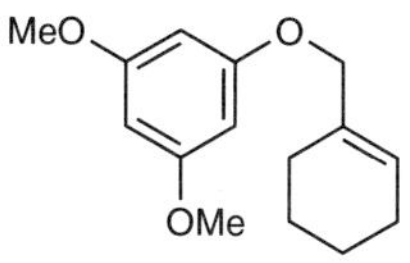	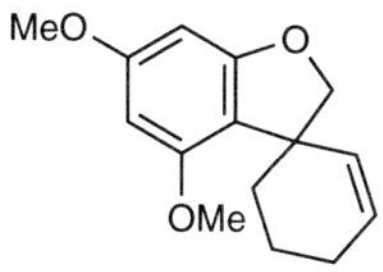	24	80	[71]
7	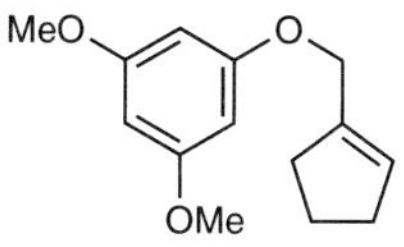	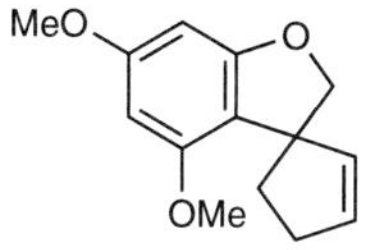	18	78	[71]
8	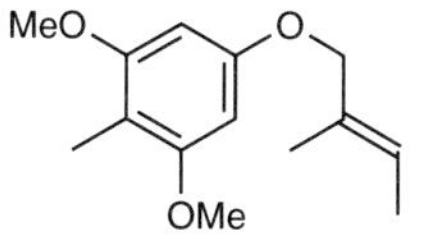	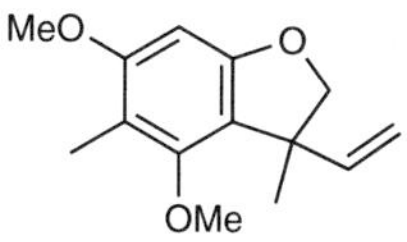	15	50	[71]
9	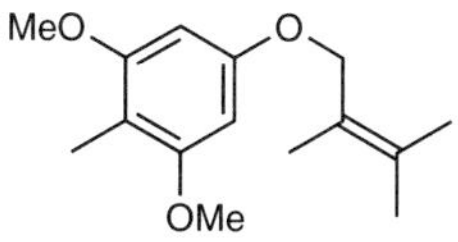	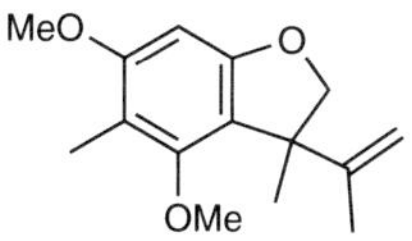	15	63	[71]
10	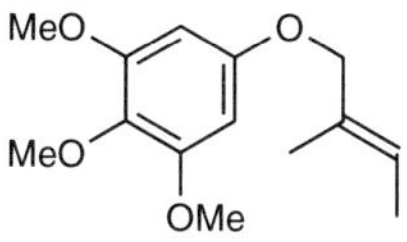	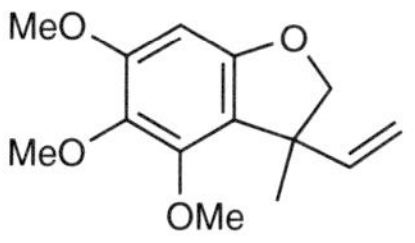	15	60	[71]
11	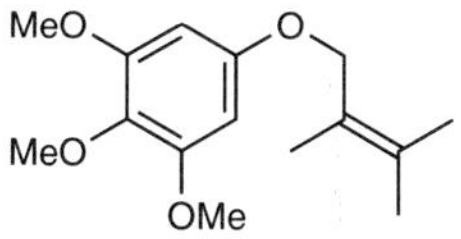	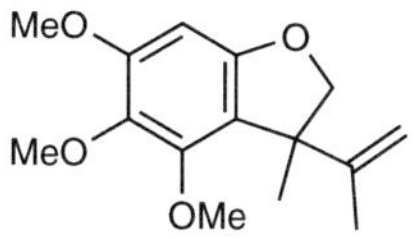	15	66	[71]

As applications of these methods, Gaunt and co-workers reported the total synthesis of rhazinicine,[73] while Stoltz and co-workers accomplished the synthesis of dragmacidin F[74] using the intramolecular oxidative coupling as the key transformation (Scheme 18).

Scheme 18 Total Syntheses of Rhazinicine and Dragmacidin F Using Intramolecular Oxidative Cyclizations[73,74]

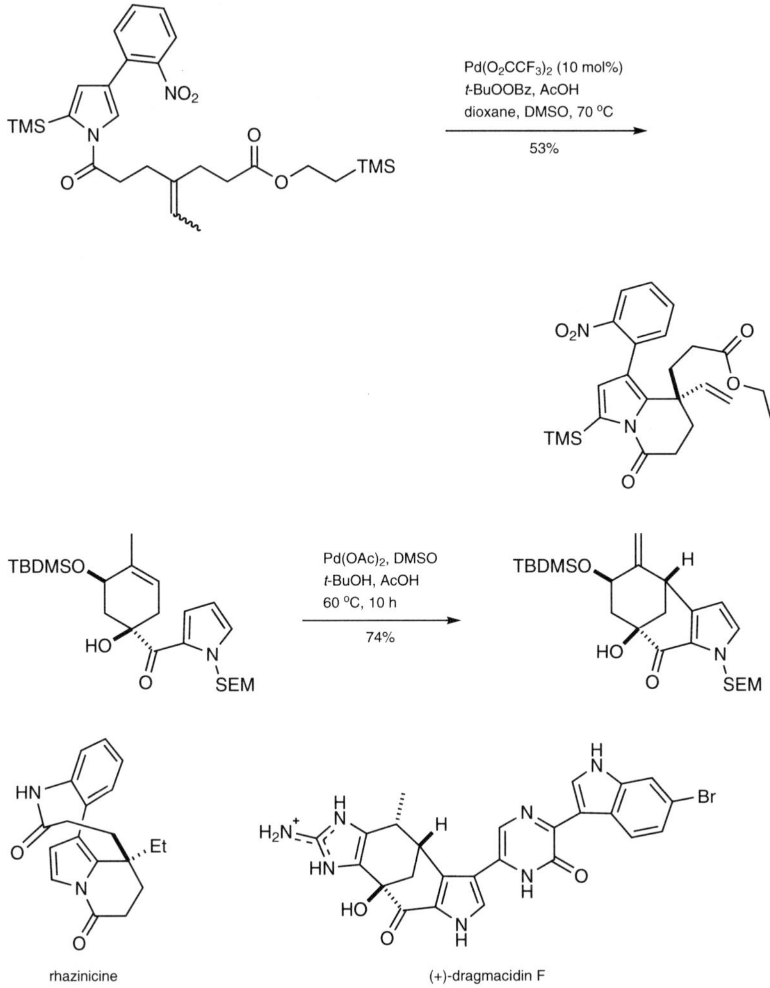

$Pd(O_2CCF_3)_2$ (10 mol%)
t-BuOOBz, AcOH
dioxane, DMSO, 70 °C
53%

$Pd(OAc)_2$, DMSO
t-BuOH, AcOH
60 °C, 10 h
74%

rhazinicine

(+)-dragmacidin F

3-Ethyl-4-methyl-3-vinyl-1,2,3,4-tetrahydrocyclopenta[*b*]indole (Table 5, Entry 2):[70]
A flame-dried, 25-mL, round-bottomed flask equipped with a magnetic stirrer bar was charged with $Pd(OAc)_2$ (17.2 mg, 0.0769 mmol), 2-methylbutan-2-ol (5.15 mL), AcOH (1.54 mL), and ethyl nicotinate (42.0 μL, 0.308 mmol, 0.400 equiv), sequentially. The flask was evacuated and backfilled with O_2 (balloon), heated to 80 °C, and stirred under O_2 (1 atm, balloon) for 10 min. A soln of 3-(3-ethylpent-3-enyl)-1-methylindole (174 mg, 0.769 mmol) in 2-methylbutan-2-ol (1.00 mL) was then added via syringe, and the mixture was stirred under O_2 for 18 h. Filtration of the mixture through a pad of silica gel, concentration, and purification of the oil by flash chromatography (hexanes/CH_2Cl_2 9:1) afforded a clear oil; yield: 123 mg (74%).

4,6-Dimethoxy-3-methyl-3-prop-1-enyl-2,3-dihydrobenzo[*b*]furan (Table 6, Entry 3):[71]
A flame-dried, 2-dram vial equipped with a magnetic stirrer bar was charged with $Pd(OAc)_2$ (11.3 mg, 0.0500 mmol), followed by ethyl nicotinate (13.8 μL, 0.100 mmol), 1,3-dimethoxy-5-(2-methylpent-2-enyloxy)benzene (118 mg, 0.500 mmol), NaOAc (8.2 mg, 0.100 mmol), and a mixture of 2-methylbutan-2-ol and AcOH (4:1; 5.00 mL). The resulting mixture was stirred at 23 °C for 2 min, and then benzo-1,4-quinone (54.1 mg, 0.500 mmol)

was added. The mixture was heated at 100 °C for 30 h and then cooled, filtered through a short plug (0.6 × 5 cm) of silica gel (eluting with Et_2O), concentrated, and purified by flash chromatography (silica gel) to afford a pale yellow oil; yield: 68 mg (58%).

1.1.2 Intramolecular Nozaki–Hiyama–Kishi Coupling Reactions

The development and application of organochromium(III) species in organic synthesis started in the 1970s. Nozaki, Hiyama, and co-workers demonstrated that by reacting allyl and vinyl halides with chromium(II) chloride under aprotic and oxygen-free conditions organochromium(III) reagents could be generated, which react as nucleophiles with aldehydes to give allylic and homoallylic alcohols (Nozaki–Hiyama reaction).[75,76] Kishi[77] and Nozaki[78] independently discovered that nickel(II) catalyzes the formation of C—Cr(III) bonds. This chromium(II)/nickel(II)-mediated "Barbier type" reaction is known as the Nozaki–Hiyama–Kishi (NHK) reaction (Scheme 19), and has been one of the most powerful tools for constructing carbon–carbon bonds both inter- and intramolecularly in organic synthesis, especially in natural product synthesis. The nucleophilic addition of organochromium(III) reagents to a carbonyl compound produces stable chromium alkoxide [O—Cr(III)] species, therefore NHK reactions normally require stoichiometric chromium(II) [>2 equiv of Cr(II) per 1 equiv of organic halide). Fürstner and co-workers developed a chromium-catalyzed process by adding a chlorosilane to release the metal salt from the chromium alkoxide, and manganese powder to reduce the released chromium(III) and regenerate chromium(II).[79–81]

Scheme 19 The Nozaki–Hiyama–Kishi Reaction

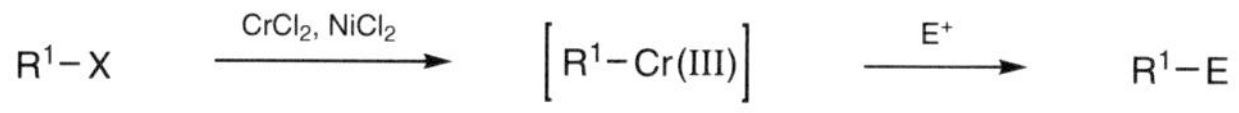

$$R^1{-}X \xrightarrow{CrCl_2,\ NiCl_2} \left[R^1{-}Cr(III)\right] \xrightarrow{E^+} R^1{-}E$$

R^1 = alkenyl, aryl, allyl, vinyl, propargyl, alkynyl, allenyl; X = Cl, Br, I, OTf

As shown in Scheme 20, the nickel(II)-catalyzed NHK reaction normally involves: (1) the reduction of nickel(II) to nickel(0); (2) oxidative addition to generate an organonickel species R^1—Ni(II)—X; (3) transmetalation with chromium(III) to form the organochromium(III) nucleophile and regenerate nickel(II); and (4) nucleophilic addition to the carbonyl compound to form the addition product (R^1—E). Alternatively, the chromium-catalyzed reaction proceeds with catalytic chromium(II) or chromium(III) to generate organochromium(III) reagents [R^1—Cr(III)X_2]. After the nucleophilic addition to the carbonyl compound, ligand exchange occurs through the σ-bond metathesis of the chlorosilane with the chromium alkoxide. The liberated chromium(III) species can be reduced to chromium(II) with manganese.

Scheme 20 General Mechanism of the Nozaki–Hiyama–Kishi Reaction

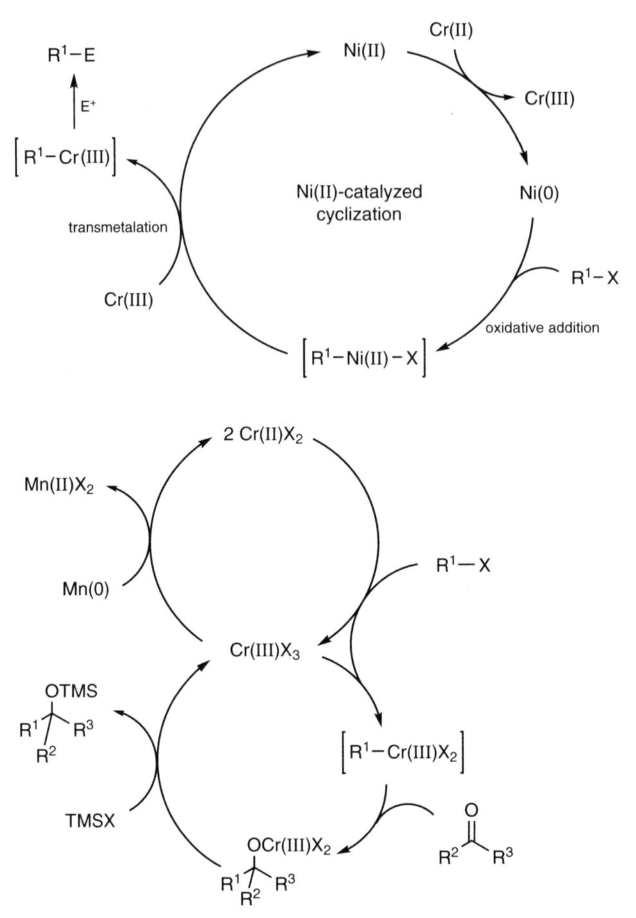

1.1.2.1 Nickel-Catalyzed Intramolecular Alkenylchromium Cyclization and Applications

Nickel-catalyzed alkenylchromium cyclizations have been used for the construction of cyclic compounds, from common rings to medium-sized and macrocyclic rings. Treatment of alkenyl iodide **29** with 5 equivalents of chromium(II) chloride in the presence of a catalytic amount of nickel(II) chloride in dimethylformamide at room temperature for 40 hours gives the cyclized product **30** in 79% yield as a single diastereomer (Scheme 21).[82] The diastereoselectivity is controlled by a chair-like transition state **31**, in which the benzyloxy and the chromium(III) alkoxide groups must be antiperiplanar to avoid an unfavorable allylic 1,3-strain between the quasi-equatorial chromium alkoxide and the alkene. Investigation of nickel-catalyzed alkenylchromium cyclization using model substrates, as shown in Table 7, demonstrated the importance of steric effects on the cyclization.[82]

Scheme 21 Nickel-Catalyzed Intramolecular Cyclization of an Alkenyl Iodide with an Aldehyde[82]

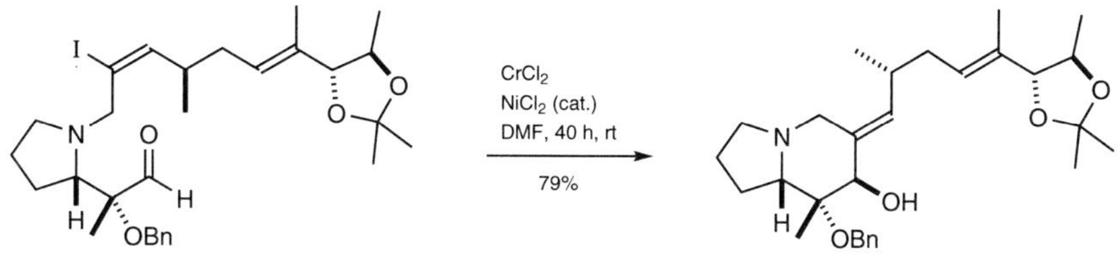

CrCl$_2$
NiCl$_2$ (cat.)
DMF, 40 h, rt
79%

29 → **30**

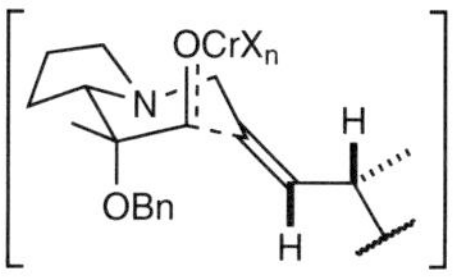

31

Table 7 Nickel-Catalyzed Intramolecular Alkenylchromium Cyclization[82]

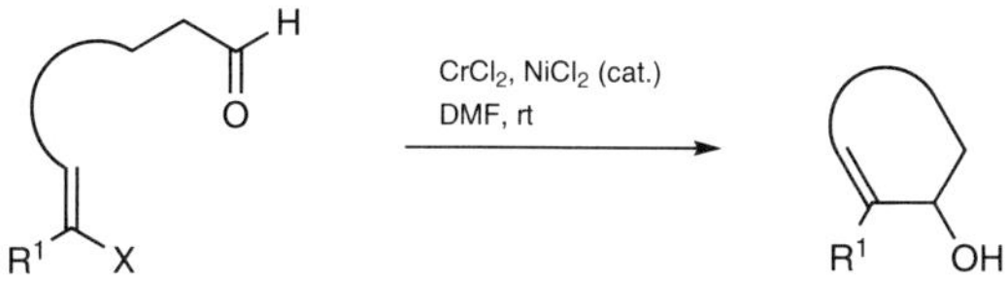

CrCl$_2$, NiCl$_2$ (cat.)
DMF, rt

Entry	Substrate	Product(s)	Yield (%)	Ref
1	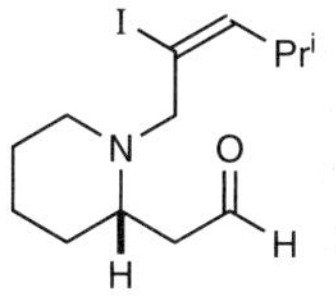	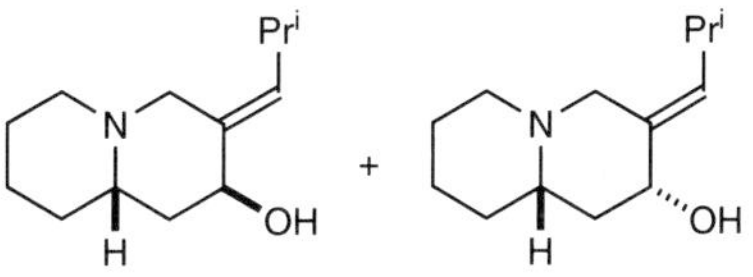 + (1.3:1)	81	[82]
2	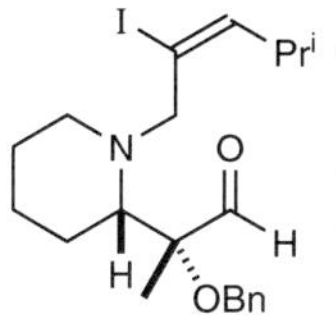	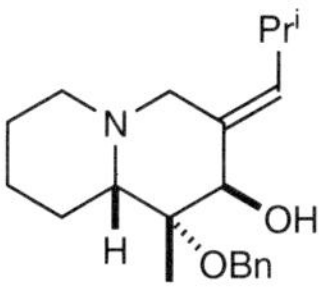	53	[82]
3	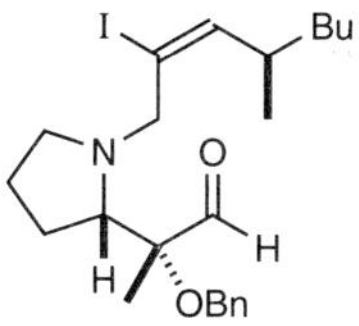	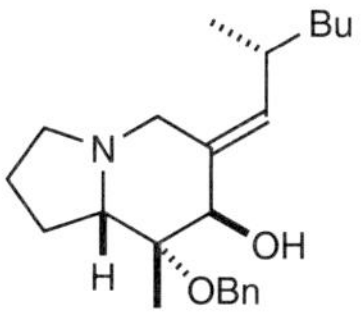	53	[82]

The nickel(II)-catalyzed alkenylchromium cyclization provides a reliable and selective ring-closure protocol for the preparation of different sized rings, which has been widely used in the synthesis of natural products. A Nozaki–Hiyama–Kishi cyclization mediated by nickel(II) chloride/chromium(II) chloride in dimethyl sulfoxide between vinyl iodide and aldehyde moieties was used in a synthesis of nigellamine A$_2$ by Ready and co-workers.[83] The 11-membered ring **32** is formed in 73% yield and with high diastereoselectivity (dr >10:1) after 6 hours of sonication at room temperature to 60 °C (Scheme 22). Curran and co-workers successfully developed a convergent approach for the total synthesis of (−)-dictyostatin.[84] It was found that the reaction of substrate **33** with excess chromium(II) chloride and 4,4′-di-*tert*-butyl-2,2′-dipyridyl (15 equiv of each) and a nickel(II) chloride complex [$NiCl_2$(dppf); 0.2 equiv] in tetrahydrofuran at room temperature gives the desired product **34** (43% yield) along with its C9 epimer (12% yield). A 22-membered macrolactone is thus efficiently constructed using this strategy (Scheme 22).

Scheme 22 Synthesis of Nigellamine A_2 and (−)-Dictyostatin[83,84]

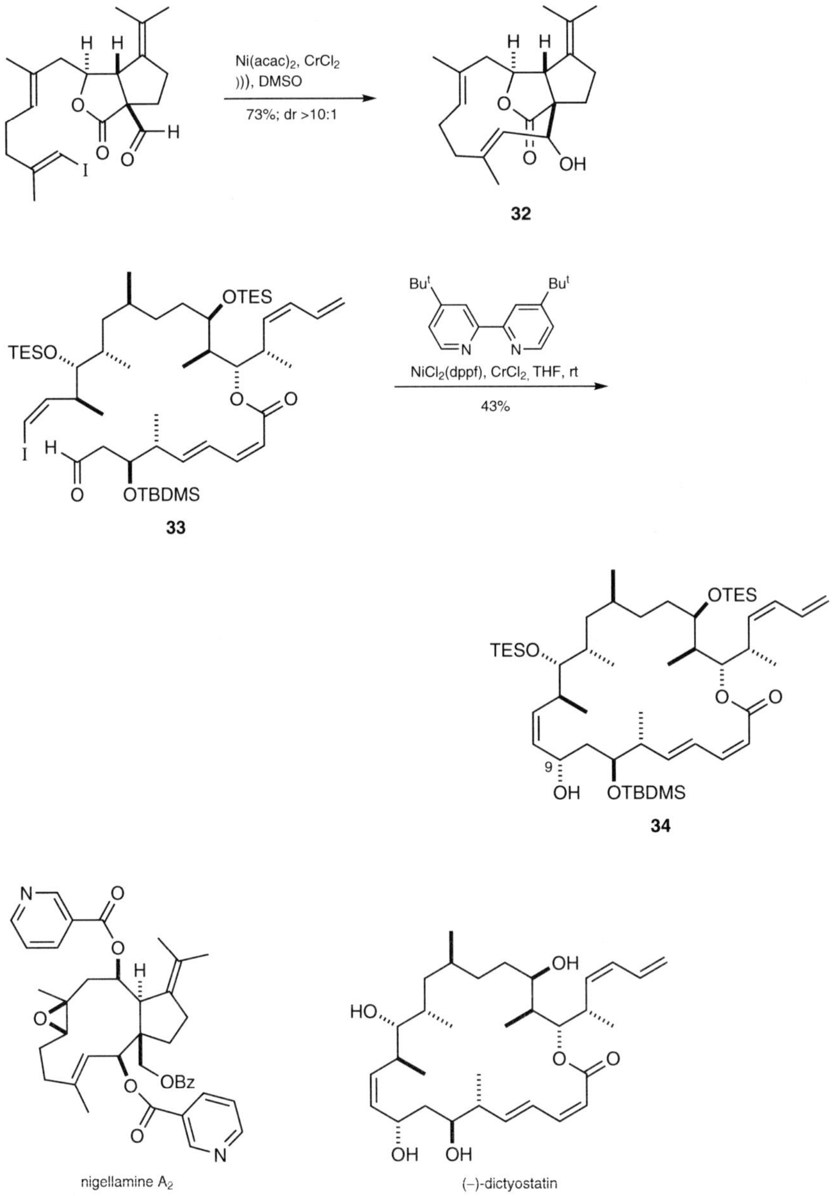

Ni(acac)$_2$, CrCl$_2$
))), DMSO
73%; dr >10:1

32

33

NiCl$_2$(dppf), CrCl$_2$, THF, rt
43%

34

nigellamine A_2

(−)-dictyostatin

(7*R*,8*R*,8a*S*,*E*)-8-(Benzyloxy)-8-methyl-6-{(*R*,*E*)-2-methyl-5-[(4*R*,5*R*)-2,2,5-trimethyl-1,3-dioxolan-4-yl]hex-4-enylidene}octahydroindolizin-7-ol (30):[82]
A suspension of $CrCl_2$ (161 mg, 1.31 mmol) and $NiCl_2$ (1 mg, 0.0077 mmol) in DMF (3 mL) was stirred at rt under argon for 10 min. To this mixture was added a soln of substrate **29** (156 mg, 0.262 mmol) in DMF (2 mL). The resulting mixture was stirred at rt for 40 h. After addition of sat. aq $NaHCO_3$ (20 mL), the mixture was extracted with EtOAc (3 × 20 mL) and the extract was washed with brine, dried ($MgSO_4$), and concentrated. The residue was purified by chromatography (silica gel); yield: 97 mg (79%).

1.1.2.2 Nickel-Catalyzed Intramolecular Alkynylchromium Cyclization and Applications

Normally, nickel-catalyzed intramolecular alkynylchromium cyclization is used for the construction of macrocyclic rings due to the highly strained cycloalkyne products. Treatment of alkynyl halides with chromium(II) chloride in the presence of an aldehyde leads to an efficient nucleophilic addition with high chemoselectivity. In close analogy to alkenyl-, aryl-, and allylchromium cyclization, a high preference for additions to aldehydes is observed, leaving unprotected ketone, ester, or cyano functions, etc. untouched. The reliability of this reaction is highly dependent on the aprotic solvent and oxygen-free conditions.

Alkynyl iodide **35** reacts with chromium(II) chloride in the presence of catalytic nickel(II) chloride, leading, after 3 hours, to the formation of a 12-membered ring in 63% yield as a mixture of two diastereomers (Scheme 23).[85] The cyclized product **36** has been used in the synthesis of phomactin A. An 11-membered ring containing a strained cycloalkyne moiety is efficiently constructed by the slow addition of the iodoalkyne **37** (1.2 mmol) in tetrahydrofuran (0.015 M) to a suspension of chromium(II) chloride (0.16 M) in tetrahydrofuran over a period of 6 hours, producing **38** in 88% yield. This product was used in the synthesis of *epi*-illudol (Scheme 23).[86]

Scheme 23 Nickel-Catalyzed Intramolecular Alkynylchromium Cyclization[85,86]

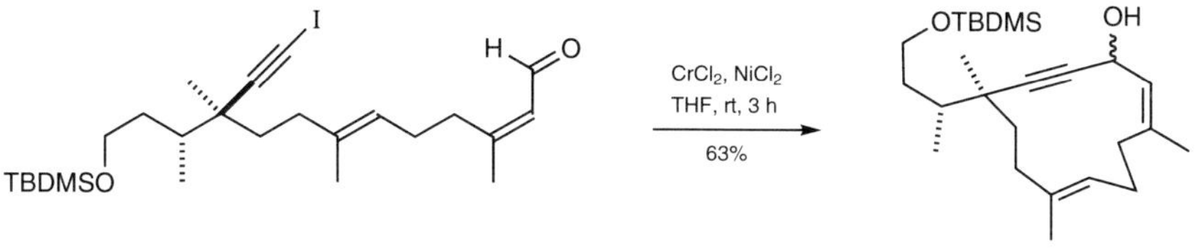

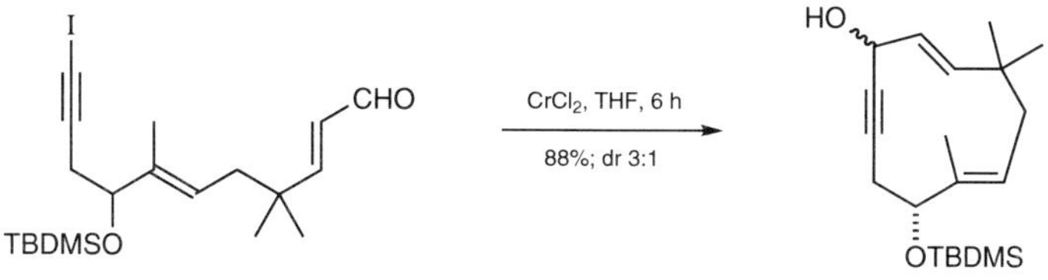

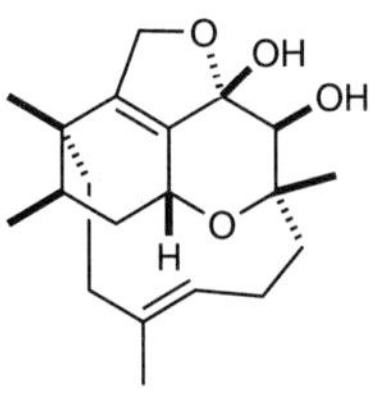

phomactin A

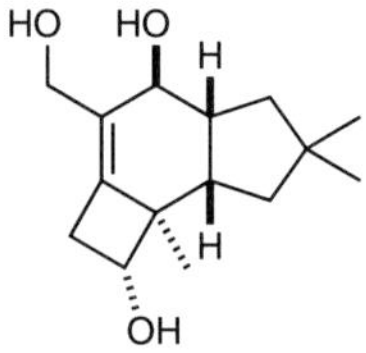

epi-illudol

(2*Z*,6*E*,10*S*)-10-[(*R*)-4-(*tert*-Butyldimethylsiloxy)butan-2-yl]-3,7,10-trimethylcyclododeca-2,6-dien-11-yn-1-ol (36):[85]

A soln of iodoalkyne **35** (295 mg, 0.556 mmol) in THF (6.3 mL) was slowly added, over 3 h, to a vigorously stirring soln of $CrCl_2$ (509 mg, 4.1 mmol) and $NiCl_2$ (0.07 mg, 0.055 mmol, as reported) in THF (44.8 mL). After approximately 3 h, the mixture was quenched with sat. NH_4Cl soln (10 mL) and extracted with Et_2O (3 × 50 mL). The extracts were washed with NaS_2O_3 (2 × 30 mL), H_2O (2 × 30 mL), and brine (2 × 30 mL), dried ($MgSO_4$), and concentrated. The resulting residue was purified by flash chromatography (silica gel, hexanes/EtOAc 95:5) to give the macrocycle **36** as a mixture of diastereomers; yield: 142 mg (63%).

1.1.2.3 Intramolecular Allylchromium Cyclization and Applications

Allylchromium species are readily available through the chromium(II) chloride induced reaction of allylic halides or 4-toluenesulfonates. The nickel-catalyzed intramolecular allylchromium cyclization provides a valuable tool for the formation of strained medium-sized and macrocyclic rings because of its chemo- and diastereoselectivity. The high stability of the organochromium(III) species serves as the thermodynamic driving force, which overrides the strain build-up during cyclization. When allyl bromide **39** is treated with a mixture of a large excess of chromium(II) chloride and nickel(II) chloride at room temperature for 12 hours, bipinnatin J (**41**) is formed in 59% yield as a major diastereomer (dr >9:1).[87] The highly diastereoselective formation of the 13-membered ring reflects the conformational rigidity of the precursor **39**, which contains a Z double bond and two 1,3-disubstituted five-membered rings of limited flexibility. This product presumably results from a chair-shaped transition state **40** and reflects the stereochemistry of the allylic bromide (Scheme 24).

Scheme 24 Intramolecular Allylchromium Cyclization[87]

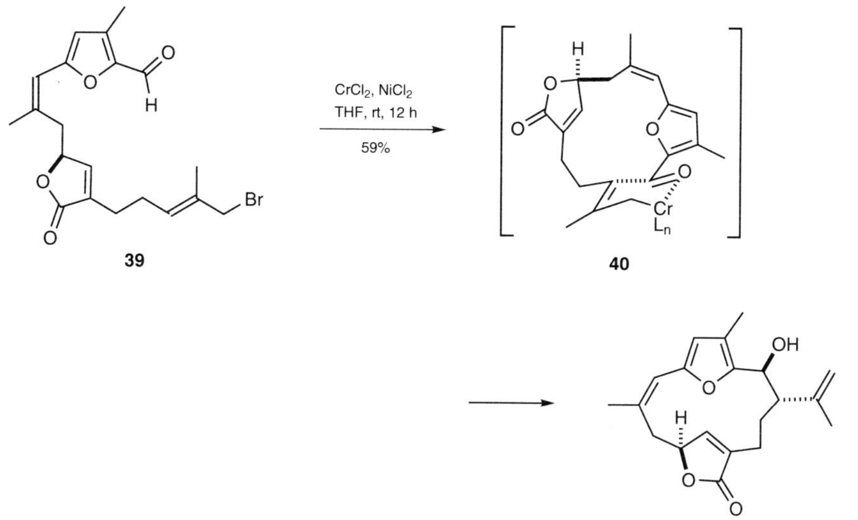

CrCl$_2$, NiCl$_2$
THF, rt, 12 h
59%

39

40

41 bipinnatin J

Bipinnatin J (41):[87]
To a suspension of CrCl$_2$ (438 mg, 3.56 mmol), NiCl$_2$•DME (196 mg, 0.891 mmol), and 4-Å molecular sieves (1.49 g) in THF (55 mL) was added a soln of allyl bromide **39** (121 mg, 0.297 mmol) in THF (5 mL) using a syringe pump over 1.5 h. After 12 h at rt, H_2O (20 mL) was added. The biphasic soln was filtered and concentrated to about 30 mL. The layers were separated, and the aqueous layer was extracted with Et_2O (25 mL). The combined organic layers were dried, filtered, and concentrated under reduced pressure. The crude oil was purified by flash column chromatography to give a white powder; yield: 58 mg (59%).

1.1.3 Copper-Catalyzed Intramolecular Coupling Reactions

Copper-promoted coupling reactions, including C—C and C—heteroatom (O, N, S, etc.) bond formations, can be traced back to the crucial work of Ullmann and Goldberg over a hundred years ago, in 1901.[88–90] These pioneering contributions laid the foundation for today's development and application of these reactions. This section is focused on the copper-catalyzed cyclization and intramolecular coupling between vinyl halides and organotin reagents or terminal alkynes to construct C—C bonds, which could be employed as beneficial complementary methods to the corresponding palladium-catalyzed Stille (Section 1.1.1.2) and Sonogashira couplings (Section 1.1.1.4). The classical Ullmann cross coupling (Ullmann and Goldberg reaction) and copper-promoted oxidative coupling normally require stoichiometric quantities of the copper reagent and will not be covered in this section. Copper-catalyzed C—N and C—O bond formations are discussed in Section 1.8.

1.1.3.1 Copper-Catalyzed Intramolecular Ene–Yne Coupling Reactions and Applications

Construction of C(sp)—C(sp^2) bonds using aryl or vinyl halides with aryl- or alkyl-substituted alkynes can be realized through a palladium/copper cocatalyzed system or a copper-promoted coupling. The original coupling normally requires stoichiometric copper(I) and high temperature due to the insolubility of the copper salts in organic solvents. Miura and co-workers modified this reaction and developed a copper/triphenylphosphine catalyst system for the coupling reaction of aryl iodides with terminal alkynes in the presence of potassium carbonate.[91] The addition of triphenylphosphine is crucial for the reaction to proceed catalytically, with the corresponding arylated alkynes obtained in excellent yields. Ma and co-workers have reported a copper(I) iodide/*N*,*N*-dimethylglycine catalyzed coupling reaction of aryl halides with terminal alkynes in dimethylformamide (100 °C), which provides a great diversity of coupling products in good to excellent yields.[92] Coleman and co-workers have applied the intramolecular coupling reaction of a vinyl iodide with a terminal alkyne using Miura's modification in the construction of the core skeleton of oximidines I and II.[93] The synthesis of the 12-membered cyclic enyne lactone was achieved, together with a 10–15% yield of a dimeric product. Related cyclic enyne lactones such as **42** can be realized by the same catalytic system in modest yield (Scheme 25). Notably, the traditional palladium/copper cocatalyzed system (Sonogashira reaction) does not give the desired coupling product.

Scheme 25 Intramolecular Coupling of Vinyl Iodides with Terminal Alkynes[93]

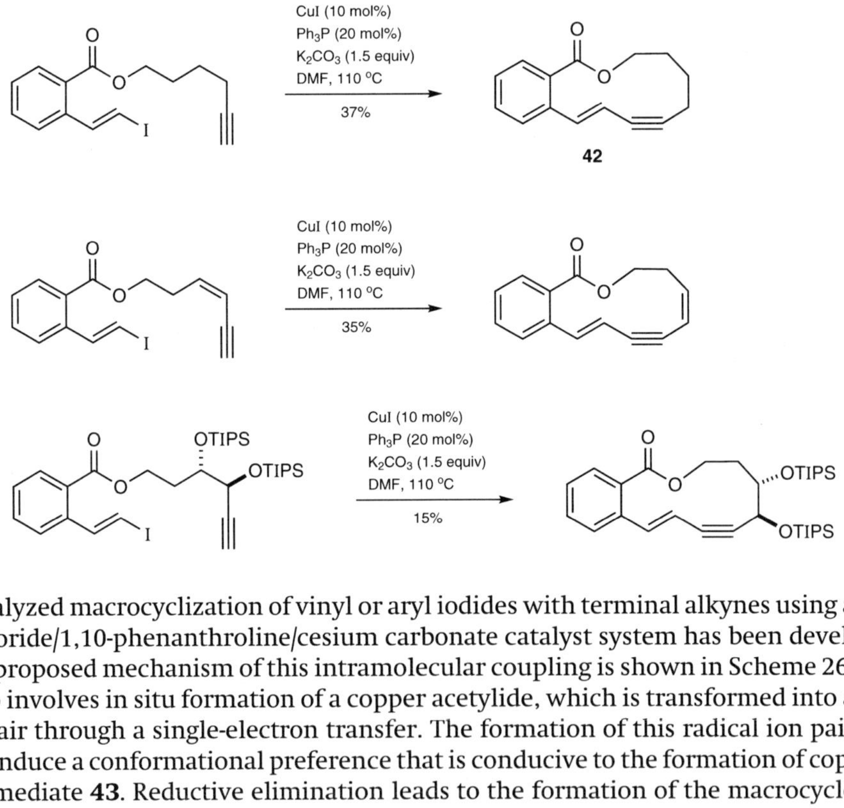

CuI (10 mol%)
Ph_3P (20 mol%)
K_2CO_3 (1.5 equiv)
DMF, 110 °C
37%
42

CuI (10 mol%)
Ph_3P (20 mol%)
K_2CO_3 (1.5 equiv)
DMF, 110 °C
35%

OTIPS
OTIPS
CuI (10 mol%)
Ph_3P (20 mol%)
K_2CO_3 (1.5 equiv)
DMF, 110 °C
15%
OTIPS
OTIPS

A copper-catalyzed macrocyclization of vinyl or aryl iodides with terminal alkynes using a copper(I) chloride/1,10-phenanthroline/cesium carbonate catalyst system has been developed.[94] The proposed mechanism of this intramolecular coupling is shown in Scheme 26. The first step involves in situ formation of a copper acetylide, which is transformed into a radical ion pair through a single-electron transfer. The formation of this radical ion pair may help to induce a conformational preference that is conducive to the formation of copper(III) intermediate **43**. Reductive elimination leads to the formation of the macrocycle and regenerates the active copper(I) catalyst. Notably, this macrocyclization can be performed at relatively high concentrations (24 mM) without a slow-addition procedure. Decreasing the concentration leads to very slow reactions and low yields of the desired macrocycle. Interestingly, the palladium/copper cocatalyzed intramolecular Sonogashira reaction requires high catalyst loadings and provides low yields. Based on this method, a range of macrocycles **44** with various ring sizes (10–25-membered rings) have been prepared using 20–30 mol% of ligand (bathophenanthroline or 1,10-phenanthroline) (Table 8).

Scheme 26 Proposed Mechanism of the Intramolecular Coupling of Vinyl or Aryl Iodides with Terminal Alkynes[94]

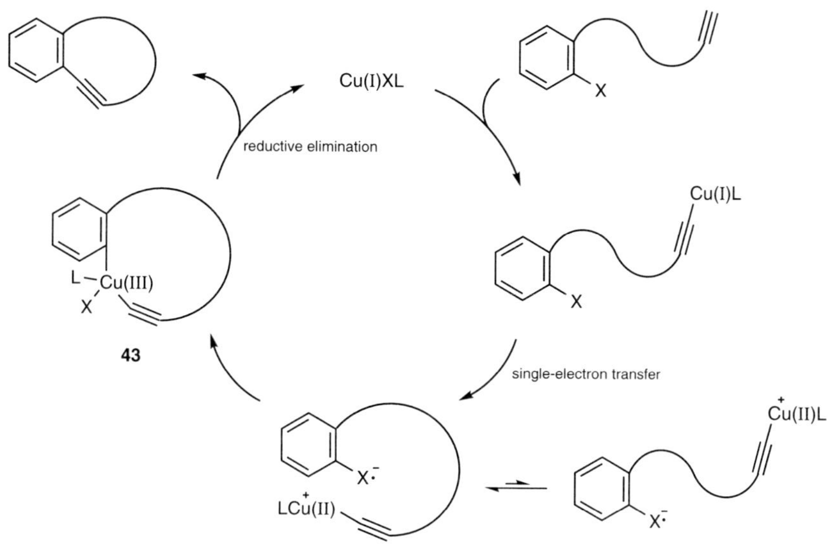

Table 8 Copper-Catalyzed Intramolecular Coupling of Aryl and Vinyl Iodides with Terminal Alkynes[94]

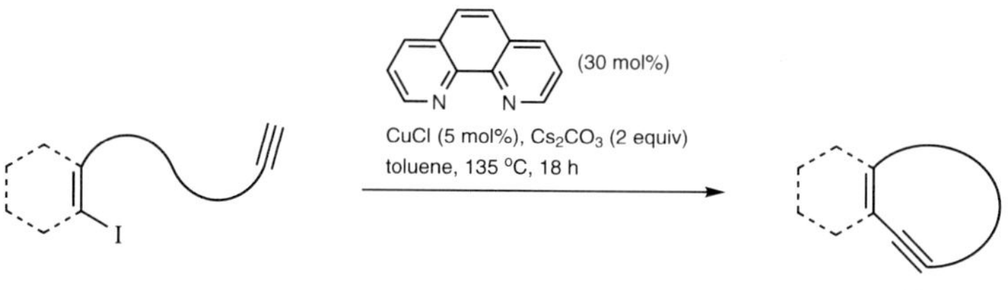

CuCl (5 mol%), Cs_2CO_3 (2 equiv), 1,10-phenanthroline (30 mol%), toluene, 135 °C, 18 h

Substrate	Product	Yield (%)	Ref
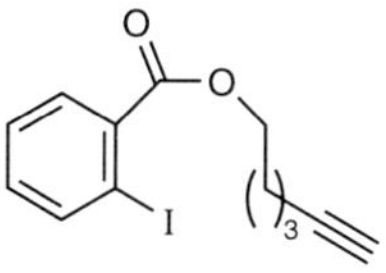	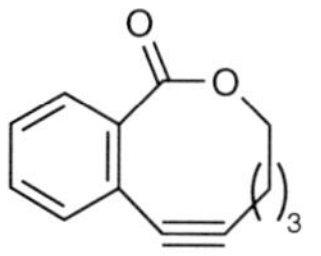	46	[94]
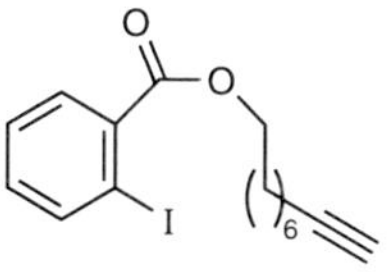	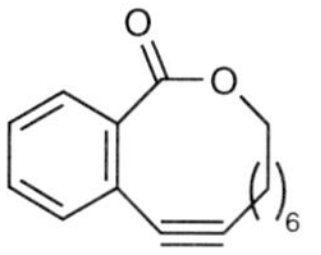	83	[94]
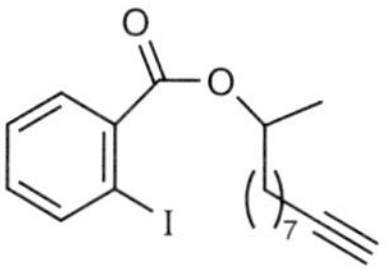	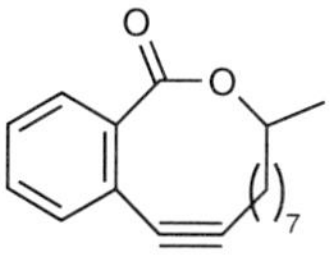	78	[94]

for references see p 51

Table 8 (cont.)

Substrate	Product	Yield (%)	Ref
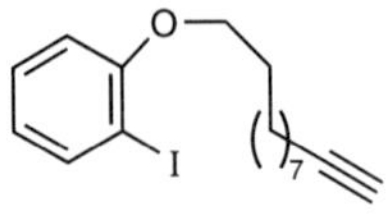	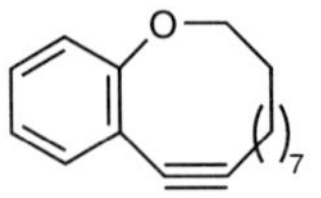	71	[94]
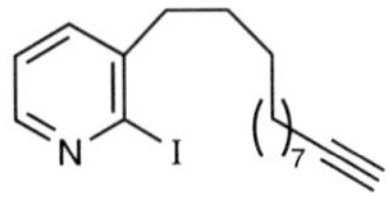	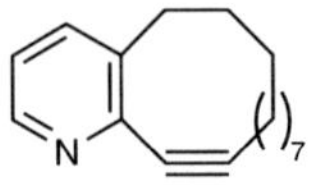	58	[94]
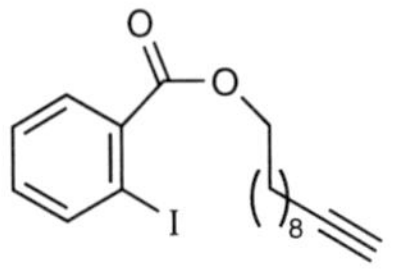	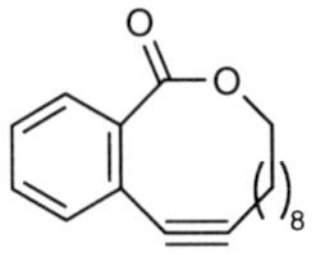	81[a]	[94]
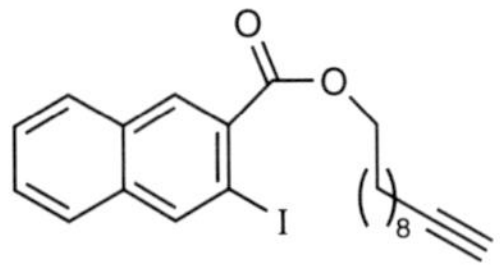	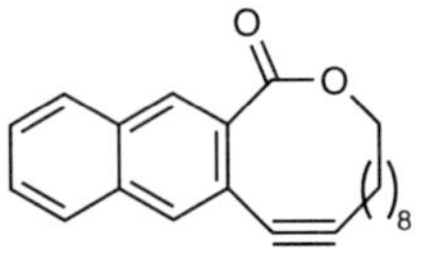	71	[94]
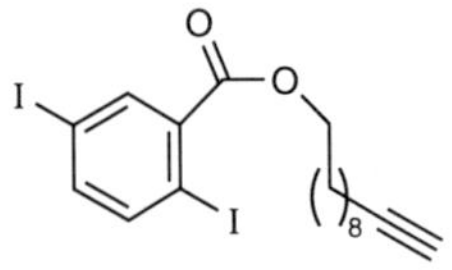	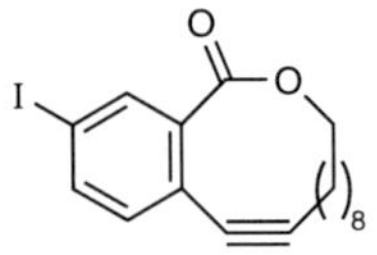	77	[94]
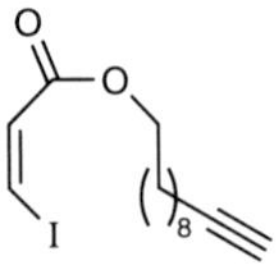	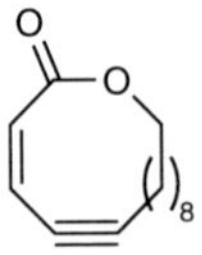	38	[94]
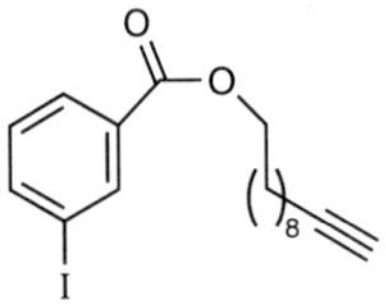	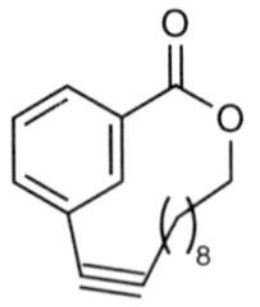	72	[94]
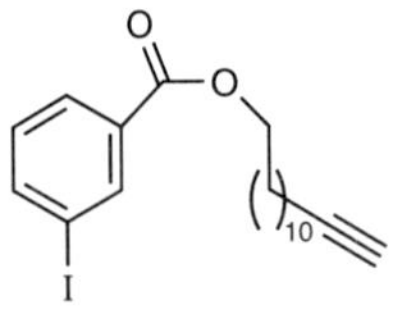	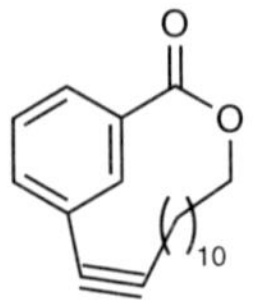	62	[94]

Table 8 (cont.)

Substrate	Product	Yield (%)	Ref
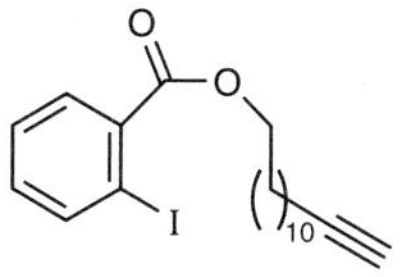	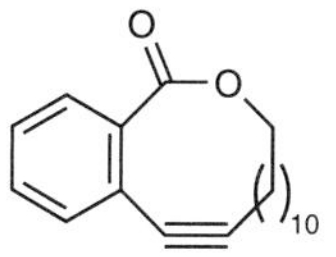	66[a]	[94]
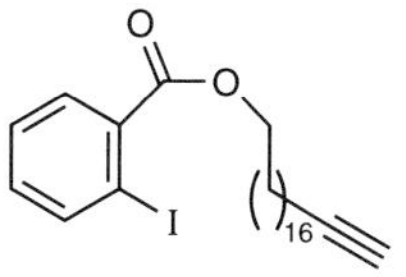	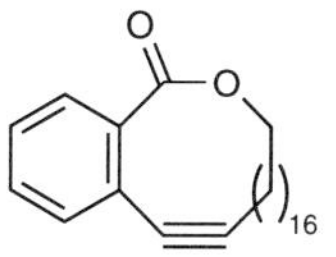	65[a]	[94]
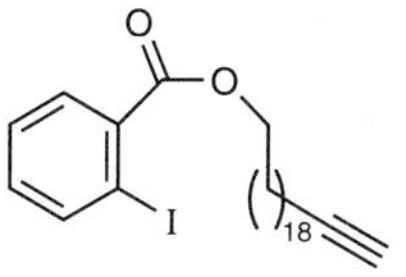	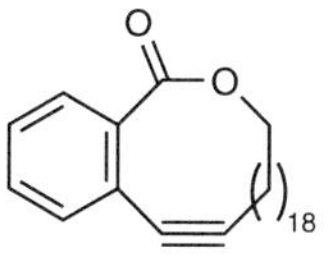	61[a]	[94]

[a] Using 20 mol% of ligand.

Cyclic Enyne Lactone 42:[93]
Hex-5-ynyl (*E*)-2-(2-iodovinyl)benzoate (1.0 mmol) was added to a mixture of CuI (0.1 mmol), Ph_3P (0.2 mmol), and K_2CO_3 (1.5 mmol) in DMF (25 mL) under N_2. The mixture was stirred at 110 °C for 26 h. Then, it was poured into H_2O (20 mL) and the mixture was extracted with Et_2O (3 × 20 mL). The combined organic extracts were dried ($MgSO_4$), filtered, and concentrated under reduced pressure. The residue was purified by flash chromatography [silica gel (5 × 20 cm), EtOAc/hexane 0:100 to 15:85]; yield: 37%.

Macrocycles 44:[94]
To a dried, sealable tube equipped with a stirrer bar was added CuCl (0.006 mmol, 5 mol%), 1,10-phenanthroline (0.024 mmol, 20 mol%), and Cs_2CO_3 (0.24 mmol, 2 equiv). A soln of the alkynyl iodide substrate (0.12 mmol, 1 equiv) in toluene (5 mL) was added to the sealable tube in one portion. The aperture of the tube was covered with a rubber septum and the mixture was bubbled with N_2 for 10 min. The septum was then replaced by a Teflon-coated screw cap and the sealed tube was placed in a preheated oil bath at 135 °C. After being stirred for 18 h, the mixture was cooled to rt and diluted with EtOAc. The resulting soln was filtered through a pad of Celite and was concentrated under reduced pressure to provide a crude mixture, which was purified by column chromatography (silica gel, hexanes/Et_2O 10:0 to 9:1).

1.1.3.2 Copper-Catalyzed Stille-Type Cross Coupling and Applications

Piers and co-workers have reported the copper(I) chloride mediated intramolecular coupling of vinylstannane and vinyl halide moieties.[95] As shown in Table 9, entry 1, when the iodo trimethylstannane is treated with 2.2 equivalents of copper(I) chloride in dimethylformamide (62 °C) for 3 minutes, the conjugated diene is obtained in 80% yield. Other sources of copper(I) (e.g., CuBr•SMe_2, CuI) are generally inferior (lower reaction rates and/or more side products) to copper(I) chloride. Using more than two equivalents of copper(I) chloride increases the reaction rate and leads to better yields. This coupling is stereospecific, with no isomerization of the alkenes observed. Additionally, the copper(I) chloride mediated reactions are generally faster and, in some instances, cleaner than the corresponding palladium(0)-catalyzed processes. This copper(I)-promoted intramolecular coupling reaction is particularly mild, with good functional-group tolerance and stereoselectivity. Notably, four-membered fused rings with a high degree of ring strain can be efficiently constructed using this method.

Table 9 Copper(I) Chloride Promoted Stille-Type Cross Coupling[95]

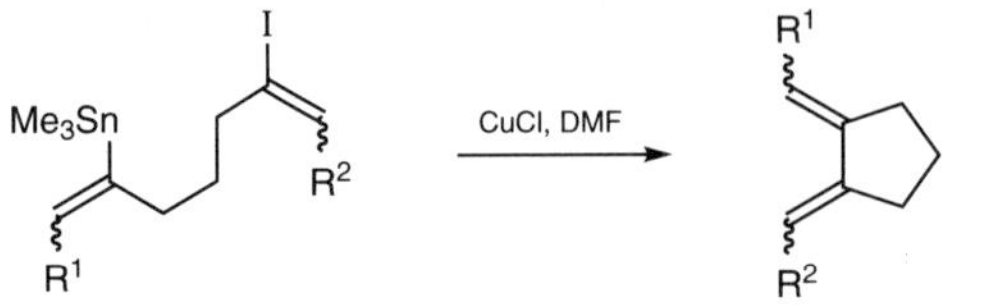

CuCl, DMF

Entry	Substrate	CuCl (Equiv)	Temp (°C)	Time (min)	Product	Yield (%)	Ref
1	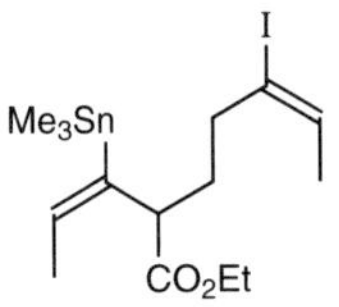	2.2	62	3	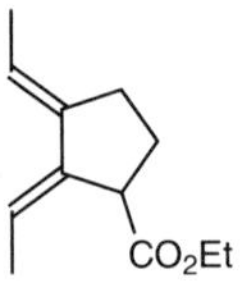	80	[95]
2	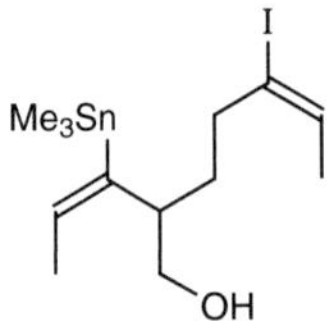	2.5	70	10	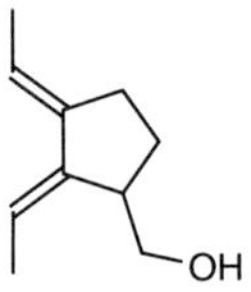	86	[95]
3	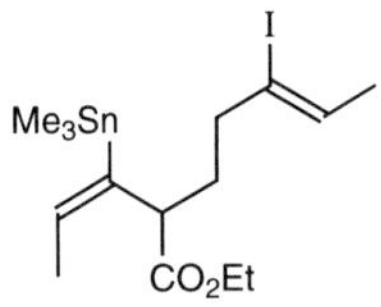	2.1	60	2	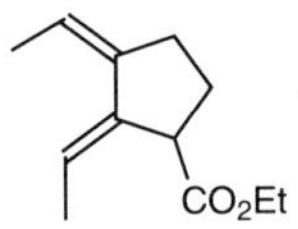	83	[95]
4	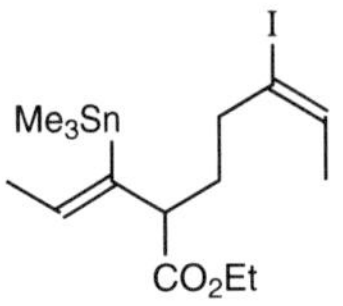	2.3	23	10	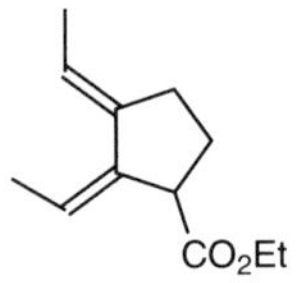	94	[95]
5	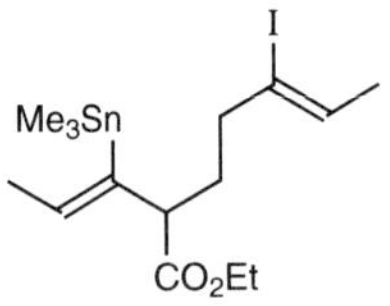	3.0	60	2	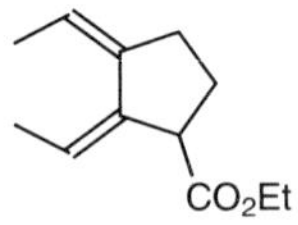	89	[95]
6	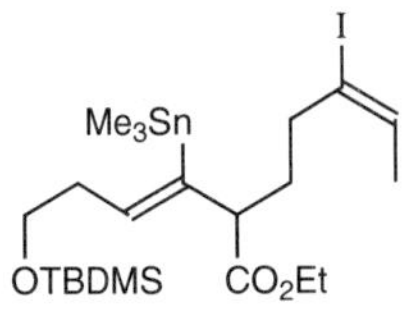	3.0	64	2	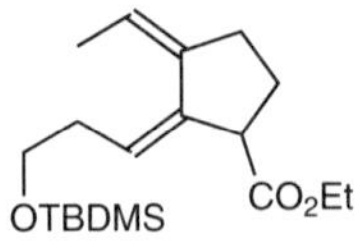	98	[95]

Table 9 (cont.)

Entry	Substrate	CuCl (Equiv)	Temp (°C)	Time (min)	Product	Yield (%)	Ref
7	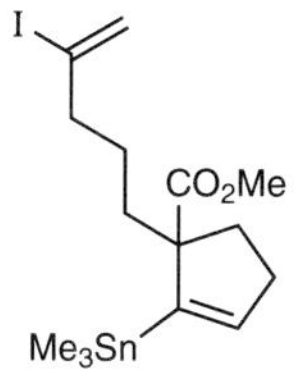 I; CO_2Me; Me_3Sn	2.5	90	5	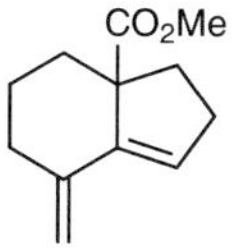 CO_2Me	75	[95]
8	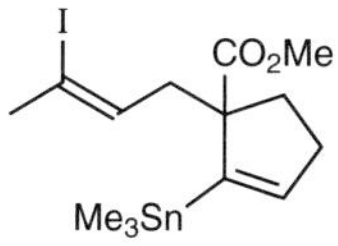 I; CO_2Me; Me_3Sn	2.1	60	5	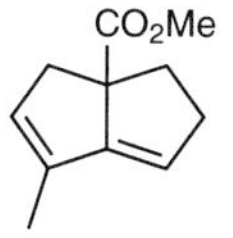 CO_2Me	93	[95]
9	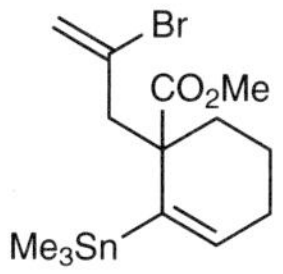 Br; CO_2Me; Me_3Sn	2.4	60	5	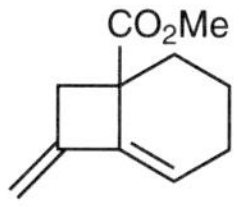 CO_2Me	78	[95]

Copper(I) thiophene-2-carboxylate (CuTC) has been proven to be an effective reagent to promote rapid Stille cross-coupling reactions under mild conditions in the absence of palladium catalysis. Paterson and co-workers developed the copper(I) thiophene-2-carboxylate promoted Stille coupling as a key cyclodimerization in the total synthesis of the macrocyclic C_2 symmetric core of elaiolide.[96] As shown in Scheme 27, treatment of a 0.01 M soln of monomer **45**, in 1-methylpyrrolidin-2-one with copper(I) thiophene-2-carboxylate (10 equiv) at room temperature for 15 min produces the required 16-membered macrocycle **46** as a white crystalline solid in 80% yield (88% based on the C7 regioisomer), accompanied by traces of other macrocycles. Dilute conditions (0.01M) are required in this reaction in order to achieve higher selectivity in favor of the cyclodimer.

Scheme 27 Synthesis of Elaiolide Using a Copper(I) Thiophene-2-carboxylate Promoted Stille Coupling[96]

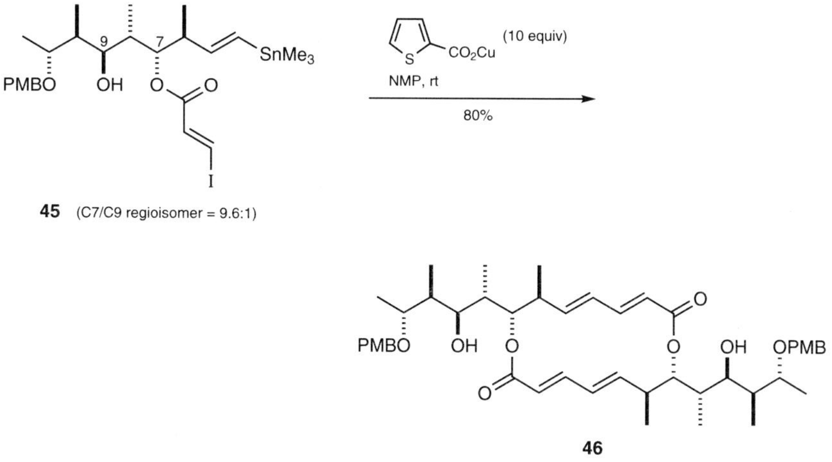

for references see p 51

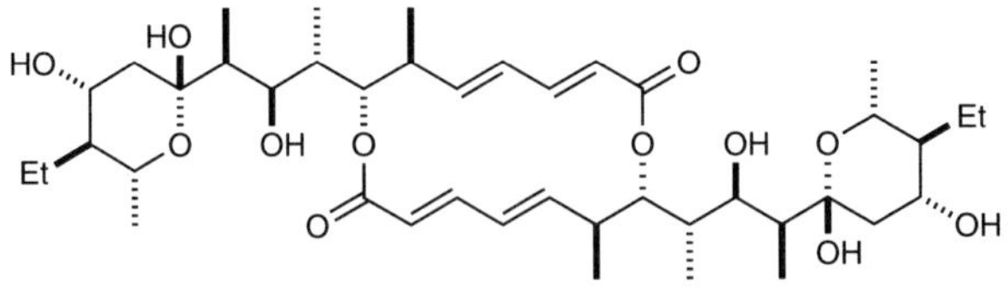

elaiolide

Ethyl (Z,Z)-2,3-Diethylidenecyclopentanecarboxylate (Table 9, Entry 5):[95]
A stirred soln of ethyl (Z)-5-iodo-2-[(Z)-1-(trimethylstannyl)prop-1-enyl]hept-5-enoate (126.5 mg, 0.26 mmol) in dry DMF (2.6 mL) was heated to 60 °C (oil bath) for 15 min. Powdered CuCl (77.8 mg, 0.786 mmol) was added and the mixture was stirred at 60 °C for 2 min. The reaction flask was removed from the oil bath, aq NH_4Cl/NH_4OH (pH 8; 7.5 mL) was added, and the mixture was opened to the atmosphere. After a short period of time, Et_2O (10 mL) was added and the mixture was stirred until the aqueous phase became deep blue (ca. 15 min). The phases were separated and the aqueous layer was extracted with Et_2O (3 × 10 mL). The combined organic extracts were washed with brine (10 mL), dried ($MgSO_4$), and concentrated. Flash chromatography [silica gel (10 g), petroleum ether/Et_2O 40:1] of the crude material and distillation (45–70 °C/0.6 Torr) of the obtained liquid afforded a colorless oil; yield: 45.2 mg (89%).

1.1.4 Nickel-Catalyzed Intramolecular Reductive Coupling between Organo Halides

The cross-coupling reactions of organometallic reagents with organo halides have been well established and applied in organic synthesis. Normally, the required organometallic reagents need to be prepared from the corresponding organic halides in a separate or one-pot operation. Alternatively, the reductive cross coupling between organic halides provides a new solution for this kind of transformation. Nickel is more nucleophilic, due to its smaller atomic radius, compared with palladium, which facilitates the oxidative addition of "unreactive" bonds such as C—H, C—O, C—X, and even C—C bonds. The nickel-catalyzed reductive cross-coupling reaction between alkyl halides and aryl halides normally involves: (1) oxidative addition of the alkyl halide (R^1—X) to nickel(0) to give an R^1Ni(II)X species; (2) reduction of the R^1Ni(II)X species by a stoichiometric metal (Zn, Mn, etc.) to give an R^1Ni(I) intermediate; (3) a second oxidative addition of the resulting R^1Ni(I) intermediate with a second halide (R^2—X) to form an R^1R^2Ni(III)X species; (4) reductive elimination to give the desired cross-coupling product R^1—R^2; (5) regeneration of the nickel(0) catalyst upon reduction (Scheme 28). Recent progress in this field has inspired extensive study by the organic chemistry community.[97–109] This section is focused on the newly developed nickel-catalyzed or -promoted cyclization and intramolecular coupling of two organo halides.

Scheme 28 The Mechanism of Nickel-Catalyzed Reductive Coupling

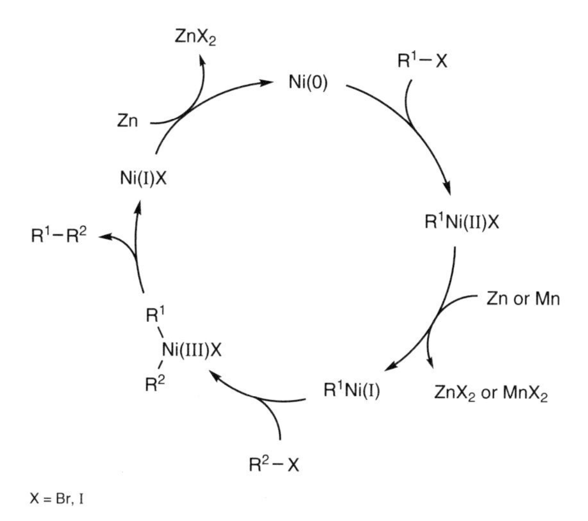

X = Br, I

In 2014, Gong and co-workers reported a nickel-catalyzed intramolecular cyclization of nitrogen- and carbon-tethered dihaloalkanes.[106] Under the optimized catalytic reaction conditions (NiI_2/2,2′-bipyridine), the cyclization of N-tethered dihaloalkane **48** smoothly produces pyrrolidine **49** in 83% yield using zinc as reductant (Table 10, entry 1). According to the general reaction mechanism (Scheme 28), a cyclic nickel(III) intermediate **47** is formed, which undergoes reductive elimination leading to the formation of the desired product. Primary/primary alkyl dihalides and secondary alkyl halides can be used in these reactions, and effectively form five-membered and six-membered nitrogen- or carbon-tethered rings in moderate yields (Table 10, entries 1–13). The construction of a seven-membered ring or a cyclopropane (entries 14 and 15) is less efficient, which may result from undesired intermolecular oligomerization or hydrodehalogenation.

for references see p 51

Table 10 Nickel-Catalyzed Intramolecular Cyclization of Dihaloalkanes[106]

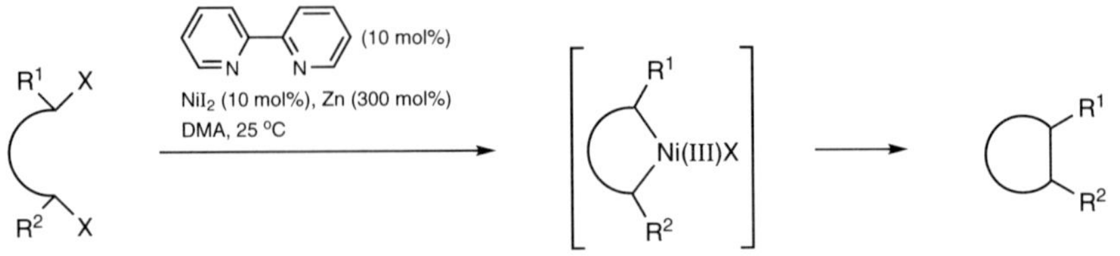

2,2'-bipyridine (10 mol%), NiI_2 (10 mol%), Zn (300 mol%), DMA, 25 °C; intermediate **47**: Ni(III)X metallacycle

Entry	Substrate	Product	Yield (%)	Ref
1	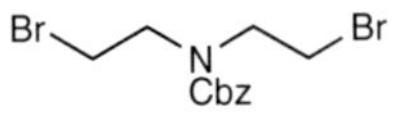 **48**	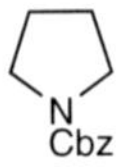 **49**	83	[106]
2	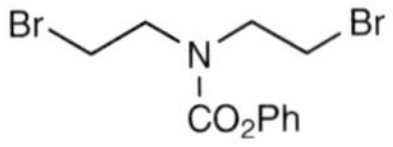	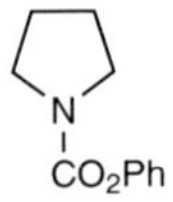	92	[106]
3	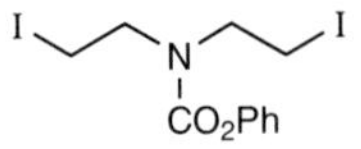	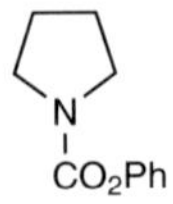	70	[106]
4	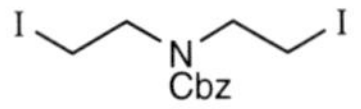	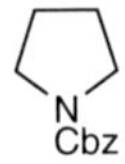	93	[106]
5	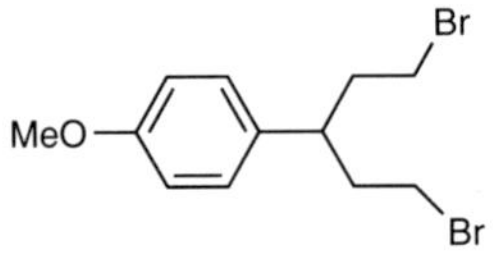	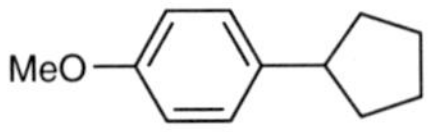	42	[106]
6	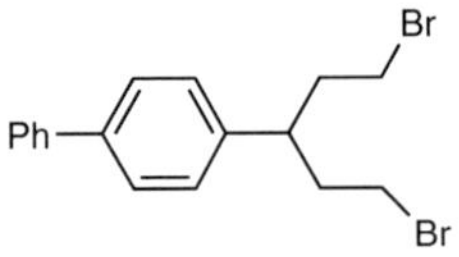	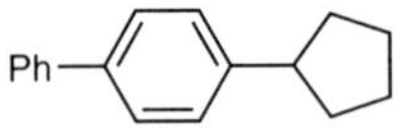	52	[106]
7	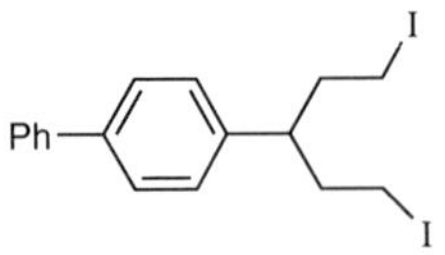	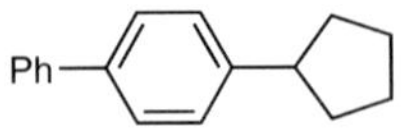	56	[106]
8	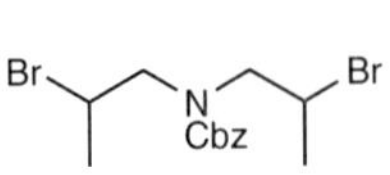	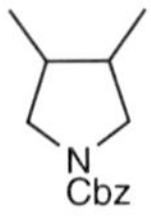	71	[106]
9	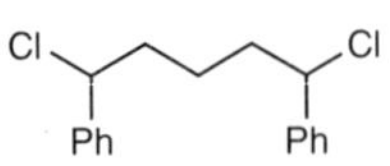	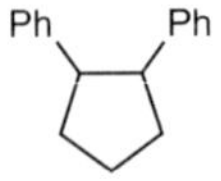	40	[106]

Table 10 (cont.)

Entry	Substrate	Product	Yield (%)	Ref
10	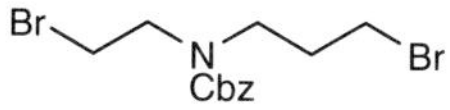 Br N Cbz Br	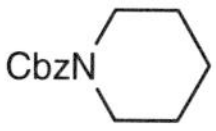 CbzN	50	[106]
11	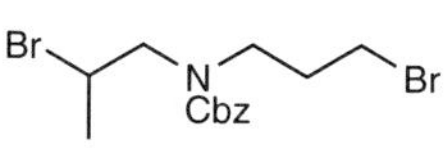 Br N Cbz Br	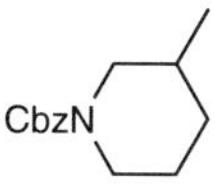 CbzN	60	[106]
12	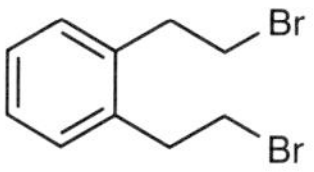 Br Br	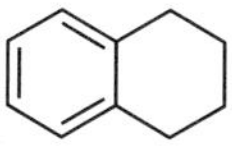	30	[106]
13	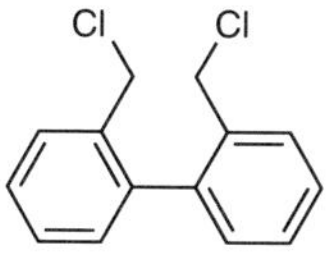 Cl Cl	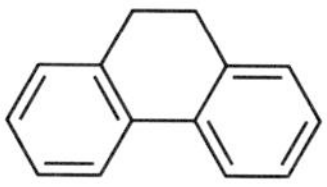	66	[106]
14	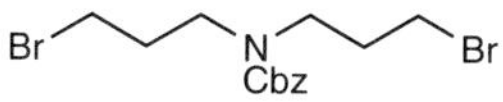 Br N Cbz Br	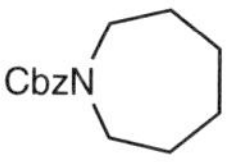 CbzN	38	[106]
15	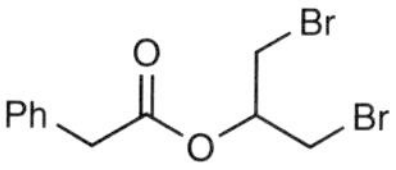 Ph O O Br Br	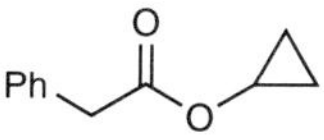 Ph O O	10	[106]

Intramolecular reductive couplings of substrates **50**, containing alkyl bromide and aryl iodide moieties, are promoted by one equivalent of nickel(0) complex [Ni(0)•py•2(ethyl crotonate)] and give various heterocycles **52** via the intermediate **51** (Scheme 29).[107–109] Dihydroindoles, tetrahydroquinolines, dihydrobenzo[*b*]furans, and benzopyrans with various functional groups can be prepared in good yield. Attempts to construct challenging medium-sized-ring compounds, such as 2,3,4,5-tetrahydro-1-benzoxepins **52** (Z = O; n = 3), lead to low yields due to competitive reduction and intermolecular homocoupling reactions.

Scheme 29 Intramolecular Reductive Coupling of Alkyl Bromides and Aryl Iodides[107]

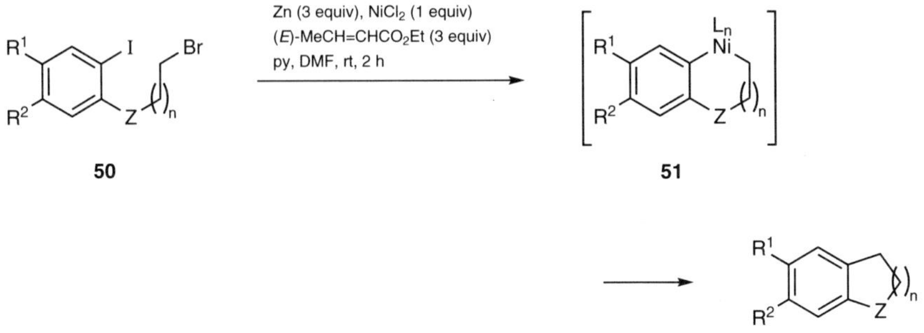

R^1	R^2	Z	n	Yield (%)	Ref
H	H	NCbz	1	41	[107]
H	H	NCbz	2	62	[107]
H	H	NCO_2Et	2	53	[107]
H	H	NTs	2	42	[107]
H	H	NMs	2	52	[107]
CO_2Me	H	NCO_2Et	2	53	[107]
H	CO_2Me	NCO_2Et	2	57	[107]
H	CF_3	NCO_2Et	2	52	[107]
H	OMe	O	2	54	[107]
H	CO_2Bu	O	1	49	[107]
H	CO_2Bu	O	2	63	[107]
H	CO_2Bu	O	3	25	[107]

Under a similar reaction system, substrates **53**, containing an alkyl bromide and a terminal alkene moiety, undergo 5-*exo-trig* cyclization and intermolecular reductive cross coupling with an aryl iodide to give products **54** (Scheme 30).[107] This reaction proceeds with good functional-group tolerance and diastereoselectivity. It is proposed that a radical species **55** is generated from the alkyl bromide by a single-electron transfer process from the nickel(0) complex. After the radical cyclization, the resulting primary radical **56** would then coordinate with previously generated nickel(I) to give a nickel(II) species, which follows a similar process to normal reductive cross couplings to give intermediate **57**. Reductive elimination then affords the desired product. Various aryl iodides, including those with electron-donating groups (OMe, SMe) or electron-withdrawing groups (CO_2Et, F), as well as 2-iodo-6-methoxynaphthalene and iodo-1-methylindoles, work well. Notably, all of the products contain linear-fused perhydrofuro[2,3-*b*]furan (or -pyran) skeletons.

Scheme 30 Tandem Cyclization–Intermolecular Cross Coupling of Alkyl Bromides with Aryl Iodides[107]

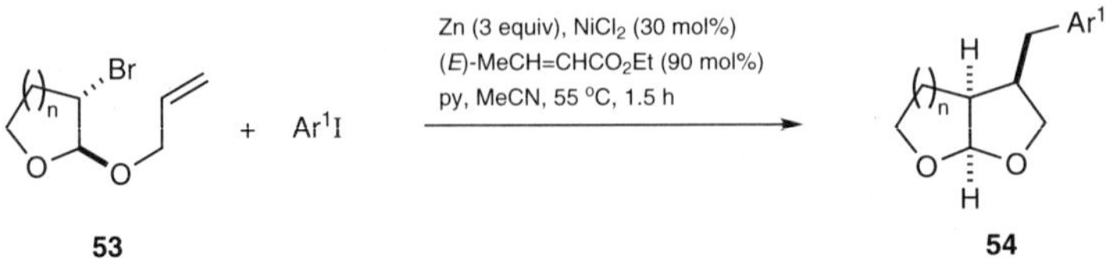

n	Ar^1	Yield (%)	Ref
1	Ph	60	[107]
1	$4\text{-}MeSC_6H_4$	63	[107]
1	$4\text{-}EtO_2CC_6H_4$	51	[107]
1	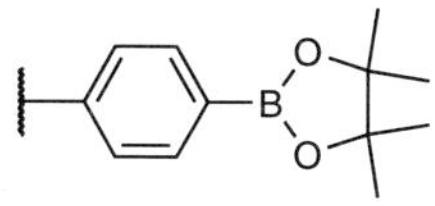 4-(pinacolatoboryl)phenyl	54	[107]
1	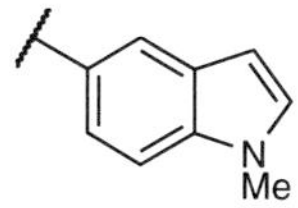 1-methylindol-5-yl	54	[107]
1	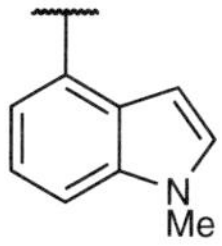 1-methylindol-4-yl	49	[107]
2	$4\text{-}FC_6H_4$	58	[107]
2	$4\text{-}MeOC_6H_4$	60	[107]
2	$4\text{-}EtO_2CC_6H_4$	53	[107]
1	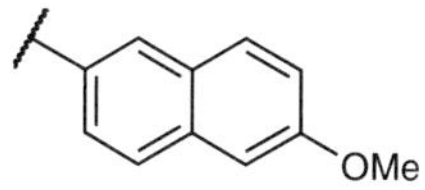 6-methoxynaphthalen-2-yl	56	[107]

Zn ZnI_2 Ni(0)L_n Ni(I)BrL_n 55 56 Ni(I)BrL_n Ni(II)BrL_n Zn $ZnBr_2$ Ni(I)L_n Ar^1I 57 Ni(III)L_n Ni(I)IL_n Ar^1

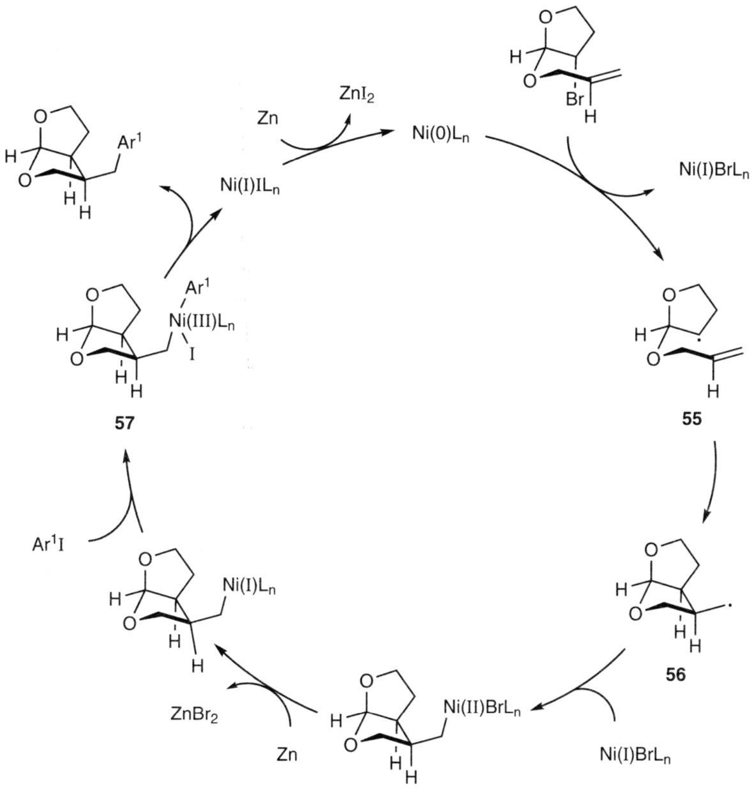

Benzyl Pyrrolidine-1-carboxylate (49):[106]
To a flame-dried Schlenk tube equipped with a stirrer bar was loaded alkyl dihalide **48** (0.15 mmol), followed by 2,2′-bipyridine (0.015 mmol, 10 mol%) and Zn power (29.6 mg, 0.45 mmol, 300 mol%). The tube was moved to a dry glovebox, after which NiI_2 (4.7 mg, 0.015 mmol, 10 mol%) was added. The tube was capped with a rubber septum, and it was moved out of the glovebox. DMA (1 mL) was then added via syringe. After the mixture had been stirred for 16 h under N_2 at 25 °C, it was directly loaded onto a silica gel column without workup. The residue in the reaction vessel was rinsed with a small amount of CH_2Cl_2. Flash column chromatography provided the product **49**; yield: 83%.

Bicyclic Heterocycles 52; **General Procedure:**[107]
Ethyl crotonate (1.9 mL, 15 mmol) was added at rt to a stirred slurry of Zn (975 mg, 15 mmol) in pyridine (5 mL). Under vigorous stirring, $NiCl_2$ (640 mg, 5 mmol) was added to the above mixture. Then, the mixture was heated to 55 °C, and stirring was continued for 15 min. The resulting red-brown Ni(0) complex was cooled to rt, and a soln of dihalide **50** (5 mmol) in DMF (50 mL) was added dropwise over 30 min. After 2 h, the mixture was filtered through a short plug of silica gel [eluting with Et_2O (80 mL)], washed with 1 M HCl, H_2O, and brine, dried ($MgSO_4$), filtered, and concentrated. The crude product was purified by flash chromatography (silica gel).

Bicyclic Heterocycles 54; **General Procedure:**[107]
Ethyl crotonate (0.12 mL, 0.9 mmol) was added to a stirred slurry of Zn (195 mg, 3 mmol) in pyridine (0.5 mL) at rt. Under vigorous stirring, $NiCl_2$ (38 mg, 0.3 mmol) was added to the above mixture. Then, the mixture was heated to 55 °C, and stirring was continued for 15 min. The resulting red-brown Ni(0) complex was cooled to rt, and a soln of the iodoarene (1.6 mmol) in MeCN (1.0 mL) was added dropwise over 2 min. Then, the mixture was heated to 55 °C, and a soln of bromide **53** (1 mmol) in MeCN (3 mL) was added slowly over 1.5 h at this temperature. After 1.5 h, the mixture was filtered through a short plug of silica gel [eluting with Et_2O (30 mL)], washed with H_2O and brine, dried ($MgSO_4$), filtered, and concentrated. The crude product was purified by flash chromatography (silica gel).

1.1.5 Conclusions and Future Perspectives

In summary, this chapter has presented palladium-, chromium-, copper-, and nickel-catalyzed or -promoted intramolecular coupling reactions for the formation of cyclic compounds with different ring sizes and functionalities. Despite considerable achievement in this field, much remains to be explored and optimized to enrich the toolbox of metal-catalyzed cyclic couplings. Instead of using the classical electrophilic and nucleophilic coupling components, the direct activation and functionalization of inert bonds, such as C—H and C—C bonds, provides novel strategies for C—C or C—X bond formations, which make such processes an essential part of current trends in metal-catalyzed reactions. Currently, this research topic is very active because of its significance in academic studies and the potential applications in pharmaceuticals. Reductive coupling of organic halides also holds the promise of extensive applications in organic synthesis, although it remains an immature method. Further studies in this field may focus on the discovery of new catalysts and ligands, which may decrease the dosage of reductants and enable a broader reaction scope in both inter- and intramolecular versions. For the chromium-mediated Nozaki–Hiyama–Kishi (NHK) coupling reactions, catalytic enantioselective coupling remains an unsolved challenge. We anticipate that explorations will continue to ensure that these methodologies and new discoveries will be powerful tools for preparing natural products and drug molecules.

References

[1] Miyaura, N.; Yanagi, T.; Suzuki, A., *Synth. Commun.*, (1981) **11**, 513.
[2] Suzuki, A., *J. Organomet. Chem.*, (1999) **576**, 147.
[3] Braga, A. A. C.; Morgon, N. H.; Ujaque, G.; Maseras, F., *J. Am. Chem. Soc.*, (2005) **127**, 9298.
[4] Miyaura, N., *J. Organomet. Chem.*, (2002) **653**, 54.
[5] Chemler, S. R.; Trauner, D.; Danishefsky, S. J., *Angew. Chem. Int. Ed.*, (2001) **40**, 4544.
[6] Miyaura, N.; Ishiyama, T.; Ishikawa, M.; Suzuki, A., *Tetrahedron Lett.*, (1986) **27**, 6369.
[7] Miyaura, N.; Ishiyama, T.; Sasaki, H.; Ishikawa, M.; Satoh, M.; Suzuki, A., *J. Am. Chem. Soc.*, (1989) **111**, 314.
[8] Miyaura, N.; Ishikawa, M.; Suzuki, A., *Tetrahedron Lett.*, (1992) **33**, 2571.
[9] Kwochka, W. R.; Damrauer, R.; Schmidt, M. W.; Gordon, M. S., *Organometallics*, (1994) **13**, 3728.
[10] Smith, B. B.; Kwochka, W. R.; Swope, R. J.; Smyth, J. R., *J. Org. Chem.*, (1997) **62**, 8589.
[11] Chemler, S. R.; Danishefsky, S. J., *Org. Lett.*, (2000) **2**, 2695.
[12] Frank, S. A.; Chen, H.; Kunz, R. K.; Schnaderbeck, M. J.; Roush, W. R., *Org. Lett.*, (2000) **2**, 2691.
[13] Uenishi, J.-i.; Beau, J.-M.; Armstrong, R. W.; Kishi, Y., *J. Am. Chem. Soc.*, (1987) **109**, 4756.
[14] Humphrey, J. M.; Aggen, J. B.; Chamberlin, A. R., *J. Am. Chem. Soc.*, (1996) **118**, 11 759.
[15] Williams, D. R.; Walsh, M. J.; Miller, N. A., *J. Am. Chem. Soc.*, (2009) **131**, 9038.
[16] Carbonnelle, A. C.; Zhu, J., *Org. Lett.*, (2000) **2**, 3477.
[17] Dufour, J.; Neuville, L.; Zhu, J., *Chem.–Eur. J.*, (2010) **16**, 10 523.
[18] Wang, Z.; Bois-Choussy, M.; Jia, Y.; Zhu, J., *Angew. Chem. Int. Ed.*, (2010) **49**, 2018.
[19] Jia, Y.; Bois-Choussy, M.; Zhu, J., *Angew. Chem. Int. Ed.*, (2008) **47**, 4167.
[20] Wu, B.; Liu, Q. S.; Sulikowski, G. A., *Angew. Chem. Int. Ed.*, (2004) **43**, 6673.
[21] Ghidu, V. P.; Wang, J. Q.; Wu, B.; Liu, Q. S.; Jacobs, A.; Marnett, L. J.; Sulikowski, G. A., *J. Org. Chem.*, (2008) **73**, 4949.
[22] Tortosa, M.; Yakelis, N. A.; Roush, W. R., *J. Am. Chem. Soc.*, (2008) **130**, 2722.
[23] Tortosa, M.; Yakelis, N. A.; Roush, W. R., *J. Org. Chem.*, (2008) **73**, 9657.
[24] Milstein, D.; Stille, J. K., *J. Am. Chem. Soc.*, (1978) **100**, 3636.
[25] Milstein, D.; Stille, J. K., *J. Am. Chem. Soc.*, (1979) **101**, 4992.
[26] Stille, J. K., *Angew. Chem. Int. Ed. Engl.*, (1986) **25**, 508.
[27] Echavarren, A. M.; Stille, J. K., *J. Am. Chem. Soc.*, (1987) **109**, 5478.
[28] Han, X.; Stoltz, B. M.; Corey, E. J., *J. Am. Chem. Soc.*, (1999) **121**, 7600.
[29] Piers, E.; Friesen, R. W.; Keay, B. A., *J. Chem. Soc., Chem. Commun.*, (1985), 809.
[30] Piers, E.; Lu, Y.-F., *J. Org. Chem.*, (1988) **53**, 926.
[31] Piers, E.; Friesen, R. W., *J. Org. Chem.*, (1986) **51**, 3405.
[32] Stille, J. K.; Tanaka, M., *J. Am. Chem. Soc.*, (1987) **109**, 3785.
[33] Nicolaou, K. C.; Xu, J. Y.; Murphy, F.; Barluenga, S.; Baudoin, O.; Wei, H.-X.; Gray, D. L. F.; Ohshima, T., *Angew. Chem. Int. Ed.*, (1999) **38**, 2447.
[34] Garg, N. K.; Hiebert, S.; Overman, L. E., *Angew. Chem. Int. Ed.*, (2006) **45**, 2912.
[35] Becker, M. H.; Chua, P.; Downham, R.; Douglas, C. J.; Garg, N. K.; Hiebert, S.; Jaroch, S.; Matsuoka, R. T.; Middleton, J. A.; Ng, F. W.; Overman, L. E., *J. Am. Chem. Soc.*, (2007) **129**, 11 987.
[36] Negishi, E.; Baba, S., *J. Chem. Soc., Chem. Commun.*, (1976), 597.
[37] Baba, S.; Negishi, E., *J. Am. Chem. Soc.*, (1976) **98**, 6729.
[38] Negishi, E.; King, A. O.; Okukado, N., *J. Org. Chem.*, (1977) **42**, 1821.
[39] King, A. O.; Okukado, N.; Negishi, E., *J. Chem. Soc., Chem. Commun.*, (1977), 683.
[40] King, A. O.; Negishi, E.; Villani, F. J., Jr.; Silveira, A., Jr., *J. Org. Chem.*, (1978) **43**, 358.
[41] Negishi, E.; Van Horn, D. E., *J. Am. Chem. Soc.*, (1977) **99**, 3168.
[42] Negishi, E., *Acc. Chem. Res.*, (1982) **15**, 340.
[43] Negishi, E.; Anastasia, L., *Chem. Rev.*, (2003) **103**, 1979.
[44] Nolasco, L.; Perez Gonzalez, M.; Caggiano, L.; Jackson, R. F. W., *J. Org. Chem.*, (2009) **74**, 8280.
[45] Sonogashira, K.; Tohda, Y.; Hagihara, N., *Tetrahedron Lett.*, (1975), 4467.
[46] Sonogashira, K., In *Comprehensive Organic Synthesis*, Trost, B. M.; Fleming, I., Eds.; Pergamon: Oxford, (1991); Vol. 3, p 521.
[47] Chinchilla, R.; Nájera, C., *Chem. Rev.*, (2007) **107**, 874.
[48] Doucet, H.; Hierso, J.-C., *Angew. Chem. Int. Ed.*, (2007) **46**, 834.
[49] Chinchilla, R.; Nájera, C., *Chem. Soc. Rev.*, (2011) **40**, 5084.
[50] Wang, D.; Gao, S., *Org. Chem. Front.*, (2014) **1**, 556.
[51] Balraju, V.; Reddy, R. S.; Periasamy, M.; Iqbal, J., *J. Org. Chem.*, (2005) **70**, 9626.

[52] Mohapatra, D. K.; Bhattasali, D.; Gurjar, M. K.; Khan, M. I.; Shashidhara, K. S., *Eur. J. Org. Chem.*, (2008), 6213.
[53] Koyama, Y.; Lear, M. J.; Yoshimura, F.; Ohashi, I.; Mashimo, T.; Hirama, M., *Org. Lett.*, (2005) **7**, 267.
[54] Yoshida, J.-i.; Tamao, K.; Takahashi, M.; Kumada, M., *Tetrahedron Lett.*, (1978), 2161.
[55] Yoshida, J.-i.; Tamao, K.; Yamamoto, H.; Kakui, T.; Uchida, T.; Kumada, M., *Organometallics*, (1982) **1**, 542.
[56] Hatanaka, Y.; Hiyama, T., *J. Org. Chem.*, (1988) **53**, 918.
[57] Hiyama, T.; Hatanaka, Y., *Pure Appl. Chem.*, (1994) **66**, 1471.
[58] Hatanaka, Y.; Hiyama, T., *Synlett*, (1991), 845.
[59] Sugiyama, A.; Ohnishi, Y.; Nakaoka, M.; Nakao, Y.; Sato, H.; Sakaki, S.; Nakao, Y.; Hiyama, T., *J. Am. Chem. Soc.*, (2008) **130**, 12975.
[60] Nakao, Y.; Hiyama, T., *Chem. Soc. Rev.*, (2011) **40**, 4893.
[61] Denmark, S. E.; Yang, S.-M., *J. Am. Chem. Soc.*, (2002) **124**, 2102.
[62] Denmark, S. E.; Yang, S.-M., *Org. Lett.*, (2001) **3**, 1749.
[63] Denmark, S. E.; Yang, S.-M., *J. Am. Chem. Soc.*, (2002) **124**, 15196.
[64] Denmark, S. E.; Yang, S.-M., *J. Am. Chem. Soc.*, (2004) **126**, 12432.
[65] Moritani, I.; Fujiwara, Y., *Synthesis*, (1973), 524.
[66] Li, C.-J., *Acc. Chem. Res.*, (2009) **42**, 335.
[67] Li, Z.; Bohle, D. S.; Li, C.-J., *Proc. Natl. Acad. Sci. U. S. A.*, (2006) **103**, 8928.
[68] Beccalli, E. M.; Broggini, G.; Martinelli, M.; Sottocornola, S., *Chem. Rev.*, (2007) **107**, 5318.
[69] Yeung, C. S.; Dong, V. M., *Chem. Rev.*, (2011) **111**, 1215.
[70] Ferreira, E. M.; Stoltz, B. M., *J. Am. Chem. Soc.*, (2003) **125**, 9578.
[71] Zhang, H.; Ferreira, E. M.; Stoltz, B. M., *Angew. Chem. Int. Ed.*, (2004) **43**, 6144.
[72] Ferreira, E. M.; Zhang, H.; Stoltz, B. M., *Tetrahedron*, (2008) **64**, 5987.
[73] Beck, E. M.; Hatley, R.; Gaunt, M. J., *Angew. Chem. Int. Ed.*, (2008) **47**, 3004.
[74] Garg, N. K.; Caspi, D. D.; Stoltz, B. M., *J. Am. Chem. Soc.*, (2004) **126**, 9552.
[75] Okude, Y.; Hirano, S.; Hiyama, T.; Nozaki, H., *J. Am. Chem. Soc.*, (1977) **99**, 3179.
[76] Okude, Y.; Hiyama, T.; Nozaki, H., *Tetrahedron Lett.*, (1977), 3829.
[77] Jin, H.; Uenishi, J.-i.; Christ, W. J.; Kishi, Y., *J. Am. Chem. Soc.*, (1986) **108**, 5644.
[78] Takai, K.; Tagashira, M.; Kuroda, T.; Oshima, K.; Utimoto, K.; Nozaki, H., *J. Am. Chem. Soc.*, (1986) **108**, 6048.
[79] Fürstner, A.; Shi, N., *J. Am. Chem. Soc.*, (1996) **118**, 12349.
[80] Fürstner, A.; Shi, N., *J. Am. Chem. Soc.*, (1996) **118**, 2533.
[81] Fürstner, A., *Chem. Rev.*, (1999) **99**, 991.
[82] Aoyagi, S.; Wang, T.-C.; Kibayashi, C., *J. Am. Chem. Soc.*, (1993) **115**, 11393.
[83] Bian, J.; Wingerden, M. V.; Ready, J. M., *J. Am. Chem. Soc.*, (2006) **128**, 7428.
[84] Zhu, W.; Jiménez, M.; Jung, W.-H.; Camarco, D. P.; Balachandran, R.; Vogt, A.; Day, B. W.; Curran, D. P., *J. Am. Chem. Soc.*, (2010) **132**, 9175.
[85] Ciesielski, J.; Gandon, V.; Frontier, A. J., *J. Org. Chem.*, (2013) **78**, 9541.
[86] Elliott, M. R.; Dhimane, A.-L.; Malacria, M., *J. Am. Chem. Soc.*, (1997) **119**, 3427.
[87] Roethle, P. A.; Trauner, D., *Org. Lett.*, (2006) **8**, 345.
[88] Ullmann, F., *Ber. Dtsch. Chem. Ges.*, (1903) **36**, 2389.
[89] Ullmann, F., *Ber. Dtsch. Chem. Ges.*, (1904) **37**, 853.
[90] Goldberg, I., *Ber. Dtsch. Chem. Ges.*, (1906) **39**, 1691.
[91] Okuro, K.; Furuune, M.; Enna, M.; Miura, M.; Nomura, M., *J. Org. Chem.*, (1993) **58**, 4716.
[92] Yu, S.; Liu, F.; Ma, D., *Tetrahedron Lett.*, (2006) **47**, 9155.
[93] Coleman, R. S.; Garg, R., *Org. Lett.*, (2001) **3**, 3487.
[94] Santandrea, J.; Bédard, A.-C.; Collins, S. K., *Org. Lett.*, (2014) **16**, 3892.
[95] Piers, E.; Wong, T., *J. Org. Chem.*, (1993) **58**, 3609.
[96] Paterson, I.; Lombart, H.-G.; Allerton, C., *Org. Lett.*, (1999) **1**, 19.
[97] Tollefson, E. J.; Erickson, L. W.; Jarvo, E. R., *J. Am. Chem. Soc.*, (2015) **137**, 9760.
[98] Durandetti, M.; Nédélec, J.-Y.; Périchon, J., *J. Org. Chem.*, (1996) **61**, 1748.
[99] Durandetti, M.; Hardou, L.; Clément, M.; Maddaluno, J., *Chem. Commun. (Cambridge)*, (2009), 4753.
[100] Everson, D. A.; Shrestha, R.; Weix, D. J., *J. Am. Chem. Soc.*, (2010) **132**, 920.
[101] Weix, D. J., *Acc. Chem. Res.*, (2015) **48**, 1767.
[102] Li, X.; Feng, Z.; Jiang, Z.-X.; Zhang, X., *Org. Lett.*, (2015) **17**, 5570.
[103] Kadunce, N. T.; Reisman, S. E., *J. Am. Chem. Soc.*, (2015) **137**, 10480.

[104] Cornella, J.; Edwards, J. T.; Qin, T.; Kawamura, S.; Wang, J.; Pan, C.-M.; Gianatassio, R.; Schmidt, M.; Eastgate, M. D.; Baran, P. S., *J. Am. Chem. Soc.*, (2016) **138**, 2174.
[105] Wang, X.; Wang, S.; Xue, W.; Gong, H., *J. Am. Chem. Soc.*, (2015) **137**, 11 562.
[106] Xue, W.; Xu, H.; Liang, Z.; Qian, Q.; Gong, H., *Org. Lett.*, (2014) **16**, 4984.
[107] Yan, C.-S.; Peng, Y.; Xu, X.-B.; Wang, Y.-W., *Chem.–Eur. J.*, (2012) **18**, 6039.
[108] Peng, Y.; Xu, X.-B.; Xiao, J.; Wang, Y.-W., *Chem. Commun. (Cambridge)*, (2014) **50**, 472.
[109] Peng, Y.; Luo, L.; Yan, C.-S.; Zhang, J.-J.; Wang, Y.-W., *J. Org. Chem.*, (2013) **78**, 10 960.

1.2 Intramolecular Heck Reactions

S. Jammi, C. Nottingham, and P. J. Guiry

General Introduction

The vinylation of aryl/vinyl/alkyl halides or pseudohalides catalyzed by palladium(0), the Heck cross-coupling reaction, has developed into one of the most widely employed C—C bond-forming reactions in organic synthesis since its discovery in 1968 by Heck.[1–3] The intramolecular version of this reaction was first realized by Mori and co-workers in 1977 through the formation of indole **1** (Scheme 1).[4] Since then the intramolecular Heck reaction has been extensively utilized in the construction of isolated, fused, bridged, or spiro rings of various sizes and as a key step in the total synthesis of many natural products.[5] Its application in cascade/domino processes has further increased the usefulness of the reaction in multiple bond formation to construct cyclic or polycyclic systems in both asymmetric and nonasymmetric modes. Additionally, tertiary or quaternary stereocenters can be established using the asymmetric intramolecular Heck reaction.[6–8]

Scheme 1 Intramolecular Heck Reaction To Form an Indole[4]

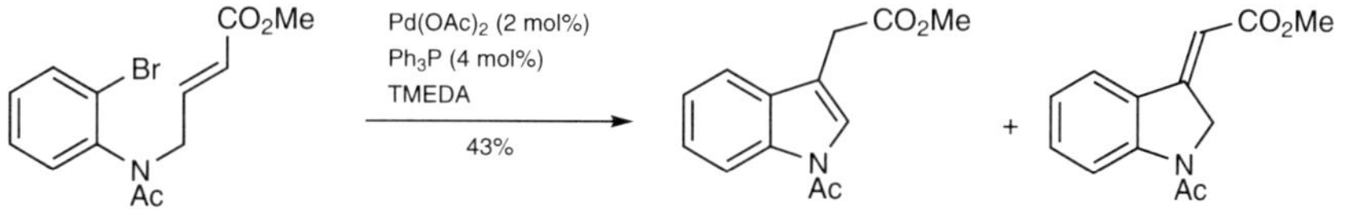

CO_2Me Br N Ac — $Pd(OAc)_2$ (2 mol%), Ph_3P (4 mol%), TMEDA, 43% → CO_2Me N Ac **1** + CO_2Me N Ac

The commonly accepted catalytic cycle for the intramolecular Heck reaction is believed to involve five major steps (Scheme 2). The first step is the oxidative addition of an aryl/vinyl/alkyl halide or pseudohalide to a palladium(0) complex **2** to form a palladium(II) complex **3**. In the second step, dissociation of this palladium(II) complex, followed by coordination of a tethered alkene, leads to the formation of a π-complex **4** or **5**. Depending on the reaction conditions, the mechanism involves the dissociation of either a halide (cationic route) or one of the ligands from the palladium(II) complex **3** (neutral route) to allow the coordination of the alkene. In the third step, *syn*-insertion of the alkene into the Pd—C bond of the π-complex occurs, providing σ-alkyl palladium(II) complexes **6**/**7** or **8**/**9**. In the fourth step, *syn*-β-hydrogen elimination gives rise to the Heck products and the hydridopalladium complex **10** or **11**. In the last step, catalytically active palladium(0) complex **2** is regenerated from the hydridopalladium complex **10** or **11** by base-assisted reductive elimination to complete the catalytic cycle.

Scheme 2 Catalytic Cycle for the Intramolecular Heck Reaction

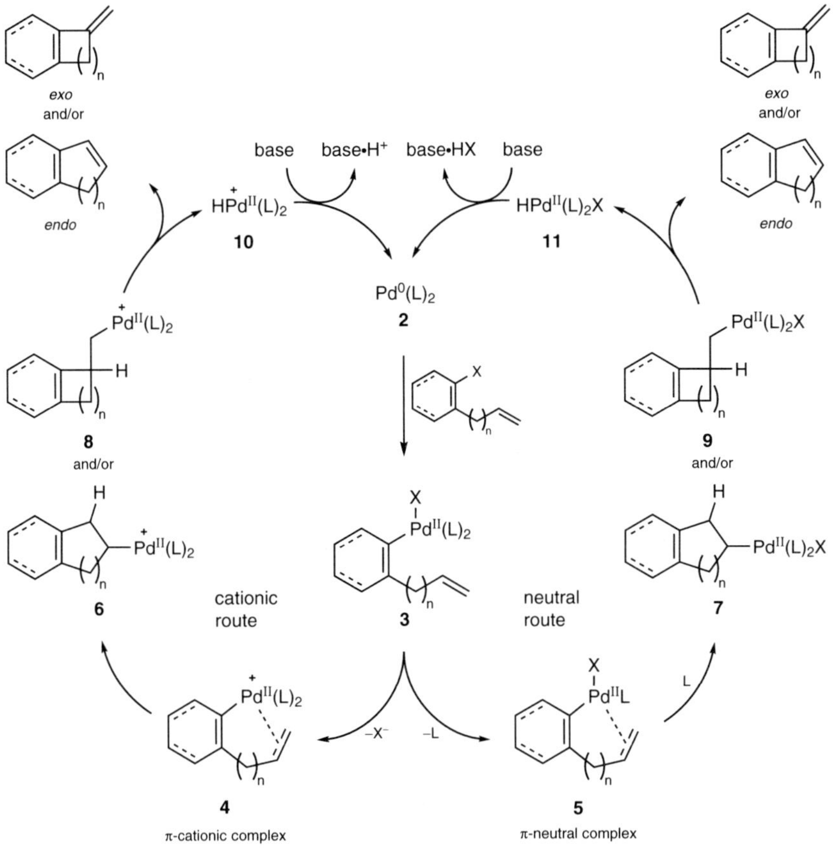

In this chapter the best methods for the construction of cyclic systems using intramolecular Heck cyclizations, including enantioselective versions, are assembled. The examples discussed herein are categorized according to Baldwin's rules.[9] These guidelines categorize the reaction by the size of the ring formed, the method of cyclization (with *exo* for the double bond external to the cyclic structure or *endo* for an internal alkene), and the geometry of the carbon undergoing cyclization (e.g., *trig* for a trigonal/sp^2 carbon).

1.2.1 Non-Enantioselective Reactions

After the initial report of the intramolecular Heck reaction in the formation of heterocyclic rings by Mori and co-workers,[4] Grigg and co-workers in 1984 reported the first formation of carbocycles using this reaction.[10] The early publications on the intramolecular Heck reaction reported low yields, and thus application in the synthesis of complex natural products was limited. Later, the momentum in this field increased and the Heck reaction is now commonly employed in cyclization steps. This is mainly due to a better understanding of the solvents, bases, additives, catalysts, and ligands necessary for optimal reaction conditions. A wide range of ring sizes can be formed by employing this reaction: small-sized rings (3 and 4; Section 1.2.1.1), common-sized rings (5 and 6; Section 1.2.1.2), medium-sized rings (7 to 11; Section 1.2.1.3), and macrocycles (>11; Section 1.2.1.4).

1.2.1.1 Formation of Small-Sized Rings (3 and 4)

The use of the intramolecular Heck cyclization for the construction of small, strained rings is not as well documented as for other ring sizes. The 3- and 4-*exo-trig* reaction pathways are favored according to Baldwin's rules, while 3- and 4-*endo-trig* processes are unknown and disfavored by Baldwin's rules.

1.2.1.1.1 3- and 4-*exo-trig* Processes

1.2.1.1.1.1 3-*exo-trig* Cyclization

The 3-*exo-trig* cyclization of enol trifluoromethanesulfonates **12** with palladium(II) acetate, triphenylphosphine, sodium carbonate, and tetraethylammonium chloride in refluxing acetonitrile gives tricyclic compounds **14** containing a fused cyclopropane ring in good yields (Scheme 3).[11] The reaction initially proceeds via a 5-*exo-trig* cyclization, generating the intermediate complex **13**, which undergoes the 3-*exo-trig* cyclization in a domino-type process. In the case of the seven-membered-ring enol trifluoromethanesulfonate **12** [X = $(CH_2)_3$], the double bond in the corresponding final product is more prone to isomerization; however, this can be suppressed by adding silver(I) or thallium(I) salts.

Scheme 3 Formation of Tricyclic Systems by 5-*exo-trig*/3-*exo-trig* Cascade Intramolecular Heck Cyclization[11]

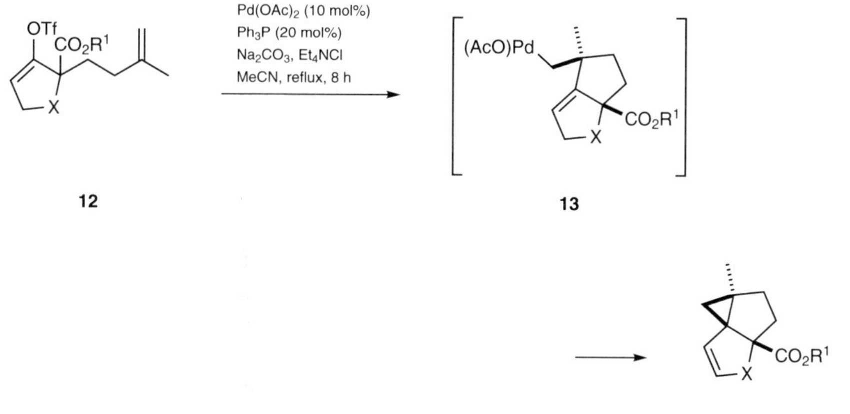

$Pd(OAc)_2$ (10 mol%)
Ph_3P (20 mol%)
Na_2CO_3, Et_4NCl
MeCN, reflux, 8 h

OTf, CO_2R^1, X — **12**

(AcO)Pd, CO_2R^1, X — **13**

CO_2R^1, X — **14**

X	R^1	Yield (%)	Ref
CH_2	Me	89	[11]
$(CH_2)_2$	Et	71	[11]
$(CH_2)_3$	Me	76[a]	[11]
$NAcCH_2$	Me	62	[11]

[a] Tl(OAc) (1.2 equiv) was used and the reaction time was 5 h.

Fused Cyclopropyl Carbo/Heterocycles 14**; General Procedure:**[11]
Enol trifluoromethanesulfonate **12** (0.87 mmol) was added to a stirred suspension of $Pd(OAc)_2$ (19.5 mg, 0.087 mmol, 10 mol%), Ph_3P (45.6 mg, 0.175 mmol, 20 mol%), Na_2CO_3 (185 mg, 1.74 mmol), and Et_4NCl (145 mg, 0.87 mmol) in MeCN (8 mL). The resulting mixture was refluxed under a N_2 atmosphere for 8 h. The solvent was then removed under reduced pressure and the residue was partitioned between H_2O (50 mL) and CH_2Cl_2

for references see p 92

(50 mL). The aqueous layer was extracted with CH_2Cl_2 (3 × 10 mL) and the combined CH_2Cl_2 extracts were dried ($MgSO_4$) and concentrated. The resulting residue was purified by flash column chromatography [silica gel, Et_2O/petroleum ether (bp 30–40 °C)].

1.2.1.1.1.2 4-*exo-trig* Cyclization

Cyclobutanes are highly valuable intermediates in total synthesis and pose a synthetic challenge due to the inherent ring strain, with a bond angle of near 90°. The *gem*-disubstituted hexa-1,5-dien-2-ol trifluoromethanesulfonates **15** and **17** have been successfully cyclized to provide functionalized methylene cyclobutanes **16** and **18** in good yields (Scheme 4).[12] Disubstitution at the α-position ($R^1 \neq H$) is crucial for promoting cyclization, with monosubstituted and unsubstituted substrates proving unreactive. In this instance, tetrakis(triphenylphosphine)palladium(0) is used as the catalyst, which reacts with the substrate to produce the palladium intermediate. Alkoxycarbonylation of this intermediate occurs in the presence of a carbon monoxide atmosphere to give cyclobutane esters **16**. Alternatively, *syn*-palladium elimination of the intermediate takes place in the presence of an argon atmosphere to produce 1,2-dimethylene cyclobutanes **18**. These products have great synthetic utility as suitable substrates for Diels–Alder transformations.

Scheme 4 Formation of Cyclobutane Systems by 4-*exo-trig* Intramolecular Heck Reactions[12]

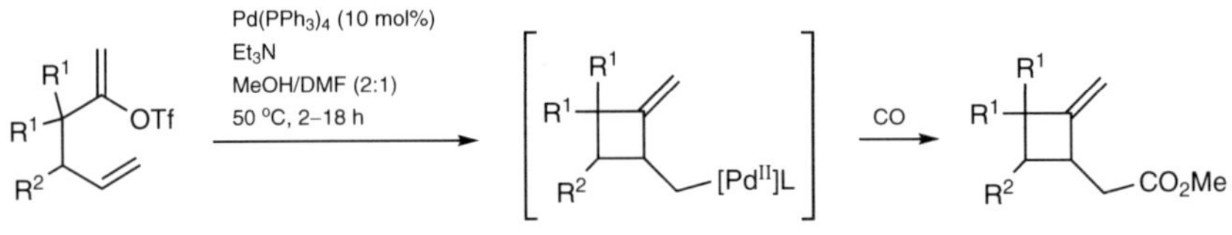

R^1	R^2	Yield (%)	Ref
Me	Me	65	[12]
Me	Et	60	[12]
Me	iPr	65	[12]

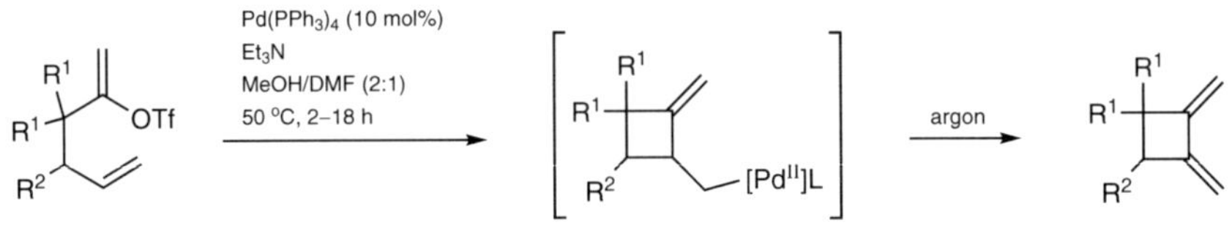

R^1	R^2	Yield (%)	Ref
Ph	Me	87	[12]
Ph	iPr	75	[12]

Methyl (2-Methylenecyclobutyl)acetates 16; General Procedure:[12]

CAUTION: *Carbon monoxide is extremely flammable and toxic, and exposure to higher concentrations can quickly lead to coma.*

To a 0.06 M soln of enol trifluoromethanesulfonate **15** (1 mmol) in MeOH/DMF (2:1) were added Et_3N (323 mg, 3.2 mmol) and $Pd(PPh_3)_4$ (115.6 mg, 0.1 mmol, 10 mol%). The resulting mixture was stirred under a CO atmosphere at 50 °C until complete consumption of starting material (TLC). 1 M HCl (5 mL) and Et_2O (30 mL) were added, the layers were separated, and the aqueous layer was extracted with Et_2O (3 × 10 mL). The combined organic layers were washed successively with sat. aq $NaHCO_3$ (5 mL), H_2O (5 mL), and brine, and then dried ($MgSO_4$), filtered, and concentrated under reduced pressure. Column chromatography [silica gel, Et_2O/petroleum ether (bp 30–40 °C)] of the residue afforded the product **16**.

1.2.1.2 Formation of Common-Sized Rings (5 and 6)

Common-sized-ring formation is well documented compared to that of small-, medium-, and large-sized rings. The most dominant mode of cyclization is the *exo* mode with the rings formed with five to seven members. *endo* Cyclization is mostly seen in the synthesis of six- and seven-membered rings as well as larger rings.

1.2.1.2.1 5- and 6-*exo-trig* Processes

1.2.1.2.1.1 5-*exo-trig* Cyclization

Substituted 2-chloroquinolin-3-yl homoallyl alcohol derivatives **19** may undergo a 5-*exo-trig* Heck cyclization to form 3-methylene-2,3-dihydro-1*H*-cyclopenta[*b*]quinoline derivatives **20** (Scheme 5).[13] The bulky phosphine ligand 2,2′-bis(diphenylphosphino)-1,1′-binaphthyl (BINAP) provides the Heck-type cyclized product exclusively in excellent yields; however, when monodentate phosphine ligands such as triphenylphosphine are used, formation of a double-bond isomerized byproduct is observed, resulting in decreased yield. Interestingly, if the benzylic alcohol in the starting material is converted into a methyl ether then both ligands give the same yield. This result suggests that with smaller monodentate phosphine ligands both the hydroxy and alkene groups in the starting material compete for palladium coordination, but with the bulky bidentate ligand BINAP only the alkene can coordinate. 2-Bromopyridyl homoallyl alcohol derivatives may also be used as substrates.

for references see p 92

Scheme 5 Intramolecular Heck Cyclization of 2-Chloroquinolin-3-yl Homoallyl Alcohols[13]

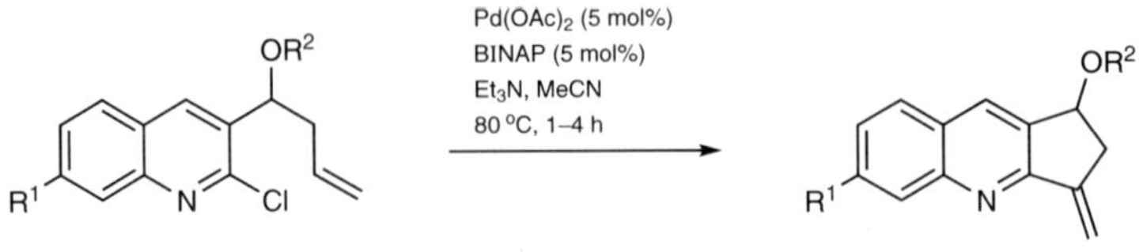

Pd(OAc)2 (5 mol%)
BINAP (5 mol%)
Et3N, MeCN
80 °C, 1–4 h

19 → **20**

R^1	R^2	Yield (%)	Ref
H	H	83	[13]
H	Me	82[a]	[13]
OMe	H	79	[13]
Me	H	80	[13]
Cl	H	89	[13]

[a] An 80% yield was obtained when Ph_3P was used as the ligand.

A range of indole, indoline, and oxindole derivatives may be synthesized from *N*-allyl-2-haloanilines via a 5-*exo-trig* intramolecular Heck cyclization.[14–16] In most cases, the 5-*exo-trig* product is formed first but the exocyclic double bond in the product isomerizes to form the more stable endocyclic double bond. If the substrates contain a 1,1-disubstituted double bond instead of 1,2 substitution or a terminal double bond then the intermediate palladium complex formed readily undergoes further polycyclizations or anion-capture reactions.

2,3-Substituted indole derivatives **23** can be prepared from 2-iodoanilines and aryl-substituted 2-diazobut-3-enoates **21** (Scheme 6). Palladium(II) chloride catalyzes the reaction in one pot, with carbenoid insertion into the N—H bond providing the Heck cyclization precursor **22**, followed by a 5-*exo-trig* cyclization. Finally, double-bond isomerization occurs to provide the more stable 5-*endo-trig* double bond, giving 2,3-substituted indole derivatives **23**.[17]

Scheme 6 N–H Bond Insertion and Intramolecular Heck Cyclization[17]

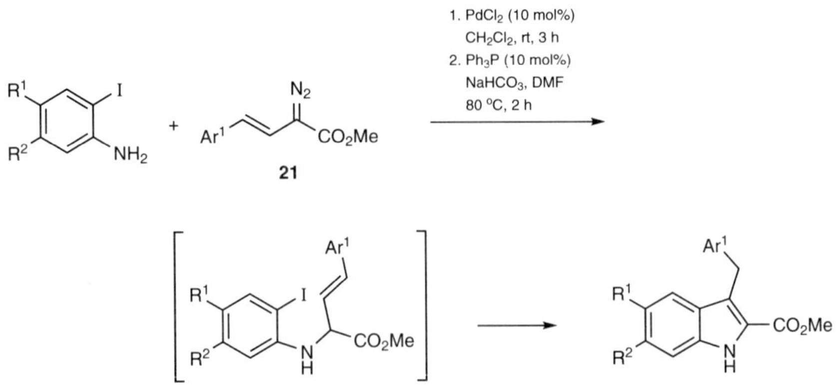

1. PdCl2 (10 mol%)
CH2Cl2, rt, 3 h
2. Ph3P (10 mol%)
NaHCO3, DMF
80 °C, 2 h

21 → [**22**] → **23**

R^1	R^2	Ar^1	Yield (%)	Ref
H	H	Ph	74	[17]
H	OMe	Ph	78	[17]
H	H	4-Tol	80	[17]
Me	H	4-ClC_6H_4	80	[17]
Me	H	2-MeOC_6H_4	67	[17]

Oxindoles are a common motif in natural products and pharmaceutically active compounds. Spiro-fused indane–oxindoles **26** may be prepared in high yields from *N*-(2-bromophenyl)acrylamides **24** using a tandem 5-*exo-trig* Heck cyclization followed by intramolecular aryl C—H bond functionalization (Scheme 7). This reaction proceeds via carbopalladation to form an alkylpalladium intermediate **25**.[18]

Scheme 7 Tandem Intramolecular Heck Reaction Followed by Intramolecular C–H Functionalization[18]

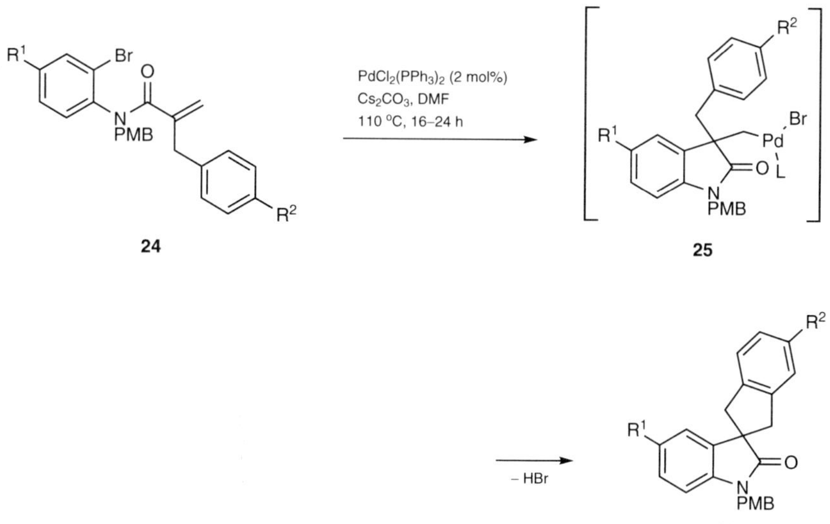

R^1	R^2	Yield (%)	Ref
H	H	65	[18]
H	Me	90	[18]
H	OMe	87	[18]
H	F	84	[18]
Me	Me	82	[18]

Using a similar approach, amides **27** initially form 5-*exo-trig* cyclized alkylpalladium intermediates with palladium(II) acetate, which then react intermolecularly with hydrazones to give oxindole-substituted alkenes **28** in high yields (Scheme 8).[19] This method tolerates various functional groups on both of the aryl rings and is also suitable for hydrazones derived from furan-2-carbaldehyde and 2-naphthaldehyde.

Scheme 8 Tandem Intramolecular Heck Reaction Followed by Intermolecular Reaction with Hydrazones[19]

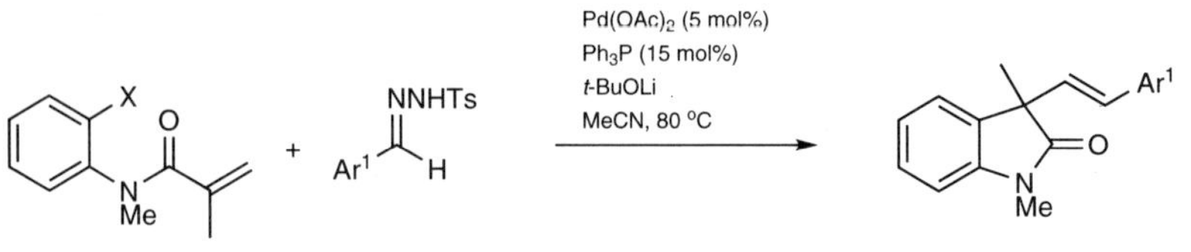

Pd(OAc)$_2$ (5 mol%)
Ph$_3$P (15 mol%)
t-BuOLi
MeCN, 80 °C

27 **28**

X	Ar^1	Yield (%)	Ref
I	Ph	98	[19]
Br	Ph	98	[19]
Br	2-Tol	99	[19]
Br	2-$MeOC_6H_4$	99	[19]
Br	4-*t*-BuC_6H_4	97	[19]
Br	4-$EtO_2CC_6H_4$	97	[19]

Pyrrolidine derivatives **29** can be synthesized using a palladium-catalyzed cascade which involves intramolecular 5-*exo-trig* cyclization of an allyl(2-bromoallyl)amine derivative followed by intermolecular cross coupling with various arylboronic acids (Scheme 9).[20] Here the intermediate alkylpalladium complex undergoes transmetalation with the arylboronic acid rather than undergoing β-hydride elimination.[20] This may be due to the intermediate palladium complex being stabilized by an oxygen of the *N*-sulfonyl group, which suppresses the usual β-hydride elimination.

Scheme 9 Intramolecular Heck Cyclization and Coupling Cascade[20]

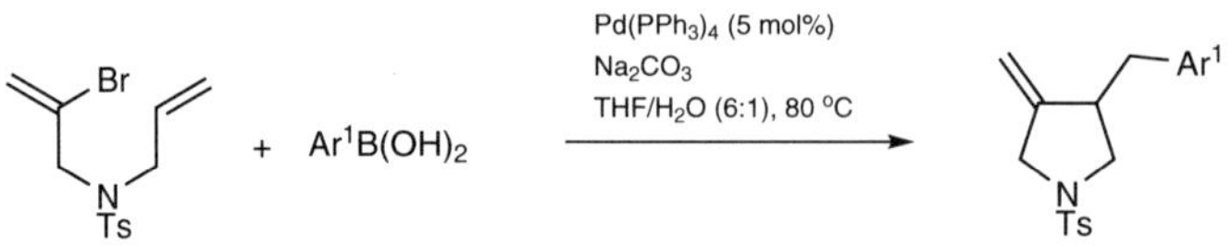

+ $Ar^1B(OH)_2$

Pd(PPh$_3$)$_4$ (5 mol%)
Na$_2$CO$_3$
THF/H$_2$O (6:1), 80 °C

29

Ar^1	Yield (%)	Ref
Ph	75	[20]
4-FC_6H_4	92	[20]
1-naphthyl	88	[20]
4-$MeOC_6H_4$	69	[20]
3-$O_2NC_6H_4$	49	[20]

In a similar approach, replacement of vinyl bromides with aryl bromide substrates facilitates the formation of indoline, oxindole, and dihydrobenzo[*b*]furan derivatives **30** in good yields (Table 1). The alkylpalladium(II) intermediate formed through initial intramolecular 5-*exo-trig* cyclization can be arylated directly with different hetarenes in a domino process to provide a variety of hetaryl-substituted products.[21]

Table 1 Domino Intramolecular Heck Cyclization and Direct Arylation[21]

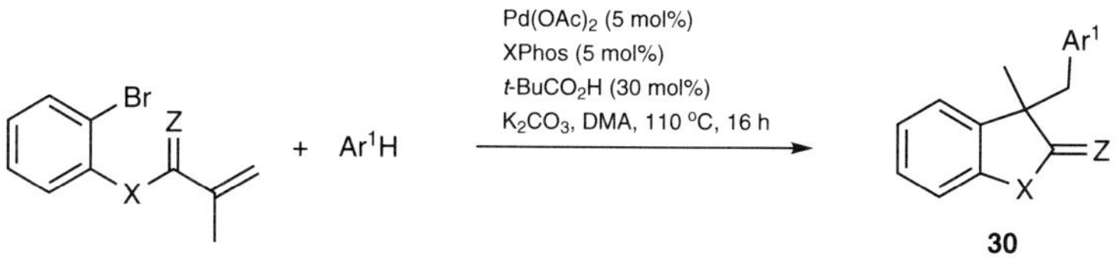

XPhos = 2-(dicyclohexylphosphino)-2',4',6'-triisopropylbiphenyl

Starting Materials		Product	Yield (%)	Ref
Aryl Bromide	Hetarene			
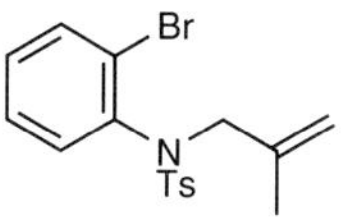 Br, N, Ts	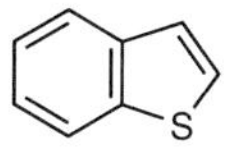 S	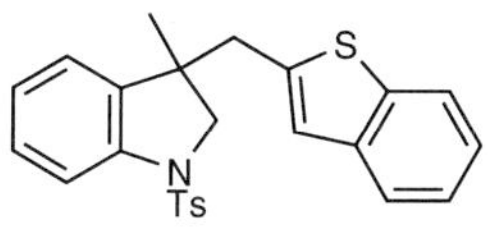 S, N, Ts	84	[21]
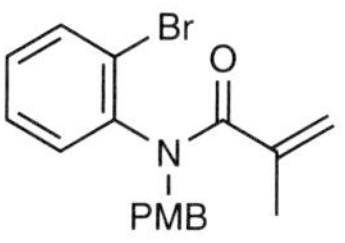 Br, O, N, PMB	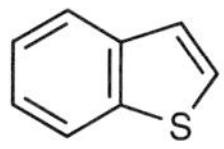 S	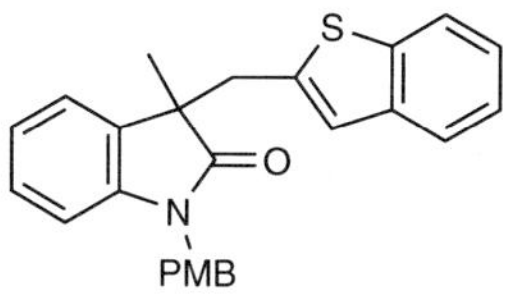 S, O, N, PMB	67	[21]
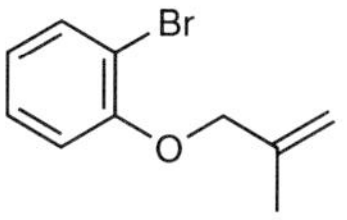 Br, O	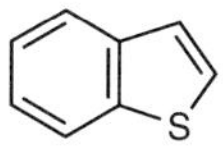 S	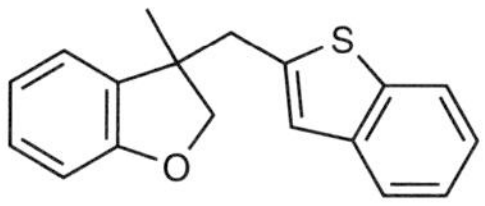 S, O	76	[21]
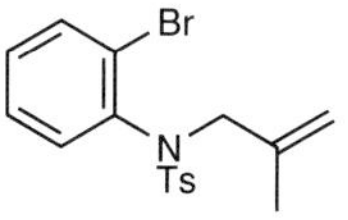 Br, N, Ts	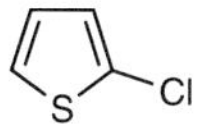 S, Cl	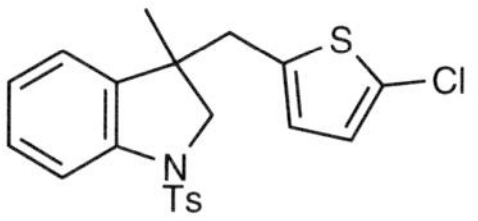 S, Cl, N, Ts	99	[21]
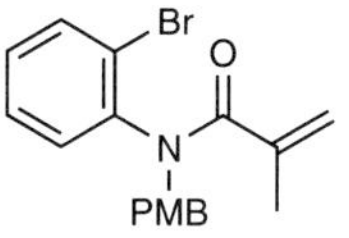 Br, O, N, PMB	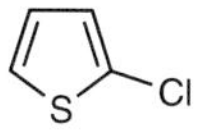 S, Cl	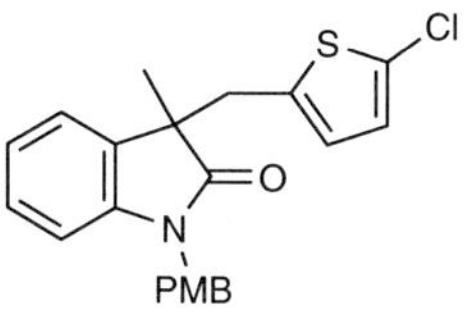 S, Cl, O, N, PMB	82	[21]
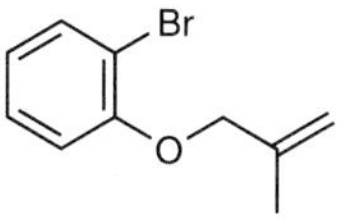 Br, O	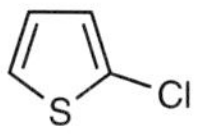 S, Cl	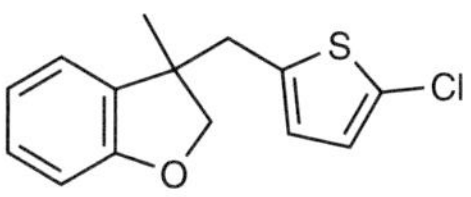 S, Cl, O	92	[21]

for references see p 92

Table 1 (cont.)

Starting Materials		Product	Yield (%)	Ref
Aryl Bromide	Hetarene			
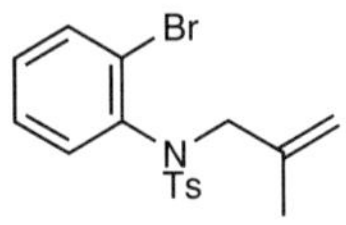	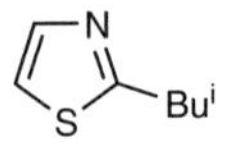	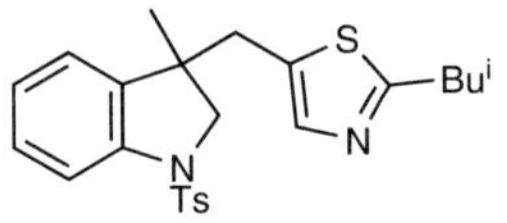	75	[21]
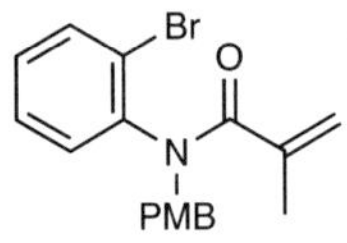	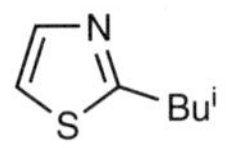	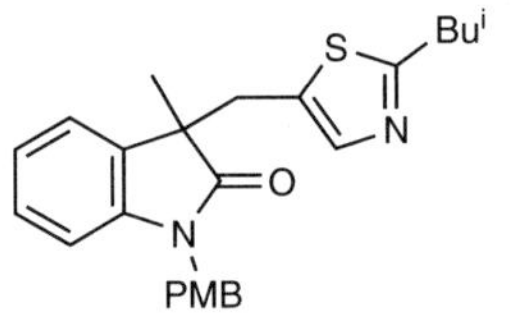	67	[21]
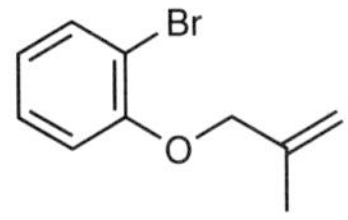	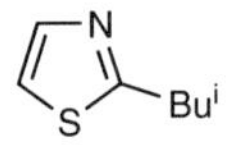	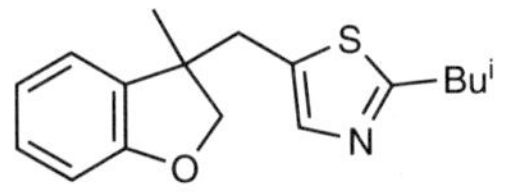	72	[21]

3-Methylene-2,3-dihydro-1*H*-cyclopenta[*b*]quinolines 20; **General Procedure:**[13]
To a suspension of $Pd(OAc)_2$ (11.22 mg, 0.05 mmol, 5 mol%), BINAP (31.13 mg, 0.05 mmol, 5 mol%), and alkenylquinoline **19** (1 mmol) in MeCN (5 mL) was added Et_3N (202 mg, 2 mmol), and the resulting mixture was stirred at 80 °C. After complete consumption of starting material (TLC), the mixture was cooled and the solvent was evaporated under reduced pressure. The crude residue was purified by column chromatography (silica gel, hexane/EtOAc).

3-Benzylindole-2-carboxylates 23; **General Procedure:**[17]
The 2-iodoaniline (0.52 mmol), 2-diazobut-3-enoate **21** (0.5 mmol), and $PdCl_2$ (8.86 mg, 0.05 mmol, 10 mol%) were dissolved in CH_2Cl_2 (2 mL). The resulting mixture was stirred at rt for 3 h and then Ph_3P (13.11 mg, 0.05 mmol, 10 mol%), $NaHCO_3$ (63 mg, 0.75 mmol), and DMF (2 mL) were added. The mixture was stirred at 80 °C for a further 2 h, and then diluted with CH_2Cl_2 (10 mL) and washed with H_2O (5 × 2 mL) and brine (5 mL). The organic layer was dried (Na_2SO_4), the solvent was evaporated under reduced pressure, and the crude mixture was purified by column chromatography (silica gel).

1,3-Dihydrospiro[indene-2,3′-indolin]-2′-ones 26; **General Procedure:**[18]
A 10-mL round-bottomed flask was charged with acrylamide **24** (1 mmol), $PdCl_2(PPh_3)_2$ (14.03 mg, 0.020 mmol, 2 mol%), and Cs_2CO_3 (814.5 mg, 2.5 mmol). The flask was purged with N_2. DMF (4 mL) was then added by syringe and the mixture was again purged with N_2. The mixture was heated in an oil bath at 110 °C for the required time and then cooled to rt. EtOAc (20 mL) and H_2O (12 mL) were added and the mixture was filtered over Celite to remove palladium. The two layers were separated and the organic layer was washed with H_2O (3 × 10 mL). The EtOAc layer was dried (Na_2SO_4) and concentrated. The residue was purified by column chromatography (silica gel, hexanes/EtOAc).

3-(2-Arylvinyl)-1,3-dimethyl-1,3-dihydro-2*H*-indol-2-ones 28; **General Procedure:**[19]
A mixture of acrylamide **27** (0.20 mmol), a hydrazone (0.40 mmol), $Pd(OAc)_2$ (2.24 mg, 0.01 mmol, 5 mol%), *t*-BuOLi (48.03 mg, 0.60 mmol), and Ph_3P (7.87 mg, 0.03 mmol, 15 mol%) in MeCN (3 mL) was stirred at 80 °C for 50 min. After complete consumption of starting material, the reaction was quenched with sat. aq NH_4Cl (3 mL). The mixture was extracted with EtOAc (2 × 5 mL), and the combined organic layers were washed with brine,

dried (Na_2SO_4), filtered, and concentrated. The residue was purified by column chromatography (silica gel, hexanes/EtOAc).

3-(Arylmethyl)-4-methylene-1-tosylpyrrolidines 29; **General Procedure:**[20]
To a stirred soln of *N*-allyl-*N*-(2-bromoallyl)-4-toluenesulfonamide (0.5 mmol) in anhyd THF (6 mL) containing $Pd(PPh_3)_4$ (28.91 mg, 0.025 mmol, 5 mol%) under N_2 was added the arylboronic acid (1 mmol) followed by 2 M aq Na_2CO_3 (1 mL, 2 mmol). The mixture was heated in an oil bath at 80 °C. After complete consumption of starting material (TLC), the mixture was cooled and poured into H_2O (30 mL), and the products were extracted with CH_2Cl_2 (3 × 30 mL). The combined extracts were dried ($MgSO_4$) and concentrated. The crude product was purified by flash chromatography (silica gel, hexanes/EtOAc).

Indoline, Oxindole, and Dihydrobenzo[*b*]furan Derivatives 30; **General Procedure:**[21]
K_2CO_3 (110.56 mg, 0.8 mmol), XPhos (9.45 mg, 0.02 mmol, 5 mol%), and $Pd(OAc)_2$ (4.5 mg, 0.02 mmol, 5 mol%) were added to a screw-cap vial equipped with a magnetic stirrer bar. The vial was purged with argon for 10 min. In a separate vial, the aryl bromide (0.4 mmol), the hetarene (1.6 mmol), and pivalic acid (12.3 mg, 0.12 mmol, 30 mol%) were dissolved in DMA (1.3 mL), and this soln was added to the catalyst mixture using a syringe. The resulting mixture was stirred at 110 °C for 16 h. Volatiles were then removed by horizontal distillation under reduced pressure, and the residual solid was purified by column chromatography (silica gel, hexane/EtOAc) without prior removal of insoluble salts.

1.2.1.2.1.2 **6-*exo-trig* Cyclization**

6-*exo-trig* intramolecular Heck cyclizations have been widely utilized in the synthesis of six-membered carbocycles. Dienyl iodide **31** cyclizes via a 6-*exo-trig* process to give cyclohexene **32** (Scheme 10).[22] Here palladium(II) acetate and triphenylphosphine catalyze the reaction in the presence of silver(I) phosphate and Proton-sponge [1,8-bis(dimethylamino)naphthalene] as additives. The reaction tolerates a free hydroxy group but with a decreased yield of 61% compared to the 93% yield obtained with the silyl-protected substrate **31**.

Scheme 10 Synthesis of a Six-Membered Carbocycle[22]

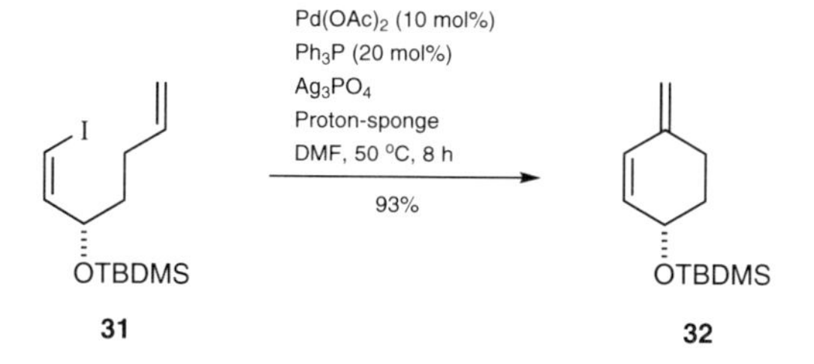

Similarly, benzo-fused carbocycles such as tetrahydronaphthalene derivatives **34** can be prepared from aryl trifluoromethanesulfonates **33** using palladium(II) acetate and 2,2′-bis(diphenylphosphino)-1,1′-binaphthyl (BINAP) (Scheme 11).[23] Here, variation of temperature and solvent have a dramatic effect on the cyclization. The use of tetrahydrofuran at 70 °C returns only starting material; changing the solvent to dimethylformamide at 90 °C affords some starting material and some hydrogen-exchanged product, whereas in dimethylacetamide at 120 °C the cyclized product is formed exclusively in high yields.

Scheme 11 Synthesis of Tetrahydronaphthalene Derivatives[23]

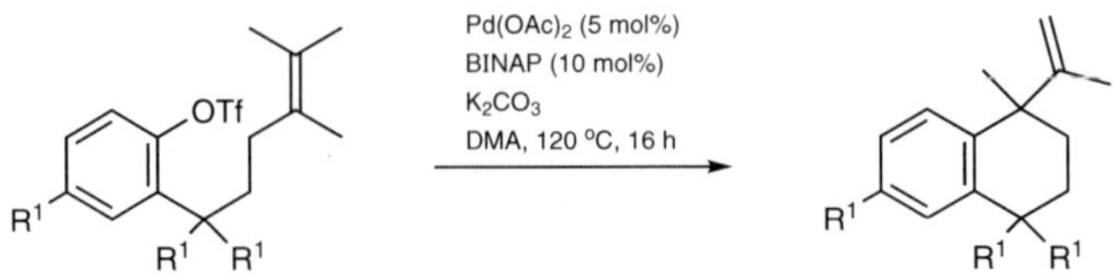

R^1	Yield (%)	Ref
H	80	[23]
Me	85	[23]

Six-membered benzo-fused heterocyclic compounds such as 1,2,3,4-tetrahydroquinoline, 1,2,3,4-tetrahydroisoquinoline, 3,4-dihydro-2*H*-1-benzopyran, and 3,4-dihydro-1*H*-2-benzopyran (isochromane) derivatives **36** can be accessed from aryl bromides **35** using a general method for palladium(II) acetate catalyzed 6-*exo-trig* intramolecular reductive Heck cyclization (Scheme 12).[24] Importantly, this method works under ligand-free conditions and can also be utilized for the synthesis of five-membered benzo-fused heterocycles such as indolines and 2,3-dihydrobenzo[*b*]furan derivatives via 5-*exo-trig* intramolecular Heck reaction. Water is essential in this reaction, with up to 5% in dimethylformamide tolerated. It is noteworthy that only dehalogenated starting material is recovered under anhydrous conditions.

Scheme 12 Synthesis of Various Heterocycles via 6-*exo-trig* Cyclization[24]

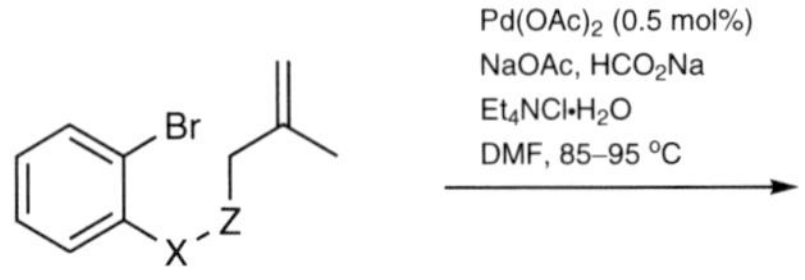

X	Z	Yield (%)	Ref
NBoc	CH_2	96	[24]
CH_2	NBoc	95	[24]
CH_2	O	91	[24]
O	CH_2	96	[24]

A diversely substituted library of isoquinoline derivatives **38** can be obtained in excellent yields in a two-step process employing an Ugi four-component reaction and palladium-catalyzed 6-*exo-trig* intramolecular Heck cyclization (Scheme 13).[25] Here the precursor for the intramolecular Heck reaction **37** is obtained from readily available starting materials such as allylamine, isonitriles, 2-halobenzaldehydes, and benzoic and alkanoic acids. A range of isoquinolinone derivatives have been prepared using the same methodology from 2-halobenzoic acids and arylaldehydes.

Scheme 13 Synthesis of Isoquinoline Derivatives[25]

X	R^1	R^2	R^3	Yield (%) of **38** from **37**	Ref
I	OMe	Ph	*t*-Bu	94	[25]
I	OMe	$4\text{-}O_2NC_6H_4$	*t*-Bu	90	[25]
I	OMe	$4\text{-}MeOC_6H_4$	Cy	92	[25]
Br	H	Me	*t*-Bu	85	[25]
Br	H	$4\text{-}MeOC_6H_4$	*t*-Bu	87	[25]

The synthesis of 3,4-dihydroisoquinolin-1-one derivatives **40** bearing an ester group has been achieved by a palladium-catalyzed carbonylative Heck cyclization (Scheme 14).[26] In this reaction, aryl iodides **39** undergo an intramolecular Heck cyclization followed by an alkoxycarbonylation. The insertion of carbon monoxide into the Pd—C bond occurs faster than β-hydride elimination from the formed intermediate σ-alkyl palladium complex during cyclization.

Scheme 14 Synthesis of 3,4-Dihydroisoquinolin-1-one Derivatives[26]

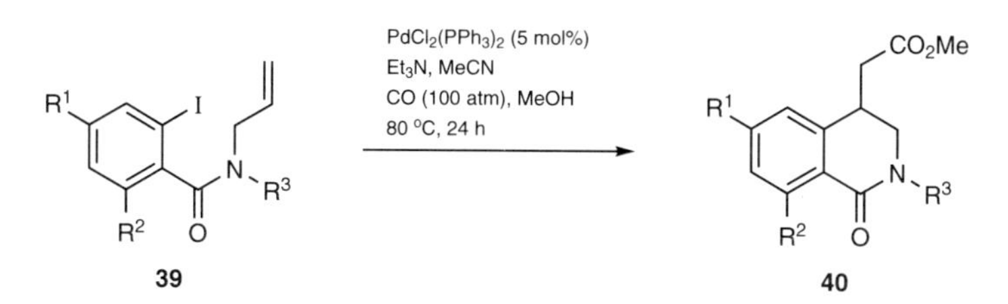

R^1	R^2	R^3	Yield (%)	Ref
H	F	$CH_2CH{=}CH_2$	86	[26]
H	F	Me	67	[26]
Cl	H	$CH_2CH{=}CH_2$	87	[26]
Cl	H	Ph	83	[26]

A range of 1,2,3-triazole-fused heterocycles and carbocycles possessing an exocyclic double bond such as **41** can be prepared via 6-*exo-trig* Heck cyclization (Scheme 15).[27] The optimized catalytic system provides exocyclic products in excellent yields, irrespective of the heteroatom present in the starting material.

Scheme 15 Synthesis of 1,2,3-Triazole-Fused Heterocycles and Carbocycles[27]

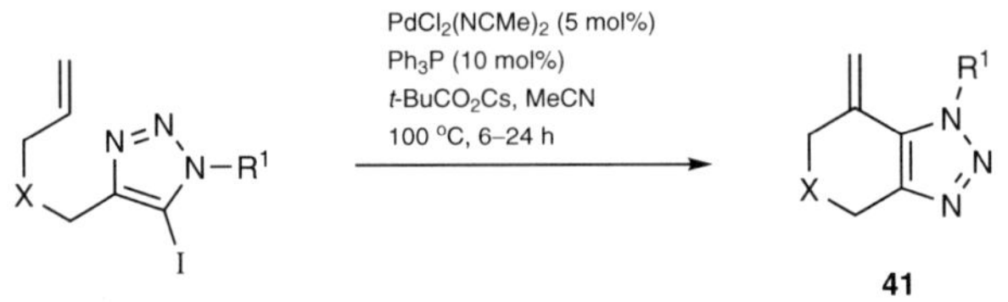

$PdCl_2(NCMe)_2$ (5 mol%)
Ph_3P (10 mol%)
t-$BuCO_2Cs$, MeCN
100 °C, 6–24 h

41

X	R^1	Yield (%)	Ref
NTs	$(CH_2)_5Me$	99	[27]
$C(CO_2Et)_2$	$(CH_2)_5Me$	89	[27]
O	Bn	99	[27]
O	PMB	92	[27]

1-(*tert*-Butyldimethylsiloxy)-4-methylenecyclohex-2-ene (32); **Typical Procedure:**[22]
A mixture of (*Z*)-1-iodohepta-1,6-diene **31** (420 mg, 1.18 mmol), $Pd(OAc)_2$ (27 mg, 0.12 mmol, 10 mol%), Ph_3P (63 mg, 0.24 mmol, 20 mol%), Ag_3PO_4 (490 mg, 1.18 mmol), and Proton-sponge (510 mg, 2.36 mmol) was dried under vacuum and maintained under an argon atmosphere. To this mixture was added DMF (5 mL) and the resulting soln was stirred at 50 °C for 8 h, diluted with H_2O (5 mL), and extracted with Et_2O (3 × 10 mL). The combined organic layers were washed with H_2O (3 × 10 mL) and brine and dried ($MgSO_4$), before being concentrated and purified by column chromatography (silica gel, hexane); yield: 0.25 g (93%).

1,4,4,6-Tetramethyl-1-(prop-1-en-2-yl)-1,2,3,4-tetrahydronaphthalene (34, **R^1 = Me); Typical Procedure:**[23]
To a soln of 2-alkenylphenyl trifluoromethanesulfonate **33** (R^1 = Me; 3.9 g, 10.3 mmol) in anhyd DMA (40 mL) was added K_2CO_3 (4.3 g, 30.9 mmol) and BINAP (0.65 g, 1.03 mmol, 10 mol%). The mixture was degassed with N_2 for 15 min, $Pd(OAc)_2$ (115 mg, 0.515 mmol, 5 mol%) was added, and the mixture was degassed with N_2 for an additional 15 min and then heated in an oil bath at 120 °C under N_2 for 16 h. The dark soln was cooled to rt and the solvent was removed under reduced pressure. The mixture was diluted with EtOAc (100 mL) and washed with H_2O (100 mL). The aqueous layer was extracted with EtOAc (2 × 100 mL). The combined organic layer was washed with brine (100 mL), dried (Na_2SO_4), filtered, and concentrated under reduced pressure. The crude product was purified by flash column chromatography (EtOAc/hexane) to provide a colorless oil; yield: 2.0 g (85%).

1,2,3,4-Tetrahydroquinoline, 1,2,3,4-Tetrahydroisoquinoline, and Dihydrobenzopyran Derivatives 36; **General Procedure:**[24]
A mixture of aryl bromide **35** (1.9 mmol), NaOAc (430 mg, 5.0 mmol), HCO_2Na (150 mg, 2.4 mmol), $Et_4NCl•H_2O$ (410 mg, 2.4 mmol), and $Pd(OAc)_2$ (2.25 mg, 0.01 mmol, 0.5 mol%) in DMF (5 mL) was degassed and refilled with N_2 (3 ×). The mixture was heated to 85–95 °C for 1–4 h. The reaction was monitored by HPLC and GC. After completion, the mixture was cooled to rt, H_2O (15 mL) was added, and the mixture was filtered through a Cel-

ite bed and washed with *t*-BuOMe (25 mL). The organic phase was separated and concentrated to give an oily residue, which was purified by column chromatography (silica gel, EtOAc/hexanes).

2-Acyl-4-methylisoquinoline-1-carboxamides 38; **General Procedure:**[25]
$Pd(OAc)_2$ (5.6 mg, 0.025 mmol, 5 mol%), Cy_3P (14.0 mg, 0.05 mmol, 10 mol%), and Ugi product **37** (0.5 mmol) were added to a dry Schlenk tube under N_2. DMA (5 mL) and Cy_2NMe (0.42 mL, 2.0 mmol) were added under N_2. The mixture was stirred at 60 or 100 °C for 4–24 h, the solvent was removed under reduced pressure, and the residue was purified by flash column chromatography (EtOAc/CH_2Cl_2/petroleum ether).

Methyl 2-(1-Oxo-1,2,3,4-tetrahydroisoquinolin-4-yl)acetates 40; **General Procedure:**[26]

CAUTION: *Carbon monoxide is extremely flammable and toxic, and exposure to higher concentrations can quickly lead to coma.*

A mixture of 2-iodobenzamide **39** (1 mmol), $PdCl_2(PPh_3)_2$ (35 mg, 0.05 mmol, 5 mol%), Et_3N (0.28 mL, 2 mmol), and MeOH (0.16 mL, 4 mmol) in MeCN (20 mL) was placed in an autoclave. It was purged 3 times with CO (20 atm) to remove residual air and then charged with CO (100 atm). The autoclave was heated at 80 °C for 24 h and then cooled to rt. After CO removal, the yellow soln was diluted with brine (20 mL) and extracted with EtOAc (3 × 20 mL). The organic phases were dried (Na_2SO_4), the solvent was evaporated, and the product was purified by column chromatography (silica gel, petroleum/EtOAc).

Annulated 1,2,3-Triazoles 41; **General Procedure:**[27]
The 5-iodo-1,2,3-triazole substrate (1 mmol), $PdCl_2(NCMe)_2$ (12.97 mg, 0.05 mmol, 5 mol%), Ph_3P (26.2 mg, 0.1 mmol, 10 mol%), and *t*-$BuCO_2Cs$ (468 mg, 2 mmol) were dissolved in MeCN (16 mL) and heated at 100 °C for 6–24 h. The solvent was removed under reduced pressure and the residue was purified by column chromatography.

1.2.1.2.2 5- and 6-*endo-trig* Processes

The 5- and 6-*endo-trig* modes of intramolecular Heck cyclization are disfavored according to Baldwin's rules. Nevertheless, there are a few results that show the possibility of 5- or 6-*endo* processes by intramolecular Heck cyclization.

1.2.1.2.2.1 5-*endo-trig* Cyclization

The 5-*endo-trig* mode of intramolecular Heck cyclization is used in the formation of a range of substituted indenones **43** (Scheme 16).[28] This reaction involves sequential intramolecular Heck cyclization followed by oxidation of the allylic alcohol. No reaction occurs in the absence of the benzylic hydroxy group in the substrate **42** or air. Similarly, conversion of vinyl bromides **44** into ketones is a useful method for the preparation of a variety of cyclopentenone derivatives **45** (Scheme 17).[29]

for references see p 92

Scheme 16 Synthesis of Indenones[28]

$Pd(OAc)_2$ (5 mol%)
Ph_3P (15 mol%)
K_2CO_3, DMF, air
80 °C, 24 h

42 → **43**

R^1	R^2	R^3	Yield (%)	Ref
OMe	4-$MeOC_6H_4$	3,5-$(MeO)_2C_6H_3$	62	[28]
OMe	H	Ph	52	[28]
OMe	Pr	4-$MeOC_6H_4$	69	[28]
H	Ph	Ph	55	[28]
H	Ph	4-$O_2NC_6H_4$	61	[28]

Scheme 17 Synthesis of Cyclopentenone Derivatives[29]

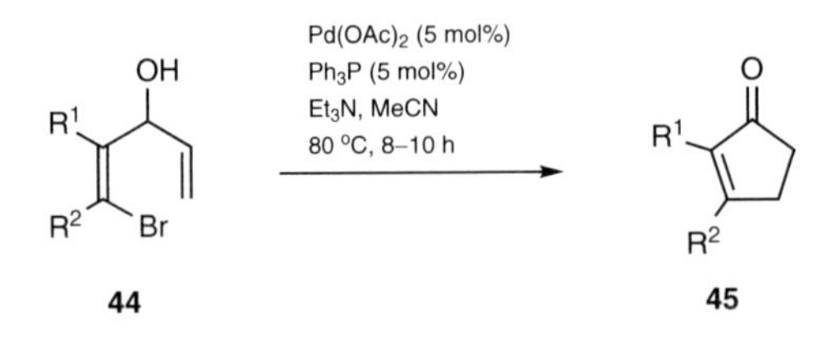

R^1	R^2	Yield (%)	Ref
$(CH_2)_4$		65	[29]
$(CH_2)_5$		70	[29]
$(CH_2)_6$		73	[29]
H	Ph	70	[29]
H	2-naphthyl	70	[29]

Among the available approaches to prepare indoles via a 5-*endo-trig* cyclization, the reaction of simple and readily available 2-bromoanilines and vinyl bromides to give variously substituted 1*H*-indoles **47** is an interesting example. This is a one-pot process that involves alkenyl amination to obtain the enamine **46** which then undergoes a 5-*endo-trig* intramolecular Heck cyclization to provide a range of substituted indole derivatives (Scheme 18).[30]

Scheme 18 Synthesis of 1*H*-Indole Derivatives[30]

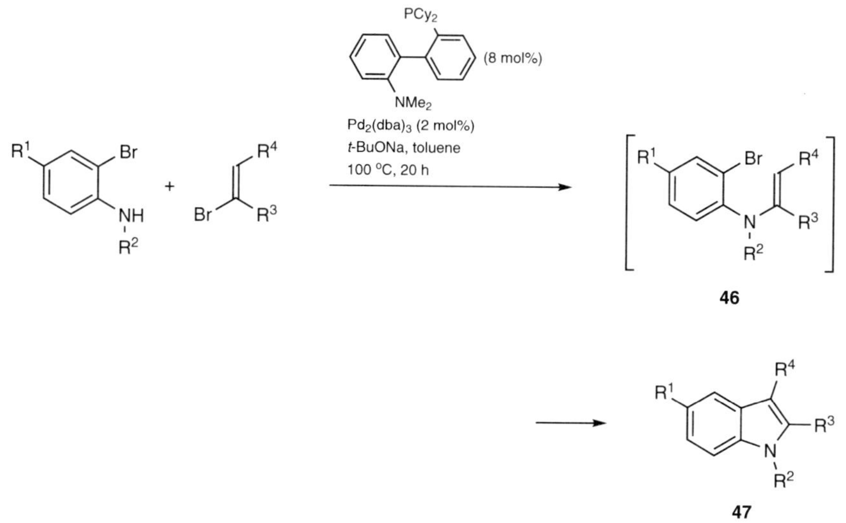

R^1	R^2	R^3	R^4	Yield (%)	Ref
H	H	Ph	H	64	[30]
Me	H	Ph	H	62	[30]
H	H	CH_2OBn	H	61	[30]
H	Me	H	Ph	70	[30]
H	Me	H	CH_2OBn	63	[30]

1*H*-Inden-1-ones 43**; General Procedure:**[28]

Aryl bromide **42** (1 mmol), K_2CO_3 (276.4 mg, 2 mmol), $Pd(OAc)_2$ (11.22 mg, 0.05 mmol, 5 mol%), and Ph_3P (39.34 mg, 0.15 mmol, 15 mol%) were dissolved in DMF (5 mL). The resulting mixture was heated to 80 °C and stirred for 24 h. The mixture was then extracted with Et_2O (3 × 20 mL), washed with brine, dried (Na_2SO_4), filtered, and concentrated under reduced pressure. The product was purified by column chromatography (silica gel, petroleum ether/EtOAc).

Cyclopent-2-en-1-one Derivatives 45**; General Procedure:**[29]

The dienyl bromide **44** (1 mmol), $Pd(OAc)_2$ (11.22 mg, 0.05 mmol, 5 mol%), Ph_3P (13.11 mg, 0.05 mmol, 5 mol%), and Et_3N (121 mg, 1.2 mmol) were dissolved in MeCN (3 mL). After being degassed with argon, the soln was heated to 80 °C for 8–10 h. The mixture was then cooled, diluted with cold H_2O (5 mL), and extracted with Et_2O (3 × 5 mL). The solvent was evaporated after drying (Na_2SO_4), and the product was purified by column chromatography (silica gel, hexanes/EtOAc).

1*H*-Indoles 47**; General Procedure:**[30]

A reaction tube under a N_2 atmosphere was charged with 2-(dicyclohexylphosphino)-2′-(*N*,*N*-dimethylamino)biphenyl (DavePhos; 0.031 g, 0.08 mmol, 8 mol%), $Pd_2(dba)_3$ (18.3 mg, 0.02 mmol, 4 mol% Pd), *t*-BuONa (0.288 g, 3 mmol), and toluene (4 mL). After 1 min, the vinyl bromide (1 mmol) and 2-bromoaniline (1 mmol) were added under N_2 and the tube was placed in a carousel block and heated to 100 °C with stirring for 20 h. The mixture was then allowed to cool to rt, was taken up in hexanes (15 mL), and filtered through Celite. The solvents were evaporated under reduced pressure. Purification by flash column chromatography (hexanes/EtOAc) afforded the product.

1.2.1.2.2.2 ***6-endo-trig* Cyclization**

The 6-*endo-trig* cyclization competes against 5-*exo-trig* cyclization, with the 5-*exo-trig* mode being favored. Nevertheless, selective formation of 6-*endo-trig* cyclization products has been reported. The cyclization of amides **48**, possessing various substituents on the alkene, gives, under standard Heck reaction conditions, 6-*endo-trig* tricyclic products **49** exclusively (Scheme 19).[31] Here the 5-*exo-trig* mode of cyclization is believed to be disfavored due to the strain involved in the approach of the palladium species to the α-position of the double bond.

Scheme 19 Synthesis of Tricyclic Systems via 6-*endo-trig* Cyclization[31]

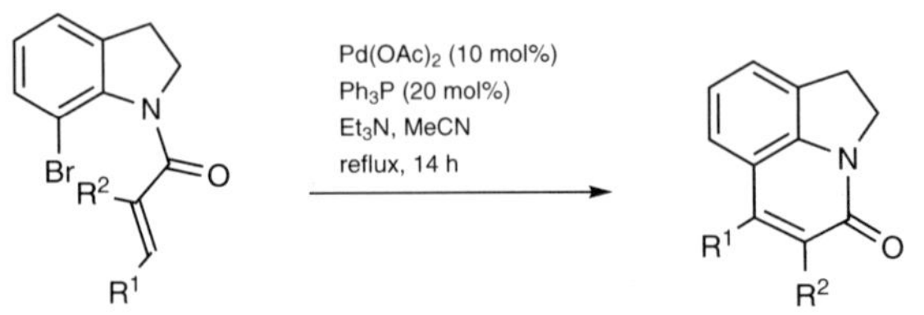

$Pd(OAc)_2$ (10 mol%)
Ph_3P (20 mol%)
Et_3N, MeCN
reflux, 14 h

48 → **49**

R^1	R^2	Yield (%)	Ref
H	H	99	[31]
Me	H	95	[31]
Ph	H	94	[31]
$CH_2N(Me)(CH_2)_2$		99	[31]
$(CH_2)_4$		88	[31]

A range of quinoline derivatives may be prepared by the reaction of 2-iodoanilines with 4-aryl-2-diazobut-3-enoates **50** catalyzed by palladium(II) chloride (Scheme 20).[17] Here, the electronic effects on both substrates play an important role in the selective formation of 6-*endo-trig* cyclized products **52** over 5-*exo-trig* products **53**. If the substrates possess strongly electron-withdrawing groups, the quinolines **52** are isolated as the major products. This cascade process involves sequential carbenoid-based N—H insertion to give intermediate **51** and an electronically controlled 5-*exo-trig* or 6-*endo-trig* intramolecular Heck cyclization. A single palladium source catalyzes both reactions in one pot and provides the products in good yields.

Scheme 20 Synthesis of Quinoline Derivatives[17]

R^1	R^2	Ratio (**52**/**53**)	Yield[a] (%)	Ref
H	NO_2	>20:1	83	[17]
CN	H	9:1	62	[17]
H	H	<1:>20	74	[17]
H	OMe	<1:>20	78	[17]

[a] Isolated yield of major product.

1,2-Dihydro-4*H*-pyrrolo[3,2,1-*ij*]quinolin-4-ones 49; **General Procedure:**[31]
The unsaturated amide **48** (1 mmol) was dissolved in MeCN (20 mL) and Et_3N (0.29 mL, 2 mmol) was added. The resulting soln was charged with Ph_3P (53 mg, 0.2 mmol) and $Pd(OAc)_2$ (23 mg, 0.1 mmol) and this mixture was refluxed for 10 h. The volatiles were removed under reduced pressure and the crude material was purified by flash chromatography (EtOAc/hexanes).

Methyl 4-Phenylquinoline-2-carboxylates 52; **General Procedure:**[17]
The 2-iodoaniline (0.52 mmol), α-diazo ester **50** (0.5 mmol), and $PdCl_2$ (4.4 mg, 0.025 mmol, 5 mol%) were dissolved in CH_2Cl_2 (2 mL). The resulting mixture was stirred at rt for 3 h, and then Ph_3P (13 mg, 0.05 mmol, 10 mol%), $NaHCO_3$ (63 mg, 0.75 mmol), and DMF (2 mL) were added. The mixture was stirred at 80 °C for a further 2 h, diluted with CH_2Cl_2 (10 mL), and washed with H_2O (5 × 2 mL) and brine (5 mL). The organic layer was dried (Na_2SO_4), the solvent was evaporated under reduced pressure, and the crude mixture was purified by column chromatography (silica gel).

1.2.1.3 Formation of Medium-Sized Rings (7 to 11)

As the ring size increases, the preference for the *endo* mode of cyclization increases. This is due to the more flexible tether in these larger rings generating a more energetically favorable transition state for *endo*, which is not possible for smaller ring sizes.

for references see p 92

1.2.1.3.1 7- and 8-*exo-trig* Processes

1.2.1.3.1.1 7-*exo-trig* Cyclization

An intramolecular 7-*exo-trig* Heck cyclization is used as a key transformation in the synthesis of many natural products and their analogues, such as faveline methyl ether and taxol. Cyclization of substrate type **54** results in the formation of azepine ring **55** (seven-membered N-heterocycle) in the synthesis of analogues of duocarmycin natural products (Scheme 21).[32] Palladium(0) catalyzes the reaction in 89% yield without the formation of any double-bond-isomerized product.

Scheme 21 Fused Azepine Synthesis[32]

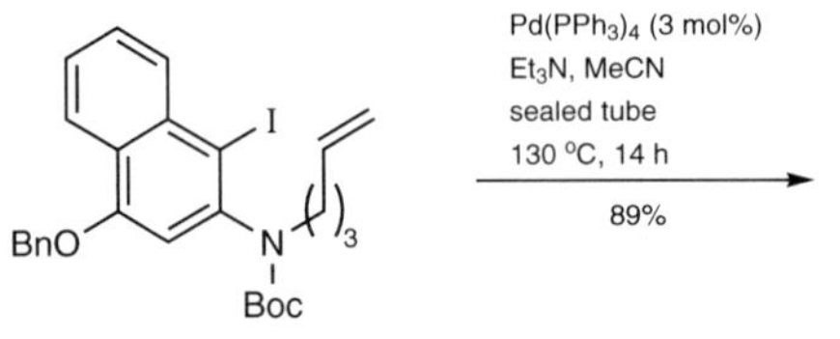

$Pd(PPh_3)_4$ (3 mol%), Et_3N, MeCN, sealed tube, 130 °C, 14 h; 89%

54 → **55**

***tert*-Butyl 7-(Benzyloxy)-1-methylene-1,2,3,4-tetrahydro-5*H*-naphtho[2,1-*b*]azepine-5-carboxylate (55); Typical Procedure:**[32]

A soln of alkenylamine **54** (503 mg, 0.926 mmol, 1 equiv) in MeCN (18 mL; degassed and purged with argon) in a thick-walled reaction tube was treated with Et_3N (0.257 mL, 1.85 mmol, 2 equiv) followed by $Pd(PPh_3)_4$ (32 mg, 0.0277 mmol, 3 mol%). The reaction vessel was sealed, and the mixture was warmed at 130 °C for 14 h. After complete consumption of starting material, concentration gave a yellow semisolid which was suspended in EtOAc/hexanes (1:19). The precipitated salts were removed by filtration and thoroughly rinsed with the solvent mixture. Concentration of the filtrate and radial chromatography (2-mm plate, silica gel, EtOAc/hexanes 1:19) gave a colorless oil, which slowly crystallized upon storage; yield: 343 mg (89%).

1.2.1.3.1.2 8-*exo-trig* Cyclization

Dibenzoazocine derivatives **57** can be prepared from acyclic precursors **56** through an 8-*exo-trig* ring-closing pathway (Scheme 22).[33] This reaction works with palladium(II) acetate as catalyst and potassium acetate as base in dimethylformamide under ligand-free conditions. The amine in **56** must be protected for the reaction to proceed. This is likely due to the unprotected amine deactivating the palladium(II) acetate through complexation.

Scheme 22 Synthesis of Dibenzoazocine Derivatives[33]

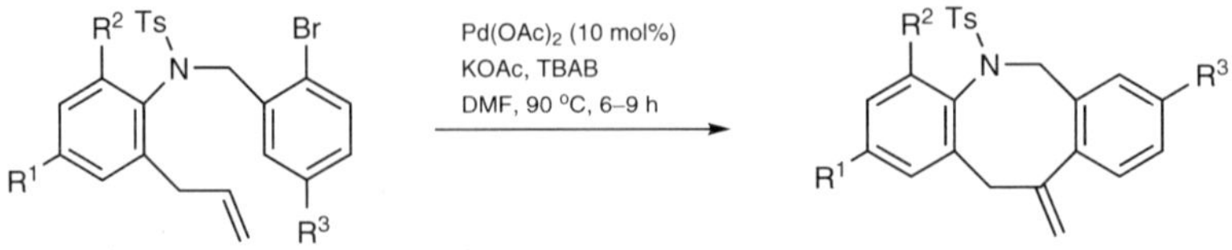

$Pd(OAc)_2$ (10 mol%), KOAc, TBAB, DMF, 90 °C, 6–9 h

56 → **57**

R^1	R^2	R^3	Yield (%)	Ref
Me	H	H	72	[33]
H	F	H	73	[33]
H	H	H	75	[33]
Me	H	OMe	79	[33]
H	F	OMe	73	[33]
H	H	OMe	79	[33]

11-Methylene-5-tosyl-5,6,11,12-tetrahydrodibenzo[*b,f*]azocines 57; **General Procedure:**[33]
A mixture of 2-allyl-*N*-tosylaniline **56** (0.212 mmol), TBAB (82 mg, 0.255 mmol), and dry KOAc (52 mg, 0.531 mmol) was dissolved in dry DMF (10 mL) under a N_2 atmosphere. $Pd(OAc)_2$ (4.75 mg, 0.021 mmol, 10 mol%) was added, and the mixture was stirred at 90 °C for the required time and then cooled to rt. H_2O (20 mL) was added, the mixture was extracted with EtOAc (3 × 30 mL), and the EtOAc extract was washed with H_2O (2 × 40 mL) and brine (30 mL). The organic layer was dried (Na_2SO_4), and the solvent was distilled off to furnish a viscous mass, which was purified by column chromatography (silica gel, EtOAc/petroleum ether).

1.2.1.3.2 7- and 8-*endo-trig* Processes

1.2.1.3.2.1 7-*endo-trig* Cyclization

The 7-*endo-trig* mode of cyclization is uncommon and is favored only when the 6-*exo-trig* cyclization leads to an intermediate that lacks β-hydrogens, or when eclipsed insertion topology is disfavored for 6-*exo-trig* cyclization by virtue of the ring system. Benzazepines and their derivatives **59** have been prepared from alkylated 2-(trimethylsilyl)ethylsulfonyl (SES) protected β-amino esters **58** in a 7-*endo-trig* Heck cyclization in excellent yields (Scheme 23).[34] The reaction requires the use of poly(ethylene glycol) (PEG-3400-OH) as the solvent and uses a microwave-assisted methodology.

Scheme 23 Synthesis of Benzazepine Derivatives[34]

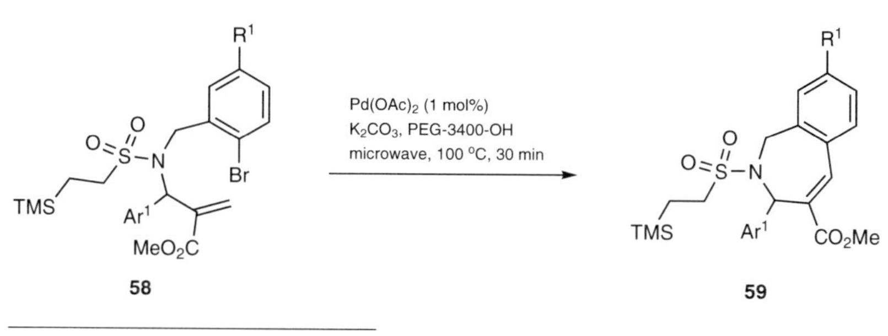

Ar^1	R^1	Yield (%)	Ref
Ph	H	95	[34]
$3-FC_6H_4$	H	91	[34]
$4-MeO_2CC_6H_4$	H	82	[34]
Ph	OMe	83	[34]
Ph	F	76	[34]

Methyl 3-Aryl-2-{[2-(trimethylsilyl)ethyl]sulfonyl}-2,3-dihydro-1*H*-benzo[*c*]azepine-4-carboxylates 59; **General Procedure:**[34]
A mixture (220 mg) of $Pd(OAc)_2$/PEG-3400-OH (1 mg/1 g) [corresponding to 0.001 mmol $Pd(OAc)_2$, 0.01 equiv] and finely powdered K_2CO_3 (41 mg, 0.3 mmol) was added to *N*-(2-bromobenzyl)-β-amino ester **58** (0.1 mmol). The mixture was heated under microwave irradiation at 100 °C (initial power 300 W) for 30 min, and then cooled, diluted with CH_2Cl_2, precipitated with Et_2O, and filtered. The solvents were removed under reduced pressure and the residue was purified by column chromatography (silica gel, Et_2O/cyclohexane).

1.2.1.3.2.2 8-*endo-trig* Cyclization

In general 7-*exo-trig* cyclization is slightly favored over 8-*endo-trig* cyclization; however, under the right conditions, 8-*endo-trig* cyclizations can proceed with excellent yields. One such example is the synthesis of various oxocin and oxathiocin derivatives **60**.[35,36] Here, substituted phenyl-based precursors are cyclized through the 8-*endo-trig* mode using palladium(II) acetate (naphthalene-based precursors have also been used). Tetrabutylammonium bromide plays an important role and the reaction does not proceed in its absence (Scheme 24).

Scheme 24 Synthesis of Oxocin and Oxathiocin Derivatives[35,36]

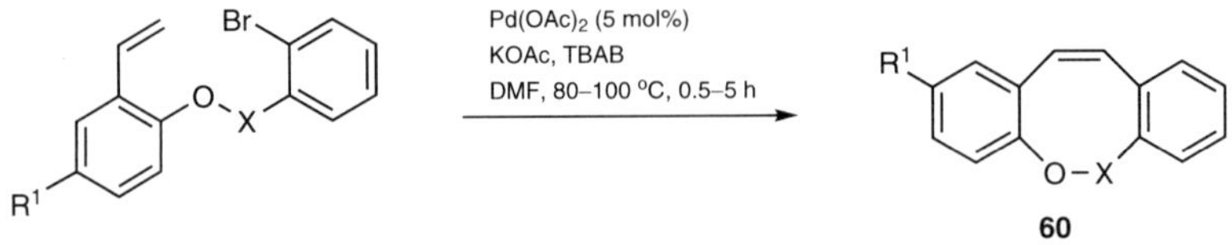

$Pd(OAc)_2$ (5 mol%)
KOAc, TBAB
DMF, 80–100 °C, 0.5–5 h

R^1	X	Yield (%)	Ref
Me	CH_2	76	[36]
Me	SO_2	91	[35]
OMe	CH_2	69	[36]
OMe	SO_2	84	[35]
H	CH_2	71	[36]
H	SO_2	88	[35]

6*H*-Dibenzo[*b*,*f*]oxocins 60 **(X = CH_2) and Dibenzo[*c*,*g*][1,2]oxathiocin 6,6-Dioxides** 60 **(X = SO_2); General Procedure:**[35,36]
A mixture of the aryl bromide (0.13 mmol), TBAB (50 mg, 0.15 mmol), and anhyd KOAc (34 mg, 0.35 mmol) was dissolved in anhyd DMF (10 mL) under a N_2 atmosphere. $Pd(OAc)_2$ (1.43 mg, 0.0065 mmol, 5 mol%) was added and the mixture was stirred at 80 °C for 60 min before being cooled. H_2O (10 mL) was added and the mixture was extracted with EtOAc (3 × 30 mL). The combined organic layers were washed with H_2O (2 × 40 mL) and then brine (30 mL). The organic layer was dried (Na_2SO_4), the solvent was distilled off, and the residue was purified by column chromatography (silica gel, EtOAc/petroleum ether).

1.2.1.4 Formation of Macrocycles (>11)

Macrocycles are most common in natural products and are even present in several marketed drugs. Normally, macrocyclization suffers from low yields and more side products due to the high functionality within the system. Recently, there has been a growing interest in the utilization of the intramolecular Heck cyclization as the key step in the later-stage synthesis of macrocycles. This is because the Heck reaction proceeds under mild reaction conditions with high functional-group tolerance and excellent yields.

An intramolecular Heck cyclization is used as the key step in the synthesis of several 17-membered functionalized macrocyclic compounds, which are of interest because several bioactive natural products contain 17-membered functionalized rings. The combination of palladium(II) acetate and tri-2-tolylphosphine catalyzes the reaction, providing good yields of the final cyclized products **62** from Heck precursors **61** (Scheme 25).[37]

Scheme 25 Synthesis of 17-Membered Macrocycles[37]

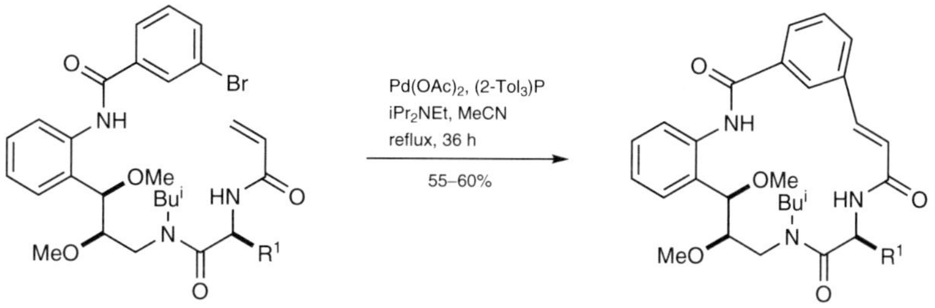

$Pd(OAc)_2$, $(2\text{-}Tol_3)P$, iPr_2NEt, MeCN, reflux, 36 h; 55–60%

61 → **62**

R^1= iPr, Bn, iBu, *s*-Bu, Me

(9*S*,13*R*,14*R*,*E*)-9-Alkyl-11-isobutyl-13,14-dimethoxy-2,8,11-triaza-1(1,2),4(1,3)-dibenzenacyclotetradecaphan-5-ene-3,7,10-triones 62; **General Procedure:**[37]

To a soln of aryl bromide **61** (0.077 mmol) in MeCN (120 mL) were added $Pd(OAc)_2$ (3.45 mg, 0.0154 mmol) and $(2\text{-}Tol)_3P$ (5 mg, 0.0154 mmol), followed by iPr_2NEt (2 mg, 0.0155 mmol). The mixture was refluxed for 36 h and then cooled to rt and diluted with CH_2Cl_2. It was then filtered through Celite, the filtrate was concentrated under reduced pressure, and the residue was dissolved in CH_2Cl_2 (15 mL). This organic soln was washed with brine, dried (Na_2SO_4), filtered, and concentrated to leave a crude oil, which was purified by column chromatography (silica gel, EtOAc/hexanes).

1.2.2 Enantioselective Reactions

The intramolecular Heck reaction is a versatile and powerful synthetic transformation that enables the construction of various spirocyclic, fused, bridged, and isolated ring systems. The conditions are mild, and the reaction has excellent functional-group tolerance and can form highly congested bonds that would be difficult to make by other means. The asymmetric variant has found extensive application in the synthesis of natural products as it allows formation of both tertiary and quaternary chiral centers during the cyclization. This section highlights intramolecular Heck processes that enable high-yielding cyclizations with control of stereochemistry at the newly formed sp^3 center. It is organized according to the substitution on the stereocenter formed (i.e., tertiary or quaternary), the type of alkene used (cyclic or acyclic), and the type of halide or pseudohalide used (aryl or vinyl).

The first asymmetric intramolecular Heck reactions were reported independently in 1989 by the groups of Overman[38] and Shibasaki.[39] The key to both approaches was maintaining the sp^3 stereocenter formed after migratory insertion; this was crucial, as uncontrolled bond rotation followed by β-hydride elimination would racemize the sp^3 center. Two different strategies were developed to overcome this problem. Overman designed a substrate whose intermediate would not have a β-hydrogen, and thus would undergo fur-

for references see p 92

ther cyclization in a tandem Heck process.[38] Shibasaki used cyclic alkenes as substrates, which led to a cyclic intermediate in which bond rotation was not possible and therefore β-hydride elimination would occur away from the stereogenic center.[39]

1.2.2.1 Formation of Tertiary Carbon Centers

1.2.2.1.1 Tertiary Centers from Acyclic Alkenes

Acyclic alkenes have found limited application in the formation of tertiary stereocenters due to the difficulty of directing β-hydride elimination away from the newly formed tertiary sp^3 center. This can be overcome by using allylsilanes as terminating alkenes, employing reactions that form endocyclic alkenes, or by employing a reductive Heck cyclization. These methods have been successfully employed with both aryl halides and pseudohalides to synthesize a range of enantioenriched benzo-fused carbo- or heterocyclic five-, six-, and seven-membered rings.

1.2.2.1.1.1 Reactions from Aryl Halides and Pseudohalides

The asymmetric cyclization of aryl iodide derivatives **66** bearing a *Z*- or *E*-allylsilane group provides the desilylated products **67** with high selectivity over the vinylsilane products (Scheme 26).[40–43] Bisphosphine **64** [(+)-TMBTP] is the optimal ligand for use with *Z*-allylsilanes, whereas ligand **65** [(*R*)-BITIANP] is optimal for *E*-allylsilanes, and both reactions employ silver(I) phosphate as the iodide scavenger. A range of enantioenriched carbo- and heterocyclic 6,6- and 6,7-ring systems have been prepared using this procedure, including 7-demethyl-2-methoxycalamene, the key intermediate for the natural product norsesquiterpene.[40]

Scheme 26 Enantioselective Cyclization of Iodoarene Derivatives with Alkenylsilane Chains To Give Vinyl-Substituted Products[40–43]

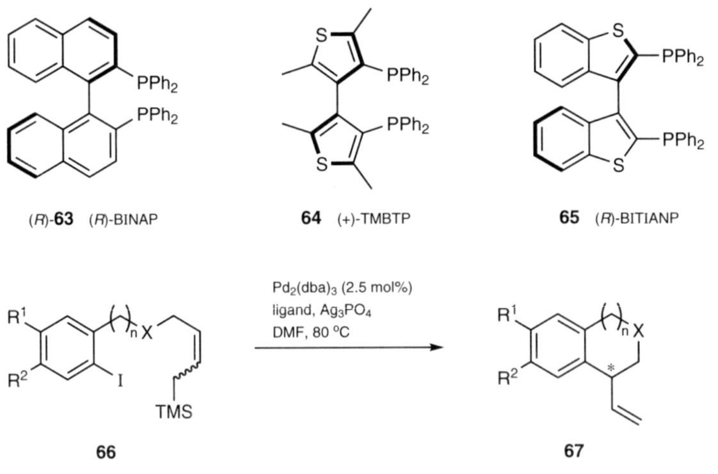

(*R*)-**63** (*R*)-BINAP

64 (+)-TMBTP

65 (*R*)-BITIANP

66 → **67**: $Pd_2(dba)_3$ (2.5 mol%), ligand, Ag_3PO_4, DMF, 80 °C

Config of **66**	R^1	R^2	n	X	Ligand	Time (h)	Config of **67**	ee[a] (%)	Yield (%)	Ref
Z	OMe	Me	1	CH_2	(*R*)-**63** (7 mol%)	48	*R*	92	91	[40,41]
Z	OMe	OMe	2	$NCOCF_3$	(*S*)-**63** (7 mol%)	40	*S*	64	72	[42]
Z	OMe	OMe	2	$NCOCF_3$	**64** (15 mol%)	68	*S*	92	71	[43]
Z	H	H	1	$NCOCF_3$	**64** (10 mol%)	65	*S*	84	80	[43]
Z	OMe	OMe	1	$NCOCF_3$	**64** (20 mol%)	64	*S*	84	73	[43]
E	OMe	OMe	2	$NCOCF_3$	**65** (10 mol%)	24	*S*	91	66	[43]

[a] Determined by HPLC analysis using a chiral stationary phase.

An efficient asymmetric reductive Heck cyclization of *ortho*-nonafluorobutylsulfonyloxy- (nonaflate) and *ortho*-trifluoromethylsulfonyloxy-substituted aryl vinyl ketones (e.g., **69**) has been developed with bisphosphine **68** [(*R*)-3,5-Xyl-MeO-BIPHEP] as the optimal chiral ligand.[44] This facilitates the synthesis of either enantiomerically enriched 3-substituted 2,3-dihydro-1*H*-inden-1-ones **70** or *exo*-methylene dihydroinden-1-ones **71** (Scheme 27). Notably, the *exo*-methylene group in **71** comes from PMP via a Mannich–Eschenmoser methylenation. This provides easy access to enantiomerically enriched *trans*- or *cis*-2-alkyl-3-aryl-substituted 2,3-dihydroinden-1-ones by subsequent conjugate addition or hydrogenation of the *exo*-methylene group in **71**, while a silyl group tolerated at the 4-position (R^1) allows even further functionalization, such as Hiyama cross coupling, iodination, or fluorination.

Scheme 27 Enantioselective Synthesis of 3-Aryl-Substituted 2,3-Dihydroinden-1-ones and *exo*-Methylene Dihydroinden-1-ones by an Asymmetric Heck Cyclization[44]

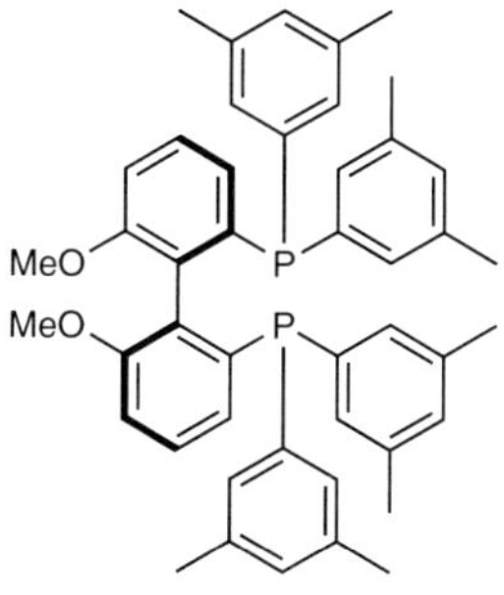

MeO
MeO
P
P
68

for references see p 92

68 (10 mol%), $Pd(OAc)_2$ (5 mol%)
Proton-sponge (2.0 equiv)
DMF, 100 °C, 12 h

69 → **70**

Proton-sponge = 1,8-bis(dimethylamino)naphthalene

R^1	R^2	X	ee[a] (%)	Yield (%)	Ref
H	H	OTf	79	88	[44]
H	H	$OSO_2(CF_2)_3CF_3$	78	90	[44]
Me	H	OTf	94	90	[44]
H	Me	OTf	94 (82% de)	82	[44]

[a] Determined by HPLC or GC analysis using a chiral stationary phase.

68 (10 mol%), $Pd(OAc)_2$ (5 mol%)
PMP, 1,4-dioxane, 100 °C, 12 h

→ **71**

PMP = 1,2,2,6,6-pentamethylpiperidine

R^1	X	PMP (Equiv)	ee[a] (%)	Yield (%)	Ref
TMS	OTf	3.0	90	46	[44]
H	$OSO_2(CF_2)_3CF_3$	2.0	80	74	[44]
Me	$OSO_2(CF_2)_3CF_3$	3.0	94	70	[44]

[a] Determined by HPLC or GC analysis using a chiral stationary phase.

Vinyl-Substituted 1,2,3,4-Tetrahydronaphthalenes, 1,2,3,4-Tetrahydroisoquinolines, and 2,3,4,5-Tetrahydro-1*H*-benzo[*d*]azepines 67**; General Procedure:**[42]
$Pd_2(dba)_3$ (23 mg, 0.025 mmol) and the appropriate amount of chiral ligand (see Scheme 26) were added to degassed DMF (20 mL) under argon. The mixture was slowly heated to 50 °C with vigorous stirring to achieve a homogeneous system (ca. 30 min). Finally, Ag_3PO_4 (460 mg, 1.1 mmol) and the alkene **66** (1 mmol) were added and the resulting mixture was heated to 80 °C. After the reaction was complete (complete consumption of starting material, as determined by TLC), the solid residue was removed by filtration and washed with Et_2O (20 mL). H_2O (20 mL) was added to the filtrate, the layers were separated, and the aqueous phase was washed with Et_2O (2 × 5 mL). The combined organic phases were washed with H_2O (10 mL) and brine (10 mL), dried ($MgSO_4$), filtered, concentrated, and purified by column chromatography (silica gel).

3-Phenyl-Substituted 2,3-Dihydro-1*H*-inden-1-ones 70 **and 2-Methylene-2,3-dihydro-1*H*-inden-1-ones** 71**; General Procedure:**[44]
$Pd(OAc)_2$ (0.025 mmol), (*R*)-3,5-Xyl-MeO-BIPHEP (**68**; 0.05 mmol), the base (Proton-sponge or 1,2,2,6,6-pentamethylpiperidine; 1.0 mmol), and the aryl sulfonate (0.5 mmol) were added to an oven-dried screw-cap test tube under an argon atmosphere. DMF or 1,4-dioxane (2 mL) was added and the mixture was stirred at 100 °C for 12 h. The mixture was then

cooled, diluted with EtOAc (10 mL), and washed 1 M aq HCl (10 mL). The organic phase was separated, washed with brine, and dried ($MgSO_4$). The solvent was evaporated under reduced pressure and the crude material was purified by column chromatography (silica gel).

1.2.2.1.2 Tertiary Centers from Cyclic Alkenes

Cyclic alkenes have found extensive application in the formation of polycyclic compounds by asymmetric Heck cyclization.[5] Chiral ligands can be used to synthesize a range of enantioenriched polycyclic compounds while cyclization of chiral substrates has become a relatively common procedure in total synthesis owing to the high-functional group tolerance and mild conditions of the reaction.[5]

1.2.2.1.2.1 Reactions from Aryl Halides and Pseudohalides

The desymmetrization of bicyclo[4.4.0]decadiene **72** by an enantioselective Heck cyclization to provide tetracyclic system **74** with three stereogenic centers in up to 78% ee has been described (Scheme 28).[45] (S_p,R)-Josiphos **73** is the most effective ligand while the addition of silver salts leads to lower yields and lower enantioselectivity. The aryl bromide, trifluoromethanesulfonate, and nonafluorobutanesulfonate (nonaflate) derivatives of **72** have also been examined as substrates; they result in decreased yields compared to iodide derivative **72**, while enantioselectivity is essentially unaffected.

Scheme 28 Enantioselective Cyclization of 8a-(2-Iodobenzyloxy)-1,5,8,8a-tetrahydronaphthalene-4a(4*H*)-carbonitrile[45]

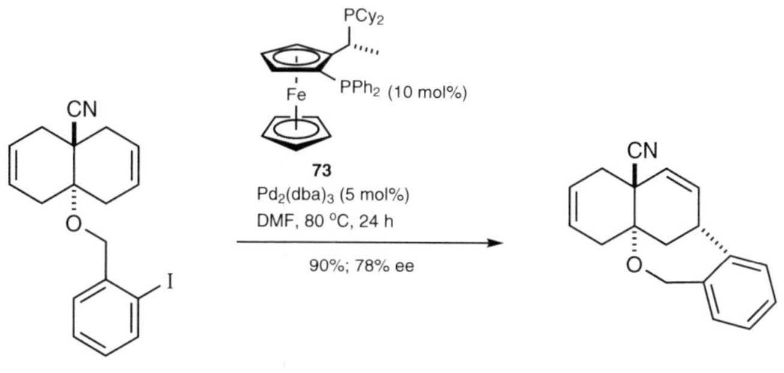

CN
O
I
72
PCy2
Fe
PPh2 (10 mol%)
73
Pd2(dba)3 (5 mol%)
DMF, 80 °C, 24 h
90%; 78% ee
CN
O
74

(4a*S*,11*R*,13a*R*)-1,4,6,11-Tetrahydro-13a*H*-4a,11-methanodibenzo[*b*,*g*]oxonine-13a-carbonitrile (74); Typical Procedure:[45]

Aryl iodide **72** (391 mg, 1 mmol), $Pd_2(dba)_3$ (46 mg, 0.05 mmol), and (S_p,R)-Josiphos **73** (73 mg, 0.1 mmol) were added to DMF (5 mL). The mixture was heated at 80 °C for 24 h and monitored by TLC. Upon completion (consumption of starting material), the mixture was cooled and Et_2O (20 mL) and H_2O (10 mL) were added. The organic phase was separated, washed with H_2O (10 mL) and brine (10 mL), dried (Na_2SO_4), and concentrated before being purified by column chromatography (silica gel); yield: 90%; 78% ee.

1.2.2.1.2.2 Reactions from Vinyl Halides and Pseudohalides

The synthesis of chiral *cis*-hydronaphthalene (*cis*-Decalin) systems by an asymmetric Heck cyclization has become a testing ground for new ligands and methodology since the initial report by Shibasaki in 1989.[39] Over the years, a series of vinyl iodides and trifluoromethanesulfonates **77** have been prepared and tested in the reaction to give *cis*-hydro-

for references see p 92

naphthalenes **78** with various ligands and additives (Scheme 29).[46–52] For the cyclization of vinyl iodides, 2,2′-bis(diphenylphosphino)-1,1′-binaphthyl (BINAP; **63**) and the analogous bisarsine **75** (BINAs) are the most efficient ligands, with 1-methylpyrrolidin-2-one as the optimal solvent. Tris(dibenzylideneacetone)dipalladium(0) and palladium(II) chloride are the best catalyst precursors for this system, while the use of silver(I) phosphate as a halide scavenger and calcium carbonate as a base have proved to be optimal. It is worth noting that the counterion of the silver salt has a significant effect on the enantioselectivity, with phosphate providing optimal results, presumably due to its low nucleophilicity.[52] In general, vinyl trifluoromethanesulfonates give better enantioselectivity, allow the omission of expensive silver salts, and enable the use of a wider range of solvents such as toluene and 1,2-dichloroethane. For these substrates BINAP has proved to be the optimal ligand, with potassium carbonate as the base of choice. Shibasaki also found that the catalytically active $Pd(0)L_n$ species is readily oxidized to $PdCl_2L_n$ in 1,2-dichloroethane but that the addition of a tertiary alcohol (such as pinacol) or potassium acetate prevents this process, resulting in a greatly accelerated reaction.[50] This methodology can be extended to substrates other than the parent Decalin, and has been applied in the synthesis of a key enone intermediate in Danishefsky's synthesis of (+)-vernolepin.[53,54]

Scheme 29 Enantioselective Synthesis of *cis*-1,2,4a,8a-Tetrahydronaphthalenes[46–51]

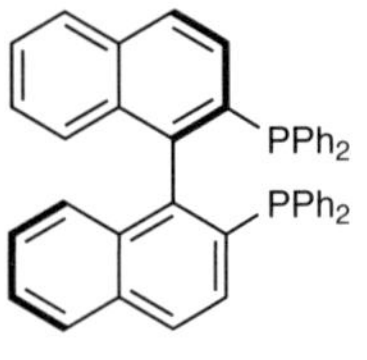

(*R*)-**63** (*R*)-BINAP 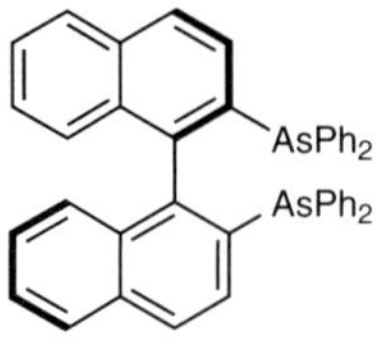 **75** (*R*)-BINAs 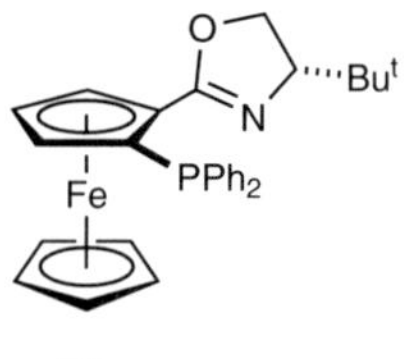 **76** (S_p,*S*)-*t*-Bu-FOX

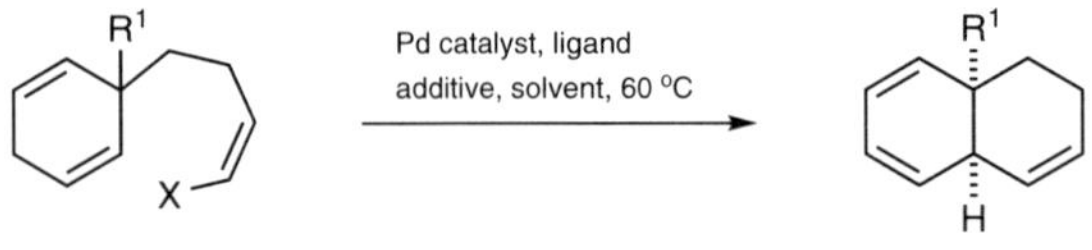

77 → Pd catalyst, ligand, additive, solvent, 60 °C → **78**

R^1	X	Conditions	ee[a] (%)	Yield (%)	Ref
$CH_2OTBDMS$	I	**75** (15 mol%), $Pd_2(dba)_3$ (5 mol%), Ag_3PO_4 (2 equiv), $CaCO_3$ (2.2 equiv), NMP, 24 h	82	90	[48]
$CH_2OTBDMS$	I	$PdCl_2$[(*R*)-**63**] (10 mol%), Ag_3PO_4 (2 equiv), $CaCO_3$ (2.2 equiv), NMP, 24 h	80	67	[48,49]
$CH_2OTBDMS$	OTf	(*R*)-**63** (10 mol%), $Pd(OAc)_2$ (5 mol%), K_2CO_3 (2 equiv), toluene, 74 h	92	35	[48,51]
$CH_2OC(O)$*t*-Bu	OTf	(*R*)-**63** (10 mol%), $Pd(OAc)_2$ (5 mol%), K_2CO_3 (2 equiv), toluene, 27 h	91	60	[50,51]
CO_2Me	OTf	**76** (10 mol%), $Pd_2(dba)_3$ (5 mol%), K_2CO_3 (2 equiv), toluene, 168 h	85	30	[46,47]
$CH_2OC(O)$*t*-Bu	OTf	(*R*)-**63** (10 mol%), $Pd(OAc)_2$ (5 mol%), K_2CO_3 (2 equiv), 1,2-dichloroethane, 106 h	92	6	[50]
$CH_2OC(O)$*t*-Bu	OTf	(*R*)-**63** (10 mol%), $Pd(OAc)_2$ (5 mol%), K_2CO_3 (2 equiv), pinacol (15 equiv), 1,2-dichloroethane, 47 h	95	78	[50]

[a] Determined by HPLC analysis using a chiral stationary phase.

The general method described for the synthesis of Decalins has also been applied to the synthesis of 6,5-bicycles such as 2,3,3a,7a-tetrahydro-1*H*-indenes (hydrindans).[55] Cyclization of the 1,1-disubstituted alkene **79** affords products **80** with good enantioselectivity under analogous reaction conditions (Scheme 30). In contrast to the pattern seen with Decalin systems, the alkenyl iodides provide cyclized products with higher enantioselective excess than the corresponding trifluoromethanesulfonates. Changing the R^1 group has very little effect on yield or enantioselectivity, while silver(I) phosphate again proves to be the most effective silver salt in this reaction.

Scheme 30 Enantioselective Synthesis of *cis*-2,3,3a,7a-Tetrahydro-1*H*-indenes[55]

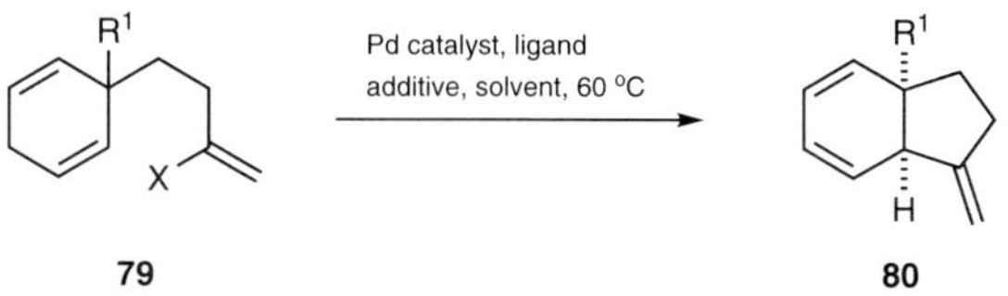

Pd catalyst, ligand
additive, solvent, 60 °C

79 **80**

R^1	X	Conditions	ee[a] (%)	Yield (%)	Ref
CO_2Me	I	$PdCl_2$[(*R*)-**63**] (10 mol%), Ag_3PO_4 (2 equiv), $CaCO_3$ (2.2 equiv), NMP, 38 h	83	73	[55]
$CH_2OTBDMS$	I	$PdCl_2$[(*R*)-**63**] (10 mol%), Ag_3PO_4 (2 equiv), $CaCO_3$ (2.2 equiv), NMP, 49 h	82	78	[55]
$CH_2OTBDMS$	OTf	(*R*)-**63** (10 mol%), $Pd(OAc)_2$ (5 mol%), K_2CO_3 (2 equiv), benzene, 64 h	73	63	[55]

[a] Determined by HPLC analysis using a chiral stationary phase.

Enantioenriched heterocyclic systems such as indolizidines may be synthesized utilizing similar methodology. Cyclization of (*Z*)-1-(3-iodoallyl)-3,4-dihydropyridin-2(1*H*)-one results in a mixture of the desired indolizidine **82** and its double-bond isomer **83** (Scheme 31).[49] The ligand of choice is chiral bis(phosphino)ferrocenyl alcohol **81** [(S_p,*R*)-BPPFOH], which greatly outperforms BINAP and other conventional ligands, presumably due to a hydrogen-bonding interaction between the hydroxy group on the ligand and the carbonyl group on the substrate.[49] Use of a silver-exchanged zeolite provides superior results to silver(I) phosphate, again hinting at the importance of the silver counterion. Treatment of the reaction mixture with a catalytic amount of palladium on carbon in methanol quantitatively transforms product **82** into its double-bond isomer **83**.

Scheme 31 Enantioselective Synthesis of the Indolizidine Skeleton[49]

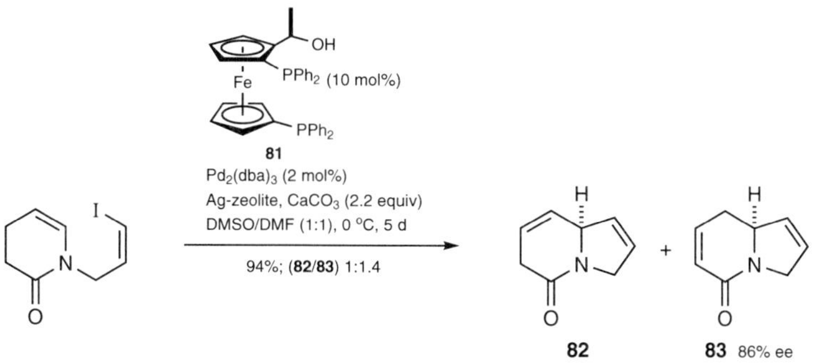

OH
Fe PPh2 (10 mol%)
PPh2
81
$Pd_2(dba)_3$ (2 mol%)
Ag-zeolite, $CaCO_3$ (2.2 equiv)
DMSO/DMF (1:1), 0 °C, 5 d
94%; (**82**/**83**) 1:1.4

82 **83** 86% ee

for references see p 92

The asymmetric Heck cyclization also allows access to enantioenriched 5,5-bicycles. The use of prochiral cyclopentadienyl systems generates a π-allylpalladium species on Heck reaction, which can subsequently be attacked by a nucleophile in a domino-style process. This allows for the construction of optically active bicyclic systems functionalized with a variety of nucleophiles such as acetates, amines, and carbanions (mainly sodium enolates) in a regio- and stereocontrolled manner. As an example, the alkenyl trifluoromethanesulfonate **84** is converted into the diquinane **86** via intermediate **85** in 89% yield and 80% ee when acetate is used as the nucleophile (Scheme 32).[56] (*S*)-BINAP [(*S*)-**63**] is the most effective ligand in all cases, while for reactions with carbanion nucleophiles the addition of 2 equivalents of sodium bromide generally enhances the levels of enantioselectivity.[57,58]

Scheme 32 An Example of the Enantioselective Synthesis of a 5,5-Bicyclic System with Capture by a Nucleophile[56]

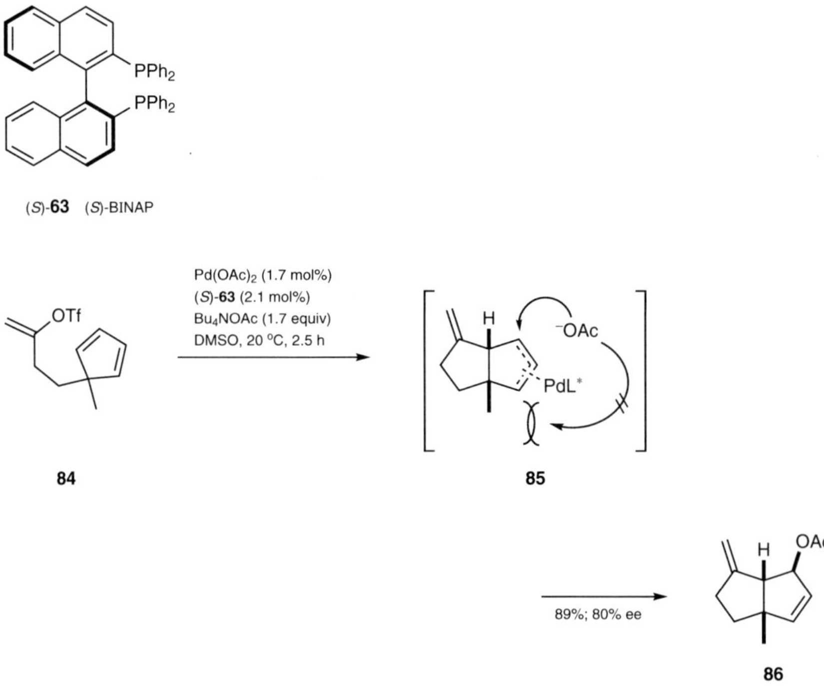

PPh_2
PPh_2
(*S*)-**63** (*S*)-BINAP

OTf
$Pd(OAc)_2$ (1.7 mol%)
(*S*)-**63** (2.1 mol%)
Bu_4NOAc (1.7 equiv)
DMSO, 20 °C, 2.5 h
H
^-OAc
PdL^*
84
85

89%; 80% ee
H
OAc
86

(4a*S*,8a*S*)-[3,8a-Dihydronaphthalen-4a(4*H*)-yl]methyl 2,2-Dimethylpropanoate [78, R^1 = $CH_2OC(O)$*t*-Bu]; Typical Procedure:[50]

To a mixture of $Pd(OAc)_2$ (1.1 mg, 0.005 mmol), (*R*)-BINAP [(*R*)-**63**; 6.2 mg, 0.01 mmol)], K_2CO_3 (27.6 mg, 0.2 mmol), and 2,3-dimethylbutane-2,3-diol (pinacol; 177 mg, 1.5 mmol) was added a soln of alkenyl trifluoromethanesulfonate **77** [R^1 = $CH_2OC(O)$*t*-Bu; X = OTf; 39.6 mg, 0.1 mmol] in 1,2-dichloroethane (1.4 mL). The mixture was degassed and stirred at 60 °C under argon until the reaction was complete (47 h). It was then diluted with Et_2O, washed with brine, dried (Na_2SO_4), and concentrated before being purified by column chromatography (silica gel); yield: 78%; 95% ee.

***cis*-1-Methylene-2,3,3a,7a-tetrahydro-1*H*-indenes** 80 **(X = I); General Procedure:**[55]

The 3-(3-iodobut-3-enyl)cyclohexa-1,4-diene **79** (1 mmol), $PdCl_2${(*R*)-**63**} (80 mg, 0.1 mmol), Ag_3PO_4 (836 mg, 2 mmol), and $CaCO_3$ (220 mg, 2.2 mmol) were added to NMP (5 mL). The resulting mixture was heated to 60 °C for the appropriate amount of time before being diluted with H_2O (10 mL) and extracted with EtOAc (3 × 10 mL). The combined organic phases were washed with brine (10 mL), dried ($MgSO_4$), and concentrated. The crude product was purified by column chromatography (silica gel).

(*S*)-6,8a-Dihydroindolizin-5(3*H*)-one (82); **Typical Procedure:**[49]
To a mixture of $Pd_2(dba)_3$•$CHCl_3$ (3.3 mg, 3.14 μmol), silver-exchanged zeolite (508 mg, ca. 0.942 mmol of Ag), $CaCO_3$ (34.6 mg, 0.345 mmol), (S_p,*R*)-BPPFOH (**81**; 9.0 mg, 15.1 μmol), and DMF (0.5 mL) were added a soln of alkenyl iodide **83** (0.157 mmol) in DMF (1.0 mL) and DMSO (1.5 mL) at rt. The mixture was degassed by freeze–pump–thaw cycles, and then stirred at 0 °C for 5 d. Upon completion of the reaction (consumption of starting material), the mixture was diluted with EtOAc/MeOH (10: 1), filtered through a short pad of silica gel, and concentrated, and the resulting residue was purified by column chromatography (silica gel).

(1*S*,3a*S*,6a*S*)-3a-Methyl-6-methylene-1,3a,4,5,6,6a-hexahydropentalen-1-yl Acetate 86; **Typical Procedure:**[56]
A mixture of trifluoromethanesulfonate **84** (240 mg, 0.99 mmol, as reported), $Pd(OAc)_2$ (3.8 mg, 16.8 μmol), (*S*)-BINAP [(*S*)-**63**; 13.2 mg, 21.2 μmol], and tetrabutylammonium acetate (512 mg, 1.70 mmol) was stirred at 20 °C for 2.5 h in degassed DMSO (4 mL). The DMSO soln was purified directly by column chromatography (silica gel).

1.2.2.2 Formation of Quaternary Carbon Centers

1.2.2.2.1 Quaternary Centers from Acyclic Alkenes

In general, the enantioselective formation of quaternary stereocenters is a difficult task. The use of an asymmetric Heck cyclization here is ideal due to the limited pathways for subsequent β-hydride elimination. Furthermore, if the σ-alkylpalladium intermediate formed after migratory insertion is attacked by a nucleophile then the reaction may proceed in a domino style, which allows for the construction of multiple carbon—carbon bonds in a one-pot sequence.

1.2.2.2.1.1 Reactions from Aryl Halides and Pseudohalides

The synthesis of 3,3-disubstituted oxindoles **88** by an asymmetric Heck cyclization of α,β-unsaturated *N*-phenylamides **87** (*Z* or *E*) has been extensively studied (Scheme 33).[59–64] The reaction proceeds under both neutral [iodide with 1,2,2,6,6-pentamethylpiperidine (PMP) as base] and cationic (iodide with added silver salt or aryl trifluoromethanesulfonate) conditions. The best enantiomeric excesses (89–95% ee) are obtained using (*R*)-2,2′-bis(diphenylphosphino)-1,1′-binaphthyl [(*R*)-BINAP, (*R*)-**63**] as the ligand. In general, for *E*-alkenes the cationic pathway is preferred whereas for *Z*-alkenes the neutral pathway is preferred; aryl trifluoromethanesulfonates can be diverted to the neutral pathway by the addition of halide salts such as tetrabutylammonium iodide. The alcohol-protecting group (R^2) has little effect on the reaction while the α-substituent (R^1) has a large effect on enantioselectivity, with small groups leading to higher levels, although substrates bearing bulky aryl or hetaryl groups can be effectively cyclized with the *E*-alkene under cationic conditions.

for references see p 92

Scheme 33 Selected Examples of the Enantioselective Synthesis of 3-Vinyl-Substituted Oxindoles[59–61]

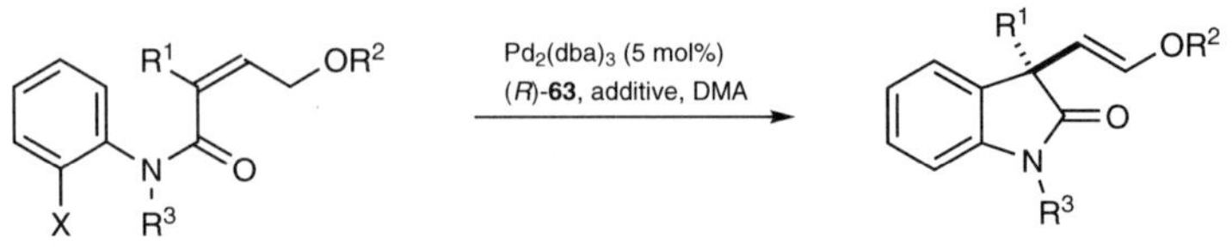

$Pd_2(dba)_3$ (5 mol%), (*R*)-**63**, additive, DMA

87 → **88**

R^1	R^2	R^3	X	BINAP (mol%)	Additive[a] (Equiv)	Temp (°C)	Time (h)	ee[b] (%)	Yield (%)	Ref
Me	TIPS	Me	I	12	PMP (4)	100	23	90	87	[60]
Me	TIPS	Me	I	12	Ag_3PO_4 (2)	100	23	80	73	[60]
Me	TBDMS	Me	OTf	11	PMP (4), TBAI (1)	100	23	90	62	[61]
Ph	Me	Bn	OTf	20	PMP (4)	80	4	84	86[c]	[59]

[a] PMP = 1,2,2,6,6-pentamethylpiperidine.
[b] Determined by HPLC analysis using a chiral stationary phase.
[c] *E*-alkene, $Pd(OAc)_2$ (10 mol%), and THF were used.

The enantioselective synthesis of tetrahydronaphthalene and tetrahydroanthracene derivatives has been accomplished by the asymmetric Heck cyclization of phenyl and naphthyl iodides and trifluoromethanesulfonates, respectively.[65–67] (*R*)-BINAP [(*R*)-**63**] is the most effective ligand for iodide derivatives, whereas for trifluoromethanesulfonates the use of the mixed phosphine/arsine BINAP analogue, 2-(diphenylarsino)-2′-(diphenylphosphino)-1,1′-binaphthyl [(*R*)-BINAPAs] increases reaction rates and yields.[65] This methodology has been applied in the synthesis of natural products, e.g. (+)-eptazocine, (+)-xestoquinone, and helenaquinone. The synthesis of tetrahydronaphthalene **90**, an intermediate in the synthesis of (+)-eptazocine, is shown as an example in Scheme 34.[67]

Scheme 34 Enantioselective Synthesis of a Tetrahydronaphthalene[67]

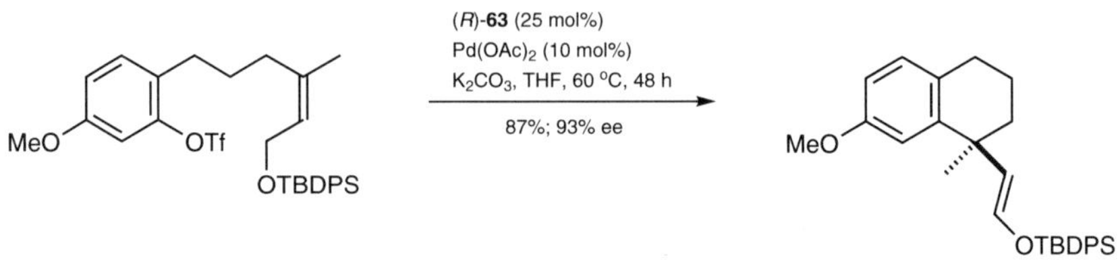

(*R*)-**63** (25 mol%), $Pd(OAc)_2$ (10 mol%), K_2CO_3, THF, 60 °C, 48 h; 87%; 93% ee

89 → **90**

(*R*)-1,3-Dimethyl-3-[(*E*)-2-(triisopropylsiloxy)vinyl]indolin-2-one (88, $R^1 = R^3 = Me$; $R^2 = TIPS$); Typical Procedure:[60]

A flask was charged with $Pd_2(dba)_3$•$CHCl_3$ (8.1 mg, 0.008 mmol, 5 mol%) and (*R*)-BINAP [(*R*)-**63**; 11.2 mg, 0.018 mmol, 12 mol%], and purged under an argon flow for 10 min. Dry DMA (0.8 mL) was added, and the mixture was stirred for 2 h before a soln of iodide **87** (0.16 mmol, 1 equiv) and 1,2,2,6,6-pentamethylpiperidine (120 μL, 100 mg, 0.66 mmol, 4 equiv) in dry DMA (0.8 mL) was added. The resulting suspension was degassed and then heated at 100 °C for 23 h under argon. After cooling to rt, the reaction was quenched with a $NaHCO_3$ soln (5 mL), and the mixture was diluted with H_2O (10 mL) and EtOAc (10 mL). The layers were separated, and the aqueous layer was extracted with EtOAc (2 × 20 mL).

The combined organic extracts were washed with brine (50 mL), dried (Na_2SO_4), filtered, and concentrated before purification by chromatography (silica gel); yield: 87%; 90% ee.

(*S*)-1-[(*E*)-2-(*tert*-Butyldiphenylsiloxy)vinyl]-7-methoxy-1-methyl-1,2,3,4-tetrahydronaphthalene (90); Typical Procedure:[67]

To a mixture of $Pd(OAc)_2$ (52 mg, 0.23 mmol, 10 mol%), (*R*)-BINAP [(*R*)-**63**; 354 mg, 0.58 mmol, 25 mol%], and K_2CO_3 (962 mg, 6.96 mmol, 3 equiv) was added a soln of trifluoromethanesulfonate **89** (1.41 g, 2.32 mmol, 1 equiv) in THF (40 mL). The resulting mixture was degassed through three freeze–pump–thaw cycles and stirred at 60 °C for 48 h. The reaction was then quenched by the addition of sat. aq NH_4Cl and this mixture was extracted with Et_2O. The organic layer was washed with brine, dried (Na_2SO_4), filtered, concentrated, and then purified by chromatography (silica gel, EtOAc/hexane 1:199) to give a colorless oil; yield: 87%; ratio (*trans*/*cis*) 21:3.

1.2.2.2.1.2 Reactions from Vinyl Halides and Pseudohalides

Vinyl halides and pseudohaldies have found limited application in the enantioselective formation of quaternary carbon centers with acyclic alkenes. The asymmetric Heck cyclization and kinetic resolution of a racemic dienyl trifluoromethanesulfonate (±)-**91** in the total synthesis of (+)-wortmannin represents one of the few examples where a high enantiomeric excess is obtained (Scheme 35).[68] The optimal conditions for this reaction employ palladium(II) acetate, (*R*)-Tol-BINAP (**92**), and potassium carbonate in toluene to provide intermediate **93** with excellent enantioselectivity (96% ee) and good diastereoselectivity (dr 11:1). The total synthesis of wortmannin was reported by the same group in 2002 (OTBDPS was replaced by OBn).[69]

Scheme 35 Synthesis of a Tetrahydronaphthalene Intermediate in the Synthesis of (+)-Wortmannin[68]

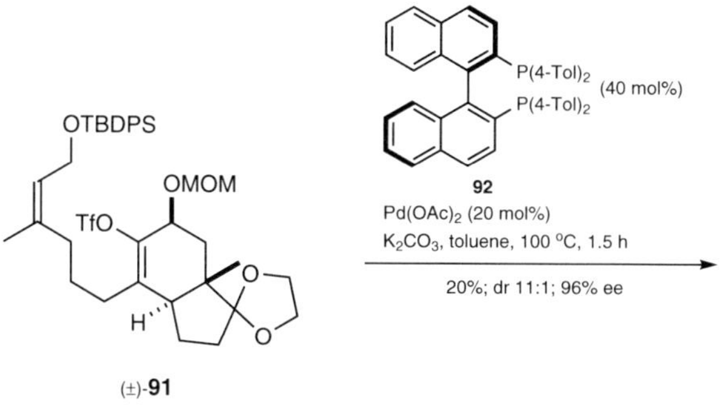

(3a*S*,5*S*,6*R*,9b*S*)-6-[2-(*tert*-Butyldiphenylsiloxy)vinyl]-5-(methoxymethoxy)-3a,6-dimethyl-1,2,3a,4,5,6,7,8,9,9b-decahydrospiro[cyclopenta[*a*]naphthalene-3,2′-[1,3]dioxolane] (93):[68]

A mixture of vinyl trifluoromethanesulfonate (±)-**91** (753 mg, 1 mmol), $Pd(OAc)_2$ (45 mg, 0.2 mmol), (*R*)-Tol-BINAP (**92**; 271 mg, 0.4 mmol), and K_2CO_3 (345 mg, 2.5 mmol) in toluene was heated at 100 °C for 1.5 h. When the reaction was complete (TLC), H_2O (10 mL) was added and the mixture was extracted with Et_2O (3 × 5 mL). The combined organic layers were washed with brine (10 mL), dried ($MgSO_4$), filtered, and concentrated before purification by chromatography (silica gel); yield: 20%; dr 11:1; 96% ee.

for references see p 92

1.2.2.2.2 Quaternary Centers from Cyclic Alkenes

The ability of the intramolecular asymmetric Heck reaction to form congested all-carbon quaternary centers with high levels of enantioselectivity is best demonstrated with cyclic alkenes. As a consequence of the cyclic substrate, this reaction has found extensive use in the synthesis of chiral spiroannulated polycyclic compounds which would be difficult to synthesize by other means.

1.2.2.2.2.1 Reactions from Aryl Halides and Pseudohalides

The synthesis of 3-spiroannulated oxindoles **97** and **98** by an asymmetric Heck cyclization of aryl halides or pseudohalides has been extensively studied (Scheme 36).[70–76] The best results are obtained using BINAP analogue (*R*)-2,2′-bis(di-2-furylphosphino)-1,1′-binaphthyl [(*R*)-TetFuBINAP; **95**] as the ligand, which provides higher enantiomeric excesses and better regioselectivity than BINAP itself.[73] Interestingly, for aryl iodides, depending upon whether the iodide scavenger is a silver salt (cationic route) or a basic tertiary amine (neutral route), either enantiomer of **97** can be formed with good selectivity using the same enantiomer of BINAP (**63**).[63,76] Regioselectivity with regards to double-bond isomers (e.g., **98**, when R^1 = H) is generally poor to moderate (75:25 optimized for BINAP). However, ligands such as (S_p,*S*)-*t*-Bu-FOX (**76**) and (*S*)-*t*-Bu-HETPHOX (**96**) provide product **97** with excellent regioselectivity (~99:1), albeit with reduced yield and enantioselectivity.[72]

Scheme 36 Enantioselective Synthesis of 3-Spiroannulated Oxindoles[70–73,76]

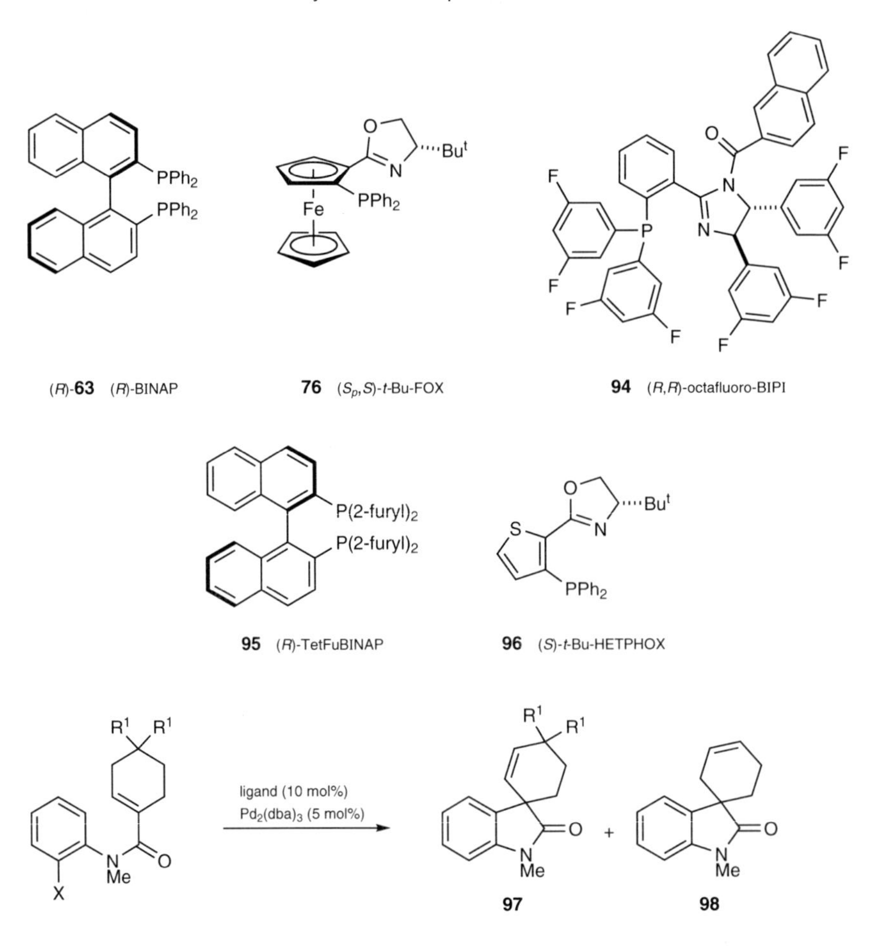

R^1	R^2	X	Ligand	Conditions[a,b]	Ratio (**97**/**98**)	ee (%)	Config	Yield (%)	Ref
H	H	I	**95**	PMP (4 equiv), DMF, 110 °C, 72 h	84:16	90	*R*	65	[73]
$O(CH_2)_2O$		I	(*R*)-**63**	Ag_3PO_4 (2 equiv), DMA, 80 °C, 26 h	**97** only	71	*S*	81	[76]
$O(CH_2)_2O$		I	(*R*)-**63**	PMP (5 equiv), DMA, 80 °C, 140 h	**97** only	66	*R*	77[c]	[76]
H	H	Br	**95**	PMP (4 equiv), DMA, 110 °C, 72 h	62:38	64	*R*	57	[73]
H	H	OTf	(*R*)-**63**	PMP (4 equiv), DMA, 110 °C, 48 h	25:75	74	*S*	90	[72]
H	H	OTf	**76**	Proton-sponge (2 equiv), DMA, 110 °C, 168 h	99:1	85	*R*	30	[72]
H	H	OTf	**96**	Proton-sponge (2 equiv), toluene, 110 °C, 168 h	98:2	76	*R*	59	[70]
Me	Me	OTf	**94**	PMP (4 equiv), Ph_2O, 95 °C, 18 h	**97** only	88	–[d]	38	[71]

[a] Proton-sponge = 1,8-bis(dimethylamino)naphthalene.
[b] PMP = 1,2,2,6,6-pentamethylpiperidine.
[c] $Pd_2(dba)_3$ (10 mol%) and ligand (20 mol%) were used.
[d] Absolute configuration was not assigned.

The enantioselective synthesis of tetracyclic diterpenoid intermediate **101** (together with the isomer **102**) has been accomplished by the asymmetric Heck cyclization of aryl trifluoromethanesulfonate **99** (Scheme 37).[77] (*R*)-SYNPHOS (**100**) provides the highest enantioselectivity while an aprotic solvent such as dimethylformamide is crucial for shortening reaction times. It has been proposed that the increased selectivity obtained with (*R*)-SYNPHOS could be due to the smaller dihedral angle about the C^1–$C^{1'}$ axis for SYNPHOS (75°), which may afford a more potent interaction between the ligand and substrate compared with BINAP (80°).

Scheme 37 Enantioselective Synthesis of a Key Intermediate of Abietane-Type Diterpenoids[77]

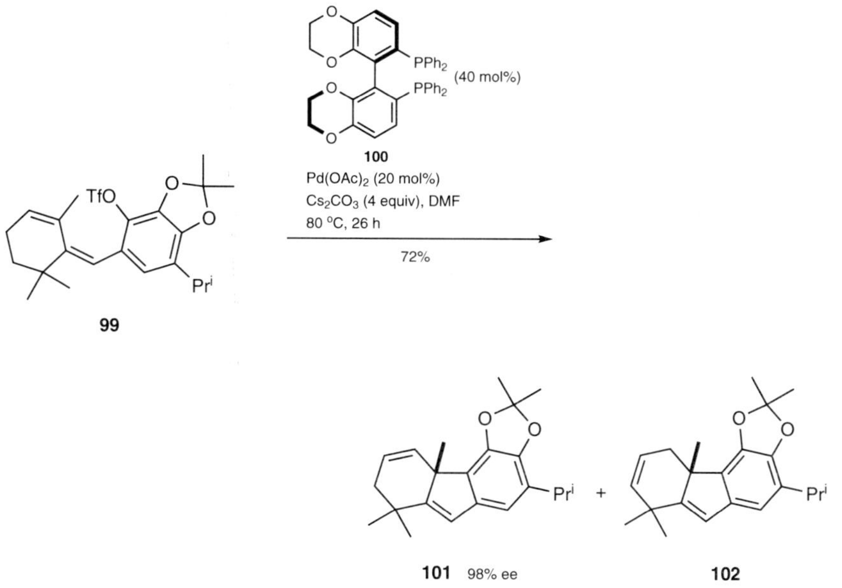

Dihydrocarbazole **105** has been synthesized with high enantiomeric purity by an asymmetric Heck cyclization of dienyl trifluoromethanesulfonate **103** in Overman's total synthesis of (+)-minfiensine (Scheme 38).[74] Optimal results are obtained using the *tert*-butyl-

for references see p 92

substituted oxazoline ligand **104** [(*S*)-*t*-Bu-PHOX] with palladium(II) acetate and 1,2,2,6,6-pentamethylpiperidine (PMP) as the base. Under standard oil-bath heating, the reaction requires 70 hours to go to completion; however, microwave heating is more efficient, promoting full conversion in just 45 minutes with no erosion of yield or enantiomeric purity.

Scheme 38 Enantioselective Synthesis of a Dihydrocarbazole Skeleton, a Key Intermediate in the Synthesis of (+)-Minfiensine[74]

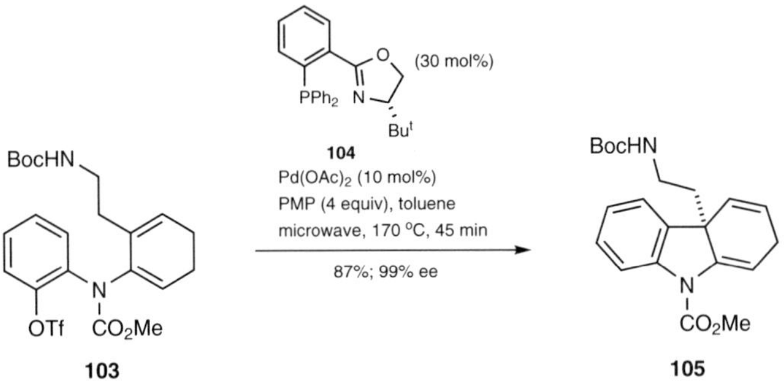

(30 mol%)
PPh2 N
Bu^t
BocHN
104
$Pd(OAc)_2$ (10 mol%)
PMP (4 equiv), toluene
microwave, 170 °C, 45 min
87%; 99% ee
OTf CO_2Me
103
BocHN
CO_2Me
105

1′-Methylspiro[cyclohexane-1,3′-indolin]-2-en-2′-ones 97 **and** 98; **General Procedure:**[72]
A mixture of $Pd_2(dba)_3$ (46 mg, 0.05 mmol) and the appropriate ligand (0.10 mmol) in the chosen solvent was stirred for 1 h. A soln of the halide or pseudohalide **95** (1 mmol) and additive in solvent (10 mL) was added and the resulting mixture was degassed by the freeze–pump–thaw method (3 ×) before being heated at the temperature indicated in Scheme 36 for the appropriate time. When the reaction was complete (TLC), H_2O (10 mL) was added and the mixture was extracted with EtOAc (3 × 5 mL). The combined organic layers were washed with brine (10 mL), dried ($MgSO_4$), filtered, and concentrated before purification by chromatography (silica gel).

(*S*)-4-Isopropyl-2,2,7,7,10a-pentamethyldihydro-7*H*-fluoreno[3,4-*d*][1,3]dioxoles 101 **and** 102:[77]
$Pd(OAc)_2$ (6.1 mg, 0.027 mmol) was added to a suspension of aryl trifluoromethanesulfonate **99** (64.0 mg, 0.135 mmol), Cs_2CO_3 (176.3 mg, 0.541 mmol), and (*R*)-SYNPHOS (**100**; 34.5 mg, 0.054 mmol) in DMF (2 mL). The mixture was stirred for 26 h at 80 °C under an argon atmosphere. When the reaction was complete (TLC), H_2O (10 mL) was added and the mixture was extracted with EtOAc (3 × 5 mL). The combined organic layers were washed with brine (10 mL), dried (Na_2SO_4), filtered, and concentrated. Purification by chromatography (silica gel, hexane/Et_2O 50:1) afforded the cyclized products as a regioisomeric mixture; yield: 72%.

Methyl (*R*)-4a-{2-[(*tert*-Butoxycarbonyl)amino]ethyl}-2,4a-dihydro-9*H*-carbazole-9-carboxylate (105):[74]
A microwave reaction tube containing $Pd(OAc)_2$ (5 mg, 23 μmol) and (*S*)-*t*-Bu-PHOX (**104**; 27 mg, 69 μmol) under argon was charged with a soln of aryl trifluoromethanesulfonate **103** (120 mg, 0.23 mmol), degassed toluene (1.5 mL), and 1,2,2,6,6-pentamethylpiperidine (170 μL, 0.92 mmol). After being stirred at rt for 10 min, the mixture was microwave-heated (CEM Discover System, 60 Hz and 300 W instrument) at 170 °C for 45 min. Following cooling, the mixture was concentrated under reduced pressure and purified by column chromatography (silica gel, EtOAc/hexanes 1:9 to 1:4) to give a colorless foam; yield: 74 mg (87%); 99% ee.

1.2.3 Conclusions

This chapter reports on the intramolecular Heck reaction, assembling the best methods for cyclization and focusing on recent advances in terms of substrate scope, reactivity, and regio- and enantioselectivity. An appreciation of reaction mechanisms/catalytic cycles is required in order to understand the concepts underpinning the significant development of these processes. The non-enantioselective cyclization examples are categorized according to Baldwin's rules and the size of the ring formed. The best methods for asymmetric intramolecular Heck reactions are organized according to the substitution on the stereocenter formed (i.e., tertiary or quaternary), the type of alkene used (cyclic or acyclic), and the type of halide or pseudohalide used (aryl or vinyl).

Although this chapter illustrates extensive research over the past number of years, this area is far from being exhausted as the formation of ring systems is crucial for the preparation of compounds that are of use in biological, medicinal, agrochemical, and material/nanoscience-related research programs. It will be of interest to follow the literature of this fascinating synthetic transformation in the future.

for references see p 92

References

[1] Heck, R. F., *J. Am. Chem. Soc.*, (1968) **90**, 5518.
[2] Heck, R. F.; Nolley, J. P., *J. Org. Chem.*, (1972) **37**, 2320.
[3] Mizoroki, T.; Mori, K.; Ozaki, A., *Bull. Chem. Soc. Jpn.*, (1971) **44**, 581.
[4] Mori, M.; Chiba, K.; Ban, Y., *Tetrahedron Lett.*, (1977), 1037.
[5] Dounay, A. B.; Overman, L. E., *Chem. Rev.*, (2003) **103**, 2945.
[6] McCartney, D.; Guiry, P. J., *Chem. Soc. Rev.*, (2011) **40**, 5122.
[7] Butler, E. M.; Doran, R.; Wilson, C. M.; Guiry, P. J., In *Comprehensive Organic Synthesis*, 2nd ed., Knochel, P.; Molander, G. A., Eds.; Elsevier: Amsterdam, (2014); Vol. 4, p 810.
[8] Coeffard, V.; Guiry, P. J., In *Science of Synthesis: Cross Coupling and Heck-Type Reactions*, Larhed, M., Ed.; Thieme: Stuttgart, (2012); Vol. 3, Section 3.1.1.4, p 303.
[9] Baldwin, J. E., *J. Chem. Soc., Chem. Commun.*, (1976), 734.
[10] Grigg, R.; Stevenson, P.; Worakun, T., *J. Chem. Soc., Chem. Commun.*, (1984), 1073.
[11] Grigg, R.; Sakee, U.; Sridharan, V.; Sukirthalingam, S.; Thangavelauthum, R., *Tetrahedron*, (2006) **62**, 9523.
[12] Innitzer, A.; Brecker, L.; Mulzer, J., *Org. Lett.*, (2007) **9**, 4431.
[13] Singh, R. M.; Chandra, A.; Sharma, N.; Singh, B.; Kumar, R., *Tetrahedron*, (2012) **68**, 9206.
[14] Caddick, S.; Kofie, W., *Tetrahedron Lett.*, (2002) **43**, 9347.
[15] Grigg, R.; Kennewell, P.; Teasdale, A. J., *Tetrahedron Lett.*, (1992) **33**, 7789.
[16] Inoue, M.; Furuyama, H.; Sakazaki, H.; Hirama, M., *Org. Lett.*, (2001) **3**, 2863.
[17] Ding, D.; Liu, G.; Xu, G.; Li, J.; Wang, G.; Sun, J., *Org. Biomol. Chem.*, (2014) **12**, 2533.
[18] Ruck, R. T.; Huffman, M. A.; Kim, M. M.; Shevlin, M.; Kandur, W. V.; Davies, I. W., *Angew. Chem. Int. Ed.*, (2008) **47**, 4711.
[19] Liu, X.; Ma, X.; Huang, Y.; Gu, Z., *Org. Lett.*, (2013) **15**, 4814.
[20] Lee, C.-W.; Oh, K. S.; Kim, K. S.; Ahn, K. H., *Org. Lett.*, (2000) **2**, 1213.
[21] René, O.; Lapointe, D.; Fagnou, K., *Org. Lett.*, (2009) **11**, 4560.
[22] Hua, D. H.; Zhao, H.; Battina, S. K.; Lou, K.; Jimenez, A. L.; Desper, J.; Perchellet, E. M.; Perchellet, J.-P. H.; Chiang, P. K., *Bioorg. Med. Chem.*, (2008) **16**, 5232.
[23] Pu, J.; Deng, K.; Butera, J.; Chlenov, M.; Gilbert, A.; Kagan, M.; Mattes, J.; Resnick, L., *Tetrahedron*, (2010) **66**, 1963.
[24] Liu, P.; Huang, L.; Lu, Y.; Dilmeghani, M.; Baum, J.; Xiang, T.; Adams, J.; Tasker, A.; Larsen, R.; Faul, M. M., *Tetrahedron Lett.*, (2007) **48**, 2307.
[25] Xiang, Z.; Luo, T.; Lu, K.; Cui, J.; Shi, X.; Fathi, R.; Chen, J.; Yang, Z., *Org. Lett.*, (2004) **6**, 3155.
[26] Ardizzoia, G. A.; Beccalli, E. M.; Borsini, E.; Brenna, S.; Broggini, G.; Rigamonti, M., *Eur. J. Org. Chem.*, (2008), 5590.
[27] Schulman, J. M.; Friedman, A. A.; Panteleev, J.; Lautens, M., *Chem. Commun. (Cambridge)*, (2012) **48**, 55.
[28] Chen, B.; Xie, X.; Lu, J.; Wang, Q.; Zhang, J.; Tang, S.; She, X.; Pan, X., *Synlett*, (2006), 259.
[29] Ray, D.; Paul, S.; Brahma, S.; Ray, J. K., *Tetrahedron Lett.*, (2007) **48**, 8005.
[30] Barluenga, J.; Fernández, M. A.; Aznar, F.; Valdés, C., *Chem.–Eur. J.*, (2005) **11**, 2276.
[31] Dankwardt, J. W.; Flippin, L. A., *J. Org. Chem.*, (1995) **60**, 2312.
[32] Boger, D. L.; Turnbull, P., *J. Org. Chem.*, (1997) **62**, 5849.
[33] Majumdar, K. C.; Chattopadhyay, B.; Samanta, S., *Tetrahedron Lett.*, (2009) **50**, 3178.
[34] Declerck, V.; Ribière, P.; Nédellec, Y.; Allouchi, H.; Martinez, J.; Lamaty, F., *Eur. J. Org. Chem.*, (2007), 201.
[35] Majumdar, K. C.; Chattopadhyay, B.; Sinha, B., *Synthesis*, (2008), 3857.
[36] Majumdar, K. C.; Chattopadhyay, B.; Sinha, B., *Tetrahedron Lett.*, (2008) **49**, 1319.
[37] Aeluri, M.; Gaddam, J.; Trinath, D. V. K. S.; Chandrasekar, G.; Kitambi, S. S.; Arya, P., *Eur. J. Org. Chem.*, (2013), 3955.
[38] Carpenter, N. E.; Kucera, D. J.; Overman, L. E., *J. Org. Chem.*, (1989) **54**, 5846.
[39] Sato, Y.; Sodeoka, M.; Shibasaki, M., *J. Org. Chem.*, (1989) **54**, 4738.
[40] Tietze, L. F.; Raschke, T., *Synlett*, (1995), 597.
[41] Tietze, L. F.; Raschke, T., *Liebigs Ann.*, (1996), 1981.
[42] Tietze, L. F.; Schimpf, R., *Angew. Chem. Int. Ed. Engl.*, (1994) **33**, 1089.
[43] Tietze, L. F.; Thede, K.; Schimpf, R.; Sannicolo, F., *Chem. Commun. (Cambridge)*, (2000), 583.
[44] Minatti, A.; Zheng, X.; Buchwald, S. L., *J. Org. Chem.*, (2007) **72**, 9253.
[45] Lormann, M. E. P.; Nieger, M.; Bräse, S., *J. Organomet. Chem.*, (2006) **691**, 2159.

[46] Kiely, D.; Guiry, P. J., *Tetrahedron Lett.*, (2003) **44**, 7377.
[47] Kiely, D.; Guiry, P. J., *J. Organomet. Chem.*, (2003) **687**, 545.
[48] Kojima, A.; Boden, C. D. J.; Shibasaki, M., *Tetrahedron Lett.*, (1997) **38**, 3459.
[49] Sato, Y.; Nukui, S.; Sodeoka, M.; Shibasaki, M., *Tetrahedron*, (1994) **50**, 371.
[50] Ohrai, K.; Kondo, K.; Sodeoka, M.; Shibasaki, M., *J. Am. Chem. Soc.*, (1994) **116**, 11737.
[51] Sato, Y.; Watanabe, S.; Shibasaki, M., *Tetrahedron Lett.*, (1992) **33**, 2589.
[52] Sato, Y.; Sodeoka, M.; Shibasaki, M., *Chem. Lett.*, (1990), 1953.
[53] Kondo, K.; Sodeoka, M.; Mori, M.; Shibasaki, M., *Tetrahedron Lett.*, (1993) **34**, 4219.
[54] Kondo, K.; Sodeoka, M.; Mori, M.; Shibasaki, M., *Synthesis*, (1993), 920.
[55] Sato, Y.; Honda, T.; Shibasaki, M., *Tetrahedron Lett.*, (1992) **33**, 2593.
[56] Kagechika, K.; Shibasaki, M., *J. Org. Chem.*, (1991) **56**, 4093.
[57] Ohshima, T.; Kagechika, K.; Adachi, M.; Sodeoka, M.; Shibasaki, M., *J. Am. Chem. Soc.*, (1996) **118**, 7108.
[58] Itano, W.; Ohshima, T.; Shibasaki, M., *Synlett*, (2006), 3053.
[59] Dounay, A. B.; Hatanaka, K.; Kodanko, J. J.; Oestreich, M.; Overman, L. E.; Pfeifer, L. A.; Weiss, M. M., *J. Am. Chem. Soc.*, (2003) **125**, 6261.
[60] Ashimori, A.; Bachand, B.; Calter, M. A.; Govek, S. P.; Overman, L. E.; Poon, D. J., *J. Am. Chem. Soc.*, (1998) **120**, 6488.
[61] Overman, L. E.; Poon, D. J., *Angew. Chem. Int. Ed. Engl.*, (1997) **36**, 518.
[62] Ashimori, A.; Matsuura, T.; Overman, L. E.; Poon, D. J., *J. Org. Chem.*, (1993) **58**, 6949.
[63] Ashimori, A.; Bachand, B.; Overman, L. E.; Poon, D. J., *J. Am. Chem. Soc.*, (1998) **120**, 6477.
[64] Oestreich, M.; Dennison, P. R.; Kodanko, J. J.; Overman, L. E., *Angew. Chem. Int. Ed.*, (2001) **40**, 1439.
[65] Cho, S. Y.; Shibasaki, M., *Tetrahedron Lett.*, (1998) **39**, 1773.
[66] Kojima, A.; Takemoto, T.; Sodeoka, M.; Shibasaki, M., *J. Org. Chem.*, (1996) **61**, 4876.
[67] Takemoto, T.; Sodeoka, M.; Sasai, H.; Shibasaki, M., *J. Am. Chem. Soc.*, (1993) **115**, 8477.
[68] Honzawa, S.; Mizutani, T.; Shibasaki, M., *Tetrahedron Lett.*, (1999) **40**, 311.
[69] Mizutani, T.; Honzawa, S.; Tosaki, S.-y.; Shibasaki, M., *Angew. Chem. Int. Ed.*, (2002) **41**, 4680.
[70] Fitzpatrick, M. O.; Coyne, A. G.; Guiry, P. J., *Synlett*, (2006), 3150.
[71] Busacca, C. A.; Grossbach, D.; Campbell, S. J.; Dong, Y.; Eriksson, M. C.; Harris, R. E.; Jones, P.-J.; Kim, J.-Y.; Lorenz, J. C.; McKellop, K. B.; O'Brien, E. M.; Qiu, F.; Simpson, R. D.; Smith, L.; So, R. C.; Spinelli, E. M.; Vitous, J.; Zavattaro, C., *J. Org. Chem.*, (2004) **69**, 5187.
[72] Kiely, D.; Guiry, P. J., *Tetrahedron Lett.*, (2002) **43**, 9545.
[73] Andersen, N. G.; McDonald, R.; Keay, B. A., *Tetrahedron: Asymmetry*, (2001) **12**, 263.
[74] Dounay, A. B.; Humphreys, P. G.; Overman, L. E.; Wrobleski, A. D., *J. Am. Chem. Soc.*, (2008) **130**, 5368.
[75] Busacca, C. A.; Grossbach, D.; So, R. C.; O'Brien, E. M.; Spinelli, E. M., *Org. Lett.*, (2003) **5**, 595.
[76] Ashimori, A.; Overman, L. E., *J. Org. Chem.*, (1992) **57**, 4571.
[77] Node, M.; Ozeki, M.; Planas, L.; Nakano, M.; Takita, H.; Mori, D.; Tamatani, S.; Kajimoto, T., *J. Org. Chem.*, (2010) **75**, 190.

1.3 Metal-Catalyzed Intramolecular Allylic Substitution Reactions

X. Zhang and S.-L. You

General Introduction

Metal-catalyzed allylic substitution reactions are one of the most powerful tools to construct carbon—carbon and carbon—heteroatom bonds. The first stoichiometric example was reported by Tsuji in 1965, demonstrating that π-allylpalladium complexes could be substituted with diethyl malonate and acetoacetate as well as a cyclohexanone-derived enamine to afford allylated products.[1] The corresponding catalytic reaction was reported in 1970[2,3] and the first palladium-catalyzed asymmetric allylic substitution reaction was realized by Trost in 1977.[4] Generally, the substitution reaction of allylic compounds with various nucleophiles via π-allylpalladium complexes is called the Tsuji–Trost reaction. As depicted in Scheme 1, the reaction usually involves the following four steps: (1) coordination of palladium to the alkene of an allylic precursor; (2) oxidative addition to generate a π-allylpalladium complex; (3) nucleophilic substitution; (4) dissociation of the product and regeneration of the palladium(0) catalyst. An asymmetric transformation can be achieved when a chiral environment is introduced. This has been realized mainly by the introduction of an appropriate chiral ligand. Since the early reports, many efforts have been devoted to this field, leading to the observation that various nucleophiles, metals, and ligands are suitable for the process.

Scheme 1 General Mechanism of the Tsuji–Trost Reaction[4]

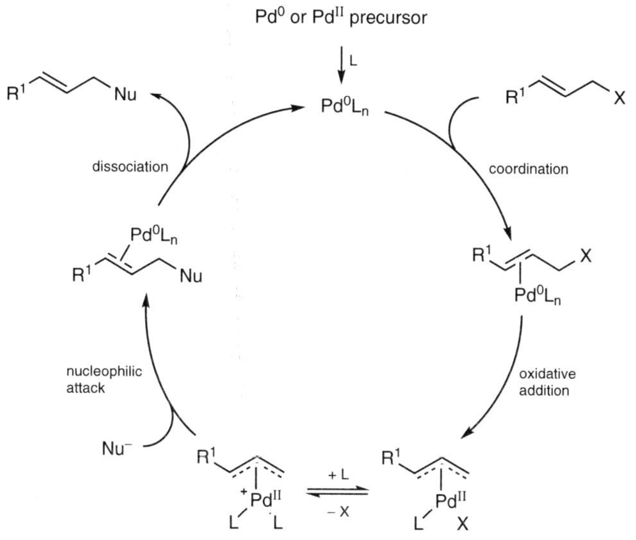

L = ligand; X = leaving group; Nu^- = soft nucleophile

Notably, an allylic substitution reaction is also responsible for the preparation of diverse rings when an intramolecular reaction is carried out (Scheme 2). In general, the intramolecular reaction tends to occur more easily, as the linkage makes the nucleophilic site closer to the electrophilic site. However, the ring strain is an unfavorable factor and an intermolecular reaction can lead to side products in some cases. In this chapter, different types of metal-catalyzed intramolecular allylic substitution reactions, including allylic alkylation, allylic amination, and allylic etherification, are discussed. The reactions are classified according to the nucleophiles employed. However, this chapter does not attempt to cover the cyclization process in a stepwise manner.

Scheme 2 Metal-Catalyzed Intramolecular Allylic Substitution Reactions

NuH LG or NuH LG metal catalyst exo-attack Nu endo-attack Nu

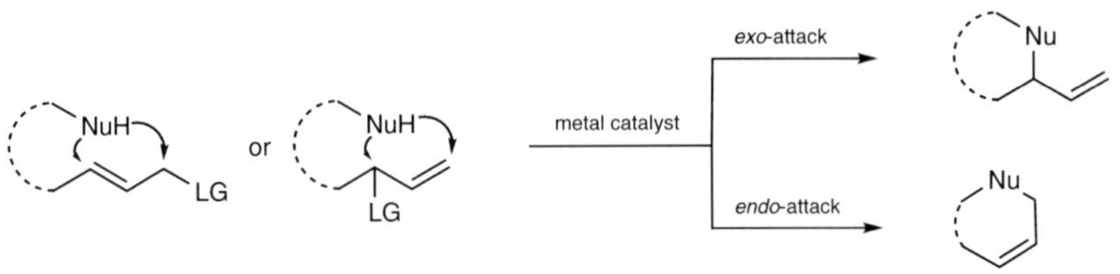

LG = leaving group; Nu = C, N, O

1.3.1 Reactions with Carbon Nucleophiles

1.3.1.1 Resonance-Stabilized Enolates

Initial attempts at intramolecular allylic substitution with carbon nucleophiles were focused on the use of resonance-stabilized enolate anions as the nucleophile to realize the cyclization due to their feasibility of formation and practical utility. Although the results obtained at the early stages in the 1980s were not satisfying, they paved the way for the development of this field. Moreover, the reactions with these soft nucleophiles are usually used as model reactions for testing new ligands.

1.3.1.1.1 Malonic Esters

The palladium-catalyzed intramolecular allylic alkylation of (*E*)-allylic carbonate **2** gives the five-membered cyclopentane derivative **3** in moderate yield with good enantioselectivity in the presence of the catalyst generated in situ from allyl(chloro)palladium(II) dimer and phosphinooxazoline ligand **1**. The results obtained strongly depend on subtle features of the reaction conditions such as nonpolar solvent, low temperature, and low substrate concentration (Scheme 3).[5]

Scheme 3 Palladium-Catalyzed Intramolecular Allylic Alkylation of a Malonic Ester To Give a Vinylcyclopentane[5]

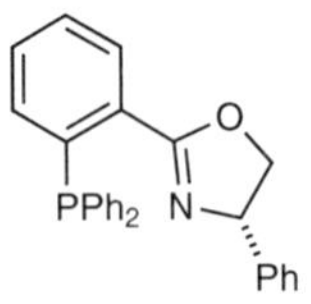

O PPh2 N Ph

1

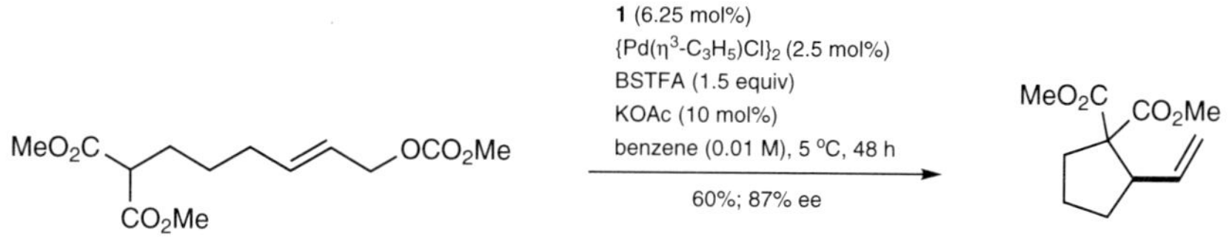

1 (6.25 mol%)
$\{Pd(\eta^3\text{-}C_3H_5)Cl\}_2$ (2.5 mol%)
BSTFA (1.5 equiv)
KOAc (10 mol%)
benzene (0.01 M), 5 °C, 48 h
60%; 87% ee

2 → **3**

BSTFA = *N*,*O*-bis(trimethylsilyl)trifluoroacetamide

In addition to palladium catalysis, this transformation has also been achieved under iridium catalysis by a complex derived from chloro(cycloocta-1,5-diene)iridium(I) dimer and phosphoramidite ligand **4** (Scheme 4).[6] Five- and six-membered carbocycles **6** are obtained in good yields and with excellent enantioselectivities from carbonates **5**. Notably, the low temperature (−78 °C) is crucial for the generation of the malonate anion and suppressing the competing non-catalyzed cyclization.

Scheme 4 Iridium-Catalyzed Intramolecular Allylic Alkylation of Malonic Esters To Give Vinylcycloalkanes[6]

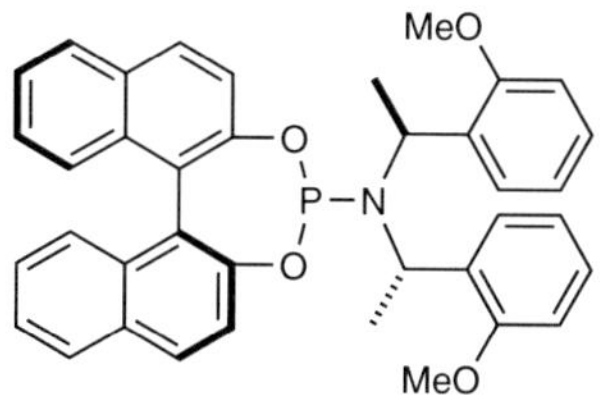

4

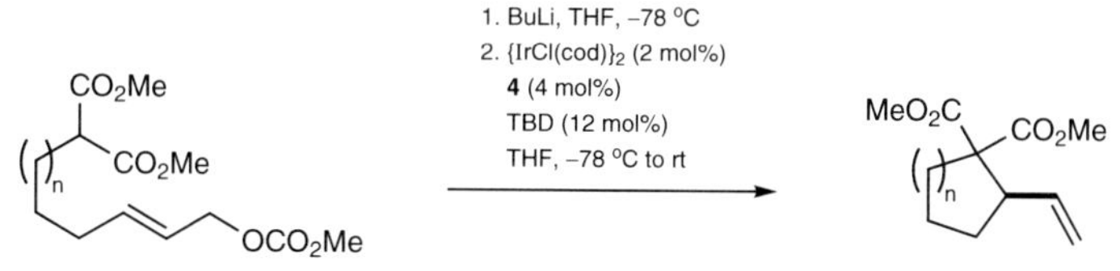

1. BuLi, THF, −78 °C
2. $\{IrCl(cod)\}_2$ (2 mol%)
4 (4 mol%)
TBD (12 mol%)
THF, −78 °C to rt

5 → **6**

TBD = 1,5,7-triazabicyclo[4.4.0]dec-5-ene

n	ee (%)	Yield (%)	Ref
1	96	77	[6]
2	97	79	[6]

In contrast to previous reports using allylic carbonates as the substrates, the newly designed chiral bidentate ligand **7** together with tris(acetonitrile)(η^5-cyclopentadienyl)ruthenium(I) hexafluorophosphate and 4-toluenesulfonic acid demonstrates high reactivity and selectivity for the direct alkylation of allylic alcohols. The substrates (e.g., **8**) undergo intramolecular C-allylation delivering the corresponding 2,3-dihydro-1*H*-indene (e.g., **9**) and 1,2,3,4-tetrahydronaphthalene derivatives in quantitative yields and excellent enantioselectivities (>98% ee) in all cases. However, the aliphatic substrate **10** without the benzene ring linker proves to be a poor substrate, leading to deterioration in both yield (57%) and enantioselectivity (76% ee). In these intramolecular cyclization reactions catalyzed by ruthenium complexes, a high substrate/catalyst (S/C) ratio of up to 1000 is realized (Table 1).[7]

Table 1 Ruthenium-Catalyzed Intramolecular Enantioselective Allylic Alkylation of Allylic Alcohols[7]

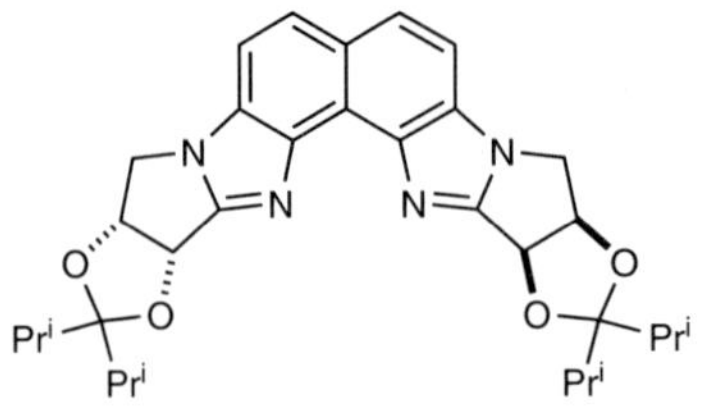

7

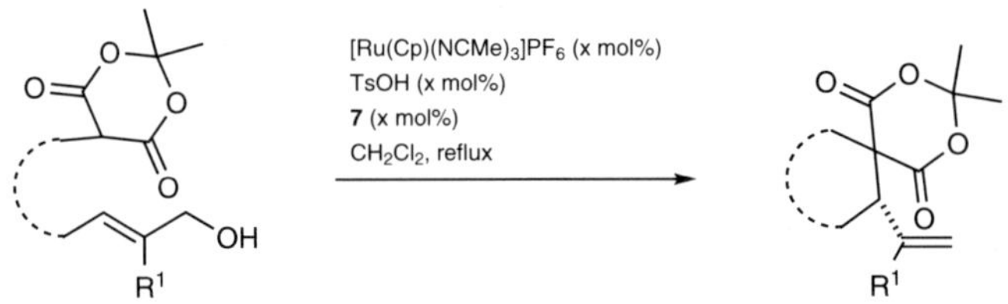

[Ru(Cp)(NCMe)$_3$]PF$_6$ (x mol%)
TsOH (x mol%)
7 (x mol%)
CH$_2$Cl$_2$, reflux

Substrate	x (mol%)	Product	ee[a] (%)	Yield[b] (%)	Ref
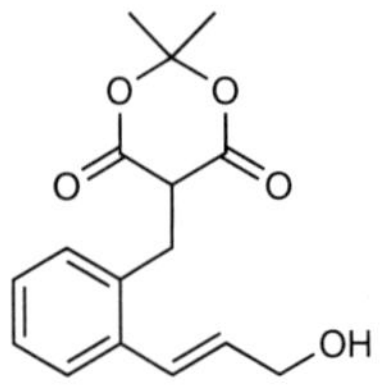 **8**	0.1	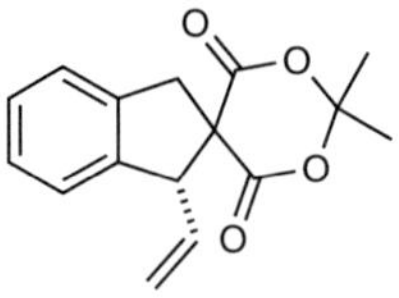 **9**	>98	98	[7]
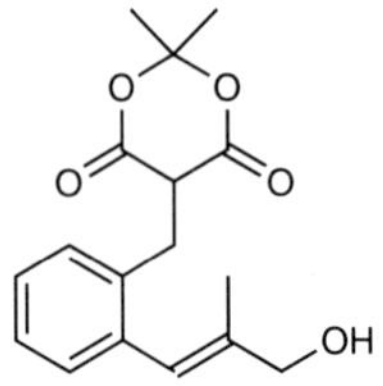	1	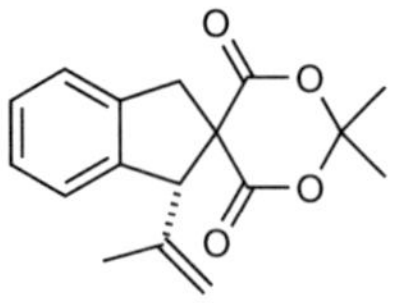	>98	98	[7]
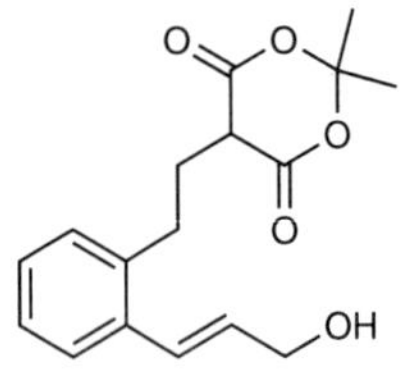	0.1	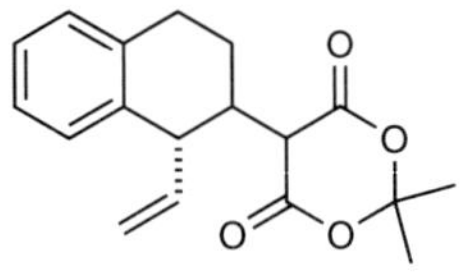	>98	97	[7]
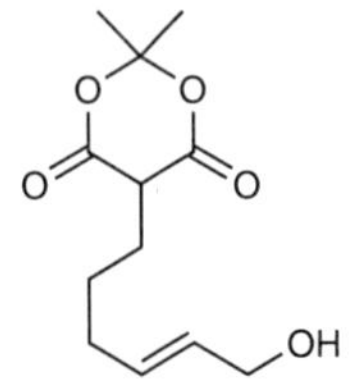 **10**	1	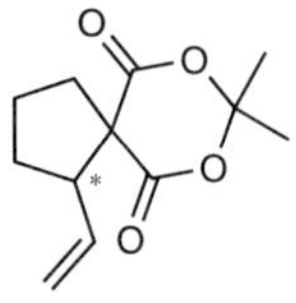	76	57	[7]

[a] Determined by HPLC analysis using a chiral stationary phase.
[b] Isolated yields after column chromatography.

Dimethyl (*R*)-2-Vinylcyclopentane-1,1-dicarboxylate (6, n = 1); General Procedure:[6]
In a flame-dried Schlenk tube under argon, $\{IrCl(cod)\}_2$ (13.4 mg, 0.02 mmol) and ligand **4** (24.0 mg, 0.04 mmol) were dissolved in dry THF (0.5 mL), forming a yellow-orange soln. After stirring for 5 min, 1,5,7-triazabicyclo[4.4.0]dec-5-ene (TBD; 17 mg, 0.12 mmol) was added and the mixture was stirred for another 2 h. In the meantime, the substrate **5** (n = 1; 1.00 mmol) was dissolved in dry THF (4.0 mL) in a second flame-dried Schlenk tube under argon. The mixture was cooled to −78 °C and 1.6 M BuLi in hexane (1.0 mmol) was added dropwise. This mixture was stirred at −78 °C for 2 h and then added to the first Schlenk tube which was also cooled to −78 °C. The resulting mixture was stirred overnight and the temperature was allowed to increase from −78 °C to rt. Then, Et_2O (5 mL) and sat. aq NH_4Cl (5 mL) were added and the phases were separated. The aqueous layer was extracted with Et_2O (2 × 10 mL) and the combined organic phases were washed with brine (10 mL). The aqueous layer was re-extracted with Et_2O (10 mL), and the combined organic phases were dried (Na_2SO_4), filtered, and concentrated under reduced pressure. The residue was purified by column chromatography (silica gel); yield: 77%; 96% ee.

2′,2′-Dimethyl-1-vinyl-1,3-dihydrospiro[indene-2,5′-[1,3]dioxane]-4′,6′-dione (9); Typical Procedure:[7]
A 150-mL Schlenk tube was charged with $[Ru(Cp)(NCMe)_3]PF_6$ (14.9 mg, 34.4 μmol), ligand **7** (18.8 mg, 34.4 μmol), and acetone (2.0 mL). The mixture was stirred at rt for 30 min and then the resulting pale yellow soln was concentrated under reduced pressure. To this residue were added CH_2Cl_2 (35.0 mL), $TsOH{\bullet}H_2O$ (6.5 mg, 34.4 μmol), and substrate **8** (10.0 g, 34.4 mmol). The mixture was stirred for 3 h at reflux temperature before cooling to rt and concentration. The residue was purified through a pad of silica gel (15 g, 3.5 cm × 3.7 cm) to afford **9**; yield: 9.18 g (98%); 99.4% ee. The white solid was recrystallized (EtOAc/hexane, 20 mL:60 mL) to give enantiomerically pure **9** as prismatic crystals; yield: 8.00 g (85%); mp 84 °C.

1.3.1.1.2 β-Keto Esters

The asymmetric cyclization of β-keto ester derivative **12** without additional base gives an (*R*)-3-vinylcyclohexanone in 83% yield and 48% ee (determined after decarboxylation of **13**) at room temperature in the presence of palladium(II) acetate and (*S*)-*N*,*N*-dimethyl-1-[(*R*)-1′,2-bis(diphenylphosphino)ferrocenyl]ethylamine (**11**) (Scheme 5).[8] C-alkylation occurs rather than O-alkylation in the case of the corresponding 8-methoxyoct-6-enoate. It can be rationalized that the formation of the six-membered ring is more thermodynamically stable and favorable in most cases.

Scheme 5 Palladium-Catalyzed Intramolecular Allylic Alkylation of a β-Keto Ester To Give a Vinylcyclohexanone[8]

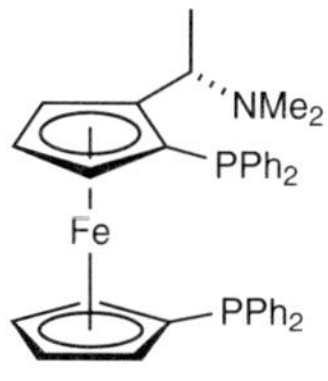

NMe2
PPh2
Fe
PPh2
11

for references see p 149

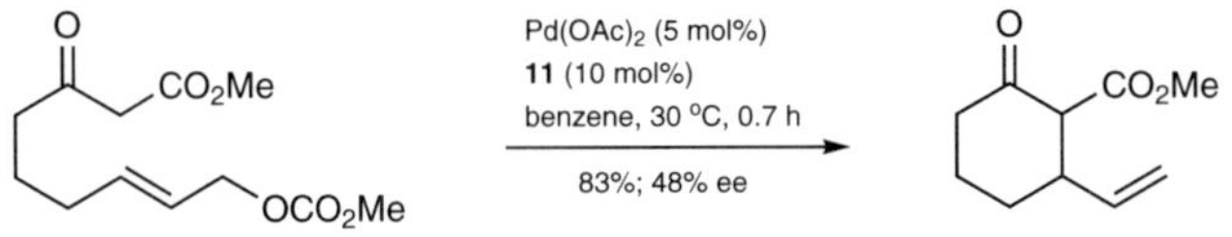

$Pd(OAc)_2$ (5 mol%)
11 (10 mol%)
benzene, 30 °C, 0.7 h
83%; 48% ee

12 → **13**

The allylic carbonate **15** is subjected to palladium-catalyzed intramolecular allylic alkylation to give quinuclidin-2-ones **16A** and **16B** with good diastereoselectivities and enantioselectivities in the absence of an exogenous base. With the addition of lanthanum fod reagents (fod = 6,6,7,7,8,8,8-heptafluoro-2,2-dimethyloctane-3,5-dionato), which can enhance the interactions between the enolate and the π-allyl system, the cyclized products are obtained in 85% yield and 1:8 diastereoselectivity favoring **16B** [68% ee in the presence of ligand (*S*,*S*)-**14**]. However, running the reaction at a more dilute concentration with (*R*,*R*)-**14** as the ligand in the absence of the lanthanum additive leads to the quinuclidinones in 84% yield and 4.6:1 diastereoselectivity favoring ester **16A** with >99% ee (Scheme 6).[9]

Scheme 6 Palladium-Catalyzed Cyclization of a β-Keto Ester Tautomer[9]

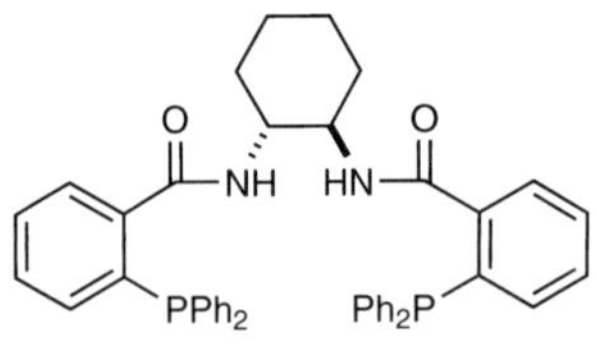

(*R*,*R*)-**14**

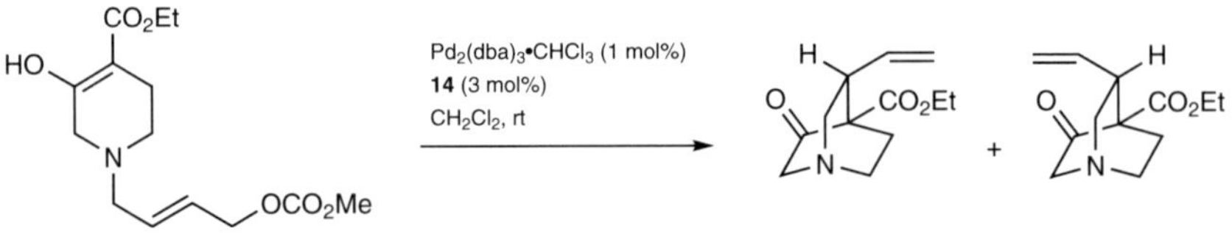

$Pd_2(dba)_3$•$CHCl_3$ (1 mol%)
14 (3 mol%)
CH_2Cl_2, rt

15 → **16A** + **16B**

Config of **14**	Concentration	Additive	dr[a] (**16A**/**16B**)	ee[b] (%)	Yield[c]	Ref
S,*S*	0.1 M	$Eu(fod)_3$ (0.1 equiv)	1:8	68 (50)[d]	85	[9]
R,*R*	0.01 M	–	4.6:1	>99	84	[9]

[a] Determined from the crude ^{1}H NMR spectra.
[b] ee of major product; determined by HPLC analysis using a chiral stationary phase.
[c] Combined yield of both diastereomers.
[d] ee (%) of **16A**.

Ethyl 3-Oxo-5-vinylquinuclidine-4-carboxylate (16A and 16B); Typical Procedure:[9]
Carbonate **15** (720 mg, 2.4 mmol), $Pd_2(dba)_3$•$CHCl_3$ (24.7 mg, 0.024 mmol), and (*R*,*R*)-**14** (50.2 mg, 0.073 mmol) were placed in a dry flask. The flask was sealed and filled with argon. Degassed CH_2Cl_2 (25 mL) was added and the mixture was stirred at rt for 43 h. After concentration under reduced pressure, the crude mixture was purified by column chromatography (silica gel, petroleum ether/EtOAc 2:1 to 1:1); yield: 84%.

1.3.1.1.3 Acetamides

The palladium-catalyzed intramolecular allylic alkylation of functionalized malonamide substrates **17** leads to the *trans*-pyrrolidin-2-one products **18** in good yields through a 5-*exo* process exclusively. This approach relies upon the concomitant generation of a π-allylpalladium appendage and stabilized acetamide enolate anion, where a strongly electron-withdrawing group [Ac, CN, SO_2Ph, or $P(O)(OEt)_2$] is essential (Scheme 7).[10]

Scheme 7 4-Vinylpyrrolidin-2-ones by Palladium-Catalyzed Intramolecular Allylic Alkylation with Stabilized Acetamide Enolate Anions[10]

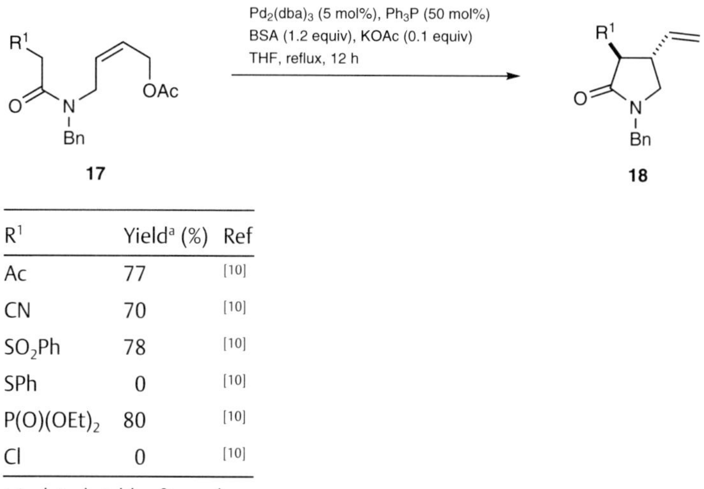

R1	Yield[a] (%)	Ref
Ac	77	[10]
CN	70	[10]
SO_2Ph	78	[10]
SPh	0	[10]
$P(O)(OEt)_2$	80	[10]
Cl	0	[10]

[a] Isolated yields after column chromatography.

The amide tether plays a key role in the complete regio- and stereoselectivity due to its preference for near-planarity resulting in a short linkage. Both the cyclic (e.g., **19**) (Scheme 8)[11] and acyclic (e.g., **20** and **21**) (Schemes 9 and 10, respectively)[12,13] allylic substrates are well tolerated. Moreover, the utility of this methodology has also been demonstrated in syntheses toward various alkaloids and kainoids of biological interest.

Scheme 8 Synthesis of Isoretronecanol[11]

Scheme 9 Synthesis of (−)-α-Kainic Acid[12]

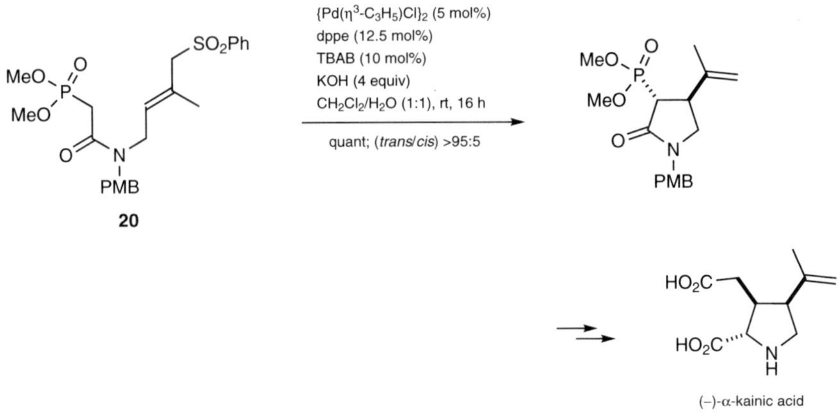

{Pd(η^3-C_3H_5)Cl}$_2$ (5 mol%)
dppe (12.5 mol%)
TBAB (10 mol%)
KOH (4 equiv)
CH_2Cl_2/H_2O (1:1), rt, 16 h

quant; (*trans*/*cis*) >95:5

20

(−)-α-kainic acid

Scheme 10 Synthesis of (−)-Trachelanthamidine[13]

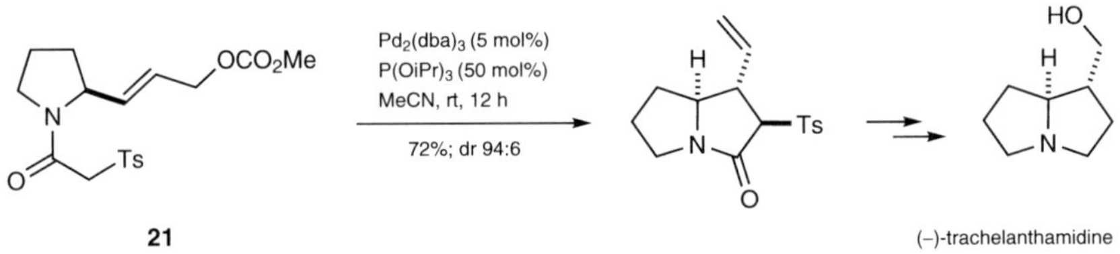

$Pd_2(dba)_3$ (5 mol%)
P(OiPr)$_3$ (50 mol%)
MeCN, rt, 12 h

72%; dr 94:6

21

(−)-trachelanthamidine

1-Benzyl-4-vinylpyrrolidin-2-ones 18; General Procedure:[10]
Under N_2, to a soln of amide **17** (1.0 mmol) in THF (20 mL) were added BSA (0.3 mL, 1.2 mmol) and KOAc (10 mg, 0.1 mmol) sequentially with stirring. In a separate flask, $Pd_2(dba)_3$ (45 mg, 0.05 mmol) and Ph_3P (130 mg, 0.5 mmol) were weighed and then added to the reaction vessel. The resulting mixture was refluxed for 12 h. Then, sat. aq NH_4Cl was added and the aqueous phase was extracted with Et_2O. The combined organic phases were dried, filtered, and concentrated under reduced pressure. The residue was purified by column chromatography (silica gel, hexanes/EtOAc) to afford the pure product as an oil.

1.3.1.2 Aliphatic Aldehydes

In contrast to cyclization of stabilized enolates, using the α-carbon of an aldehyde as a suitable nucleophile remains challenging in intramolecular allylic alkylation due to the weak electron-withdrawing effect of the aldehyde and its potential side reactions. The combined use of transition-metal catalysis and organocatalysis, falling into the concept of dual catalysis, has overcome this limitation. A catalytic intramolecular stereoselective α-allylic alkylation of aldehydes **24** has been realized by merging aminocatalysis (catalyst **23**) with gold(I) catalysis (catalyst **22**). The corresponding 3,4-disubstituted pyrrolidines **25** are obtained in good yields, diastereoselectivities, and enantioselectivities (Scheme 11).[14] The addition of benzoic acid has an accelerating effect on the reaction, a fact that can be attributed to assistance in enamine formation. Notably, piperidine **26** can also be isolated in 34% yield with 8:1 *trans*/*cis* diastereoselectivity and 98% ee for the *trans*-isomer (Scheme 12).[14]

Scheme 11 3-Vinylpyrrolidines and Vinylcyclopentanes by Cocatalyzed Enantioselective Intramolecular α-Allylic Alkylation of Aldehydes[14]

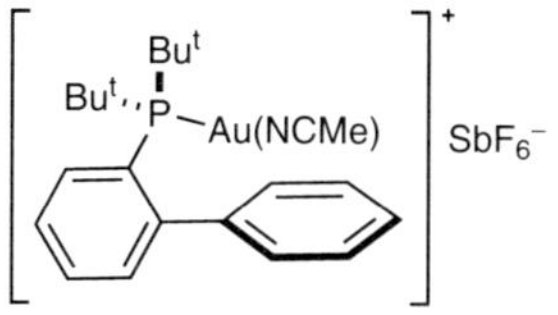

22 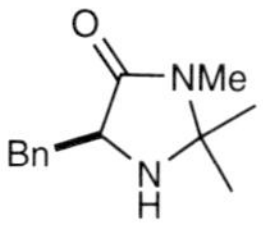 23

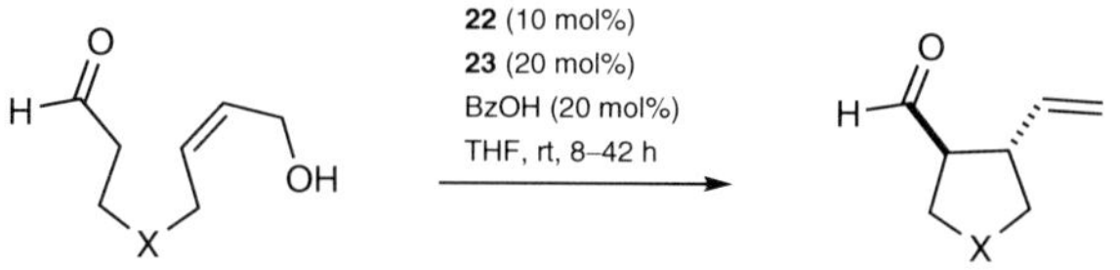

22 (10 mol%)
23 (20 mol%)
BzOH (20 mol%)
THF, rt, 8–42 h

24 25

X	dr[a] (*trans*/*cis*)	ee[b] (%)		Yield[c] (%)	Ref
		trans	*cis*		
NSO_2Mes	5.3:1	94	88	53	[14]
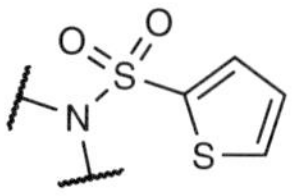	19:1	84	65	76	[14]
NCbz	16:1	97	85	83	[14]
NCO_2Me	12:1	98	–	53	[14]
$C(CO_2Me)_2$	2.1:1	84	82	78	[14]
$C(CO_2Et)_2$	2.3:1	96	85	71	[14]
$C(CO_2Et)_2$	4.9:1	90	92	41[d]	[14]
$C(CO_2t\text{-}Bu)_2$	2.2:1	95	81	90	[14]
$C(CO_2Bn)_2$	2:1	89	83	79	[14]

[a] Determined by GC and ^1H NMR analysis of the crude reaction mixture.
[b] Determined by HPLC analysis using a chiral stationary phase after derivatization.
[c] Isolated yields after column chromatography.
[d] Substrate with *E* configuration was used instead.

Scheme 12 Enantioselective Synthesis of a Piperidine via Cocatalyzed Intramolecular α-Allylic Alkylation of an Aldehyde[14]

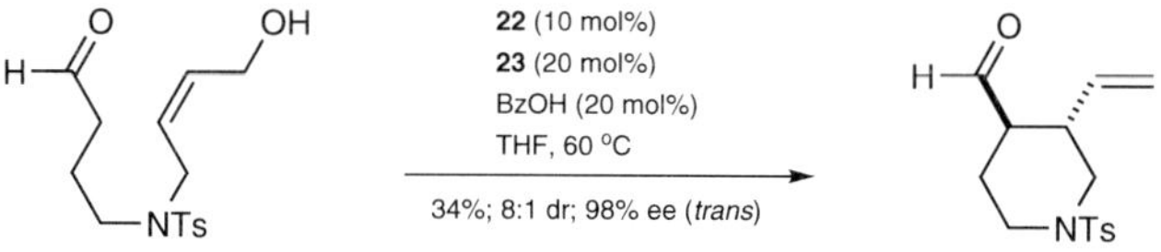

22 (10 mol%)
23 (20 mol%)
BzOH (20 mol%)
THF, 60 °C

34%; 8:1 dr; 98% ee (*trans*)

26

4-Vinylpyrrolidine-3-carbaldehydes 25**; General Procedure:**[14]
In a screw-capped vial, with no exclusion of moisture or O_2, substrate **24** (0.07 mmol, 1 equiv) was dissolved in THF and gold catalyst **22** (0.007 mmol, 10 mol%), the organo-

for references see p 149

catalyst **23** (0.014 mmol, 20 mol%), and BzOH (0.014 mmol, 20 mol%) were added in sequence. The reaction was monitored by TLC and the mixture was stirred at rt until satisfactory conversion of the starting material was obtained. The crude mixture was directly submitted to column chromatography (silica gel, cyclohexane/EtOAc typically 8:2). The dr was determined by GC or NMR analysis of the crude mixture; the ee was determined by chiral HPLC analysis after suitable product derivatization.

1.3.1.3 Ketones

A palladium-catalyzed intramolecular asymmetric allylic alkylation of ketones **28** has been developed with the SIOCPhox ligand **27**. 2,3-Disubstituted indan-1-ones **29**, bearing two vicinal chiral centers, are obtained with moderate to good yields and good to excellent diastereoselectivities and enantioselectivities (Scheme 13).[15]

Scheme 13 3-Vinylindanones by Palladium-Catalyzed Intramolecular Asymmetric Allylic Alkylation of Ketones[15]

27 (5 mol%)
$\{Pd(\eta^3\text{-}C_3H_5)Cl\}_2$ (2.5 mol%)
K_3PO_4 (2 equiv), TBAF (10 mol%)
t-BuOH/$MeO(CH_2)_2OH$ (1:1), −20 °C

R^1	R^2	R^3	Time (h)	dr (*anti*/*syn*)[a]	ee[b] (%)	Yield[c] (%)	Ref
Me	H	H	7	91:9	88	70	[15]
iPr	H	H	9	>95:5	87	79	[15]
CH_2CO_2Me	H	H	5	93:7	87	70	[15]
Me	H	Me	33	>95:5	89	46	[15]
Me	Br	H	5	92:8	86	85	[15]
Ph	H	H	3.5	95:5	67	69[d]	[15]
$(CH_2)_2CH{=}CH_2$	H	H	12	95:5	80	85	[15]
$(CH_2)_2OBn$	H	H	12	94:6	83	84	[15]

[a] Determined by GC and ^1H NMR analysis of the crude reaction mixture.
[b] Determined by HPLC analysis using a chiral stationary phase.
[c] Isolated yields after column chromatography.
[d] $MeO(CH_2)_2OH$/DME (1:1) was used as the solvent and TBAB was used as the additive.

(2*S*,3*S*)-2-Methyl-3-vinyl-2,3-dihydro-1*H*-inden-1-one (29, R^1 = Me; R^2 = R^3 = H); Typical Procedure:[15]
To a flame-dried Schlenk tube were added $\{Pd(\eta^3\text{-}C_3H_5)Cl\}_2$ (1.8 mg, 0.005 mmol), SIOC-Phox ligand **27** (6.4 mg, 0.010 mmol), freshly distilled anhyd *t*-BuOH (1 mL), and $MeO(CH_2)_2OH$ (1 mL). The resulting mixture was allowed to stir for 30 min before it was cooled to −20 °C. Then, compound **28** (R^1 = Me; R^2 = R^3 = H; 65.3 mg, 0.2 mmol), K_3PO_4 (84.9 mg, 0.4 mmol), and TBAF (15 wt% on Al_2O_3; 34.8 mg, 0.02 mmol) were added. The resulting mixture was stirred at −20 °C for 7 h, filtered through a pad of Celite, and concentrated under reduced pressure to afford a crude oil. The *anti*/*syn* ratio was determined by ^{1}H NMR analysis of the crude product. Then, the residue was purified by column chromatography (silica gel, petroleum ether/EtOAc 10:1) to give a colorless oil; yield: 70%; 88% ee.

1.3.1.4 Aliphatic Nitro Compounds

Nitro compounds are very feasible and attractive substrates for use in intramolecular allylic alkylation reactions. This is mainly attributed to the electron-withdrawing nitro group, which not only results in the strong acidity of the α-hydrogen, similar to that of doubly activated methylene compounds, but also provides access to further versatile transformations. One representative example of the application of nitro compounds is in the synthesis of ergot alkaloids, such as (−)-chanoclavine I, in which the key step relies on the palladium-catalyzed allylic alkylation of aliphatic nitro compound **30**. The desired product **31** is obtained in 57% yield and with excellent diastereo- and enantiostereoselectivity at room temperature (Scheme 14).[16]

Scheme 14 Synthesis of (−)-Chanoclavine I Using a Palladium-Catalyzed Intramolecular Allylic Alkylation of an Aliphatic Nitro Compound[16]

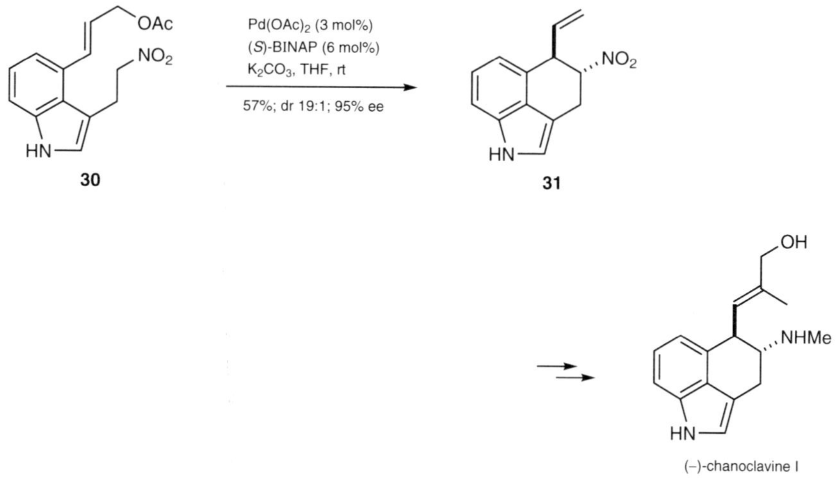

OAc
NO_2
HN
30
$Pd(OAc)_2$ (3 mol%)
(*S*)-BINAP (6 mol%)
K_2CO_3, THF, rt
57%; dr 19:1; 95% ee
NO_2
HN
31
OH
NHMe
HN
(−)-chanoclavine I

(4*R*,5*R*)-4-Nitro-5-vinyl-1,3,4,5-tetrahydrobenz[*cd*]indole (31); Typical Procedure:[16]
Under argon, a mixture of K_2CO_3 (9.58 g, 69.4 mmol), $Pd(OAc)_2$ (0.187 g, 0.83 mmol), and (*S*)-BINAP (1.048 g, 1.66 mmol) in dry THF (30 mL) was stirred at rt. The color of the mixture turned from light orange to dark red. Then, a soln of nitro compound **30** (8 g, 27.7 mmol) in dry THF (60 mL) was added dropwise and the resulting mixture was stirred for another 6 h. After that, the mixture was filtered through silica gel and the crude residue was washed with THF. The solvent was removed under reduced pressure. Column chromatography (cyclohexane/EtOAc 9:1) gave a white solid; yield: 3.6 g (57%); mp 210 °C (MeOH).

for references see p 149

1.3.1.5 Indoles and Pyrroles

1.3.1.5.1 Friedel–Crafts-Type Reactions

1.3.1.5.1.1 Indoles with an Allylic Linkage at the C2-Position

Given the high reactivity of the C3-position of indoles, indol-2-yl allyl carbonates **33** are treated with $Pd_2(dba)_3$•$CHCl_3$ and Trost ligand (*R*,*R*)-**32** to give the corresponding products **34** in up to 98% yield and 97% ee through an intramolecular Friedel–Crafts-type process. The nature of the leaving group does not affect the stereochemical outcome. The reaction features a broad range of substrates with good tolerance toward different steric and electronic effects. In particular, substrates bearing an electron-donating group, such as a methyl or methoxy group, deliver the cyclized compounds in relatively higher enantiomeric excesses (90–97% ee). Notably, quaternary stereocenters are accessible with excellent stereoinduction when a substrate with a trisubstituted alkene is employed. A pyrrolyl-based substrate is also applicable to this transformation (Scheme 15).[17] Palladium catalytic systems generally provide linear-selective products; however, with an intramolecular design, branched-selective compounds are obtained due to the favorable formation of six-membered rings.

Scheme 15 Tetrahydrocarbolines by Palladium-Catalyzed Intramolecular Friedel–Crafts-Type Alkylation of Indol-2-yl-Substituted Allyl Carbonates[17]

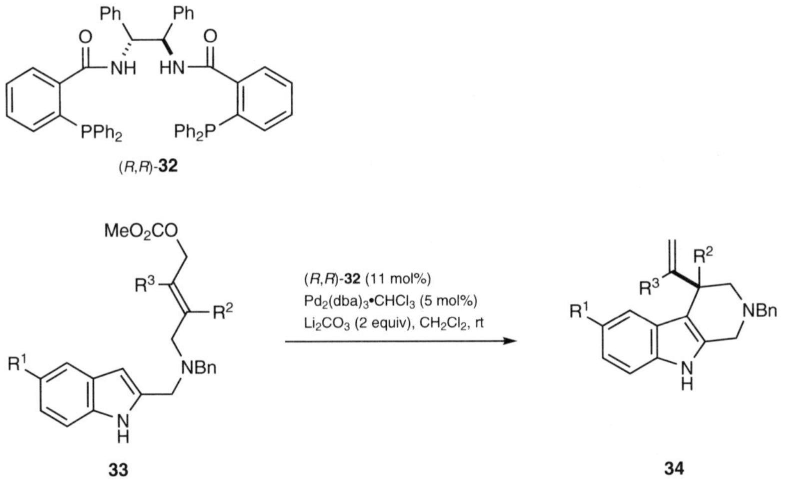

(*R*,*R*)-**32**

(*R*,*R*)-**32** (11 mol%)
$Pd_2(dba)_3$•$CHCl_3$ (5 mol%)
Li_2CO_3 (2 equiv), CH_2Cl_2, rt

33 → **34**

R^1	R^2	R^3	ee[a] (%)	Yield[b] (%)	Ref
H	H	H	92	88	[17]
OMe	H	H	90	95	[17]
OMe	H	H	94	60[c]	[17]
Cl	H	H	82	80	[17]

R^1	R^2	R^3	ee[a] (%)	Yield[b] (%)	Ref
Me	H	H	97	98	[17]
H	Me	H	90	45[d]	[17]
H	H	Me	94	49[e]	[17]

[a] Determined by HPLC analysis using a chiral stationary phase.
[b] Isolated yields after column chromatography.
[c] Reaction at 0 °C.
[d] Reaction was carried out under reflux.
[e] (*R*,*R*)-**14** was used as the ligand.

MeO_2CO … NBn … (*R*,*R*)-**32** (11 mol%), $Pd_2(dba)_3$•$CHCl_3$ (5 mol%), Li_2CO_3 (2 equiv), CH_2Cl_2, rt; 85%; 95% ee → NBn

This approach has also been realized under an iridium-catalyzed system using the easily accessible ligand **35**. Various substituents on the nitrogen linkage, such as benzyl, methyl, and allyl groups, have been investigated and, in all cases, the corresponding alkylated products are delivered with excellent enantioselectivities (98 to >99% ee) (Scheme 16).[18] However, the iridium catalytic system does not tolerate a trisubstituted alkene substrate. Furthermore, the generation of branched-selective products is not only because of the formation of the favorable six-membered ring, but is also a result of the nature of the iridium catalyst that generally favors branched selectivity.

Scheme 16 Tetrahydrocarbolines by Iridium-Catalyzed Intramolecular Friedel–Crafts-Type Alkylation of Indol-2-yl-Substituted Allyl Carbonates[18]

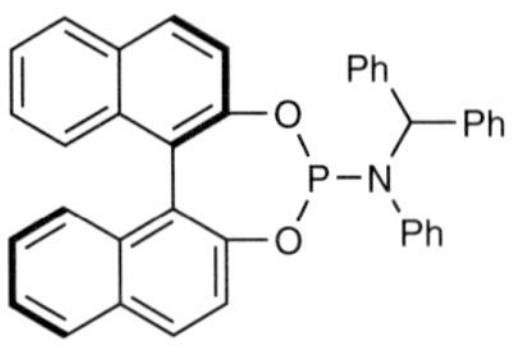

35

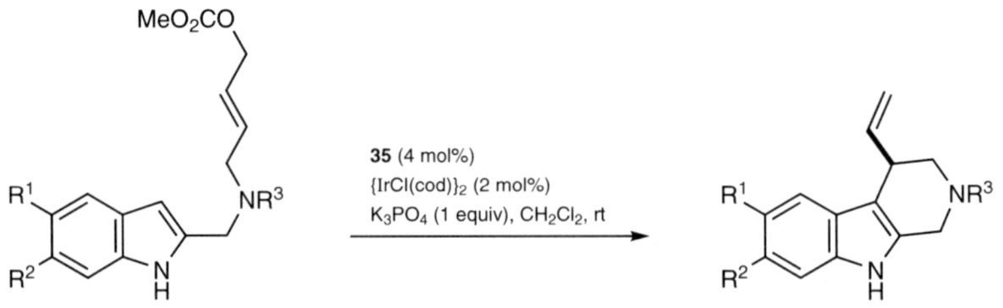

MeO_2CO

35 (4 mol%)
$\{IrCl(cod)\}_2$ (2 mol%)
K_3PO_4 (1 equiv), CH_2Cl_2, rt

R^1	R^2	R^3	Time (h)	ee[a] (%)	Yield[b] (%)	Ref
H	H	Bn	2	>99	80	[18]
H	H	Me	32	98	30	[18]
H	H	$CH_2CH=CH_2$	18	>99	65	[18]
Me	H	Bn	2	>99	75	[18]
OMe	H	Bn	6	>99	70	[18]
Br	H	Bn	5	>99	77	[18]
H	Cl	Bn	5	98	76	[18]

[a] Determined by HPLC analysis using a chiral stationary phase.
[b] Isolated yields after column chromatography.

Under gold catalysis, both indol-3-yl- and indol-2-yl-substituted allyl alcohols lead to the corresponding polycyclic compounds in good yields and with good enantioselectivities (Schemes 17 and 18).[19] Trifluoromethanesulfonate is found to be the best counterion for the gold catalyst. A wide range of substrates bearing either an electron-withdrawing or electron-donating group on the indole core have been investigated, furnishing the desired tetrahydrocarbazoles in good yields and with good enantioselectivities. Various R^1 groups on the malonate tether can be tolerated, giving excellent results. However, when a substrate bearing a methyl group at the N1-position is employed, the reaction does not occur at all, likely due to the detrimental steric hindrance between the methyl group and the cyclization site (C2-position) (Scheme 17). If the position of the linkage is changed from the C3-position to the C2-position of the indole, N—Me substituted substrates successfully undergo this transformation (Scheme 18). This method showcases the capability of gold(I) catalysts to activate allyl alcohols toward aromatic C—H bond functionalization in a highly enantioselective fashion.

Scheme 17 Tetrahydrocarbazoles by Gold-Catalyzed Intramolecular Friedel–Crafts-Type Alkylation of Indol-3-yl-Substituted Allyl Alcohols[19]

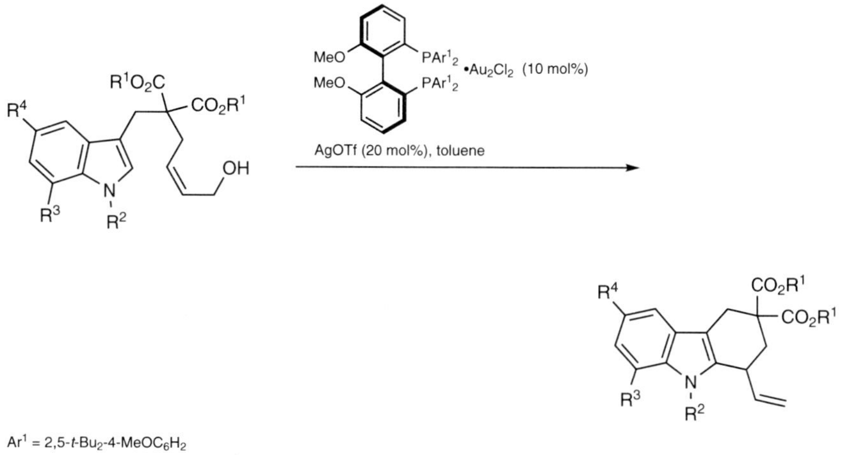

MeO, MeO, PAr^1_2, PAr^1_2 •Au_2Cl_2 (10 mol%)
AgOTf (20 mol%), toluene

Ar^1 = 2,5-*t*-Bu_2-4-$MeOC_6H_2$

R^1	R^2	R^3	R^4	Temp	Time (h)	ee[a] (%)	Yield[b] (%)	Ref
Et	H	H	Br	rt	48	86	60	[19]
Et	H	H	OMe	rt	24	84	79	[19]
Et	H	H	OMe	0°C	48	96	55	[19]
Et	H	Me	H	0°C	48	83	91	[19]
Et	Me	H	H	0°C	48	–	–[c]	[19]
Me	H	H	H	0°C	24	85	74	[19]
t-Bu	H	H	H	0°C	48	92	53	[19]
Et	H	H	OBn	rt	24	80	78	[19]
Et	H	H	OBn	0°C	48	82	69	[19]

[a] Determined by HPLC analysis using a chiral stationary phase.
[b] Isolated yields after column chromatography.
[c] Unchanged substrate was recovered (80%).

Scheme 18 Tetrahydrocarbazoles by Gold-Catalyzed Intramolecular Friedel–Crafts-Type Alkylation of Indol-2-yl-Substituted Allyl Alcohols[19]

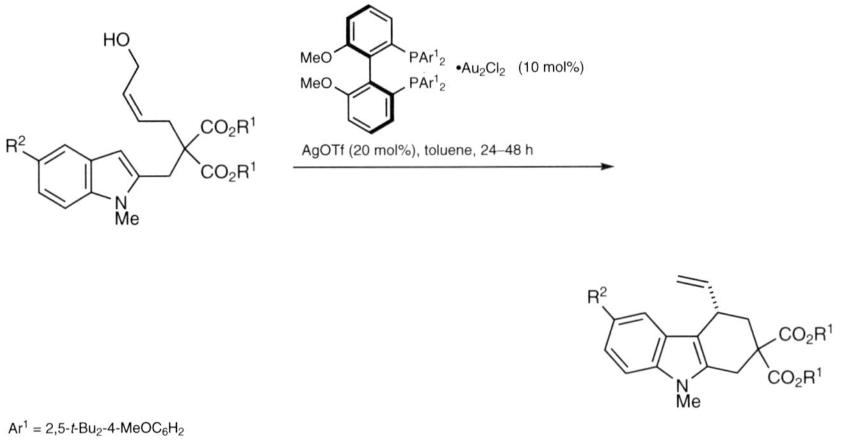

MeO, MeO, PAr^1_2, PAr^1_2 •Au_2Cl_2 (10 mol%)

AgOTf (20 mol%), toluene, 24–48 h

Ar^1 = 2,5-t-Bu_2-4-$MeOC_6H_2$

R^1	R^2	Temp	ee[a] (%)	Yield[b] (%)	Ref
Et	H	0°C	86	79	[19]
t-Bu	H	rt	80	80	[19]
Me	H	0°C	74	87	[19]
Et	OMe	0°C	83	55	[19]
Et	Me	0°C	80	87	[19]

[a] Determined by HPLC analysis using a chiral stationary phase.

[b] Isolated yields after column chromatography.

4-Vinyl-2,3,4,9-tetrahydro-1*H*-pyrido[3,4-*b*]indoles 34; **General Procedure:**[17]
Under N_2, a soln of $Pd_2(dba)_3$•$CHCl_3$ (3.6 μmol, 5 mol%) and Trost ligand (*R*,*R*)-**32** (11 mol%) in anhyd CH_2Cl_2 (1.0 mL) was stirred at rt until the soln turned from deep red to orange (about 30 min). Then, carbonate **33** (0.07 mmol, 1 equiv) dissolved in CH_2Cl_2 (0.5 mL) and Li_2CO_3 (2 equiv) were added. The resulting mixture slowly turned yellow and was stirred overnight (monitored by TLC). The reaction was then quenched with H_2O (4 mL) and extracted with EtOAc. The combined organic phases were dried (Na_2SO_4), filtered, and concentrated under reduced pressure. The crude mixture was purified by column chromatography (silica gel, CH_2Cl_2/EtOAc 4:1).

1.3.1.5.1.2 Indoles with an Allylic Linkage at the C4-Position

Introducing an allyl carbonate linkage at the C4-position of an indole, the potential cyclization patterns become diverse, such as C4-to-C3 or C4-to-C5, and depend upon various reaction parameters. In the presence of a palladium catalyst, the cyclization of substrate **36** affords an indole-based nine-membered ring **37** in 70% yield, which is in accordance with the fact that the nucleophile (C3 of indole) tends to attack the less-substituted π-allyl terminus due to steric factors (Scheme 19).[20] However, when using an iridium catalyst with ligand **38**, the highly enantioenriched seven-membered rings **39** are obtained as a result of a preference for branched selectivity. Notably, the substrate with a carbon linkage is not applicable in this approach (Scheme 20).[20] Interestingly, when there is a substituent at the C3-position of indole derivatives **40**, the allylic alkylation reaction proceeds at the C5-position, giving the corresponding products **41** in 40–80% yield and 56–97% ee (Scheme 21).[20]

Scheme 19 Fused Indole by Palladium-Catalyzed Intramolecular Friedel–Crafts-Type Alkylation of an Indole through C4–C3[20]

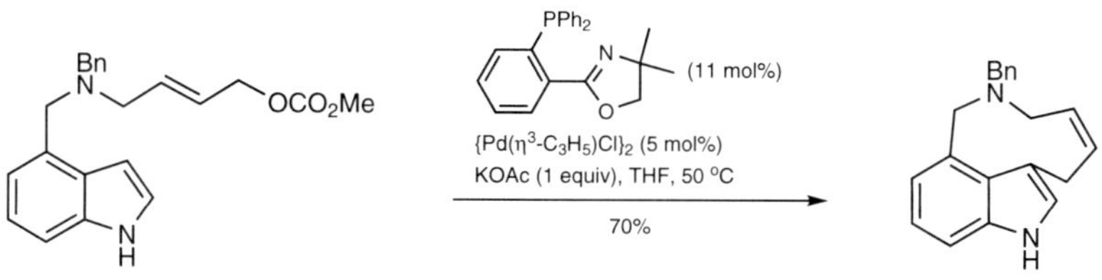

Scheme 20 Fused Indoles by Iridium-Catalyzed Intramolecular Friedel–Crafts-Type Alkylation of Indoles through C4–C3[20]

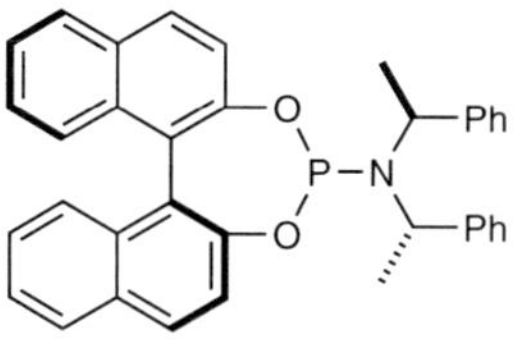

X	R[1]	R[2]	ee[a] (%)	Yield[b] (%)	Ref
NBn	H	H	94	60	[20]
$NCH_2CH=CH_2$	H	H	94	51	[20]
NBn	Me	H	94	46	[20]
NBn	H	Ph	94	78	[20]
NBn	H	$CH_2CH=CH_2$	97	60	[20]
$C(CO_2Me)_2$	H	H	–	–[c]	[20]

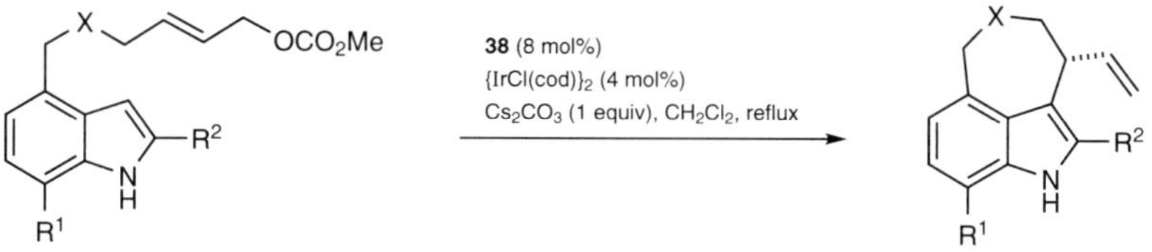

[a] Determined by HPLC analysis using a chiral stationary phase.
[b] Isolated yields after column chromatography.
[c] No reaction.

Scheme 21 Fused Indoles by Iridium-Catalyzed Intramolecular Friedel–Crafts-Type Alkylation of Indoles through C4–C5[20]

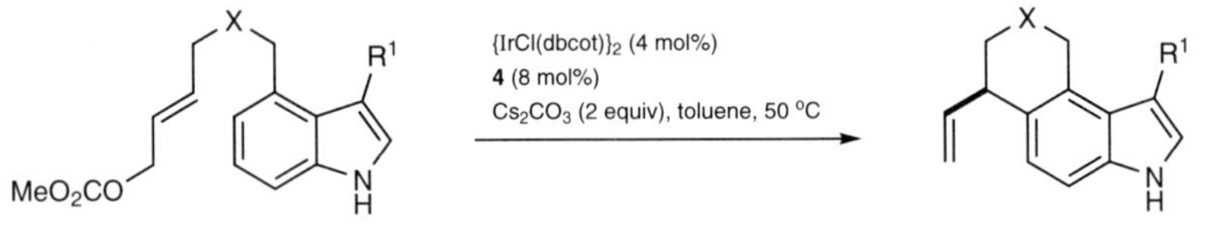

$\{IrCl(dbcot)\}_2$ (4 mol%)
4 (8 mol%)
Cs_2CO_3 (2 equiv), toluene, 50 °C

40 → **41**

dbcot = dibenzo[*a*,*e*]cyclooctatetraene

X	R[1]	ee[a] (%)	Yield[b] (%)	Ref
NBn	$CH_2CH{=}CH_2$	83	48	[20]
$NCH_2CH{=}CH_2$	$CH_2CH{=}CH_2$	90	40	[20]
NBn	Me	79	56	[20]
NBn	Bn	86	63	[20]
NBn	$4\text{-}ClC_6H_4$	56	52	[20]

[a] Determined by HPLC analysis using a chiral stationary phase.
[b] Isolated yields after column chromatography.

1.3.1.5.2 Dearomatization Reactions

1.3.1.5.2.1 Allylic Dearomatization Reactions

The iridium-catalyzed asymmetric allylic dearomatization of indole derivatives **43** has been achieved in an intramolecular fashion, delivering the corresponding spiroindolenines **44** in excellent yields, diastereoselectivities, and enantioselectivities (Scheme 22).[21] The introduction of phosphoramidite ligand **42** (Me-THQphos) is crucial for the high yields and selectivities. In contrast to Feringa-type ligands where the active iridium catalyst is formed via C(sp^3)—H activation, the active catalyst derived from Me-THQphos is formed via the cleavage of a C(sp^2)—H bond. Notably, the substrates utilized in this methodology are restricted to those bearing an electron-donating group, such as benzyl, allyl, or methyl, on the nitrogen linkage. Interestingly, a ruthenium catalytic system has been reported which features a broader relative substrate scope, where substrates bearing either an electron-donating or electron-withdrawing group on the nitrogen tether and carbon-tethered allylic carbonates are well tolerated (Scheme 23).[22] Compared with the iridium catalytic system, several features of the ruthenium catalytic system should be noted: (1) an inexpensive and easily prepared catalyst; (2) high reactivity; (3) broad substrate scope; (4) insensitivity to water; and (5) mild conditions. In addition, iridium-catalyzed allylic dearomatization of pyrrole derivatives **45** has been realized in excellent yields, diastereoselectivities, and enantioselectivities (Scheme 24).[23]

Scheme 22 Spiro-Fused Indole Derivatives by Iridium-Catalyzed Intramolecular Asymmetric Allylic Dearomatization of Indoles[21]

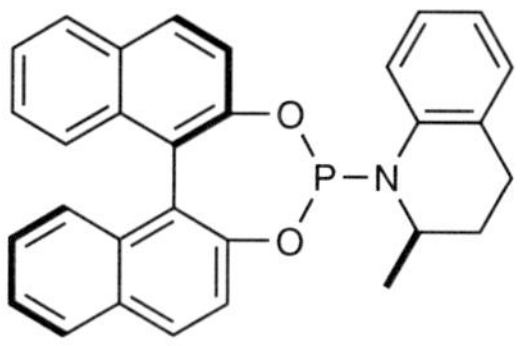

42

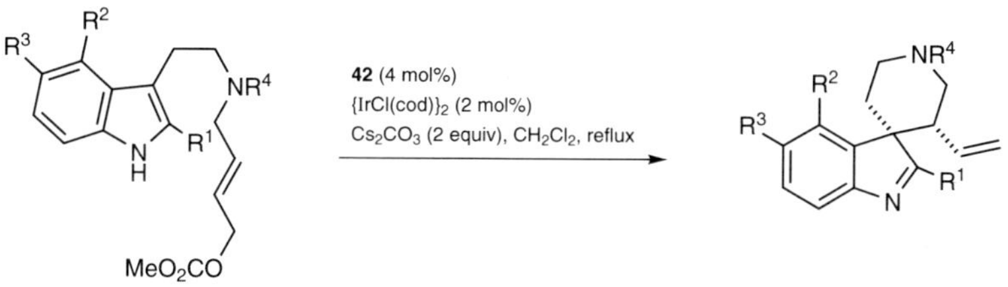

43

42 (4 mol%)
$\{IrCl(cod)\}_2$ (2 mol%)
Cs_2CO_3 (2 equiv), CH_2Cl_2, reflux

44

R^1	R^2	R^3	R^4	dr[a]	ee[b] (%)	Yield[c] (%)	Ref
H	H	H	Bn	>99:1	96	95	[21]
H	H	H	Me	>99:1	96	95	[21]
H	H	H	$CH_2CH{=}CH_2$	96:4	96	93	[21]
H	Me	H	Bn	>99:1	93	97	[21]
H	H	F	Bn	97:3	88	93	[21]
Ph	H	H	Bn	75:25	93 (91)[d]	68 (20)[d]	[21]

[a] Determined by ^{1}H NMR analysis of the crude reaction mixture.
[b] Determined by HPLC analysis using a chiral stationary phase.
[c] Isolated yields after column chromatography.
[d] Value for minor diastereomer.

Scheme 23 Spiro-Fused Indole Derivatives by Ruthenium-Catalyzed Intramolecular Allylic Dearomatization of Indoles[22]

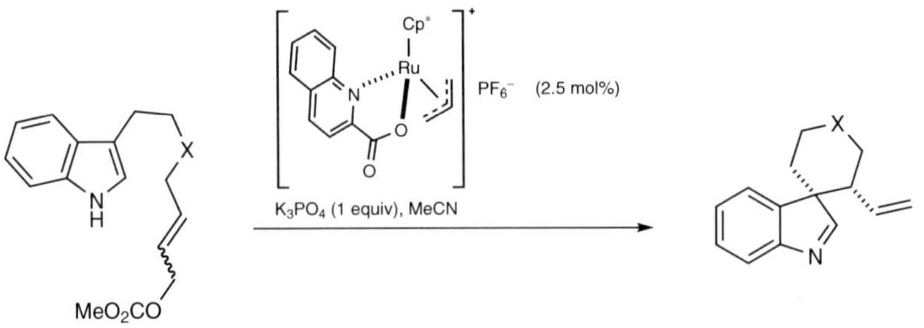

Cp*; Ru; PF_6^- (2.5 mol%); K_3PO_4 (1 equiv), MeCN; MeO_2CO

X	Config[a]	Temp	dr[b]	Yield[c] (%)	Ref
NBn	*E*	rt	81:19	87	[22]
NTs	*E*	rt	87:13	87	[22]
NBoc	*E*	reflux	64:36	77	[22]
$C(CO_2Me)_2$	*E*	reflux	90:10	71	[22]
NTs	*Z*	reflux	68:32	91	[22]
NBoc	*Z*	reflux	76:24	77	[22]

[a] Config of starting alkene.
[b] Determined by ^1H NMR analysis of the crude reaction mixture.
[c] Isolated yields after column chromatography.

Scheme 24 Spiro-Fused Pyrrole Derivatives by Iridium-Catalyzed Intramolecular Asymmetric Allylic Dearomatization of Pyrroles[23]

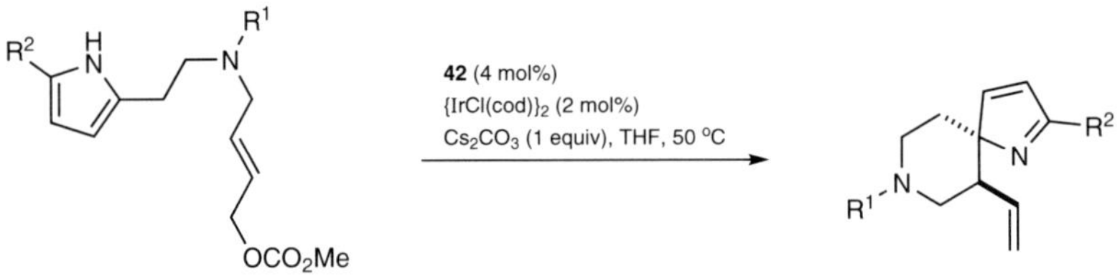

42 (4 mol%); $\{IrCl(cod)\}_2$ (2 mol%); Cs_2CO_3 (1 equiv), THF, 50 °C; OCO_2Me; **45**

R^1	R^2	dr[a]	ee[b] (%)	Yield[c] (%)	Ref
Bn	Ph	99:1	93	80	[23]
$CH_2CH{=}CH_2$	Ph	95:5	86	85	[23]
Me	Ph	>99:1	94	77	[23]
Bn	Et	90:10	96	83	[23]
Bn	H	97:3	96	61	[23]

[a] Determined by ^1H NMR analysis of the crude reaction mixture.
[b] Determined by HPLC analysis using a chiral stationary phase.
[c] Isolated yields after column chromatography.

By installing the side chain of the allylic moiety at the C4-position of the indole core, the C4-to-C3 cyclization can be achieved via palladium-catalyzed allylic dearomatization of

C3-substituted indoles **46**. However, only moderate enantioselectivities of the products **47** are obtained in the presence of a planar chiral ferrocene-based Phox-type ligand (Scheme 25).[20]

Scheme 25 Fused Indole Derivatives by Palladium-Catalyzed Allylic Dearomatization of Indoles through C4–C3[20]

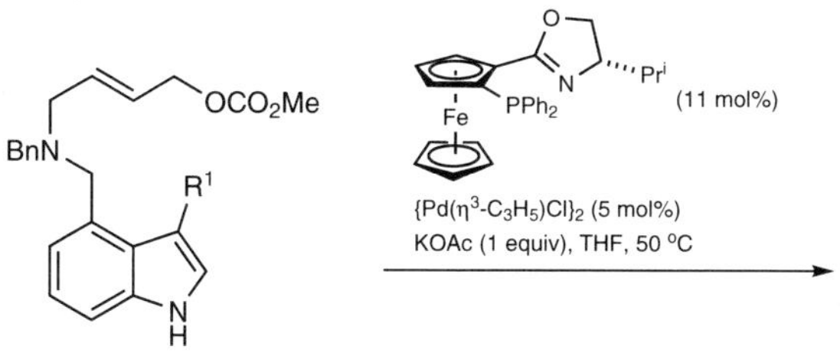

BnN, OCO_2Me, R^1, N H, **46**; ligand: Fe, PPh_2, O, N, Pr^i (11 mol%); $\{Pd(\eta^3\text{-}C_3H_5)Cl\}_2$ (5 mol%), KOAc (1 equiv), THF, 50 °C; BnN, R^1, N, **47**

R^1	ee (%)	Yield (%)	Ref
$CH_2CH{=}CH_2$	78	74	[20]
Me	74	62	[20]
Bn	74	68	[20]

Spiroindolenines 44; General Procedure:[21]
A flame-dried Schlenk tube was cooled to rt and filled with argon. To this flask were added $\{IrCl(cod)\}_2$ (2.7 mg, 0.004 mmol, 2 mol%), phosphoramidite ligand **42** (4.3 mg, 0.008 mmol, 4 mol%), THF (0.5 mL), and $PrNH_2$ (0.5 mL). The mixture was heated at 50 °C for 30 min and then the volatile solvents were removed in vacuo to give a pale yellow solid. At this point, allylic carbonate **43** (0.20 mmol) dissolved in CH_2Cl_2 (2.0 mL) and Cs_2CO_3 (0.40 mmol, 2 equiv) were added and the mixture was refluxed for 1.5 h. After the reaction was complete (monitored by TLC), the crude mixture was filtered through a pad of Celite which was then washed with EtOAc. The solvents were removed from the filtrate under reduced pressure and the dr was determined by ^1H NMR analysis of the crude mixture. Then, the residue was purified by column chromatography (silica gel, petroleum ether/EtOAc 2:1).

1.3.1.5.2.2 Allylic Dearomatization/Migration Reactions

Cyclopentane-1,3′-dihydroindoles **49** can be obtained through an iridium-catalyzed allylic dearomatization of all-carbon-tethered indolyl allylic carbonates **48** with subsequent reduction by sodium cyanoborohydride. Interestingly, upon treatment with a catalytic amount of 4-toluenesulfonic acid, the stereospecific allylic migration of the spiroindolenine intermediates is observed, delivering the corresponding tetrahydrocarbazoles **50** in excellent yields and enantioselectivities (Scheme 26).[24]

for references see p 149

Scheme 26 Spiro-Fused Indole and Tetrahydrocarbazole Derivatives by Iridium-Catalyzed Intramolecular Asymmetric Allylic Dearomatization of Indoles[24]

1. $\{IrCl(cod)\}_2$ (2 mol%), **38** (4 mol%) Cs_2CO_3 (2 equiv), dioxane, 50 °C
2. $NaBH_3CN$, MeOH, 0 °C to rt

48 → **49**

R^1	dr	ee (%)	Yield (%)	Ref
H	16:1	98	93	[24]
OMe	14:1	95	91	[24]
Br	13:1	96	92	[24]

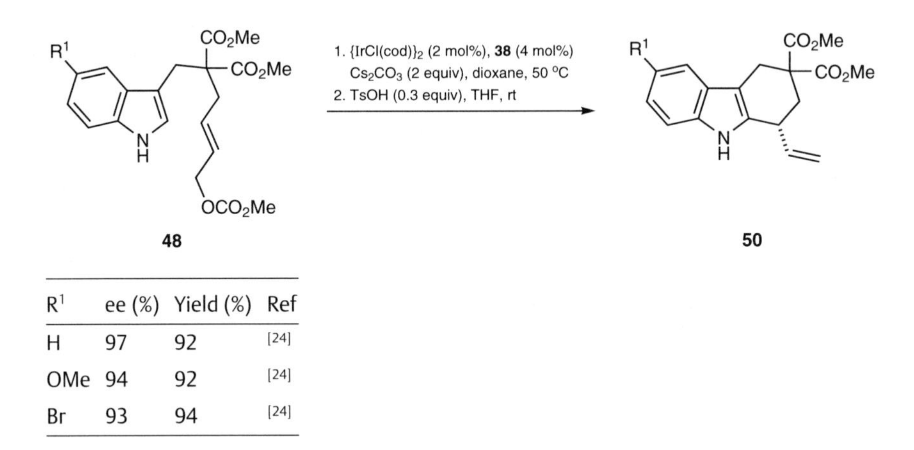

R^1	ee (%)	Yield (%)	Ref
H	97	92	[24]
OMe	94	92	[24]
Br	93	94	[24]

By utilizing a substrate with a benzylamino-tethered linkage (e.g., **51**) instead of a carbon linkage (as in **48**), enantioenriched polycyclic indoles **52** can be obtained through iridium-catalyzed allylic dearomatization and an unprecedented in situ methylene migration process, resulting in the substituent originally at the C3-position of the substrate transferring to the C2-position in the product (Scheme 27).[25] Under slightly modified conditions, the analogous dearomatization/migration phenomenon is also observed for C2-substituted pyrroles **53** giving tetrahydro-1*H*-pyrrolo[3,2-*c*]pyridines **54** (Scheme 28).[25] Notably, this approach has also been realized by a ruthenium catalytic system.[26]

Scheme 27 Vinyltetrahydrocarbolines by Iridium-Catalyzed Enantioselective Functionalization of Indoles via an In Situ Formed Spiro Intermediate[25]

$\{IrCl(cod)\}_2$ (2 mol%)
35 (4 mol%)
K_3PO_4 (2 equiv)
THF, reflux, 48 h

51 → **52**

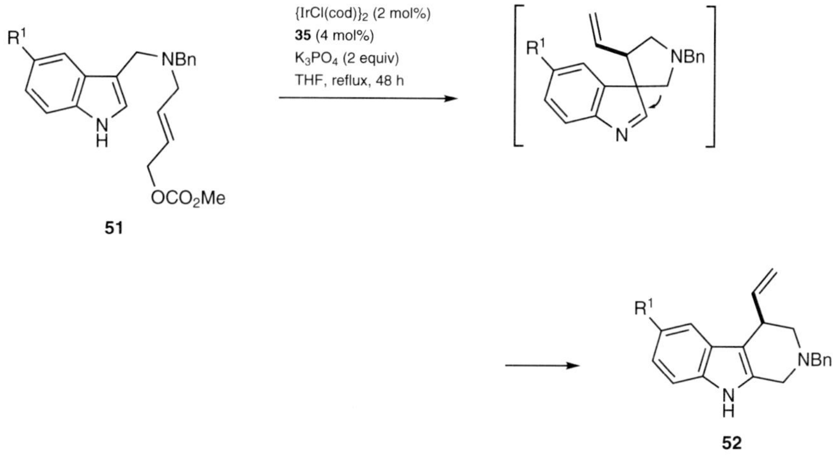

R^1	ee (%)	Yield (%)	Ref
H	96	80	[25]
F	96	88	[25]
Br	94	74	[25]
OMe	94	80	[25]

Scheme 28 7-Vinyl-4,5,6,7-tetrahydro-1*H*-pyrrolo[3,2-*c*]pyridines by Iridium-Catalyzed Enantioselective Functionalization of Pyrroles via an In Situ Formed Spiro Intermediate[25]

$\{IrCl(cod)\}_2$ (2 mol%)
35 (4 mol%)
K_3PO_4 (2 equiv), 50 °C

53 → **54**

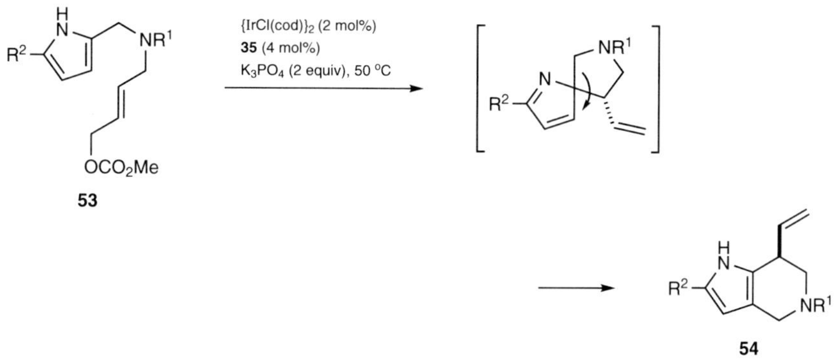

R^1	R^2	Time (h)	ee[a] (%)	Yield[b] (%)	Ref
Bn	H	34	98	80	[25]
$CH_2CH{=}CH_2$	4-Tol	58	98	57	[25]
Bn	Ph	34	99	91	[25]
Bn	4-ClC_6H_4	33	99	88	[25]
Bn	Et	33	97	89	[25]

[a] Determined by HPLC analysis using a chiral stationary phase.
[b] Isolated yields after column chromatography.

for references see p 149

1.3.1.6 Phenols

1.3.1.6.1 Friedel–Crafts-Type Reactions

Within transition-metal-catalyzed allylic substitution reactions, there are only limited reports of Friedel–Crafts-type allylic alkylation of phenols, mainly due to the strong competition from O-allylation. However, relocating the allyl carbonate tether to the *meta*-position relative to the hydroxy group allows the intramolecular allylic alkylation of phenols **55** in the presence of an iridium catalyst, affording the corresponding tetrahydroisoquinolines **56** and **57** in high yields, regioselectivities, and enantioselectivities (Scheme 29).[27] Notably, when a substituent is introduced at the 2-position of the phenol (i.e., $R^1 \neq H$), a single regioisomer **57** is obtained with excellent enantioselectivity as a result of reaction at the *para*-position. On the other hand, if the substrate bears a substituent at the 4- or 5-position of the phenol ring (i.e., R^2 and/or $R^3 \neq H$), the reaction leads to product **56** exclusively in good to excellent yield and enantioselectivity.

Scheme 29 Vinyltetrahydroisoquinolines and Vinyltetrahydronaphthalenes by Iridium-Catalyzed Intramolecular Friedel–Crafts-Type Alkylation of Phenols[27]

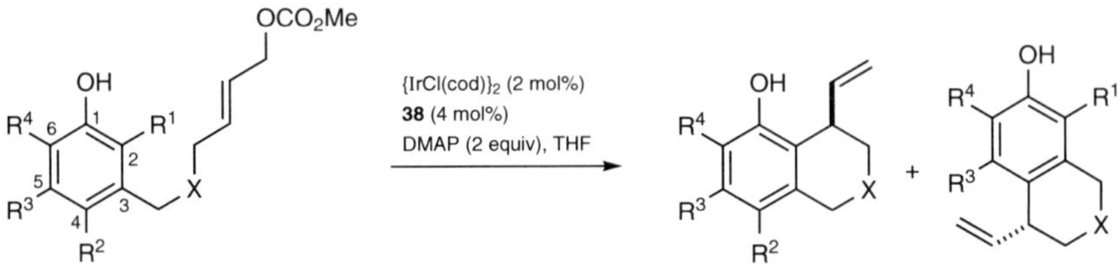

OCO2Me; OH; R4; R1; R3; R2; X; 55; {IrCl(cod)}2 (2 mol%), 38 (4 mol%), DMAP (2 equiv), THF; 56; + ; 57

X	R^1	R^2	R^3	R^4	Temp	Ratio[a] (**56**/**57**)	ee[b] (%)	Yield[c] (%)	Ref
NBn	H	H	H	H	rt	4.8:1	92	84	[27]
$NCH_2CH=CH_2$	H	H	H	H	rt	6.0:1	92	88	[27]
NBn	OMe	H	H	H	rt	**57** only	93	72	[27]
NBn	Cl	H	H	H	rt	**57** only	94	80	[27]
NBn	H	H	H	OMe	rt	4.0:1	95	82	[27]
NBn	H	H	H	NO_2	50 °C	**56** only	93	25	[27]
NBn	H	Br	H	H	50 °C	**56** only	89	78	[27]
NBn	Br	H	H	OMe	50 °C	**57** only	96	50	[27]
NBn	H	Br	H	OMe	50 °C	**56** only	92	80	[27]
NBn	H	H	OH	H	rt	**56** only	86	90	[27]
NBn	H	H	OH	OMe	rt	**56** only	88	86	[27]
$C(CO_2Me)_2$	H	H	H	H	50 °C	4.0:1	91	87[d]	[27]

[a] Determined by ^1H NMR analysis of the crude reaction mixture.
[b] Determined by HPLC analysis using a chiral stationary phase.
[c] Isolated yields of **56** and **57** after column chromatography.
[d] The reaction was carried out with Cs_2CO_3 (2 equiv) in dioxane (2 mL) at 50 °C.

Tetrahydroisoquinolines 56 **and** 57**; General Procedure:**[27]
$\{IrCl(cod)\}_2$ (2.7 mg, 0.004 mmol, 2 mol%), phosphoramidite ligand **38** (4.3 mg, 0.008 mmol, 4 mol%), THF (0.5 mL), and $PrNH_2$ (0.3 mL) were added to a flame-dried Schlenk tube under argon. The mixture was heated at 50 °C for 30 min and then the volatile solvents were removed under reduced pressure to give a yellow solid. A soln of allyl

carbonate **55** (0.20 mmol) in THF (2.0 mL) and DMAP (25.0 mg, 0.20 mmol, 100 mol%) were added successively and the mixture was stirred at the required temperature. After the reaction was complete (monitored by TLC), the crude mixture was filtered through a pad of Celite, which was then washed with EtOAc. The solvents were removed under reduced pressure and the ratio of **56/57** was determined by ^1H NMR analysis of the crude mixture. Then, the residue was purified by column chromatography (silica gel, petroleum ether/EtOAc 8:1) to afford **56** and **57**.

1.3.1.6.2 Dearomatization Reactions

As exemplified above, the strategy of intramolecular design has the advantage of avoiding competitive O-alkylation of phenols. Interestingly, with a substituent at the *para*-position, phenol derivatives **58** undergo a desirable dearomatization reaction to give spirocyclohexadienone derivatives **59** (Scheme 30).[28] This racemic process has been achieved by palladium catalysis. Generally, the utilization of monodentate phosphorus ligands increases the reactivity. Among them, triphenylphosphine is chosen for practical reasons. The current method tolerates a broad range of substrates including secondary alcohol derivatives as well as substrates bearing varied linkages such as dimethyl acetal, *N*—tosyl, and oxygen moieties. In general, the corresponding spiro[4.5]cyclohexadienones with an *E*-alkene unit are obtained in good yields. Products with two contiguous chiral centers are accessible when substrates with a methyl or chloride group at the *meta*-position (R^2) of the phenol are used. The initial attempt at an asymmetric transformation was realized by the utilization of chiral ligand **60**, providing the major diastereomer in 80% yield and 89% ee (Scheme 31).[28] Almost at the same time, the highly enantioselective construction of all-carbon quaternary spirocenters was disclosed, using iridium-catalyzed allylic dearomatization of *para*-substituted phenol derivatives tethered with an allylic carbonate, forming either five- or six-membered rings in excellent yields and enantioselectivities (Scheme 32).[29] Notably, the linkage between the phenol and allyl carbonate is important for the cyclization process. Cyclohexadienone products with a six-membered spiro ring can be formed in good yields and excellent enantiomeric excess from substrates bearing a benzyl or 4-bromobenzyl group on the linking nitrogen atom, while five-membered ring formation is achieved with good results with a tosyl-substituted nitrogen or carbon-tethered substrates.

Scheme 30 Spiro-Fused Cyclohexadienones by Palladium-Catalyzed Allylic Dearomatization of Phenols[28]

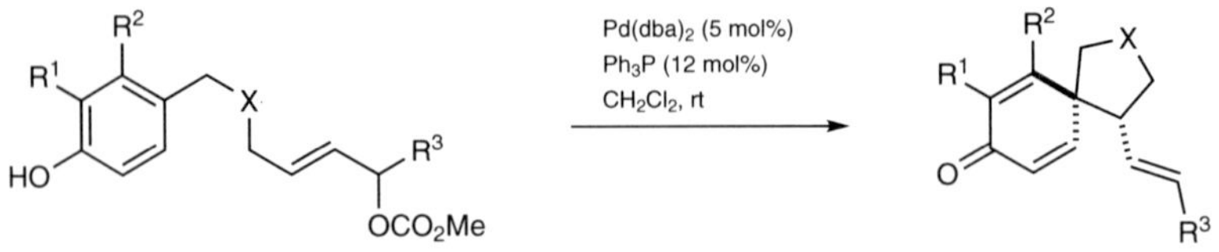

X	R^1	R^2	R^3	Time (h)	dr[a]	Yield[b] (%)	Ref
$C(CO_2Me)_2$	H	H	H	3	–	94	[28]
$C(CO_2Me)_2$	H	H	Me	6	–	84	[28]
$C(CO_2Me)_2$	H	H	Ph	6	–	89	[28]
$C(OMe)_2$	H	H	H	6	–	93	[28]
NTs	H	H	H	6	–	86	[28]
O	H	H	H	6	–	63[c]	[28]
$C(CO_2Me)_2$	H	Me	H	6	13.4:1	97	[28]
$C(CO_2Me)_2$	H	Cl	H	8	11.0:1	92	[28]
NTs	(CH=CH)$_2$		H	24	3.0:1	97[d]	[28]

[a] Determined by ^1H NMR analysis of the crude reaction mixture.
[b] Isolated yields after column chromatography.
[c] Isolated yield using triphenyl phosphite as the ligand.
[d] (*R*)-Monophos [(*R*)-(−)-(3,5-dioxa-4-phosphacyclohepta[2,1-a:3,4-a]dinaphthalen-4-yl)dimethylamine] was used as the ligand and acetonitrile as the solvent.

Scheme 31 Synthesis of a Spiro-Fused Cyclohexadienone by Palladium-Catalyzed Asymmetric Allylic Dearomatization of a Phenol[28]

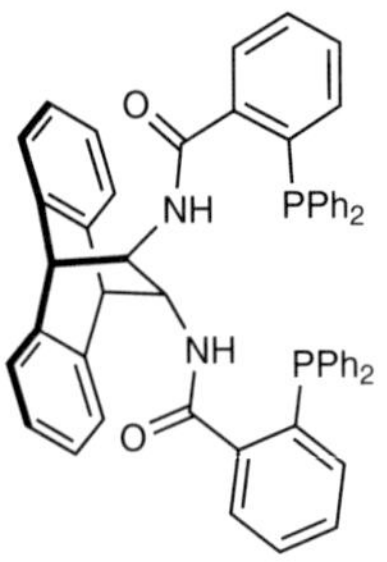

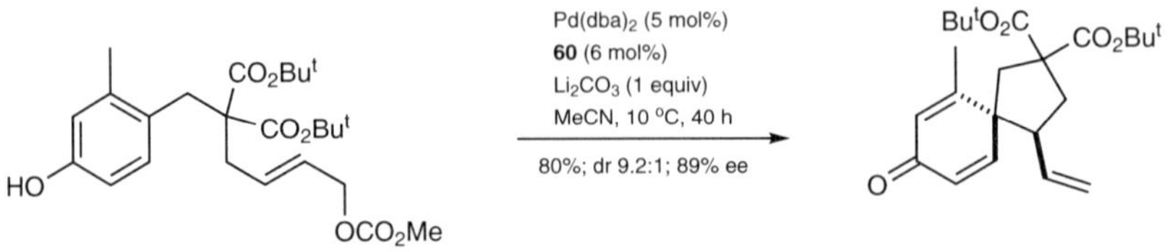

Scheme 32 Spiro-Fused Cyclohexadienones by Iridium-Catalyzed Asymmetric Allylic Dearomatization of Phenols[29]

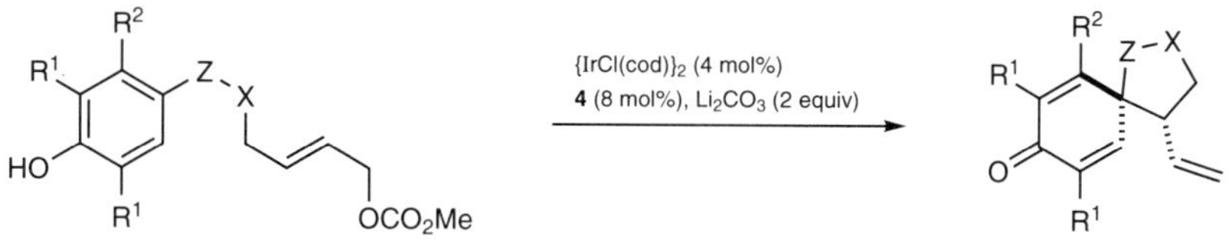

R^1	R^2	X	Z	Conditions	ee[a] (%)	Yield[b] (%)	Ref
H	H	NBn	$(CH_2)_2$	THF, 50 °C	96	68	[29]
H	H	$(4\text{-}BrC_6H_4CH_2)N$	$(CH_2)_2$	THF, 50 °C	91	60	[29]
iPr	H	NTs	CH_2	THF, 50 °C	95	92	[29]
OMe	H	NTs	CH_2	THF, 50 °C	88	65	[29]
H	H	$C(CO_2Me)_2$	CH_2	dioxane, reflux	97 (95)[c]	95 (92)[c]	[29]
H	H	$C(CO_2t\text{-}Bu)_2$	CH_2	dioxane, reflux	86	75	[29]
H	Me	$C(CO_2Me)_2$	CH_2	dioxane, reflux	93[d]	95	[29]

[a] Determined by HPLC analysis using a chiral stationary phase.
[b] Isolated yields after column chromatography.
[c] The results in parentheses were obtained for a 1-mmol scale reaction.
[d] dr 11:1; minor diastereomer was obtained in 80% ee.

Dimethyl 8-Oxo-4-vinylspiro[4.5]deca-6,9-diene-2,2-dicarboxylate [59, X = $C(CO_2Me)_2$; $R^1 = R^2 = R^3$ = H]; Typical Procedure:[28]

$Pd(dba)_2$ (8.6 mg, 0.015 mmol), Ph_3P (9.4 mg, 0.036 mmol), and phenol **58** [X = $C(CO_2Me)_2$; $R^1 = R^2 = R^3$ = H; 110.0 mg, 0.30 mmol] were dissolved in CH_2Cl_2 (1.5 mL) and the resulting soln was stirred at rt. After 3 h, the reaction was quenched with sat. aq NH_4Cl and the mixture was extracted with EtOAc. The organic layer was washed with brine, dried (Na_2SO_4), and concentrated under reduced pressure. Then, the residue obtained was purified by column chromatography (silica gel, hexane/EtOAc 3:1); yield: 94%.

1.3.1.7 Simple Arenes

Metal-catalyzed allylic alkylation involving simple arenes has also been developed despite the relatively low nucleophilicity exhibited compared to heteroaromatic substrates. The ring closure of allylic carbonate **61** is facilitated to a comparable extent by the employment of a molybdenum catalyst [$\{MoBr_2(CO)_4\}_2$]. Although the catalyst is air-sensitive, the reaction proceeds smoothly even in the presence of air and with reagent-grade 1,2-dichloroethane. In addition, the configuration of the double bond does not seem to have an influence on this transformation. The starting materials **61** containing mixtures of *Z*/*E*-diastereomers work well, giving the cyclized products **62** in good yields (Scheme 33).[30]

Scheme 33 Diethyl 4-Vinyl-3,4-dihydronaphthalene-2,2(1*H*)-dicarboxylates by Molybdenum(II)-Catalyzed Intramolecular Allylic Alkylation of Arenes[30]

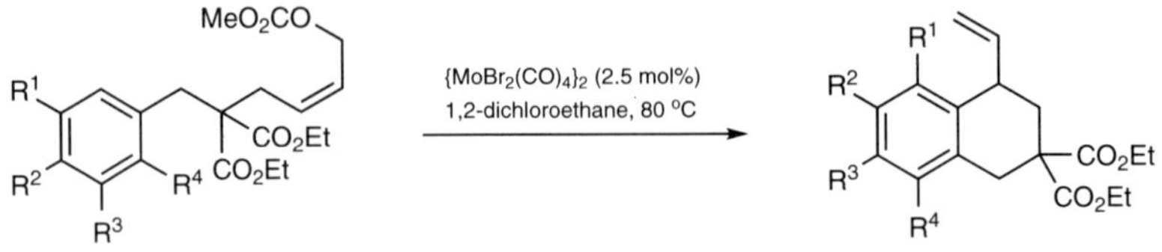

$\{MoBr_2(CO)_4\}_2$ (2.5 mol%)
1,2-dichloroethane, 80 °C

61 → **62**

R^1	R^2	R^3	R^4	Ratio[a] (*Z*/*E*) of **61**	Yield[b] (%)	Ref
OMe	H	OMe	H	90:10	72	[30]
H	OMe	H	H	75:25	85	[30]
H	Me	H	H	>98:2	90	[30]
H	H	H	H	71:29	92	[30]
(CH=CH)$_2$		H	H	88:12	96	[30]
H	Ph	H	H	97:3	84	[30]
H	SMe	H	H	87:13	70	[30]
(CH=CH)$_2$		(CH=CH)$_2$		98:2	93	[30]

[a] Determined by ^{1}H NMR and GC/MS analysis.
[b] Isolated yields after column chromatography.

The intramolecular allylic alkylation of arenes has also been realized under silver catalysis using *Z*-allylic alcohols **63** as the substrates, affording the functionalized polycyclic aromatic compounds **64** in good yields and regioselectivities (Scheme 34).[31] However, the substrates in this study are limited to electron-rich or electron-neutral arenes. It is only with the introduction of iron(III) catalysts that arenes with electron-withdrawing groups are compatible with this cyclization process. Allyl alcohols **65** with either *Z* or *E* configuration of the double bond undergo nucleophilic allylic alkylation, leading to the corresponding tetrahydronaphthalenes **66** in excellent yields (Scheme 35).[32]

Scheme 34 1-Vinyltetrahydronaphthalenes by Silver(I)-Catalyzed Intramolecular Allylic Alkylation of Arenes with Allylic Alcohols[31]

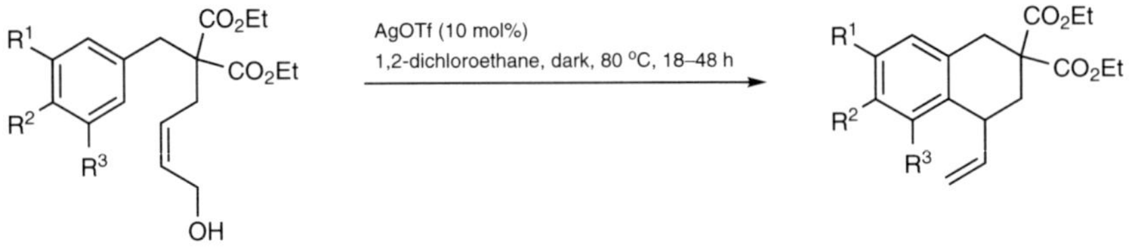

AgOTf (10 mol%)
1,2-dichloroethane, dark, 80 °C, 18–48 h

63 → **64**

R^1	R^2	R^3	Yield[a] (%)	Ref
OMe	OMe	H	90	[31]
OMe	H	OMe	70	[31]
H	OMe	H	77	[31]

R^1	R^2	R^3	Yield[a] (%)	Ref
H	H	H	73	[31]
$(CH{=}CH_2)_2$		H	80	[31]
H	Me	H	76	[31]

[a] Isolated yields after column chromatography.

Scheme 35 1-Vinyltetrahydronaphthalenes by Iron(III)-Catalyzed Intramolecular Allylic Alkylation of Electron-Deficient Arenes with π-Activated Alcohols[32]

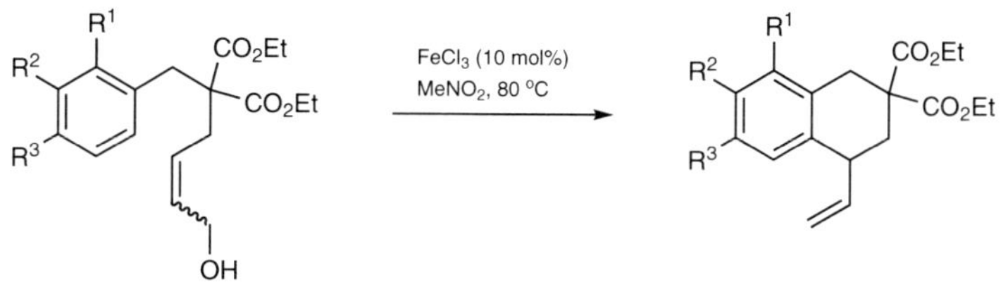

R^1	R^2	R^3	Config of **65**	Time (min)	Yield[a] (%)	Ref
H	H	F	*Z*	30	88	[32]
H	H	F	*E*	15	85	[32]
H	F	H	*Z*	15	78[b]	[32]
H	H	CN	*Z*	15	71	[32]
H	H	CO_2Me	*Z*	15	75	[32]
H	H	CO_2Et	*Z*	60	75	[32]
NO_2	H	H	*Z*	60	43	[32]

[a] Isolated yields after column chromatography.
[b] 2:1 ratio of **66** and regioisomeric diethyl 5-fluoro-4-vinyltetrahydronaphthalene-2,2-dicarboxylate; determined by GC/MS and ^1H NMR analysis of the crude reaction mixture.

Iridium catalysis opens up access to the highly enantioselective allylic alkylation of arenes. With the combination of an iridium complex with ligand **67** and a Lewis acid, the asymmetric polyene cyclization of branched, racemic allylic alcohols **68** gives the corresponding polycyclic rings **69** as single diastereomers in good yields and excellent enantioselectivities. Both alkoxy-substituted electron-rich arenes and heterocycles such as pyrroles, indoles, and furans can be used successfully in this transformation (Table 2). Notably, the reaction can be scaled up to 1.0 mmol without any erosion in yield or selectivity.[33]

Table 2 Iridium-Catalyzed Enantioselective Polyene Cyclization[33]

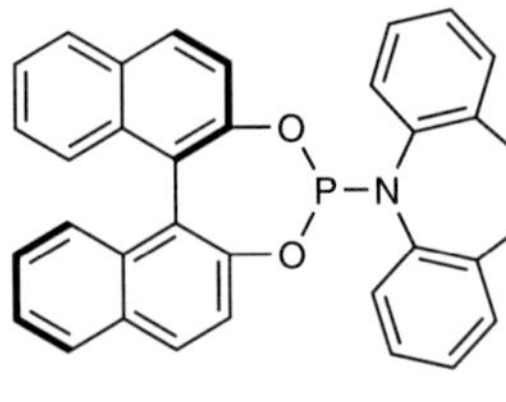

67

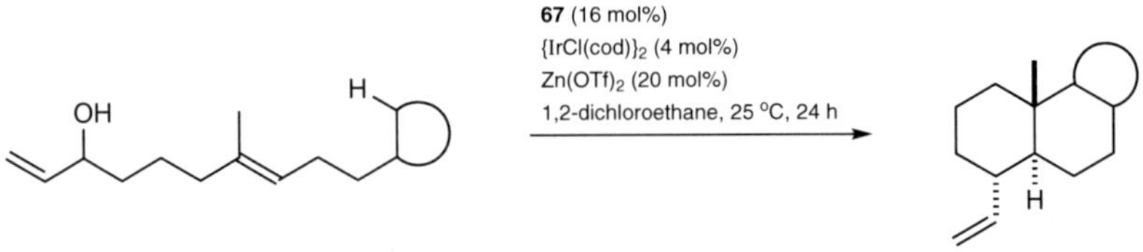

67 (16 mol%)
$\{IrCl(cod)\}_2$ (4 mol%)
$Zn(OTf)_2$ (20 mol%)
1,2-dichloroethane, 25 °C, 24 h

68 → **69**

Substrate	Product	ee[a] (%)	Yield[b] (%)	Ref
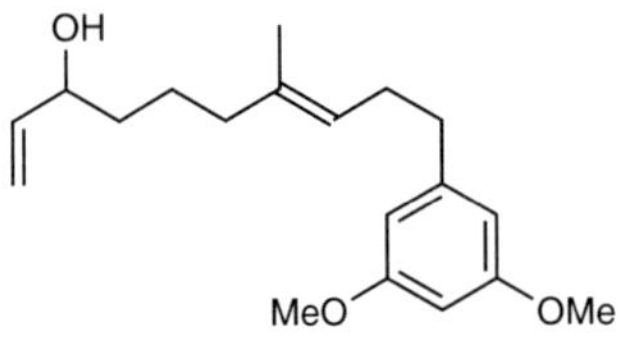	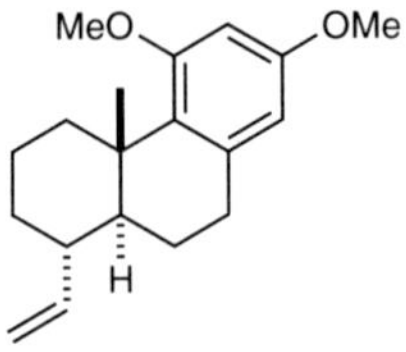	>99.5[c]	90[c]	[33]
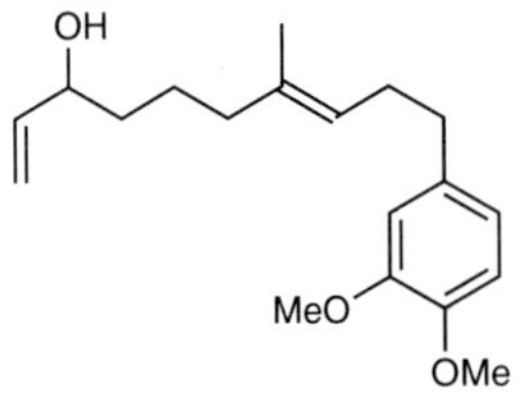	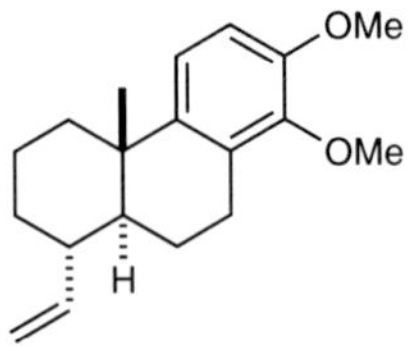	>99.5	71	[33]
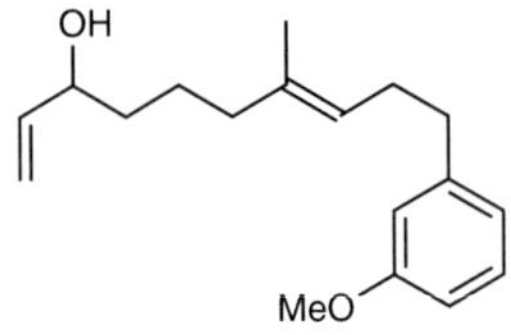	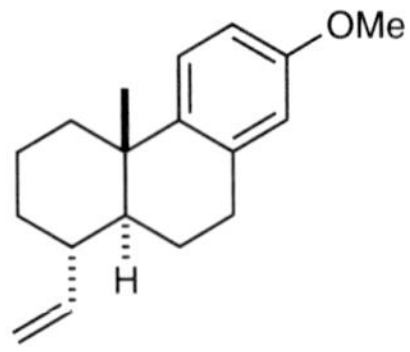	>99.5	69	[33]
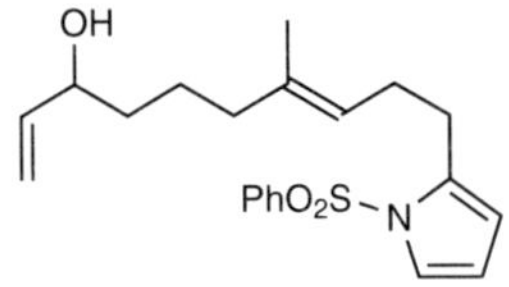	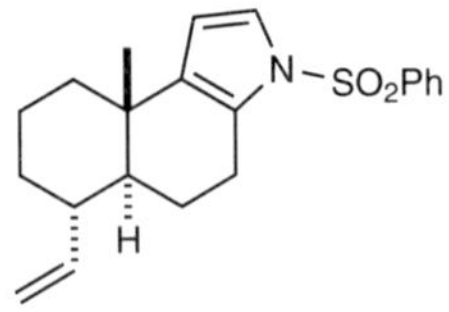	>99.5	90	[33]

Table 2 (cont.)

Substrate	Product	ee[a] (%)	Yield[b] (%)	Ref
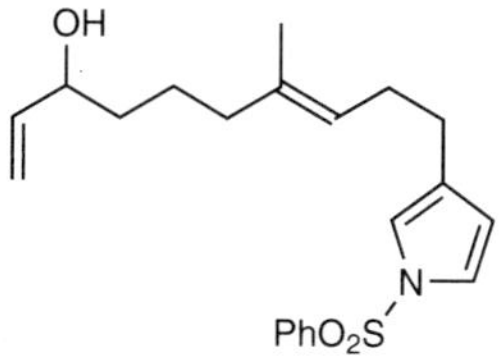	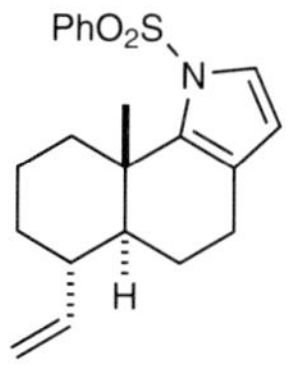	>99.5	93	[33]
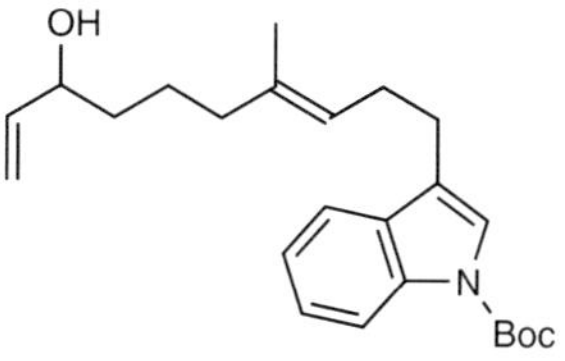	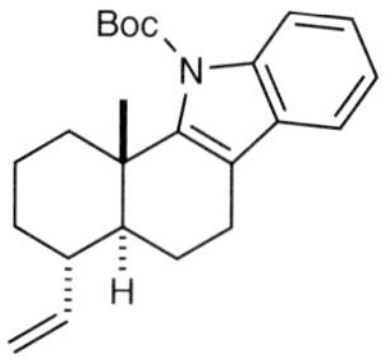	>99.5	86	[33]
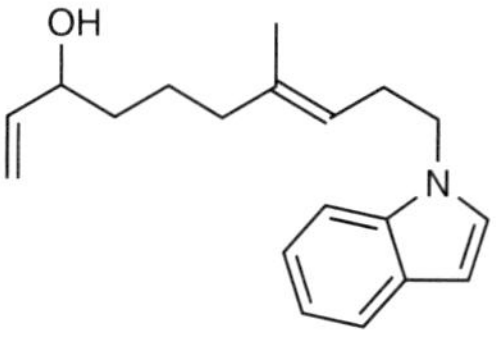	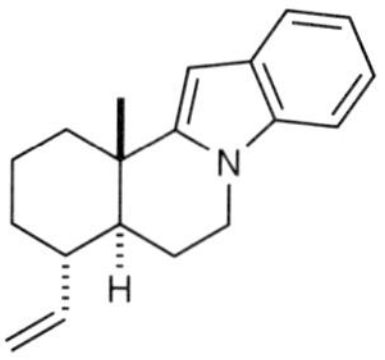	99	71	[33]
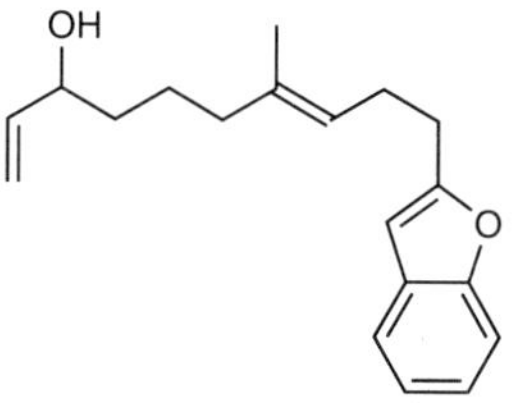	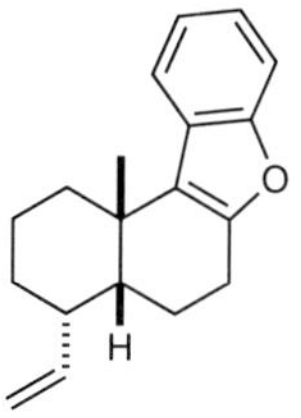	99.5	89	[33]

[a] Determined by HPLC analysis using a chiral stationary phase.
[b] Isolated yields after column chromatography.
[c] Reaction was run on a 1.0-mmol scale.

Diethyl 4-Vinyl-3,4-dihydronaphthalene-2,2(1*H*)-dicarboxylates 62; **General Procedure:**[30]
In a dry, two-necked, round-bottomed, 25-mL flask equipped with a condenser, substrate **61** (0.2 mmol) was dissolved in reagent grade 1,2-dichloroethane (2 mL). Then, $\{MoBr_2(CO)_4\}_2$ (1.9 mg, 2.5 mol%) was added and the mixture was stirred at 80 °C for 16 h. The volatiles were removed under reduced pressure and the residue was purified by column chromatography (cyclohexane/EtOAc, typically 4:1).

Diethyl 6,7-Dimethoxy-4-vinyl-3,4-dihydronaphthalene-2,2(1*H*)-dicarboxylate (64, **R^1 = R^2 = OMe; R^3 = H); Typical Procedure:**[31]
To a flame-dried Schlenk tube equipped with a magnetic stirrer bar, anhyd 1,2-dichloroethane (1 mL) and AgOTf (1.3 mg, 5 μmol) were added. The tube was covered with aluminum foil and a soln of substrate **63** (R^1 = R^2 = OMe; R^3 = H) in 1,2-dichloroethane (50 μmol, 0.05 M) was added at once and the mixture was refluxed for 18 h in the dark. The volatiles were removed under reduced pressure and the residue was purified by column chromatography (silica gel, cyclohexane/EtOAc, typically 4:1) to afford a white wax; yield: 90%.

for references see p 149

Diethyl 4-Vinyl-3,4-dihydronaphthalene-2,2(1*H*)-dicarboxylates 66; **General Procedure:**[32]
A Schlenk tube was charged with alcohol **65** (60 μmol) dissolved in $MeNO_2$ (1 mL) and hydrated $FeCl_3$ (1 mg, 10 mol%) dissolved in $MeNO_2$ (1 mL). The yellow-orange mixture was heated to 80 °C for the indicated time and then cooled to rt. The volatiles were removed under reduced pressure and the residue was directly purified by column chromatography (cyclohexane/EtOAc, typically 95:5).

Polycycles 69; **General Procedure:**[33]
$\{IrCl(cod)\}_2$ (13.4 mg, 20.0 μmol, 0.04 equiv) and ligand **67** (40.8 mg, 80.0 μmol, 0.16 equiv) were placed in a screw-capped vial (5.0 mL) or flask with a magnetic stirrer bar. Commercial grade 1,2-dichloroethane (3 mL) was added and the reaction vessel was quickly purged with N_2 and closed. The contents were stirred vigorously for 15 min, during which time the soln turned dark red. Allylic alcohol **68** (0.5 mmol, 1.0 equiv) and $Zn(OTf)_2$ (36.4 mg, 0.1 mmol, 0.2 equiv) were added to the mixture, resulting in an orange soln which was stirred at rt for 24 h. The solvent was removed under reduced pressure and the residue was purified by column chromatography (silica gel, hexane/CH_2Cl_2).

1.3.2 Reactions with Nitrogen Nucleophiles

Metal-catalyzed allylic amination reactions are particularly important as they provide a significant way to construct a C—N bond in a straightforward and convenient manner. Successful examples of intramolecular allylic amination reactions are mainly focused on the use of protected amines such as alkylamines, amides, and acetimidates. It is only very recently that electron-deficient pyridines and pyrazines have been successfully employed in these reactions.

1.3.2.1 Protected Amines

1.3.2.1.1 Palladium Catalysis

Palladium-catalyzed intramolecular allylic amination of branched allylic ester **70** in the presence of Trost ligand (*R*,*R*)-**14** affords the seven-membered ring **71** in 84% yield and 92% ee (Scheme 36).[34] Similarly, bicyclic tosylamide **74** is formed in 90% yield and 88% ee when amido carbonate **73** is subjected to a catalyst derived from ligand **72** in dichloromethane at 0 °C (Scheme 37).[35] Extending the nitrogen nucleophile to a carboxamide, the palladium-mediated cyclization of allylic ester **76** delivers the corresponding 1-vinyltetrahydroisoquinoline derivative **77A** in 89% yield and 88% ee using chiral 2-(phosphinophenyl)pyridine **75** as the ligand. (*R*)-(+)-Carnegine can be easily accessed through subsequent transformation of the key intermediate **77A** (Scheme 38).[36]

Scheme 36 Synthesis of a 2-Vinylazepane by Palladium-Catalyzed Intramolecular Allylic Amination[34]

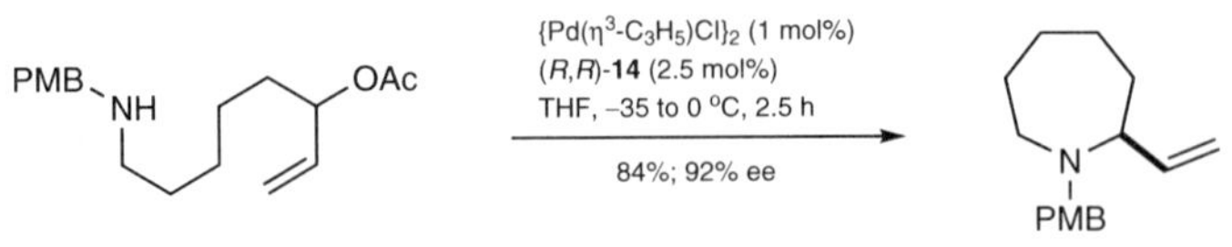

PMB-NH ... OAc (**70**) → $\{Pd(\eta^3-C_3H_5)Cl\}_2$ (1 mol%), (*R*,*R*)-**14** (2.5 mol%), THF, –35 to 0 °C, 2.5 h; 84%; 92% ee → **71** (N-PMB)

Scheme 37 Synthesis of a Bridged Nitrogen Heterocycle by Palladium-Catalyzed Intramolecular Allylic Amination with an Amido Carbonate[35]

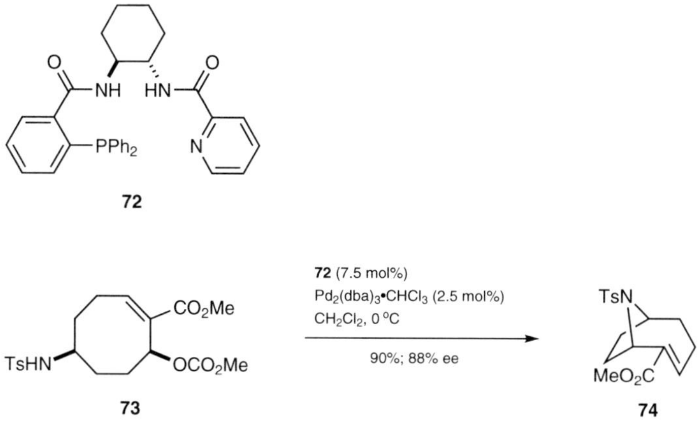

Scheme 38 Synthesis of (*R*)-(+)-Carnegine[36]

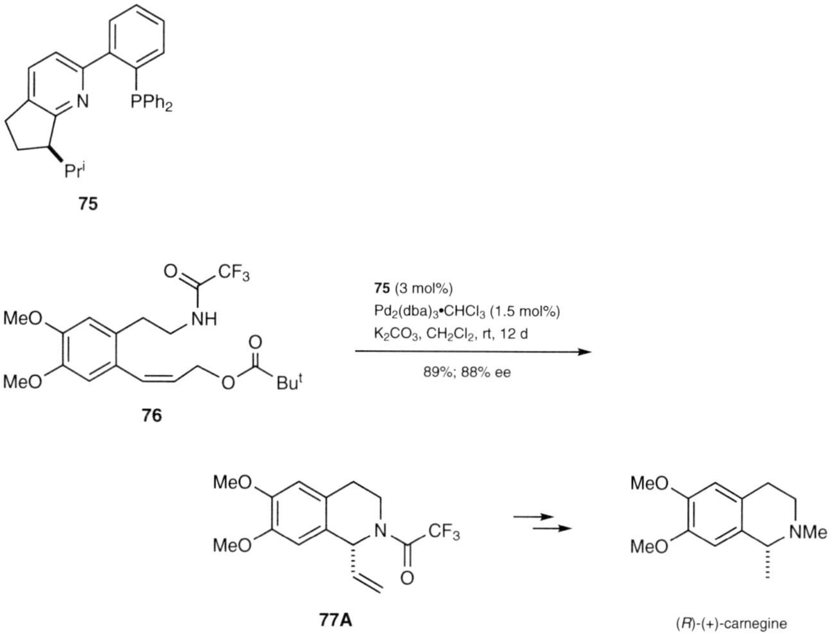

(*R*)-1-[6,7-Dimethoxy-1-vinyl-3,4-dihydroisoquinolin-2(1*H*)-yl]-2,2,2-trifluoroethanone (77A); Typical Procedure:[36]

Under N_2, $Pd_2(dba)_3$•$CHCl_3$ (1.9 mg, 1.8 μmol) and ligand **75** (1.5 mg, 3.6 μmol) were placed in a flask and CH_2Cl_2 (0.36 mL) was added. After being stirred for 30 min at rt, compound **76** (50 mg, 0.12 mmol) in CH_2Cl_2 (0.24 mL) and K_2CO_3 (49.8 mg, 0.36 mmol) were added successively and the mixture was stirred at rt for 12 d, quenched with H_2O, and extracted with CH_2Cl_2. The extract was dried ($MgSO_4$) and concentrated. Column chromatography of the residue (silica gel, hexane/EtOAc 9:1) gave an oil; yield: 33.7 mg (89%).

for references see p 149

1.3.2.1.2 Iridium Catalysis

In the presence of a well-developed catalyst generated from $\{IrCl(cod)\}_2$ and phosphoramidite ligand **38**, an intramolecular allylic amination reaction with carbonate **78** affords the corresponding *N*-benzyl-2-vinyl-piperidine **79** in 85% isolated yield and 92% ee. Notably, a catalytic amount of 1,5,7-triazabicyclo[4.4.0]undec-5-ene (TBD) is used for the in situ activation of the catalyst (Scheme 39).[37]

Scheme 39 Synthesis of a 2-Vinylpiperidine by Iridium-Catalyzed Intramolecular Allylic Amination with an Amine[37]

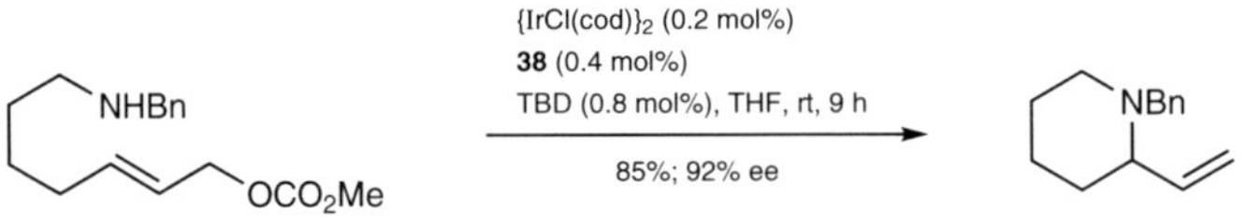

NHBn OCO$_2$Me **78** → $\{IrCl(cod)\}_2$ (0.2 mol%), **38** (0.4 mol%), TBD (0.8 mol%), THF, rt, 9 h; 85%; 92% ee → NBn **79**

TBD = 1,5,7-triazabicyclo[4.4.0]undec-5-ene

A new approach to the synthesis of chiral nitrogen-containing heterocycles such as tetrahydroisoquinolines has been developed relying on iridium-catalyzed intramolecular asymmetric allylic amidation. Desired products including five-, six-, and seven-membered chiral heterocycles (e.g., **77B**) are produced in excellent yields and enantioselectivities from the corresponding allylic carbonates (e.g., **80**) (Table 3).[38] The utility of this methodology is further demonstrated by application in the synthesis of the core structure of almorexant (**81**) (Scheme 40).[39]

Table 3 Iridium-Catalyzed Intramolecular Allylic Amidation[38]

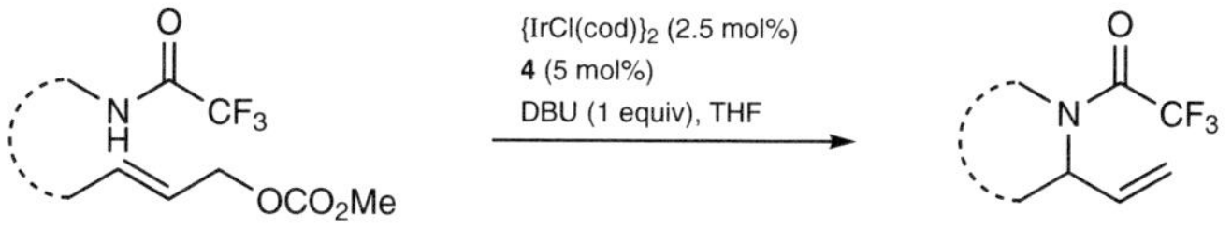

$\{IrCl(cod)\}_2$ (2.5 mol%)
4 (5 mol%)
DBU (1 equiv), THF

Substrate	Product	Temp	ee[a] (%)	Yield[b] (%)	Ref
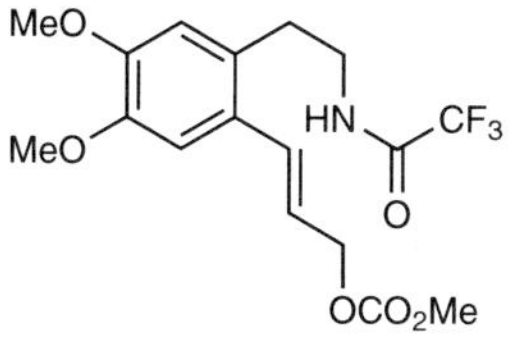 80	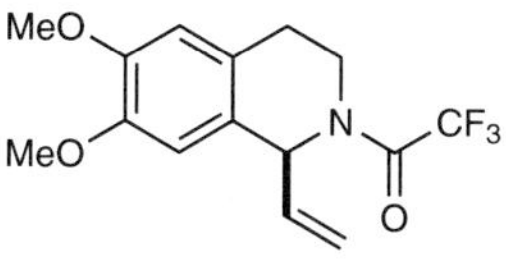 77B	rt	95	97	[38]
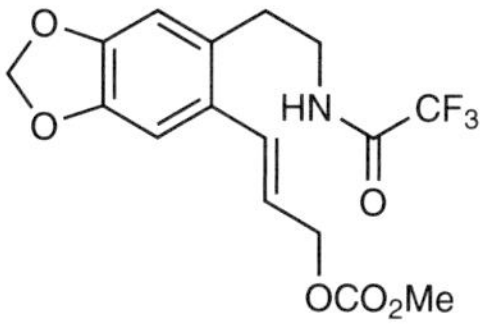	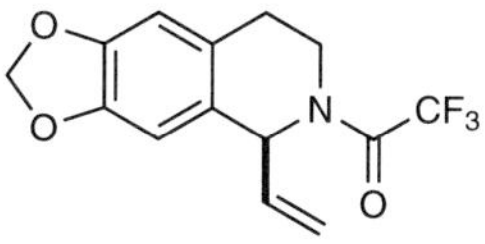	rt	94	89	[38]
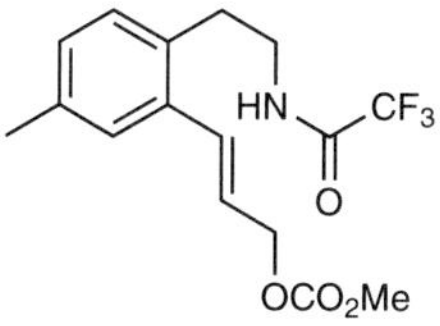	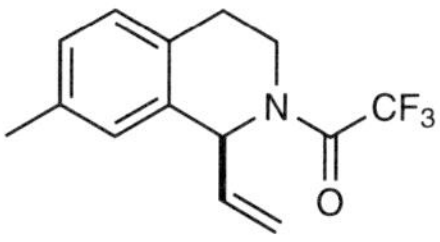	rt	91	92	[38]
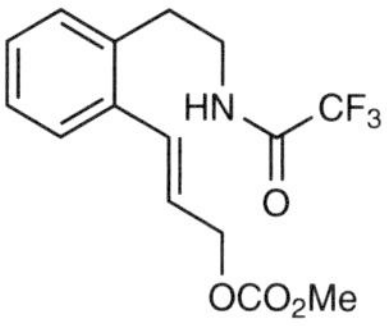	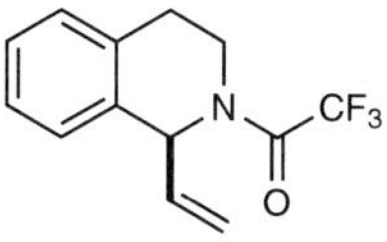	rt	94	78	[38]

for references see p 149

Table 3 (cont.)

Substrate	Product	Temp	ee[a] (%)	Yield[b] (%)	Ref
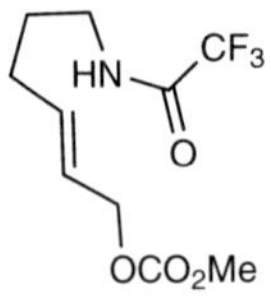 HN, CF3, O, OCO2Me	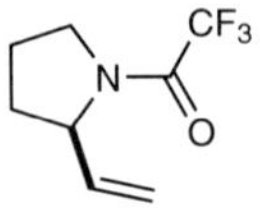 N, CF3, O	50 °C	96	56[c]	[38]
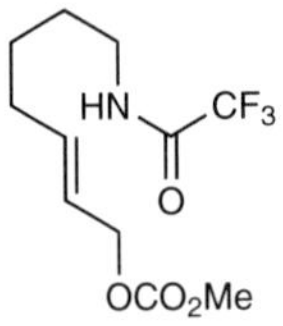 HN, CF3, O, OCO2Me	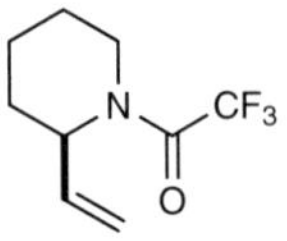 N, CF3, O	50 °C	88	68[c]	[38]
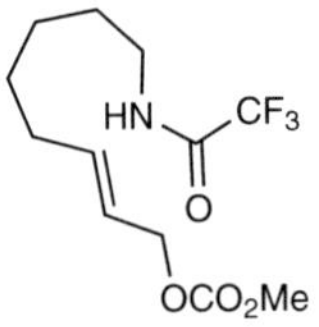 HN, CF3, O, OCO2Me	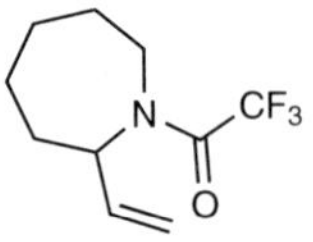 N, CF3, O	50 °C	92	25[c]	[38]

[a] Determined by HPLC analysis using a chiral stationary phase.
[b] Isolated yields after column chromatography.
[c] Products are volatile.

Scheme 40 Synthesis of Almorexant[39]

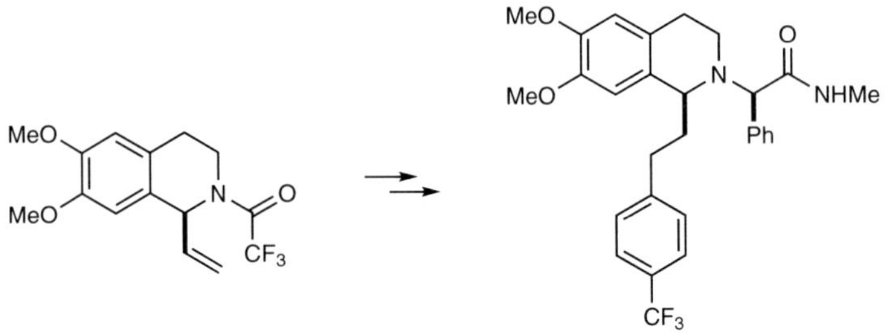

MeO, MeO, N, O, CF3

77B

MeO, MeO, N, O, NHMe, Ph, CF3

81 almorexant

(*S*)-1-[6,7-Dimethoxy-1-vinyl-3,4-dihydroisoquinolin-2(1*H*)-yl]-2,2,2-trifluoroethanone (77B); Typical Procedure:[38]

Under N_2, $\{IrCl(cod)\}_2$ (2.5 mol%) and ligand **4** (5 mol%) were dissolved in dry THF (1.0 mL per 0.2 mmol). Then, DBU (1 equiv) was added and the mixture was heated at 50 °C for 30 min and brought to rt. Amido carbonate **80** (1 equiv) was added, the mixture was stirred until TLC showed full conversion, and then volatiles were removed under reduced pressure to yield the crude product. Purification by column chromatography (silica gel, pentane/EtOAc 3:1) afforded a colorless oil which was a mixture of two conformers in a 3.6:1 ratio (determined by ^1H NMR analysis at 20 °C); yield: 97%; 95% ee.

1.3.2.1.3 Ruthenium Catalysis

In the presence of a catalyst derived from ligand **82** and tris(acetonitrile)(η^5-cyclopentadienyl)ruthenium(I) hexafluorophosphate, an intramolecular allylic amination with N-protected ω-amino and aminocarbonyl allylic alcohols (e.g., **83**) leads to the corresponding N-heterocycles including pyrrolidines, piperidines, and azepanes and their benzo-fused variants (e.g., **84**) in good yields and enantioselectivities. Not only are both *E*- and *Z*-isomers suitable substrates, but allylic alcohols bearing an alkyl substituent as C3 are also found to be applicable to this transformation (Table 4).[40]

Table 4 Ruthenium-Catalyzed Intramolecular Asymmetric Allylic Amination[40]

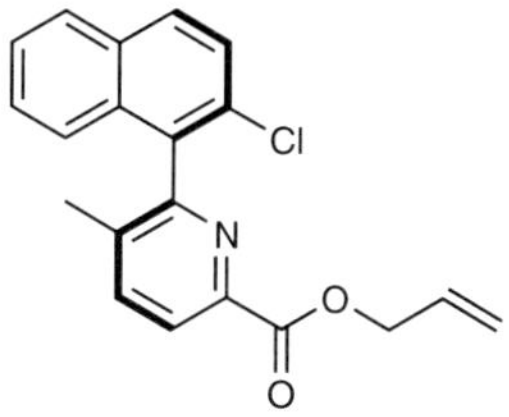

82

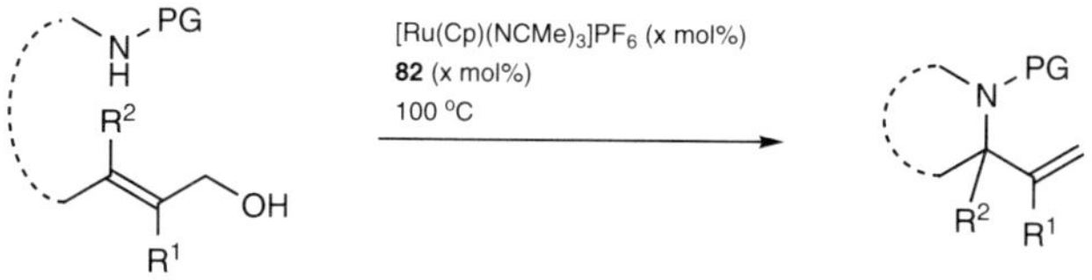

$[Ru(Cp)(NCMe)_3]PF_6$ (x mol%)
82 (x mol%)
100 °C

PG = protecting group

Substrate	x (mol%)	Solvent	Time (h)	Product	ee[a] (%)	Yield[b] (%)	Ref
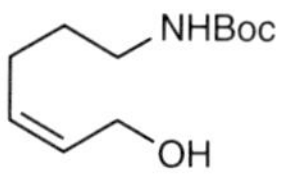	1	DMA	1	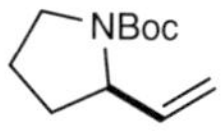	88	92	[40]
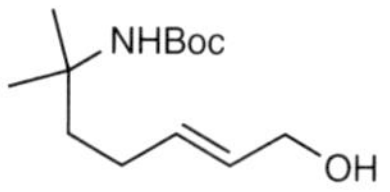	0.1	DMA	6	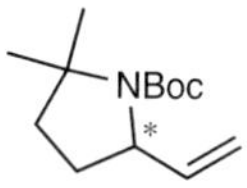	94	92	[40]
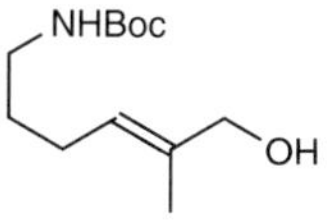	0.1	DMA	3	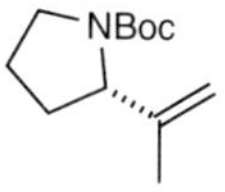	96	96	[40]
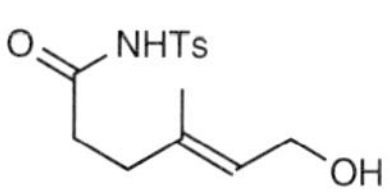	1	*t*-BuOH/DMA (10:1)	3	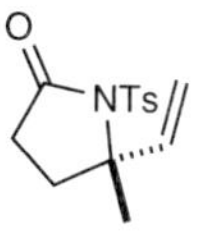	86	92	[40]
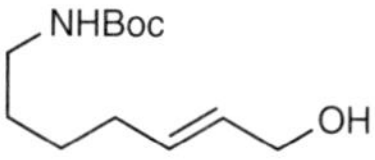	1	DMA	3	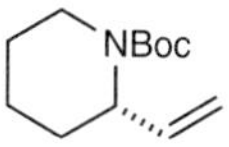	94	90	[40]

for references see p 149

Table 4 (cont.)

Substrate	x (mol%)	Solvent	Time (h)	Product	ee[a] (%)	Yield[b] (%)	Ref
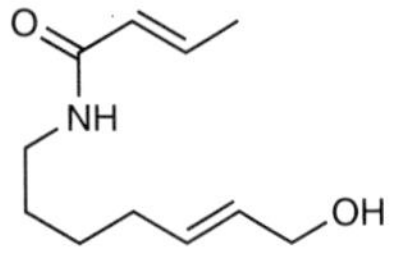	1	*t*-BuOH/DMA (10:1)	24	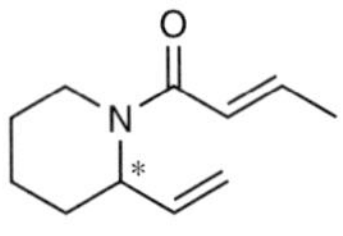	88	96	[40]
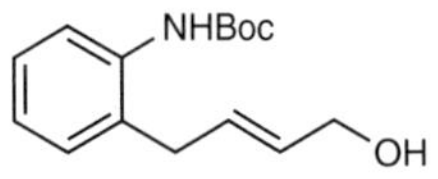 83	0.1	*t*-BuOH/DMA (10:1)	6	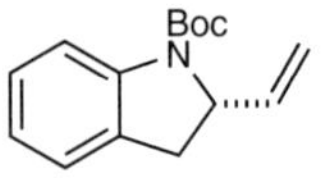 84	>98	98	[40]
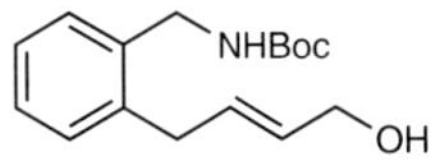	1	*t*-BuOH/DMA (10:1)	3	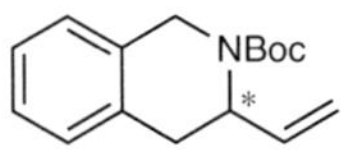	92	90	[40]
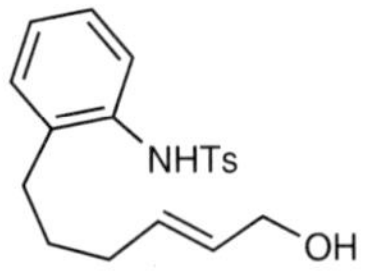	1	*t*-BuOH/DMA (10:1)	3	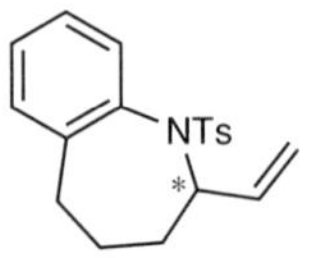	98	99	[40]

[a] Determined by HPLC analysis on a chiral stationary phase.
[b] Isolated yields after column chromatography.

1-*tert*-Butoxycarbonyl-2-vinyl-2,3-dihydroindole (84)**; Typical Procedure:**[40]
To an argon-filled 5-mL Schlenk flask containing [Ru(Cp)(NCMe)$_3$]PF$_6$ (8.25 mg, 19.0 μmol) was added a 10 mM soln of ligand **82** (1.90 mL, 19.0 μmol) at rt. The resulting yellow soln was transferred into an argon-filled 100-mL Schlenk flask using a cannula under a slightly positive argon pressure. After concentration of the soln, a soln of carbamate **83** (5.00 g, 19.0 mmol) in *t*-BuOH/DMA (10:1; 38.0 mL) was added. The whole system was heated to 100 °C under argon and then sealed by a tap. After 6 h at 100 °C, the volatiles were removed under reduced pressure. The resulting brown oil (5 g) was purified by column chromatography (silica gel, hexane/EtOAc 2:1) to give a colorless oil; yield: 4.54 g (98%).

1.3.2.1.4 Gold Catalysis

Treatment of allylic alcohols **86** and **88** bearing a secondary alkylamine with a catalyst generated from gold complex **85** and silver(I) hexafluoroantimonate results in the substituted pyrrolidine **87** and piperidine **89**, respectively, in good yields. Notably, the cyclization of (*R*,*Z*)-8-(benzylamino)oct-3-en-2-ol (**88**; 96% ee) under gold catalysis delivers (*R*,*E*)-1-benzyl-2-(prop-1-enyl)piperidine (**89**) in 99% yield and 96% ee, which is in accordance with the net *syn* addition of the amine relative to the departing hydroxy group (Scheme 41).[41]

Scheme 41 2-Vinylpiperidines and 2-Vinylpyrrolidines by Gold(I)-Catalyzed Intramolecular Amination of Allylic Alcohols with Alkylamines[41]

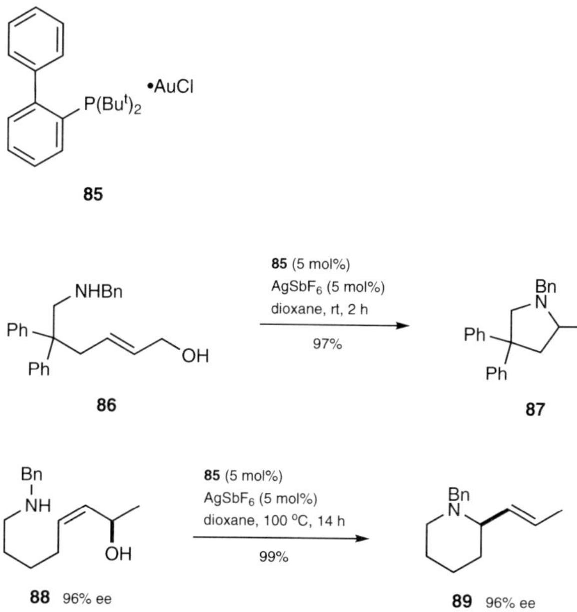

$P(Bu^t)_2$ •AuCl

85

NHBn, Ph, Ph, OH

86

85 (5 mol%)
$AgSbF_6$ (5 mol%)
dioxane, rt, 2 h
97%

Bn, N, Ph, Ph

87

Bn, NH, OH

88 96% ee

85 (5 mol%)
$AgSbF_6$ (5 mol%)
dioxane, 100 °C, 14 h
99%

Bn, N

89 96% ee

Soon after, the enantioselective version of this process was achieved, forming five- and six-membered nitrogen-containing heterocycles with up to 95% ee. However, the protecting group on the nitrogen atom is crucial to the stereoinduction of the reaction. The intramolecular enantioselective amination of a benzylamino-containing allylic alcohol leads to 2-vinylpyrrolidine in quantitative yield, but with only 29% ee. After careful manipulation of the nitrogen nucleophile, Fmoc-protected ω-amino allylic alcohol (*E*)-**91** in dioxane at room temperature for 48 hours gives the cyclized product **92** in 95% yield and 91% ee under the optimized catalyst system comprising gold complex **90** (2.5 mol%) and silver(I) perchlorate (5 mol%) (Scheme 42).[42] Notably, this transformation is limited to the (*E*)-alkene; substrate (*Z*)-**91** is converted into **92** in 99% yield but with <5% ee.

Scheme 42 A 2-Vinylpyrrolidine by Gold(I)-Catalyzed Enantioselective Intramolecular Amination of an Allylic Alcohol with a Carbamate[42]

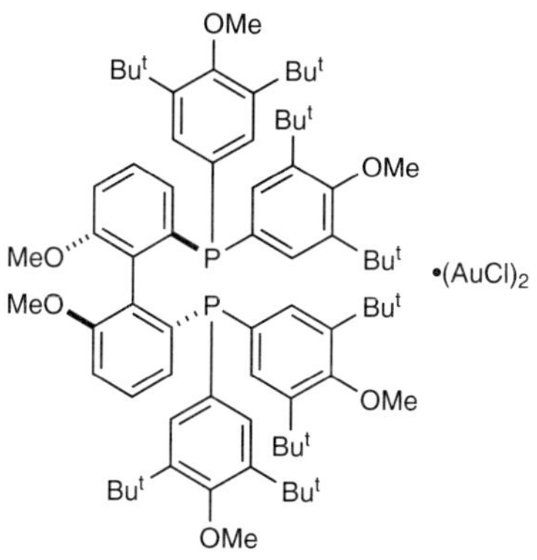

OMe, Bu^t, Bu^t, Bu^t, OMe, MeO, P, Bu^t, $•(AuCl)_2$, MeO, P, Bu^t, OMe, Bu^t, Bu^t, Bu^t, OMe

90

for references see p 149

NHFmoc Ph Ph OH (E)-91 → 90 (2.5 mol%), $AgClO_4$ (5 mol%), dioxane, rt, 48 h; 95%; 91% ee → Fmoc N Ph Ph 92

1-Benzyl-4,4-diphenyl-2-vinylpyrrolidine (87); Typical Procedure:[41]
A suspension of complex **85** (2.6 mg, 5 μmol), $AgSbF_6$ (1.7 mg, 5 μmol), and alcohol **86** (36 mg, 0.10 mmol) in 1,4-dioxane (0.34 mL) in a sealed tube was stirred at rt for 2 h. The resulting suspension was concentrated under reduced pressure and the residue was chromatographed (hexane/EtOAc 40:1 to 10:1) to give a pale yellow oil; yield: 33 mg (97%).

1.3.2.1.5 Catalysis by Other Metals

The combination of the chiral ligand (*R*)-BINAPHANE (**93**) with mercury(II) trifluoromethanesulfonate serves as a highly efficient catalyst in the enantioselective cyclization of anilino sulfonamide allyl alcohol derivatives **94**, resulting in the formation of 2-vinyl-2,3-dihydroindoles **95**. The protecting group on the nitrogen atom dramatically affects the enantioselectivity. The substrates bearing a benzenesulfonamide group on the nitrogen afford a low enantiomeric excess while substrates with an alkanesulfonamide substituent are found to afford a high enantiomeric excess. In this regard, the generality of the reaction has been investigated based on the *tert*-butyl-substituted sulfonamide derivatives. Substituent methyl groups at different positions on the aromatic ring also dramatically affect the enantioselectivity (Scheme 43).[43] The 3-methyl-substituted derivative induces the highest enantiomeric excess (99% ee), while substrates with a methyl group introduced at other positions give only moderate enantioselectivity (72%, 79%, and 84% ee, respectively). It is likely that the introduction of a 3-methyl group has a beneficial effect on the transition state.

Scheme 43 2-Vinyl-2,3-dihydroindoles by Mercury(II) Trifluoromethanesulfonate–(*R*)-BINAPHANE Catalyzed Cyclization of Sulfonamide Allyl Alcohols[43]

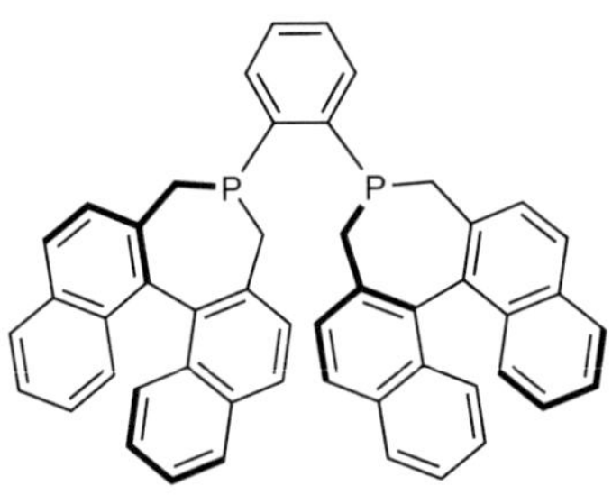

P P

93 (*R*)-BINAPHANE

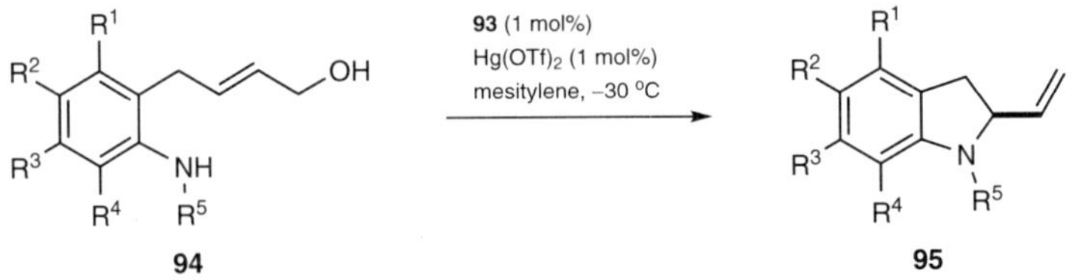

R1 R2 R3 R4 OH NH R5 94 → 93 (1 mol%), $Hg(OTf)_2$ (1 mol%), mesitylene, −30 °C → R1 R2 R3 R4 N R5 95

R^1	R^2	R^3	R^4	R^5	ee[a] (%)	Yield[b] (%)	Ref
H	H	H	H	Ts	74	99	[43]
H	H	H	H	SO_2t-Bu	78	99	[43]
H	H	H	Me	SO_2t-Bu	72	91	[43]
H	H	Me	H	SO_2t-Bu	79	98	[43]
H	Me	H	H	SO_2t-Bu	84	95	[43]
Me	H	H	H	SO_2t-Bu	99	96	[43]
H	Me	Me	H	SO_2t-Bu	83	93	[43]
Me	H	Me	H	SO_2t-Bu	93	99	[43]

[a] Determined by HPLC analysis using a chiral stationary phase.
[b] Isolated yields after column chromatography.

Easy access to the substituted 1,2-dihydroquinolines **97** is provided by an iron-catalyzed intramolecular allylic amination reaction of protected anilines **96**. The quinoline **98** can also be obtained using a different workup procedure (treating the reaction mixture with sodium hydroxide). The reaction can be performed with the flask open to ambient air (Scheme 44).[44]

Scheme 44 Quinolines and Dihydroquinolines by Iron(III) Chloride Hexahydrate Catalyzed Intramolecular Allylic Amination[44]

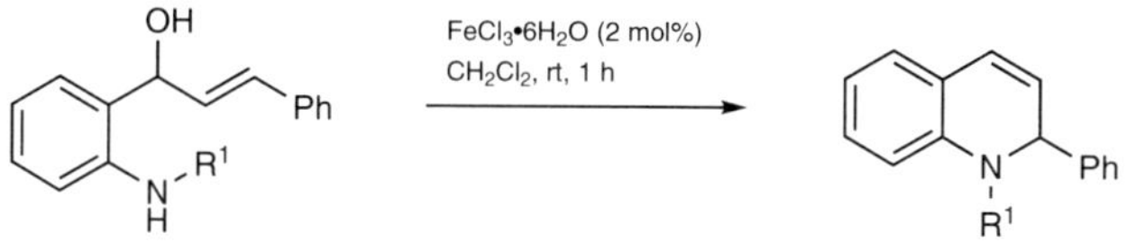

OH Ph N-R¹ H **96** → $FeCl_3•6H_2O$ (2 mol%), CH_2Cl_2, rt, 1 h → N Ph R¹ **97**

R^1	Yield (%)	Ref
Ts	96	[44]
Ms	90	[44]

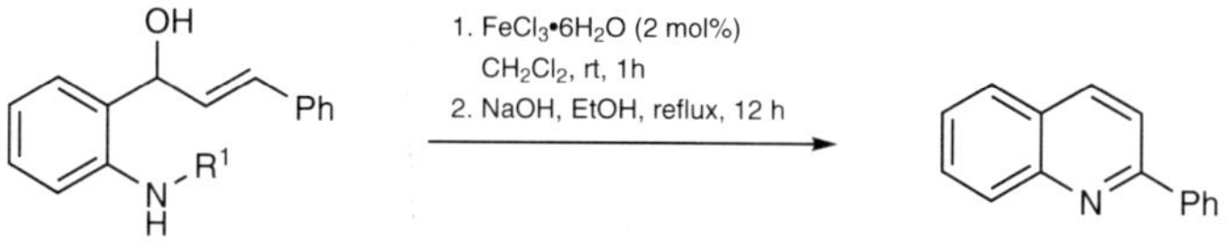

OH Ph N-R¹ H **96** → 1. $FeCl_3•6H_2O$ (2 mol%), CH_2Cl_2, rt, 1h; 2. NaOH, EtOH, reflux, 12 h → N Ph **98**

R^1	Yield (%)	Ref
Ts	93	[44]
Ms	90	[44]

1-Tosyl-2-vinyl-2,3-dihydroindole (95, **$R^1 = R^2 = R^3 = R^4 =$ H; $R^5 =$ Ts); Typical Procedure:**[43]
A 0.1 M soln of $Hg(OTf)_2$ in MeCN (30 μL, 3.0 μmol) was added to a dried two-necked flask under argon and the MeCN was removed under reduced pressure. Then, mesitylene (3.0 mL) and ligand **93** (2.0 mg, 3.0 μmol) were added. After stirring for 5 min at rt, sulfon-

for references see p 149

amide **94** (100 mg, 315 μmol) was added to the mixture, which was then stirred for 40 h at −30 °C. The mixture was immediately subjected to column chromatography (silica gel, hexane/EtOAc 5:1) to give a white solid; yield: 93 mg (99%); 74% ee.

2-Phenyl-1,2-dihydroquinolines 97; **General Procedure:**[44]
To a soln of protected aniline **96** (0.5 mmol) in CH_2Cl_2 was added $FeCl_3$•$6H_2O$ (0.01 mmol). The mixture was stirred at rt for 1 h (monitored by TLC). The resulting mixture was purified by column chromatography (silica gel, petroleum ether/EtOAc).

2-Phenylquinoline (98); **General Procedure:**[44]
To a soln of protected aniline **96** (0.5 mmol) in CH_2Cl_2 was added $FeCl_3$•$6H_2O$ (0.01 mmol). After the mixture was stirred at rt for 1 h (monitored by TLC), NaOH (2.5 mmol) and EtOH (0.1 mL) were added and the reaction was continued for 12 h under reflux (monitored by TLC). The resulting mixture was purified by column chromatography (silica gel, petroleum ether/EtOAc).

1.3.2.2 Pyridines and Pyrazines

Apart from protected amines, the nitrogen atoms of electron-deficient pyridines, e.g. **99**, and pyrazines, e.g. **101**, can also act as nucleophiles in iridium-catalyzed intramolecular asymmetric allylic substitution reactions. The reaction tolerates various substituents on the pyridine or pyrazine core, delivering the dearomatized 2,3-dihydroindolizine (e.g., **100**) and 6,7-dihydropyrrolo[1,2-*a*]pyrazine derivatives (e.g., **102**) in excellent yields and enantioselectivities. Notably, an electron-withdrawing group is usually introduced on the linking substituent to stabilize the dearomatized products (Schemes 45 and 46).[45]

Scheme 45 3-Vinyl-2,3-dihydroindolizines by Iridium-Catalyzed Intramolecular Asymmetric Allylic Dearomatization of Pyridines[45]

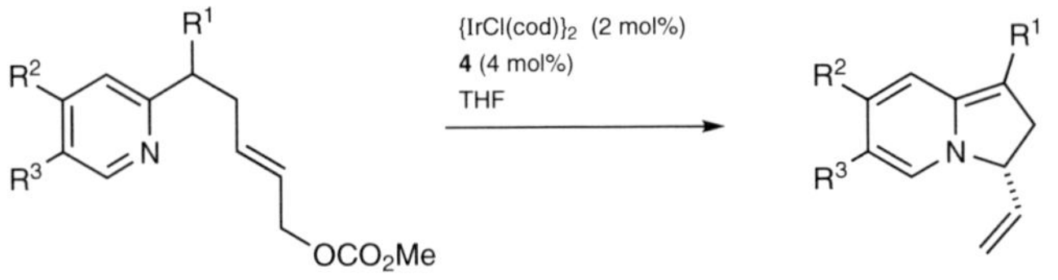

$\{IrCl(cod)\}_2$ (2 mol%)
4 (4 mol%)
THF

R1, R2, R3, N, OCO_2Me — **99** → R1, R2, R3, N — **100**

R^1	R^2	R^3	Temp	Time (h)	ee[a] (%)	Yield[b] (%)	Ref
CO_2Me	H	H	rt	4	98	99	[45]
CO_2Et	H	H	rt	4	97	93	[45]
CO_2Et	H	OMe	rt	2	95	95	[45]
CO_2Et	H	Ph	rt	4	98	99	[45]
CO_2Et	Me	H	rt	4	95	93	[45]
CO_2Et	Cl	H	rt	4	99[c]	>95[c]	[45]
Ac	H	H	50 °C	12	93	82	[45]
Bz	H	H	50 °C	3	99	94	[45]
SO_2Ph	H	H	50 °C	12	83	82	[45]

[a] Determined by HPLC analysis using a chiral stationary phase.
[b] Isolated yields after column chromatography.
[c] Determined after Diels–Alder reaction with diethyl acetylenedicarboxylate.

Scheme 46 6-Vinyl-6,7-dihydropyrrolo[1,2-*a*]pyrazines by Iridium-Catalyzed Intramolecular Asymmetric Allylic Dearomatization of Pyrazines[45]

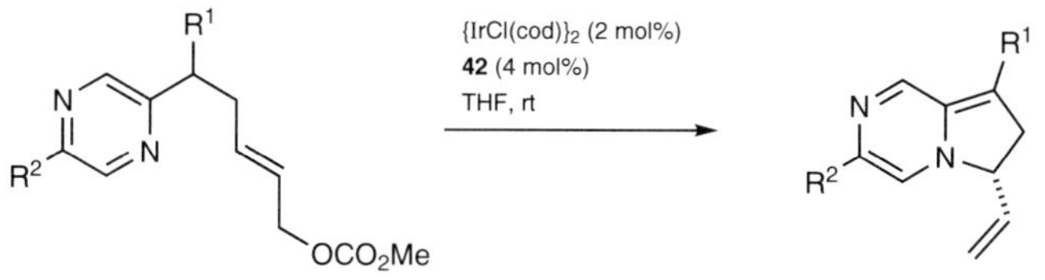

{IrCl(cod)}$_2$ (2 mol%)
42 (4 mol%)
THF, rt

101 → **102**

R^1	R^2	Time (h)	ee[a] (%)	Yield[b] (%)	Ref
CO_2Et	H	18	94	88	[45]
CO_2*t*-Bu	H	17	92	68	[45]
Bz	H	12	97	95	[45]
CO_2Me	4-FC_6H_4	36	96	75	[45]
Bz	Me	1.5	91	64	[45]

[a] Determined by HPLC analysis using a chiral stationary phase.
[b] Isolated yields after column chromatography.

6-Vinyl-6,7-dihydropyrrolo[1,2-*a*]pyrazines 102; **General Procedure:**[45]
A flame-dried Schlenk tube was cooled to rt and filled with argon. To this flask were added {IrCl(cod)}$_2$ (2.7 mg, 0.004 mmol, 2 mol%), phosphoramidite ligand **42** (4.8 mg, 0.008 mmol, 4 mol%), THF (0.5 mL), and $PrNH_2$ (0.5 mL). The mixture was heated at 50 °C for 30 min and then the volatile solvents were removed under reduced pressure to give a pale yellow solid. After that, allylic carbonate **101** (0.20 mmol) was added and the mixture was stirred at rt until the starting material had been consumed (monitored by TLC). The solvents were removed under reduced pressure and the residue was purified by column chromatography (neutral alumina, EtOAc/MeOH 10:1).

1.3.3 Reactions with Oxygen Nucleophiles

The power of metal-catalyzed intramolecular allylic substitution reactions has also been well demonstrated in the construction of carbon—oxygen bonds, thereby offering an efficient strategy to prepare oxygen-containing heterocycles. However, compared with the extensive studies on intramolecular allylic substitution reactions with soft carbon nucleophiles, reactions with oxygen nucleophiles have been less explored. To date, only phenols, aliphatic alcohols, esters, and acetimidates have been successfully utilized as oxygen nucleophiles in the intramolecular allylic cyclization reaction.

1.3.3.1 Phenols

The palladium-catalyzed asymmetric allylic substitution reaction of phenol-derived allylic carbonates in an intramolecular fashion allows access to chiral dihydrobenzopyrans (e.g., **104**) with up to 98% ee. The addition of acetic acid plays an important role in the dramatic increase in the enantioselectivity. Moreover, alkene geometry in the substrates has a profound influence on the absolute configuration and enantioselectivity. The reaction features a good tolerance of both disubstituted and trisubstituted alkene substrates (e.g., **103**) leading to the enantioselective construction of tertiary and quaternary stereocenters, respectively. This protocol has been applied in the synthesis of (+)-clusifoliol (Schemes 47, 48, and 49).[46]

for references see p 149

Scheme 47 2-Vinyl-3,4-dihydrobenzopyrans by Palladium-Catalyzed Asymmetric Allylic Substitution of (*E*)-Allylic Carbonates[46]

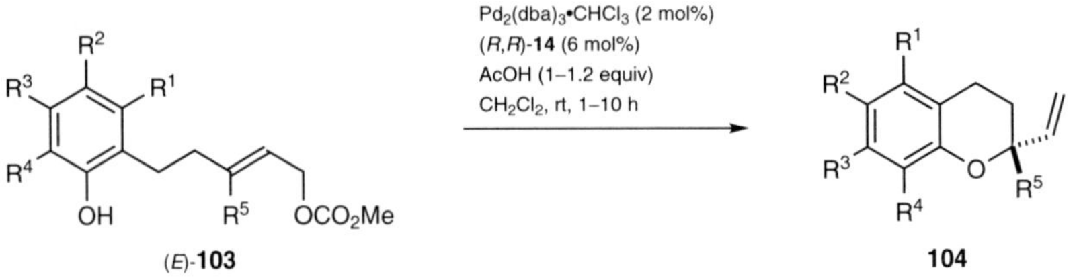

$Pd_2(dba)_3$•$CHCl_3$ (2 mol%)
(*R*,*R*)-**14** (6 mol%)
AcOH (1–1.2 equiv)
CH_2Cl_2, rt, 1–10 h

(*E*)-**103** → **104**

R^1	R^2	R^3	R^4	R^5	ee[a] (%)	Yield[b] (%)	Ref
OMe	H	Me	H	Me	84	94	[46]
OMe	Me	Me	OMe	Me	80	99	[46]
H	H	H	H	Me	73	62	[46]
H	F	H	H	H	80	88	[46]
H	H	H	H	H	84	99	[46]
H	Me	H	H	H	87	97	[46]
H	OMe	H	H	H	89	82	[46]
H	Me	H	OMe	H	82	82	[46]

[a] Determined by HPLC analysis using a chiral stationary phase.
[b] Isolated yields after column chromatography.

Scheme 48 2-Vinyl-3,4-dihydrobenzopyrans by Palladium-Catalyzed Asymmetric Allylic Substitution of (*Z*)-Allylic Carbonates[46]

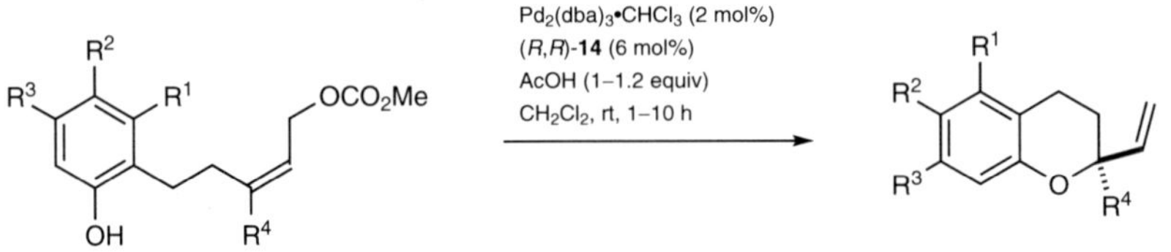

$Pd_2(dba)_3$•$CHCl_3$ (2 mol%)
(*R*,*R*)-**14** (6 mol%)
AcOH (1–1.2 equiv)
CH_2Cl_2, rt, 1–10 h

R^1	R^2	R^3	R^4	ee[a] (%)	Yield[b] (%)	Ref
H	H	H	Me	95	68	[46]
OMe	H	Me	Me	97[c]	79	[46]
H	Me	H	H	48	72	[46]
H	H	H	H	34	67	[46]
H	F	H	H	57	93	[46]

[a] Determined by HPLC analysis using a chiral stationary phase.
[b] Isolated yields after column chromatography.
[c] Ligand (*S*,*S*)-**14** was used to give the opposite configuration at the quaternary center.

Scheme 49 Synthesis of (+)-Clusifoliol[46]

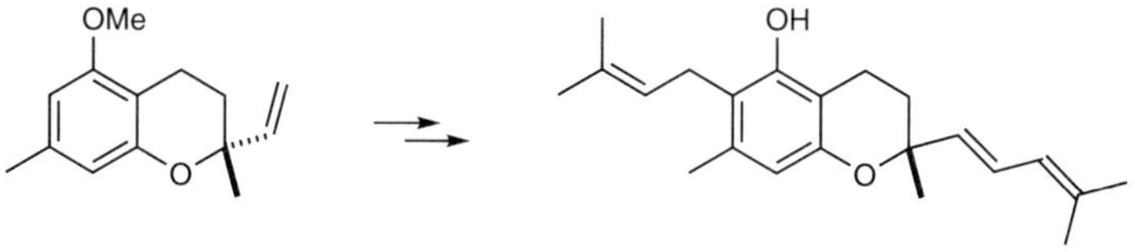

OMe

OH

97% ee

(+)-clusifoliol 97% ee

2-Vinyl-3,4-dihydrobenzopyrans 104**; General Procedure:**[46]
To a degassed mixture of $Pd_2(dba)_3$•$CHCl_3$ (2 mol%) and chiral ligand **14** (6 mol%) was added CH_2Cl_2. The soln was stirred for 10 min under argon to yield a yellow soln. To this soln was added AcOH (1–1.2 equiv). After 5 min, a soln of carbonate **103** in CH_2Cl_2 was added. The mixture was stirred at rt for 1–10 h and the volatiles were removed under reduced pressure. The residue was purified by column chromatography (silica gel, petroleum ether/Et_2O 11:1).

1.3.3.2 Aliphatic Alcohols

Aliphatic alcohols **105** can serve as oxygen nucleophiles in an intramolecular substitution reaction, giving the five- and six-membered oxygen heterocycles **106** in good yields and enantioselectivities under catalysis by a ruthenium complex of bidentate ligand **7**. The reaction even proceeds with a substrate/catalyst (S/C) ratio of 1000 to 5000. However, in the case where the substrate has an aliphatic chain, the product is obtained in quantitative yield but with poor enantioselectivity (20% ee) (Table 5).[7]

Table 5 Ruthenium-Catalyzed Intramolecular Asymmetric Allylic Etherification[7]

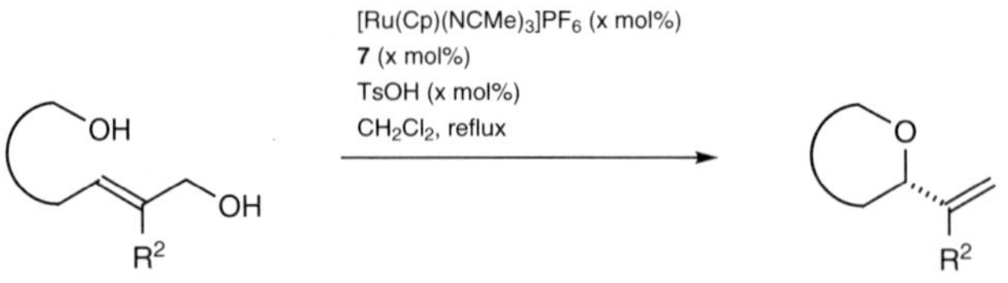

[Ru(Cp)(NCMe)$_3$]PF$_6$ (x mol%)
7 (x mol%)
TsOH (x mol%)
CH$_2$Cl$_2$, reflux

105 → **106**

Substrate	x (mol%)	Product	ee[a] (%)	Yield[b] (%)	Ref
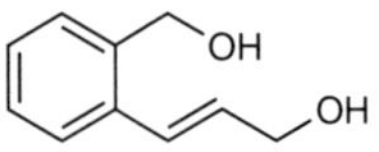	0.1	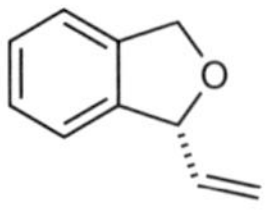	98	89	[7]
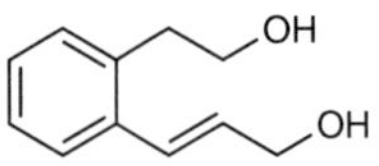	0.1	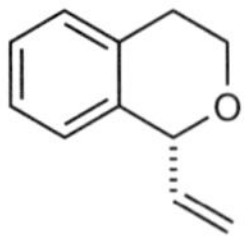	>98	98	[7]
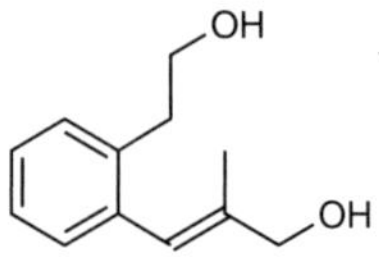	0.1	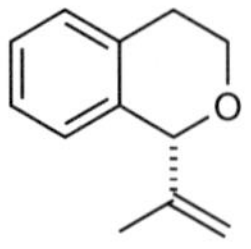	98	94	[7]
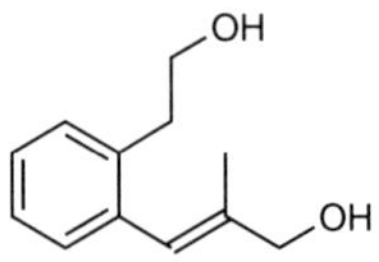	0.02	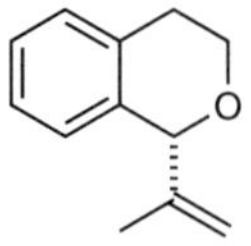	88	93	[7]
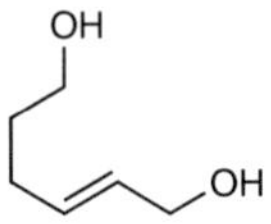	0.2	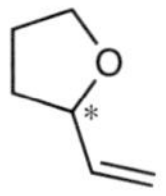	20	–[c]	[7]

[a] Determined by HPLC analysis using a chiral stationary phase.
[b] Isolated yields after column chromatography.
[c] Not isolated; quantitative by GC analysis.

In addition to the ruthenium catalyst, substituted tetrahydropyrans **108** can also be obtained through the intramolecular allylic etherification of alcohols **107** using an eco-friendly iron catalyst. Thermodynamic equilibration of the product leads to the more stable *cis*-isomers preferentially (Scheme 50).[47]

Scheme 50 Iron(III) Chloride Catalyzed Synthesis of *cis*-2,6-Disubstituted Tetrahydropyrans from Hydroxyalkyl Allylic Alcohols and Derivatives[47]

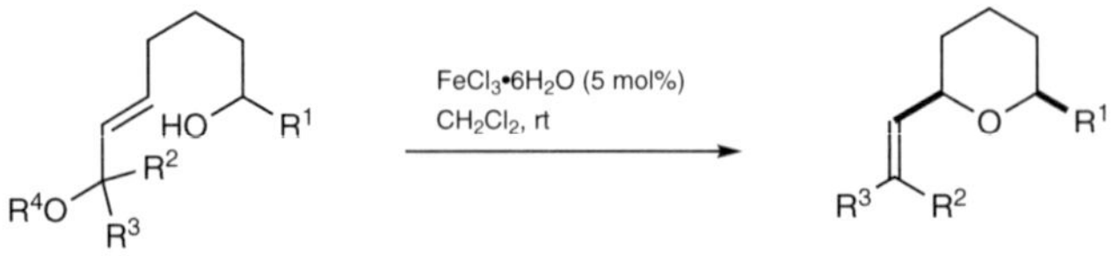

FeCl$_3$•6H$_2$O (5 mol%)
CH$_2$Cl$_2$, rt

107 → **108**

R^1	R^2	R^3	R^4	Time (h)	dr[a]	Yield[b]	Ref
H	Ph	H	H	0.5	–	81	[47]
$(CH_2)_4Me$	Ph	H	H	2	97:3	83	[47]
$(CH_2)_4Me$	Mes	H	H	1	95:5	89	[47]
iPr	Mes	H	H	1	95:5	82	[47]
Me	Mes	H	H	1	98:2	51	[47]
CO_2Et	Ph	H	Ac	48	90:10	quant[c]	[47]
$(CH_2)_4Me$	Me	Me	H	0.25	91:9	87	[47]

[a] Determined by ^1H NMR or by GC/MS analysis.
[b] Isolated yields after column chromatography.
[c] Using 15 mol% $FeCl_3•6H_2O$.

Tetrahydro-2*H*-pyrans 108; **General Procedure:**[47]
To a soln of substrate **107** in CH_2Cl_2 (0.1 M) at rt was added $FeCl_3•6H_2O$ (5 mol%). After the required time, the resulting mixture was directly filtered through a pad of silica gel (CH_2Cl_2 as the eluent) and the volatiles were removed under reduced pressure to yield the corresponding cyclized product **108** without any further purification.

1.3.3.3 Esters

Synthetically useful vinylbutyrolactones **111** are obtained via gold(I)–N-heterocyclic carbene (NHC) complex **110** catalyzed intramolecular direct etherification of allylic alcohols with esters. Good isolated yields are obtained for a range of malonyl derivatives **109** with diverse functional groups at the methylene carbon atom. Notably, there is no appreciable difference in reaction rate between the substrates bearing labile and non-labile ester alkyl groups. In addition, the monoester analogs **112** and **113** have also been subjected to cyclization, leading to the corresponding lactones in good to excellent yields. However, decomposition of the starting material is observed when using β-keto ester **114** as the substrate under these conditions (Schemes 51 and 52).[48]

Scheme 51 Synthesis of Five-Membered Lactones by Gold(I)-Catalyzed Intramolecular Allylic Alkylation with Malonates[48]

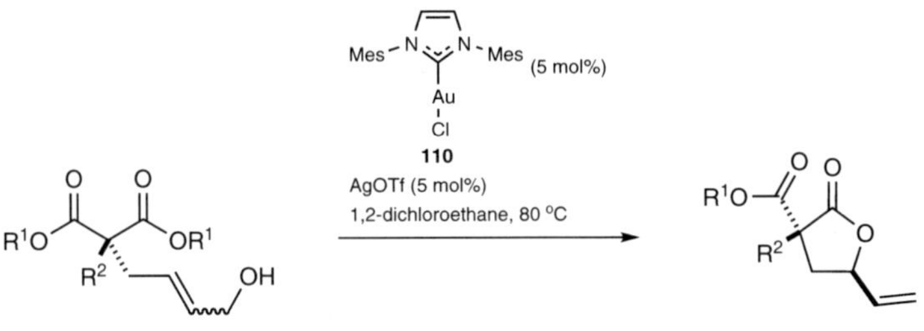

110 (5 mol%), AgOTf (5 mol%), 1,2-dichloroethane, 80 °C; 109 → 111

R^1	R^2	Config of **109**	Ratio[a] (*trans*/*cis*)	Yield[b] (%)	Ref
Me	Me	*E*	1.3:1	72	[48]
Et	H	*Z*	1:1	95	[48]
t-Bu	H	*Z*	1.1:1	66	[48]
Me	H	*Z*	1:1	85	[48]
Me	$CH_2CH{=}CH_2$	*Z*	1.5:1	94	[48]
Et	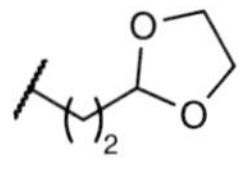 $(CH_2)_2$-1,3-dioxolan-2-yl	*Z*	1.4:1	45	[48]

[a] Determined by GC of the crude reaction mixture.
[b] Isolated yield after column chromatography.

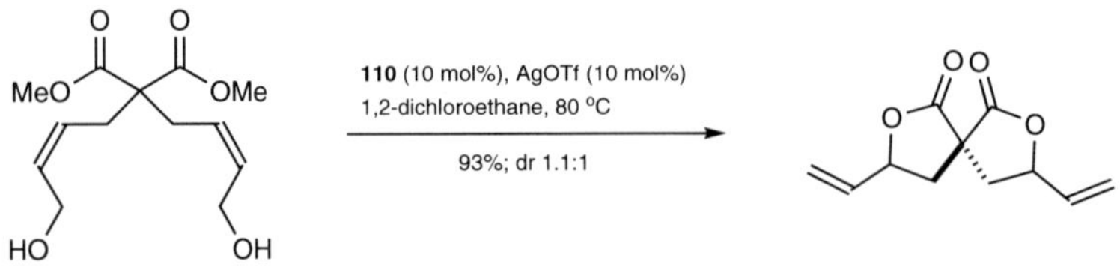

110 (10 mol%), AgOTf (10 mol%), 1,2-dichloroethane, 80 °C; 93%; dr 1.1:1

Scheme 52 Gold(I)-Catalyzed Synthesis of Five-Membered Lactones from Unsaturated Monoesters[48]

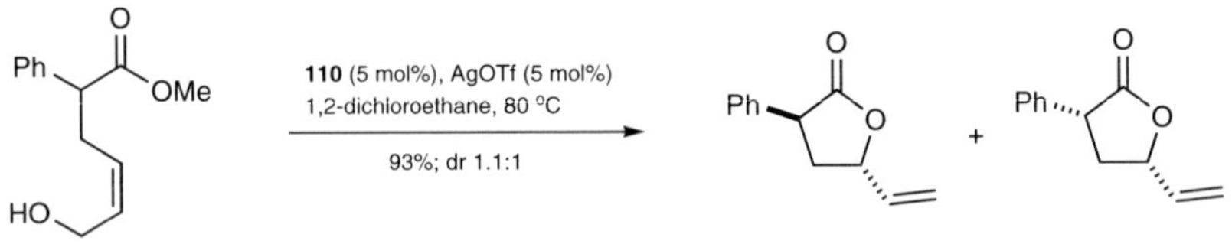

112: 110 (5 mol%), AgOTf (5 mol%), 1,2-dichloroethane, 80 °C; 93%; dr 1.1:1

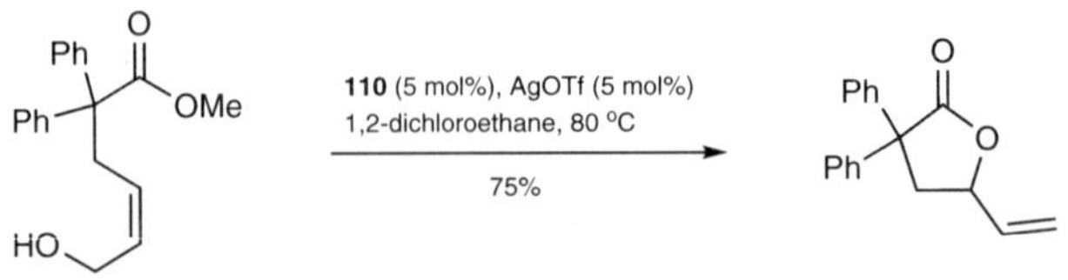

113: 110 (5 mol%), AgOTf (5 mol%), 1,2-dichloroethane, 80 °C; 75%

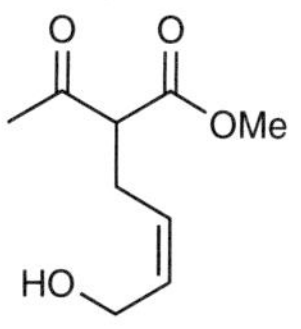

114

2-Oxo-5-vinyltetrahydrofuran-3-carboxylates 111; **General Procedure:**[48]
In a screw-capped vial, under air, gold complex **110** (0.05 equiv) and AgOTf (0.05 equiv) were dissolved in reagent-grade 1,2-dichloroethane (300 μL) and the mixture was stirred for 30 min at rt in the dark. Then, malonate **109** (1 equiv) was added and the mixture was stirred at 80 °C until complete consumption of the starting material. The crude product was purified by column chromatography (silica gel, cyclohexane/EtOAc, typically 9:1).

1.3.3.4 Acetimidates

Embedding the ambident amide nucleophile **116** in the cyclization process, a series of N,O-heterocyclic derivatives such as dihydrooxazoles, oxazines, and benzoxazines **117** are prepared in high yields and excellent enantioselectivities using a catalyst generated from chloro(cycloocta-1,5-diene)iridium(I) dimer and tetrahydroquinoline-derived phosphoramidite ligand **115** (Scheme 53).[49]

Scheme 53 4-Vinyl-4*H*-benzo[*d*][1,3]oxazines by Iridium-Catalyzed Intramolecular Allylic Substitution with Nucleophilic Attack of the Amide Oxygen Atom[49]

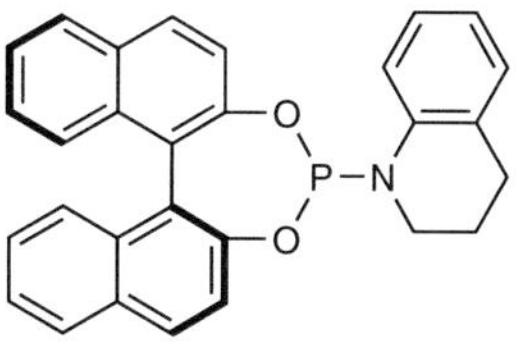

115

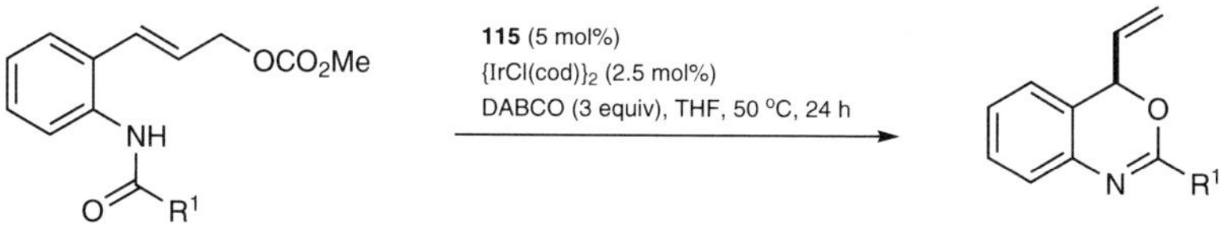

116 → 117: **115** (5 mol%), $\{IrCl(cod)\}_2$ (2.5 mol%), DABCO (3 equiv), THF, 50 °C, 24 h

R^1	ee[a] (%)	Yield[b] (%)	Ref
Ph	97	81	[49]
4-Tol	94	93	[49]
2-furyl	97	87	[49]
$4-O_2NC_6H_4$	92	93	[49]
$4-ClC_6H_4$	95	90	[49]
$3-ClC_6H_4$	83	67	[49]
$2-ClC_6H_4$	97	67	[49]
t-Bu	86	63	[49]

[a] Determined by HPLC analysis using a chiral stationary phase.
[b] Isolated yields after column chromatography.

for references see p 149

4-Vinyl-4*H*-benzo[*d*][1,3]oxazines 117; **General Procedure:**[49]
Under N_2, to a suspension of {IrCl(cod)}$_2$ (3.3 mg, 2.5 mol%) and ligand **115** (4.47 mg, 5 mol%) in dry THF (2 mL) was added DABCO (72.6 mg, 0.6 mmol). Then the mixture was heated at 50 °C for 30 min to generate the catalyst. The allylic carbonate **116** (0.2 mmol) was added and the mixture was stirred until TLC showed full conversion (24 h). All volatiles were removed under reduced pressure and the residue was purified by column chromatography (silica gel, pentane/EtOAc 10:1).

1.3.4 Applications in the Syntheses of Natural Products and Drug Molecules

1.3.4.1 Allylic Alkylation

Metal-catalyzed intramolecular allylic alkylation reactions have found wide application in the synthesis of natural products and drug molecules. As shown in Schemes 53–58, the synthesis usually involves a key step which relies on the allylic cyclization of certain carbon nucleophiles such as malonic esters (Scheme 54),[50] aldehydes (Scheme 55),[51] phenols (Scheme 56),[52] allylic silanes (Scheme 57),[53] and electron-rich arenes (Scheme 58).[54]

Scheme 54 Synthesis of *ent*-Nigellamine A_2[50]

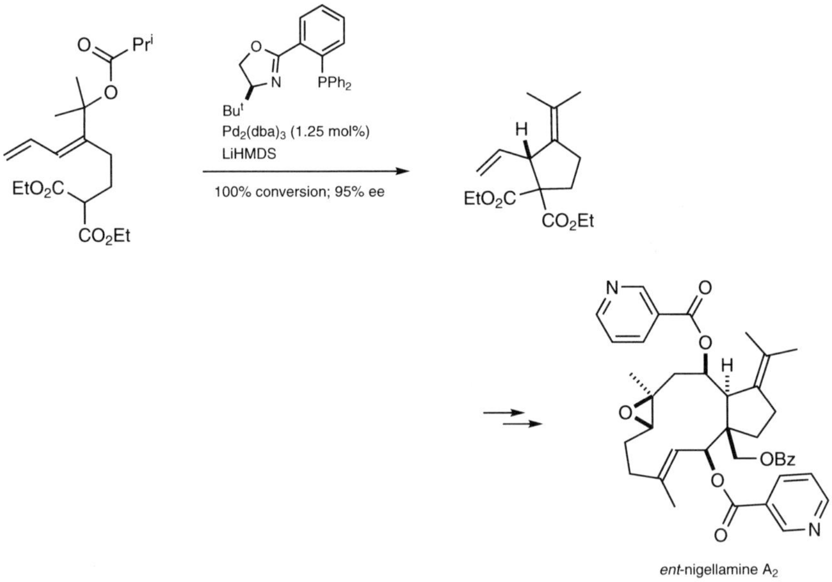

O Pr^i O EtO$_2$C CO$_2$Et

O N PPh$_2$ Bu^t
Pd$_2$(dba)$_3$ (1.25 mol%)
LiHMDS
100% conversion; 95% ee

H EtO$_2$C CO$_2$Et

N O O H O OBz O O N

ent-nigellamine A_2

Scheme 55 Synthesis of (+)-Allokainic Acid[51]

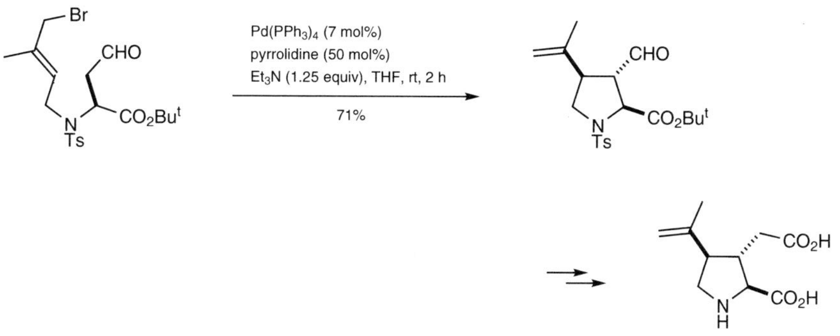

Pd(PPh_3)$_4$ (7 mol%)
pyrrolidine (50 mol%)
Et_3N (1.25 equiv), THF, rt, 2 h
71%

(+)-allokainic acid

Scheme 56 Synthesis of (−)-Cedrelin A[52]

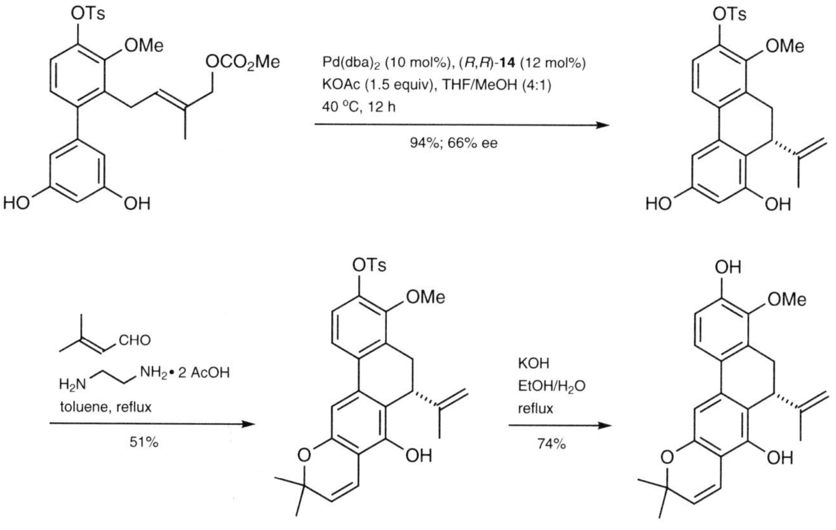

Pd(dba)$_2$ (10 mol%), (*R*,*R*)-**14** (12 mol%)
KOAc (1.5 equiv), THF/MeOH (4:1)
40 °C, 12 h
94%; 66% ee

H_2NCH$_2$CH$_2$$NH_2$• 2 AcOH
toluene, reflux
51%

KOH
EtOH/H_2O
reflux
74%

(−)-cedrelin A

Scheme 57 Synthesis of Asperolide C[53]

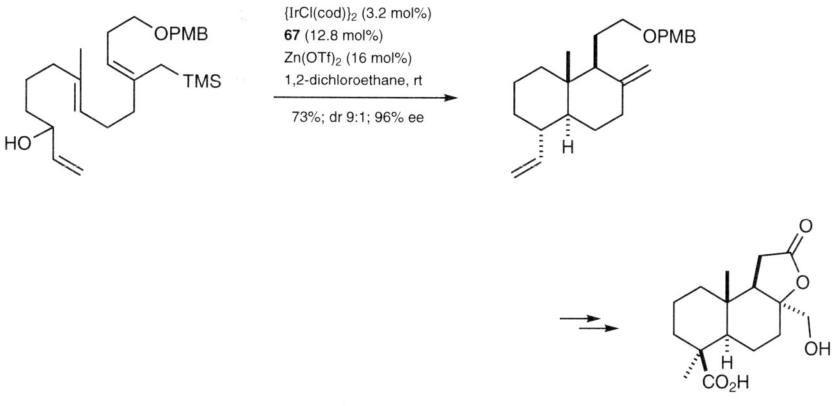

{IrCl(cod)}$_2$ (3.2 mol%)
67 (12.8 mol%)
Zn(OTf)$_2$ (16 mol%)
1,2-dichloroethane, rt
73%; dr 9:1; 96% ee

asperolide C

for references see p 149

Scheme 58 Synthesis of Taiwaniaquinone F[54]

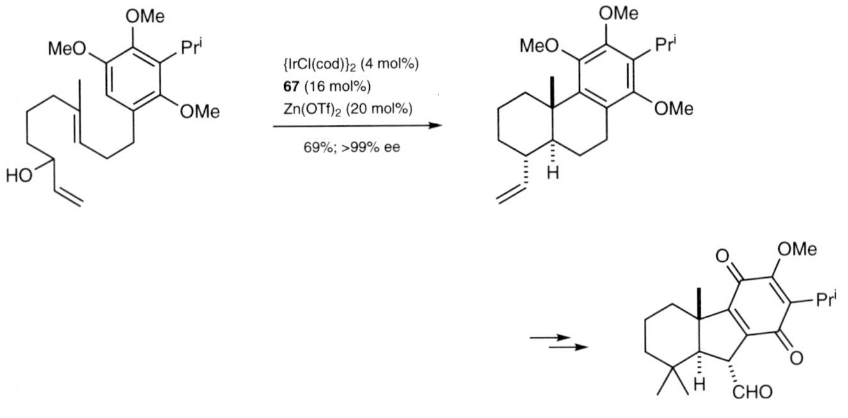
OMe
MeO
Pri
OMe
HO
{IrCl(cod)}2 (4 mol%)
67 (16 mol%)
Zn(OTf)2 (20 mol%)
69%; >99% ee
H
CHO
O

taiwaniaquinone F

1.3.4.2 Allylic Amination

Elegant syntheses of (−)-aurantioclavine (Scheme 59)[55] and alkaloid (+)-241D (Scheme 60)[56] have been reported by Takemoto and Helmchen, respectively. The key for both studies is the formation of nitrogen-containing heterocycles which are based on intramolecular allylic amination reactions.

Scheme 59 Synthesis of (−)-Aurantioclavine[55]

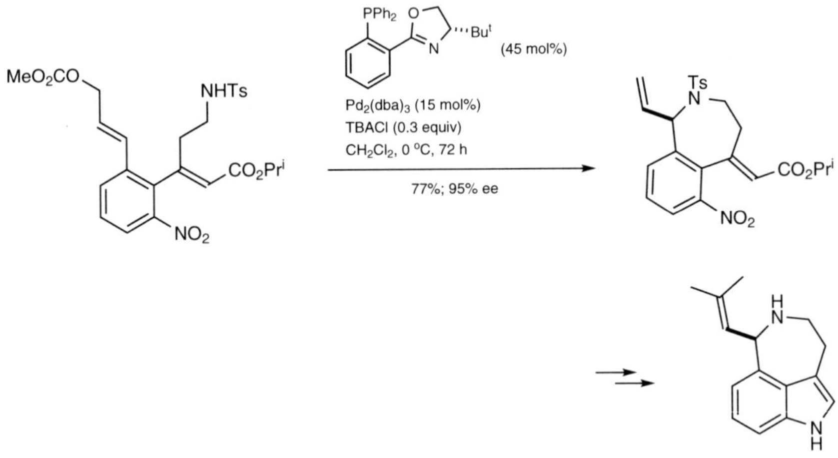
PPh2
But
(45 mol%)
MeO2CO
NHTs
CO2Pri
NO2
Pd2(dba)3 (15 mol%)
TBACl (0.3 equiv)
CH2Cl2, 0 °C, 72 h
77%; 95% ee
Ts
N
H

(−)-aurantioclavine

Scheme 60 Synthesis of Alkaloid (+)-241D[56]

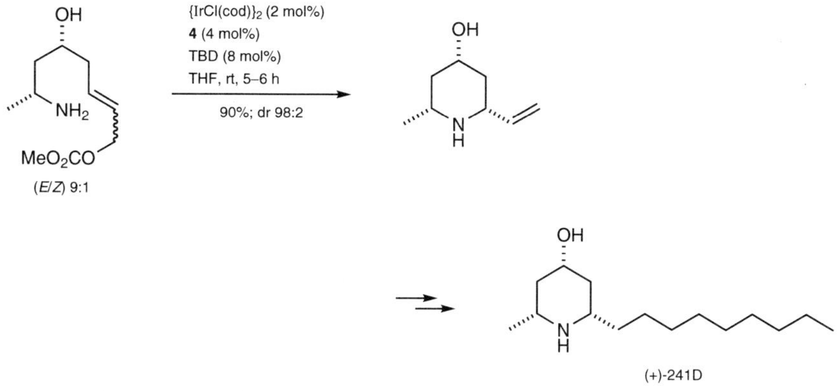

{IrCl(cod)}$_2$ (2 mol%)
4 (4 mol%)
TBD (8 mol%)
THF, rt, 5–6 h
90%; dr 98:2

OH
NH_2
MeO_2CO
(*E*/*Z*) 9:1

OH
N H

OH
N H
(+)-241D

TBD = 1,5,7-triazabicyclo[4.4.0]dec-5-ene

1.3.4.3 Allylic Etherification

The dihydrobenzopyran skeleton is a popular scaffold found in many natural products. A novel strategy to construct the structure relies on the palladium-catalyzed intramolecular allylic etherification of phenol derivatives, as exemplified in Schemes 61 and 62. The cyclization process is followed by several further transformations to complete the total syntheses of (+)-clusifoliol (Scheme 61)[57] and (−)-siccanin (Scheme 62).[58]

Scheme 61 Synthesis of (+)-Clusifoliol[57]

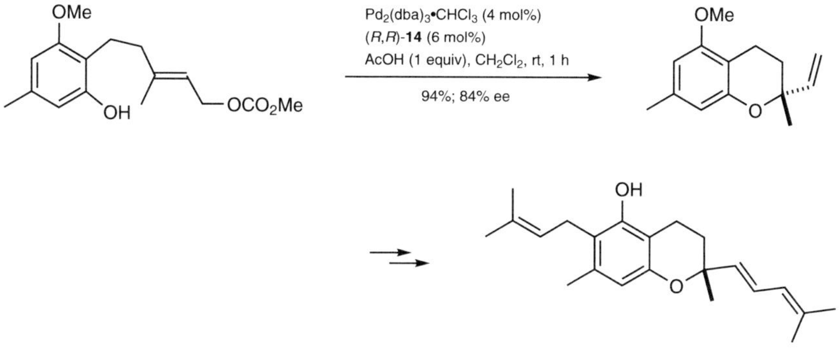

OMe
OH
OCO_2Me

$Pd_2(dba)_3{\cdot}CHCl_3$ (4 mol%)
(*R*,*R*)-**14** (6 mol%)
AcOH (1 equiv), CH_2Cl_2, rt, 1 h
94%; 84% ee

OMe
O

OH
O
(+)-clusifoliol

for references see p 149

Scheme 62 Synthesis of (–)-Siccanin[58]

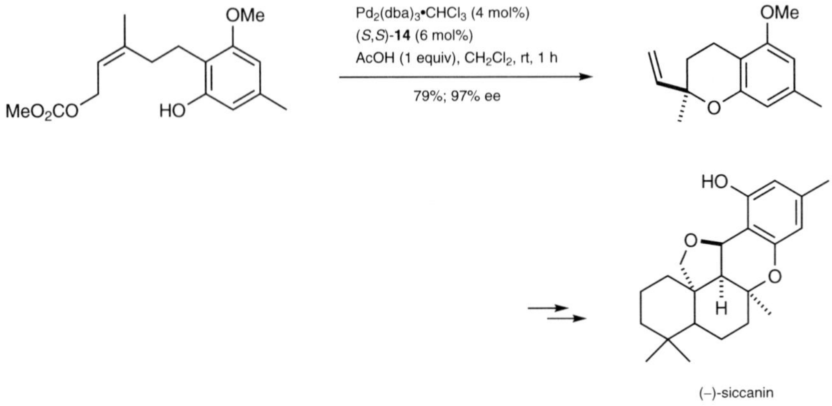

OMe
MeO_2CO
HO
$Pd_2(dba)_3{\bullet}CHCl_3$ (4 mol%)
(*S*,*S*)-**14** (6 mol%)
AcOH (1 equiv), CH_2Cl_2, rt, 1 h
79%; 97% ee
OMe
O
HO
O
O
H
(–)-siccanin

1.3.5 Conclusions and Future Perspectives

Metal-catalyzed intramolecular allylic substitution has been developed into a powerful synthetic methodology to construct a variety of rings in a highly efficient manner. The ability to tune the catalyst by varying metals and/or ligands allows for the formation of diverse bond types including C—C, C—N, and C—O bonds. The synthetic applications of this methodology have been demonstrated successfully as key steps in the synthesis of a broad range of natural products and biologically active compounds. Although great progress has been made in this field, some remaining limitations should be noted: (1) the nucleophiles employed in this strategy are restricted to those based on carbon, nitrogen, and oxygen; (2) five- or six-membered rings are preferred in most cases; (3) the metal catalysts that enable the cyclization reaction are mainly limited to palladium, iridium, ruthenium, and gold complexes; (4) the catalyst loading is generally high, which represents a big hurdle for potential practical applications; (5) for some of the reactions, only the racemic version has been reported and efforts toward the development of enantioselective cyclization processes are still in great demand. In this regard, future studies in this area should be focused on the exploration of novel nucleophiles (such as sulfur, silicon, and phosphorus), development of efficient catalysts, and practical applications of intramolecular allylic cyclization reactions.

References

[1] Tsuji, J.; Takahashi, H.; Morikawa, M., *Tetrahedron Lett.*, (1965), 4387.
[2] Atkins, K. E.; Walker, W. E.; Manyik, R. M., *Tetrahedron Lett.*, (1970), 3821.
[3] Hata, G.; Takahashi, K.; Miyake, A., *J. Chem. Soc. D.*, (1970), 1392.
[4] Trost, B. M., *Tetrahedron*, (1977) **33**, 2615.
[5] Koch, G.; Pfaltz, A., *Tetrahedron: Asymmetry*, (1996) **7**, 2213.
[6] Streiff, S.; Welter, C.; Schelwies, M.; Lipowsky, G.; Miller, N.; Helmchen, G., *Chem. Commun. (Cambridge)*, (2005), 2957.
[7] Miyata, K.; Kutsuna, H.; Kawakami, S.; Kitamura, M., *Angew. Chem.*, (2011) **123**, 4745; *Angew. Chem. Int. Ed.*, (2011) **50**, 4649.
[8] Yamamoto, K.; Tsuji, J., *Tetrahedron Lett.*, (1982) **23**, 3089.
[9] Trost, B. M.; Sacchi, K. L.; Schroeder, G. M.; Asakawa, N., *Org. Lett.*, (2002) **4**, 3427.
[10] Giambastiani, G.; Pacini, B.; Porcelloni, M.; Poli, G., *J. Org. Chem.*, (1998) **63**, 804.
[11] Lemaire, S.; Giambastiani, G.; Prestat, G.; Poli, G., *Eur. J. Org. Chem.*, (2004), 2840.
[12] Thuong, M. B. T.; Sottocornola, S.; Prestat, G.; Broggini, G.; Madec, D.; Poli, G., *Synlett*, (2007), 1521.
[13] Craig, D.; Hyland, C. J. T.; Ward, S. E., *Synlett*, (2006), 2142.
[14] Chiarucci, M.; di Lillo, M.; Romaniello, A.; Cozzi, P. G.; Cera, G.; Bandini, M., *Chem. Sci.*, (2012) **3**, 2859.
[15] Li, X.-H.; Zheng, B.-H.; Ding, C.-H.; Hou, X.-L., *Org. Lett.*, (2013) **15**, 6086.
[16] Kardos, N.; Genet, J.-P., *Tetrahedron: Asymmetry*, (1994) **5**, 1525.
[17] Bandini, M.; Melloni, A.; Piccinelli, F.; Sinisi, R.; Tommasi, S.; Umani-Ronchi, A., *J. Am. Chem. Soc.*, (2006) **128**, 1424.
[18] Xu, Q.-L.; Zhuo, C.-X.; Dai, L.-X.; You, S.-L., *Org. Lett.*, (2013) **15**, 5909.
[19] Bandini, M.; Eichholzer, A., *Angew. Chem.*, (2009) **121**, 9697; *Angew. Chem. Int. Ed.*, (2009) **48**, 9533.
[20] Xu, Q.-L.; Dai, L.-X.; You, S.-L., *Chem. Sci.*, (2013) **4**, 97.
[21] Wu, Q.-F.; He, H.; Liu, W.-B.; You, S.-L., *J. Am. Chem. Soc.*, (2010) **132**, 11418.
[22] Zhang, X.; Liu, W.-B.; Wu, Q.-F.; You, S.-L., *Org. Lett.*, (2013) **15**, 3746.
[23] Zhuo, C.-X.; Liu, W.-B.; Wu, Q.-F.; You, S.-L., *Chem. Sci.*, (2012) **3**, 205.
[24] Wu, Q.-F.; Zheng, C.; You, S.-L., *Angew. Chem.*, (2012) **124**, 1712; *Angew. Chem. Int. Ed.*, (2012) **51**, 1680.
[25] Zhuo, C.-X.; Wu, Q.-F.; Zhao, Q.; Xu, Q.-L.; You, S.-L., *J. Am. Chem. Soc.*, (2013) **135**, 8169.
[26] Yang, Z.-P.; Zhuo, C.-X.; You, S.-L., *Adv. Synth. Catal.*, (2014) **356**, 1731.
[27] Xu, Q.-L.; Dai, L.-X.; You, S.-L., *Org. Lett.*, (2012) **14**, 2579.
[28] Nemoto, T.; Ishige, Y.; Yoshida, M.; Kohno, Y.; Kanematsu, M.; Hamada, Y., *Org. Lett.*, (2010) **12**, 5020.
[29] Wu, Q.-F.; Liu, W.-B.; Zhuo, C.-X.; Rong, Z.-Q.; Ye, K.-Y.; You, S.-L., *Angew. Chem.*, (2011) **123**, 4547; *Angew. Chem. Int. Ed.*, (2011) **50**, 4455.
[30] Bandini, M.; Eichholzer, A.; Kotrusz, P.; Umani-Ronchi, A., *Adv. Synth. Catal.*, (2008) **350**, 531.
[31] Bandini, M.; Eichholzer, A.; Kotrusz, P.; Tragni, M.; Troisi, S.; Umani-Ronchi, A., *Adv. Synth. Catal.*, (2009) **351**, 319.
[32] Bandini, M.; Tragni, M.; Umani-Ronchi, A., *Adv. Synth. Catal.*, (2009) **351**, 2521.
[33] Schafroth, M. A.; Sarlah, D.; Krautwald, S.; Carreira, E. M., *J. Am. Chem. Soc.*, (2012) **134**, 20276.
[34] Trost, B. M.; Krische, M. J.; Radinov, R.; Zanoni, G., *J. Am. Chem. Soc.*, (1996) **118**, 6297.
[35] Trost, B. M.; Oslob, J. D., *J. Am. Chem. Soc.*, (1999) **121**, 3057.
[36] Ito, K.; Akashi, S.; Saito, B.; Katsuki, T., *Synlett*, (2003), 1809.
[37] Welter, C.; Koch, O.; Lipowsky, G.; Helmchen, G., *Chem. Commun. (Cambridge)*, (2004), 896.
[38] Teichert, J. F.; Fañanás-Mastral, M.; Feringa, B. L., *Angew. Chem.*, (2011) **123**, 714; *Angew. Chem. Int. Ed.*, (2011) **50**, 688.
[39] Fañanás-Mastral, M.; Teichert, J. F.; Fernández-Salas, J. A.; Heijnen, D.; Feringa, B. L., *Org. Biomol. Chem.*, (2013) **11**, 4521.
[40] Seki, T.; Tanaka, S.; Kitamura, M., *Org. Lett.*, (2012) **14**, 608.
[41] Mukherjee, P.; Widenhoefer, R. A., *Org. Lett.*, (2011) **13**, 1334.
[42] Mukherjee, P.; Widenhoefer, R. A., *Angew. Chem.*, (2012) **124**, 1434; *Angew. Chem. Int. Ed.*, (2012) **51**, 1405.
[43] Yamamoto, H.; Ho, E.; Namba, K.; Imagawa, H.; Nishizawa, M., *Chem.–Eur. J.*, (2010) **16**, 11271.

[44] Wang, Z.; Li, S.; Yu, B.; Wu, H.; Wang, Y.; Sun, X., *J. Org. Chem.*, (2012) **77**, 8615.
[45] Yang, Z.-P.; Wu, Q.-F.; You, S.-L., *Angew. Chem.*, (2014) **126**, 7106; *Angew. Chem. Int. Ed.*, (2014) **53**, 6986.
[46] Trost, B. M.; Shen, H. C.; Dong, L.; Surivet, J.-P., *J. Am. Chem. Soc.*, (2003) **125**, 9276.
[47] Guérinot, A.; Serra-Muns, A.; Gnamm, C.; Bensoussan, C.; Reymond, S.; Cossy, J., *Org. Lett.*, (2010) **12**, 1808.
[48] Chiarucci, M.; Locritani, M.; Cera, G.; Bandini, M., *Beilstein J. Org. Chem.*, (2011) **7**, 1198.
[49] Zhao, D.; Fañanás-Mastral, M.; Chang, M.-C.; Otten, E.; Feringa, B. L., *Chem. Sci.*, (2014) **5**, 4216.
[50] Bian, J.; Wingerden, M. V.; Ready, J. M., *J. Am. Chem. Soc.*, (2006) **128**, 7428.
[51] Vulovic, B.; Gruden-Pavlovic, M.; Matovic, R.; Saicic, R. N., *Org. Lett.*, (2014) **16**, 34.
[52] Suzuki, Y.; Matsuo, N.; Nemoto, T.; Hamada, Y., *Tetrahedron*, (2013) **69**, 5913.
[53] Jeker, O. F.; Kravina, A. G.; Carreira, E. M., *Angew. Chem.*, (2013) **125**, 12 388; *Angew. Chem. Int. Ed.*, (2013) **52**, 12 166.
[54] Deng, J.; Zhou, S.; Zhang, W.; Li, J.; Li, R.; Li, A., *J. Am. Chem. Soc.*, (2014) **136**, 8185.
[55] Suetsugu, S.; Nishiguchi, H.; Tsukano, C.; Takemoto, Y., *Org. Lett.*, (2014) **16**, 996.
[56] Gnamm, C.; Krauter, C. M.; Brödner, K.; Helmchen, G., *Chem.–Eur. J.*, (2009) **15**, 2050.
[57] Trost, B. M.; Shen, H. C.; Dong, L.; Surivet, J.-P.; Sylvain, C., *J. Am. Chem. Soc.*, (2004) **126**, 11 966.
[58] Trost, B. M.; Shen, H. C.; Surivet, J.-P., *J. Am. Chem. Soc.*, (2004) **126**, 12 565.

1.4 Metal-Catalyzed Intramolecular Cyclizations Involving Cyclopropane and Cyclopropene Ring Opening

X. Tang and M. Shi

General Introduction

Cyclopropanes and cyclopropenes, as the smallest carbocycles, are readily available starting materials and are versatile building blocks in organic synthesis. Their special electronic properties and ring strain usually give them unique reactivities that give rise to a wide range of interesting transformations. In particular, the ring-opening process offers a strong driving force to break the three-membered ring to achieve molecular complexity easily with high atom efficiency. Chemistry in this area is very rich and has drawn much attention over the last few decades. Without doubt, ring-opening reactions of three-membered carbocycles catalyzed by transition metals, Lewis acids, or Brønsted acids have become one of the most quickly developed research areas of synthetic organic chemistry. In this chapter, we primarily focus on transition-metal and metallic Lewis acid catalyzed intramolecular cyclizations involving opening of cyclopropane and cyclopropene rings, because such transformations are extremely useful for the rapid construction of carbocycles and heterocycles.

1.4.1 Intramolecular Cyclizations of Cyclopropanes

1.4.1.1 Methylenecyclopropanes

Methylenecyclopropanes contain C=C bond and cyclopropane moieties, both of which are very reactive owing to ring strain. In the presence of transition-metal catalysts, the double bond can be activated by coordination or insertion of a metal into the three-membered ring to generate a metallocyclic species.

1.4.1.1.1 Palladium-Catalyzed Cycloisomerization of Methylenecyclopropanes

In 2004, Ma reported the elegant synthesis of 4*H*-pyrans **2** through the cycloisomerization of acyl-substituted methylenecyclopropanes **1** catalyzed by palladium(II).[1] The reactions are very fast and can be performed under very mild conditions. If substrate **3**, possessing a tetrasubstituted double bond, is treated under the standard reaction conditions, 2*H*-pyran **4** is formed as the sole product (Scheme 1).

Scheme 1 Palladium(II)-Catalyzed Ring-Opening Cycloisomerization of Methylenecyclopropanes[1]

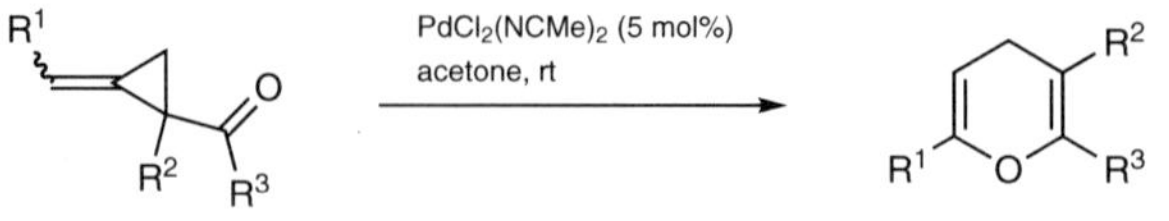

$PdCl_2(NCMe)_2$ (5 mol%)
acetone, rt

R^1	R^2	R^3	Yield (%)	Ref
$(CH_2)_6Me$	CO_2Et	Me	80	[1]
$(CH_2)_6Me$	CO_2Et	Ph	96	[1]
Bu	SO_2Ph	Me	91	[1]

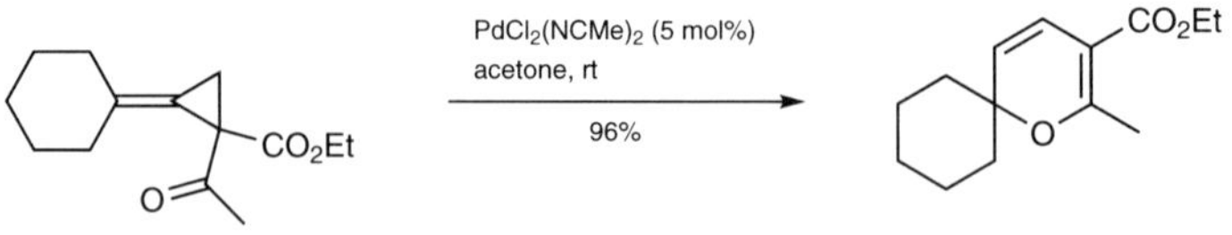

$PdCl_2(NCMe)_2$ (5 mol%)
acetone, rt
96%

A plausible mechanism is proposed on the basis of control experiments (Scheme 2). First, regioselective chloropalladation of the methylenecyclopropane with palladium(II) chloride generates cyclopropyl palladium intermediate **5**; this is followed by β-carbon elimination to give ring-opened palladium enolate **6**. After *endo*-addition to the C=C bond, homoallyl chloride **7** is formed. An alternative pathway leading to the homoallyl chloride may involve coordination of palladium(II) chloride to the *exo*-methylene double bond of the starting methylenecyclopropane, from which oxypalladation is facilitated to form cationic bicyclic palladium species **10**. Subsequent tandem ring opening of the cyclopropane and rearrangement also deliver homoallyl chloride **7**. Next, β-hydride elimination/hydropalladation takes place, for which the regioselectivity depends on the nature of the R^4 substituent. If R^4 is hydrogen, then the new C=C bond will be formed after β-hydride elimination of R^4, because enol ether **8** is more thermodynamically favored. Hydropalladation of **8** furnishes intermediate **9**, which gives rise to the final product through β-dechloropalladation. For the synthesis of the final product ($R^4 \neq H$), β-hydride elimination of this site is blocked but takes place at the other side to generate intermediate **11**, and following the same process from **8** to the 4*H*-pyran product, β-dechloropalladation of **12** produces the 2*H*-pyran product.

Scheme 2 Mechanism of the Bis(acetonitrile)dichloropalladium(II)-Catalyzed Cycloisomerization of Acyl-Substituted Methylenecyclopropanes[1]

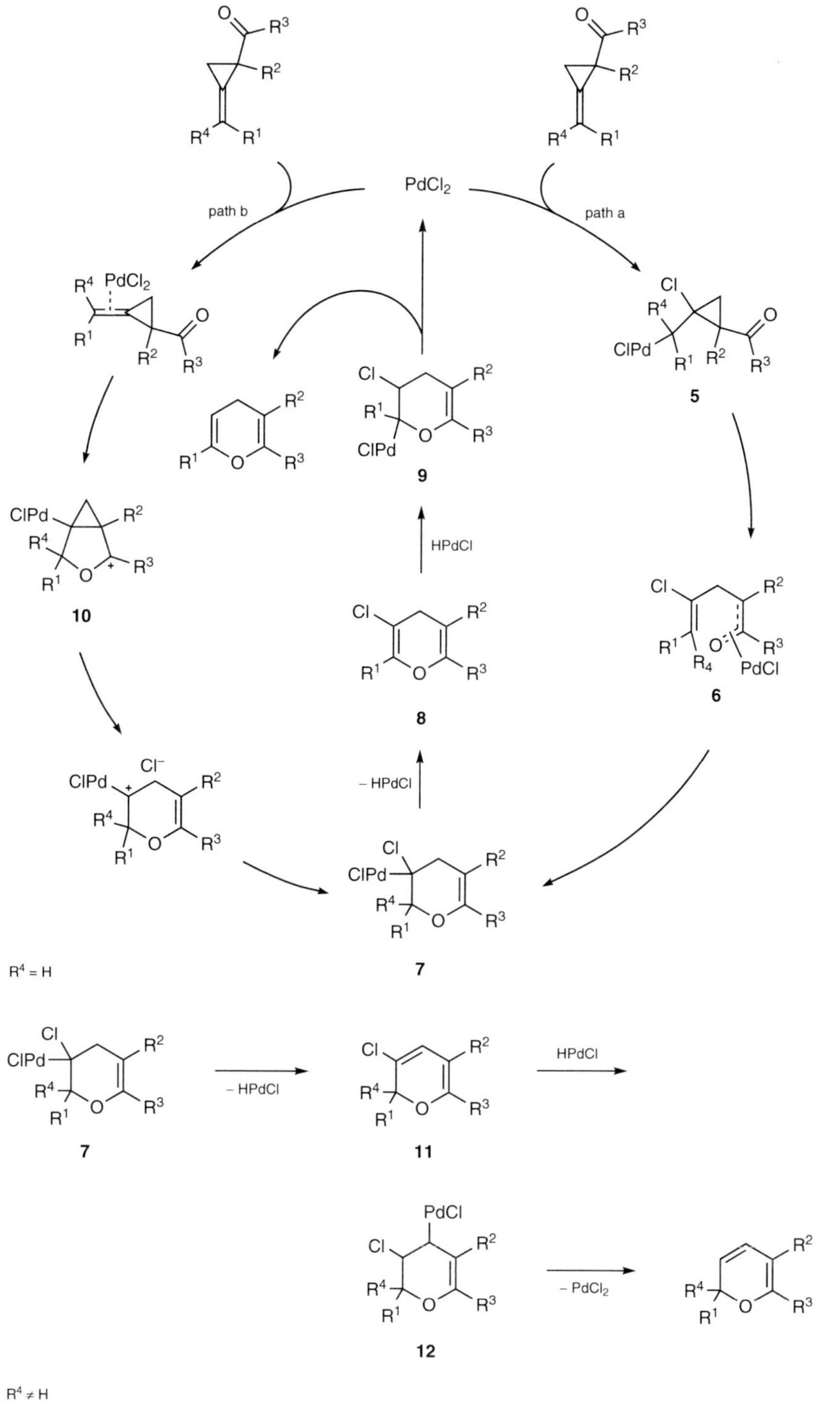

$PdCl_2$

path a

path b

HPdCl

– HPdCl

R^4 = H

– HPdCl

HPdCl

– $PdCl_2$

R^4 ≠ H

for references see p 256

Interestingly, if the cycloisomerization of acyl-substituted methylenecyclopropanes is performed in the presence of sodium iodide (2.0 equiv) by heating at reflux in acetone, furan derivatives **13** are obtained (Scheme 3).

Scheme 3 Palladium(II)-Catalyzed Cycloisomerization of Acyl-Substituted Methylenecyclopropanes To Give Furans[1]

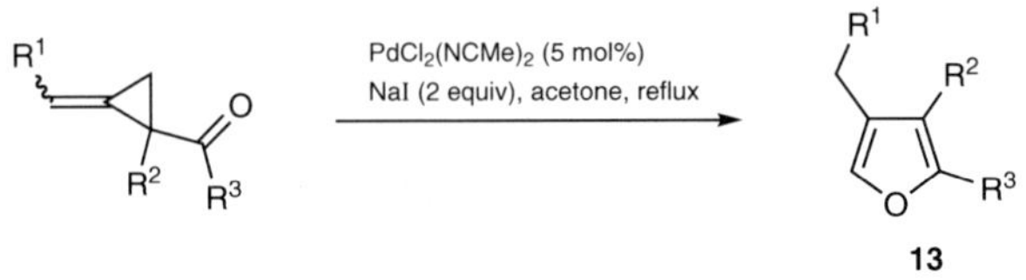

$PdCl_2(NCMe)_2$ (5 mol%)
NaI (2 equiv), acetone, reflux

13

R^1	R^2	R^3	Yield (%)	Ref
$(CH_2)_6Me$	CO_2Et	Me	74	[1]
Bn	CO_2Et	Me	78	[1]
$(CH_2)_6Me$	CO_2Et	Ph	88	[1]
H	H	$4\text{-}FC_6H_4$	84	[1]

Yamamoto[2] reported the intramolecular hydroamination of benzylidenecyclopropanes **14** bearing an *ortho*-amino group to provide an efficient method for the construction of tetrahydroquinolines **15** (Scheme 4). *syn*-Hydropalladation at the double bond of the methylenecyclopropane is believed to occur, followed by opening of the cyclopropane ring to furnish the final product. In addition, for alkyl-chain-tethered methylenecyclopropane–amine substrate **16**, the reaction also proceeds smoothly to give seven-membered nitrogen heterocycle derivative **17** (Scheme 4).

Scheme 4 Intramolecular Hydroamination of Methylenecyclopropanes To Give Tetrahydroquinolines and Azepanes[2]

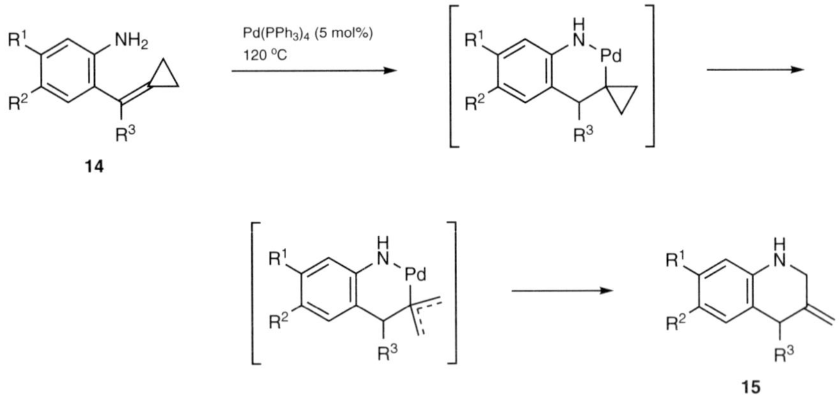

$Pd(PPh_3)_4$ (5 mol%)
120 °C

14

15

R^1	R^2	R^3	Yield (%)	Ref
H	H	Ph	86	[2]
Cl	H	Ph	71	[2]
Cl	H	$2\text{-}ClC_6H_4$	70	[2]
H	Me	Ph	86	[2]

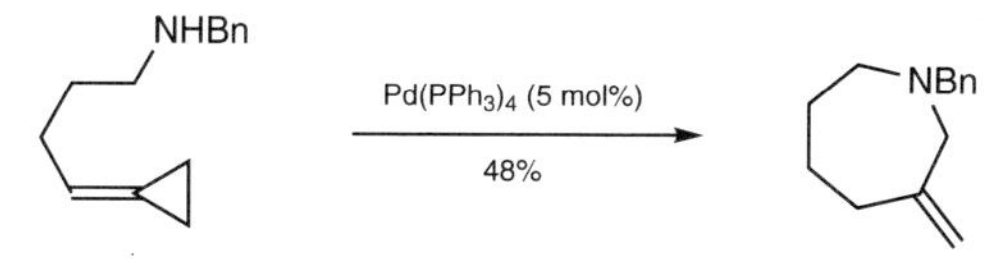

NHBn

Pd(PPh3)4 (5 mol%)

48%

NBn

16 17

The intramolecular cycloadditions of unsaturated carbon–carbon bonds with methylenecyclopropanes promoted by transition-metal catalysts have been extensively investigated by Mascareñas and co-workers.[3] Reactions catalyzed by palladium or nickel are thought to go through common trimethylenemethane intermediate **18** and bicyclic derivatives, e.g. **21**, are delivered by formal (3+2) cycloaddition of compounds **19** or **20** (Scheme 5). Interestingly, in the presence of the first-generation Grubbs carbene complex, the reaction of methylenecyclopropanes **22** also gives bicyclic compounds **23** (Scheme 5). The mechanism of this transformation is not clear. However, on the basis of the outcome of certain research, a non-carbenoid ruthenium species, which is generated during the reaction process, might be the real catalyst.[4] In continuation of their research interest, Mascareñas and co-workers further investigated the intramolecular cycloaddition of methylenecyclopropanes to conjugated dienes. The corresponding cyclohepta[*c*]furan and -pyrrole derivatives **24** are obtained with high diastereoselectivities (Scheme 5).[5]

Scheme 5 Intramolecular Cycloaddition of Yne– or Ene–Methylenecyclopropanes To Give Cycloalkenes[3–5]

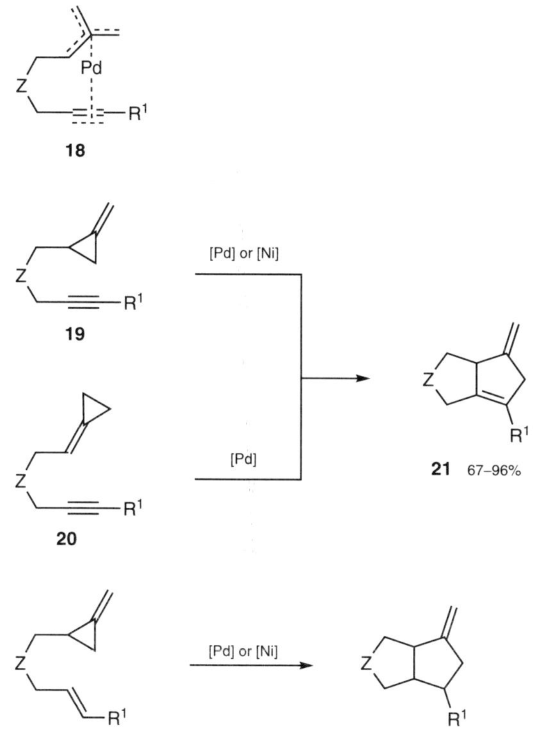

Pd

Z

R1

18

[Pd] or [Ni]

Z

R1

19

Z

R1

21 67–96%

[Pd]

Z

R1

20

Z

[Pd] or [Ni]

Z

R1

R1

Z = O, C(CO2Et)2; R1 = H, Me, CH2OH, CH2OTBDMS, TMS

for references see p 256

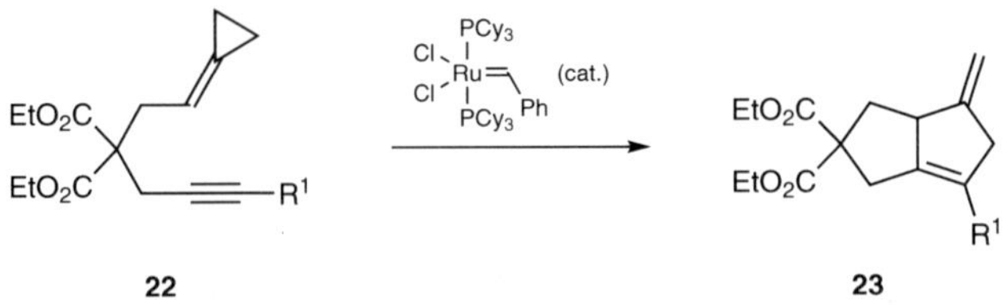

R^1	Yield (%)	Ref
Me	78	[4]
H	47	[4]
$CH_2OTBDMS$	62	[4]
CH_2OH	26	[4]

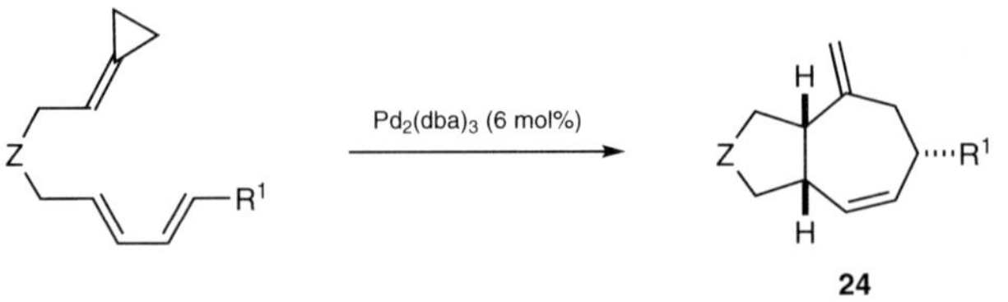

R^1	Z	Yield (%)	Ref
CO_2Et	O	74	[5]
H	NBn	60	[5]

In 2006, Shi reported the palladium-catalyzed ring expansion of methylenecyclopropanes **25** to provide an efficient approach to cyclobutenes **26** (Scheme 6).[6] The mechanism suggested by Shi involves bromopalladation of the methylenecyclopropane with palladium(II) bromide, which is formed in situ by the reaction of palladium(II) acetate with copper(II) bromide. The resulting cyclopropylpalladium intermediate **27** undergoes reversible β-hydride elimination to afford η^2-palladium intermediate **28**, and another hydropalladation delivers cyclopropylpalladium intermediate **29**, which is a regioisomer of **27**. Subsequent α-elimination of bromide gives rise to palladium carbene **30**, which initiates ring expansion of the methylenecyclopropane to the final cyclobutane product. Given that synthetic protocols toward cyclobutenes are very limited, this method is of great value and may have potential application.

Scheme 6 Shi's Palladium-Catalyzed Cyclobutene Synthesis[6]

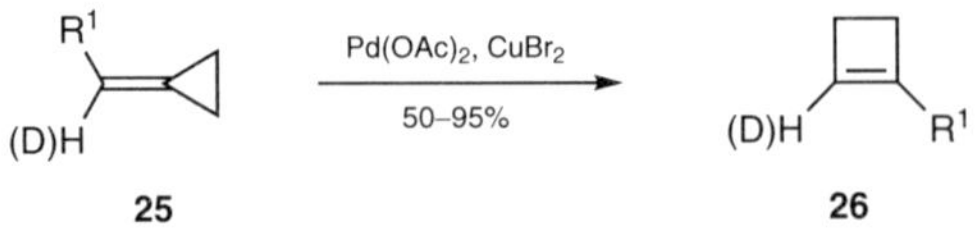

R^1 = alkyl, aryl

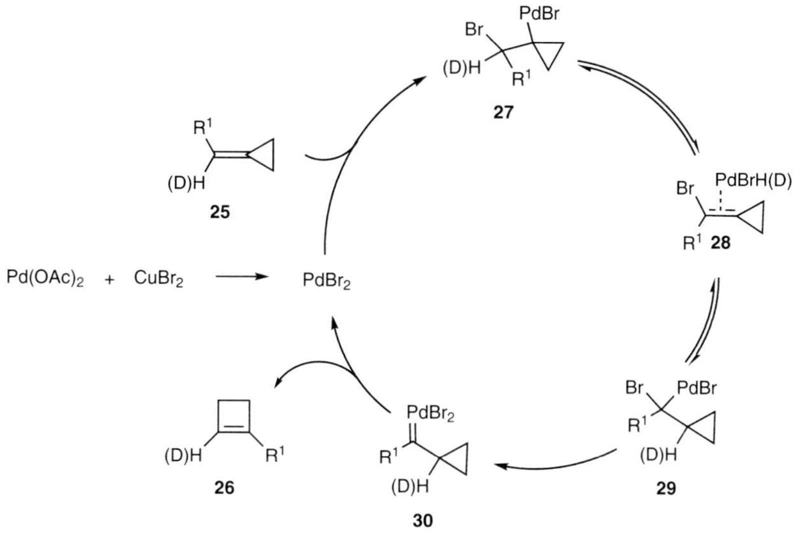

PdBr
Br
(D)H
R^1
27
R^1
(D)H
25
Br
PdBrH(D)
R^1 **28**
$Pd(OAc)_2$ + $CuBr_2$ ⟶ $PdBr_2$
Br
PdBr
R^1
(D)H
29
(D)H
R^1
26
$PdBr_2$
R^1
(D)H
30

Ring enlargement of (2-methylenecyclopropyl)carbinols **31** in the presence of a palladium(II) complex to afford cyclobutenyl carbinols **32** and **33** has also been developed by Shi (Scheme 7).[7] Palladium carbenes **34** and **35** are believed to be the key intermediates in the reaction process, and they deliver the corresponding products **32** (major product) and **33** (minor product) in a ring-expansion process. According to a control experiment, if the hydroxy group is replaced by an *N*-benzylamino group, the reaction does not take place, which reveals that an interaction between the hydroxy group and the palladium catalyst may exist.

Scheme 7 Cyclobutenes by Ring-Opening Cycloisomerization of (2-Methylenecyclopropyl)carbinols[7]

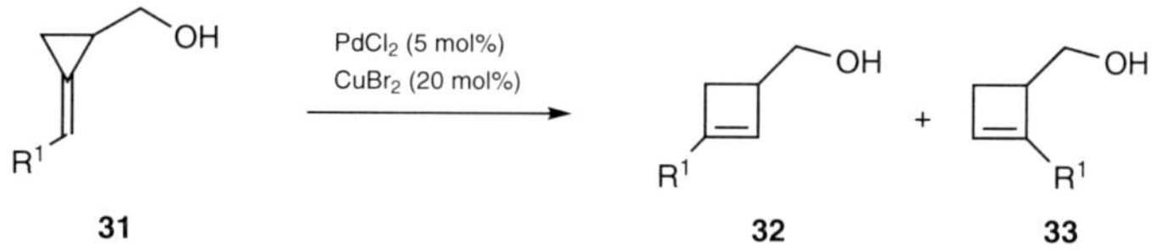

OH
R^1
31
$PdCl_2$ (5 mol%)
$CuBr_2$ (20 mol%)
OH
R^1
32
+
OH
R^1
33

R^1	Ratio (**32**/**33**)	Yield (%)	Ref
4-$MeOC_6H_4$	11:1	86	[7]
3-$BnOC_6H_4$	1:0	75	[7]
2,5-$(MeO)_2C_6H_3$	13:1	89	[7]
4-Tol	17:1	70	[7]
4-BrC_6H_4	1:0	61	[7]

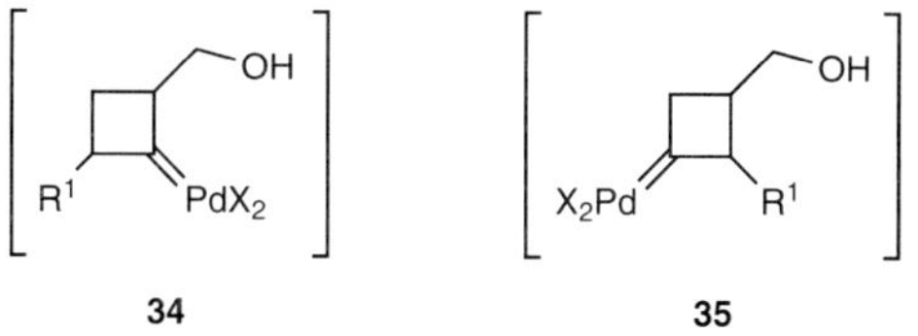

OH
R^1
PdX_2
34
OH
X_2Pd
R^1
35

for references see p 256

Ethyl 6-Heptyl-2-methyl-4*H*-pyran-3-carboxylate [2, $R^1 = (CH_2)_6Me$; $R^2 = CO_2Et$; $R^3 = Me$]; Typical Procedure:[1]
$PdCl_2(NCMe)_2$ (6 mg, 5 mol%) was added to a soln of acyl-substituted methylenecyclopropane **1** (133 mg, 0.50 mmol) in acetone (2 mL) under an atmosphere of argon. The mixture was then stirred at rt for 15 min. Evaporation and chromatography (silica gel, petroleum ether/Et_2O 100:1) under an atmosphere of argon afforded the title compound as an air-sensitive liquid; yield: 106 mg (80%).

Ethyl 2-Methyl-1-oxaspiro[5.5]undeca-2,4-diene-3-carboxylate (4); Typical Procedure:[1]
$PdCl_2(NCMe)_2$ (6 mg, 5 mol%) was added to a soln of methylenecyclopropane **3** (0.50 mmol) in acetone (2 mL) under an atmosphere of argon. The mixture was then stirred at rt for 15 min. Evaporation and chromatography (silica gel, petroleum ether/Et_2O 100:1) under an atmosphere of argon afforded the title compound; yield: 123 mg (96%).

Ethyl 2-Methyl-4-octylfuran-3-carboxylate [13, $R^1 = (CH_2)_6Me$; $R^2 = CO_2Et$; $R^3 = Me$]; Typical Procedure:[1]
A soln of the acyl-substituted methylenecyclopropane (356 mg, 1.3 mmol), NaI (400 mg, 2.6 mmol), and $PdCl_2(NCMe)_2$ (17 mg, 0.065 mmol) in acetone (5 mL) was heated at reflux for 10 h. Evaporation and flash chromatography (silica gel, petroleum ether/Et_2O 100:1) afforded the title compound as a liquid; yield: 265 mg (74%).

3-Methylene-4-phenyl-1,2,3,4-tetrahydroquinoline (15, $R^1 = R^2 = H$; $R^3 = Ph$); Typical Procedure:[2]
Methylenecyclopropane **14** (110.5 mg, 0.5 mmol) was added to a mixture of $Pd(PPh_3)_4$ (28.9 mg, 0.025 mmol) and Bu_3PO (8.4 mg, 0.075 mmol) under an argon atmosphere in a screw-capped Wheaton microreactor. After heating at 120 °C for 2 d, the mixture was filtered through a short Florisil column (EtOAc). Separation by passing through a Florisil column (hexane/EtOAc) and purification by medium-pressure liquid column chromatography (RP-18, EtOAc) afforded the title compound; yield: 95 mg (86%).

Diethyl 4-Methylene-6-(trimethylsilyl)-3,3a,4,5-tetrahydropentalene-2,2(1*H*)-dicarboxylate [21, $R^1 = TMS$; $Z = C(CO_2Et)_2$]; Typical Procedure:[3]
$P(OiPr)_3$ (90 μL of a soln of 100 μL in 1 mL of dioxane) and $Pd_2(dba)_3$ (10 mg, 6 mol%) were added to a soln of methylenecyclopropane **19** (66 mg, 0.18 mmol) in dioxane (3.6 mL). The resulting mixture was heated at reflux for 30 min by using an oil bath preheated at 140 °C. The crude mixture was filtered through a short pad of Celite (Et_2O). The filtrate was concentrated and purified by flash chromatography (silica gel, 0 to 2% EtOAc in hexanes) to afford the title compound as a slightly yellow oil; yield: 63 mg (95%).

Benzyl 2-(Cyclobut-1-enyl)phenyl Ether (26, $R^1 = 2\text{-}BnOC_6H_4$); Typical Procedure:[6]
$Pd(OAc)_2$ (2.0 mg, 3 mol%) and $CuBr_2$ (7.0 mg, 10 mol%) were added to a soln of methylenecyclopropane **25** (71 mg, 0.3 mmol) in 1,2-dichloroethane (2.0 mL). The mixture was stirred at rt for 3 h (monitored by TLC), and then the solvent was removed under reduced pressure. The residue was subjected to flash column chromatography to give the title compound as a white solid; yield: 66 mg (93%).

[3-(4-Methoxyphenyl)cyclobut-2-enyl]methanol (32, $R^1 = 4\text{-}MeOC_6H_4$) and [2-(4-Methoxyphenyl)cyclobut-2-enyl]methanol (33, $R^1 = 4\text{-}MeOC_6H_4$); Typical Procedure:[7]
Compound **31** (0.30 mmol), $PdCl_2$ (0.015 mmol), $CuBr_2$ (0.06 mmol), and $NaHCO_3$ (0.15 mmol) were weighed into an oven-dried test tube that was sealed with a rubber cap. The tube was then evacuated and backfilled with argon. Then, CH_2Cl_2 (1.0 mL) was added to the mixture by syringe. The mixture was stirred at 0 °C. Upon completion of the transformation, as monitored by TLC, the mixture was directly purified by chromatog-

raphy (silica gel, petroleum ether/EtOAc) to afford the title compounds; total yield: 86%; ratio (**32/33**) 11:1. These two products are not very stable under air at rt and should therefore be stored at <0 °C.

1.4.1.1.2 Platinum-Catalyzed Cycloisomerization of Methylenecyclopropanes

Fürstner[8] found that the ring expansion of methylenecyclopropanes to cyclobutenes could be realized in the presence of platinum(II) chloride. The difference between the reports of Fürstner and Shi (see Section 1.4.1.1.1) is that Fürstner's method evokes a platinum(II) chloride/carbon monoxide system as the catalyst, whereas Shi's approach applies palladium(II) acetate/copper(II) bromide as the precatalyst. According to the mechanism proposed by Fürstner, the platinum catalyst first initiates electrophilic addition to the double bond of the methylenecyclopropane **36** to give zwitterionic intermediate **38**, which undergoes ring expansion to provide zwitterionic intermediate **39** or its platinum carbenoid resonance structure **40**. Subsequent β-hydride elimination from **41** and release of the catalyst affords cyclobutenes **37** (Scheme 8).

Scheme 8 Fürstner's Platinum-Catalyzed Cyclobutene Synthesis[8]

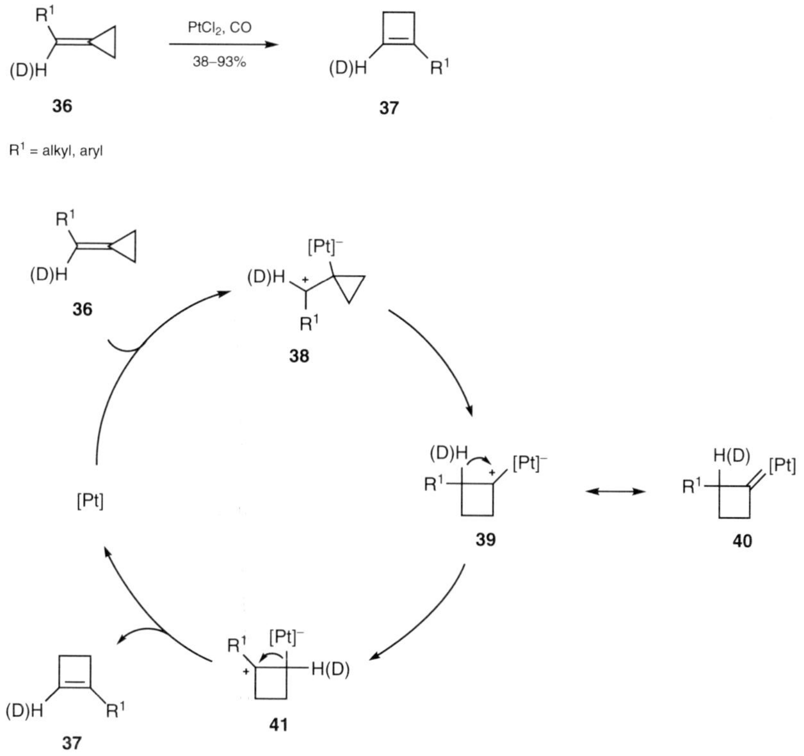

R^1 = alkyl, aryl

4-(Cyclobut-1-enyl)biphenyl (37, R^1 = 4-PhC$_6$H$_4$); Typical Procedure:[8]

CAUTION: *Carbon monoxide is extremely flammable and toxic, and exposure to higher concentrations can quickly lead to a coma.*

$PtCl_2$ (1 mg, 0.0038 mmol) was added to a soln of methylenecyclopropane **36** (90 mg, 0.437 mmol) in toluene (4.3 mL). CO was bubbled into the soln by needle for ca. 30 s, and the resulting mixture was then stirred at 80 °C under a CO atmosphere (1 atm). For workup, the mixture was filtered through a short pad of silica gel under an atmosphere of argon, and the filtrate was concentrated to give the title compound as a white, air-sensitive solid; yield: 69 mg (77%).

1.4.1.1.3 Rhodium-Catalyzed Cycloisomerization of Methylenecyclopropanes

Transition-metal-catalyzed cycloisomerizations of ene– or yne–methylenecyclopropanes have been extensively investigated, whereas examples of the intramolecular cyclization of (het)aryl–methylenecyclopropanes are very rare. A representative example is the rhodium(I)-catalyzed cycloisomerization of nitrogen-tethered indole–alkylidenecyclopropanes **42** as an efficient method for the synthesis of polycyclic indoles **43**.[9] A six- or seven-membered ring is formed along with the opening of the cyclopropane ring (Scheme 9).

Scheme 9 Pyrido- and Azepino[3,4-*b*]indoles by Rhodium(I)-Catalyzed Cycloisomerization of Nitrogen-Tethered Indolyl–Methylenecyclopropanes[9]

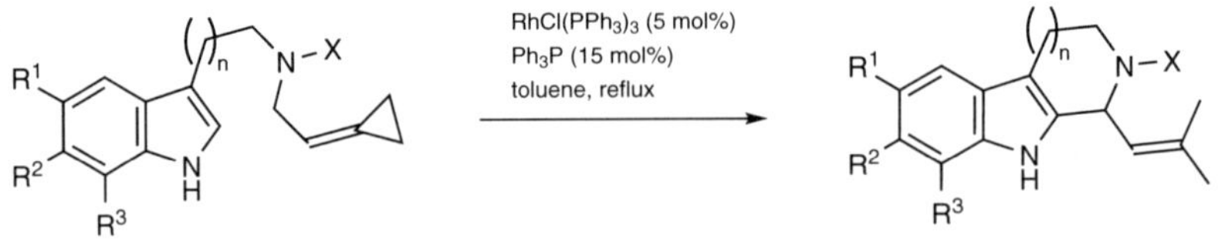

n	R1	R2	R3	X	Yield (%)	Ref
1	OMe	H	H	Ts	95	[9]
1	Me	H	H	Ts	82	[9]
1	Br	H	H	Ts	76	[9]
1	H	F	H	Ts	52	[9]
1	H	H	OBn	Ts	88	[9]
1	H	H	H	4-$O_2NC_6H_4SO_2$	85	[9]
1	H	H	H	Ms	73	[9]
2	H	H	H	Ts	68	[9]

To probe the mechanism, several deuterium-labeling experiments were conducted. As shown in Scheme 10, substrate **44** bearing a deuterium atom at the C2-position of the indole ring undergoes the reaction smoothly to give product **45** without the incorporation of deuterium (Scheme 10). Treatment of substrate **46** possessing two deuterium atoms at the allylic position under the standard reaction conditions results in the formation of **47** in 74% yield with >65% deuterium incorporation at the methyl carbon atom and 80% deuterium incorporation at the ring carbon (Scheme 10). The reactions of **46** and diene **48** in the presence of deuterium oxide (50 equiv) give deuterated product **47** in yields of 53 (>90% D at the methyl carbon atom) and 86% (>80% D at the methyl carbon atom), respectively (Scheme 10).

Scheme 10 Isotopic Labeling Experiments for the Rhodium(I)-Catalyzed Cycloisomerization of Nitrogen-Tethered Indolyl–Methylenecyclopropanes[9]

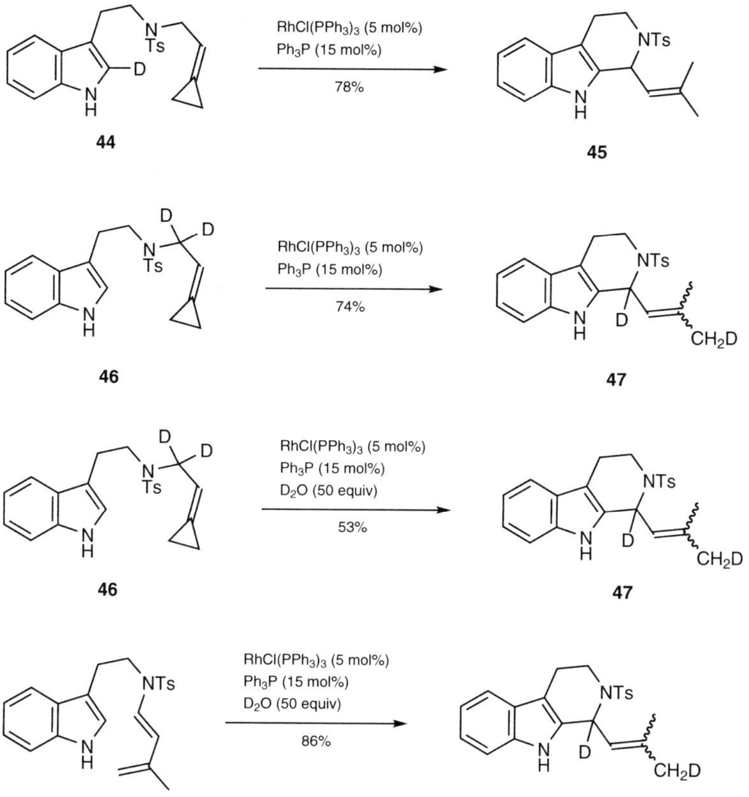

44 → 45: $RhCl(PPh_3)_3$ (5 mol%), Ph_3P (15 mol%), 78%

46 → 47: $RhCl(PPh_3)_3$ (5 mol%), Ph_3P (15 mol%), 74%

46 → 47: $RhCl(PPh_3)_3$ (5 mol%), Ph_3P (15 mol%), D_2O (50 equiv), 53%

48 → 47: $RhCl(PPh_3)_3$ (5 mol%), Ph_3P (15 mol%), D_2O (50 equiv), 86%

On the basis of control experiments, a plausible mechanism is proposed in Scheme 11. First, the insertion of rhodium(I) into the cyclopropane with cleavage of the distal C—C bond gives cyclorhodium intermediate **49**, which undergoes isomerization to deliver rhodacyclic intermediate **50**. Diene **51** is formed by subsequent β-hydride elimination. After reductive elimination of the rhodium catalyst, diene intermediate **47** is formed. There may be a trace amount of protons existing in the reaction medium; therefore, diene **47** readily initiates a Friedel–Crafts-type reaction from either the C3- or C2-position of the indole backbone. Attack at C3 forms spiro intermediate **52**, which leads to cationic intermediate **53** after 1,2-alkyl shift; this intermediate is the same species that results from attack at C2. Product **54** is obtained after an aromatization process. The driving force for the Friedel–Crafts-type reaction is not very clear at this stage; however, such a reaction can also be promoted by using a Brønsted acid (e.g., trifluoromethanesulfonic acid).

Scheme 11 Proposed Mechanism for the Rhodium(I)-Catalyzed Cycloisomerization of Nitrogen-Tethered Indolyl–Methylenecyclopropanes[9]

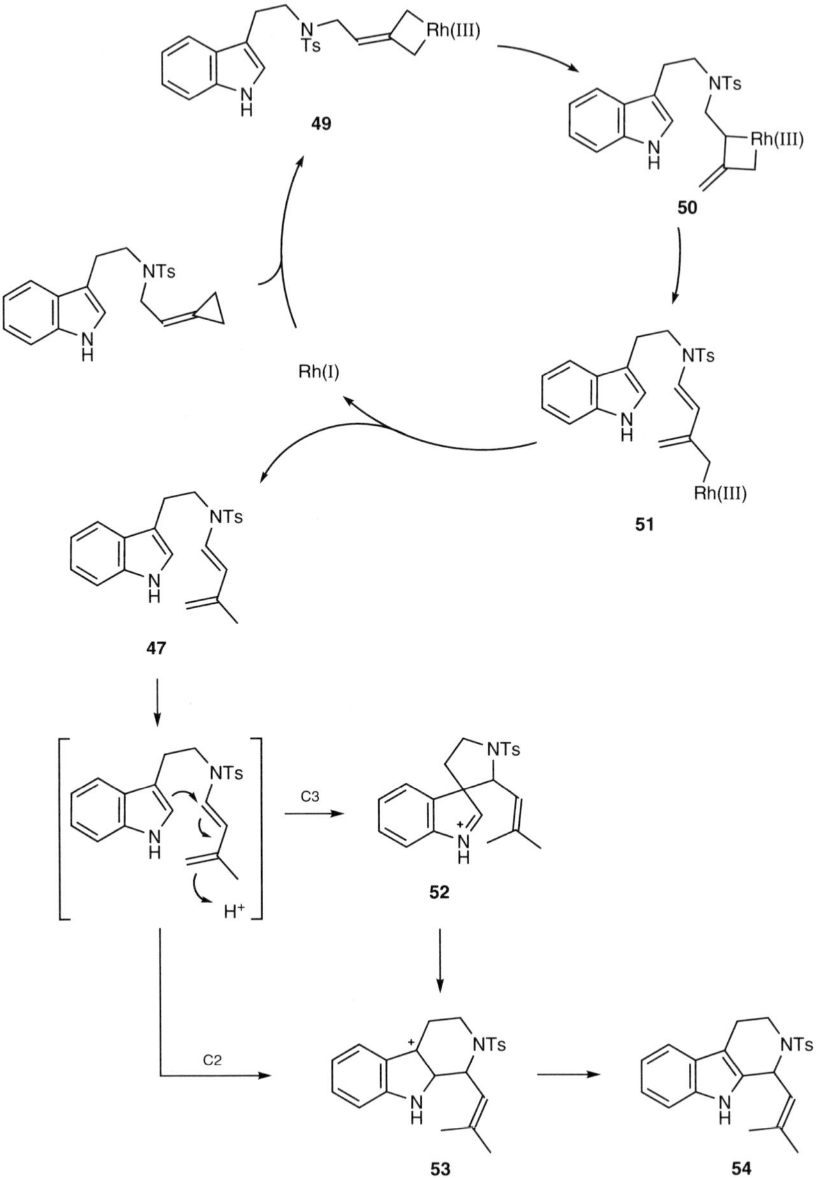

Through C—H activation using pyridine as a directing group, Fürstner has established a protocol to strategically guide how the metal adds to the methylenecyclopropane.[10] For example, cyclopropane-containing 1,5-diene **55** can be easily converted into the corresponding cycloheptene derivative **56**. In a second scenario with the use of chlorobis(cyclooctene)rhodium(I) dimer [$Rh_2Cl_2(coe)_4$; coe = cyclooctene] as the catalyst, metal insertion into the C(=O)—H bond of **57** is believed to take place, and the resulting rhodium species can initiate opening of the three-membered ring to yield ketone **58** (Scheme 12).

Scheme 12 Rhodium-Catalyzed Ring Opening of Methylenecyclopropanes through a C–H Activation Strategy[10]

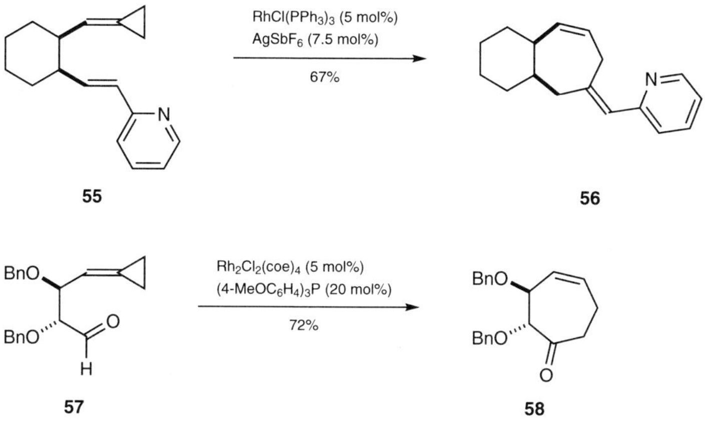

The intramolecular reaction of methylenecyclopropanes with in situ generated carbenes has been reported by Tang and Shi.[11] From *N*-sulfonyl-1,2,3-triazole-tethered methylenecyclopropanes **59**, the formation of an azavinyl carbene intermediate easily takes place through rhodium(II)-catalyzed denitrogenative ring opening of the triazole. Subsequent cyclization and ring expansion gives imines **60**, which undergo a novel intermolecular aza-Diels–Alder cycloaddition to provide polycyclic nitrogen heterocycles **61** (Scheme 13).

Scheme 13 Rhodium(II)-Catalyzed Intramolecular Cycloisomerization of Methylenecyclopropane–Triazoles[11]

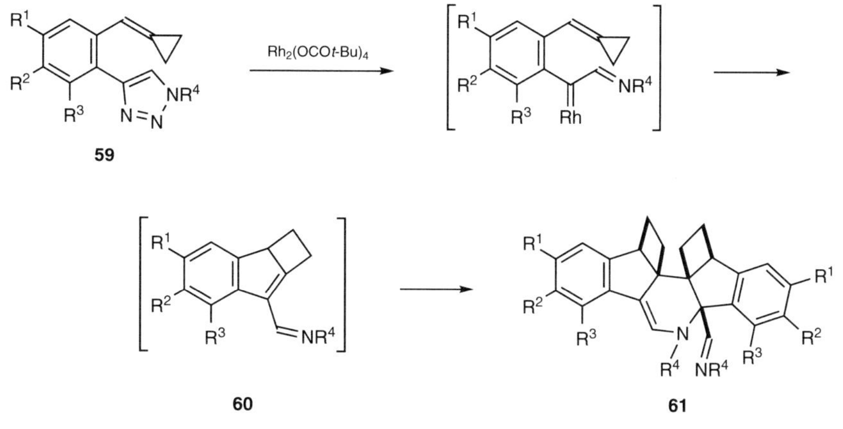

R^1	R^2	R^3	R^4	Yield (%)	Ref
Me	H	H	Ts	74	[11]
H	CF_3	H	Ts	79	[11]
H	Me	H	Ts	77	[11]
H	OMe	H	Ts	71	[11]
H	H	F	Ts	73	[11]
H	H	H	$4\text{-}BrC_6H_4SO_2$	67	[11]
H	H	H	Ms	81	[11]

for references see p 256

To investigate the substrate scope further, several substrates with different spacers were synthesized and treated under the standard reaction conditions (Scheme 14). The reactions of thiophene- and cyclohexene-tethered methylenecyclopropane–triazoles proceed smoothly to give dihydroindole derivatives **62** and **63**. If the alkene of the methylenecyclopropane is fully substituted, such as in substrate **64**, the reaction delivers piperidine derivative **66** via key intermediate **65**. The reaction of an *N*-tosyl-tethered triazole gives spirocyclopropane **67** after cyclopropanation of the alkene with the carbene. Treatment of carbon-tethered triazole **68** with the rhodium(II) catalyst results in the formation of cyclobutane **69**. The outcomes of these reactions demonstrate the versatility of the intramolecular cycloisomerization of methylenecyclopropane–triazoles.

Scheme 14 Rhodium(II)-Catalyzed Intramolecular Cycloisomerizations of Various Methylenecyclopropane–Triazoles[11]

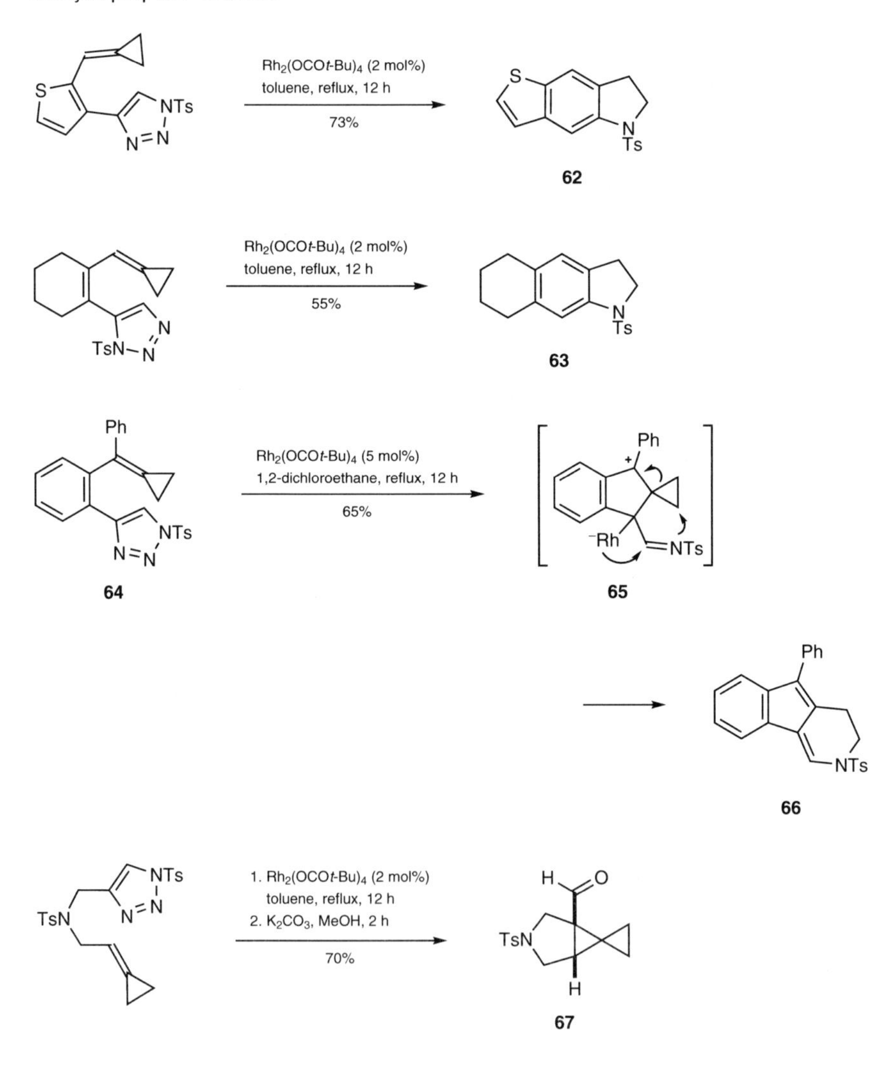

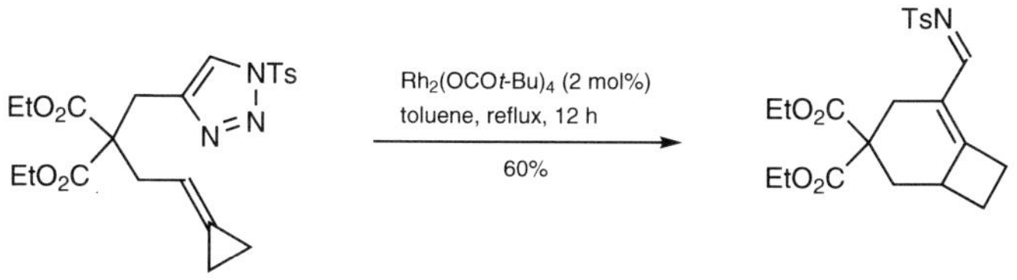

Pyrido[3,4-*b*]indoles 43 **(n = 1) and Azepino[3,4-*b*]indoles** 43 **(n = 2); General Procedure:**[9]
Under an argon atmosphere, indole–alkylidenecyclopropane **42** (0.1 mmol, 1.0 equiv) was dissolved in toluene (8.0 mL, 0.0125 M) in a Schlenk tube, and then $RhCl(PPh_3)_3$ (5 mol%) and Ph_3P (15 mol%) were added. Then, the mixture was stirred at 110 °C until the reaction was complete. The solvent was removed under reduced pressure, and the residue was purified by flash column chromatography (silica gel) to give the title compound.

2-[(*E*)-(4a*R,9a*R**)-1,2,3,4,4a,5,7,9a-Octahydro-6*H*-benzo[7]annulen-6-ylidenemethyl]pyridine** (56)**; Typical Procedure:**[10]
A Teflon-screwed Schlenk tube equipped with a magnetic stirrer bar was charged with compound **55** (24 mg, 0.1 mmol), $RhCl(PPh_3)_3$ (4.6 mg, 0.005 mmol), $AgSbF_6$ (2.6 mg, 0.0075 mmol), and THF (2 mL). The flask was sealed and immersed into a preheated oil bath (120 °C bath temperature). After stirring for 6 h at that temperature, the mixture was allowed to cool before it was filtered through a short pad of silica gel, which was carefully rinsed with Et_2O. The combined filtrate was concentrated, and the residue was purified by flash chromatography (silica gel, hexanes/Et_2O 20:1 to 10:1) to give the title compound as a colorless oil; yield: 16 mg (67%).

(2*R,3*S**)-2,3-Bis(benzyloxy)cyclohept-4-en-1-one** (58)**; Typical Procedure:**[10]
A Teflon-screwed Schlenk flask equipped with a small stirrer bar was charged with compound **57** (39 mg, 0.12 mmol), $Rh_2Cl_2(coe)_4$ (4 mg, 0.005 mmol), $(4\text{-}MeOC_6H_4)_3P$ (7 mg, 0.02 mmol), and 1,2-dichloroethane (2 mL) under an atmosphere of argon. Ethene was bubbled through the soln using a needle for 60 s before the flask was sealed and immersed into a preheated oil bath (120 °C bath temperature). After stirring for 2.5 h at that temperature, the mixture was allowed to cool before it was diluted with Et_2O (5 mL). Filtration through a short pad of silica gel and concentration of the filtrate followed by purification by flash chromatography (hexanes/EtOAc 20:1) afforded the title compound as a colorless oil; yield: 28 mg (72%).

Fused Nitrogen Heterocycles 61**; General Procedure:**[11]
A soln of compound **59** (0.1 mmol) and $Rh_2(OCO\mathit{t}\text{-}Bu)_4$ (1.2 mg, 0.002 mmol) in dry toluene (1 mL) was stirred at 110 °C under an atmosphere of N_2 overnight. The mixture was cooled to rt and then purified by flash column chromatography (silica gel, petroleum ether/EtOAc 15:1) to afford the title compound.

1.4.1.1.4 Nickel-Catalyzed Cycloisomerization of Methylenecyclopropanes

Nickel(0) catalysts are also able to insert into the cyclopropane ring to yield a nickelacycle species.[12] For example, substrates **70** undergo intramolecular cycloaddition to alkynes catalyzed by bis(cyclooctadiene)nickel(0) to furnish cyclopenta[*a*]indene derivatives **73**. Owing to geometric requirements, the insertion of nickel(0) occurs with cleavage of the proximal bond, which results in intermediate **71**. This intermediate generates six-membered nickelacycle **72** through intramolecular insertion of the C≡C bond; reductive elimination affords **73** (Scheme 15).

Scheme 15 Cyclopenta[*a*]indenes by Nickel-Catalyzed Intramolecular Cycloaddition of Methylenecyclopropanes to Alkynes[12]

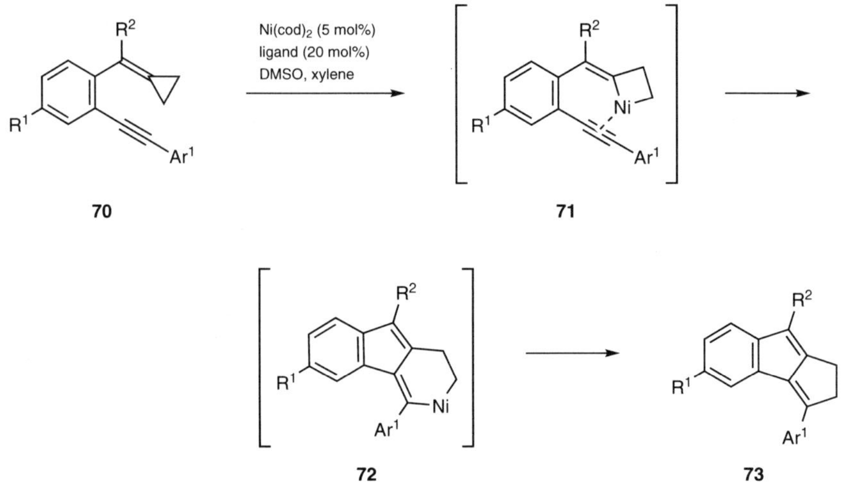

$Ni(cod)_2$ (5 mol%)
ligand (20 mol%)
DMSO, xylene

70 71 72 73

R^1	R^2	Ar^1	Ligand	Yield (%)	Ref
H	H	Ph	Ph_3P	71	[12]
H	H	4-$MeOC_6H_4$	Ph_3P	58	[12]
H	H	4-ClC_6H_4	Ph_3P	43	[12]
Me	H	Ph	Ph_3P	62	[12]
H	Me	Ph	cod	46	[12]

Cyclopenta[*a*]indenes 73; **General Procedure:**[12]

Under a N_2 atmosphere, an oven-dried Schlenk tube equipped with a magnetic stirrer bar was charged with substrate **70** (0.3 mmol), $Ni(cod)_2$ (4 mg, 0.0015 mmol), Ph_3P (16 mg, 0.06 mmol), DMSO (4.5 mL), and xylene (0.5 mL). The resulting mixture was stirred at 120 °C for 12 h. After cooling to rt, the mixture was filtered through a pad of Celite, and the solvent was removed under reduced pressure. The residue was subjected to flash column chromatography (silica gel) to afford the title compound.

1.4.1.1.5 Copper-Catalyzed Cycloisomerization of Methylenecyclopropanes

In 2011, Wu reported the intramolecular rearrangement of in situ formed methylenecyclopropane–triazoles **75** derived from the copper-catalyzed alkyne–azide cycloaddition of methylenecyclopropanes **74** with sulfonyl azides.[13] Unlike the reaction described in Scheme 13 (Section 1.4.1.1.3), intermediate **75** derived from click chemistry can be readily transformed into reactive ketenimine **76** through a denitrogenative ring-opening process. Subsequent 6π-electrocyclization then takes place to produce ring-closed bicyclic imines **77**, which give the corresponding dihydroindoles **78** after ring expansion (Scheme 16).

Scheme 16 2,3-Dihydrobenzo[*f*]indoles by Copper-Catalyzed Intramolecular Cycloisomerizations of Various Methylenecyclopropane–Triazoles[13]

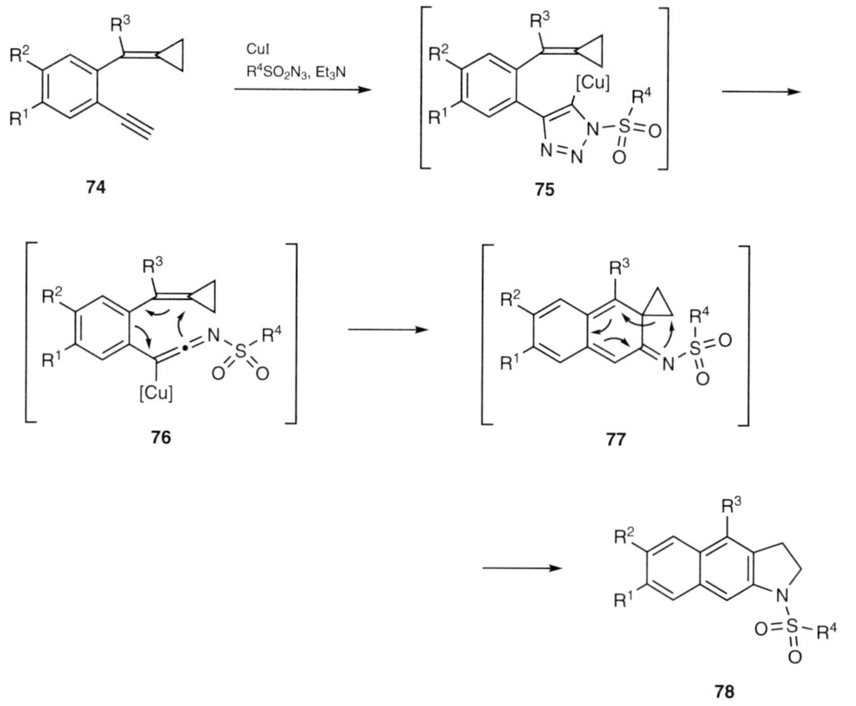

R^1	R^2	R^3	R^4	Yield (%)	Ref
H	H	H	Ts	67	[13]
OMe	H	H	Ts	70	[13]
H	Cl	H	Ts	60	[13]
Me	H	H	Ph	71	[13]
H	H	Me	Ts	46	[13]

2,3-Dihydrobenzo[*f*]indoles 78; **General Procedure:**[13]

Et_3N (0.42 mmol) was added to a soln of 1-(cyclopropylidenemethyl)-2-ethynylbenzene **74** (0.3 mmol), sulfonyl azide (0.36 mmol), and CuI (5 mol%) in 1,4-dioxane (1.0 mL). The resulting mixture was stirred at rt. Upon completion of the reaction, as indicated by TLC, the mixture was diluted with CH_2Cl_2 (10 mL) and filtered through a thin layer of silica gel. The solvent was evaporated, and the residue was purified by column chromatography (silica gel, hexane/EtOAc 6:1) to provide the title compound.

1.4.1.1.6 Magnesium Chloride Catalyzed Cycloisomerization of Methylenecyclopropanes

Methylenecyclopropanes bearing an additional nucleophile are attractive because their transformations usually lead to diverse heterocycles. In the synthesis of cyclic diazadienes in the presence of magnesium chloride, Lautens[14] reports that methylenecyclopropane hydrazones **79** undergo ring expansion via zwitterionic intermediate **80**. The corresponding dienamines **81** are obtained in good to excellent yields with excellent selectivities (Scheme 17).

Scheme 17 1,2-Dihydropyridazines by Lewis Acid Catalyzed Ring Expansion of Methylene-cyclopropane–Hydrazones[14]

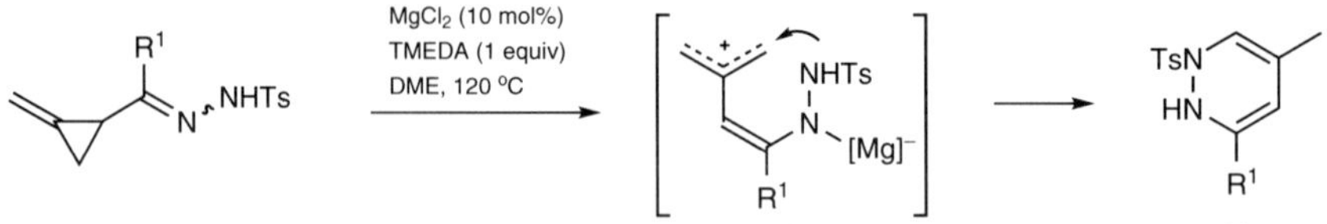

R^1	Yield (%)	Ref
Me	86	[14]
Et	77	[14]
Bn	70	[14]
$CH_2CH=CH_2$	87	[14]
Ph	93	[14]

1,2-Dihydropyridazines 81**; General Procedure:**[14]
A sealed, oven-dried vial purged with argon was charged with the methylenecyclopropane–hydrazone, DME (0.01 M), and TMEDA (redistilled, 1 equiv) under a cone of argon with stirring. $MgCl_2$ (10 mol%) was then added, and the vial was immediately sealed and heated at 120 °C until the reaction was complete (TLC). The resulting soln was evaporated, and the residue was directly purified by flash chromatography (silica gel) to afford the title compound.

1.4.1.1.7 Gold- or Silver-Catalyzed Cycloisomerization of Methylenecyclopropanes

In contrast to palladium and nickel catalysts,[3] gold catalysts, as powerful soft Lewis acids that can easily activate carbon–carbon multiple bonds, exhibit different activities; they are also very attractive in organic synthesis because unprecedented results are usually obtained. For example, in 2008 Toste reported the novel cycloisomerization of enyne–methylenecyclopropanes **82** for the rapid construction of complex polycyclic ring systems (Scheme 18).[15] The methylenecyclopropane moiety plays an important role in the regiocontrol of the reaction with latent ring-strain reactivity. First, the gold catalyst initiates a 6-*exo-dig* addition of methylenecyclopropane **82** to produce cationic intermediate **83** or resonance structure **84**. Ring expansion of **83** gives cyclobutane intermediate **85** or gold–carbene intermediate **86**. If the alkyne substituent is a butyl chain, carbene-induced 1,2-hydrogen shift readily occurs and conjugated dienes **87** and **88** are obtained as a pair of isomers. If there is no α-hydrogen atom adjacent to the gold–carbene (the alkyne substituent is an aryl group), tetracycle **89** is afforded as a single diastereomer through Nazarov-type electrocyclization. Alternatively, if the alkyne substrate bears a distant hydroxy group [$R^1 = (CH_2)_4OH$], carbene intermediate of type **84** undergoes O—H bond insertion to produce pyran derivative **90** in 60% yield (Scheme 18).

Scheme 18 Rearrangements of Enyne–Methylenecyclopropanes under Gold Catalysis[15]

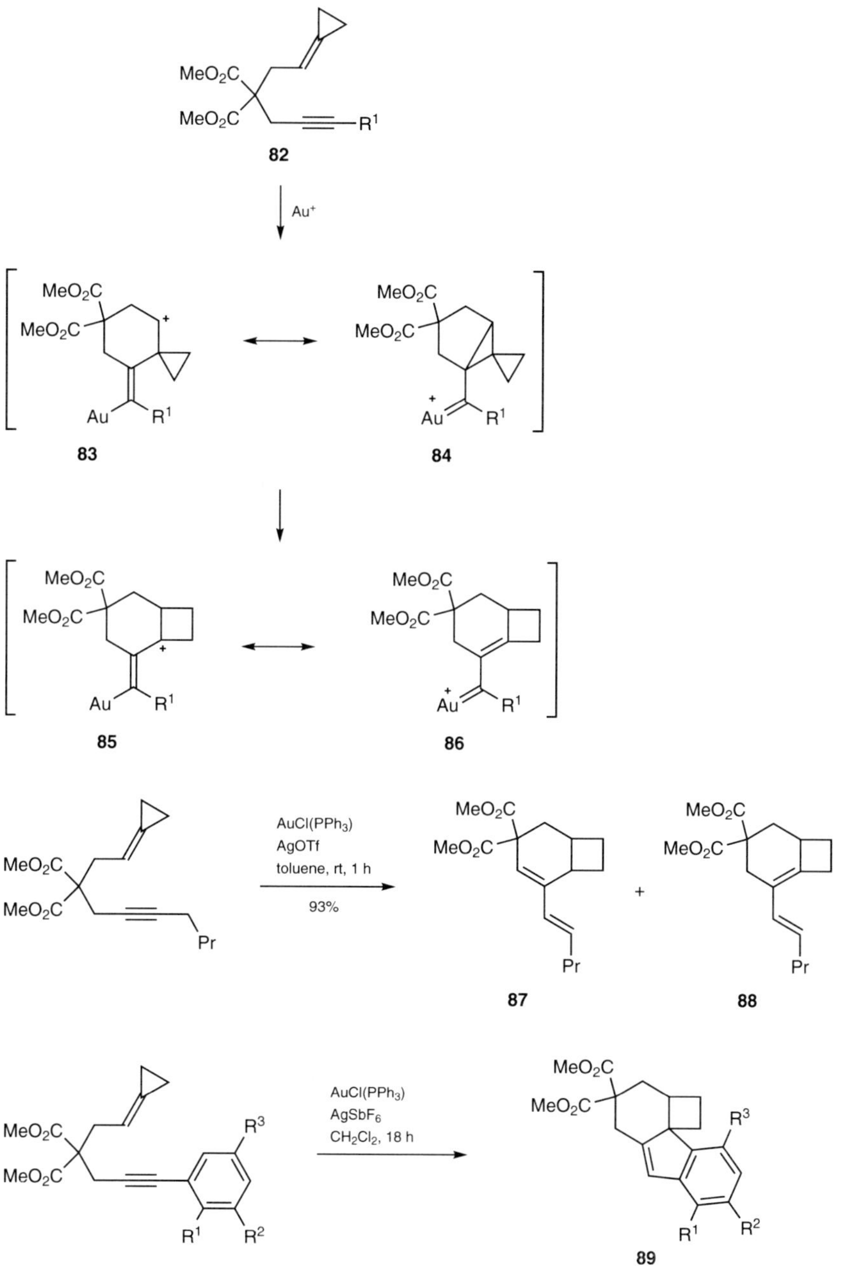

R^1	R^2	R^3	Yield (%)	Ref
H	H	H	75	[15]
Me	H	H	75	[15]
I	H	H	91	[15]
H	Me	Me	44	[15]
H	Cl	Cl	86	[15]

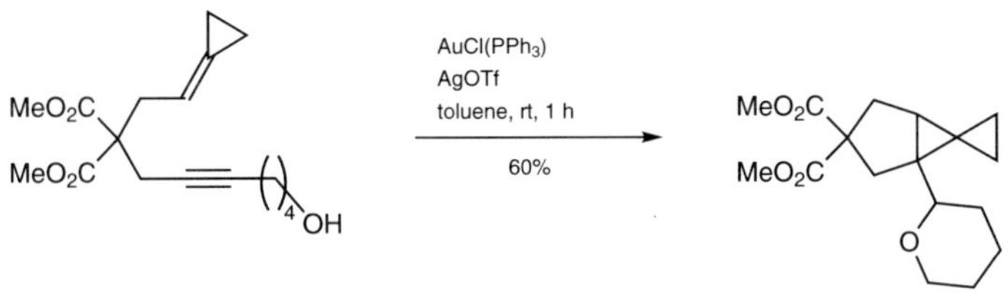

AuCl(PPh_3)
AgOTf
toluene, rt, 1 h
60%

90

The intramolecular rearrangement of propargylic alcohol tethered methylenecyclopropanes **91** to stereoselectively synthesize allenylcyclobutanols **92** and 1-vinyl-3-oxabicyclo-[3.2.1]octan-8-ones **93** under silver and gold catalysis has also been developed by Shi.[16] Investigation into this reaction revealed that **92** can also be converted into **93** under gold-catalyzed reaction conditions, which indicates that the gold catalyst may be more active toward the allene moiety than the silver catalyst (Scheme 19).

Scheme 19 Gold(I)- and Silver(I)-Catalyzed Cycloisomerization of Propargylic Alcohol Tethered Methylenecyclopropanes[16]

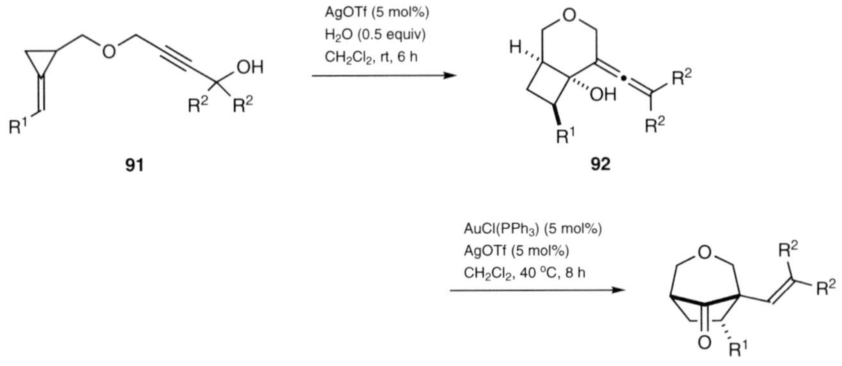

AgOTf (5 mol%)
H_2O (0.5 equiv)
CH_2Cl_2, rt, 6 h

91 **92**

AuCl(PPh_3) (5 mol%)
AgOTf (5 mol%)
CH_2Cl_2, 40 °C, 8 h

93

R^1	R^2	Yield (%) of **92**	Yield (%) of **93**	Ref
Ph	Ph	50	65	[16]
4-Tol	Ph	53	74	[16]
3-BnC_6H_4	Ph	40	60	[16]
Ph	4-ClC_6H_4	34	77	[16]
Ph	4-$MeOC_6H_4$	40	85	[16]
$(CH_2)_6Me$	Ph	32	52	[16]

The proposed mechanism is outlined in Scheme 20. Upon treatment of **91** with silver(I) trifluoromethanesulfonate, allenylic cation **94** is formed. Subsequent intramolecular nucleophilic attack from the double bond of the methylenecyclopropane gives cyclopropyl cation **95**, which then undergoes cation-induced ring expansion to deliver cyclobutane intermediate **96**. Once the reaction is quenched by ambient water in the reaction system, allenylcyclobutanol **92** is formed. With regard to gold catalysis, the allene moiety of intermediate **92** is further activated by the gold complex. A Wagner–Meerwein-type rearrangement of **97** is then trigged by release of ring strain to afford vinylgold ketone **98**. Finally, deprotonation and protodeauration give rise to 1-vinyl-3-oxabicyclo[3.2.1]octan-8-one **93**.

Scheme 20 Proposed Mechanism for the Gold(I)- and Silver(I)-Catalyzed Cycloisomerization of Propargylic Alcohol Tethered Methylenecyclopropanes[16]

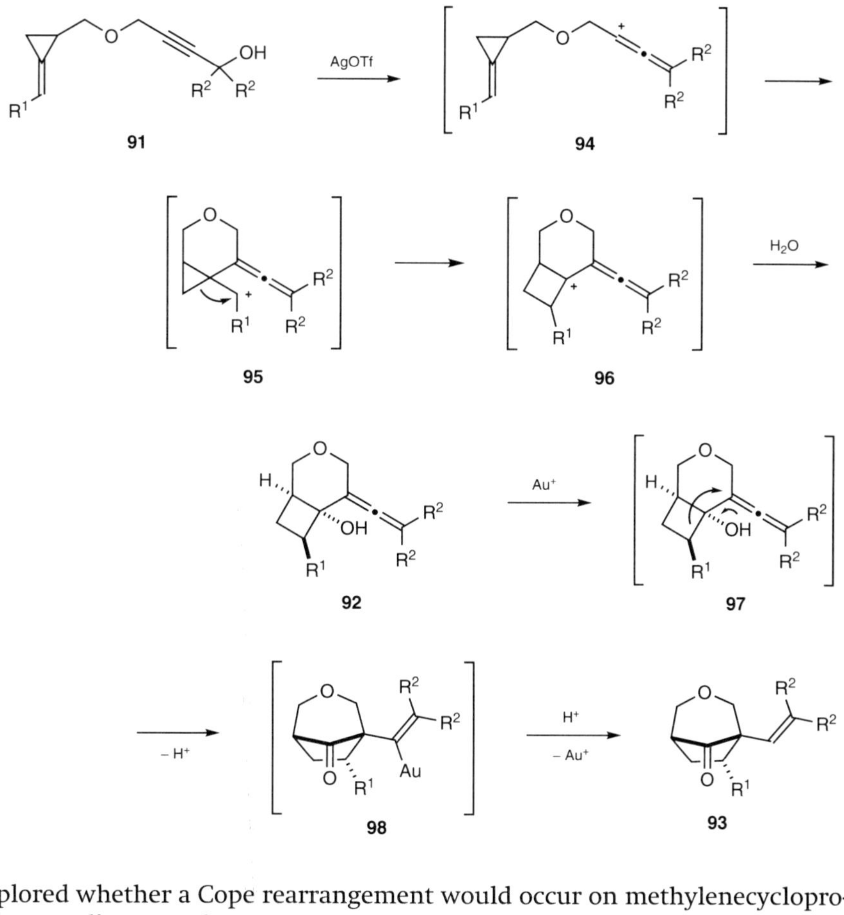

Gagné has explored whether a Cope rearrangement would occur on methylenecyclopropane-containing 1,5-dienes such as **99**. However, the desired reaction path is not followed, and tricyclic compounds **100**, which feature a bicyclo[4.2.0]oct-1-ene core, are obtained (Scheme 21).[17] The only drawback of this method is that it suffers from a very narrow substrate scope, as it operates only if R^1 is a methyl or phenyl group.

for references see p 256

Scheme 21 Unprecedented Rearrangement of Methylenecyclopropane-Containing 1,5-Dienes[17]

$Au(NTf_2)(PPh_3)$ (10 mol%)
CH_2Cl_2

99 **100**

R^1	Yield (%)	Ref
Me	76	[17]
Ph	88	[17]

Recently, the gold-catalyzed diastereoselective cycloisomerization of methylenecyclopropane-bearing 1,6-diynes **101** was developed by Gagné (Scheme 22).[18] This tandem rearrangement enables rapid access to a variety of fused norbornanes **102**, which are useful building blocks for the synthesis of abiotic and sesquiterpene core skeletons.

Scheme 22 Fused Norbornanes by Cycloisomerization of Methylenecyclopropane-Bearing 1,6-Diynes[18]

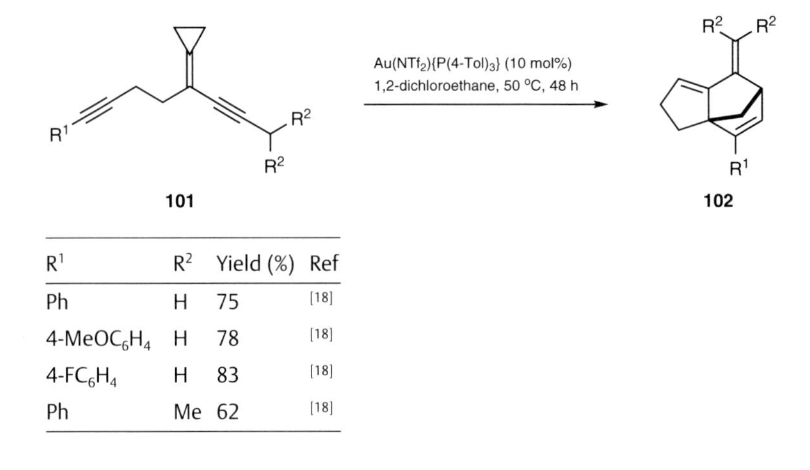

R^1	R^2	Yield (%)	Ref
Ph	H	75	[18]
4-$MeOC_6H_4$	H	78	[18]
4-FC_6H_4	H	83	[18]
Ph	Me	62	[18]

The mechanism of this cycloisomerization is complicated, but has been probed by ^{13}C-labeling experiments. The 6-*endo-dig* cyclization of 1,6-diyne **101** gives rise to cationic intermediate **103**, which initiates ring expansion to generate diene intermediate **104**. Then, the remaining alkyne is activated by gold to facilitate intramolecular nucleophilic attack of the diene unit to yield bridged carbocycle **105**. Subsequent 1,2-alkyl migration and proton elimination deliver intermediate **106**, which undergoes tandem protonation and ring expansion of the cyclobutane ring to produce intermediate **107**. Final product **102** is obtained after regeneration of the gold catalyst (Scheme 23).

Scheme 23 Proposed Mechanism for the Cycloisomerization of Methylenecyclopropane-Bearing 1,6-Diynes[18]

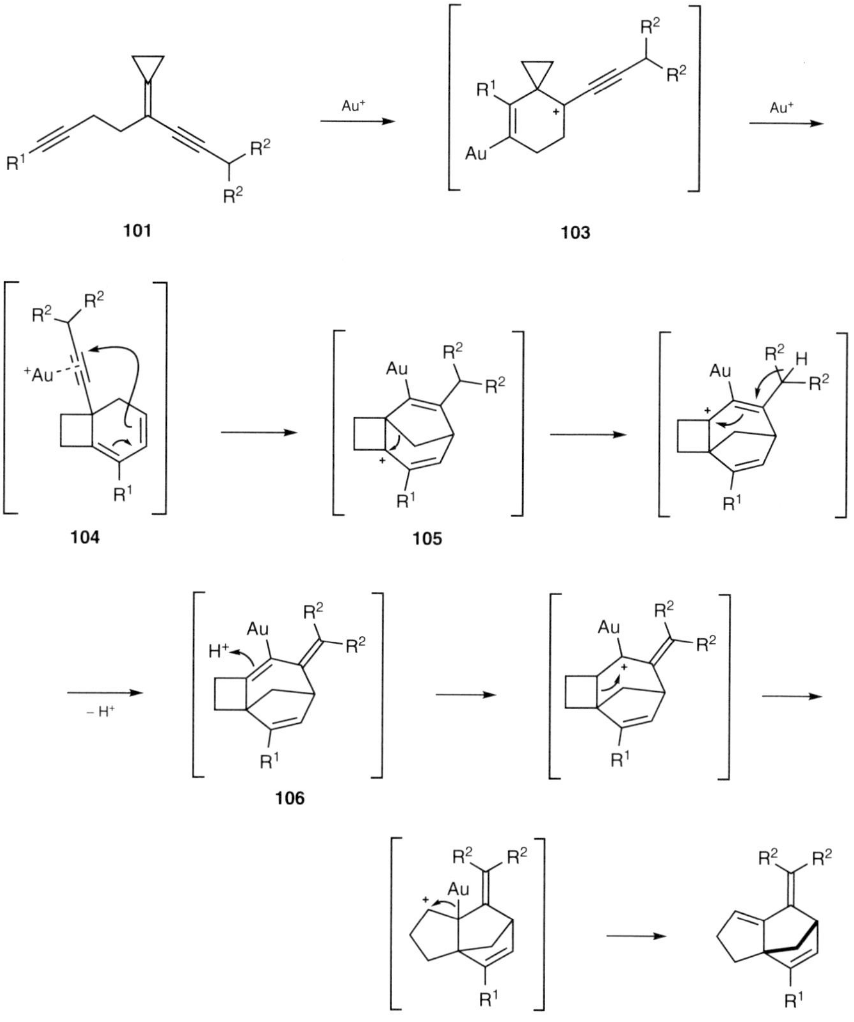

O-{[2-(Arylmethylene)cyclopropyl]methyl}hydroxylamines **108**, as demonstrated by Shi, can undergo a ring-opening reaction in the presence of a gold catalyst to furnish a variety of 4-substituted isoxazolidine derivatives **109** in good to high yields with high regioselectivities (Scheme 24).[19]

Scheme 24 Isoxazolidines by Ring-Opening Hydroamination of *O*-{[2-(Arylmethylene)cyclopropyl]methyl}hydroxylamines[19]

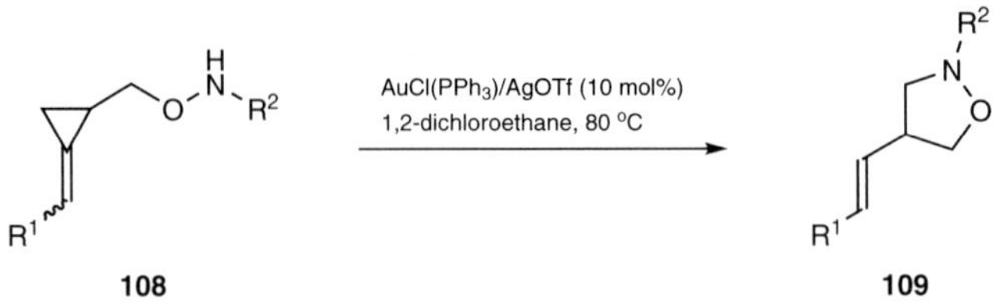

R^1	R^2	Yield (%)	Ref
Ph	Ts	79	[19]
$4\text{-}ClC_6H_4$	Ts	86	[19]
$4\text{-}MeOC_6H_4$	Ts	67	[19]
$4\text{-}BrC_6H_4$	$4\text{-}O_2NC_6H_4SO_2$	67	[19]

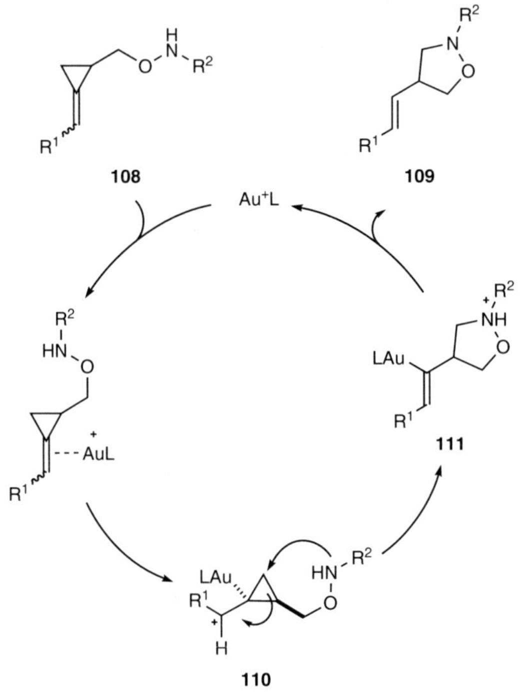

A plausible mechanism for the formation of five-membered N,O-heterocycle **109** is also depicted in Scheme 24: The cationic gold(I) complex first coordinates to the alkene moiety of **108** to produce carbocationic intermediate **110**. Subsequent nucleophilic attack from the nitrogen group results in the formation of more stable *E*-intermediate **111**, which gives rise to **109** after protodeauration.

Fensterbank and Malacria have reported an interesting structure-dependent rearrangement of methylenecyclopropanes **112** (Scheme 25).[20] The length of the carbon chain plays an important role in controlling the reaction pathway in the presence of a gold catalyst. If n = 1, electron depletion of the methylenecyclopropane moiety induced by gold facilitates elimination of an allylic proton, and this is followed by tandem rearrangement to form more stable α,β,γ,δ-unsaturated ketone **113** (distal bond cleavage) driven by relief of the ring strain. With a longer tether (n = 2 or 3), the hydroxy group attacks the double bond or opens the cyclopropane ring (proximal bond cleavage), which might be faster than proton elimination, and this results in tricyclic ketone **114** or spiro compound **115**, respectively.

Scheme 25 Gold(I)-Catalyzed Cycloisomerization of Hydroxyalkyl-Tethered Methylenecyclopropanes[20]

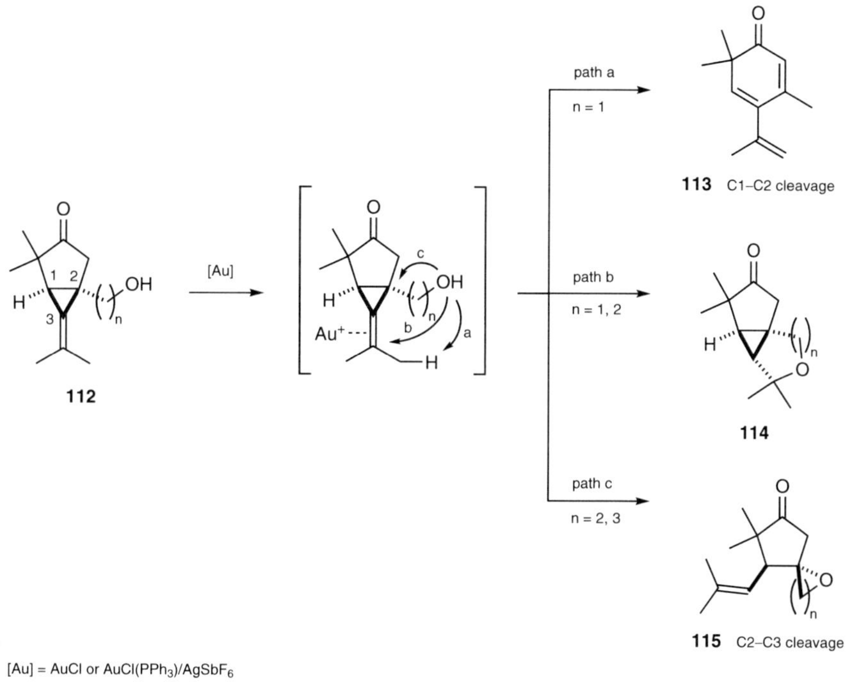

[Au] = AuCl or AuCl(PPh$_3$)/AgSbF$_6$

Dienes 87/88 **or Pyrans** 90**; General Procedure:**[15]
A 0.45 M soln of the substrate in toluene was added to a mixture of 0.25 M AuCl(PPh$_3$) and AgOTf in toluene. The mixture was stirred at rt for 1 h and was then purified directly by chromatography (silica gel, hexanes/EtOAc 9:1) to afford the title compound.

Fused Tetracycles 89**; General Procedure:**[15]
A 0.45 M soln of the substrate in CH_2Cl_2 was added to a mixture of 0.25 M AuCl(PPh$_3$) and AgSbF$_6$ in CH_2Cl_2. The mixture was stirred at rt until the reaction was complete (about 18 h), and then the mixture was purified by chromatography (silica gel, hexanes/EtOAc 9:1) to afford the title compound.

(1*S,6*R**,7*R**)-5-(Diphenylvinylidene)-7-phenyl-3-oxabicyclo[4.2.0]octan-6-ol** (92, **R^1 = R^2 = Ph); Typical Procedure:**[16]
(*E*)-4-[(2-Benzylidenecyclopropyl)methoxy]-1,1-diphenylbut-2-yn-1-ol (**91**, R^1 = R^2 = Ph; 76 mg, 0.2 mmol) and AgOTf (3 mg, 5 mol%) were dissolved in CH_2Cl_2 (2.0 mL). H_2O (1.8 mg, 0.1 mmol) was then added to this mixture by using a 25-μL microsyringe. The mixture was stirred at rt (25 °C) for 6 h. The solvent was removed under reduced pressure, and the residue was purified by flash column chromatography (silica gel, petroleum ether/EtOAc 4:1) to give the title compound as a light yellow oil; yield: 38 mg (50%).

for references see p 256

(1*S,5*R**,7*R**)-7-Phenyl-1-(2,2-diphenylethenyl)-3-oxabicyclo[3.2.1]octan-8-one** (93, $R^1 = R^2 =$ Ph)**; Typical Procedure:**[16]
Compound **92** (76 mg, 0.2 mmol) was added to a stirring suspension of $AuCl(PPh_3)$ (5 mg, 5 mol%) and AgOTf (3 mg, 5 mol%) in CH_2Cl_2 (2 mL), and the mixture was stirred for 8 h at 40 °C. After removal of the solvent under reduced pressure, purification by flash chromatography (silica gel, EtOAc/petroleum ether 1:9) afforded the title compound as a colorless crystalline solid; yield: 50 mg (65%).

6,7-Dimethyltricyclo[5.3.1.0^{1,4}]undec-4-ene (100, R^1 = Me)**; Typical Procedure:**[17]
A 1-dram vial equipped with a stirrer bar was charged with $Au(NTf_2)(PPh_3)$ (0.011 g, 0.0142 mmol, 0.10 equiv) followed by CH_2Cl_2 (0.5 mL). The mixture was stirred briefly before **99** (0.025 g, 0.142 mmol, 1.00 equiv) was added. The mixture was then stirred for 12 h and then concentrated under reduced pressure. The crude product was dissolved in a small amount of CH_2Cl_2 and was then purified by chromatography by using a pipet (silica gel, hexanes) to afford the title compound as a colorless oil; yield: 0.019 g (76%).

Fused Norbornenes 102**; General Procedure:**[18]
$Au(NTf_2)\{P(4\text{-Tol})_3\}$ (7.8 mg, 0.01 mmol) was added to a soln of 1,6-diyne **101** (0.1 mmol) in 1,2-dichloroethane (1.0 mL) at rt. The resulting soln was stirred at 50 °C for 48 h. Upon evaporation of the solvent under reduced pressure, the residue was purified by column chromatography (silica gel, hexanes) to afford the title compound.

4-Vinylisoxazolidines 109**; General Procedure:**[19]
Under an argon atmosphere, sulfonamide-substituted hydroxylamine **108** (0.2 mmol, 1.0 equiv) was dissolved in 1,2-dichloroethane (2.0 mL) in a Schlenk tube, and $AuCl(PPh_3)$/AgOTf (10 mol%) were added. Then, the mixture was stirred at 80 °C until the reaction was complete. The solvent was removed under reduced pressure, and the residue was purified by flash column chromatography (silica gel) to give the title compound.

Ketones 113–115**; General Procedure:**[20]
A soln of methylenecyclopropane **112** (0.45 mmol) in CH_2Cl_2 (9 mL) was added to a soln of $AuCl(PPh_3)$ (2 mol%) and $AgSbF_6$ (2 mol%) in CH_2Cl_2. The mixture was stirred under reflux for 2 h. After cooling to rt, the mixture was filtered through a short pad of silica gel and concentrated under reduced pressure to give the title compounds.

1.4.1.2 Donor–Acceptor Cyclopropanes

Scientists have paid much attention to the research of intermolecular annulations of donor–acceptor cyclopropanes for many decades. Whereas the intramolecular versions of such transformations did not initially draw much attention, this field has developed very quickly over recent years. In principle, intramolecular cyclization allows the construction of increased molecular complexity owing to the ability to form polycyclic derivatives in a single operation.

1.4.1.2.1 Metal Trifluoromethanesulfonate Catalyzed Cycloisomerization of Donor–Acceptor Cyclopropanes

In 2008, Kerr realized the stereodivergent synthesis of complex pyrrolidines through the intramolecular (3+2) cycloaddition of (*E*)-oxime-containing cyclopropane 1,1-diesters **116** (Scheme 26).[21] In most cases, the reactions give bicyclic products **118** as single diastereomers (*trans*) in high yields. An intramolecular ring-opening reaction gives zwitterionic intermediate **117**, and a subsequent Mannich-type reaction results in ring closure. Interestingly, if (Z)-oxime diester **119** is treated under the standard reaction conditions, *cis*-diastereomer **120** is formed exclusively.

Scheme 26 Pyrrolo[1,2-*b*]isoxazolidines by Intramolecular Cycloaddition of Oximes with Cyclopropane 1,1-Diesters[21]

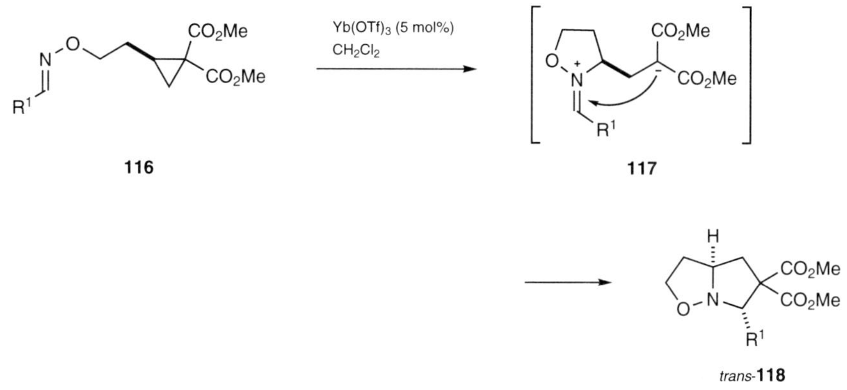

R^1	Yield (%)	Ref
Ph	98	[21]
(*E*)-CMe=CHPh	85	[21]
t-Bu	75	[21]

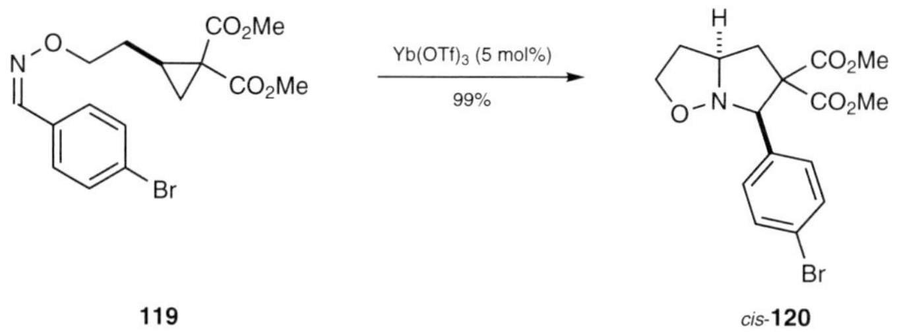

Kerr realized the homodipolar cycloaddition of nitrones with donor–acceptor cyclopropanes for the efficient synthesis of bridged tetrahydro-1,2-oxazines.[22] From cyclopropyl aldehyde **121**, the reaction with a hydroxylamine delivers cyclopropane-containing nitrones **122**. The authors envisage that the in situ formed nitrone derived from the aldehyde and the hydroxylamine might initiate nucleophilic ring opening in the presence of ytterbium(III) trifluoromethanesulfonate to provide intermediate **123**; a subsequent Mannich-type reaction furnishes the bridged heterocycles **124** (Scheme 27). The generality of this method is broad. For various carbon-chain- and aryl-ring-tethered substrates, the reactions proceed smoothly to give the desired products **125–129** in good to excellent yields.

for references see p 256

Scheme 27 Bridged Tetrahydro-1,2-oxazines by Intramolecular Cycloaddition of Nitrones with Cyclopropane 1,1-Diesters[22]

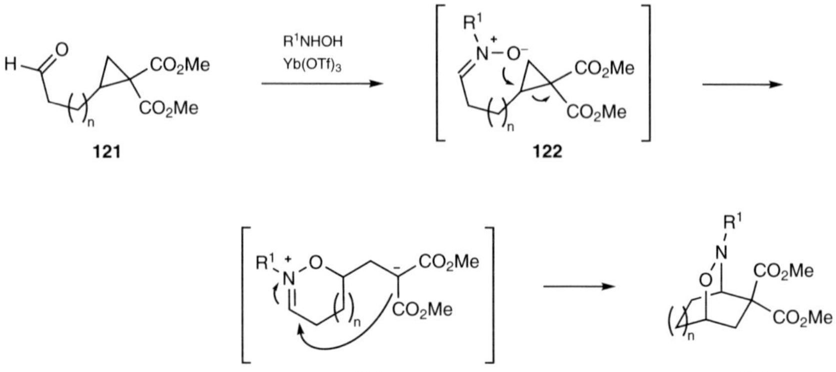

R^1	Yield (%)	Ref
Ph	92	[22]
4-Tol	92	[22]
Bn	89	[22]
PMB	84	[22]
$(CH_2)_6Me$	82	[22]

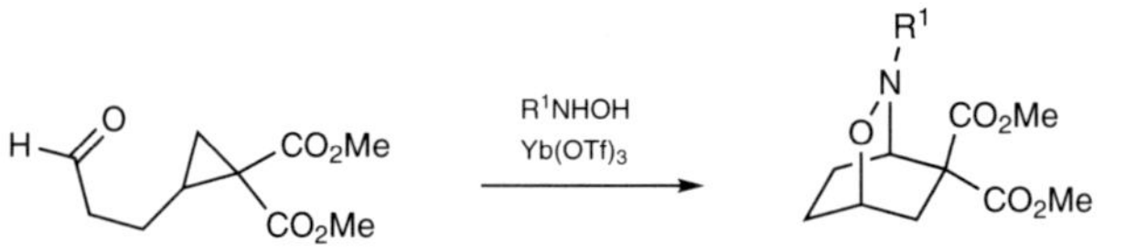

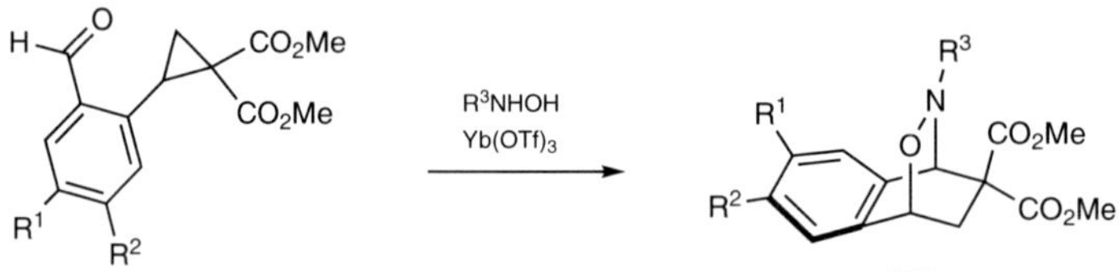

R^1	R^2	R^3	Yield (%)	Ref
H	H	Bn	80	[22]
H	H	PMB	85	[22]
H	H	Me	83	[22]
H	H	Ph	89	[22]
H	H	$(CH_2)_6Me$	91	[22]
NO_2	H	PMB	98	[22]
H	NO_2	PMB	91	[22]

PMBNHOH, Yb(OTf)3, 63% → 127

PMBNHOH, Yb(OTf)3, 89% → 128

PMBNHOH, Yb(OTf)3, 62% → 129

Ns = 4-nitrophenylsulfonyl

Wang discovered a new type of (3 + 2)-cycloaddition reaction of cyclopropane 1,1-diesters in 2010.[23] Previously reported examples had all been related to the intramolecular parallel-cycloaddition reaction to construct bicyclic systems (Scheme 28), as demonstrated by Kerr[21] and Snider.[24] In this new strategy, termed the intramolecular cross-cycloaddition reaction, the electronic properties of the dipolarophile are reversed, and this results in a different route through which the cycloaddition proceeds to afford bridged bicyclic derivatives (Scheme 28).

Scheme 28 Two Types of Intramolecular (3 + 2) Cycloadditions of Cyclopropane 1,1-Diesters[21,23,24]

intramolecular parallel-cycloaddition reaction

Lewis acid

X = N; Y = C

intramolecular cross-cycloaddition reaction

Lewis acid

X = C; Y = N, O

Study of the substrate scope revealed that a variety of functional groups are tolerated if ytterbium(III) trifluoromethanesulfonate is used as the catalyst. As shown in Scheme 29, aryl-ring-tethered substrates give the corresponding products **130** and **131** in moderate to excellent yields, whereas alkyl-tethered substrates give products **132** in varying yields depending on the length of the tether. For indole and pyrrole tethers, the corresponding oxygen-bridged products **133** and **134** are formed in high yields, and even if a substrate with a longer chain tether is used, eight-membered ring product **135** is still delivered in 42% yield.

The reaction can also be extended for the synthesis of aza-bridged bicycles. For various imines derived from aldehyde **136** and amines, the reactions all proceed smoothly to give the corresponding 8-azabicyclo[3.2.1]octanes **137**. The challenge of the desired reaction is the direct intramolecular reaction of the aldehyde with the cyclopropane. By adding 4-Å molecular sieves to remove water and by performing the reaction in toluene instead of 1,2-dichloroethane, the side reaction can be suppressed and **137** is formed as the sole product. Reactions of other imines such as **139** and **142** derived from the condensation of aldehydes **138** and **141** with *p*-toluidine successfully afford 9-azabicyclo[4.2.1]nonane **140** and diaza[3.2.1]octane **143** in yields of 55 and 61%, respectively (Scheme 29). Notably, this methodology has been successfully utilized in the formal total synthesis of platensimycin and also for the construction of acetal [n.2.1] skeletons.[25]

Scheme 29 Intramolecular (3 + 2) Cross-Cycloadditions of Cyclopropane 1,1-Diesters To Give Bridged [n.2.1] Heterocycles[23]

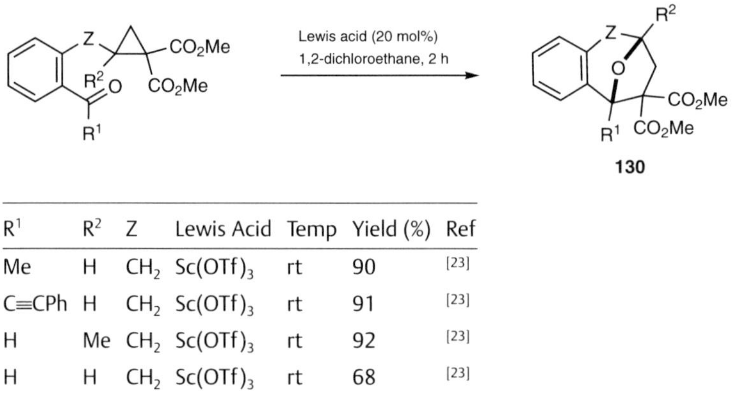

Z, CO_2Me, R^2, O, CO_2Me, R^1 → Lewis acid (20 mol%), 1,2-dichloroethane, 2 h → R^2, Z, O, CO_2Me, R^1, CO_2Me

130

R1	R2	Z	Lewis Acid	Temp	Yield (%)	Ref
Me	H	CH_2	$Sc(OTf)_3$	rt	90	[23]
C≡CPh	H	CH_2	$Sc(OTf)_3$	rt	91	[23]
H	Me	CH_2	$Sc(OTf)_3$	rt	92	[23]
H	H	CH_2	$Sc(OTf)_3$	rt	68	[23]
H	H	O	$SnCl_4$	60°C	75	[23]
Me	H	O	$SnCl_4$	60°C	35	[23]
H	H	S	$SnCl_4$	60°C	85	[23]

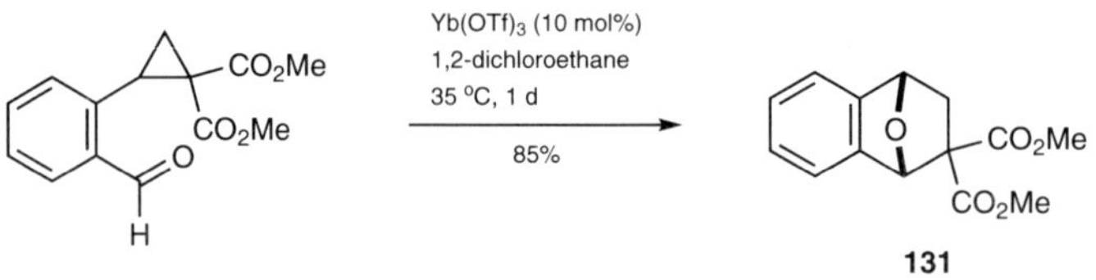

CO_2Me, CO_2Me, O, H → $Yb(OTf)_3$ (10 mol%), 1,2-dichloroethane, 35 °C, 1 d; 85% → O, CO_2Me, CO_2Me

131

Yb(OTf)$_3$ (10 mol%)
1,2-dichloroethane
35 °C, 1 d

132

n	Yield (%)	Ref
2	47	[23]
1	27	[23]

Sc(OTf)$_3$ (20 mol%)
1,2-dichloroethane
rt, 2 h
87%

133

Sc(OTf)$_3$ (20 mol%)
1,2-dichloroethane
rt, 2 h
90%

134

$SnCl_4$ (20 mol%)
1,2-dichloroethane
60 °C, 2 h
42%

135

136 + R^1NH_2

Sc(OTf)$_3$ (10 mol%)
4-Å molecular sieves
toluene

137

R^1	Yield (%)	Ref
4-BrC$_6$H$_4$	80	[23]
4-Tol	75	[23]
Bn	79	[23]

for references see p 256

$SnCl_4$ (10 mol%)
4-Å molecular sieves
toluene

138 + Ar^1NH_2 → [139] → 140 (55%)

Ar^1 = 4-Tol

$Sc(OTf)_3$ (10 mol%)
4-Å molecular sieves
toluene

141 + Ar^1NH_2 → [142] → 143 (61%)

Ar^1 = 4-Tol

An intramolecular (3 + 2)-cross-cycloaddition reaction strategy using alkene–cyclopropane substrates to synthesize bridged [n.2.1] carbocycles has been realized by Wang.[26] Previously, the cycloadditions between cyclopropane and alkenes were restricted to transition-metal-catalyzed reactions. Wang has reported the first examples of such reactions catalyzed by a Lewis acid. The reaction of substrates **144** gives rise to desired cycloadducts **145**, whereas byproducts **146**, which are probably derived from proton elimination, are also formed in some cases. Compounds **146** cannot be converted into **145** under the standard reaction conditions, which indicates that they are not an intermediate for the synthesis of **145** (Scheme 30). By catalyzing the reaction with scandium(III) trifluoromethanesulfonate in 1,2-dichloroethane, the formation of byproducts can be suppressed in most cases. The substrate scope of this method is broad and various benzo-fused [n.2.1] carbocycles **145** and **147–150** are obtained in moderate to good yields (Scheme 30).

Scheme 30 Bridged [n.2.1] Carbocycles by Intramolecular (3 + 2)-Cross-Cycloaddition Reaction and Ring Opening of Alkene–Cyclopropanes[26]

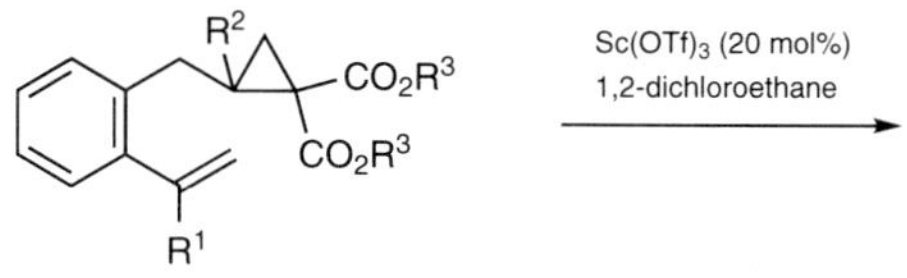

Sc(OTf)$_3$ (20 mol%)
1,2-dichloroethane

144 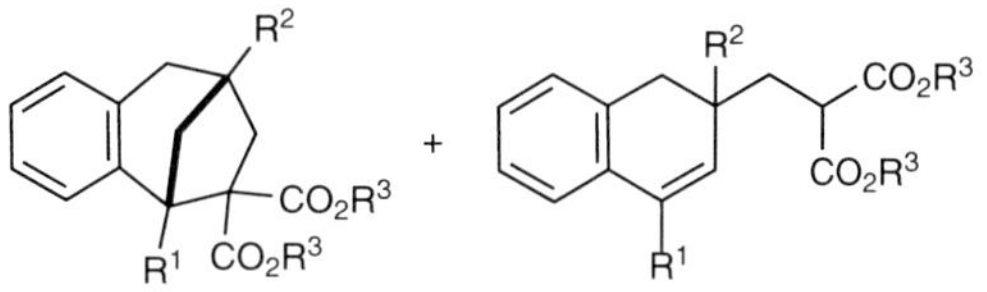 → **145** + **146**

R^1	R^2	R^3	Yield (%)		Ref
			145	**146**	
H	H	Me	78	15	[26]
H	H	Et	70	–	[26]
Me	H	Me	82[a]	14	[26]
H	Me	Me	78	–	[26]
C≡CPh	H	Me	71	16	[26]

[a] 30 mol% of catalyst was used.

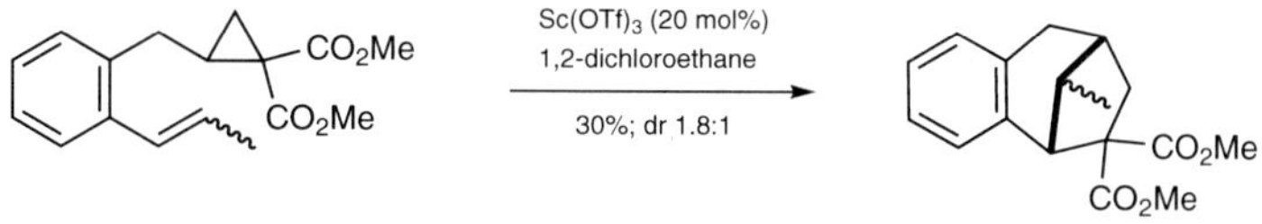

(Z/E) 3:1 → Sc(OTf)$_3$ (20 mol%), 1,2-dichloroethane; 30%; dr 1.8:1 → **147**

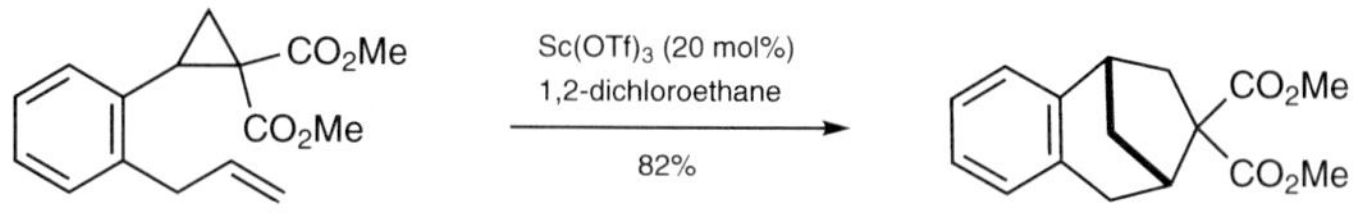

Sc(OTf)$_3$ (20 mol%), 1,2-dichloroethane; 82% → **148**

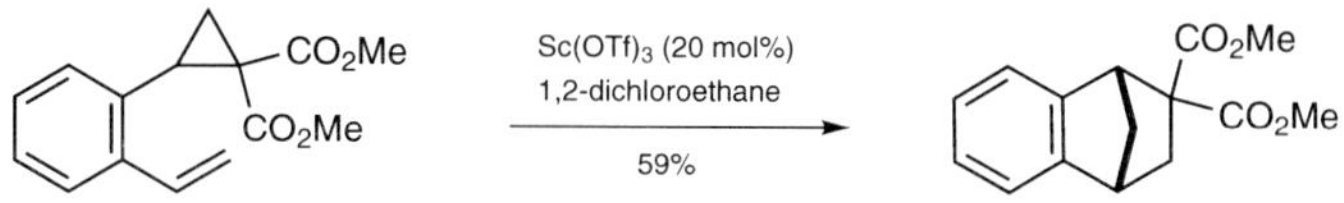

Sc(OTf)$_3$ (20 mol%), 1,2-dichloroethane; 59% → **149**

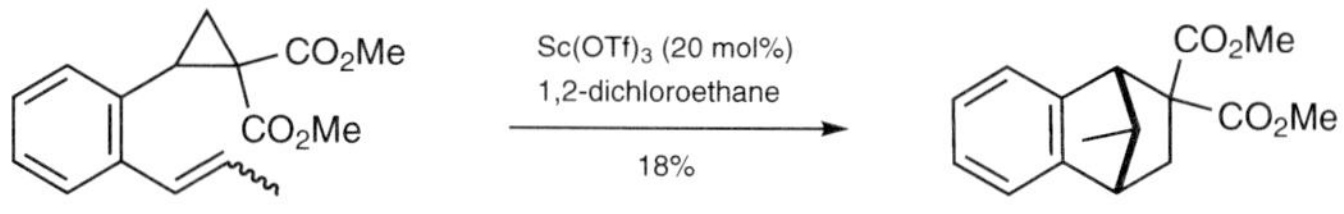

(Z/E) 2.7:1 → Sc(OTf)$_3$ (20 mol%), 1,2-dichloroethane; 18% → **150**

Saturated acyclic and cyclic carbon-chain-tethered substrates **151** are less reactive (Table 1), presumably because the benzene ring can stabilize the cationic intermediate that is formed in the reaction. In the case of entry 6, the product is formed in only 36% yield. Fortunately, the cation can also be stabilized by alkenes, whereas the alkyl chain cannot. Therefore, for most diene-containing substrates, the corresponding products (entries 1–5) are obtained in relatively high yields. Such scaffolds exist in a number of natural products. To further demonstrate the potential of this strategy, it has been applied successfully as the key step in the total synthesis of the natural products (±)-phyllocladanol and (±)-phyllocladene.

Table 1 Synthesis of Other [n.2.1] Carbocycles toward Natural Product Synthesis[26]

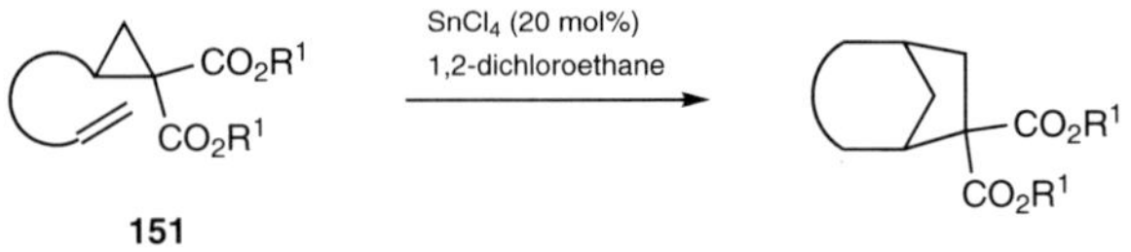

CO_2R^1 CO_2R^1 **151** → ($SnCl_4$ (20 mol%), 1,2-dichloroethane) → CO_2R^1 CO_2R^1

Entry	Substrate	Product	Yield (%)	Ref
1	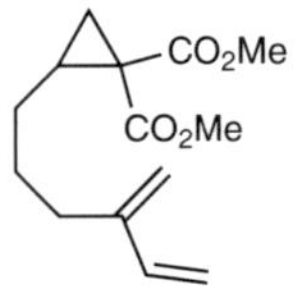	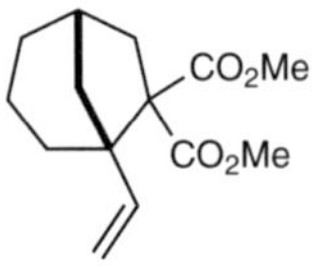	77[a]	[26]
2	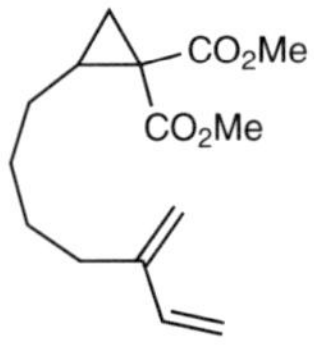	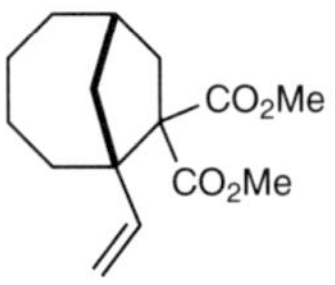	30	[26]
3	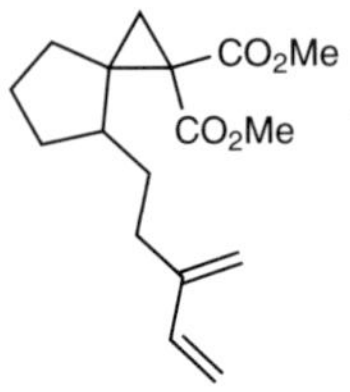	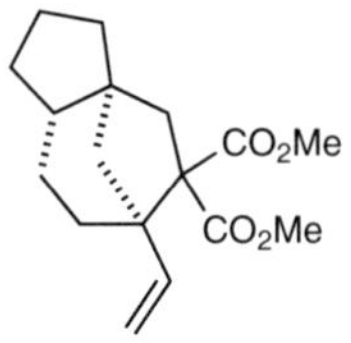	82	[26]

Table 1 (cont.)

Entry	Substrate	Product	Yield (%)	Ref
4	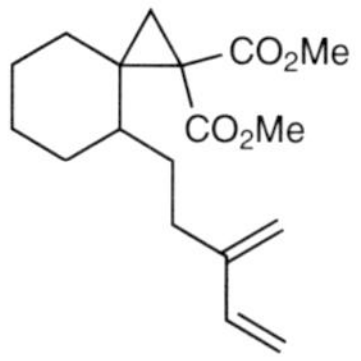 CO_2Me, CO_2Me	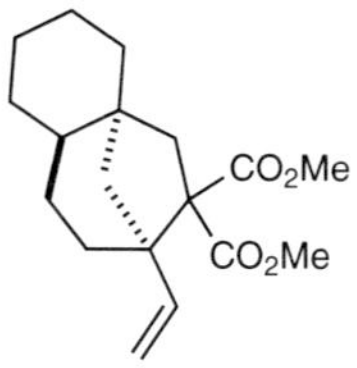 CO_2Me, CO_2Me	74	[26]
5	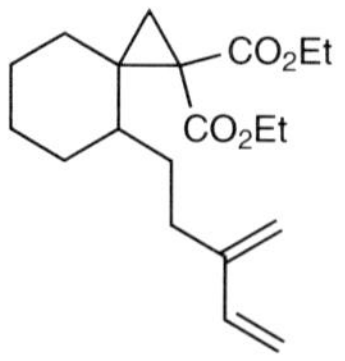 CO_2Et, CO_2Et	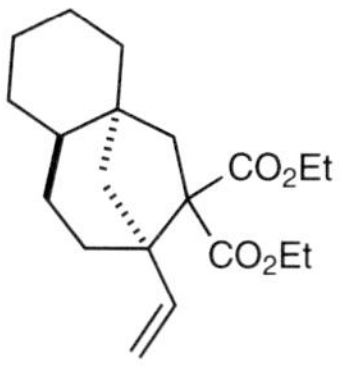 CO_2Et, CO_2Et	50	[26]
6	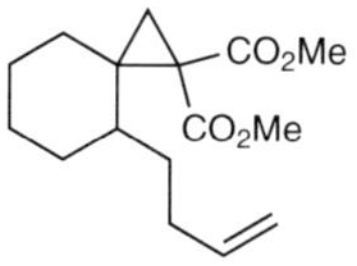 CO_2Me, CO_2Me	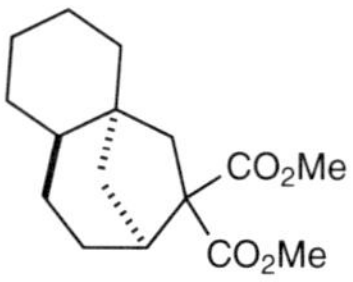 CO_2Me, CO_2Me	36	[26]

[a] $Sc(OTf)_3$ (20 mol%) was used as catalyst.

The catalyst-dependent divergent synthesis of [3.2.1]octanes and [4.3.0]nonanes can be achieved through the intramolecular (3 + 2)-cross-cycloaddition reaction and the intramolecular (3 + 2)-parallel-cycloaddition reaction of cyclopropane 1,1-diesters with allenes.[27] The internal and terminal double bonds of allenes **152** selectively take part in the cycloaddition with the cyclopropane ring, depending on the Lewis acid used. Scandium(III) trifluoromethanesulfonate facilitates the intramolecular (3+2)-parallel-cycloaddition reaction, and the corresponding cycloadducts **153** are afforded in yields of 24–98%. The reaction preferentially proceeds through an intramolecular (3+2) cross-cycloaddition if promoted by ytterbium(III) trifluoromethanesulfonate (1.0 equiv) and [3.2.1]octanes **154** are obtained in yields of 76–91% in addition to the [4.3.0]nonanes (Scheme 31).

Scheme 31 Divergent Synthesis of [4.3.0]Nonanes and [3.2.1]Octanes from Cyclopropane 1,1-Diesters[27]

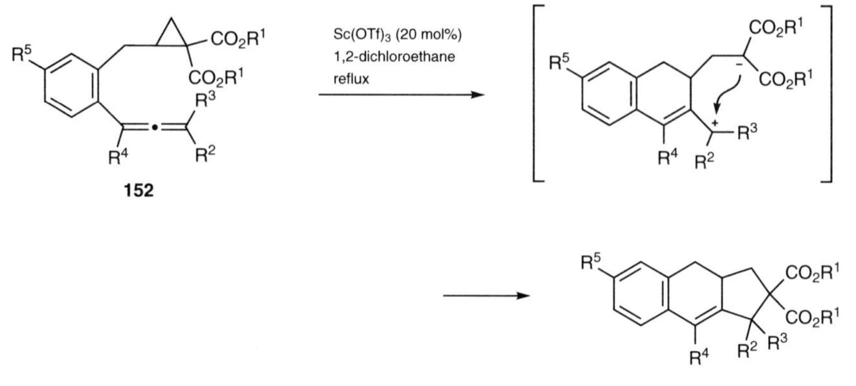

$Sc(OTf)_3$ (20 mol%), 1,2-dichloroethane, reflux

152 → 153

R^1	R^2	R^3	R^4	R^5	Yield (%)	Ref
Me	H	H	H	H	81	[27]
Et	H	H	H	H	78	[27]
Me	Me	Me	H	H	98	[27]
Me	Ph	Me	H	H	24[a]	[27]
Me	H	H	Me	H	95	[27]
Me	H	H	H	OMe	78	[27]

[a] dr 1.4:1; the product of cross-cycloaddition is also produced in 66% yield [ratio (*Z*/*E*) 1:1.3].

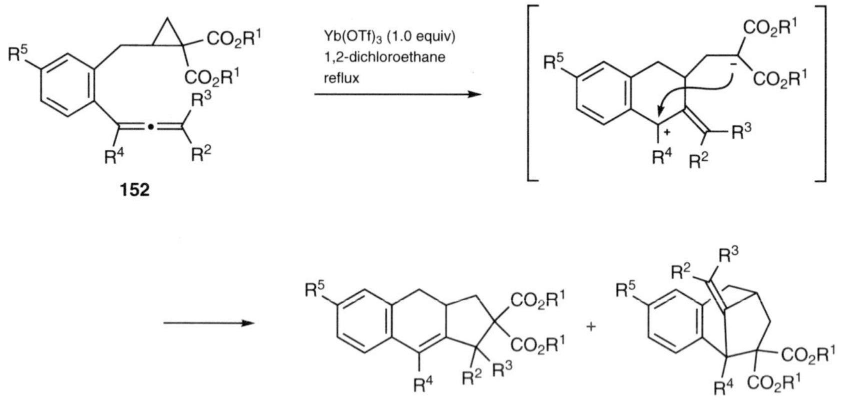

$Yb(OTf)_3$ (1.0 equiv), 1,2-dichloroethane, reflux

152 → 153 + 154

R^1	R^2	R^3	R^4	R^5	Ratio (**153**/**154**)	Overall Yield (%)	Ref
Me	H	H	H	H	1:2.2	80	[27]
Et	H	H	H	H	1:2.1	76	[27]
Me	Me	Me	H	H	1:0.76[a]	93	[27]

R^1	R^2	R^3	R^4	R^5	Ratio (**153**/**154**)	Overall Yield (%)	Ref
Me	Ph	Me	H	H	1:1.4	87[b]	[27]
Me	H	H	Me	H	1:2[c]	93	[27]
Me	H	H	H	OMe	1.7:1	–[d]	[27]

[a] Reaction was performed at 45 °C.
[b] Ratio (*Z*/*E*) for **154** was 1:1.5.
[c] Reaction was performed at 50–55 °C.
[d] Reaction was performed at 60–65 °C; the reaction was very slow, and the product was not isolated.

Alkynes can also be used in intramolecular (3 + 2) cycloadditions with donor–acceptor cyclopropanes.[28] For example, Liang and co-workers have reported the synthesis of cyclopenta[*c*][1]benzopyrans **156** from cyclopropyl diesters **155** in moderate to excellent yields. A substrate linked with an *N*-tosyl group (**157**) affords the corresponding 1*H*-cyclopenta[*c*]quinoline product **158** in 83% yield (Scheme 32). A variety of aryl substituents are tolerated on the alkyne; however, reactions of aliphatic and terminal alkynes are not successful.

Scheme 32 Intramolecular Parallel Cycloaddition of Alkyne-Tethered Cyclopropanes To Give Cyclopenta[*c*][1]benzopyrans and 1*H*-Cyclopenta[*c*]quinolines[28]

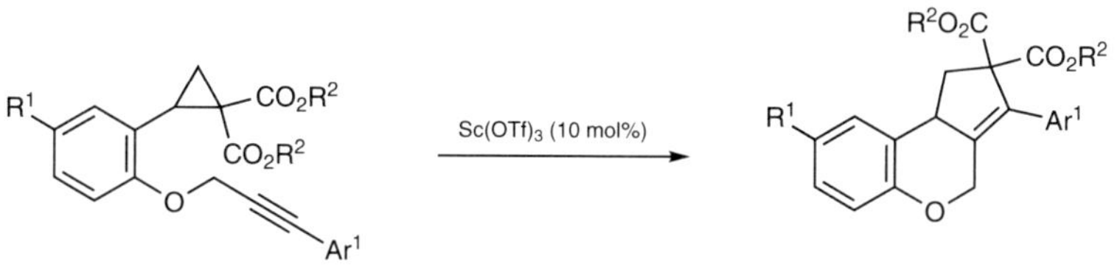

R^1 CO_2R^2 CO_2R^2 O Ar^1 **155** — $Sc(OTf)_3$ (10 mol%) → R^2O_2C CO_2R^2 R^1 Ar^1 O **156**

R^1	R^2	Ar^1	Yield (%)	Ref
Me	Me	Ph	69	[28]
Cl	Et	Ph	65	[28]
H	Me	2-thienyl	77	[28]

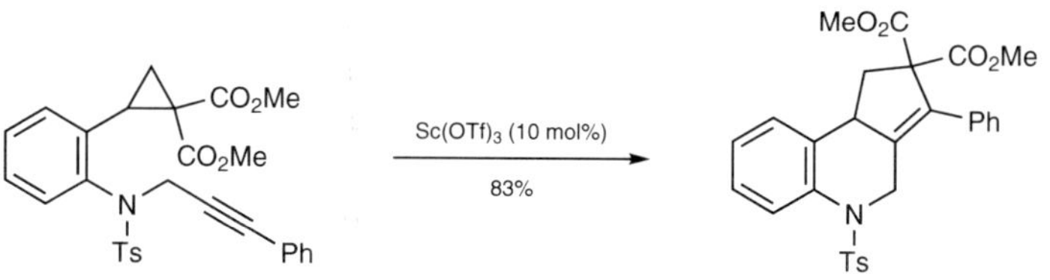

CO_2Me CO_2Me N Ts Ph **157** — $Sc(OTf)_3$ (10 mol%), 83% → MeO_2C CO_2Me Ph N Ts **158**

In 2010, France reported an efficient homo-Nazarov cyclization of alkenyl cyclopropyl ketones catalyzed by indium(III) trifluoromethanesulfonate[29] to afford six-membered carbocycles, instead of the five-membered rings derived from traditional Nazarov cyclizations. Cation **160** is proposed as the key intermediate, and it is generated from a tandem ring-opening/recyclization process of cyclopropane β-oxo esters **159**. Subsequent β-hydride elimination of proton H^a or H^b gives two different carbocycles (Scheme 33). In some cases, both products **161** (elimination of H^a) and **162** (elimination of H^b) can be obtained in high total yields. Notably, for cyclopropane β-oxo ester **159** in which the Ar[1] group is 4-FC_6H_4, **161** (Ar[1] = 4-FC_6H_4) is obtained exclusively; similarly, for a pyran-substituted substrate, **163** is obtained exclusively (via loss of H^b).

Scheme 33 Indium-Catalyzed Homo-Nazarov Cyclizations of Cyclopropane β-Oxo Esters To Give Cyclohexanols and Cyclohexanones[29]

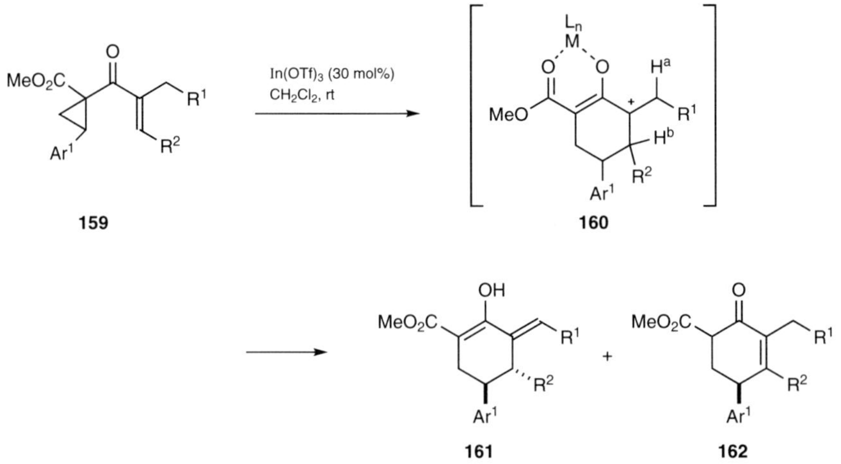

R[1]	R[2]	Ar[1]	Time	Yield (%) of **161**	Yield (%) of **162**	Ref
H	Me	4-$MeOC_6H_4$	3 h	45	30	[29]
$(CH_2)_3$		4-$MeOC_6H_4$	40 min	45	30	[29]
H	H	4-$MeOC_6H_4$	40 min	46	31	[29]
H	H	4-FC_6H_4	30 h	55	0	[29]

France also found that by replacing the alkene by an aromatic ring, such as in alkylidenecyclopropane β-oxo esters **164**, formal homo-Nazarov-type cyclization takes place for the facile construction of functionalized arenes and heteroaromatics (Scheme 34).[30] There are two pathways that can be followed for the ring-closing step, and they lead to two different isomers. In the presence of ytterbium(III) trifluoromethanesulfonate, the ring-opening reaction of **164** gives allylic cationic intermediate **165**. If nucleophilic attack to form the aromatic ring takes place at the carbon atom bearing the R[1] substituent, phenol **166** is formed (Scheme 34, path a). Reaction at the other carbon atom gives **167** (Scheme 34, path b). Generally, the R[1] substituent plays an important role in stabilizing the carbon cation, which accounts for the fact that **166** is obtained as the sole or major product.

Scheme 34 Formal Homo-Nazarov-Type Cyclizations of Alkylidene–Cyclopropanes[30]

Yb(OTf)$_3$ (15 mol%)

164 165 166 major 167 minor

R^1 = alkyl; R^2 = alkyl, aryl

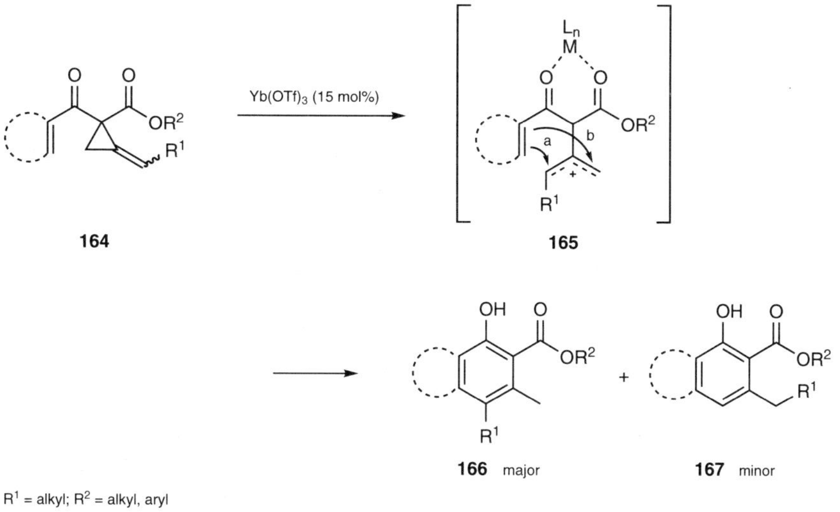

If methylenecyclopropanes such as **168** possessing two substituents at the terminal alkene position are treated with a Lewis acid, the outcome of the reaction is quite different from that shown in Scheme 34. In this scenario, the derived cationic intermediate undergoes Friedel–Crafts-type reaction to furnish product **169** in 63% yield. The aromatization process is blocked by the *gem*-dimethyl groups. Interestingly, in the case of **170**, in which the alkylidene is substituted by methyl and trimethylsilyl groups, the reaction presumably goes through a silyl-stabilized methyl allyl cationic intermediate; subsequent Friedel–Crafts-type cyclization and 1,3-silicon migration affords benzo[*b*]furan **171** in 68% yield (Scheme 35).

Scheme 35 Substituent Effects of the Alkylidene Group in Homo-Nazarov-Type Cyclizations of Alkylidene–Cyclopropanes[30]

Yb(OTf)$_3$ (15 mol%)

168

63%

169

for references see p 256

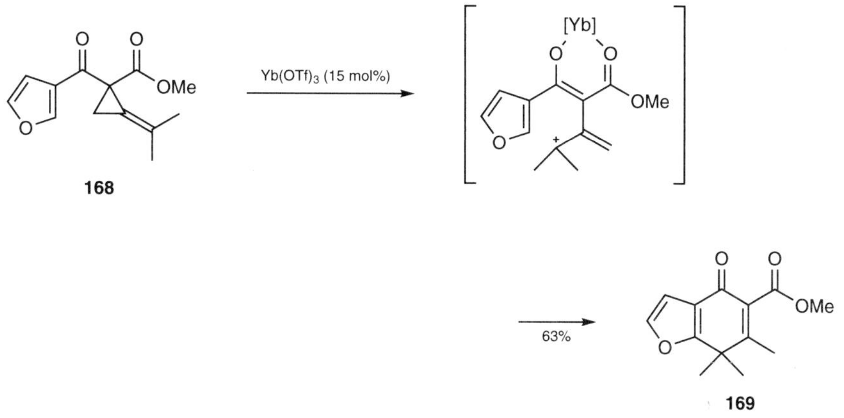

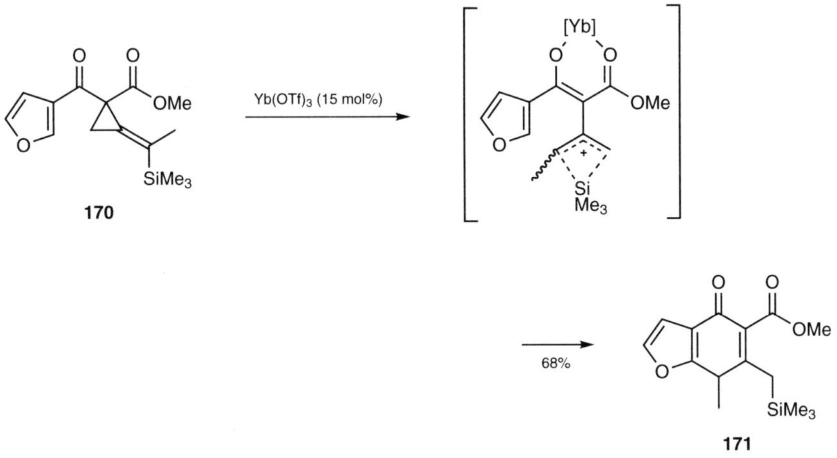

Yb(OTf)$_3$ (15 mol%)

170 → [intermediate] → 68% → 171

Moreover, France has also developed an efficient approach to indole-fused lactams **175** through a Lewis acid catalyzed tandem ring-opening/Friedel–Crafts alkylation sequence.[31] Upon treatment of cyclopropyl amido esters **172** with indium(III) trifluoromethanesulfonate (30 mol%), the corresponding products **175** are obtained in 31–99% yield. With the assistance of a Lewis acid, the ring-opening reaction of **172** gives cationic intermediate **173**, which is stabilized by the R^1 and R^2 substituents. After Friedel–Crafts alkylation, lactam **174** is formed, and the final product is furnished after an aromatization process (Scheme 36).

Scheme 36 Indole-Fused Lactams by Tandem Ring-Opening/Friedel–Crafts Alkylation of Cyclopropyl Amido Esters[31]

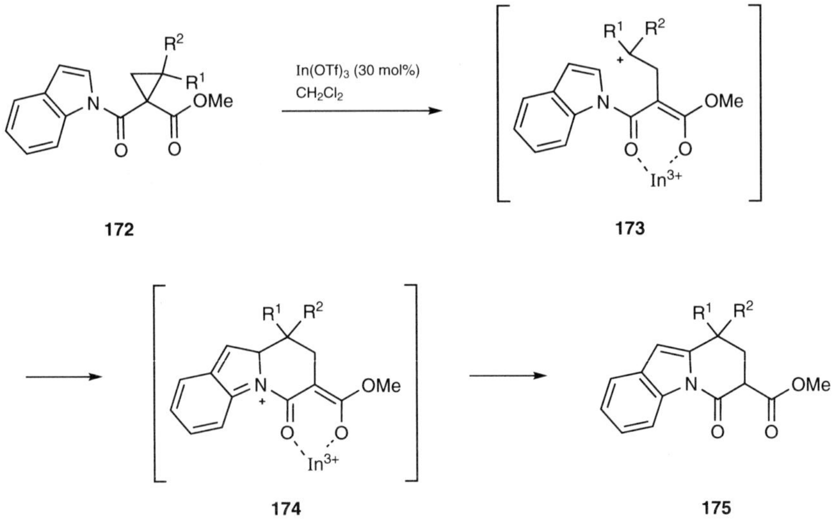

In(OTf)$_3$ (30 mol%), CH_2Cl_2

172 → 173 → 174 → 175

R^1	R^2	Yield (%)	Ref
H	4-MeOC$_6$H$_4$	86	[31]
H	4-FC$_6$H$_4$	48	[31]
Me	Ph	94	[31]
H	SPh	81	[31]

Werz has developed the facile synthesis of [n,5]spiroketals from readily available cyclopropyl aldehydes, which can be generated in situ by oxidation of the corresponding (hydroxymethyl)cyclopropanes **176**. Activated by ytterbium(III) trifluoromethanesulfonate, the cyclopropyl aldehydes are transformed into zwitterionic intermediates **177** after opening of the cyclopropane ring; subsequent recyclization gives spiroketals **178** (Table 2).[32] Synthetic approaches to spiroketals traditionally involve hetero-Diels–Alder reactions and cross couplings. This method stands as an excellent complementary example.

Table 2 Synthesis of [n,5]Spiroketals from Cyclopropyl Aldehydes[32]

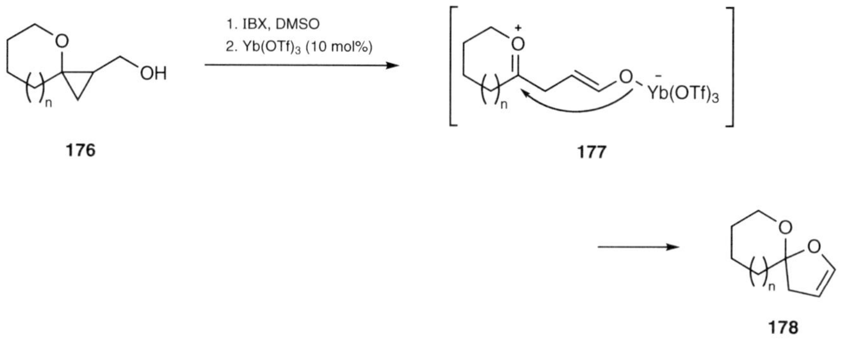

1. IBX, DMSO
2. $Yb(OTf)_3$ (10 mol%)

176 → [**177**] → **178**

Entry	Substrate	Product	Yield (%)	Ref
1	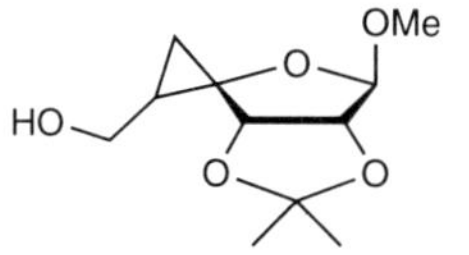	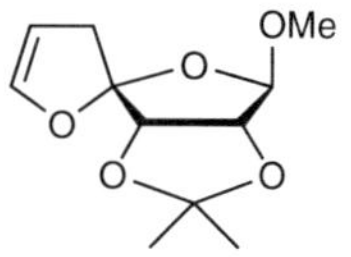	55	[32]
2	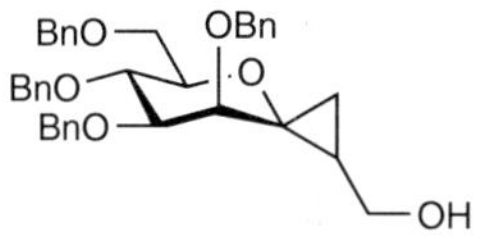	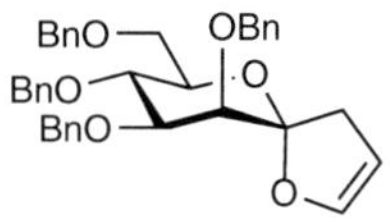	26	[32]
3	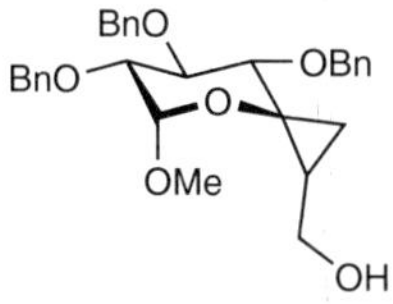	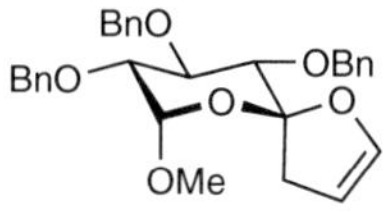	87	[32]

***trans*-Pyrrolo[1,2-*b*]isoxazolidines** 118**; General Procedure:**[21]

Oxime ether **116** (1.0 equiv) was dissolved in dry CH_2Cl_2 (0.1 M) and $Yb(OTf)_3$•nH_2O (5 mol%) was added. The mixture was stirred overnight at rt (some reactions were heated to 40 °C if not complete after 16 h at rt), and the progress of the reaction was monitored by TLC. Upon completion of the reaction, the mixture was diluted with CH_2Cl_2, washed with H_2O (1 ×) and brine (1 ×), dried ($MgSO_4$), and concentrated under reduced pressure. The crude product was preadsorbed onto silica gel and purified by flash column chromatography (EtOAc/hexanes) to afford the title compound.

Dimethyl 3-Phenyl-2-oxa-3-azabicyclo[2.2.2]octane-5,5-dicarboxylate (125, **R**^{1} **= Ph); Typical Procedure:**[22]

Toluene (10 mL) was added to a flask containing the aldehyde cyclopropane 1,1-diester (750 mg, 3.50 mmol) and freshly ground 4-Å molecular sieves (1.5 g). A soln of *N*-phenylhy-

for references see p 256

droxylamine (382 mg, 3.50 mmol) in toluene (10 mL + 10 mL rinse) was then added by cannula. The mixture was then stirred at rt for 30–40 min. $Yb(OTf)_3$ (217 mg, 0.35 mmol) was added, and the soln was stirred until the reaction was complete (TLC). The mixture was then filtered through Celite, washing with CH_2Cl_2, and the solvent was removed under reduced pressure. The crude material was purified by column chromatography (silica gel, EtOAc/hexanes 1:9) to give the title compound as an off-white solid; yield: 983 mg (92%).

Bridged [n.2.1] Heterocycles, e.g. 130–135; **General Procedure:**[23]
Using scandium(III) trifluoromethanesulfonate: $Sc(OTf)_3$ (28.5 mg, 0.058 mmol, 20 mol%) was added to a soln of the cyclopropane substrate (0.29 mmol) dissolved in 1,2-dichloroethane (4 mL) at rt under an argon atmosphere. The mixture was stirred for 2 h. After filtration through Celite, the organic phase was concentrated under reduced pressure, and the residue was purified by flash column chromatography (silica gel, petroleum ether/EtOAc 20:1) to afford the title compound.

Using ytterbium(III) trifluoromethanesulfonate: $Yb(OTf)_3$ (18 mg, 0.029 mmol, 10 mol%) was added to a soln of the cyclopropane substrate (0.29 mmol) dissolved in 1,2-dichloroethane (40 mL) under an argon atmosphere. The mixture was stirred at 60 °C overnight. After filtration through Celite, the organic phase was concentrated under reduced pressure, and the residue was purified by flash column chromatography (silica gel, petroleum ether/EtOAc 20:1) to afford the title compound.

Using tin(IV) chloride: $SnCl_4$ (15.1 mg, 0.058 mmol, 20 mol%) was added to a soln of the cyclopropane substrate (0.29 mmol) dissolved in 1,2-dichloroethane (4 mL) under an argon atmosphere. The mixture was stirred at 60 °C for 2 h. After filtration through Celite, the organic phase was concentrated under reduced pressure, and the residue was purified by flash column chromatography (silica gel, petroleum ether/EtOAc) to afford the title compound.

8-Azabicyclo[3.2.1]octanes, e.g. 137; **General Procedure:**[23]
A mixture of cyclopropane aldehyde **136** (0.29 mmol) and the amine (0.44 mmol, 1.5 equiv) in toluene (4.0 mL) were stirred in the presence of 4-Å molecular sieves at rt under argon for 2 h, then $Sc(OTf)_3$ (10 mol%) was added. The mixture was stirred for another 2 h. After filtration through Celite, the organic phases were concentrated under reduced pressure and the residue was purified by flash column chromatography (silica gel, petroleum ether/EtOAc 20:1) to afford the title compound.

9-Azabicyclo[4.2.1]nonanes, e.g. 140; **General Procedure:**[23]
A mixture of cyclopropane aldehyde **138** (0.29 mmol) and aniline (0.44 mmol, 1.5 equiv) in toluene (4.0 mL) were stirred in the presence of 4-Å molecular sieves at rt under argon for 2 h, then $SnCl_4$ (10 mol%) was added. The mixture was stirred for another 2 h. After filtration through Celite, the organic phases were concentrated under reduced pressure and the residue was purified by flash column chromatography (silica gel, petroleum ether/EtOAc 20:1) to afford the title compound.

Bridged [3.2.1] Carbocycles 145; **General Procedure:**[26]
$Sc(OTf)_3$ (30 mg, 0.06 mmol, 0.2 equiv) was added to a soln of the cyclopropane (0.30 mmol) in 1,2-dichloroethane (6 mL) under an argon atmosphere at rt. The mixture was stirred at 60–65 °C for 24 h. After the mixture had cooled to rt, the solvent was removed under reduced pressure, and the residue was purified by flash column chromatography (silica gel, petroleum ether/EtOAc) to afford the title compound.

[4.3.0]Nonanes 153; **General Procedure:**[27]

$Sc(OTf)_3$ (0.2 equiv) was added to a soln of freshly prepared cyclopropane **152** (1.0 equiv) in dry 1,2-dichloroethane (0.02 M) under an argon atmosphere at rt. The mixture was immersed in a preheated oil bath (85–90 °C) and stirred at reflux for 12 h or until the reaction was complete, as judged by TLC or ^{1}H NMR spectroscopy. The mixture was filtered through a pad of silica gel, the organic phase was concentrated under reduced pressure, and the residue was purified by flash column chromatography (silica gel, petroleum ether/EtOAc 20:1) to afford the title compound.

[3.2.1]Octanes 154; **General Procedure:**[27]

$Yb(OTf)_3$ (1.0 equiv) was added to a soln of freshly prepared cyclopropane **152** (1.0 equiv) in dry 1,2-dichloroethane (0.02 M) under an argon atmosphere at rt. The mixture was immersed in a preheated oil bath and was stirred at a set temperature until the reaction was completed, as judged by TLC or ^{1}H NMR spectroscopy. After filtering through a pad of silica gel, the organic phase was concentrated under reduced pressure, and the residue was purified by flash column chromatography (silica gel, petroleum ether/EtOAc 20:1) to afford the title compound.

1*H*-Cyclopenta[*c*][1]benzopyrans, e.g. 156 **or 1*H*-Cyclopenta[*c*]quinolines, e.g.** 158; **General Procedure:**[28]

A test tube was charged with the cyclopropane 1,1-diester (0.2 mmol), $Sc(OTf)_3$ (0.02 mmol), 4-Å molecular sieves (50 mg), and 1,2-dichloroethane (2 mL), and then the mixture was stirred at 75 °C. Upon completion of the reaction, as determined by TLC, EtOAc (20 mL) and H_2O (20 mL) were added to the mixture. The organic layer was extracted with EtOAc (2 × 20 mL). The combined organic phase was washed with brine and dried (Na_2SO_4). The residue was purified by flash chromatography (silica gel) to afford the title compound.

Cyclohexanols 161 **and Cyclohexenones** 162; **General Procedure:**[29]

The alkenyl cyclopropyl ketone (0.36 mmol, 1.0 equiv) was added to a soln of $In(OTf)_3$ (41 mg, 0.3 equiv) in anhyd CH_2Cl_2 (2 mL) at rt. The mixture was stirred for a set amount of time. The mixture was quenched with H_2O (5 mL) and extracted with CH_2Cl_2 (3 × 10 mL). The combined organic layer was washed with brine (3 × 10 mL) and dried ($MgSO_4$), and the solvent was removed under reduced pressure. Column chromatography (silica gel, hexanes/EtOAc) afforded the title compounds.

2-Hydroxybenzoates 166 **and** 167; **General Procedure:**[30]

A 0.1 M soln of the cyclopropane in CH_2Cl_2 (or 1,2-dichloroethane) was added to a dried round-bottomed flask containing $Yb(OTf)_3$ and 4-Å molecular sieves (1–2-mm beads, 0.6–0.8 g). The mixture was heated to reflux, and progress of the reaction was monitored by TLC. Upon complete disappearance of the starting material, the reaction was quenched with H_2O. The mixture was extracted with CH_2Cl_2, dried (Na_2SO_4), and purified by column chromatography (silica gel, hexanes/EtOAc 20:1) to afford the title compounds.

Lactams 175; **General Procedure:**[31]

Cyclopropyl amido ester **172** (1.0 equiv) was added to a soln of $In(OTf)_3$ (0.30 equiv) in anhyd CH_2Cl_2 (2 mL) at rt. Upon completion of the reaction, the mixture was quenched with H_2O, and the product was extracted from the aqueous phase with CH_2Cl_2. The combined organic layer was washed with brine, dried ($MgSO_4$), and concentrated. The residue was purified by flash column chromatography (silica gel, hexanes/EtOAc) to afford the title compound.

for references see p 256

Spiroketals 178; **General Procedure:**[32]

A 10-mL flask was charged with a soln of the (hydroxymethyl)cyclopropane (0.26 mmol, 1.0 equiv) in dry DMSO (3 mL). 2-Iodoxybenzoic acid (IBX; 88 mg, 0.31 mmol, 1.2 equiv) and $Yb(OTf)_3$ (16 mg, 0.03 mmol, 0.1 equiv) were sequentially added, and the resulting mixture was stirred at rt for 8 h. The reaction was stopped by the addition of H_2O (5 mL). Et_2O (10 mL) was added, and the layers were separated. The aqueous layer was extracted with Et_2O (3 × 5 mL), and the combined organic phase was washed with brine (10 mL), dried (Na_2SO_4), filtered, and concentrated under reduced pressure. Purification by column chromatography (silica gel, pentane/Et_2O 6:1) afforded the title compound.

1.4.1.2.2 Gold- or Nickel-Catalyzed Cycloisomerization of Alkynylcyclopropyl Ketones

In 2008, Zhang[33] developed an intermolecular (4 + 2)-annulation reaction with 1-(alk-1-ynyl)cyclopropyl ketones as all-carbon 1,4-dipoles for the easy formation of carbo- and heterocycles. Gold catalysts are used as strong π-Lewis acids to activate the unsaturated π-systems. The intramolecular version, also called the intramolecular (4 + 2)-cross-cycloaddition reaction, was discovered by Wang in 2012 (Scheme 37).[34] This type of reaction affords bridged [n.3.1] skeletons, which are found in a number of natural products with a broad range of biological activities. The generality of the substrates has been studied. For substrates **179**, the reactions all proceed smoothly to give oxa[3.2.1] and oxa[3.3.1] derivatives **180** in good to excellent yields through intramolecular (4 + 2) cross cycloaddition (Scheme 38).

Scheme 37 Inter- and Intramolecular (4 + 2) Cycloadditions[34]

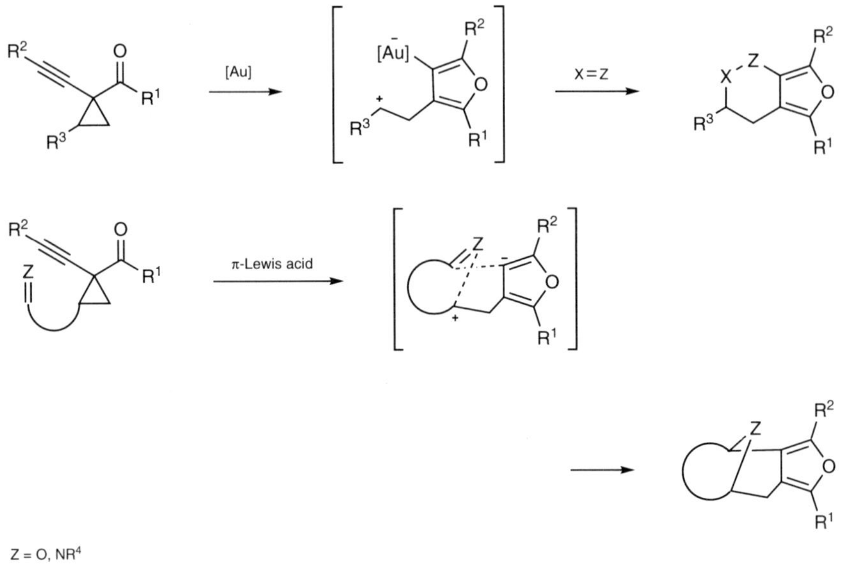

[Au]

X=Z

π-Lewis acid

Z = O, NR[4]

Scheme 38 Generality of the Intramolecular (4 + 2)-Cross-Cycloaddition Reaction[34]

AuCl(PPh$_3$) (10 mol%)
AgOTf (10 mol%)
1,2-dichloroethane
60 °C, 40 min

179 → 180

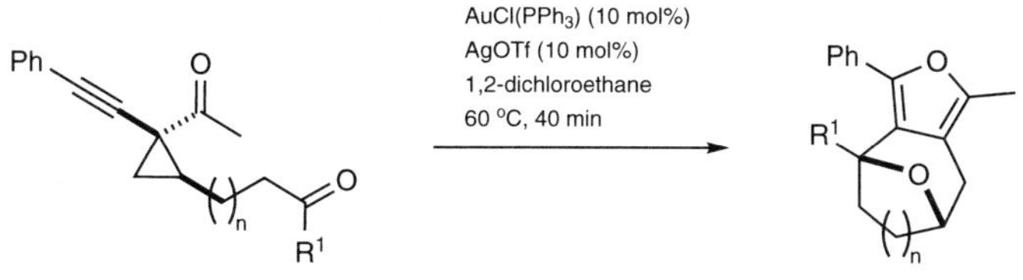

R^1	n	Yield (%)	Ref
H	2	86	[34]
Ph	2	91	[34]
H	1	76	[34]

Furthermore, by using other σ-Lewis acids such as nickel(II) perchlorate hexahydrate to activate the ketone, a conventional intramolecular (3 + 2)-cross-cycloaddition reaction takes place. The challenge is the chemoselective control of these two competitive cycloadditions. Among various catalysts, nickel(II) perchlorate hexahydrate exhibits the highest selectivity toward the intramolecular (3 + 2)-cross-cycloaddition reaction. This method has also been tested for various substrates **181** and gives versatile oxa[2.2.1] and oxa[3.2.1] skeletons **182** (Scheme 39). Considering that the two functional groups of the alkyne and ketone remain untouched, it is possible to derivatize these products further.

Scheme 39 Generality of the Intramolecular (3 + 2)-Cross-Cycloaddition Reaction[34]

Ni(ClO$_4$)$_2$•6H$_2$O (10 mol%)

181 → 182

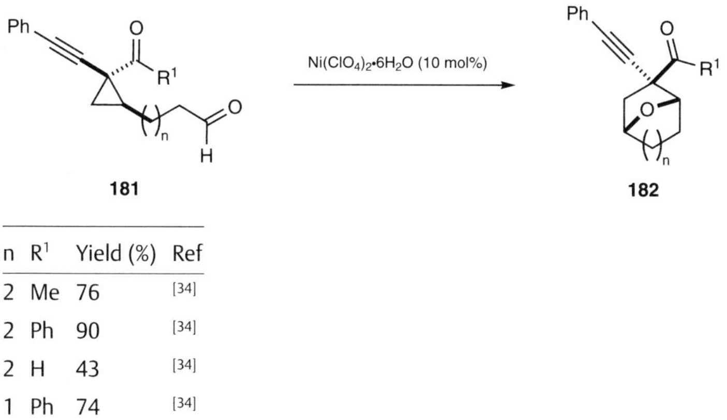

n	R^1	Yield (%)	Ref
2	Me	76	[34]
2	Ph	90	[34]
2	H	43	[34]
1	Ph	74	[34]

Oxatricyclo[3.2.1]octanes 180 **(n = 1) and Oxatricyclo[3.3.1]nonanes** 180 **(n = 2); General Procedure:**[34]

An oven-dried, 25-mL flask filled with argon gas was charged with AuCl(PPh$_3$) (16 mg, 0.032 mmol, 10 mol%) and AgOTf (8 mg, 0.031 mmol, 10 mol%) and then 1,2-dichloroethane (10 mL) was added. The resulting soln was stirred for 10 min, and then **179** (80 mg, 0.31 mmol) was added. After stirring at 60 °C for 40 min, the soln was cooled to rt and passed through a short silica gel column. The product was then purified by column chromatography (silica gel, petroleum ether/EtOAc 20:1) to give the title compound.

Oxatricyclo[2.2.1]heptanes 182 **(n = 1) and Oxatricyclo[3.2.1]octanes** 182 **(n = 2); General Procedure:**[34]

An oven-dried, 25-mL flask filled with argon gas was charged with $Ni(ClO_4)_2 \cdot 6H_2O$ (11.2 mg, 0.031 mmol, 10 mol%) and 1,2-dichloroethane (10 mL). Then **181** (80 mg, 0.31 mmol) was added, and the resulting soln was stirred at 60 °C for 100 min. The soln was cooled to rt and passed through a short diatomaceous column. The product was then purified by column chromatography (silica gel, petroleum ether/EtOAc 20:1) to give the title compound.

1.4.1.2.3 Copper-Catalyzed Cycloisomerization of Donor–Acceptor Cyclopropanes

In 2014, Tang and Houk developed a copper-catalyzed intramolecular cycloaddition of donor–acceptor cyclopropanes with indoles to provide a facile diastereodivergent synthesis of tetracyclic spiroindolines using ligand **183**.[35] The challenge of this transformation is the control of the diastereoselectivity. Remote ester groups are critical to the reaction results: whereas isopropyl esters highly favor the formation of the *trans*-diastereomer, 2-adamantyl esters are the best choice for the formation of the *cis*-diastereomer. The substrate scope is broad; for various cyclopropane 1,1-diesters **184** and **186** the corresponding tetracyclic spiroindolines **185** and **187** are obtained in moderate to excellent yields. Notably, in most cases, high diastereoselectivities are achieved (Scheme 40).

Scheme 40 Diastereodivergent Synthesis of Tetracyclic Spiroindolines[35]

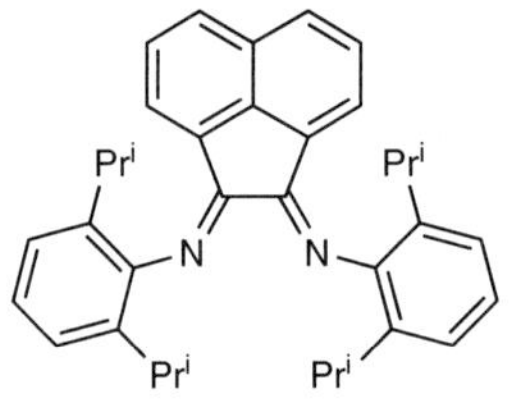

$Cu(SbF_6)_2$ (10 mol%)
183 (12 mol%)

183 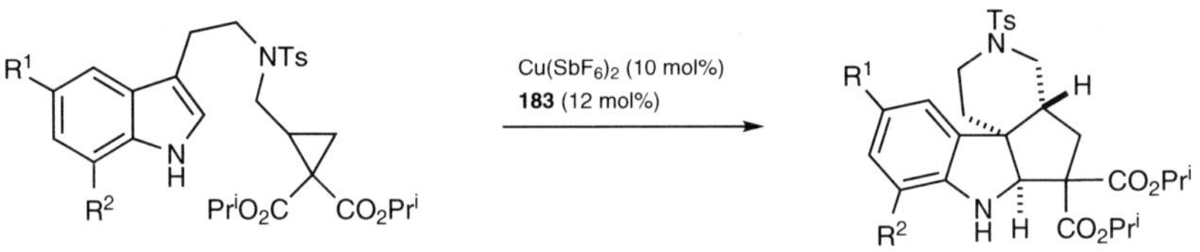 **184** **185**

R^1	R^2	dr	Yield (%)	Ref
H	H	9:1	77	[35]
Me	H	15.7:1	71	[35]
H	Br	15.7:1	84	[35]

R^3 = 2-adamantyl

Cu(SbF$_6$)$_2$ (10 mol%), **183** (13 mol%); **186** → **187**

R^1	R^2	dr	Yield (%)	Ref
H	H	19:1	79	[35]
Me	H	24:1	90	[35]
H	Cl	15.7:1	91	[35]

Diisopropyl (4a*R,6a*S**,11b*R**)-3-Tosyl-1,2,3,4,4a,5,6a,7-octahydro-6*H*-pyrido[4′,3′:2,3]cyclopenta[1,2-*b*]indole-6,6-dicarboxylate** 185 **($R^1 = R^2 = H$); Typical Procedure:**[35]
A mixture of $CuBr_2$ (11.2 mg, 0.05 mmol), $AgSbF_6$ (34.4 mg, 0.10 mmol), and ligand **183** (30.0 mg, 0.06 mmol) in dry 1,2-dichloroethane (8.0 mL) was stirred at rt for about 10–12 h under a N_2 atmosphere to yield the active [Cu(**183**)(SbF_6)$_2$] catalyst as a clear black soln. Then, another oven-dried Schlenk reaction tube containing a magnetic stirrer bar was charged with substrate **184** (135.2 mg, 0.25 mmol) under a N_2 atmosphere, and this was followed by the addition of the above catalyst soln (4.0 mL). The resulting mixture was stirred until the substrate was completely consumed (monitoring by TLC, hexane/EtOAc 4:1). The mixture was then diluted with sat. aq NH_4Cl and extracted with CH_2Cl_2 (3 × 30 mL). The combined extract was washed with brine and dried (Na_2SO_4). Filtration and concentration under reduced pressure gave the crude mixture, and the dr was determined by ^1H NMR spectroscopy (dr 90:10). Purification by column chromatography (silica gel, hexane/EtOAc 10:1 to 7:1) gave the title compound as a white solid; yield: 104.1 mg (77%).

1.4.1.2.4 Pentacarbonyltungsten-Catalyzed Cycloisomerization of Alkynylcyclopropyl Ketones or Oximes

Sarpong and co-workers have disclosed the tungsten-catalyzed hetero-cycloisomerization of 1-(alk-1-ynyl)cyclopropyl ketones or oximes for the facile construction of 4,5-dihydrobenzo[*b*]furans and 4,5-dihydroindoles (Scheme 41).[36] Upon treatment of substrates **188** with pentacarbonyl(tetrahydrofuran)tungsten (10 mol%), 4,5-dihydrobenzo[*b*]furans and 4,5-dihydroindoles **189** are obtained in yields of 57–98%. If bicyclo[5.1.0] substrates such as **190** are employed as the starting materials, the reactions are not affected by the seven-membered carbocycles and cycloheptene-fused furans **191** are delivered in good to excellent yields. The derived products can be easily functionalized; for example, simple dehydrogenation of **189** gives the corresponding benzo[*b*]furans and indoles.

Scheme 41 4,5-Dihydrobenzo[*b*]furans and 4,5-Dihydroindoles by Tungsten-Catalyzed Hetero-cycloisomerization of Donor–Acceptor Cyclopropanes[36]

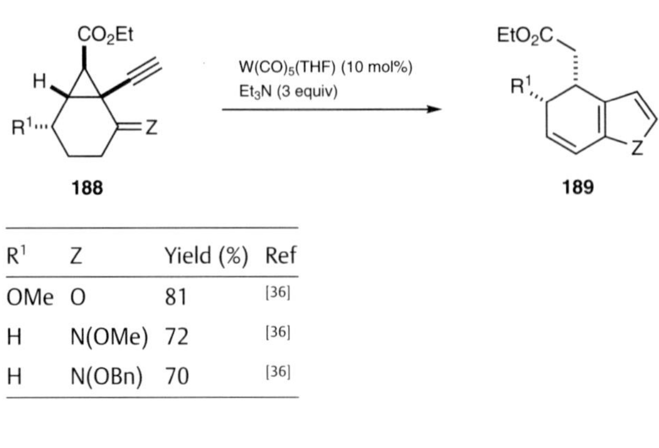

R^1	Z	Yield (%)	Ref
OMe	O	81	[36]
H	N(OMe)	72	[36]
H	N(OBn)	70	[36]

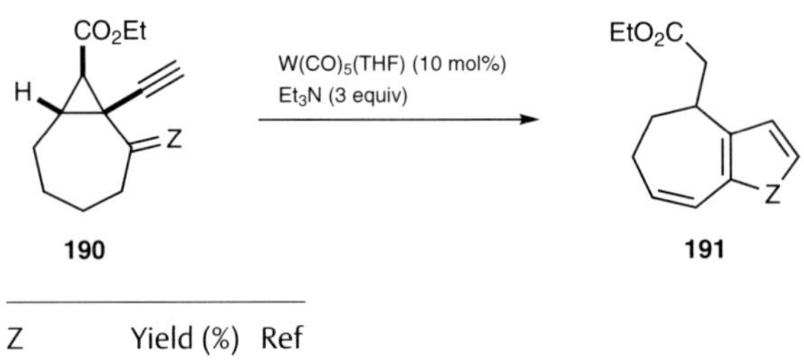

Z	Yield (%)	Ref
O	90	[36]
N(OMe)	78	[36]

The mechanism is provided in Scheme 42. In the presence of the tungsten catalyst, bicycle **188** is first converted into metallovinylidene **192**, which is susceptible to nucleophilic attack from the proximal heteroatom (N or O) to generate zwitterionic intermediate **193**. After tandem deprotonation and protodemetalation, fused tricycle **194** is formed. At this stage, a ring-opening process triggered by relief of ring strain delivers zwitterionic intermediate **195**; this is followed by proton transfer and concomitant aromatization to afford desired dihydrobenzo[*b*]furans or 4,5-dihydrobenzo[*b*]indoles **189**.

Scheme 42 Proposed Catalytic Cycle for the Tungsten-Catalyzed Hetero-cycloisomerization of Donor–Acceptor Cyclopropanes[36]

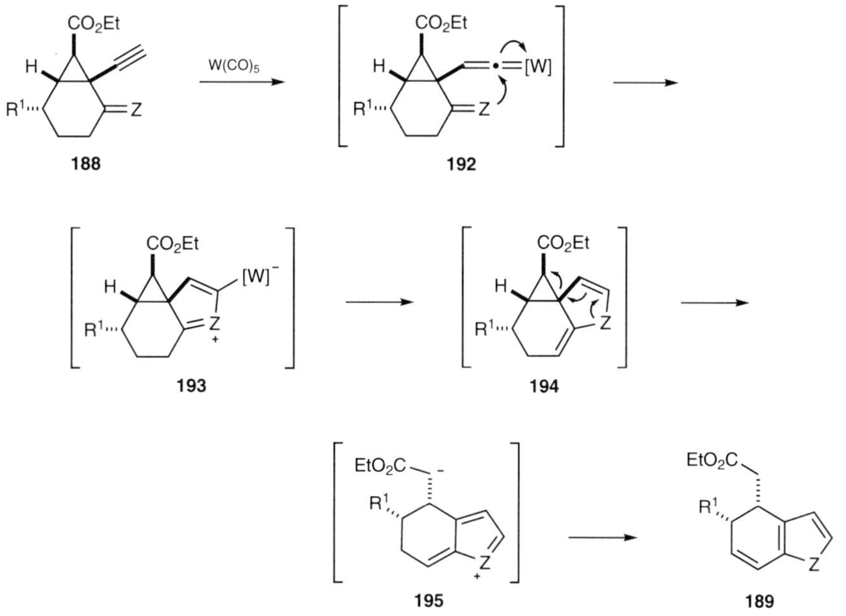

Fused Furan and Pyrrole Derivatives, e.g. 189 and 191; General Procedure:[36]

A dry vial was charged with $W(CO)_6$ (10 mol%) and a stirrer bar; it was then sealed, evacuated, and placed under a N_2 atmosphere. Dry THF (4.6 mL/mmol substrate) was added, and the resulting soln was irradiated at 350 nm for 2 h. A soln of the cycloisomerization substrate and Et_3N (3 equiv) in THF (4.6 mL/mmol substrate) was added to the contents of the vial. The yellow soln turned dark orange/red and was stirred at rt for 12 h, at which time the reaction was judged to be complete by TLC. The solvents were removed under reduced pressure, and the residue was purified by chromatography (silica gel, hexanes/EtOAc) to afford the title compound.

1.4.1.3 Vinylidenecyclopropanes

The cyclopropane rings in vinylidenecyclopropanes are even more strained than those in methylenecyclopropanes. With both allene and cyclopropane moieties (or methylenecyclopropane and C=C moieties), vinylidenecyclopropanes can behave as allenes and also as methylenecyclopropanes in many reactions, and this makes them excellent resources to study the chemistry of allenes and cyclopropanes.

1.4.1.3.1 Tin-Catalyzed Cycloisomerization of Vinylidenecyclopropanes

In 2005, Shi reported the Lewis acid catalyzed rearrangement of multisubstituted vinylidenecyclopropanes for the synthesis of benzo[*c*]fluorene and phenyl-1*H*-indene derivatives.[37] The outcome of the reaction is greatly dependent on the substituents on the substrates. For instance, for substrates **196** in which R^2 is an aryl ring and R^4 is an alkyl group, the reaction leads to highly functionalized benzo[*c*]fluorenes **197** in 47–97% yield. Interestingly, the reaction of vinylidenecyclopropane **198** results in indene derivative **199** in 95% yield (Scheme 43). All reactions proceed through the same type of zwitterionic intermediate, but the reaction pathways are greatly affected by the substituents.

Scheme 43 Benzo[c]fluorenes and Phenyl-1*H*-indenes by Lewis Acid Catalyzed Rearrangements of Vinylidenecyclopropanes[37]

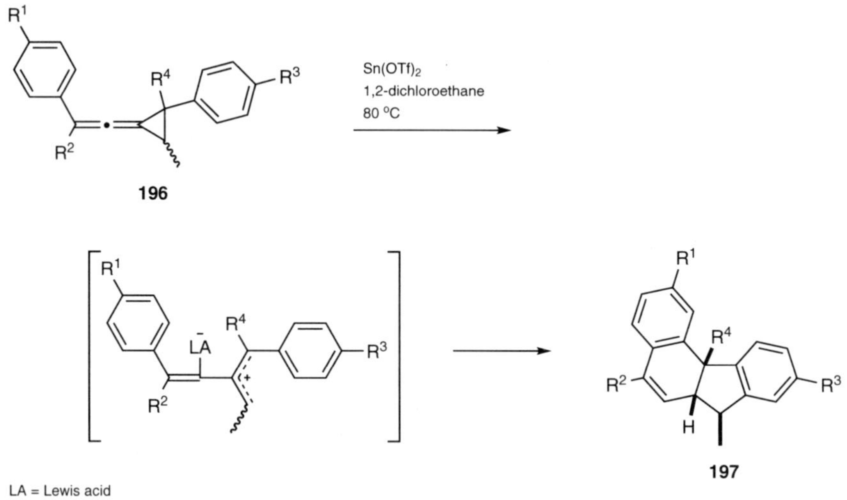

LA = Lewis acid

R^1	R^2	R^3	R^4	Yield (%)	Ref
H	Ph	Cl	Me	91	[37]
H	Ph	H	Et	96	[37]
F	4-FC_6H_4	H	Me	97	[37]

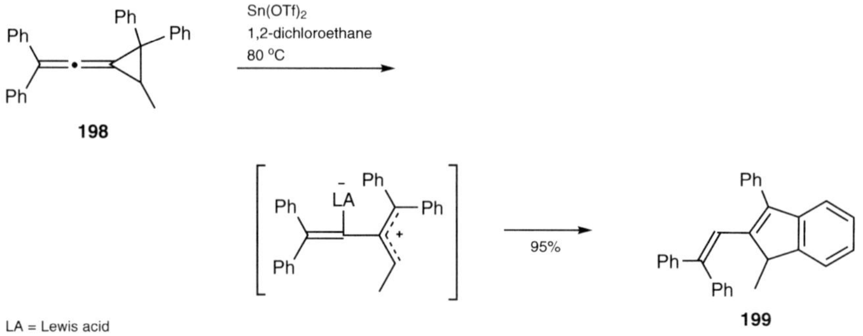

LA = Lewis acid

Benzo[c]fluorenes 197**; General Procedure:**[37]
$Sn(OTf)_2$ (0.02 mmol) was added to a soln of 2-(2,2-diarylvinylidene)cyclopropane **196** (0.2 mmol) in 1,2-dichloroethane (2.0 mL), and the mixture was stirred for 5 h at 80 °C (monitored by TLC). After the starting material was consumed, the solvent was removed under reduced pressure, and the residue was subjected to flash column chromatography.

1.4.1.3.2 Gold-Catalyzed Cycloisomerization of Vinylidenecyclopropanes

Interestingly, as powerful soft Lewis acids, gold catalysts exhibit different reactivities toward vinylidenecyclopropanes compared to traditional Lewis acids. For example, treatment of vinylidenecyclopropanes **200** with a gold catalyst gives functionalized 1,2-dihydronaphthalenes **201** in moderate to excellent yields. According to the proposed mechanism, gold coordinates with the vinylidenecyclopropane to activate the allene moiety, and this is followed by opening of the cyclopropane ring to deliver allylic cation **202**. Subsequent Friedel–Crafts reaction generates another cationic intermediate **203**. After tandem deprotonation and demetalation, the final product is obtained (Scheme 44).[38]

Scheme 44 Functionalized 1,2-Dihydronaphthalenes by Gold-Catalyzed Rearrangement of Vinylidenecyclopropanes[38]

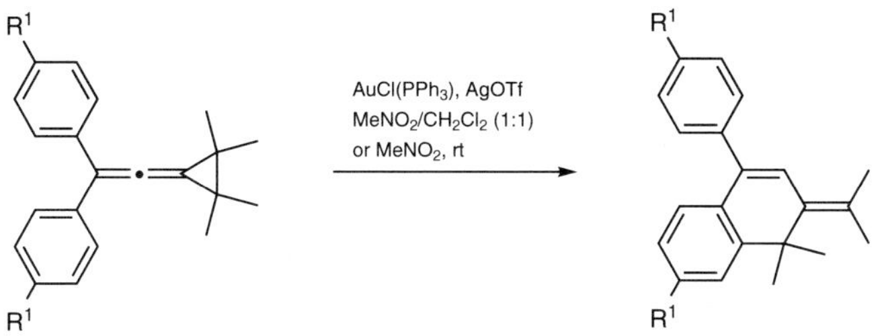

$AuCl(PPh_3)$, AgOTf
$MeNO_2/CH_2Cl_2$ (1:1)
or $MeNO_2$, rt

R^1	Yield (%)	Ref
Me	80	[38]
Cl	99	[38]

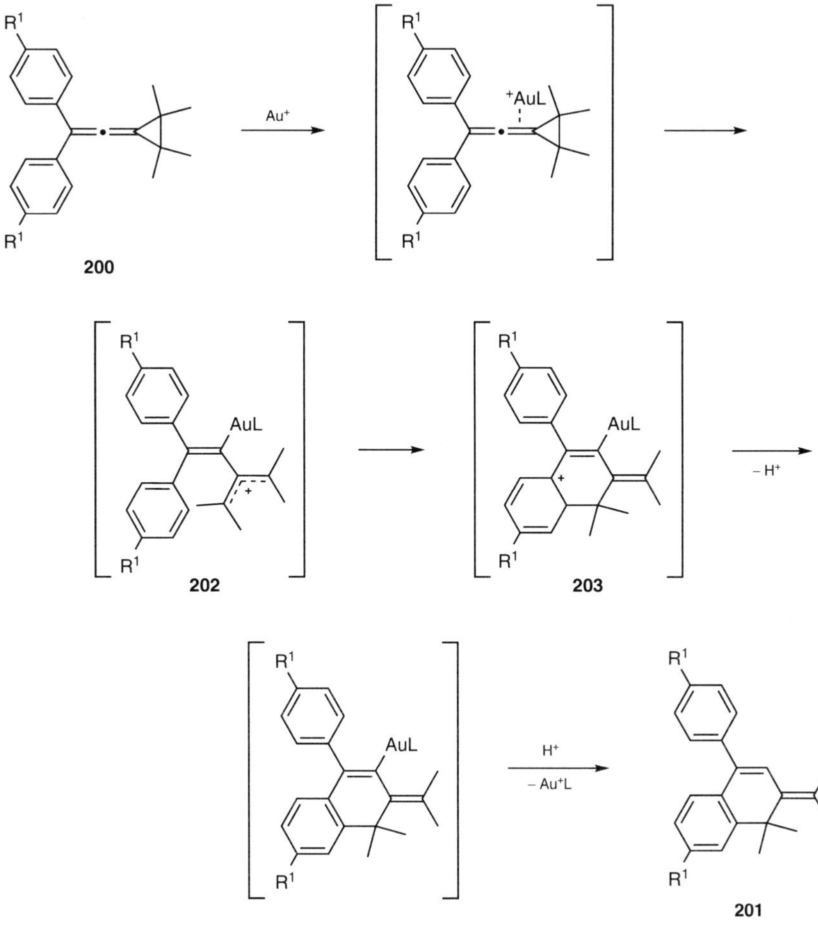

for references see p 256

In the presence of a gold complex as a catalyst, the reactivity of functionalized vinylidenecyclopropanes tethered with nucleophilic moieties is quite different from that of unfunctionalized vinylidenecyclopropanes. For instance, vinylidenecyclopropane-tethered alcohols **204** undergo a nucleophilic ring-opening reaction catalyzed by a gold complex for the facile construction of allene-containing tetrahydropyran derivatives **206** in moderate yields (Scheme 45).[39] Two possible pathways are suggested: either the allene functionality is activated by a gold cation or a gold(I) species catalyzes the opening of the cyclopropane ring. Both reaction pathways lead to tetrahydropyran **205** as the common intermediate through intramolecular nucleophilic addition of the hydroxy group. Subsequent protodemetalation generates the desired product.

Scheme 45 Tetrahydropyrans by Nucleophilic Ring Opening of Vinylidenecyclopropane-Tethered Alcohols Catalyzed by a Gold Complex[39]

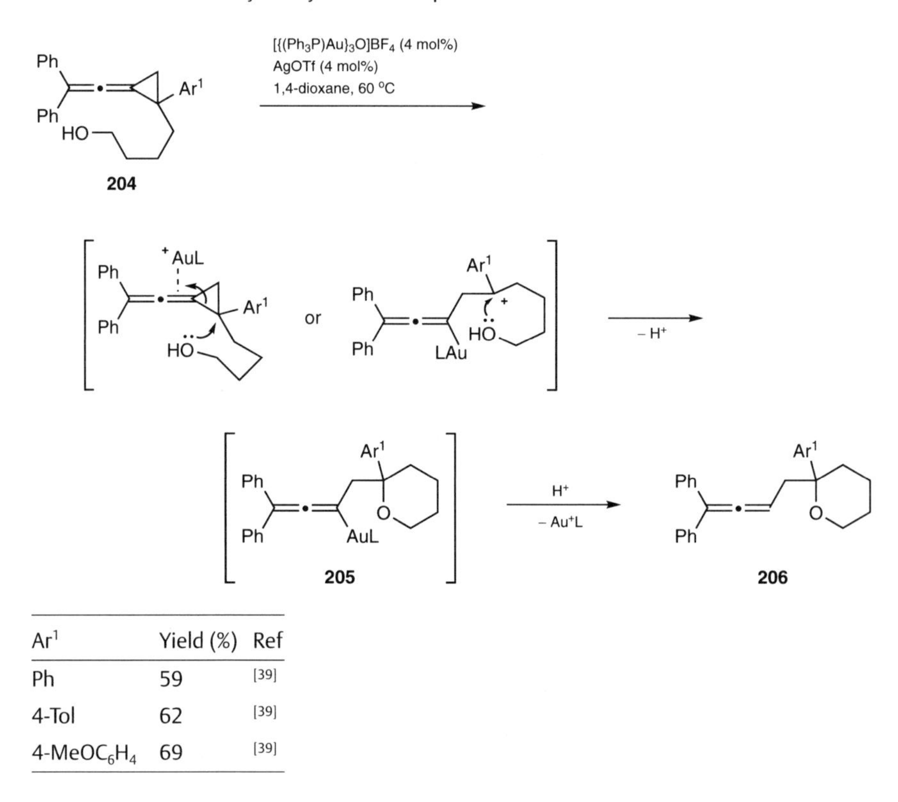

Ar1	Yield (%)	Ref
Ph	59	[39]
4-Tol	62	[39]
4-MeOC$_6$H$_4$	69	[39]

Changing the length of the spacer in the vinylidenecyclopropane-tethered alcohols is also possible, and Shi has reported a different nucleophilic ring-opening reaction under such circumstances.[40] In this case, functionalized vinylidenecyclopropanes **207** are synthesized, in which the cyclopropane moiety bears an adjacent secondary alcohol group. Upon activation by a gold(I) catalyst, the vinylidenecyclopropane undergoes intramolecular nucleophilic addition to form five-membered ring **208**, and this is followed by C—O bond cleavage to give cationic intermediate **209**. This intermediate is stabilized by the electron-donating R^1 substituent and the cyclopropyl group. Subsequent cation-induced opening of the cyclopropane ring gives ketone intermediate **210**. Finally, deprotonation and protodeauration give dienone derivative **211** under mild conditions (Scheme 46). This reaction mechanism has been proved by ^{18}O-labeling experiments.

Scheme 46 Dienones by Gold-Catalyzed Intramolecular Rearrangement of Vinylidenecyclopropane-Tethered Alcohols[40]

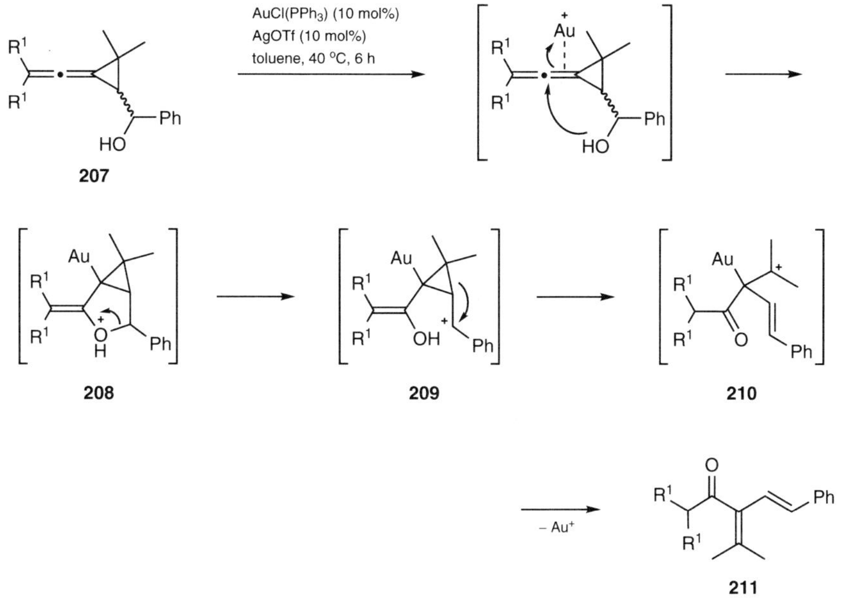

R1	Yield (%)	Ref
Me	80	[40]
Bu	84	[40]

In addition, Shi reported that hydroxylamine-tethered vinylidenecyclopropane 1,1-diesters **212** can undergo an interesting domino hydroamination and ring-opening reaction (Scheme 47).[41] In the presence of an in situ formed gold complex, the allene moiety of the vinylidenecyclopropane is activated and then intramolecular hydroamination occurs to form N,O-heterocycle **213**. After opening of the cyclopropane ring, the corresponding alkyne **214** is obtained. Furthermore, upon hydrolysis of the C≡C bond catalyzed by a gold(I) catalyst, the final ketone derivative **215** is afforded through classical alkyne hydration with ambient water.

Scheme 47 Gold(I)-Catalyzed Domino Intramolecular Hydroamination and Ring-Opening Reaction of Hydroxylamine-Tethered Vinylidenecyclopropane 1,1-Diesters[41]

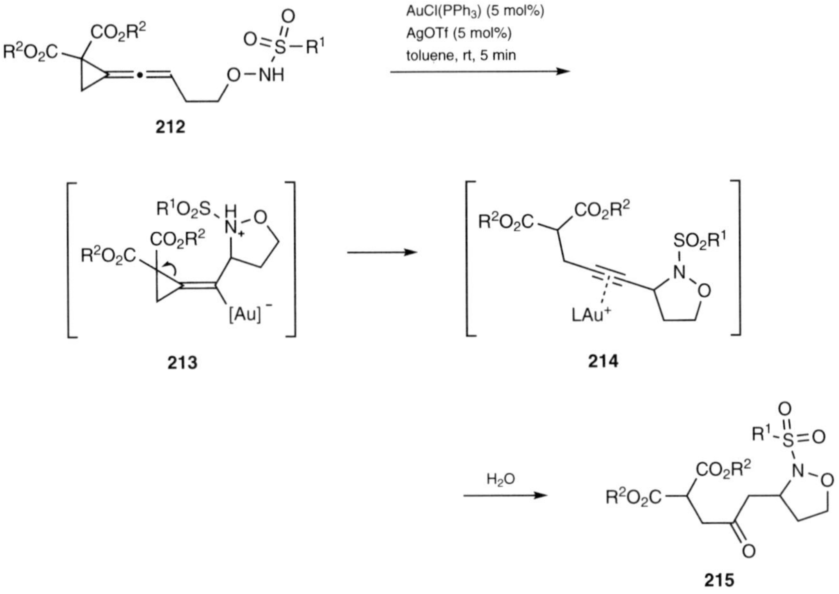

R^1	R^2	Yield (%)	Ref
4-$MeOC_6H_4$	Me	69	[41]
4-$O_2NC_6H_4$	Me	65	[41]
4-Tol	Bn	66	[41]

Homogeneous gold catalysis is widely applied in activating unsaturated carbon–carbon bonds, especially for alkyne activation. Novel yne–vinylidenecyclopropanes **217** and **220** can be used in cycloisomerizations catalyzed by gold catalyst **216**.[42] In these transformations, the vinylidenecyclopropanes serve as three-carbon-atom synthons for an intramolecular (3 + 2)-cycloaddition reaction to furnish ring-fused [4.3.0] and [5.3.0] bicyclic ring systems. The outcome of the reaction is largely dependent on the length of the carbon chain in the vinylidenecyclopropanes. If **217** is used, the reactions give [4.3.0] bicyclic derivatives **218** as the major products in yields of 25–85%, together with minor amounts of allene byproducts **219** in some cases. If the carbon chain is longer (e.g., **220**), allene products **221** are obtained as the major products, and cycloadducts **222** are also formed as minor byproducts (Scheme 48).

Scheme 48 Gold-Catalyzed Cycloisomerization of Yne–Vinylidenecyclopropanes[42]

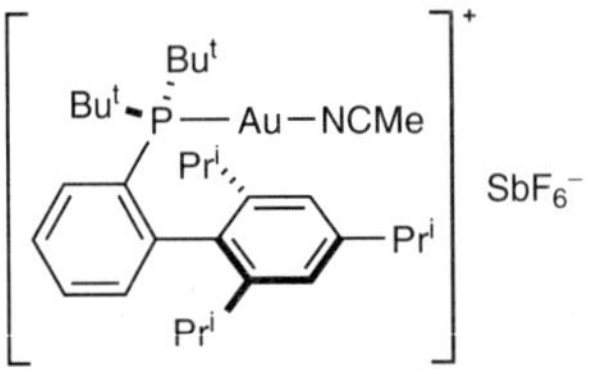

216, 1,2-dichloroethane

217 → **218** major + **219** minor

R^1	Z	Yield (%) 218	Yield (%) 219	Ref
Me	NTs	80	–[a]	[42]
Me	$4\text{-BrC}_6H_4SO_2N$	67	–[a]	[42]
Bn	NTs	56	14	[42]

[a] Yield not reported.

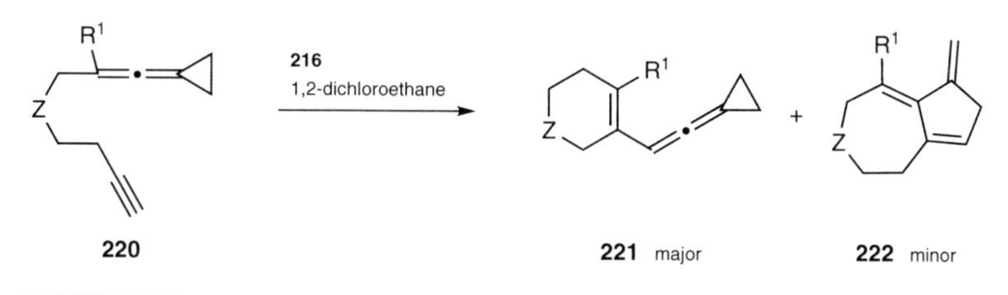

R^1	Z	Yield (%) 221	Yield (%) 222	Ref
Me	NTs	56	–[a]	[42]
Me	$4\text{-BrC}_6H_4SO_2N$	55	25	[42]
Bn	NTs	47	25	[42]

[a] Yield not reported.

1,2-Dihydronaphthalenes 201; **General Procedure:**[38]

CAUTION: *Nitromethane is flammable, a shock- and heat-sensitive explosive, and an eye, skin, and respiratory tract irritant.*

AgOTf (8 mg, 0.01 mmol) was added to a soln of 2-(2,2-diarylvinylidene)cyclopropane **200** (0.2 mmol) and $AuCl(PPh_3)$ (5 mg, 0.01 mmol) in $CH_2Cl_2/MeNO_2$ (1:1; 2.0 mL), and the mixture was stirred for 10 min at rt (monitored by TLC). After the starting material was consumed, the solvent was removed under reduced pressure, and the residue was subjected to flash column chromatography (silica gel) to give the title compound as a colorless liquid.

Tetrahydro-2*H*-pyrans 206; **General Procedure:**[39]

The 2-(2,2-diarylvinylidene)cyclopropane **204** (0.1 mmol) was dissolved in 1,4-dioxane (3.0 mL), and then $[\{(Ph_3P)Au\}_3O]BF_4$ (6.0 mg, 4 mol%) and AgOTf (1.0 mg, 4 mol%) were added at 60 °C. The mixture was stirred for another 10–30 min at this temperature. The solvent was removed under reduced pressure, and the residue was purified by flash column chromatography (silica gel, petroleum ether/EtOAc 80:1) to afford the title compound.

Dienones 211; **General Procedure:**[40]

Under an argon atmosphere, a Schlenk tube was charged with the vinylidenecyclopropane (0.1 mmol), $AuCl(PPh_3)$ (0.01 mmol), AgOTf (0.01 mmol), and toluene. The mixture

for references see p 256

was stirred at 40 °C until the reaction was complete. Then, the solvent was removed under reduced pressure, and the residue was purified by flash column chromatography (silica gel) to afford the title compound.

Dialkyl 2-[3-(Isoxazolidin-3-yl)-2-oxopropyl]malonates 215; **General Procedure:**[41]
Under an argon atmosphere, a Schlenk tube was charged with the vinylidenecyclopropane **212** (0.1 mmol) and $AuCl(PPh_3)$ (5 mol%), and then toluene (1.0 mL) was added. The mixture was stirred at rt (20 °C) for 5 min and AgOTf (5 mol%) was added followed by H_2O (0.1 mmol, 1.0 equiv). After the substrate was consumed, the solvent was removed under reduced pressure, and the residue was then purified by flash column chromatography (silica gel) to afford the title compound.

[4.3.0]Bicycles 218/**[5.3.0]Bicycles** 222 **and Allenes** 219/221; **Typical Procedure:**[42]
A flame-dried tube was charged with the vinylidenecyclopropane **217** or **220** (0.1 mmol), the gold catalyst (5 mol%), and anhyd 1,2-dichloroethane (1.0 mL) under an atmosphere of argon. Then, the resulting soln was stirred at rt for 0.5 h. Then, the mixture was evaporated to dryness, and the residue was purified by flash column chromatography (silica gel, petroleum ether/EtOAc 20:1) to give the title compound.

1.4.1.3.3 Titanium-Catalyzed Cycloisomerization of Vinylidenecyclopropanes

Huang and co-workers have described a novel protocol for facile access to medium- and large-sized fused naphthalenes **223** through ring expansion of bicyclic vinylidenecyclopropanes.[43] This reaction is performed under very mild conditions with readily available starting materials, and the procedure is also very simple. Therefore, this protocol is attractive for the construction of common to large-sized carbocycles (Scheme 49).

Scheme 49 Synthesis of Medium- to Large-Sized Carbocycles by Ring Opening of Vinylidenecyclopropanes[43]

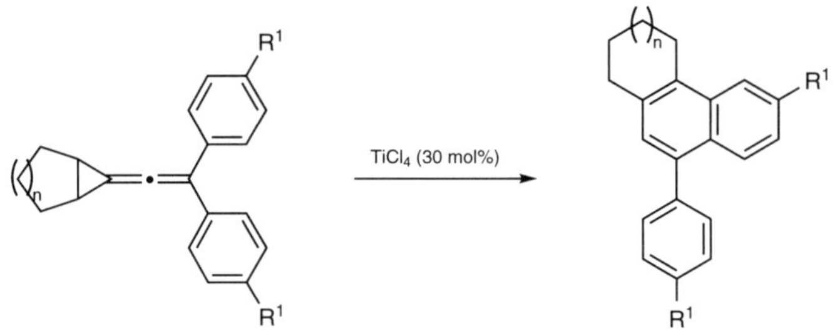

R^1	n	Yield (%)	Ref
H	1	82	[43]
Cl	1	83	[43]
H	2	85	[43]
H	3	78	[43]

The mechanism is shown in Scheme 50. First, the vinylidenecyclopropane is activated by the Lewis acid to generate cationic intermediate **224**. Triggered by a ring-opening process, allylic intermediate **225** is formed. Subsequent Friedel–Crafts cyclization delivers tricyclic cationic intermediate **226**. After deprotonation and demetalation, the corresponding tricarbocycle **227** is obtained. The final product is formed after 1,3-hydrogen shift.

Scheme 50 Proposed Mechanism[43]

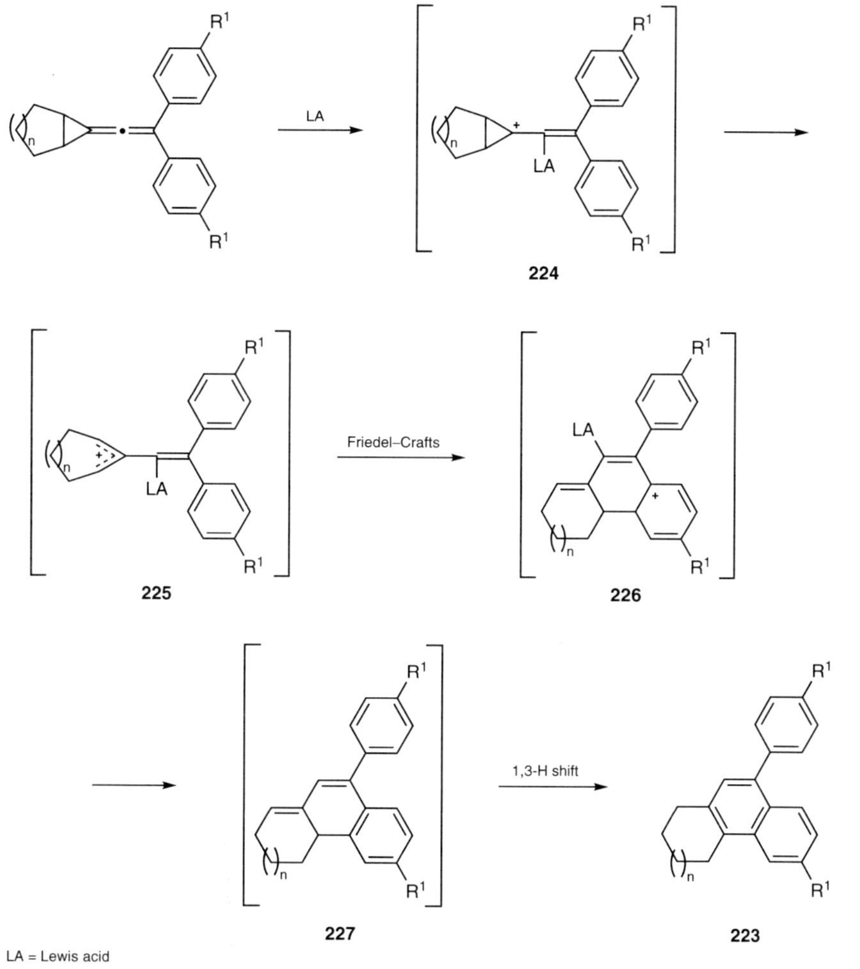

LA = Lewis acid

Fused Naphthalenes 223; General Procedure:[43]

$TiCl_4$ (28.4 mg, 0.15 mmol) was added to a soln of the bicyclic vinylidenecyclopropane (0.5 mmol) in anhyd CH_2Cl_2 (3 mL) under a N_2 atmosphere. The mixture was stirred at rt for 2 h. Upon completion of the reaction, the mixture was concentrated, and the residue was purified by flash chromatography (silica gel) to afford the title compound.

1.4.1.3.4 Rhodium-Catalyzed Cycloisomerization of Vinylidenecyclopropanes

An interesting intramolecular cycloaddition of vinylidenecyclopropanes **228** was disclosed by Lu and Shi (Scheme 51).[44] The eight-membered nitrogen heterocycles **229** derived from this reaction are useful building blocks in organic synthesis. The exocyclic double bonds allow these molecules to be very easily modified.

Scheme 51 5,6-Dimethylenehexahydroazocines by Rhodium-Catalyzed Intramolecular Cycloaddition of Vinylidenecyclopropanes[44]

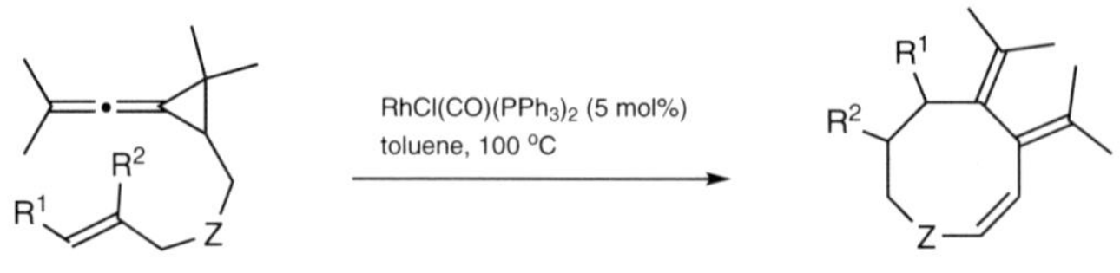

R2
R1
Z
228

$RhCl(CO)(PPh_3)_2$ (5 mol%)
toluene, 100 °C

R1
R2
Z
229

R^1	R^2	Z	Yield (%)	Ref
Me	H	NTs	38	[44]
H	Me	NTs	31	[44]
Me	H	$4\text{-}BrC_6H_4SO_2N$	37	[44]
H	Me	$4\text{-}BrC_6H_4SO_2N$	29	[44]

A plausible mechanism is outlined in Scheme 52. Vinylidenecyclopropane **228** undergoes C—C bond insertion or cyclopropyl C—H bond insertion by the rhodium catalyst to generate four-membered intermediate **230** or cyclopropylmethyl metallic intermediate **231**. Next, either β-hydride elimination or β-carbon elimination takes place to yield rhodium species **232**, which then produces conjugated triene intermediate **233** through 1,3-migration of the rhodium catalyst. Subsequent addition to the alkene delivers ring-closed nitrogen heterocycle **234**, which gives final product **229** by reductive elimination.

Scheme 52 Mechanism for the Synthesis of 5,6-Dimethylenehexahydroazocines[44]

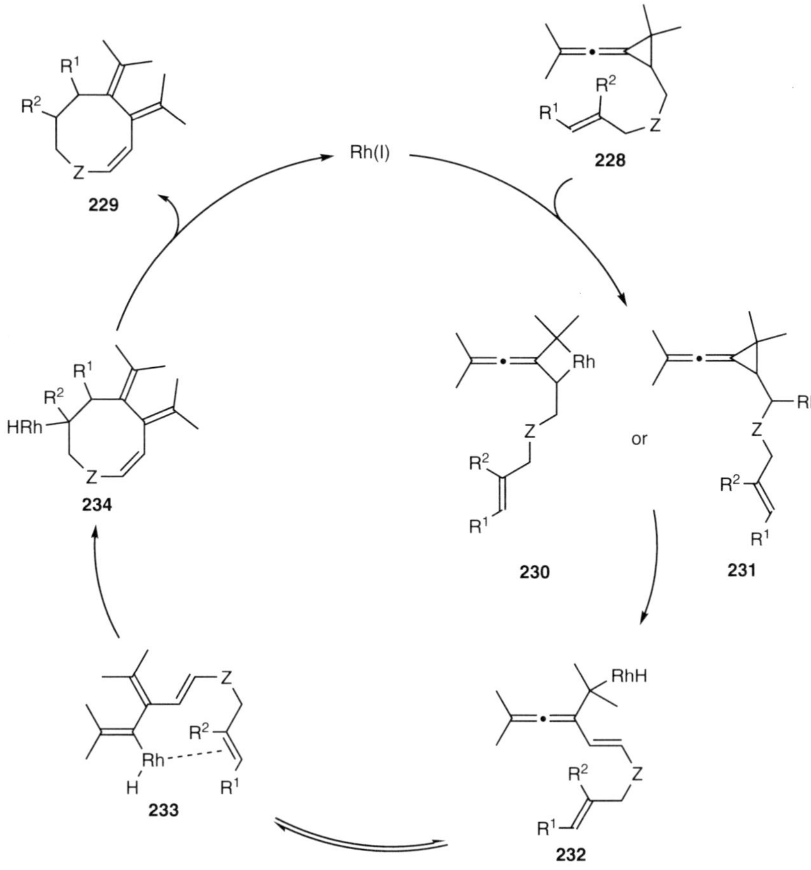

5,6-Dimethylenehexahydroazocines 229**; General Procedure:**[44]
Under an argon atmosphere, $RhCl(CO)(PPh_3)_2$ (10 mg, 0.015 mmol), the substrate (0.15 mmol), and toluene (6.0 mL) were charged into a Schlenk tube. The mixture was stirred at 100 °C until the reaction was complete. Then, the solvent was removed under reduced pressure, and the residue was purified by flash column chromatography (silica gel, EtOAc/petroleum ether 1:30) to afford the title compound.

1.4.1.3.5 Copper-Catalyzed Cycloisomerization of Vinylidenecyclopropanes

As shown in Scheme 53, 3-cyclopropylideneprop-2-en-1-ones are converted into benzo[*b*]furan-7(3a*H*)-one derivatives **236** by a tandem ring-opening and intermolecular (4 + 2)-cycloaddition process.[45] Previously, it was reported that furan-fused cyclobutene intermediate **235** could also be formed under palladium catalysis; however, no further cycloaddition was observed.[46] In this case, copper(I) chloride catalyzes the formation of **235**, and then also acts as a Lewis acid to further activate the cyclobutene intermediate, thereby facilitating the intermolecular cycloaddition.

Scheme 53 Tandem Ring Opening and Cycloaddition of 3-Cyclopropylideneprop-2-en-1-ones[45]

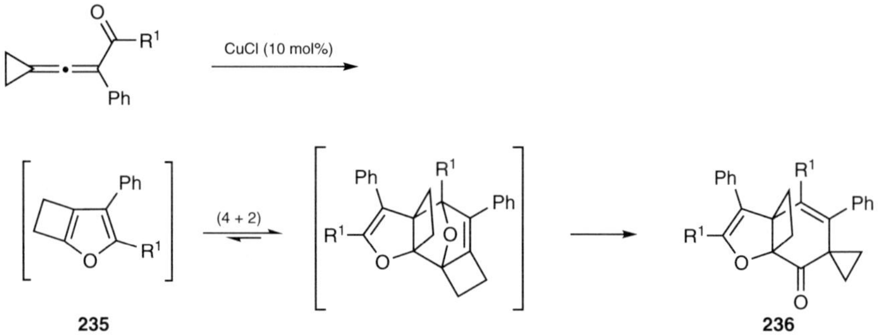

R^1	Yield (%)	Ref
4-Tol	87	[45]
Ph	83	[45]
$4\text{-FC}_6\text{H}_4$	61	[45]
$4\text{-MeOC}_6\text{H}_4$	91	[45]
$3,4,5\text{-(MeO)}_3\text{C}_6\text{H}_2$	64	[45]
2-furyl	86	[45]
2-thienyl	63	[45]

3′,5′-Diphenyl-2′,4′-di-4-tolyl-7′*H*-spiro[cyclopropane-1,6′-[3a,7a]ethanobenzofuran]-7′-one (236, R^1 = 4-Tol); Typical Procedure:[45]
Under an atmosphere of dry N_2, CuCl (1.5 mg, 0.015 mmol, 10 mol%) was added to a soln of the vinylidenecyclopropane (39 mg, 0.150 mmol) in anhyd THF (3 mL) at 50 °C. After stirring for 10–12 h (monitored by TLC), the reaction was quenched with H_2O (5 mL), and the mixture was extracted with EtOAc (3 × 10 mL). The combined organic layer was dried ($MgSO_4$). After filtration and removal of the solvent under reduced pressure, the residue was purified by flash chromatography (silica gel, petroleum ether/EtOAc 15:1) to afford the title compound as a white solid; yield: 34 mg (87%).

1.4.1.4 Other Cyclopropanes

Other functionalized cyclopropanes such as cyclopropanols and alkynylcyclopropanes have been used widely as readily accessible three-carbon-atom synthons in organic synthesis.

1.4.1.4.1 Gold-Catalyzed Cycloisomerization of Functionalized Cyclopropanes

Transition-metal-catalyzed ring expansions of cyclopropanols have been intensively investigated by many groups. Iwasawa has shown that 1-(alk-1-ynyl)cyclopropanols can be converted into cyclopent-2-enones in the presence of hexacarbonyldicobalt(0).[47] Interestingly, gold(I) catalysts show different reactivity toward such cyclopropanols, and cyclobutanones **237** are formed instead of the cyclopentenones (Scheme 54).[48] Notably, this reaction stereoselectively provides a single *E*-alkene isomer and is stereospecific with regard to the substituents on the ring. As suggested by Toste, there are two possible mechanisms for this rearrangement. In path a, activation of the alkyne moiety by the gold catalyst induces a 1,2-alkyl shift. Path b involves gold activation of the cycloalkanol and then β-carbon elimination to yield the alkylgold species. The (*E*)-alkene geometry of the resulting products reveals that path a is more probable.

Scheme 54 Cyclobutanones by Gold-Catalyzed Ring Expansion of 1-(Alk-1-ynyl)cyclopropanols[48]

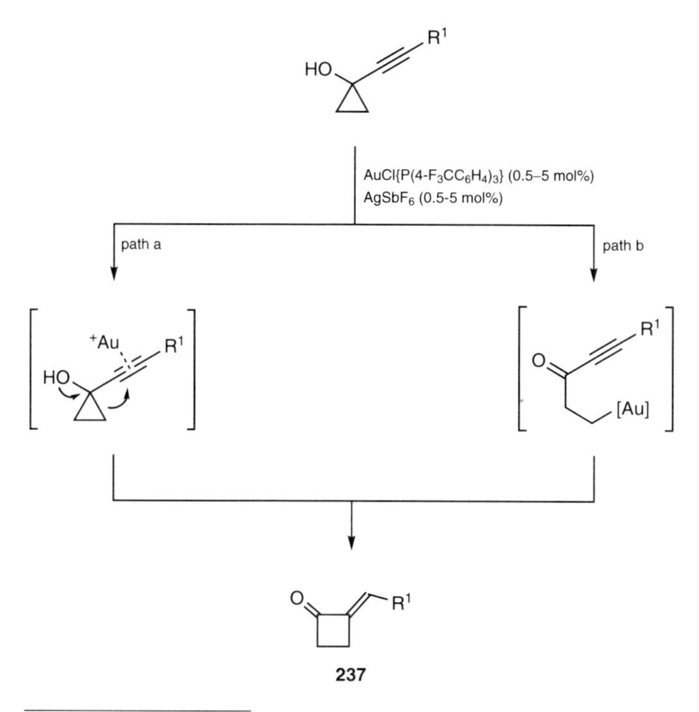

R^1	Yield (%)	Ref
t-Bu	99	[48]
4-ClC_6H_4	94	[48]
TMS	95	[48]
H	90	[48]
I	88	[48]

Alkynylcyclopropanes are another useful starting material that can be used to investigate the chemistry of cyclopropanes. In 2011, Barluenga reported the gold-catalyzed rearrangement of 6-alkynylbicyclo[3.1.0]hex-2-enes **238** to afford 6-alkynylcyclohexa-1,3-dienes **239** and **240** (Scheme 55).[49] There are two pertinent features of this reaction. First, the distal cyclopropane C—C bond is selectively broken; second, the alkyne functional group remains unreacted. Interestingly, exchange of the R^1 and R^2 substituents is sometimes observed in this reaction. Moreover, alkynylcyclopropanes possessing a primary alkyl substituent (i.e., R^2 = alkyl) give a completely different reaction outcome, and the bicyclic products **241** and **242** are formed.

Scheme 55 6-Alkynylcyclohexa-1,3-dienes by Gold-Catalyzed Rearrangement of Alkynylcyclopropanes[49]

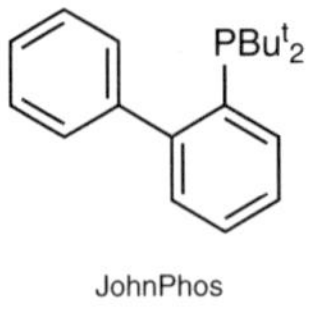

PBu$^t_2$

JohnPhos

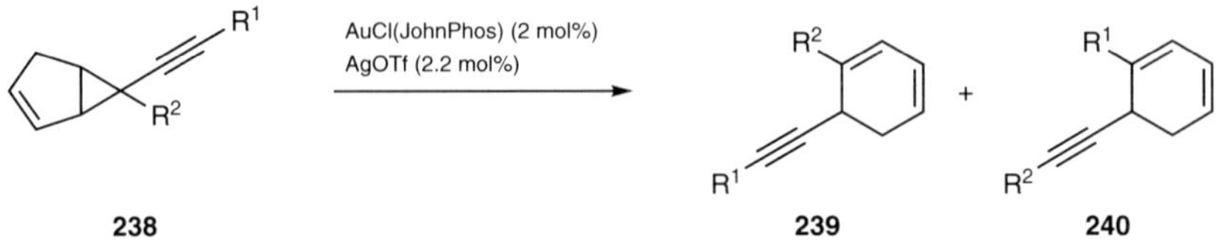

238 → AuCl(JohnPhos) (2 mol%), AgOTf (2.2 mol%) → 239 + 240

R^1	R^2	Ratio (**239**/**240**)	Overall Yield (%)	Ref
Ph	Ph	–	75	[49]
4-ClC_6H_4	4-ClC_6H_4	–	88	[49]
4-ClC_6H_4	Ph	1.4:1	70	[49]
4-ClC_6H_4	4-$MeOC_6H_4$	3:1	62	[49]
Ph	cyclopropyl	1:0	54	[49]
Ph	*t*-Bu	1:0	47	[49]

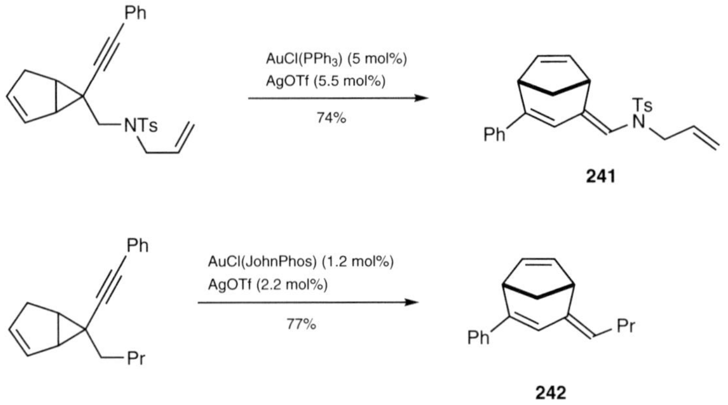

AuCl(PPh_3) (5 mol%), AgOTf (5.5 mol%), 74% → 241

AuCl(JohnPhos) (1.2 mol%), AgOTf (2.2 mol%), 77% → 242

A plausible mechanistic rationale for exchange of the substituents and the formation of these different products is outlined in Scheme 56. Upon coordination of alkyne **238** with gold to give activated complex **243**, nucleophilic attack from either the alkene (path a) or the cyclopropane (path b) is likely to occur. The derived cyclopropyl and allyl cationic intermediates formed through path a stay in equilibrium, from which exchange of the substituents can be rationalized by reversible cleavage A or B, and **244** is formed. For substrates with a primary alkyl substituent as R^2, elimination of a proton yields the corresponding bicyclic derivatives **241**/**242**. Cleavage of the cyclopropane ring in **243** or **244** gives newly formed cyclopropane intermediate **245** or **246**, which delivers final product **239** or **240** through a ring-opening process (path b).

Scheme 56 Mechanism for the Exchange of the Substituents and the Formation of the Products in the Gold-Catalyzed Rearrangement of Alkynylcyclopropanes[49]

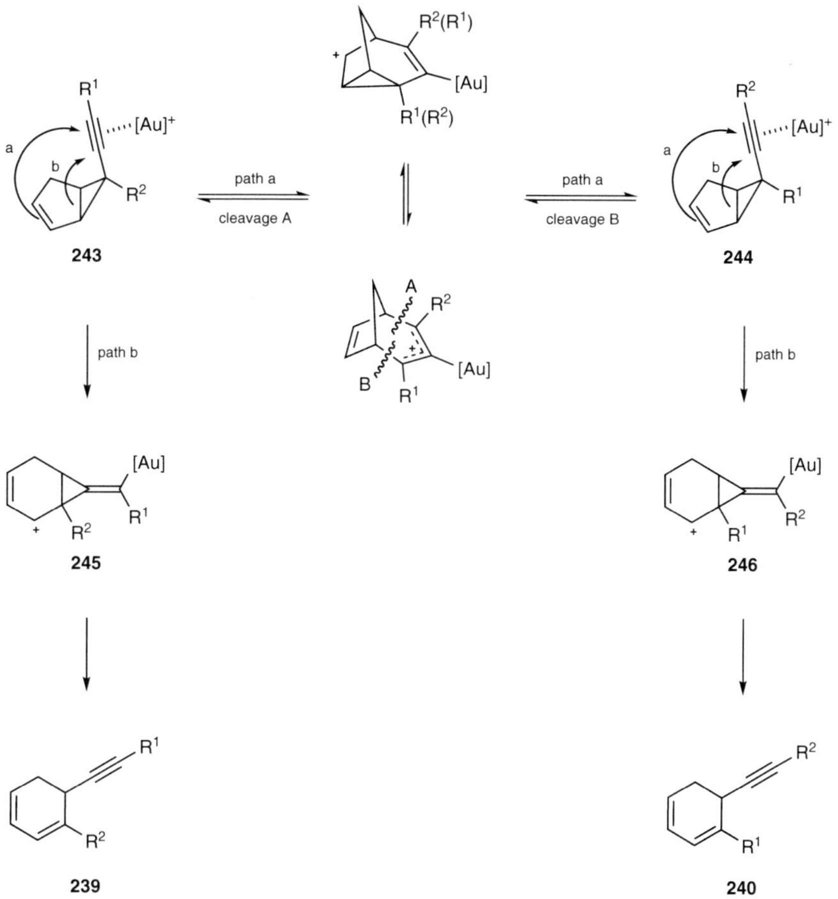

The gold-catalyzed intramolecular cycloisomerization of 1-(alk-1-ynylcyclopropyl)-alkanols **247** has been reported by Shi and co-workers; using gold catalysts the reaction leads to highly functionalized carbazoles through ring expansion or ring opening of cyclopropanes (Scheme 57).[50] Unlike Jiao's example (see Section 1.4.1.4.4),[51] indolyl substituents introduced into the substrates take part in the reaction. The outcome of the reaction is critically dependent on two factors: first, only gold catalysts with sterically bulky ligands work well, and others usually lead to complex mixtures; second, the reaction intermediates are very sensitive to other nucleophiles, and if different solvents are used, a ring-expansion or nucleophilic ring-opening reaction is observed. By performing the reaction in dichloromethane, cyclobutane-fused carbazoles **248** are obtained in yields of 34–87%; if toluene is used as the solvent, carbazole ethers **249** are formed as the major products in moderate yields; if alcohols are used, the corresponding ether products **250** are delivered in yields of 44–99% after nucleophilic opening of the cyclopropane ring by the alcohol.

Scheme 57 Cycloisomerization of Indolyl–1-(Alk-1-ynylcyclopropyl)alkanols[50]

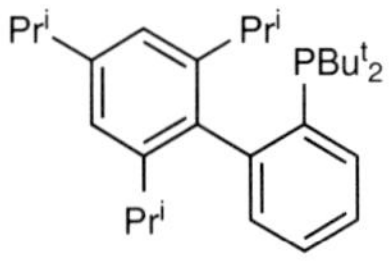

t-Bu-XPhos

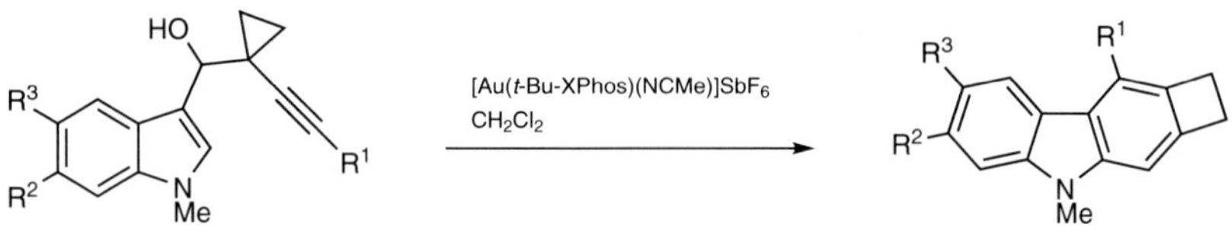

247 → 248: [Au(t-Bu-XPhos)(NCMe)]SbF6, CH_2Cl_2

R^1	R^2	R^3	Yield (%)	Ref
4-Tol	H	H	60	[50]
4-$MeOC_6H_4$	H	H	34	[50]
Ph	F	H	87	[50]

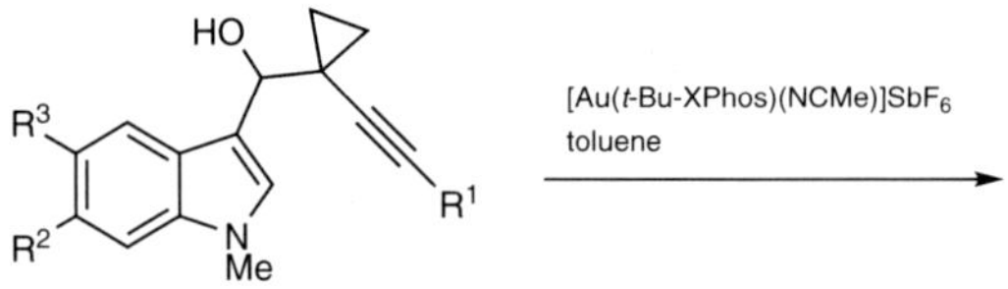

247 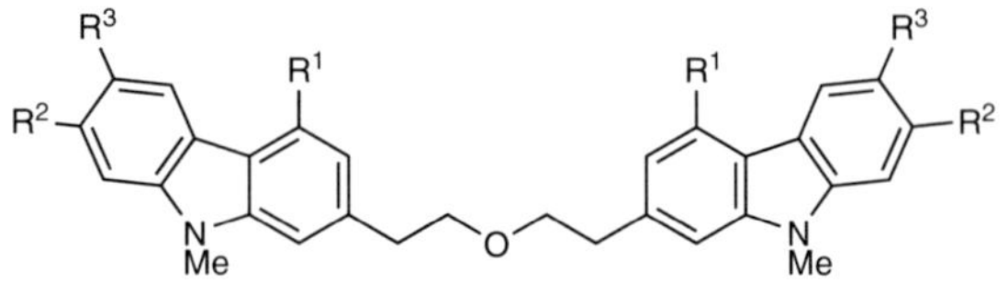 → 249: [Au(t-Bu-XPhos)(NCMe)]SbF6, toluene

R^1	R^2	R^3	Yield (%)	Ref
4-Tol	H	H	63	[50]
Ph	F	H	76	[50]

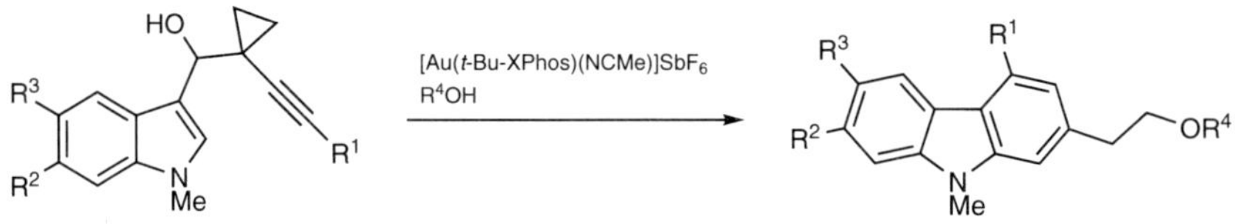

→ 250: [Au(t-Bu-XPhos)(NCMe)]SbF6, R^4OH

R^1	R^2	R^3	R^4	Yield (%)	Ref
Ph	H	H	Me	94	[50]
Ph	H	OMe	Et	81	[50]
4-$MeOC_6H_4$	H	H	Me	87	[50]

A proposed mechanism is provided in Scheme 58. First, the gold(I) complex activates the starting material by coordination with the alkyne moiety to give **251**. The following Friedel–Crafts-type reaction delivers spirocyclic intermediate **252**, which rearranges into cationic intermediate **253** through 1,2-migration. After proton elimination and release of one molecule of water, gold carbene **254** is formed. Cyclobutane-fused carbazole **255** is afforded after carbene-induced ring expansion if dichloromethane is used as the solvent, and deauration of **255** furnishes **248**. If water or alcohol is present in the reaction system, these nucleophiles will attack carbene intermediate **254** to initiate a ring-opening reaction to generate the corresponding alcohols **250** and ether derivatives **249** via intermediate **256**.

Scheme 58 Proposed Catalytic Cycle for the Cycloisomerization of Indolyl-1-(Alk-1-ynylcyclopropyl)alkanols[50]

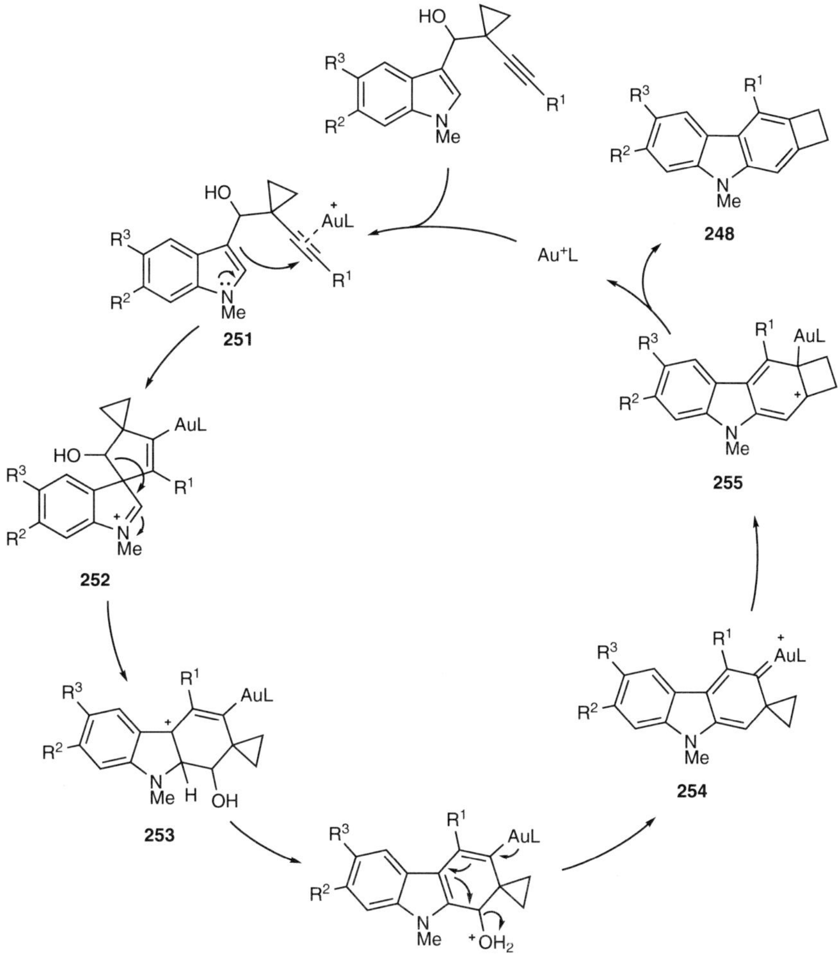

HO R3 R2 N Me R1

Au^+L

251 **252** **253** **254** **255** **248**

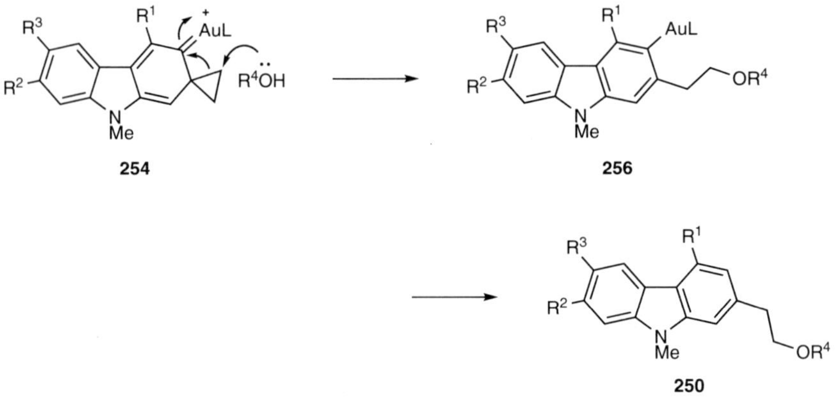

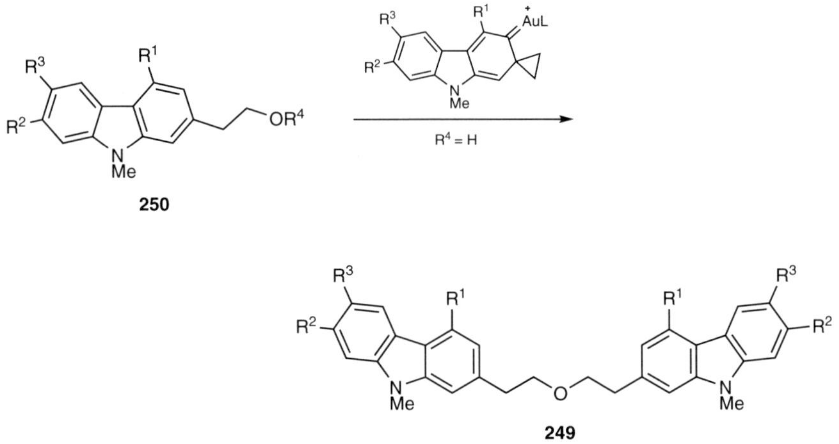

(*E*)-2-(2,2-Dimethylpropylidene)cyclobutan-1-one (237, $R^1 = t$-Bu)**; Typical Procedure:**[48]
1-(3,3-Dimethylbut-1-ynyl)cyclopropan-1-ol (1.00 mmol) was added to a stirring suspension of AuCl{P(4-$F_3CC_6H_4)_3$} (0.005 mmol) and $AgSbF_6$ (0.005 mmol) in CH_2Cl_2 (1 mL), and the mixture was stirred at rt for 6 h. After removal of the solvent under reduced pressure, purification by flash chromatography (silica gel, EtOAc/hexanes 1:9) afforded the title compound as a white crystalline solid; yield: 136 mg (99%).

[(2-Phenylcyclohexa-2,4-dienyl)ethynyl]benzene (239, $R^1 = R^2 =$ Ph)**; Typical Procedure:**[49]
A prepared stock soln containing AuCl(JohnPhos) (3.2 mg, 2.0 mol%) and AgOTf (1.7 mg, 2.2 mol%) in 1,2-dichloroethane (1.0 mL) was added to a soln of the alkynylcyclopropane (77 mg, 0.30 mmol) in 1,2-dichloroethane (2.0 mL) under an atmosphere of argon at ambient temperature. The resulting mixture was immediately placed into a preheated oil bath at 70 °C. After 10 min, silica gel was added and the solvent was removed under reduced pressure. The remaining residue was purified by column chromatography (silica gel, hexanes/CH_2Cl_2 10:1) to yield the title compound as a colorless oil; yield: 57 mg (75%).

Carbazoles 248 **and** 250**; General Procedure:**[50]
The gold catalyst (5 mol%) was added to a stirred soln of the cyclopropyl alcohol in CH_2Cl_2 or alcohol (0.1 M) under an air atmosphere at rt, and the mixture was stirred until the reaction was complete (monitored by TLC). Then, the mixture was concentrated under reduced pressure, and the residue was purified by flash column chromatography (silica gel, petroleum ether/EtOAc 10:1) to give the title compound.

1.4.1.4.2 Ruthenium-Catalyzed Cycloisomerization of Functionalized Cyclopropanes

Trost has demonstrated that the ring expansion of 1-alkynylcyclopropanols is highly dependent on the substituents on the alkyne.[52] For silyl-substituted 1-alkynylcyclopropanols **256**, ring expansion catalyzed by a ruthenium(II) complex gives cyclobutanone derivatives **257**, for which the *Z*-isomers are the major products (Scheme 59). The reaction of electron-deficient 1-alkynylcyclopropanols **258** also delivers cyclobutanones **259**, and in all the reported cases, high chemoselectivity is achieved (*Z*/*E* ratio >20:1). For alkyl-substituted 1-alkynylcyclopropanols **260**, the reaction might go through C—C bond insertion to form a ruthenacyclohexenone intermediate, which then delivers cyclopentenones **261**.

Scheme 59 Cyclobutanones and Cyclopentenones by Ruthenium-Catalyzed Ring Expansion of 1-Alkynylcyclopropanols[52]

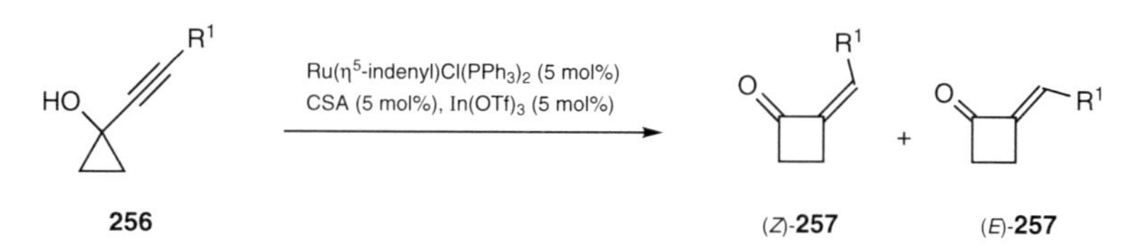

R^1	Ratio [(*Z*)-**257**/(*E*)-**257**]	Total Yield (%)	Ref
TMS	5.7:1	98	[52]
$SiMe_2Bn$	6.0:1	94	[52]
$SiMe_2Ph$	6.0:1	96	[52]
TIPS	>20:1	87	[52]

O
R1
HO
Ru(η5-indenyl)Cl(PPh3)2 (5 mol%)
CSA (5 mol%), In(OTf)3 (5 mol%)
R1
O
O
258
259 (*Z*/*E*) >20:1

R^1	Yield (%)	Ref
Cy	88	[52]
OEt	68	[52]
OBn	81	[52]
$4-O_2NC_6H_4O$	85	[52]

for references see p 256

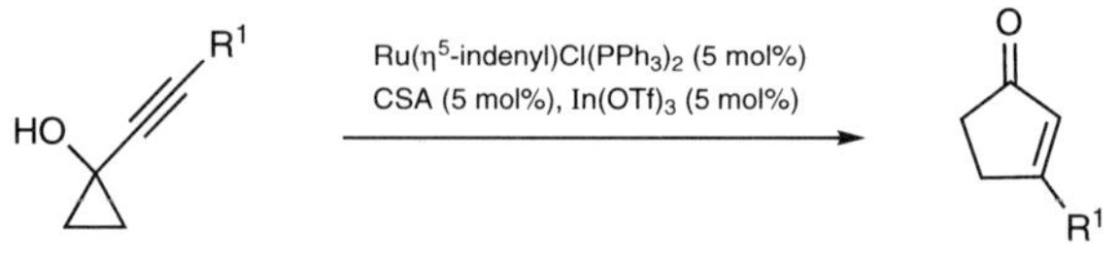

Ru(η^5-indenyl)Cl(PPh_3)$_2$ (5 mol%)
CSA (5 mol%), In(OTf)$_3$ (5 mol%)

260 → 261

R^1	Yield (%)	Ref
Bn	81	[52]
Cy	88	[52]
$(CH_2)_5Me$	78	[52]
$(CH_2)_3OBn$	76	[52]
$(CH_2)_4OBn$	68	[52]

Tang and Shi have developed a convenient method for the easy construction of cyclobutane-fused pyrroles **263** from alkynyl imines **262** (Scheme 60).[53] The substrate scope is broad. It is noteworthy that potential byproducts derived from Wagner–Meerwein-type 1,2-alkyl shift (as shown in **264**) are not observed, whereas carbene-initiated 1,2-shift (as shown in **265**) does occur. This reaction can also be extended to other cyclic and noncyclic alkynyl imines.

Scheme 60 Cycloisomerization of Cyclopropyl-Tethered Alkynyl Imines for the Synthesis of Pyrroles[53]

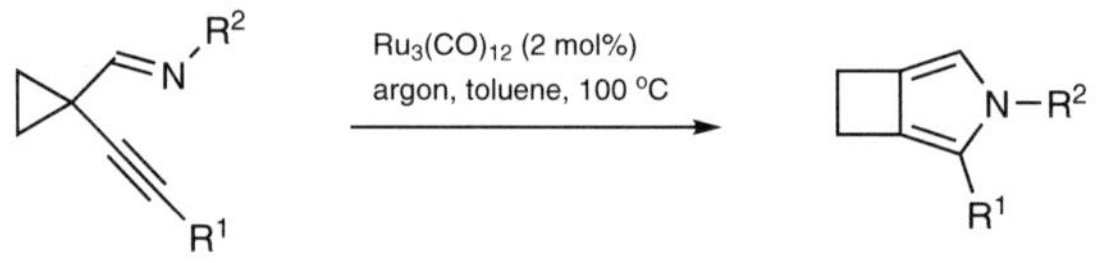

$Ru_3(CO)_{12}$ (2 mol%)
argon, toluene, 100 °C

262 → 263

R^1	R^2	Yield (%)	Ref
Ph	Pr	97	[53]
Ph	iPr	83	[53]
Ph	Bu	99	[53]
Ph	iBu	95	[53]
Ph	Bn	91	[53]
Ph	PMB	94	[53]
Ph	$(CH_2)_2Ph$	89	[53]
4-PhC_6H_4	PMB	94	[53]
4-Tol	Bu	97	[53]

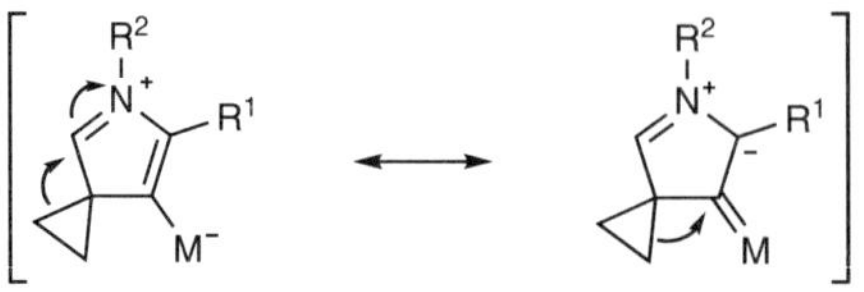

264 ⟷ 265

A mechanism involving the formation of a carbene has been proposed (Scheme 61). Upon coordination of imine **262** with a ruthenium(0) or rhodium(I) catalyst, metallic complex **266** is formed. Rearrangement furnishes carbene **267** or zwitterionic intermediate **268**. Then, carbene-induced ring expansion takes place to form cyclobutane intermediate **269**, which after aromatization then affords the final product **263**.

Scheme 61 Proposed Mechanism for the Ruthenium(0)- or Rhodium(I)-Catalyzed Cycloisomerization of Cyclopropyl-Tethered 3-Alkynyl Imines[53]

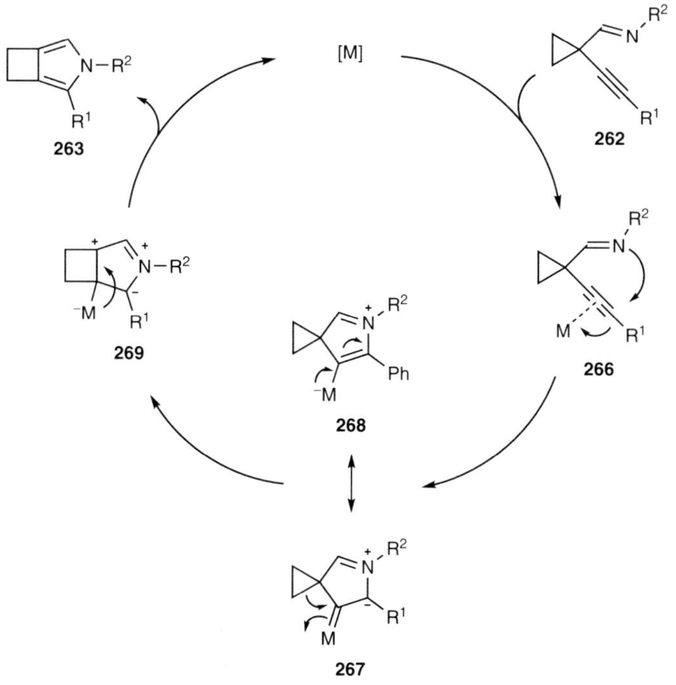

M = Rh(I) or Ru(0)

Cyclopropanes substituted with a single acyl group have been less explored than activated cyclopropanes such as methylenecyclopropanes and vinylidenecyclopropanes because such substrates are not very active. Unlike methylenecyclopropanes and vinylidenecyclopropanes, there are no reports describing the insertion of a transition metal into the C—C bond of such cyclopropanes; Lewis acids are more commonly used, but a catalyst used in stoichiometric amount is usually required. However, radical ring opening of such substrates seems very effective. In 2011, Yoon reported the intramolecular (3 + 2) cycloaddition of aryl cyclopropyl ketones with alkenes to generate highly substituted cyclopentanes **272**. Using a photocatalytic system comprising tris(2,2′-bipyridyl)ruthenium(II) chloride, one-electron reduction of the aryl cyclopropyl ketone generates the corresponding radical anion **270**, which is stabilized by the Lewis acid lanthanum(III) trifluoromethanesulfonate. Subsequent opening of the cyclopropane ring gives alkyl radical species **271**, which leads to the final products after tandem rearrangement (Scheme 62).[54]

for references see p 256

Scheme 62 (3 + 2) Cycloaddition of Aryl Cyclopropyl Ketones by Visible-Light Photocatalysis[54]

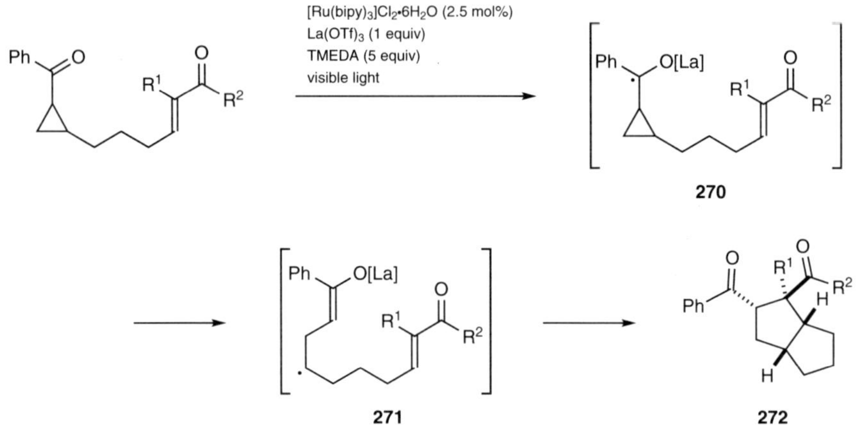

$[Ru(bipy)_3]Cl_2{\cdot}6H_2O$ (2.5 mol%)
$La(OTf)_3$ (1 equiv)
TMEDA (5 equiv)
visible light

270

271

272

R^1	R^2	dr	Yield (%)	Ref
Me	OEt	6:1	83	[54]
H	OEt	2.3:1	67	[54]
Me	O*t*-Bu	5:1	86	[54]
Et	SEt	10:1	70	[54]
Me	*t*-Bu	4:1	82	[54]

(*Z*)-2-[(Trimethylsilyl)methylene]cyclobutanone [(*Z*)-257, R^1 = TMS] and (*E*)-2-[(Trimethylsilyl)methylene]cyclobutanone [(*E*)-257, R^1 = TMS]; Typical Procedure:[52]
A mixture of the ruthenium catalyst (31.0 mg, 0.040 mol), $In(OTf)_3$ (22.5 mg, 0.040 mol), CSA (9.3 mg, 0.040 mmol), and 1-alkynylcyclopropanol **256** (123.4 mg, 0.80 mmol) in THF (16 mL) was heated at reflux for 2 h. Purification by flash chromatography (silica gel, Et_2O/petroleum ether 4:96) afforded (*Z*)-**257** as a colorless oil; yield: 102.4 mg (83%); and (*E*)-**257** as a colorless oil; yield 18.5 mg (15%); ratio (*Z*/*E*) 5.7:1. The yield and *Z*/*E* ratio were determined through analysis of the protons of the vinyl group on the cyclobutanone by ^1H NMR spectroscopy with the use of mesitylene (12.2 mg, 0.10 mmol) as an internal standard.

3-Hexylcyclopent-2-en-1-one [261, R^1 = $(CH_2)_5Me$]; Typical Procedure:[52]
THF (2.0 mL) was added to a flame-dried flask charged with the ruthenium catalyst (23.3 mg, 0.03 mmol), $In(OTf)_3$ (16.9 mg, 0.03 mmol), and CSA (7.0 mg, 0.03 mmol). The mixture was stirred for 10 min at rt, and then a soln of cyclopropanol **260** (100 mg, 0.60 mmol) in THF (1.0 mL) was added. The soln was heated to reflux. After stirring at reflux for 4 h, the soln was cooled to rt and concentrated under reduced pressure. Purification of the residue by flash chromatography (silica gel, Et_2O/petroleum ether 5:95 to 15:85) gave the title compound as a liquid; yield: 78 mg (78%).

3-Butyl-2-phenyl-3-azabicyclo[3.2.0]hepta-1,4-diene (263, R^1 = Ph; R^2 = Bu); Typical Procedure:[53]
A 25-mL Schlenk tube was flame-dried under vacuum and was then charged with **262** (45 mg, 0.2 mmol) and $Ru_3(CO)_{12}$ (2 mg, 2 mol%). The tube was evacuated and backfilled with argon (5 ×). Then, toluene (2.0 mL) was added, and the mixture was stirred for 10 h in an oil bath heated at 100 °C. The solvent was removed under reduced pressure, and

the residue was eluted through a short column (basic alumina, petroleum ether/EtOAc 60:1) to afford the title compound as a colorless oil; yield: 45 mg (99%).

Fused Cyclopentanes 272; General Procedure:[54]
A 25-mL Schlenk flask containing $MgSO_4$ (2 wt% with respect to the substrate) was flame-dried under vacuum and cooled under an atmosphere of dry N_2. The flask was then charged with $[Ru(bipy)_3]Cl_2{\bullet}6H_2O$ (2.5 mol%), $La(OTf)_3$ (1 equiv), the substrate (1 equiv), MeCN (0.05 M), and TMEDA (5 equiv). The soln was then degassed through three freeze–pump–thaw cycles under an atmosphere of N_2. The reaction flask was placed in a water bath, and the mixture was stirred in front of a 23-W (1380 lumen) compact fluorescent lamp at a distance of 30 cm. Upon consumption of the starting material, the mixture was diluted with Et_2O and passed through a short pad of silica gel (Et_2O). The filtrate was concentrated under reduced pressure, and the residue was purified by column chromatography (silica gel, hexanes/EtOAc) to give the title compound.

1.4.1.4.3 Palladium-Catalyzed Cycloisomerization of Cyclopropane Aminals

Tsuritani has reported a synthetic method for the construction of 3,4-dihydro-2(1*H*)-quinolinones **275** from readily available cyclopropane aminals **273**.[55] In some cases, 2-alkoxy-3,4-dihydroquinolines **274** are formed, but such products might be intermediates of the quinolinones, as hydrolysis affords the desired products **275** (Scheme 63).

Scheme 63 3,4-Dihydro-2(1*H*)-quinolinones by Palladium-Catalyzed Ring Opening of Cyclopropane Aminals[55]

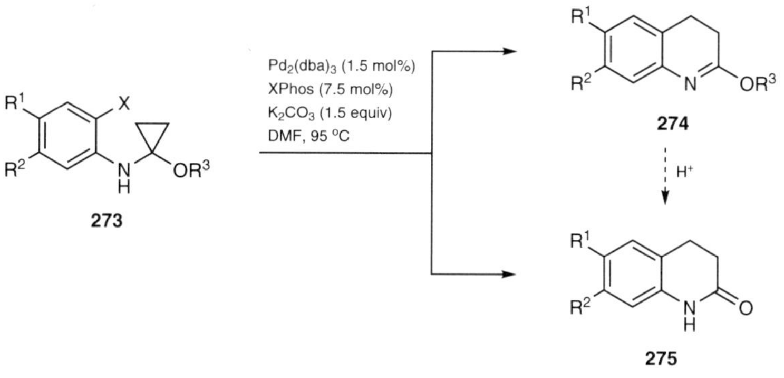

Pd2(dba)3 (1.5 mol%)
XPhos (7.5 mol%)
K2CO3 (1.5 equiv)
DMF, 95 °C

273 → 274 (H+) → 275

R^1	R^2	R^3	X	Yield (%) of **274**	Yield (%) of **275**	Ref
H	H	Me	I	88	–[a]	[55]
H	H	Me	Br	90	–[a]	[55]
H	H	Me	Cl	70	–[a]	[55]
F	H	Me	Br	–[a]	87[b]	[55]
CN	H	Et	Br	–[a]	58[b]	[55]
H	OMe	Et	Br	–[a]	82[b]	[55]

[a] Product not formed.
[b] Reaction was quenched with 1 M HCl instead of H_2O.

A plausible reaction mechanism is shown in Scheme 64. Oxidative addition of aryl halide **273** to palladium(0) and intramolecular ligand exchange give rise to four-membered azapalladacycle intermediate **276**. Trigged by relief of ring strain, β-carbon elimination readily takes place to furnish seven-membered azapalladacycle intermediate **277**. After reductive elimination, dihydroquinoline **274** is formed and the catalyst is regenerated; hydrolysis of **274** furnishes **275**. Notably, in some cases, dihydroquinolines **274** resist hydrolysis and are obtained as the final products.

Scheme 64 Proposed Mechanism for the Palladium-Catalyzed Ring Opening of Cyclopropane Aminals[55]

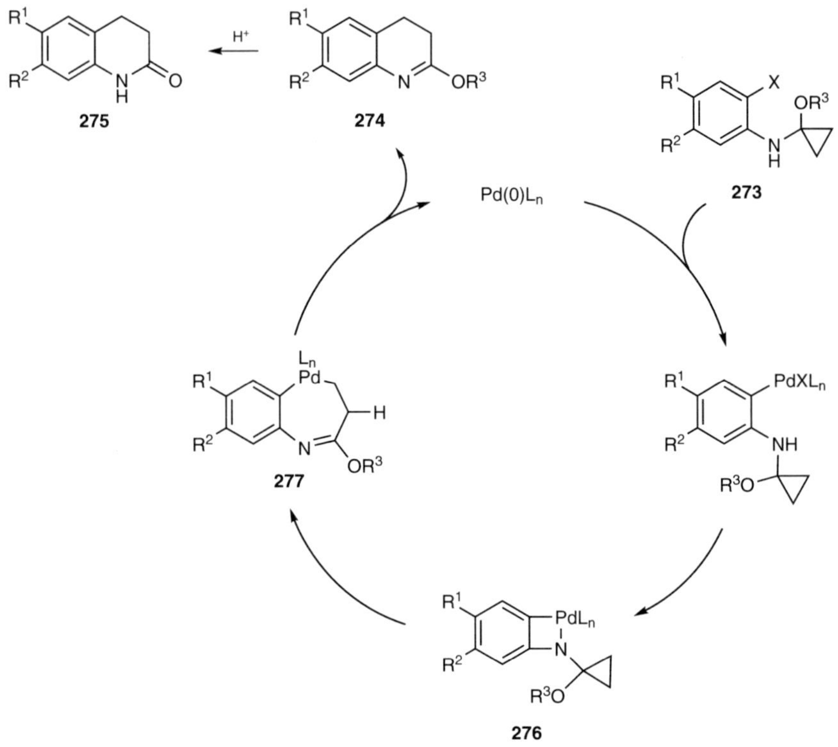

3,4-Dihydro-2(1*H*)-quinolinones 275; General Procedure:[55]

A mixture of the *N*-cyclopropyl-2-haloaniline **273** (1.0 mmol, 0.2 M), K_2CO_3 (1.5 equiv), $Pd_2(dba)_3$ (1.5 mol%), and XPhos (7.5 mol%) in DMF (5.0 mL) was stirred at 95 °C under a N_2 atmosphere in a round-bottomed flask for the time specified. The mixture was then poured into aq 1 M HCl, and the resulting mixture was extracted with EtOAc (3 ×). The combined organic extract was washed with brine (2 ×). The obtained soln was dried (Na_2SO_4) and concentrated under reduced pressure. The residue was purified by flash chromatography to give the title compound.

1.4.1.4.4 Iron-Catalyzed Cycloisomerization of 1-(Alkynylcyclopropyl)alkanols

1-(Alkynylcyclopropyl)alkanols can undergo ring expansion to cyclobutanols *trans*-**278** (major) and *cis*-**278** (minor) in the presence of iron(II) chloride and oxygen (Scheme 65).[51] It is postulated that iron(III) is generated in situ through oxidation by oxygen to promote the reaction. The R^2 substituent controls the stereoselectivity, and *trans*-products are favored.

Scheme 65 Cyclobutanols by Iron-Catalyzed Selective Ring Expansion of 1-(Alkynylcyclopropyl)alkanols[51]

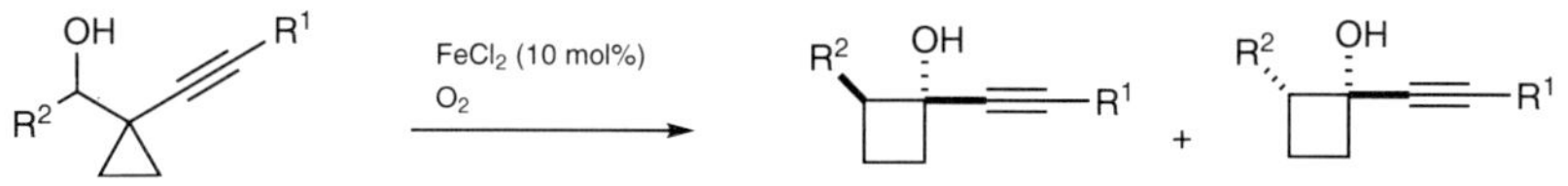

FeCl2 (10 mol%)
O2

trans-**278** cis-**278**

R^1	R^2	Ratio (*trans*/*cis*)	Yield (%)	Ref
Ph	Me	9:1	78	[51]
Ph	Et	6:1	88	[51]
Ph	Pr	4:1	57	[51]
2-Tol	Me	3:1	62	[51]
$(CH_2)_5Me$	Me	4:1	54	[51]

favored → trans   disfavored → cis

(1*R,2*R**)-2-Methyl-1-(phenylethynyl)cyclobutanol (*trans*-278, R^1 = Ph; R^2 = Me) and (1*R**,2*S**)-2-Methyl-1-(phenylethynyl)cyclobutanol (*cis*-278, R^1 = Ph; R^2 = Me); Typical Procedure:**[51]

CAUTION: *Nitromethane is flammable, a shock- and heat-sensitive explosive, and an eye, skin, and respiratory tract irritant.*

1-[1-(Phenylethynyl)cyclopropyl]ethanol (37.2 mg, 0.2 mmol) was added to a mixture of $FeCl_2$ (2.5 mg, 0.02 mmol, 10 mol%) and acetone/$MeNO_2$ (1:1; 2 mL). The mixture was stirred at rt for 48 h. The resulting mixture was concentrated and purified by flash chromatography (silica gel, petroleum ether/Et_2O 5:1) to afford a mixture of the title compounds as a liquid; yield: 25 mg (78%, based on the conversion); ratio (*trans*/*cis*) 9:1; and recovered 1-[1-(phenylethynyl)cyclopropyl]ethanol; yield: 4.3 mg (12%).

1.4.1.4.5 Silver-Catalyzed Cycloisomerization of Cyclopropanes

Zhang reported the ring expansion of alkynylcyclopropanes for the synthesis of naphthalen-2(1*H*)-ones **281** (Scheme 66).[56] This transformation is very special owing to the fact that the silver catalyst does not coordinate with the alkyne but with the ketone group. Upon treatment with the silver catalyst, the ketone is activated and nucleophilic addition takes place from the alkyne to form cyclopropyl cation **279**, which then undergoes ring expansion to yield cyclobutane intermediate **280**. Subsequent rearrangement and 1,2-migration of the R^3 substituent produces the final naphthalene derivatives **281**.

for references see p 256

Scheme 66 Naphthalen-2(1*H*)-ones by Silver-Catalyzed Ring Expansion of Alkynylcyclopropanes[56]

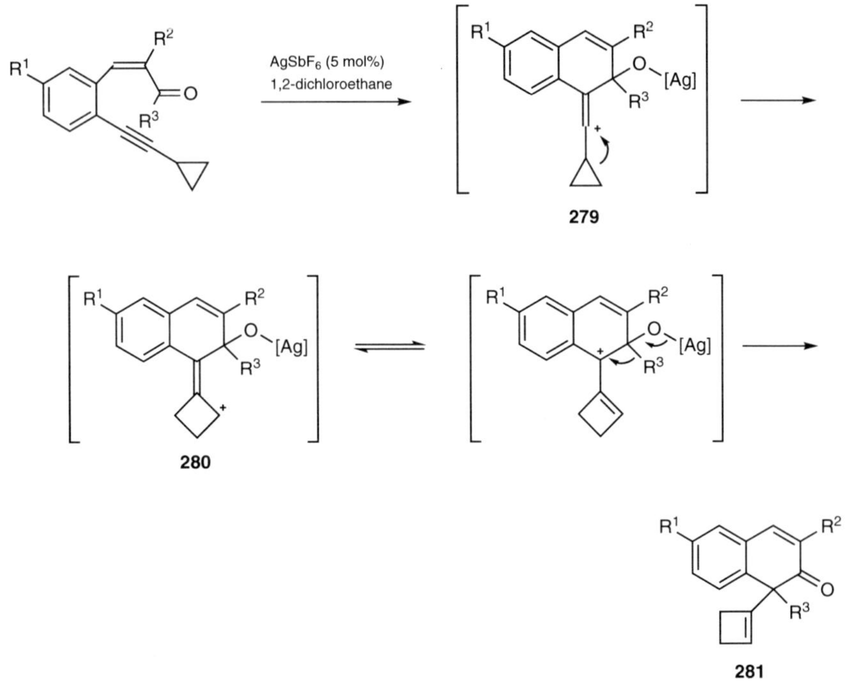

$AgSbF_6$ (5 mol%)
1,2-dichloroethane

279

280

281

R^1	R^2	R^3	Yield (%)	Ref
H	Bz	Ph	93	[56]
H	CO_2Me	Ph	80	[56]
H	CO_2Et	Ph	84	[56]
F	Bz	Ph	90	[56]
H	CO_2Me	4-$MeOC_6H_4$	86	[56]
H	CO_2Me	4-Tol	82	[56]

It is well known that diazo compounds and triazoles can generate carbenes in the presence of transition-metal catalysts. Therefore, it is possible that an in situ formed cyclopropyl carbene can initiate a ring-expansion reaction. For example, Tang has reported the carbene-induced ring expansion of cyclopropane-containing diazo compounds **282** for the easy construction of synthetically valuable cyclobutenes **283** (Scheme 67).[57] Moreover, silver azavinyl carbenes **285** derived from triazoles **284** can also be converted into the corresponding cyclobutene derivatives according to two different protocols: hydrolysis of the imine intermediate gives aldehyde products **286**, whereas amine derivatives **287** are obtained after one-pot reduction (Scheme 67).[58]

Scheme 67 Carbene-Induced Ring Expansion of Cyclopropanes[57,58]

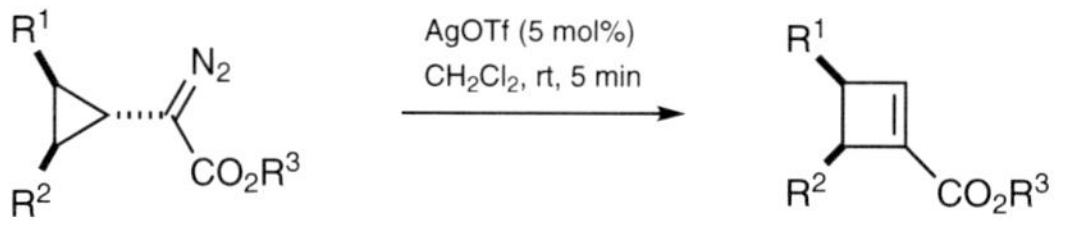

AgOTf (5 mol%), CH_2Cl_2, rt, 5 min: 282 → 283

R^1	R^2	R^3	Yield (%)	Ref
H	H	$(CH_2)_2Ph$	91	[57]
Pr	Pr	Et	71	[57]
$(CH_2)_4$		Et	72	[57]
$(CH_2)_6$		Et	90	[57]
CO_2Me	H	Et	70	[57]

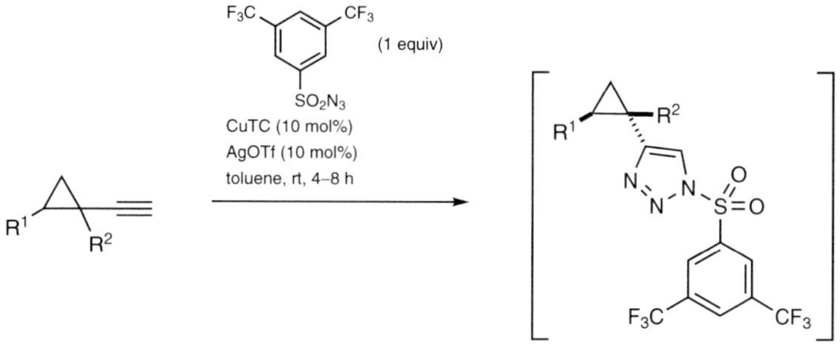

3,5-$(F_3C)_2C_6H_3SO_2N_3$ (1 equiv), CuTC (10 mol%), AgOTf (10 mol%), toluene, rt, 4–8 h → [284] → [285] → H_2O, alumina → 286

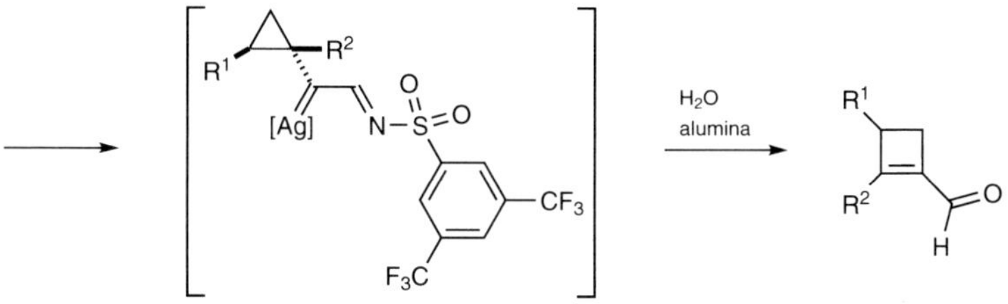

CuTC = copper(I) thiophene-2-carboxylate

R^1	R^2	Yield (%)	Ref
Ph	Ph	85	[58]
$4-BrC_6H_4$	Ph	80	[58]
$4-FC_6H_4$	Ph	80	[58]
Ph	$4-FC_6H_4$	82	[58]

for references see p 256

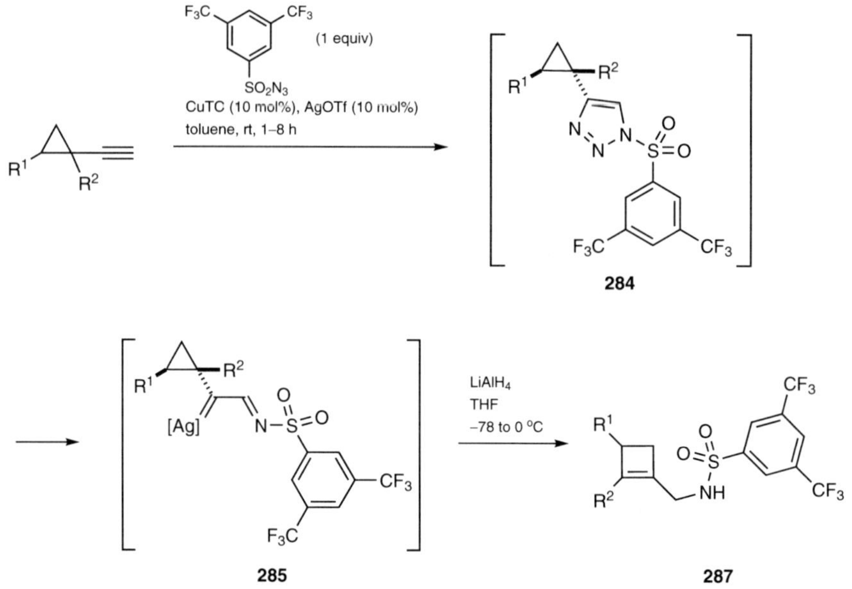

CuTC = copper(I) thiophene-2-carboxylate

R^1	R^2	Yield (%)	Ref
Ph	Ph	81	[58]
H	H	85	[58]
H	$CH_2OTBDMS$	80	[58]
Ph	Me	76	[58]

3-Benzoyl-1-(cyclobut-1-enyl)-1-phenylnaphthalen-2(1*H*)-one (281, $R^1 = H$; $R^2 = Bz$; $R^3 = Ph$)**; Typical Procedure:**[56]

Under a N_2 atmosphere, the alkynylcyclopropane (112.8 mg, 0.3 mmol) was added to a soln of $AgSbF_6$ (5.2 mg, 0.015 mmol) in 1,2-dichloroethane (3 mL). The mixture was stirred for 12 h at 80 °C, at which point the alkynylcyclopropane was completely consumed, as determined by TLC analysis. After concentration under reduced pressure, the residue was purified by column chromatography (silica gel, hexanes/EtOAc 6:1; R_f 0.41) to give the title compound as a white solid; yield: 105.2 mg (93%).

Cyclobut-1-enecarbaldehydes 286**; General Procedure:**[58]

A 4-mL vial was equipped with a stirrer bar, the azide (0.2 mmol), and the alkyne (0.2 mmol). Copper(I) thiophene-2-carboxylate (0.02 mmol), AgOTf (0.02 mmol), and toluene (2 mL) were then added. The mixture was stirred in a capped vial at rt for 4–8 h until the starting material was completely consumed, as determined by TLC. H_2O (several drops) and alumina (20 mg) were added to the mixture. The resulting suspension was stirred for 1 h until hydrolysis of the imine was complete. The solvents were removed under reduced pressure. The residue was purified by flash column chromatography (silica gel, EtOAc/hexane 1:10 to 1:4) to give the title compound.

N-(Cyclobut-1-enylmethyl)-3,5-bis(trifluoromethyl)benzenesulfonamides 287; **General Procedure:**[58]

CAUTION: *Solid lithium aluminum hydride reacts vigorously with a variety of substances, and can ignite on rubbing or vigorous grinding.*

A 4-mL vial was equipped with a stirrer bar, the azide (0.2 mmol), and the alkyne (0.2 mmol). Copper(I) thiophene-2-carboxylate (0.02 mmol), AgOTf (0.02 mmol), and toluene (2 mL) were then added. The mixture was stirred in a capped vial at rt for 1–8 h until the starting material was completely consumed, as determined by TLC. The mixture was cooled to −78 °C, and a soln of 0.2 M $LiAlH_4$ in THF (2 mL, 0.4 mmol) was added slowly. The mixture was stirred at the same temperature for 1 h and then warmed to 0 °C. The reaction was quenched by the careful addition of MeOH followed by H_2O. The solid was removed by filtration through a thin silica gel pad. The resulting soln was concentrated under reduced pressure to afford an oily residue, which was purified by column chromatography (silica gel) to give the title compound.

1.4.2 Intramolecular Cyclizations of Cyclopropenes

Metal catalysts can either coordinate with the double bonds of cyclopropenes or undergo insertion into the C—C single bonds; therefore, the chemistry of cyclopropenes is very rich. Generally, Lewis acids tend to activate double bonds, whereas transition-metal catalysts prefer to undergo oxidative addition, from which metal vinyl carbenes are usually formed.

1.4.2.1 Rhodium-Catalyzed Cycloisomerization of Cyclopropenes

Nefedov first reported the transition-metal-catalyzed cycloisomerization of cyclopropenes into furan derivatives, for which a carbene intermediate is proposed.[59] Later, Davies found that vinylcyclopropenes undergo cycloisomerization into multisubstituted furans.[60] However, the mechanistic rationale suggested by Davies involves the formation of a zwitterionic intermediate. The different behavior may be due to the vinyl substituent of the cyclopropene ring.

Padwa's discovery of the transition-metal-catalyzed acylcyclopropene-to-furan rearrangement further supports the vinyl carbenoid mechanism. Disubstituted furan **289**, derived from carbonyl–cyclopropene **288**, readily undergoes intermolecular cyclopropanation with the vinyl rhodium carbene species to generate oxabicyclohexene intermediate **290**, which is not very stable and thus quickly rearranges into oxabicyclic compound **291** as the major product (Scheme 68).[61] Müller and Doyle have also described the formation of vinyl carbenes.[62]

Scheme 68 Rhodium-Catalyzed Cycloisomerization of a Cyclopropene to a Furan and an 8-Oxa[3.2.1]octa-2,6-diene[61]

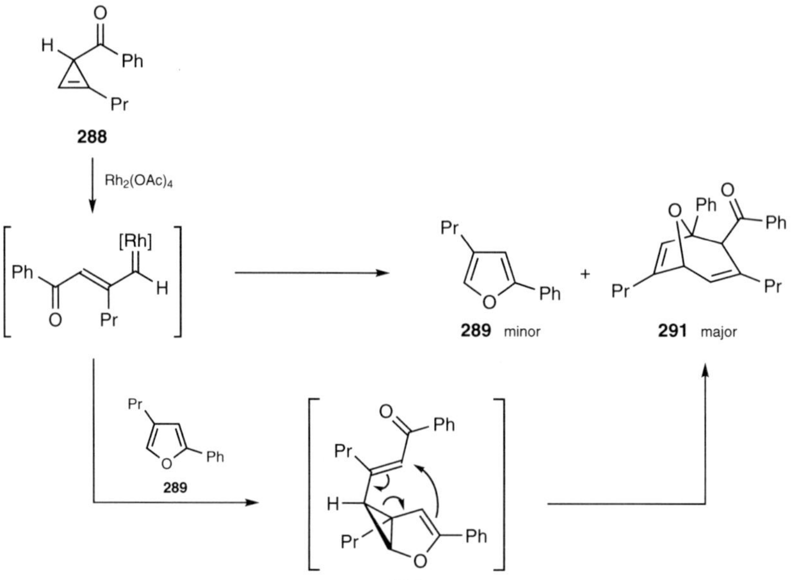

The intramolecular addition of α-diazo ketones to acetylenes via in situ formed vinyl carbenes has been intensively investigated by Padwa.[63] Several alkynyl-substituted diazo ketones were examined. For substrate **292** bearing a pentyl group, the vinyl carbene intermediate is generated from opening of the in situ formed cyclopropene ring. Subsequent carbene-induced 1,2-hydrogen shift or C(sp^3)—H functionalization leads to the corresponding conjugated diene **293** or cyclopentane **294**. Treatment of substrate **295**, however, which bears a tertiary carbon center, gives final product **296** after 1,2-migration (Scheme 69).

Scheme 69 Rearrangement of In Situ Formed Cyclopropenes[63]

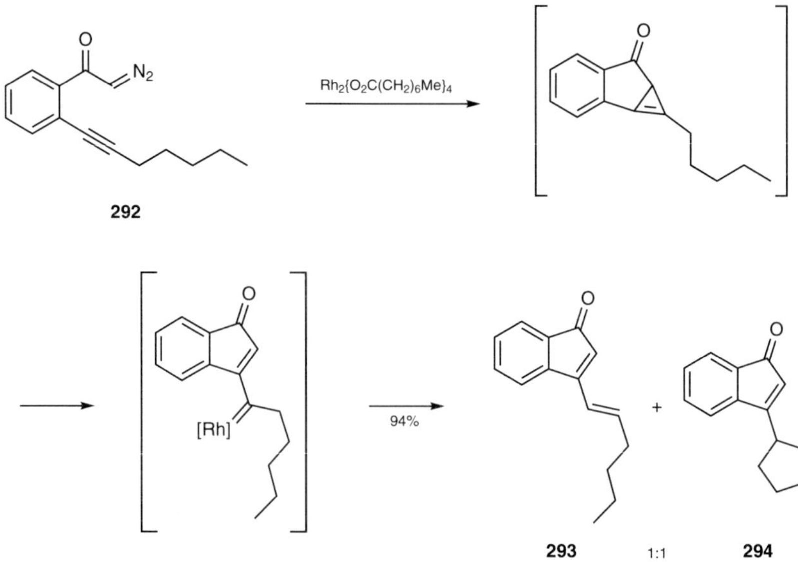

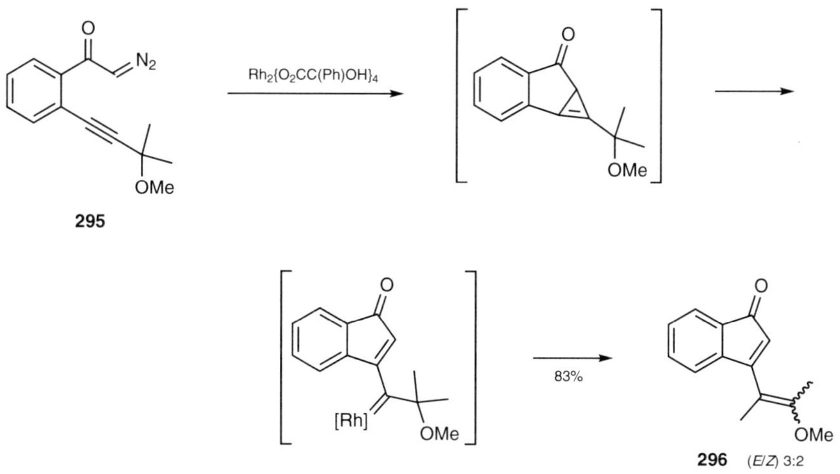

Rh2{O2CC(Ph)OH}4

295

83%

296 (E/Z) 3:2

Padwa has also designed a tandem process involving carbene transfer/cycloisomerization/intramolecular Diels–Alder cycloaddition in the stereoselective synthesis of complex polycyclic scaffolds using dirhodium(II) tetrakis(perfluorobutanoate) {$Rh_2(pfb)_4$; pfb = heptafluorobutanoate} as the catalyst. Thus, from diazo compound **297**, classic cyclopropenation takes place to give cyclopropene intermediate **298**, which then undergoes carbene-induced rearrangement to furnish furan intermediate **299**. Finally, after intramolecular Diels–Alder cycloaddition and a ring-opening process, the corresponding polycycle **300** is delivered (Scheme 70).[64]

Scheme 70 Construction of Polycyclic Ring Systems through Ring Opening of In Situ Formed Cyclopropenes[64]

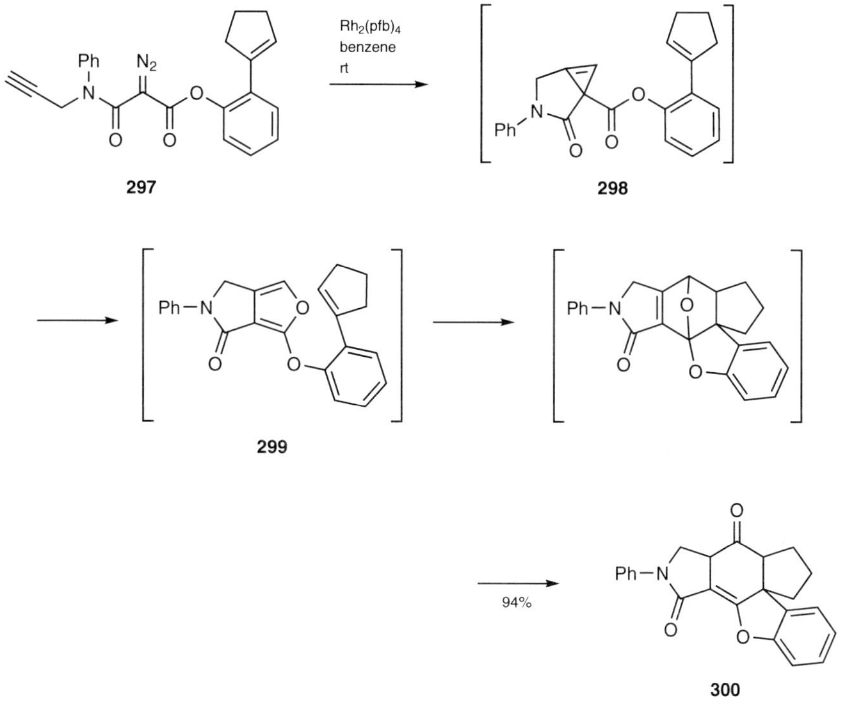

$Rh_2(pfb)_4$, benzene, rt

297

298

299

94%

300

pfb = $O_2C(CF_2)CF_3$

for references see p 256

Gevorgyan has demonstrated a highly efficient regiodivergent method for the synthesis of nitrogen-fused heterocycles via metal vinyl carbenes by using dirhodium(II) tetrakis-{(2*S*)-1-[(4-dodecylphenyl)sulfonyl]pyrrolidine-2-carboxylate} (Rh_2\{(*S*)-DOSP\}$_4$; (*S*)-DOSP = (2*S*)-1-[(4-dodecylphenyl)sulfonyl]pyrrolidine-2-carboxylate).[65] Starting from readily available triazoles **301** and alkynes, intermolecular cyclopropenation gives the corresponding pyridine-substituted cyclopropenes **302** in moderate to excellent yields. Interestingly, upon treatment of **302** with a rhodium(I) catalyst, products **303** are obtained with high regioselectivity in good to excellent yields (Scheme 71).

Scheme 71 Rhodium(I)-Catalyzed Rearrangement of 2-(Cycloprop-2-enyl)pyridines[65]

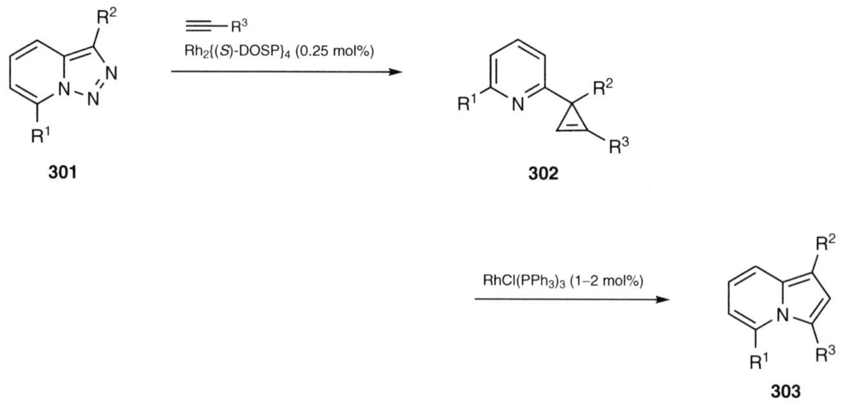

R^1	R^2	R^3	Yield (%) of **303**	Ref
Cl	Ph	Ph	85	[65]
Br	Ph	Ph	87	[65]
Cl	Ph	4-$MeOC_6H_4$	91	[65]
Br	4-$MeOC_6H_4$	Ph	81	[65]

The proposed mechanism is shown in Scheme 72. First, opening of the cyclopropene ring generates vinyl carbene **304** with cleavage of the more-substituted C—C bond of the cyclopropene. Then, nucleophilic addition or a pericyclic reaction takes place to yield zwitterion **305** or six-membered cyclic intermediate **306**. Subsequent rearrangement or reductive elimination gives **303**.

Scheme 72 Proposed Mechanism for the Rhodium(I)-Catalyzed Cycloisomerization of Cyclopropenes[65]

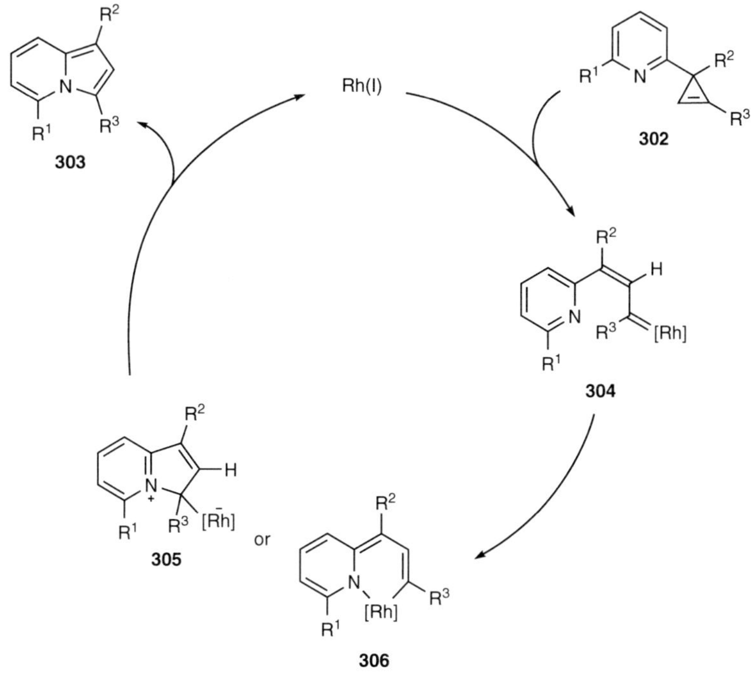

Rh(I)

302

304

305 or 306

303

Most of the carbenes appearing in the literature can be classified as donor–acceptor carbenes, because electron-deficient substituents are usually present. Owing to hazards associated with the preparation and handling of unstabilized diazo precursors, studies on donor rhodium carbenes are largely unexplored. Meyer and Cossy have reported a facile protocol to generate purely donor carbenes **308** from electron-rich cyclopropenes **307**; carbenes **308** can trigger highly efficient and diastereoselective intramolecular C(sp^3)—H insertions leading to various functionalized carbocycles/oxygen heterocycles (e.g., **309**) (Scheme 73).[66] This protocol nicely illustrates that 3,3-dimethylcyclopropenes can be used as good alternatives to α-diazo ketones in C—H insertion reactions. Moreover, it is certain that this study opens new opportunities for the investigation of donor carbenes.

for references see p 256

Scheme 73 Tetrahydropyrans by C(sp^3)–H Insertions with Donor Rhodium Carbenes[66]

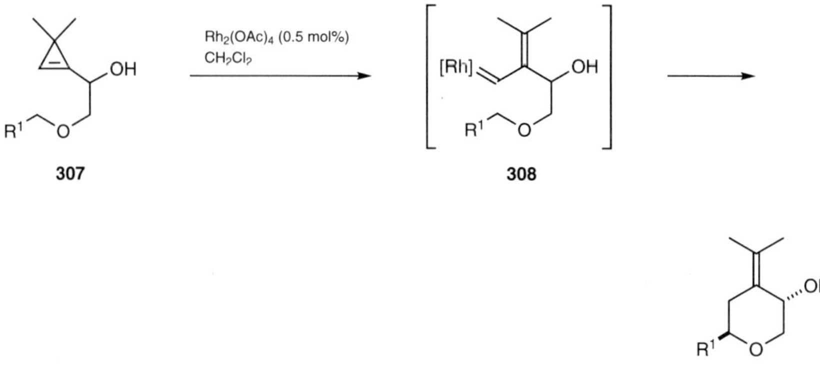

R^1	Yield (%)	Ref
(CH$_2$)$_6$Me	46	[66]
3-MeOC$_6$H$_4$	99	[66]
Ph	98	[66]
4-FC$_6$H$_4$	90	[66]

For cyclopropenes containing a cyclopropane moiety, the transformation under rhodium(I) catalysis is very interesting.[67] Rhodium can undergo C—C bond insertion into cyclopropenes **310** to give cyclic metallospecies **312**. Then, β-carbon elimination takes place and seven-membered cyclic intermediates **313** are formed. After reductive elimination of the catalyst, cyclohexadienes **314** are produced. The silyl enolate of cyclohexadiene **314** can be hydrolyzed to afford the corresponding ketone **311** (Scheme 74), or be further oxidized to afford a phenol derivative. The requirement of a silyl group in the substrate is presumably due to steric hindrance, which may prevent the rhodium(I) catalyst from inserting into the C—C bond of the cyclopropanol.

Scheme 74 Rhodium(I)-Catalyzed Rearrangement of [1,1′-Bi(cyclopropan)]-2′-en-1-ols To Give Cyclohexenones, and Catalytic Cycle[67]

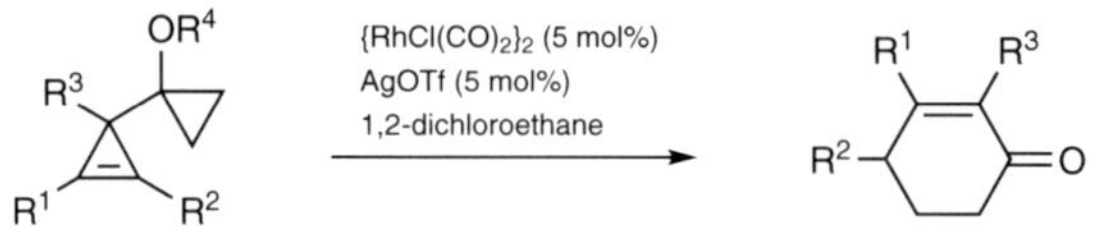

R^1 = Bu, Cy, Ph; R^2 = Me, Bu, Bn, Cy, Ph; R^3 = H, Ph; R^4 = TMS, TES, TIPS

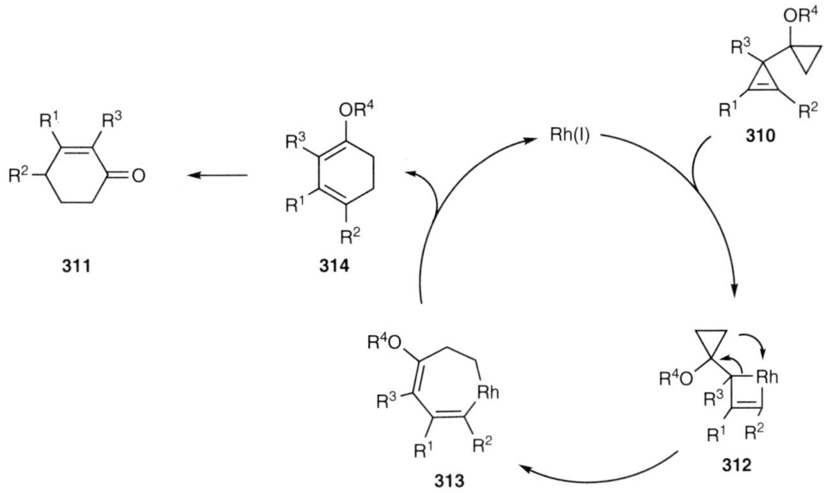

10-Phenyl-5,6,7,7a,9,10-hexahydro-8*H*-benzofuro[2,3-*e*]cyclopenta[*f*]isoindole-8,11(8a*H*)-dione (300); **Typical Procedure:**[64]

$Rh_2(pfb)_4$ (2 mg) was added to a soln of **297** (0.1 g, 0.27 mmol) in dry benzene (8 mL) (**CAUTION:** *carcinogen*) at 25 °C. The mixture was stirred for 10 min at rt, the solvent was removed under reduced pressure, and the residue was subjected to chromatography (silica gel) to give the title compound as a white solid; yield: 0.09 g (94%).

5-Chloro-1,3-diphenylindolizine (303, **R^1 = Cl; R^2 = R^3 = Ph); Typical Procedure:**[65]

An oven-dried, 3-mL Wheaton vial was charged with 2-chloro-6-(1,2-diphenylcycloprop-2-enyl)pyridine (**302**, R^1 = Cl; R^2 = R^3 = Ph; 0.152 g, 0.5 mmol) and Wilkinson's catalyst (9.2 mg, 0.01 mmol) under a N_2 atmosphere. Dry DMF (1 mL) was then added, and the mixture was stirred for 24 h, at which point the reaction was judged complete by TLC analysis. The mixture was quenched with sat. aq NH_4Cl (50 mL) and thoroughly extracted with hexanes. The organic phase was washed with brine and dried ($MgSO_4$). After removal of the solvents under reduced pressure, the residue was purified by flash chromatography (silica gel, Et_3N/hexanes 1:99) to afford the title compound as a yellow oil; yield: 0.129 g (85%).

(3*S,6*S**)-6-Heptyl-4-(propan-2-ylidene)tetrahydro-2*H*-pyran-3-ol** [309, **R^1 = $(CH_2)_6Me$]; Typical Procedure:**[66]

To a soln of cyclopropene **307** (124 mg, 0.517 mmol) in CH_2Cl_2 (5 mL) at rt, was added $Rh_2(OAc)_4$ (1.1 mg, 2.6 μmol, 0.5 mol%). After 7 h, the reaction was not complete and more $Rh_2(OAc)_4$ was added (2.2 mg, 5.2 μmol, 1 mol%). After a further 1 h stirring at rt, the mixture was concentrated under reduced pressure and analysis of the residue by ^1H NMR spectroscopy confirmed the formation of tetrahydropyran **309** as a single diastereomer. Purification by flash chromatography (silica gel, petroleum ether/EtOAc 90:10 to 80:20) afforded the title compound as a colorless oil; yield: 57.3 mg (46%).

Cyclohex-2-enones 311; **General Procedure:**[67]

A 10-mL, oven-dried Schlenk flask was charged with the Rh(I) catalyst (0.01 mmol, 5 mol%) under an atmosphere of argon. Then, a soln of silylated substrate **310** (0.2 mmol) in 1,2-dichloroethane (4 mL) was added, and the mixture was stirred at 70 or 110 °C under an atmosphere of argon. Upon completion of the reaction, the solvent was removed under reduced pressure, and the residue was purified by flash column chromatography (silica gel, petroleum ether or petroleum ether/EtOAc 30:1) to give the title compound.

1.4.2.2 Ruthenium-Catalyzed Cycloisomerization of Cyclopropenes

In 2010, Ma reported an interesting divergent synthesis of furan derivatives through catalyst-controlled ring-opening cycloisomerization of cycloprop-2-enecarboxylates **315** (Scheme 75).[68] A similar method had been previously reported by Ma, in which the reaction was conducted in the presence of bis(acetonitrile)dichloropalladium(II) (5 mol%) or copper(I) iodide (5 mol%).[69] In that case, cyclopropenes were found to undergo tandem rearrangement to form various furan derivatives with high regioselectivity. In the case of cycloprop-2-enecarboxylates **315**, high regioselectivity can be achieved by using a ruthenium(II) catalyst. The substrate scope is broad, and the corresponding 2,3,5-trisubstituted furans **316** are obtained in good yields.

Scheme 75 Ruthenium-Catalyzed Synthesis of Functionalized Furans[68]

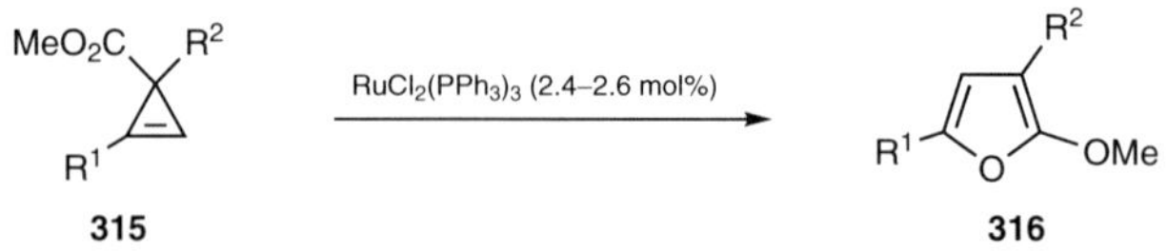

MeO_2C, R^2, R^1 — **315** —[$RuCl_2(PPh_3)_3$ (2.4–2.6 mol%)]→ R^1, R^2, O, OMe — **316**

R^1	R^2	Yield (%)	Ref
Bu	CO_2Me	95	[68]
$(CH_2)_7Me$	CO_2Me	87	[68]
Ph	CO_2Me	89	[68]
Ph	SO_2Ph	78	[68]

A rationale involving metal vinyl carbenes is provided in Scheme 76. Chlorometalation of **315** takes place, with the chloride anion attacking the less-substituted double bond. The chloride **317** derived from this reaction gives rise to vinyl carbene intermediate **318** after rearrangement. Subsequent nucleophilic attack of the carbonyl group at the carbene generates zwitterionic intermediate **319**, which results in furan derivatives **316** after elimination of the catalyst.

Scheme 76 Proposed Mechanism for Ruthenium(II)-Catalyzed Cycloisomerizations[68]

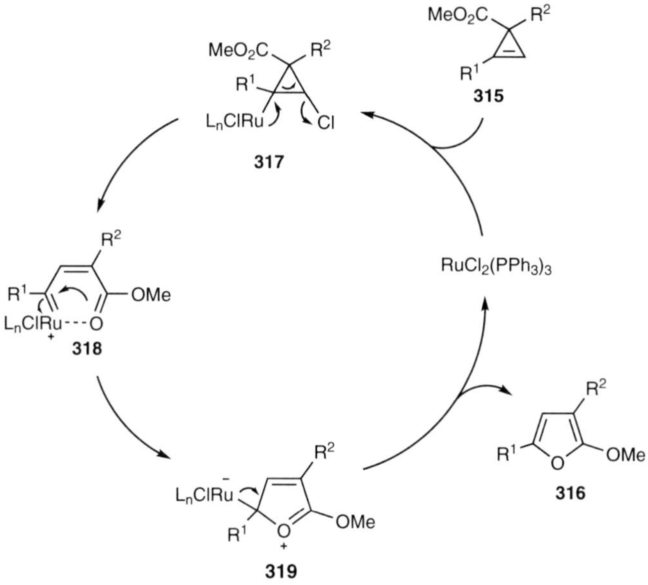

Methyl 5-Butyl-2-methoxyfuran-3-carboxylate (316, R^1 = Bu; R^2 = CO_2Me); Typical Procedure:[68]

$RuCl_2(PPh_3)_3$ (5 mg, 0.0052 mmol), cycloprop-2-ene-1,1-dicarboxylate **315** (42 mg, 0.20 mmol), and THF (2 mL) were added sequentially to a Schlenk reaction tube that had been evacuated and backfilled with argon. The resulting mixture was stirred at rt. After 11 h, the reaction was complete, as determined by TLC. The mixture was concentrated under reduced pressure, and purification of the residue by column chromatography (silica gel, petroleum ether/EtOAc 5:1) afforded the title compound as an oil; yield: 40 mg (95%).

1.4.2.3 Zinc-Catalyzed Cycloisomerization of Cyclopropanol-Substituted Cyclopropenes

Cyclopropenes and cyclopropanols are important, highly strained small-ring compounds that possess unique structures as well as chemical properties. Wang and co-workers envisaged that if the high reactivities of cyclopropene and cyclopropanol were merged together, very interesting reaction outcomes might be obtained if those substrates were treated with transition-metal or Lewis acid catalysts. Therefore, they synthesized a novel type of silylated alcohol **320** bearing connected cyclopropene and cyclopropanol moieties, and investigated their performance in the presence of a catalytic amount of zinc(II) iodide. Intriguing carbon skeletal reorganizations of the substrates with high regioselectivities are observed. Cyclohexanone derivatives **321** are obtained in moderate to good yields (Scheme 77).[67]

Scheme 77 Rearrangement of Silylated 1,1′-Bi(cyclopropan)-2′-en-1-ols[67]

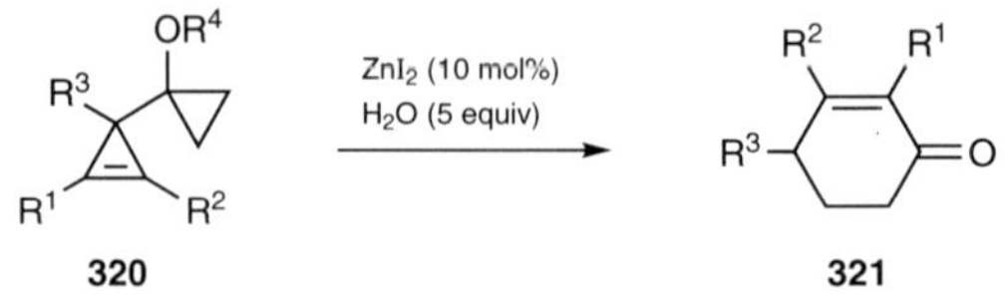

R^1	R^2	R^3	R^4	Yield (%)	Ref
Ph	Ph	H	TIPS	95	[67]
Ph	Ph	Ph	TMS	83	[67]
Bu	Bu	H	TIPS	94	[67]
Ph	Me	H	TIPS	82	[67]

The proposed mechanism of the zinc(II) iodide catalyzed rearrangement is shown in Scheme 78. Initially, cyclopropanol **320** is activated by the zinc(II) iodide Lewis acid to form zwitterionic complex **322**, from which cleavage of either the O—Si or C—O bond is possible. Cleavage of the O—Si bond gives ring-opened ketone **324** through zinc intermediate **323**. During optimization experiments, the formation of a ketone byproduct was observed in most cases. However, cleavage of the C—O bond is much more complicated. Cation-induced ring expansion furnishes spirocarbocycle **325**, which leads to the formation of bicyclo[2.2.0] skeleton **326** after expansion of the cyclopropane ring. Subsequent nucleophilic attack from a siloxy anion and rearrangement deliver cyclohexadiene intermediate **327**, which gives final cyclohexenone **321** after hydrolysis of the silyl enolate.

Scheme 78 Catalytic Cycle for the Zinc(II) Iodide Catalyzed Rearrangement of Silylated [1,1′-Bi(cyclopropan)]-2′-en-1-ols[67]

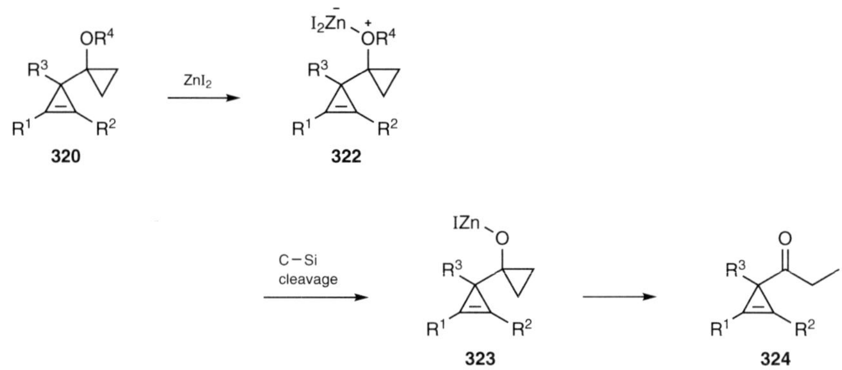

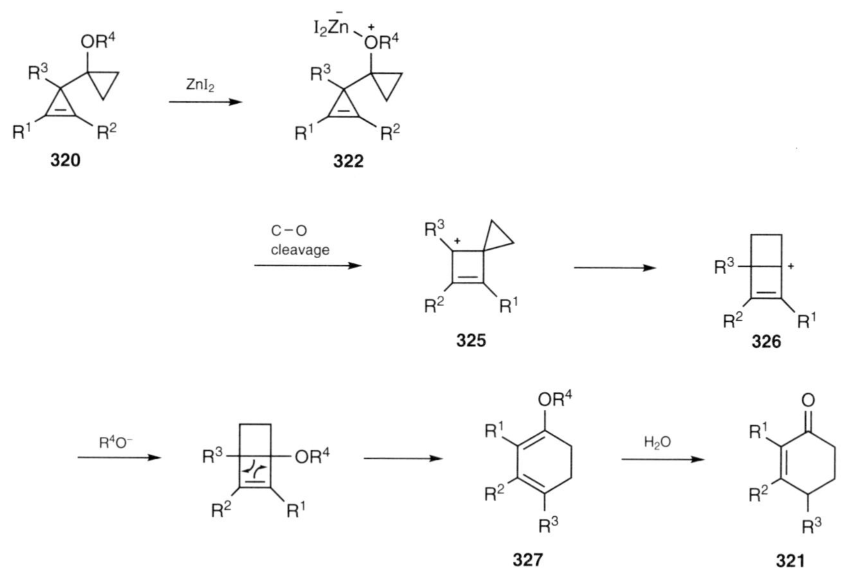

Cyclohex-2-enones 321; **General Procedure:**[67]

A 10-mL, oven-dried Schlenk flask was charged with ZnI_2 (6.4 mg, 0.02 mmol, 10 mol%) under an atmosphere of argon. Then, a soln of the silylated 1,1′-bi(cyclopropan)-2′-en-1-ol substrate (0.2 mmol) in 1,2-dichloroethane (4 mL) was added, and the mixture was stirred at 70 °C under an atmosphere of argon. Upon completion of the reaction, H_2O (18 μL, 5 equiv) was added, and the mixture was stirred at 70 or 110 °C under an atmosphere of argon for an additional 2 h. The solvent was then removed under reduced pressure, and the residue was purified by flash column chromatography (silica gel, petroleum ether/EtOAc 30:1) to afford the title compound.

1.4.2.4 Copper-Catalyzed Cycloisomerization of Cyclopropenes

The reaction of vinyl cyclopropenes **328** in the presence of copper(II) trifluoromethanesulfonate leads to indenes **329** with high regioselectivity (Scheme 79).[70] A mechanism is proposed based on control experiments. As can be seen from Scheme 79, copper coordinates with the cyclopropene to generate the corresponding cationic intermediates **330**. Triggered by a ring-opening process, vinyl cations **331** are formed. Subsequent Friedel–Crafts-type reaction at the aryl ring delivers the corresponding indene derivatives **329**. Steric hindrance may account for the high regioselectivity; the copper catalyst is small, so it prefers to form an aryl-ring-stabilized cation, which is more stable.

for references see p 256

Scheme 79 Indenes by Copper-Catalyzed Cycloisomerization of Vinyl Cyclopropenes[70]

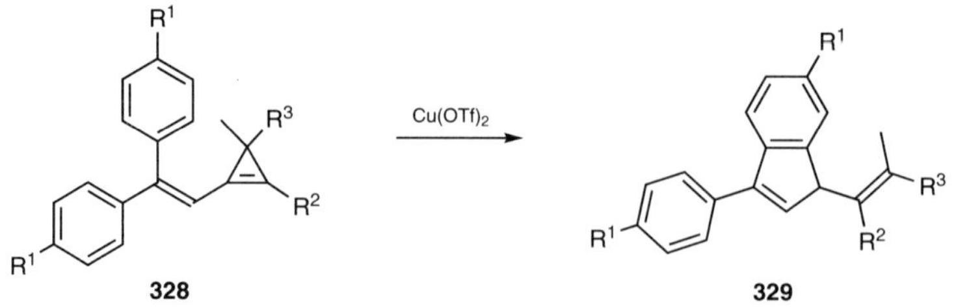

R^1	R^2	R^3	Yield (%)	Ref
H	Ph	Me	98	[70]
F	Ph	Me	68	[70]
H	Ph	H	84	[70]
Cl	4-ClC_6H_4	H	87	[70]

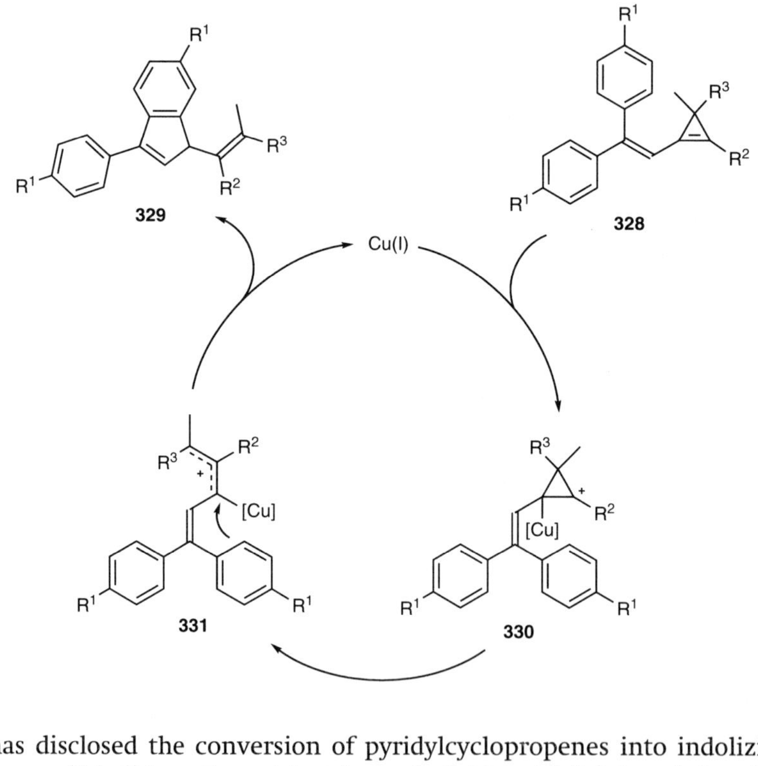

Gevorgyan has disclosed the conversion of pyridylcyclopropenes into indolizines **332** upon using copper(I) iodide as the catalyst instead of chlorotris(triphenylphosphine)rhodium(I) (see Section 1.4.2.1, Scheme 71).[65] In this case, copper coordinates with the cyclopropene at the less-substituted carbon of the C=C bond, and then vinyl carbene **333** is formed. Ensuing nucleophilic attack from the nitrogen atom or a pericyclic reaction takes place to give intermediate **334** or **335**, which delivers final product **332** (Scheme 80).

Scheme 80 Copper-Catalyzed Synthesis of Indolizines by Rearrangement of 2-(Cycloprop-2-enyl)pyridines[65]

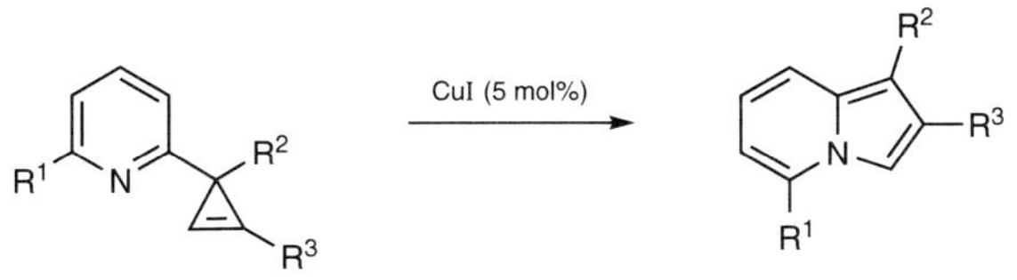

R^1	R^2	R^3	Yield (%)	Ref
Cl	Ph	Ph	75	[65]
Cl	CO_2Me	4-Tol	83	[65]
Cl	CO_2Me	4-$MeOC_6H_4$	95	[65]
Br	CO_2Me	Bu	71	[65]

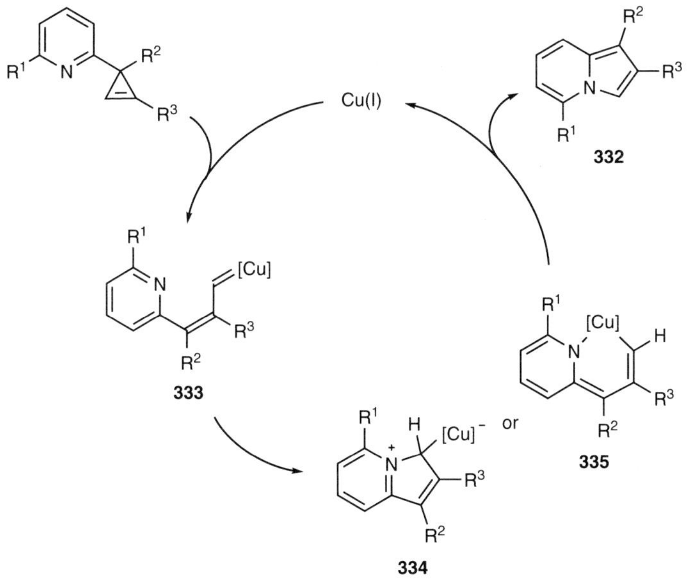

Ma has reported that in the reaction of cycloprop-2-enecarboxylates, copper(II) acetylacetonate exhibits catalytic activity that is different from that of dichlorotris(triphenylphosphine)ruthenium(II) (see Section 1.4.2.2, Scheme 75), with furan derivatives **336** formed in moderate to excellent yields.[68] According to the proposed mechanism, it is believed that the copper catalyst coordinates with the less-substituted side of the double bond to give vinyl copper carbene intermediate **337**. Subsequent nucleophilic attack and elimination afford the final furan derivatives (Scheme 81).

Scheme 81 Copper-Catalyzed Synthesis of Furans[68]

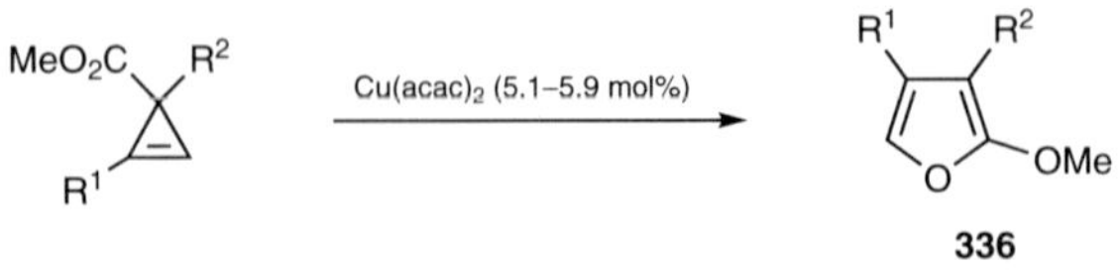

$Cu(acac)_2$ (5.1–5.9 mol%)

R^1	R^2	Yield (%)	Ref
Bu	CO_2Me	88	[68]
$(CH_2)_7Me$	CO_2Me	76	[68]
Ph	CO_2Me	96	[68]
Ph	SO_2Ph	67	[68]

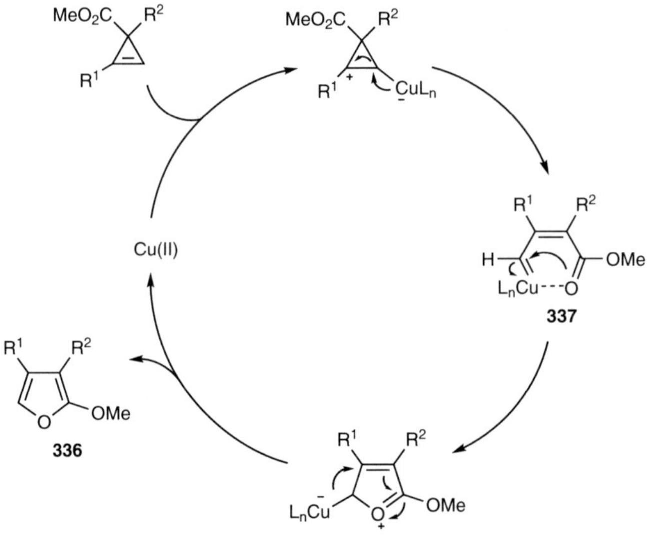

Indenes 329; **General Procedure:**[70]
$Cu(OTf)_2$ (10 mol%) was added to a 1,2-dichloroethane (5.0 mL) soln containing vinylcyclopropene **328** (0.2 mmol). The mixture was vigorously stirred at 50 °C for about 5 h. Flash column chromatography (silica gel) of the resulting mixture gave the title compound.

Indolizines 332; **General Procedure:**[65]
An oven-dried, 3-mL Wheaton vial was charged with 2-(cycloprop-2-enyl)pyridine (0.5 mmol) and CuI (4.8 mg, 0.025 mmol) under a N_2 atmosphere. Dry DMF (1 mL) was then added, and the mixture was stirred for 10 h, at which point the reaction was judged complete by TLC. The reaction was quenched with sat. aq NH_4Cl (50 mL) and was thoroughly extracted with hexanes. The organic phase was washed with brine and dried ($MgSO_4$). After removal of the solvents under reduced pressure, the residue was purified by flash chromatography (silica gel, Et_3N/hexanes 1:99) to afford the title compound.

Methyl 4-Butyl-2-methoxyfuran-3-carboxylate (336, **R^1 = Bu; R^2 = CO_2Me); Typical Procedure:**[68]
$Cu(acac)_2$ (3 mg, 0.011 mmol), dimethyl 2-butylcycloprop-2-ene-1,1-dicarboxylate (41 mg, 0.19 mmol), and MeCN (2 mL) were added sequentially to a Schlenk tube fitted with a screw cap that had been evacuated and backfilled with argon. The resulting mixture was

heated at reflux at 100 °C. After 48 h the reaction was complete, as judged by TLC. Concentration of the mixture and column chromatography of the residue (silica gel, petroleum ether/EtOAc 10:1) afforded the title compound as an oil; yield: 36 mg (88%).

1.4.2.5 Gold-Catalyzed Cycloisomerization of Cyclopropenes

If compounds **338** are treated with a gold catalyst instead of copper(II) trifluoromethanesulfonate (see Section 1.4.2.4, Scheme 79), the reaction outcome is quite different. Gold catalysts with ligands are much more sterically bulky; thus, gold coordinates to the less-hindered site of a cyclopropene to give a new type of indene derivative **339** (Scheme 82).[71]

Scheme 82 Gold-Catalyzed Synthesis of Indene Derivatives[71]

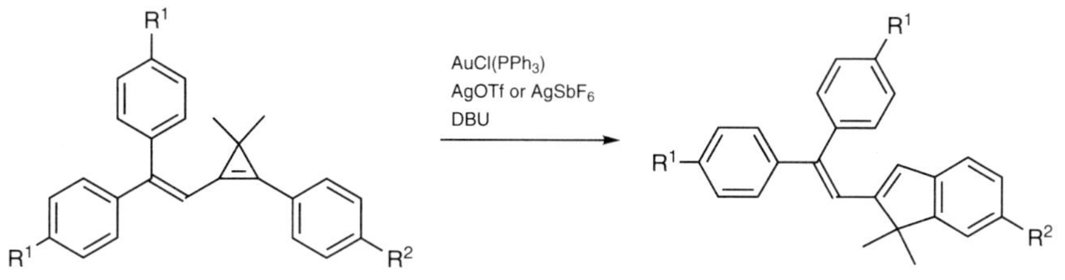

R1 R1 R1 R2 **338** $AuCl(PPh_3)$ AgOTf or $AgSbF_6$ DBU R1 R1 R2 **339**

R^1	R^2	Yield (%)	Ref
H	Me	99	[71]
Me	H	94	[71]
H	Cl	89	[71]
Cl	H	90	[71]

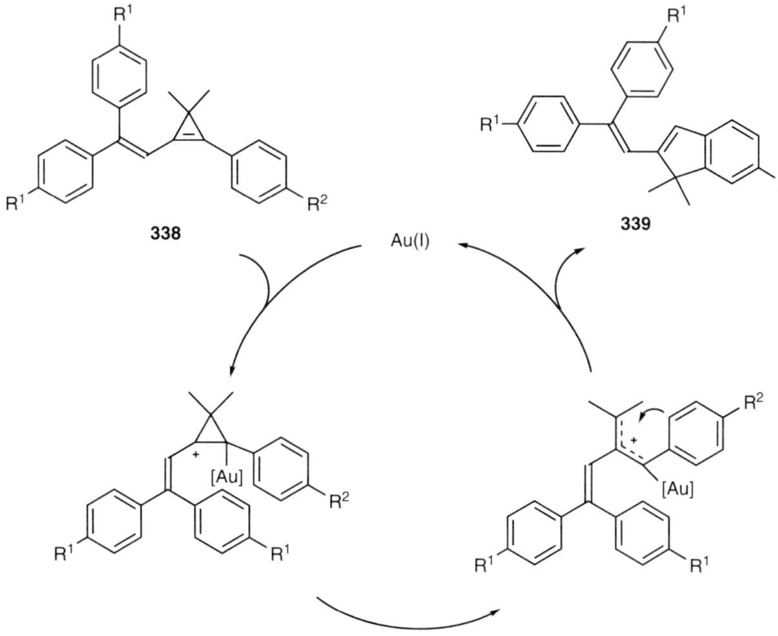

R1 R1 R2 **338** Au(I) R1 R1 R2 **339** [Au] R1 R1 R2 [Au] R1 R1 R2

for references see p 256

Meyer and Cossy have reported the gold-catalyzed cycloisomerization of allyl (3,3-dimethylcyclopropenyl)methyl ethers **340** (Z = O) or sulfonamides **340** (Z = NTs) to give 3-oxa- and 3-azabicyclo[4.1.0]heptanes **342** (Z = O, NTs) in excellent yields and with high diastereoselectivities (Scheme 83).[72,73] The formation of a cyclopropane indicates that vinyl carbene intermediate **341** is generated through classic rearrangement of a cyclopropene, as previously reported.

Scheme 83 3-Oxa- and 3-Azabicyclo[4.1.0]heptanes by Intramolecular Cyclopropanation via In Situ Formation of a Gold Carbene[72,73]

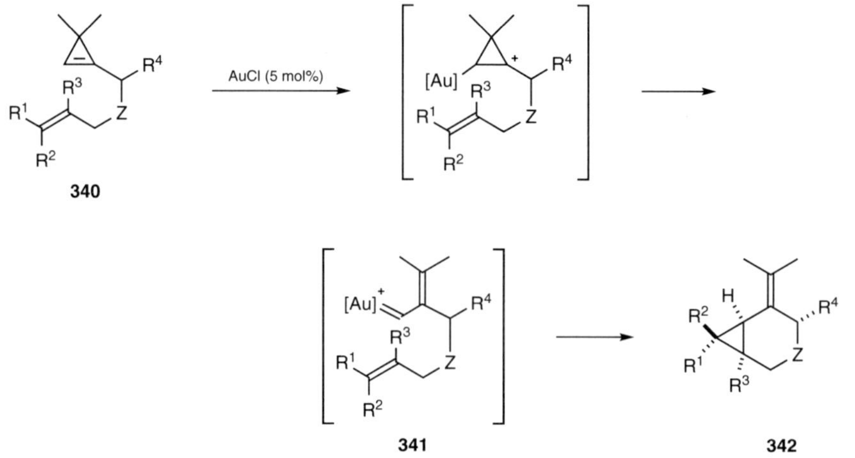

AuCl (5 mol%)

340 **341** **342**

Z	R^1	R^2	R^3	R^4	dr	Yield (%)	Ref
O	H	H	Me	CH_2OBn	6.7:1	72	[72]
O	Ph	H	H	CH_2OBn	>24:1	93	[72]
O	Me	Me	H	CH_2OBn	>24:1	98	[72]
NTs	H	H	H	Ph	–[a]	99	[72]
NTs	H	H	Me	Ph	11.5:1	99	[72]

[a] Single diastereomer.

The rearrangement of fully substituted cyclopropenes has been examined previously by Müller[62] using rhodium(II) catalysts; however, a long reaction time is required (around 48 h). Wang has reported that if substrates **343** are treated with a gold catalyst the reaction is complete within one hour to give indene products **345** in high yields (Scheme 84);[74] this indicates that the gold catalyst is more efficient than the corresponding rhodium(II) catalyst in the cycloisomerization of such cyclopropenes. A gold-stabilized cationic intermediate is proposed in the reaction mechanism. However, the formation of vinyl gold carbene **344** should also be possible.

Scheme 84 Indenes by Gold-Catalyzed Cycloisomerization of Cyclopropenes[74]

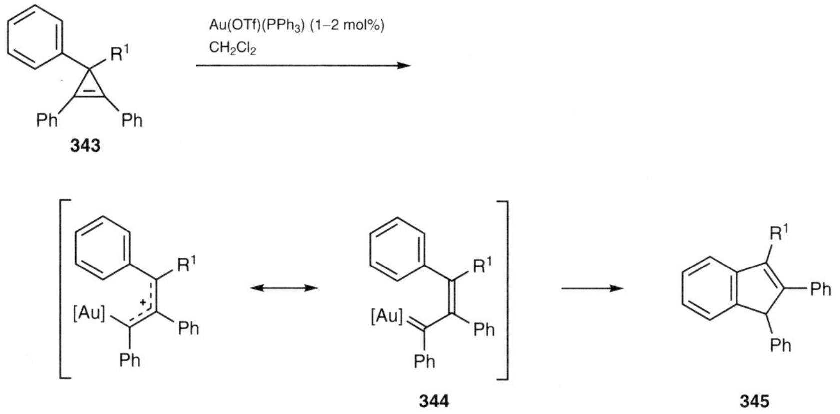

Au(OTf)(PPh$_3$) (1–2 mol%)
CH_2Cl_2

343 344 345

R^1	Yield (%)	Ref
H	99	[74]
Me	97	[74]
Ph	98	[74]
$(CH_2)_3Ph$	99	[74]

Cyclopropenes **346** bearing a protected hydroxymethyl group at C3 have also been tested.[74] Only acetates undergo clean reaction to furnish 2-substituted 1-methylene-1*H*-indenes **349**. The yields can be improved by adding 1,8-diazabicyclo[5.4.0]undec-7-ene. Electrophilic activation of the cyclopropene occurs regioselectively, probably as a result of steric hindrance. In situ formed gold-stabilized cations **347** or gold carbene intermediates **348** then rearrange into the indene products, which may not be stable in the presence of acetic acid; this is the reason why 1,8-diazabicyclo[5.4.0]undec-7-ene is added (Scheme 85).

for references see p 256

Scheme 85 1-Methylene-1*H*-indenes by Gold-Catalyzed Cycloisomerization of O-Protected 3-(Hydroxymethyl)cyclopropenes[74]

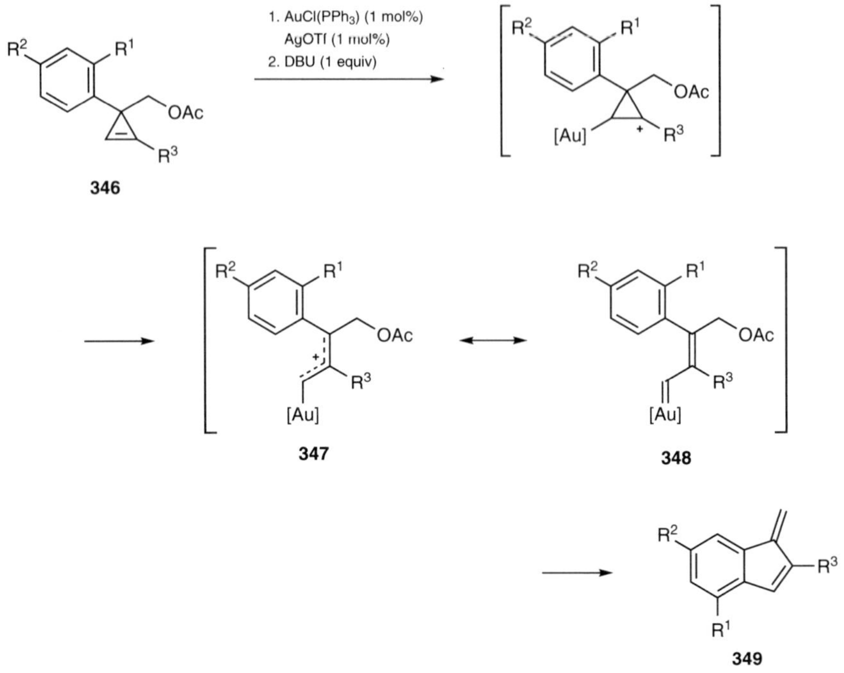

1. $AuCl(PPh_3)$ (1 mol%)
$AgOTf$ (1 mol%)
2. DBU (1 equiv)

346 → **347** ↔ **348** → **349**

R^1	R^2	R^3	Yield (%)	Ref
H	H	Ph	80	[74]
F	H	Ph	82	[74]
H	Me	Ph	82	[74]
H	H	1-naphthyl	85	[74]
H	H	$(CH_2)_4Me$	90	[74]

The cycloisomerization of yne-containing cyclopropenes **350** to construct functionalized phenols **353** has been intensively investigated by Wang.[75] In the presence of a gold complex, the reaction of **350** first produces gold intermediate **351** after classic enyne cycloisomerization. A subsequent ring-opening process yields gold carbene intermediate **352**, which then induces 1,2-migration to give the final products. The reaction scope is quite broad: secondary propargylic alcohols, tertiary alcohols, and *O*-trimethylsilyl cyanohydrins are all tolerated under the standard reaction conditions (Scheme 86).

Scheme 86 Functionalized Phenols by Gold-Catalyzed Cycloisomerization of Yne–Cyclopropenes[75]

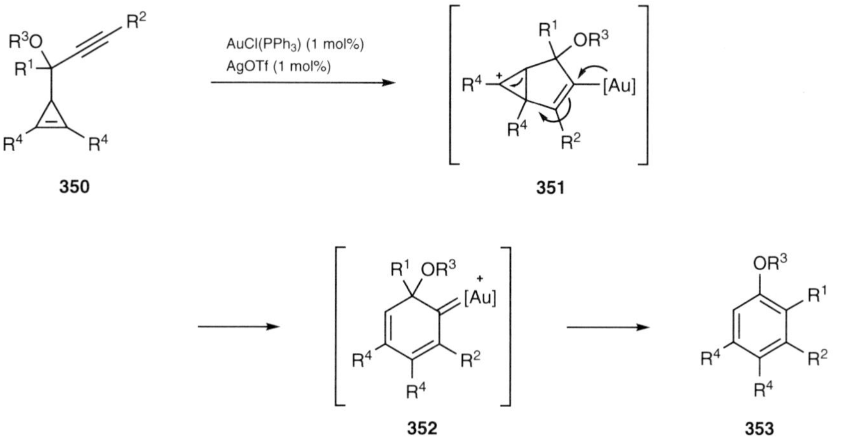

R^1	R^2	R^3	R^4	Yield (%)	Ref
Me	Ph	H	Ph	97	[75]
Bu	Ph	H	Ph	96	[75]
C≡CPh	Ph	H	Ph	82	[75]
Me	Ph	H	Ph	90	[75]
CN	Ph	TMS	Ph	89	[75]
H	Ph	Ac	Ph	96	[75]
H	CO_2Et	H	Ph	97	[75]
C≡CBu	Bu	H	Bu	74	[75]

Abnormal results are observed in some special cases. A pair of unsymmetrical and symmetrical benzene derivatives **355** and **356** is obtained in high yields in a 1:1 ratio if terminal or alkyl substituted alkynes **354** are employed as substrates, which is indicative of complete cleavage of both the cyclopropene double bond and the alkyne triple bond in the reaction (Scheme 87). Further investigation has revealed that the substituents on the cyclopropene moiety play a more important role in switching the reaction pathway. If substrates **357** with butyl groups on the cyclopropene ring are used, symmetrical phenol derivatives **358** are afforded exclusively through the double-cleavage process. These results encouraged the authors to investigate the influence of the substituents on the cyclopropene ring further. For instance, for substrates **359**, in which there are two different substituents (butyl and trimethylsilyl) on the double bond, the reaction also delivers the double-cleavage products exclusively. Notably, from the two diastereomers in the substrates, the same phenol products **360** are obtained. The trimethylsilyl group is situated between the R^1 and R^2 groups, which indicates that the cationic intermediate is greatly stabilized by silicon (β-cation stabilizing effect of silicon for carbon cations).

Scheme 87 Symmetrical and Unsymmetrical Phenol Synthesis[75]

R^1	Ratio (**355**/**356**)	Combined Yield (%)	Ref
H	1:1	99	[75]
$(CH_2)_4Me$	1:1	91	[75]

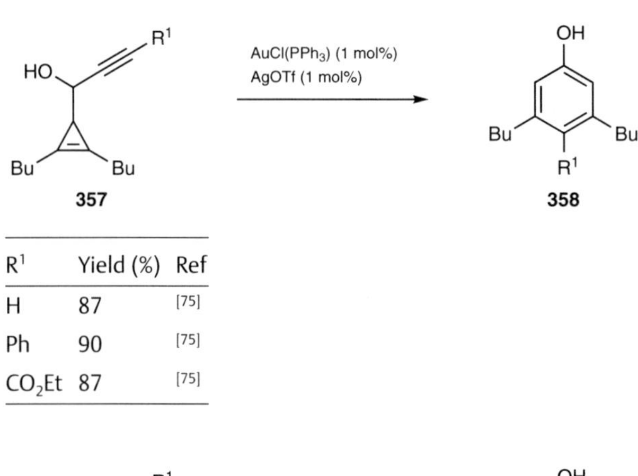

R^1	Yield (%)	Ref
H	87	[75]
Ph	90	[75]
CO_2Et	87	[75]

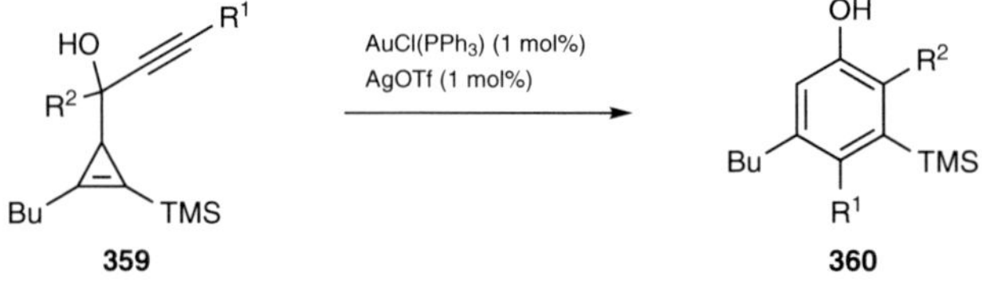

R^1	R^2	Yield (%)	Ref
H	H	84	[75]
Ph	H	93	[75]
CO_2Et	H	86	[75]
Ph	Me	89	[75]

A plausible mechanism for the synthesis of double-cleavage products is outlined in Scheme 88. The enyne cycloisomerization of the substrate gives cyclopropyl cation **361**, which accepts back donation of electrons from the gold center to furnish gold carbene **362**. Subsequent 1,2-alkyl migration delivers Dewar-benzene-type intermediate **363**, from which six-membered cyclic derivative **364** is obtained after a ring-opening process. Finally, the desired benzene derivative is formed by cation-induced 1,2-alkyl migration.

Scheme 88 Proposed Mechanism for the Synthesis of Double-Cleavage Products[75]

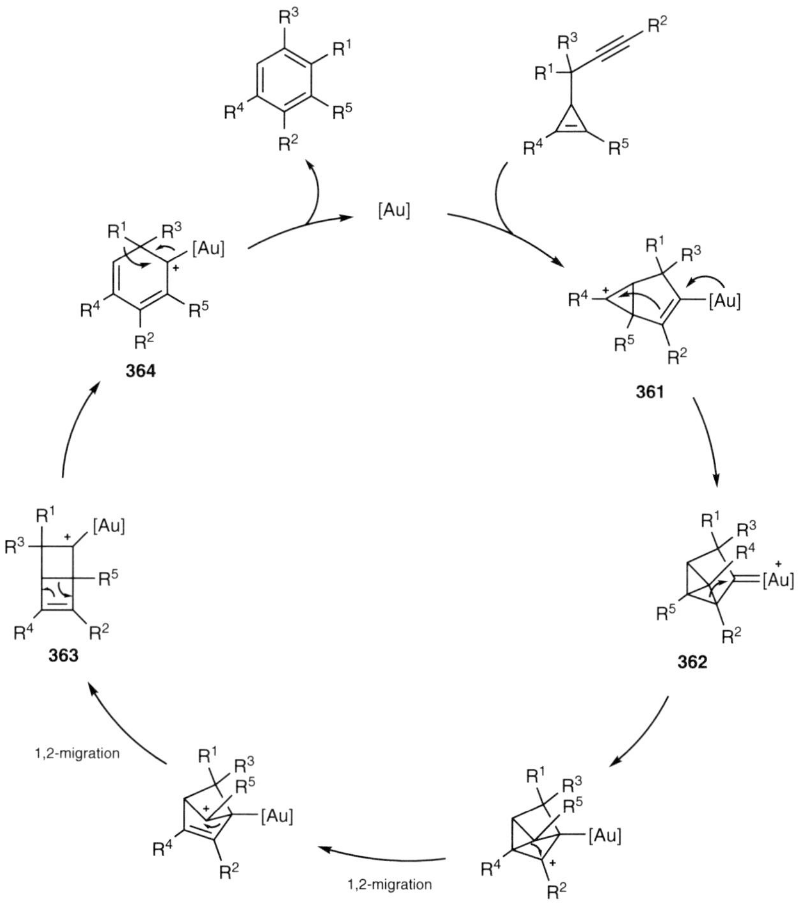

If 1,6-enyne system **365** is used in gold catalysis, tricyclic compounds **368** are obtained through a similar reaction pathway. The reaction starts with rearrangement of the 1,6-enyne to give cyclopropyl cation **366**, which then delivers the corresponding gold carbene intermediate **367**. Friedel–Crafts cyclization gives the final tricyclic products (Scheme 89).

Scheme 89 1,3-Dihydro-2*H*-fluorene-2,2-dicarboxylates by Gold-Catalyzed Cycloisomerization of 1,6-Enynes[75]

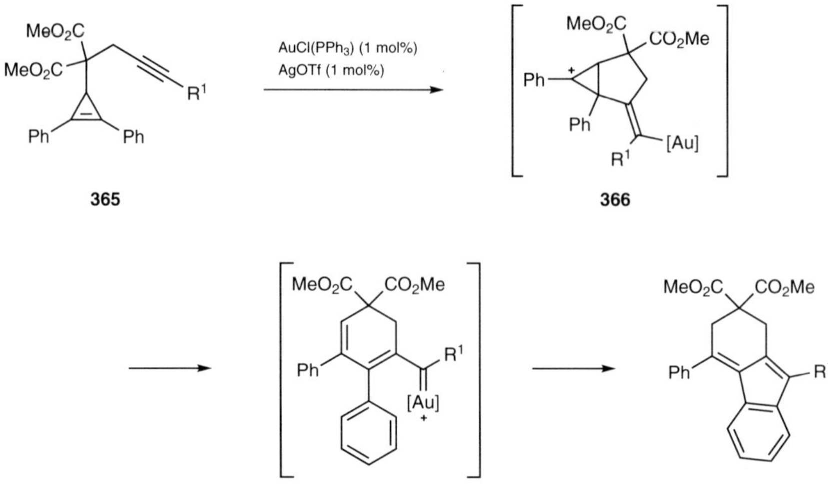

R^1	Yield (%)	Ref
H	80	[75]
Ph	74	[75]

1,7-Enyne system **369** has also been tested, and the corresponding tetrahydroindeno[2,1-*c*]azepine **370** is obtained in 88% yield through carbene intermediate formation and C—H functionalization. Moreover, if the alkyne is replaced by an alkene or allene, substrates **371** undergo tandem rearrangement and Friedel–Crafts cyclization to afford indene products **372** in good yields (Scheme 90).

Scheme 90 Indenes by Gold-Catalyzed Cycloisomerization of Cyclopropene–Ynes, Cyclopropene–Enes, and Cyclopropene–Allenes[75]

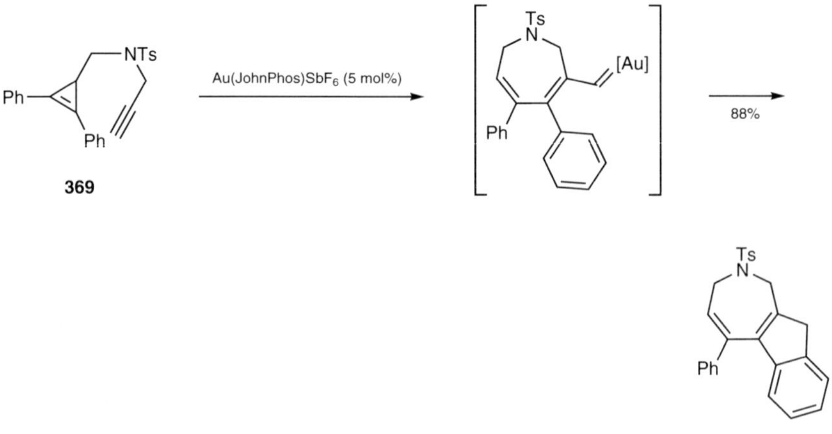

JohnPhos = biphenyl-2-yldi-*tert*-butylphosphine

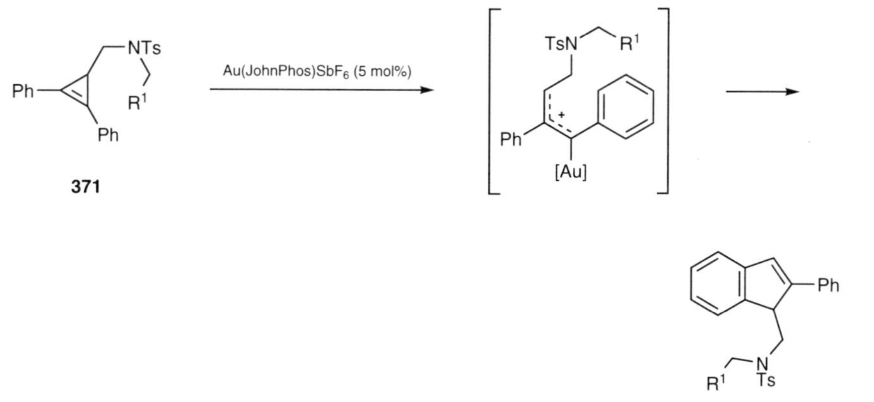

R^1	Yield (%)	Ref
CH=CH$_2$	80	[75]
CH=C=CH$_2$	78	[75]

Indenes 339; **General Procedure:**[71]
Under an argon atmosphere, aryl(vinyl)cyclopropene **338** (0.2 mmol), $AgSbF_6$ (0.01 mmol), $AuCl(PPh_3)$ (0.01 mmol), DBU (0.1 mmol), and 1,2-dichloroethane (1.0 mL) were added into a Schlenk tube. The mixture was stirred at 50 °C until the reaction was complete. Then, the solvent was removed under reduced pressure, and the residue was purified by flash column chromatography (silica gel) to afford the title compound.

3-Oxa- and 3-Azabicyclo[4.1.0]heptanes 342 **(Z = O, NTs); General Procedure:**[72]
AuCl (5 mol%) was added to a 0.05 M soln of cyclopropene **340** (0.122 mmol) in CH_2Cl_2 at 0 °C. After 15 min at 0 °C, the mixture was filtered through a short pad of Celite (CH_2Cl_2), and the filtrate was concentrated under reduced pressure. Analysis of the crude material by ^{1}H NMR spectroscopy indicated the formation of a single diastereomer. The crude material was purified by flash chromatography (silica gel, petroleum ether/EtOAc) to afford the title compound.

1,2-Diphenyl-3-(3-phenylpropyl)-1*H*-indene [345, **R^1 = $(CH_2)_3Ph$**]; **Typical Procedure:**[74]
CH_2Cl_2 (2 mL) was added to a flask containing $AuCl(PPh_3)$ (3.3 mg, 0.0067 mmol) and AgOTf (1.7 mg, 0.0067 mmol). The resulting mixture was stirred for 5 min at rt. A soln of 1,2,3-triphenyl-3-(3-phenylpropyl)cyclopropene [**343**, R^1 = $(CH_2)_3Ph$; 130 mg, 0.34 mmol] dissolved in CH_2Cl_2 (5 mL) was then added. The soln immediately turned yellow. The reaction was monitored by TLC. Upon complete consumption of the substrate, the solvent was removed under reduced pressure and the crude product was purified by column chromatography (silica gel, petroleum ether) to give the title compound as a yellow oil; yield: 129 mg (99%).

Substituted Phenols 353; **General Procedure:**[75]
$AuCl(PPh_3)$ (0.01 equiv), AgOTf (0.01 equiv), and CH_2Cl_2 (2 mL) were added to an open, 25-mL round-bottomed flask. The mixture was stirred at rt for 5 min and a white precipitate appeared. A soln of the 3-propargylic cyclopropene (1.0 equiv) in CH_2Cl_2 (2 mL) was added to the catalyst system, and the soln turned yellow. After 5 min, TLC showed that the starting material had disappeared. The solvent was removed under reduced pressure, and the residue was purified by flash column chromatography (silica gel, petroleum ether/EtOAc 10:1) to afford the title compound.

1.4.3 Applications in the Syntheses of Natural Products and Drug Molecules

The total synthesis of (−)-α-kainic acid through ene–cycloisomerization of methylenecyclopropanes has been accomplished by the Evans group (Scheme 91).[76] A key and striking feature of this protocol is that the alkene geometry does not have an impact on the efficiency or the diastereocontrol: (*E*)- and (*Z*)-allylic substrates **373** furnish the corresponding aldehydes **374** with similar efficiencies and selectivities. In addition, similarly derived diene products **375** are also very versatile building blocks in organic synthesis.

Scheme 91 Rhodium-Catalyzed Ene–Cycloisomerization of Methylenecyclopropanes[76]

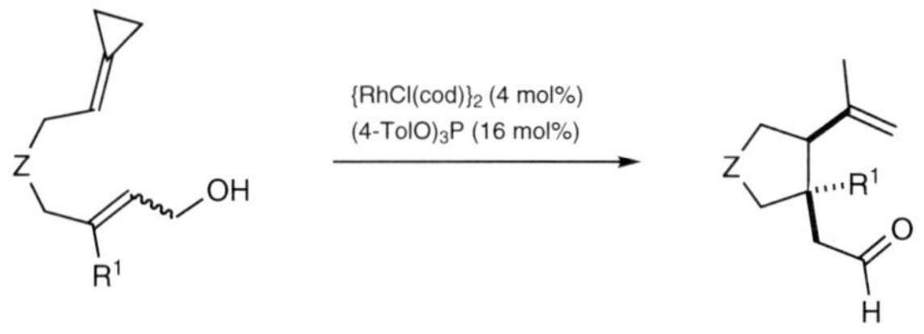

{RhCl(cod)}$_2$ (4 mol%)
(4-TolO)$_3$P (16 mol%)

373 → **374**

Z	R^1	Config of **373**	Yield (%)	Ref
O	H	*Z*	97	[76]
O	Me	*E*	93	[76]
NTs	Me	*E*	80	[76]

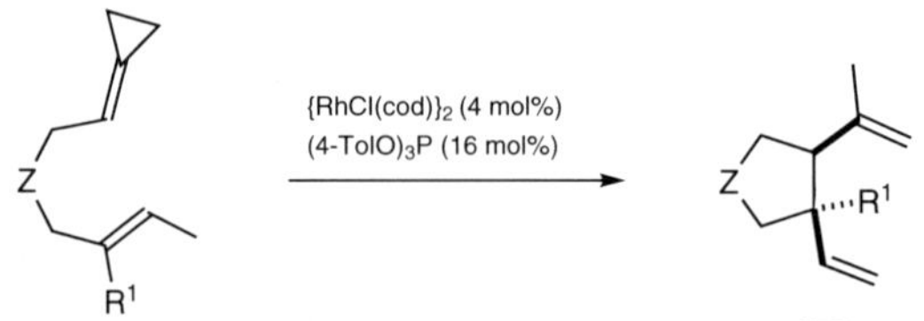

{RhCl(cod)}$_2$ (4 mol%)
(4-TolO)$_3$P (16 mol%)

→ **375**

Z	R^1	Yield (%)	Ref
O	H	80	[76]
O	Me	89	[76]
C(CO$_2$Me)$_2$	Me	93	[76]

This methodology has been applied in the total synthesis of (−)-α-kainic acid (Scheme 92).[76] From substrate **376**, the key rhodium-catalyzed ene–cycloisomerization reaction proceeds smoothly to furnish *anti,syn*-2,3,4-trisubstituted pyrrolidine skeleton **377** in 69% yield with excellent diastereoselectivity (ds >19:1). Subsequent oxidation using pyridinium dichromate gives intermediate **378**, which is converted into alcohol **379** upon hydrolysis under basic conditions. After several other transformations, (−)-α-kainic acid is obtained. Notably, this total synthesis represents one of the most concise and efficient developments that may be suitable for industrial application.

Scheme 92 Total Synthesis of (−)-α-Kainic Acid[76]

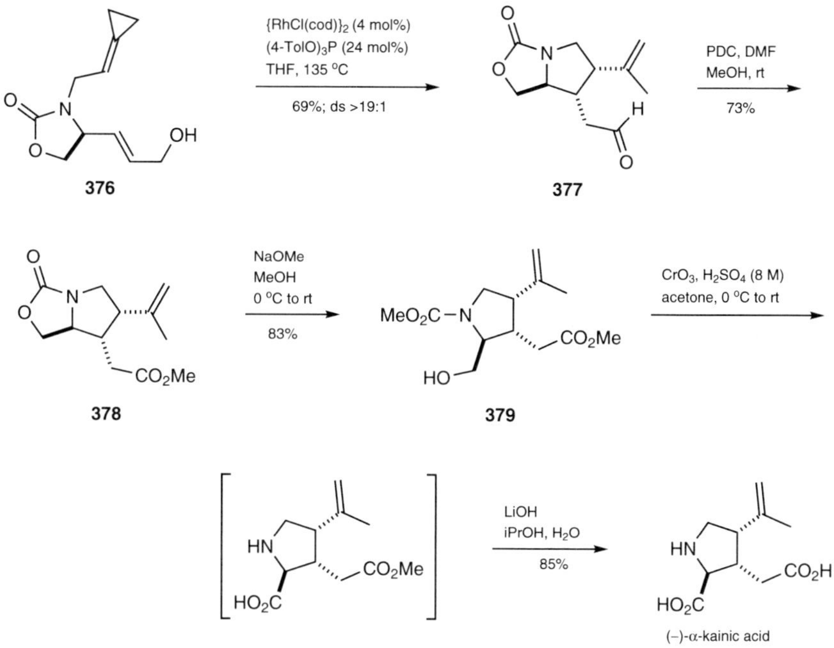

{RhCl(cod)}$_2$ (4 mol%)
(4-TolO)$_3$P (24 mol%)
THF, 135 °C
69%; ds >19:1

PDC, DMF
MeOH, rt
73%

376
377

NaOMe
MeOH
0 °C to rt
83%

CrO$_3$, H$_2$SO$_4$ (8 M)
acetone, 0 °C to rt

378
379

LiOH
iPrOH, H$_2$O
85%

(−)-α-kainic acid

Jung has accomplished the enantiospecific formal total synthesis of (+)-fawcettimine using a nucleophilic ring-opening of a donor–acceptor cyclopropane (Scheme 93).[77] The reaction of optically pure cyclopropane **380** and ketone **381** in the presence of bis(trifluoromethane)sulfonimide gives the corresponding cyclopropyl ketone **382**, and subsequent Wittig reaction furnishes vinyl cyclopropane **383**. Then, nucleophilic ring opening catalyzed by a Lewis acid leads to the formation of tricyclic compound **384**. Removal of one of the ester groups by Krapcho decarboxylation under microwave irradiation gives monoester **385**, which was identical in all respects to a previously reported fawcettimine intermediate.[78] Although there are still eight steps left toward the target natural product, this method provides facile access to functionalized bicyclic ketones.

Scheme 93 Formal Total Synthesis of (+)-Fawcettimine[77]

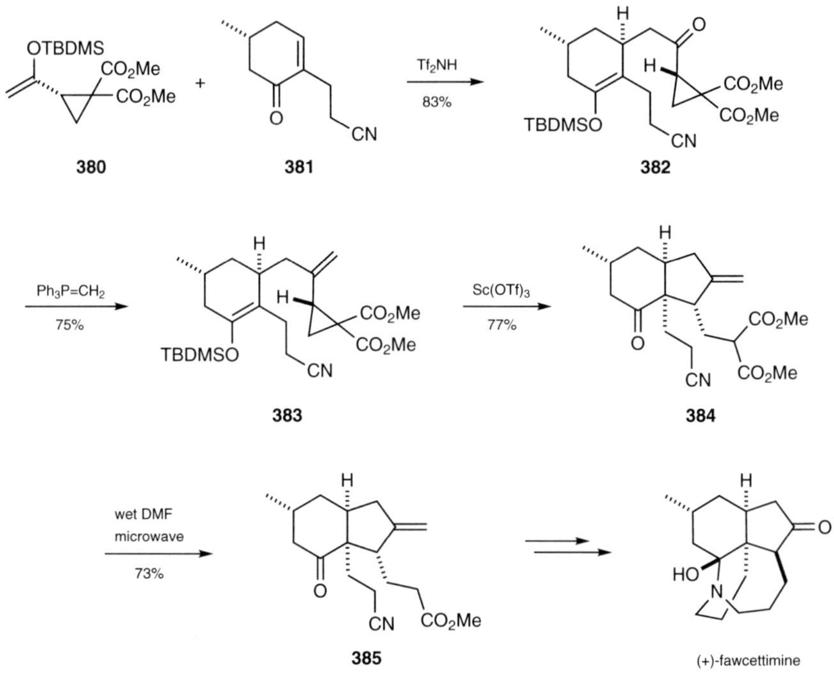

In the total synthesis of (±)-vincorine,[79] Qin has used a nucleophilic ring-opening annulation strategy to construct ring-fused indoline frameworks. From diazo compound **386**, cyclopropanation of the indole gives intermediate **387**, which delivers **388** from a ring-opening process. Mannich-type reaction of **388** results in key intermediate **389**, which then leads to the first total synthesis of the *akuammiline* alkaloid (±)-vincorine (Scheme 94).

Scheme 94 Total Synthesis of (±)-Vincorine[79]

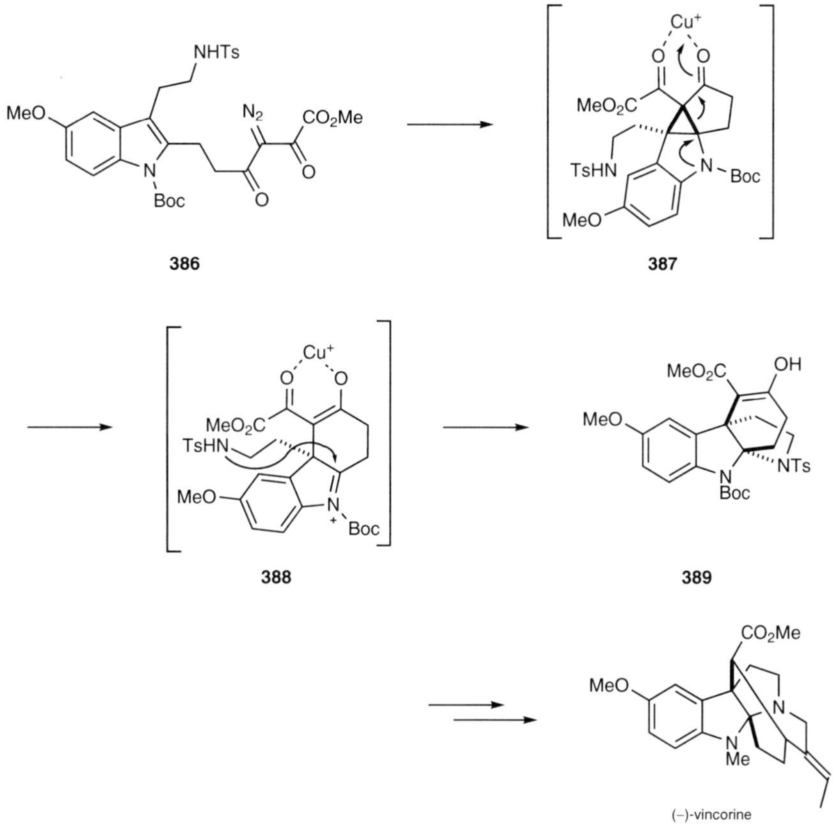

Using an intramolecular cross-cycloaddition reaction strategy, Wang successfully completed a formal total synthesis of platensimycin.[23] Starting material **391** can be easily synthesized in high yield by oxidative cleavage of alkene **390**. Then, in the presence of scandium(III) trifluoromethanesulfonate, intramolecular cross-cycloaddition readily takes place to construct the desired oxa-bridged heterocycle **392**. Decarboxylation delivers monoester **393**, which is the key intermediate in the synthesis of platensimycin (Scheme 95).

Scheme 95 Formal Total Synthesis of Platensimycin[23]

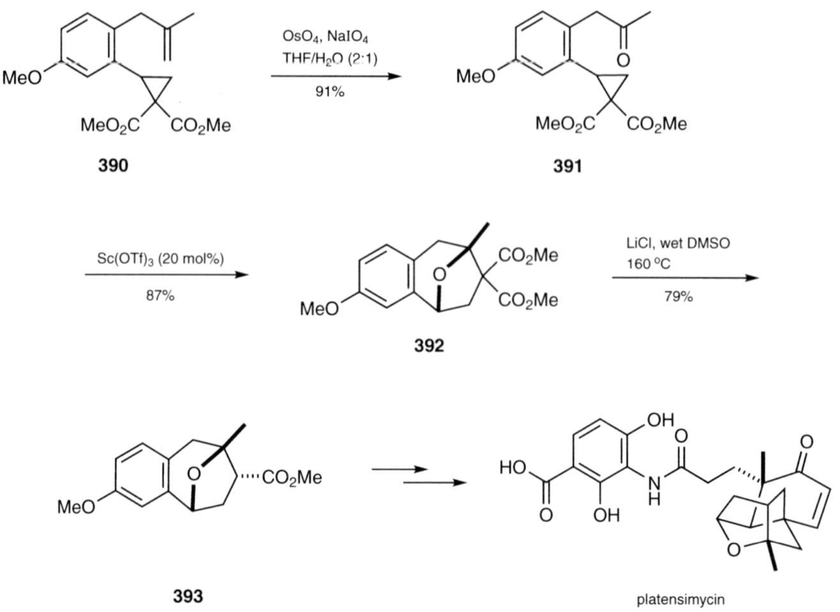

MeO, MeO$_2$C, CO$_2$Me, **390**; OsO_4, $NaIO_4$, THF/H_2O (2:1), 91%; **391**; $Sc(OTf)_3$ (20 mol%), 87%; **392**; LiCl, wet DMSO, 160 °C, 79%; **393**; platensimycin

In a total synthesis of ventricosene, Toste has reported that cyclopropanol **394**, with cyclic vinyl and propargylic substituents in a *cis* relationship, can be converted into ring-expanded cyclobutanone derivative **396**.[15] A classic enyne cyclization is thought to take place, generating key cyclopropyl cationic intermediate **395**, which induces ring expansion of the cyclopropanol. The resulting cyclobutanone **396** is further converted into the natural product ventricosene (Scheme 96).

Scheme 96 Ring-Expanding Cycloisomerization: Total Synthesis of Ventricosene[15]

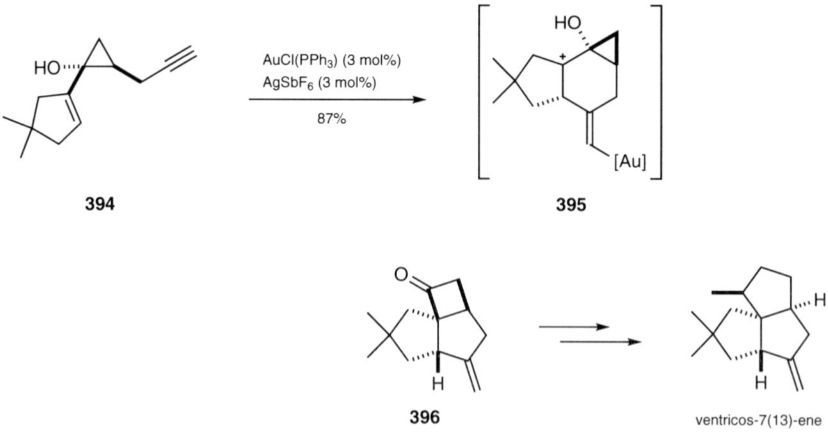

394; AuCl(PPh$_3$) (3 mol%), $AgSbF_6$ (3 mol%), 87%; **395** [Au]; **396**; ventricos-7(13)-ene

The ring-opening of cyclopropyl diazo compound **397** has been elegantly applied by Tang in the total syntheses of members of the intriguing family of natural products with a cyclobutane core, such as piperchabamide G, the proposed structure of which is shown in Scheme 97.[80] Key intermediate **398** is simply constructed by carbine-induced ring expansion of the starting cyclopropane.

Scheme 97 Total Synthesis of Piperchabamide G[80]

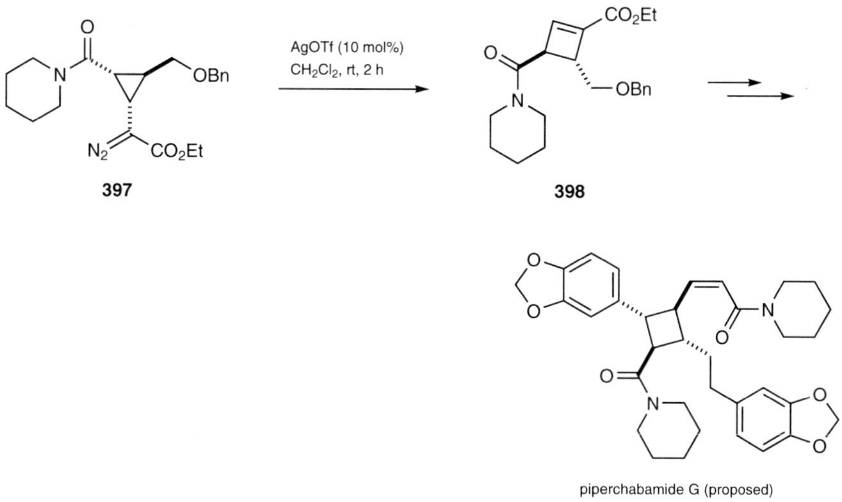

AgOTf (10 mol%)
CH_2Cl_2, rt, 2 h

397

398

piperchabamide G (proposed)

for references see p 256

References

[1] Ma, S.; Lu, L.; Zhang, J., *J. Am. Chem. Soc.*, (2004) **126**, 9645.
[2] Siriwardana, A. I.; Kamada, M.; Nakamura, I.; Yamamoto, Y., *J. Org. Chem.*, (2005) **70**, 5932.
[3] Delgado, A.; Rodrígues, J. R.; Castedo, L.; Mascareñas, J. L., *J. Am. Chem. Soc.*, (2003) **125**, 9282.
[4] López, F.; Delgado, A.; Rodrígues, J. R.; Castedo, L.; Mascareñas, J. L., *J. Am. Chem. Soc.*, (2004) **126**, 10 262.
[5] Gulías, M.; Durán, J.; López, F.; Castedo, L.; Mascareñas, J. L., *J. Am. Chem. Soc.*, (2007) **129**, 11 026.
[6] Shi, M.; Liu, L.-P.; Tang, J., *J. Am. Chem. Soc.*, (2006) **128**, 7430.
[7] Tian, G.-Q.; Yuan, Z.-L.; Zhu, Z.-B.; Shi, M., *Chem. Commun. (Cambridge)*, (2008), 2668.
[8] Fürstner, A.; Aïssa, C., *J. Am. Chem. Soc.*, (2006) **128**, 6306.
[9] Zhang, D.-H.; Tang, X.-Y.; Wei, Y.; Shi, M., *Chem.–Eur. J.*, (2013) **19**, 13 668.
[10] Aïssa, C.; Fürstner, A., *J. Am. Chem. Soc.*, (2007) **129**, 14 836.
[11] Chen, K.; Zhu, Z.-Z.; Tang, X.-Y.; Shi, M., *Angew. Chem. Int. Ed.*, (2014) **53**, 6645.
[12] Yao, B.; Li, Y.; Liang, Z.; Zhang, Y., *Org. Lett.*, (2011) **13**, 640.
[13] Li, S.; Luo, Y.; Wu, J., *Org. Lett.*, (2011) **13**, 3190.
[14] Scott, M. E.; Bethuel, Y.; Lautens, M., *J. Am. Chem. Soc.*, (2007) **129**, 1482.
[15] Sethofer, S. G.; Staben, S. T.; Hung, O. Y.; Toste, F. D., *Org. Lett.*, (2008) **10**, 4315.
[16] Yao, L.-F.; Wei, Y.; Shi, M., *J. Org. Chem.*, (2009) **74**, 9466.
[17] Felix, R. J.; Gutierrez, O.; Tantillo, D. J.; Gagné, M. R., *J. Org. Chem.*, (2013) **78**, 5685.
[18] Zheng, H.; Adduci, L. L.; Felix, R. J.; Gagné, M. R., *Angew. Chem. Int. Ed.*, (2014) **53**, 7904.
[19] Zhang, D.-H.; Du, K.; Shi, M., *Org. Biomol. Chem.*, (2012) **10**, 3763.
[20] Zriba, R.; Gandon, V.; Aubert, C.; Fensterbank, L.; Malacria, M., *Chem.–Eur. J.*, (2008) **14**, 1482.
[21] Jackson, S. K.; Karadeolian, A.; Driega, A. B.; Kerr, M. A., *J. Am. Chem. Soc.*, (2008) **130**, 4196.
[22] Dias, D. A.; Kerr, M. A., *Org. Lett.*, (2009) **11**, 3694.
[23] Xing, S.; Pan, W.; Liu, C.; Ren, J.; Wang, Z., *Angew. Chem. Int. Ed.*, (2010) **49**, 3215.
[24] Beal, R. B.; Dombroski, M. A.; Snider, B. B., *J. Org. Chem.*, (1986) **51**, 4391.
[25] Xing, S.; Li, Y.; Li, Z.; Liu, C.; Ren, J.; Wang, Z., *Angew. Chem. Int. Ed.*, (2011) **50**, 12 605.
[26] Zhu, W.; Fang, J.; Liu, Y.; Wang, Z., *Angew. Chem. Int. Ed.*, (2013) **52**, 2032.
[27] Wang, Z.; Ren, J.; Wang, Z., *Org. Lett.*, (2013) **15**, 5682.
[28] Xia, X.-F.; Song, X.-R.; Liu, X.-Y.; Liang, Y.-M., *Chem.–Asian J.*, (2012) **7**, 1538.
[29] Patil, D. V.; Phun, L. H.; France, S., *Org. Lett.*, (2010) **12**, 5684.
[30] Aponte-Guzmán, J. A.; Taylor, J. E.; Tillman, J. E.; France, S., *Org. Lett.*, (2014) **16**, 3788.
[31] Patil, D. V.; Cavitt, M. A.; Grzybowski, P.; France, S., *Chem. Commun. (Cambridge)*, (2011) **47**, 10 278.
[32] Brand, C.; Rauch, G.; Zanoni, M.; Dittrich, B.; Werz, D. B., *J. Org. Chem.*, (2009) **74**, 8779.
[33] Zhang, G.; Huang, X.; Li, G.; Zhang, L., *J. Am. Chem. Soc.*, (2008) **130**, 1814.
[34] Bai, Y.; Tao, W.; Ren, J.; Wang, Z., *Angew. Chem. Int. Ed.*, (2012) **51**, 4112.
[35] Zhu, J.; Liang, Y.; Wang, L.; Zheng, Z.-B.; Houk, K. N.; Tang, Y., *J. Am. Chem. Soc.*, (2014) **136**, 6900.
[36] Fisher, E. L.; Wilkerson-Hill, S. M.; Sarpong, R., *J. Am. Chem. Soc.*, (2012) **134**, 9946.
[37] Xu, G.-C.; Liu, L.-P.; Lu, J.-M.; Shi, M., *J. Am. Chem. Soc.*, (2005) **127**, 14 552.
[38] Shi, M.; Wu, L.; Lu, J.-M., *J. Org. Chem.*, (2008) **73**, 8344.
[39] Li, W.; Yuan, W.; Pindi, S.; Shi, M.; Li, G., *Org. Lett.*, (2010) **12**, 920.
[40] Lu, B.-L.; Wei, Y.; Shi, M., *Chem.–Eur. J.*, (2010) **16**, 10 975.
[41] Wu, L.; Shi, M., *Chem.–Eur. J.*, (2011) **17**, 13 160.
[42] Yuan, W.; Tang, X.-Y.; Wei, Y.; Shi, M., *Chem.–Eur. J.*, (2014) **20**, 3198.
[43] Huang, X.; Su, C.; Liu, Q.; Song, Y., *Synlett*, (2008), 229.
[44] Lu, B.-L.; Shi, M., *Angew. Chem. Int. Ed.*, (2011) **50**, 12 027.
[45] Miao, M.; Cao, J.; Zhang, J.; Huang, X.; Wu, L., *J. Org. Chem.*, (2013) **78**, 2687.
[46] Miao, M.; Cao, J.; Zhang, J.; Huang, X.; Wu, L., *Org. Lett.*, (2012) **14**, 2718.
[47] Iwasawa, N.; Matsuo, T.; Iwamoto, M.; Ikeno, T., *J. Am. Chem. Soc.*, (1998) **120**, 3903.
[48] Markham, J. P.; Staben, S. T.; Toste, F. D., *J. Am. Chem. Soc.*, (2005) **127**, 9708.
[49] Barluenga, J.; Tudela, E.; Vicente, R.; Ballesteros, A.; Tomás, M., *Angew. Chem. Int. Ed.*, (2011) **50**, 2107.
[50] Zhang, Z.; Tang, X.-Y.; Xu, Q.; Shi, M., *Chem.–Eur. J.*, (2013) **19**, 10 625.
[51] Chen, A.; Lin, R.; Liu, Q.; Jiao, N., *Chem. Commun. (Cambridge)*, (2009), 6842.
[52] Trost, B. M.; Xie, J.; Maulide, N., *J. Am. Chem. Soc.*, (2008) **130**, 17 258.
[53] Chen, G.; Zhang, X.-N.; Wei, Y.; Tang, X.-Y.; Shi, M., *Angew. Chem. Int. Ed.*, (2014) **53**, 8492.

[54] Lu, Z.; Shen, M.; Yoon, T. P., *J. Am. Chem. Soc.*, (2011) **133**, 1162.
[55] Tsuritani, T.; Yamamoto, Y.; Kawasaki, M.; Mase, T., *Org. Lett.*, (2009) **11**, 1043.
[56] Liu, L.; Zhang, J., *Angew. Chem. Int. Ed.*, (2009) **48**, 6093.
[57] Xu, H.; Zhang, W.; Shu, D.; Werness, J. B.; Tang, W., *Angew. Chem. Int. Ed.*, (2008) **47**, 8933.
[58] Liu, R.; Zhang, M.; Winston-McPherson, G.; Tang, W., *Chem. Commun. (Cambridge)*, (2013) **49**, 4376.
[59] Tomilov, Yu. V.; Shapiro, E. A.; Protopopova, M. N.; Ioffe, A. I.; Dolgii, I. E.; Nefedov, O. M., *Izv. Akad. Nauk SSSR, Ser. Khim.*, (1985), 631; *Bull. Acad. Sci. USSR, Div. Chem. Sci. (Engl. Transl.)*, (1985) **34**, 576.
[60] Davies, H. M. L.; Romines, K. R., *Tetrahedron*, (1988) **44**, 3343.
[61] Padwa, A.; Kassir, J. M.; Xu, S. L., *J. Org. Chem.*, (1991) **56**, 6971.
[62] Müller, P.; Pautex, N.; Doyle, M. P.; Bagheri, V., *Helv. Chim. Acta.*, (1990) **73**, 1233.
[63] Padwa, A.; Chiacchio, U.; Garreau, Y.; Kassir, J. M.; Krumpe, K. E.; Schoffstall, A. M., *J. Org. Chem.*, (1990) **55**, 414.
[64] Padwa, A.; Straub, C. S., *Org. Lett.*, (2000) **2**, 2093.
[65] Chuprakov, S.; Gevorgyan, V., *Org. Lett.*, (2007) **9**, 4463.
[66] Archambeau, A.; Meige, F.; Meyer, C.; Cossy, J., *Angew. Chem. Int. Ed.*, (2012) **51**, 11 540.
[67] Zhang, H.; Li, C.; Xie, G.; Wang, B.; Zhang, Y.; Wang, J., *J. Org. Chem.*, (2014) **79**, 6286.
[68] Chen, J.; Ma, S., *Chem.–Asian J.*, (2010) **5**, 2415.
[69] Ma, S.; Zhang, J., *J. Am. Chem. Soc.*, (2003) **125**, 12 386.
[70] Shao, L.-X.; Zhang, Y.-P.; Qi, M.-H.; Shi, M., *Org. Lett.*, (2007) **9**, 117.
[71] Zhu, Z.-B.; Shi, M., *Chem.–Eur. J.*, (2008) **14**, 10 219.
[72] Miege, F.; Meyer, C.; Cossy, J., *Org. Lett.*, (2010) **12**, 4144.
[73] Miege, F.; Meyer, C.; Cossy, J., *Chem.–Eur. J.*, (2012) **18**, 7810.
[74] Li, C.; Zeng, Y.; Wang, J., *Tetrahedron Lett.*, (2009) **50**, 2956.
[75] Li, C.; Zeng, Y.; Zhang, H.; Fang, J.; Zhang, Y.; Wang, J., *Angew. Chem. Int. Ed.*, (2010) **49**, 6413.
[76] Evans, P. A.; Inglesby, P. A., *J. Am. Chem. Soc.*, (2012) **134**, 3635.
[77] Jung, M. E.; Chang, J. J., *Org. Lett.*, (2010) **12**, 2962.
[78] Heathcock, C. H.; Blumenkopf, T. A.; Smith, K. M., *J. Org. Chem.*, (1989) **54**, 1548.
[79] Zhang, M.; Huang, X.; Shen, L.; Qin, Y., *J. Am. Chem. Soc.*, (2009) **131**, 6013.
[80] Liu, R.; Zhang, M.; Wyche, T. P.; Winston-McPherson, G. N.; Bugni, T. S.; Tang, W., *Angew. Chem. Int. Ed.*, (2012) **51**, 7503.

1.5 Cyclization Reactions of Alkenes and Alkynes

L. Zhang

General Introduction

Alkenes and alkynes are two common classes of substrates of ubiquitous usage in metal catalysis, be it involving soft late transition, hard early transition, or main group metals. Intramolecular reactions of these π-systems bearing a diverse range of appropriately tethered functional groups serve as essential tools for studying reactivities of metal complexes, and are a main avenue for developing synthetically versatile methodologies. A majority of these reactions belong to metal-catalyzed cyclization reactions, which are typically kinetically more facile than the corresponding intermolecular reactions due to decreased entropy penalty and provide efficient access to synthetically versatile ring structures. To avoid overlapping with other chapters, this chapter is focused on metal-catalyzed cyclization reactions involving polar intermediates with the exception of the intramolecular Heck and the oxidative Heck reactions (Section 1.2), intramolecular allylation reactions (Section 1.3), cycloisomerization reactions with multiple unsaturated bonds (Section 1.7), amination and C—O bond-forming reactions (Section 1.8), and cycloaddition reactions (Sections 2.4–2.8). Cyclizations involving radical intermediates are discussed in Section 2.9. The reactions discussed here are grouped into two types: one where an unactivated carbon–carbon double/triple bond reacts as nucleophile to attack tethered electrophiles (Section 1.5.1), and the other where the π-system is activated by a metal-based π-acid and subsequently attacked by carbon nucleophiles (Section 1.5.2).

1.5.1 Unactivated Alkenes and Alkynes as Nucleophiles

Electron-rich and electron-neutral alkenes and alkynes are broadly used as nucleophiles for various transformations, many of which are cyclization reactions. Among the ring-formation reactions, many are catalyzed by Brønsted acids, and others are promoted by stoichiometric amounts of metal salts or complexes. The metal-catalyzed cases, which in general avoid strongly acidic conditions in comparison to those using Brønsted acids, or avoid the use of large amounts of a metal and are thus more environmentally friendly, offer versatile methodologies for organic synthesis. Notably, in most cases a metal complex behaves as a hard Lewis acid to facilitate the generation of a cationic electrophile and hence does not interact with the π-nucleophiles. This section is devoted to covering this aspect of metal catalysis using less nucleophilic, electronically unactivated alkynes and alkenes.

1.5.1.1 Oxocarbenium-Based Electrophiles

The Prins cyclization[1] and intramolecular Lewis acid catalyzed carbonyl-ene reactions[2] are two types of reaction that fall into this category.

for references see p 314

1.5.1.1.1 Prins Cyclization between Homoallylic Alcohols and Aldehydes

The Prins cyclization between a homoallylic alcohol and an aldehyde involves a key step where the cyclization of an alkene with a tethered oxocarbenium generates tetrahydropyran **1** in a one-step process or a stepwise fashion via the carbocation **2** (Scheme 1).[1,3] A large majority of examples are promoted by superstoichiometric amounts of Brønsted acids or stoichiometric amounts of metal salts, where the conjugate bases (acid counter-anions) or metal counteranions (often halides) end up at the C4-position of the tetrahydropyran ring. Notwithstanding, several practically useful metal-catalyzed cases have been realized since the early report using scandium(III) trifluoromethanesulfonate as catalyst.[4]

Scheme 1 Prins Cyclization of a Homoallylic Alcohol and an Aldehyde[1,3]

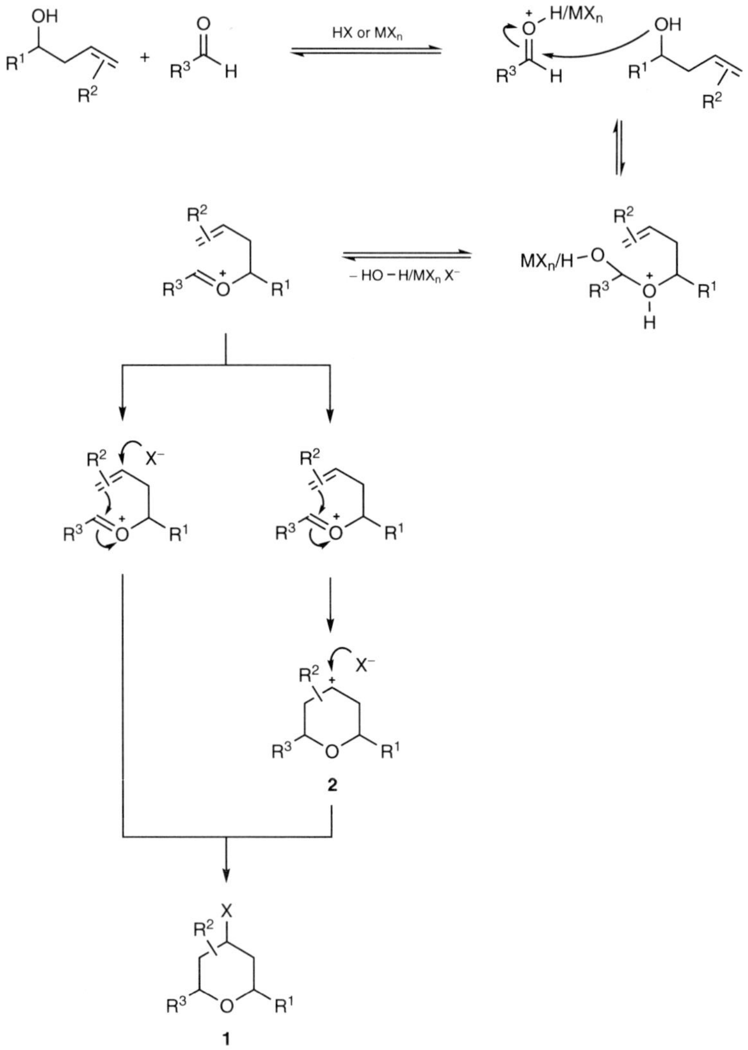

OH, R^1, R^2, +, O, R^3, H, HX or MX_n, H/MX_n, MX_n/H–O, – HO – H/MX_n X^-, X^-, X, **2**, **1**

1.5.1.1.1.1 Highly Stereoselective Iron(III)-Catalyzed Synthesis of *cis*-2-Alkyl-4-halotetrahydropyrans

In many cases of the metal-promoted Prins reaction, the metal counteranions, being halides or carboxylates, are incorporated into the products. To make the process catalytic, one approach is to supply the requisite counteranion with an inexpensive reagent. In the case of halides, the use of a halotrimethylsilane additive (1.0 equiv) facilitates a tris(acetylacetonato)iron(III)-catalyzed Prins cyclization (Scheme 2).[5] It is notable that the reaction tolerates functionalized aldehydes, which is not the case in the stoichiometric version, although the explored scope is limited. Both 4-bromo- and 4-chlorotetrahydropyrans **3** can be formed.

Scheme 2 Efficient Tris(acetylacetonato)iron(III)-Catalyzed Synthesis of *cis*-2-Alkyl-4-halotetrahydropyrans[5]

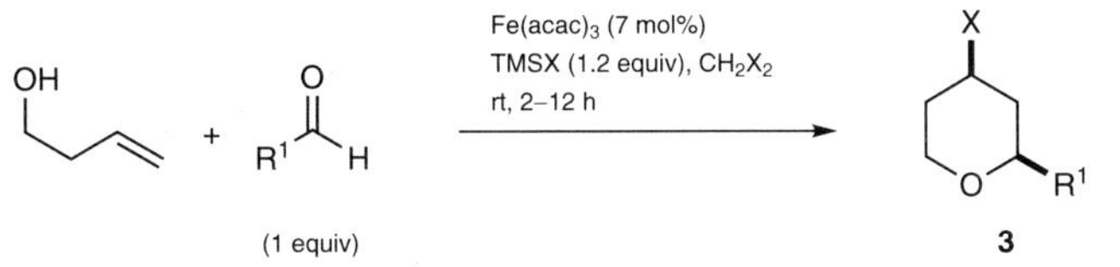

OH + R^1CHO (1 equiv) → Fe(acac)$_3$ (7 mol%), TMSX (1.2 equiv), CH_2X_2, rt, 2–12 h → **3**

R^1	X	Yield (%)	Ref
iBu	Br	88	[5]
4-$O_2NC_6H_4$	Br	78	[5]
$(CH_2)_2CH{=}CH_2$	Cl	70	[5]
$(CH_2)_3OAc$	Cl	60	[5]
$(CH_2)_2OBn$	Cl	86	[5]
$(CH_2)_7OTBDMS$	Cl	71[a]	[5]

[a] 30% protected as the TBDMS ether, and 41% as the free alcohol.

***cis*-2-Substituted 4-Halotetrahydropyrans 3; General Procedure:**[5]
To a soln of but-3-en-1-ol (1.0 equiv) in anhyd CH_2X_2 (0.1 M) was added Fe(acac)$_3$ (0.07 equiv). The resulting soln was cooled to 0 °C using an ice bath followed by the addition of TMSX (1.2 equiv). After 5 min, an aldehyde (1.0 equiv) was then added dropwise. The mixture was stirred until TLC analysis showed complete formation of product. The reaction was then quenched by addition of H_2O with stirring and the mixture was extracted with CH_2Cl_2. The combined organic layers were dried ($MgSO_4$), and the solvent was removed under reduced pressure. The residue was purified by flash column chromatography (silica gel, hexane/EtOAc).

1.5.1.1.1.2 Highly Stereoselective Rhenium(VII)-Catalyzed Synthesis of Tetrasubstituted Tetrahydro-2*H*-pyran-4-ols

An alternative approach toward catalyzed Prins cyclizations is to use a Lewis acid that does not form strong M—O bonds. As a result, the hydroxide that is formally formed during the generation of the oxocarbenium intermediate can also act as the anion that is delivered to the C4-position, thereby affording tetrahydro-2*H*-pyran-4-ols (see Scheme 1). Such Lewis acids can be scandium(III) trifluoromethanesulfonate[4] or triphenylsilyl perrhenate [$ReO_3(OSiPh_3)$].[6] In the latter case, the reaction offers highly stereoselective access to tetrasubstituted tetrahydropyrans **5** with a free hydroxy group at the C4-position (Scheme 3). The reaction mostly proceeds smoothly with aromatic aldehydes, regardless of the electronic nature of the arene ring; a reported exception is with the highly electron-rich 2-hydroxy-3,5-dimethoxy-4-methylbenzaldehyde (30% yield). Due to complications caused by oxonia-Cope induced side-chain exchanges, saturated aliphatic aldehydes do not work well. As a viable solution to this problem, the corresponding α,β-unsaturated aldehydes can be used; the C=C bond of the products can subsequently be hydrogenated. Notably, the use of a chiral homoallylic alcohol (99% ee) led to little erosion of the stereopurity in two selected cases, where both tetrahydrofuran products were obtained with 97% ee. Mechanistically, the perruthenate ion is believed to trap the carbocation generated upon alkene nucleophilic cyclization to form the perruthenate **4**, which would then undergo transesterification to release the observed product **5**.

Scheme 3 Highly Stereoselective Triphenylsilyl Perrhenate Catalyzed Synthesis of Tetrahydro-2*H*-pyran-4-ols[6]

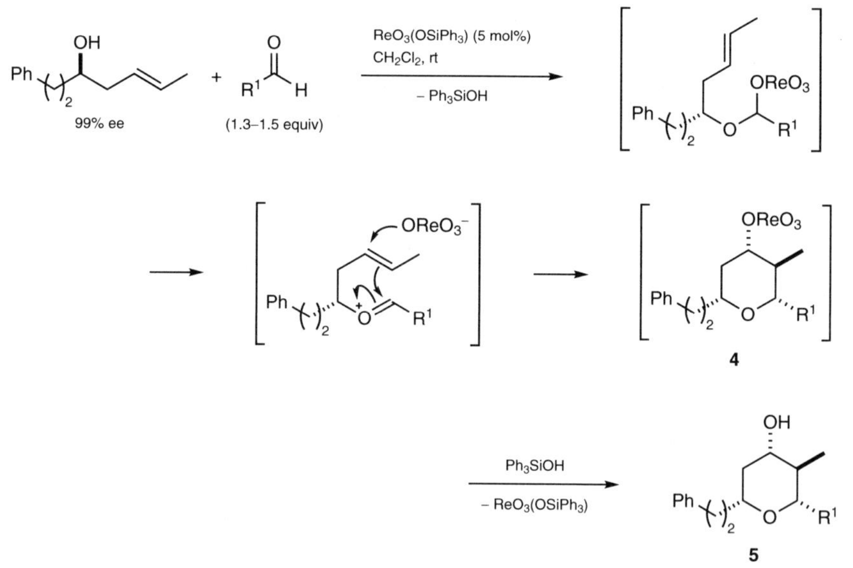

99% ee; (1.3–1.5 equiv); $ReO_3(OSiPh_3)$ (5 mol%), CH_2Cl_2, rt; – Ph_3SiOH; **4**; Ph_3SiOH, – $ReO_3(OSiPh_3)$; **5**

R^1	Time (h)	Yield (%)	Ref
Ph	24	82	[6]
4-$MeOC_6H_4$	25	88[a,b]	[6]
4-HOC_6H_4	27	75	[6]
4-FC_6H_4	26	88[a,b]	[6]
2,4-$(MeO)_2C_6H_3$	50	58[a]	[6]
2-$MeOC_6H_4$	18	84	[6]
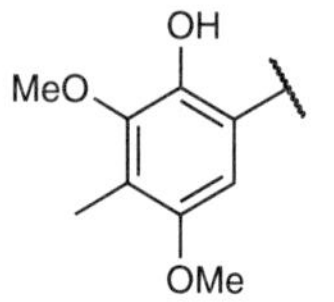 2-hydroxy-3,5-dimethoxy-4-methylphenyl (structure: OH, MeO, OMe)	168	30[c]	[6]

R^1	Time (h)	Yield (%)	Ref
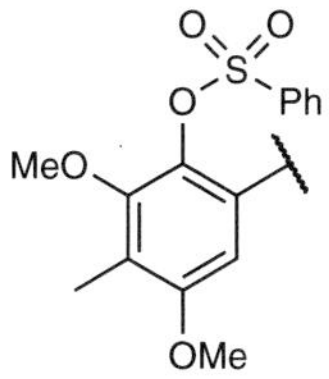 (aryl structure: 2-(phenylsulfonyloxy)-3-methoxy-5-methoxy-4-methylphenyl)	23	83[d]	[6]
4-$F_3CC_6H_4$	24	73	[6]
4-NCC_6H_4	24	55	[6]
(*E*)-CH=CHPr	42	83	[6]
(*E*)-CH=CHPh	48	70	[6]
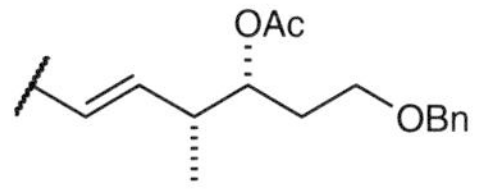 (alkenyl structure with OAc and OBn)	24	55	[6]

[a] The *S*-alcohol (99% ee) was used as the starting material.
[b] 97% ee by HPLC analysis.
[c] Starting material was also recovered in 11% yield.
[d] 10 mol% of triphenylsilyl perrhenate was used.

(2*R*,3*R*,4*S*,6*R*)-2-Alkenyl- and (2*R*,3*R*,4*S*,6*R*)-2-Aryl-3-methyl-6-(2-phenylethyl)tetrahydro-2*H*-pyran-4-ols 5; General Procedure:[6]
To an oven-dried flask were added (*R*,*E*)-1-phenylhept-5-en-3-ol (0.40 mmol, 1 equiv) and CH_2Cl_2 (4 mL, 0.1 M). An aldehyde (0.52–0.60 mmol, 1.3–1.5 equiv) and $ReO_3(OSiPh_3)$ (0.02 mmol, 5 mol%) were added and the mixture was stirred at rt under argon until TLC analysis indicated complete disappearance of the homoallylic alcohol. The solvent was removed under reduced pressure, and the crude product was purified by flash chromatography (Et_2O/hexanes 2:3 to 3:2).

1.5.1.1.1.3 Indium(III) Chloride Catalyzed Synthesis of Spirotetrahydropyrans

An alternative approach to achieve catalysis by Lewis acids is to trap the carbocation intermediate (e.g., **2**; Scheme 1) or such a developing electrophilic carbon center efficiently with a tethered nucleophile,[7–9] or remove it via E1-type elimination.[10] In the former approach, the use of phenols **6** gives tricyclic spirotetrahydropyrans **7** in the presence of 10 mol% indium(III) chloride (Scheme 4).[7] In cases where Z = O, the reactions are highly diastereoselective; a diastereomeric ratio of 24:1 is detected in the case where R^1 = 4-BrC_6H_4. For 2-aminophenol-based substrates **6** (Z = NTs), the reactions are best run at −20 °C to achieve good diastereoselectivity without compromising the reaction yield too much. In all cases, the product **7A**, with the phenolic oxygen *cis* to R^1, is the major isomer. The reaction works well with a variety of aromatic aldehydes, although with the electron-rich 3,4-dimethoxybenzaldehyde a moderate yield and low diastereoselectivity are obtained.

for references see p 314

Scheme 4 Indium(III) Chloride Catalyzed Synthesis of Spirotetrahydropyrans[7]

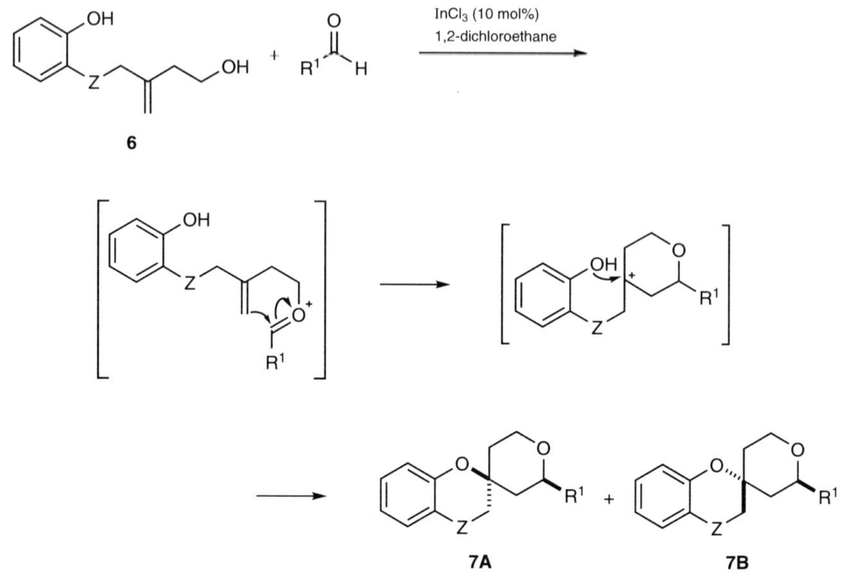

$InCl_3$ (10 mol%)
1,2-dichloroethane

6

7A 7B

Z	R^1	Time (h)	Temp	Yield (%)	Ratio[a] (**7A**/**7B**)	Ref
O	4-BrC_6H_4	4	0 °C to rt	81	24:1	[7]
O	Ph	6	0 °C to rt	75	n.r.	[7]
O	4-$O_2NC_6H_4$	5	0 °C to rt	70	n.r.	[7]
O	2-naphthyl	8	0 °C to rt	70	n.r.	[7]
O	iPr	4	0 °C to rt	72	n.r.	[7]
NTs	4-ClC_6H_4	3	−20 °C	85	91:9	[7]
NTs	4-Tol	4	−20 °C	82	93:7	[7]
NTs	4-$O_2NC_6H_4$	6	−20 °C	78	92:8	[7]
NTs	iPr	4	−20 °C	75	90:10	[7]
NTs	3,4-$(MeO)_2C_6H_3$	3	−20 °C	65	63:37	[7]

[a] n.r. = not reported.

(2*R,2′*S**)-2′-(4-Chlorophenyl)-4-tosyl-2′,3,3′,4,5′,6′-hexahydrospiro[benzo[*b*][1,4]oxazine-2,4′-pyran]** (7A, **Z = NTs; R^1 = 4-ClC_6H_4); Typical Procedure:**[7]
To a mixture of *N*-(4-hydroxy-2-methylenebutyl)-*N*-(2-hydroxyphenyl)-4-toluenesulfonamide (**6**; 0.5 mmol) and 4-chlorobenzaldehyde (0.6 mmol) in anhyd 1,2-dichloroethane (5 mL) was added $InCl_3$ (10 mol%) at −20 °C. The resulting mixture was stirred at the same temperature under N_2 for 3 h. The reaction was quenched with aq $NaHCO_3$ and the mixture was extracted with CH_2Cl_2. The organic phase was dried (Na_2SO_4) and concentrated on a rotary evaporator. The crude product was purified by column chromatography [silica gel (100–200 mesh), EtOAc/hexane gradient] to afford the desired product as the major isomer of an inseparable mixture of diastereomers; yield: 85%; dr 91:9.

1.5.1.1.1.4 Indium(III) Bromide Catalyzed Synthesis of 3-Oxaterpenoids

Appropriately tethered arenes and/or alkenes can likewise trap a carbocation intermediate in an indium(III) bromide catalyzed Prins-cyclization-initiated polyene cyclization (Scheme 5).[9] The reaction tolerates a broad range of aldehydes such as aryl, hetaryl, alkenyl, and aliphatic ones (including sterically hindered pivalaldehyde) and even ketones. Moreover, the homoallylic alcohol reacting partner can have a tethered phenyl ring with various substituents including a mildly deactivating chloride and electron-donating methoxy group(s), electron-rich heteroaromatic rings, or additional C=C bonds. In addition to the generally good efficiencies that are observed, the reactions also exhibit, similar to related polyene cyclizations, high diastereoselectivities in the formed products **8**. The mild reaction conditions are of particular practical importance as a chiral aldehyde does not suffer epimerization.

Scheme 5 Indium(III) Bromide Catalyzed Synthesis of 3-Oxaterpenoids[9]

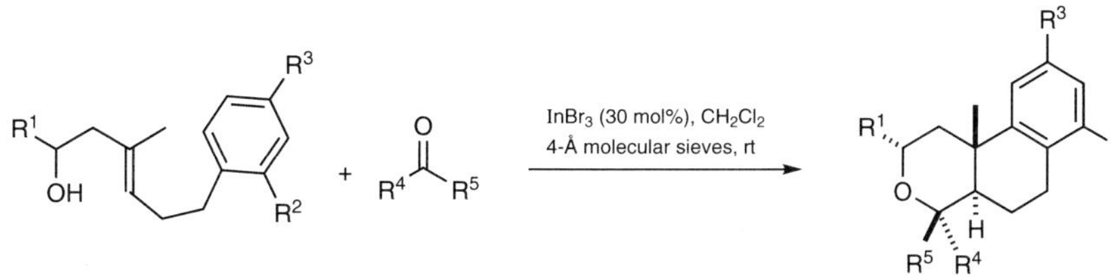

$InBr_3$ (30 mol%), CH_2Cl_2
4-Å molecular sieves, rt

8

R^1	R^2	R^3	R^4	R^5	Yield (%)	Ref
H	H	H	Ph	H	88	[9]
H	H	H	2-furyl	H	79	[9]
H	H	H	2-thienyl	H	71	[9]
H	H	H	*t*-Bu	H	79	[9]
H	H	H	CH_2Br	H	58[a]	[9]
H	H	H	Me	Me	73	[9]
H	OMe	H	Ph	H	94	[9]
H	H	Cl	Ph	H	81	[9]
H	H	H	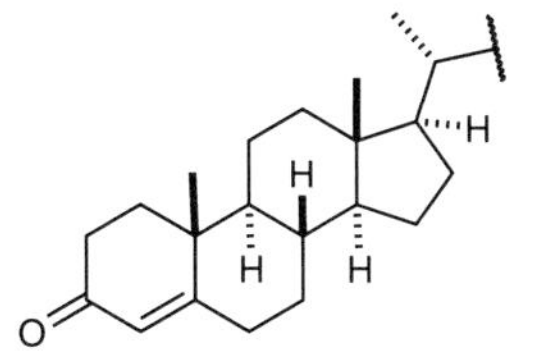 (steroid substituent)	H	74	[9]
Et	H	H	Ph	H	75	[9]

[a] 2-Bromo-1,1-diethoxyethane was used instead of the corresponding aldehyde.

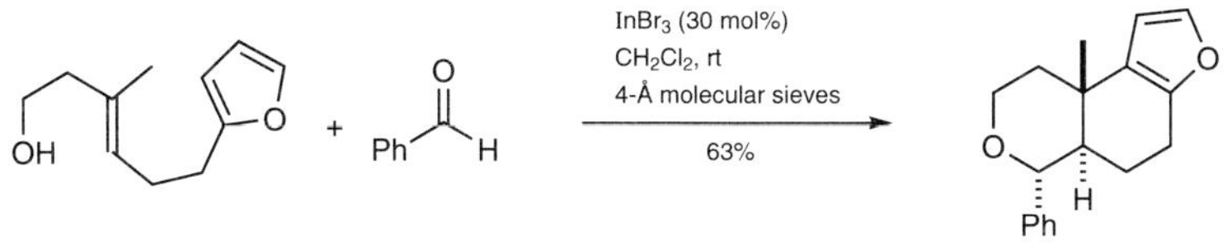

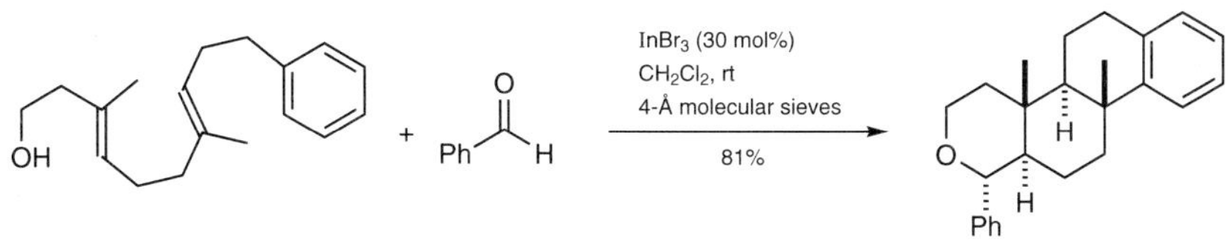

$InBr_3$ (30 mol%)
CH_2Cl_2, rt
4-Å molecular sieves
63%

$InBr_3$ (30 mol%)
CH_2Cl_2, rt
4-Å molecular sieves
81%

for references see p 314

OH + Ph(C=O)H

$InBr_3$ (30 mol%)
CH_2Cl_2, rt
4-Å molecular sieves
59%

(4*R,4a*R**,10b*S**)-10b-Methyl-4-phenyl-1,4,4a,5,6,10b-hexahydro-2*H*-benzo[*f*]-2-benzopyran (8, $R^1 = R^2 = R^3 = R^5 = H$; $R^4 = Ph$); Typical Procedure:**[9]

An oven-dried round-bottomed flask (10 mL) equipped with a magnetic stirrer bar was charged with 4-Å molecular sieves (100 mg) and $InBr_3$ (0.06 mmol), and sealed with a rubber septum. (*E*)-3-Methyl-6-phenylhex-3-en-1-ol (0.20 mmol) and benzaldehyde (0.24 mmol, 1.2 equiv) were dissolved in anhyd CH_2Cl_2 (2 mL) and added via syringe. The soln was stirred at rt for 24–96 h. The reaction was quenched with sat. aq $NaHCO_3$ (5 mL) and the mixture was extracted with CH_2Cl_2 (3 × 20 mL). The combined organic layers were washed with brine (30 mL), dried (Na_2SO_4), and concentrated under reduced pressure. The crude product was purified by flash column chromatography (silica gel, hexane/EtOAc mixture) to provide the product as a white solid; yield: 88%.

1.5.1.1.1.5 Highly Stereoselective Iron(III) Chloride Catalyzed Synthesis of Tetrahydro-2*H*-pyran-4-ols with Complete OH Selectivity

Chiral homoallylic alcohols **10** bearing an aryl group and with R^1 = OEt or aryl can be synthesized by a highly enantioselective nickel(II)/*N*,*N*′-dioxide ligand **9** catalyzed ene reaction.[11] These products serve as excellent substrates for a subsequent iron(III) chloride catalyzed highly stereoselective Prins cyclization to afford tetrahydro-2*H*-pyran-4-ols **11** with good to excellent diastereomeric ratios and excellent enantiomeric excess (Scheme 6).[12] The R^1 group is essential for success as it stabilizes the tetrahydropyran carbocation intermediate (see Scheme 1) and thereby facilitates its formation, and also, most importantly, retards the oxonia-Cope rearrangement. This stereopurity eroding rearrangement is also much discouraged by the presence of the carbonyl group α to the hydroxy group in homoallylic alcohol **10**. With glyoxal (R^1 = H), no reaction is detected. *tert*-Butyl(chloro)dimethylsilane helps to suppress the formation of dihydropyran products. This two-step reaction has a broad scope with regard to aldehydes, including benzaldehydes of varying electronic nature, heteroaromatic aldehydes, aliphatic aldehydes, and enals. It is, however, mostly limited to arylalkenes where the aryl group is not electron-rich, as the second step does not occur in the case where Ar^1 = 4-MeOC_6H_4, and the α-oxoaldehydes are likely limited to those without β-hydrogens.

Even though iron(III) chloride is the catalyst, no corresponding 4-chlorotetrahydropyrans are detected. The high selectivity toward the 4-hydroxy products is rationalized by a mechanism differing from the Prins cyclization, which is supported by mechanistic studies and DFT calculations. Nevertheless, the reaction is formally a Prins cyclization and offers a two-step, three-component access to functionalized tetrahydropyrans with high to excellent stereochemical control.

Scheme 6 Iron(III) Chloride Catalyzed Highly Stereoselective Synthesis of Tetrahydro-2*H*-pyran-4-ols[12]

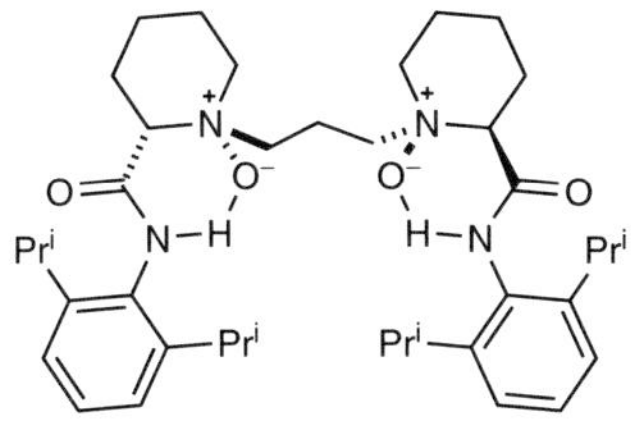

Ar^1	R^1	R^2	Yield[a] (%)	dr	ee (%)	Ref
Ph	OEt	Ph	74	90:10	98	[12]
Ph	OEt	2-MeOC_6H_4	72	90:10	96	[12]
Ph	OEt	2-naphthyl	92	>95:5	98	[12]
Ph	OEt	3-BrC_6H_4	70	>95:5	99	[12]
Ph	OEt	iPr	84	99:1	98	[12]
4-MeOC_6H_4	OEt	Et	–	–	–	[12]
2-Tol	OEt	4-BrC_6H_4	50	99:1	97	[12]
4-ClC_6H_4	OEt	Et	72	99:1	98	[12]
Ph	Ph	2-naphthyl	76	99:1	97	[12]
Ph	4-MeOC_6H_4	Et	50	99:1	98	[12]

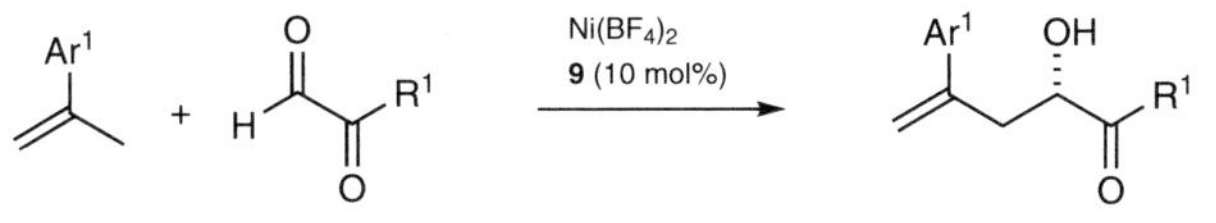

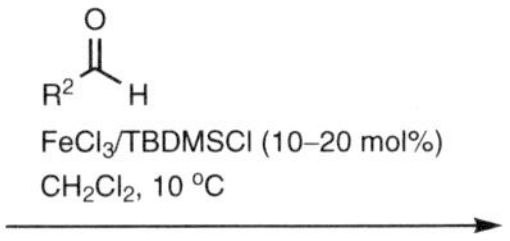

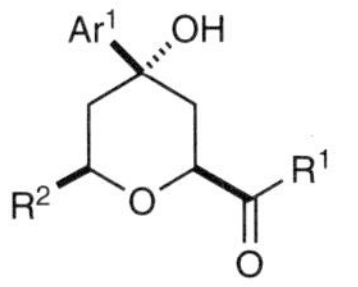

[a] Two-step overall yield.

Ethyl (2*S*,4*S*,6*R*)-4-Hydroxy-6-(naphthalen-2-yl)-4-phenyltetrahydro-2*H*-pyran-2-carboxylate (11, Ar^1 = Ph; R^1 = OEt; R^2 = 2-Naphthyl); Typical Procedure:[12]

A mixture of *N*,*N*′-dioxide ligand **9** (6.4 mg, 0.02 mmol) and $Ni(BF_4)_2\cdot 6H_2O$ (3.4 mg, 0.02 mmol) in CH_2Cl_2 (1.0 mL) was stirred at 30 °C for 30 min. Then, ethyl glyoxylate (0.2 mmol) and 2-phenylpropene (2.0 equiv) were added at 35 °C and the resulting mixture was stirred for 48 h. After flash column chromatography or filtration, the crude ethyl (*S*)-2-hydroxy-4-phenylpent-4-enoate (**10**, R^1 = Ph; R^2 = OEt) was added to a mixture of 2-naphthaldehyde (0.24 mmol, 1.2 equiv), $FeCl_3$ (20 mol%), and TBDMSCl (20 mol%) in CH_2Cl_2 (1.0 mL) at 10 °C. After the mixture had been stirred at the same temperature for 12 h, it was directly purified by column chromatography (silica gel, EtOAc/petroleum ether 1:7) to afford the product as a colorless liquid; yield: 92%; dr >95:5; 98% ee (for major diastereomer, determined by chiral HPLC analysis).

1.5.1.1.2 Prins Cyclization Involving Nucleophilic Alkynes

1.5.1.1.2.1 Iron(III)-Catalyzed Cyclization of Alkynyl Aldehyde Acetals

With a tethered C≡C bond as the nucleophile, Prins-type cyclizations can be likewise promoted by Lewis acids.[5,13] The use of a stoichiometric amount of an acetyl halide enables an iron(III) halide catalyzed Prins cyclization of alkynyl aldehyde acetals to give five-membered carbo- and heterocycles **12** (Scheme 7).[14] The reaction is generally applicable to substrates with oxygen- or nitrogen-containing backbones and with methyl or aryl groups of varying substitution patterns at the alkyne terminus. When the combination of iron(III) chloride hexahydrate (5 mol%) and acetyl chloride (1.2 equiv) is used, the reaction yields are mostly good to excellent, with the exception of the substrate where R^1 = Me. In contrast, using the combination of iron(III) bromide (5 mol%) and acetyl bromide (1.1 equiv), the yields are typically lower; moreover, in some cases dibromides are formed and the reaction is further complicated by double-bond isomers.

Scheme 7 Iron(III)-Catalyzed Efficient Synthesis of Five-Membered Carbo- and Heterocycles[14]

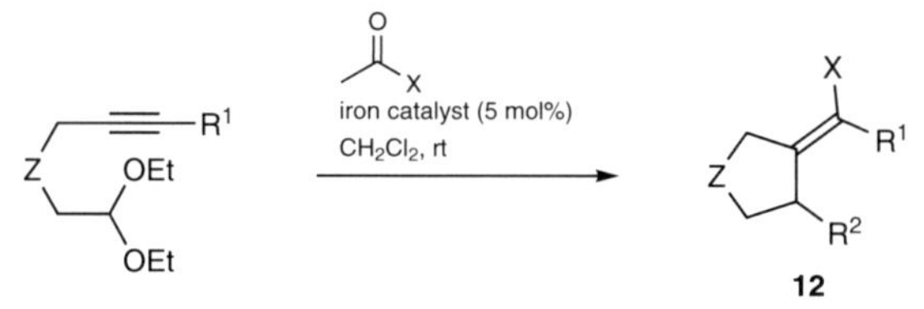

Z	R^1	R^2	X	Catalyst	Yield (%)	Ref
O	Ph	OEt	Cl	$FeCl_3•6H_2O$	83	[14]
O	4-BnO-2-MeC_6H_3	OEt	Cl	$FeCl_3•6H_2O$	80	[14]
O	3-MeOC_6H_4	OEt	Cl	$FeCl_3•6H_2O$	76	[14]
O	4-$O_2NC_6H_4$	OEt	Cl	$FeCl_3•6H_2O$	80	[14]
NTs	Me	OEt	Cl	$FeCl_3•6H_2O$	48	[14]
NTs	Ph	OEt	Cl	$FeCl_3•6H_2O$	81	[14]
NTs	Me	OEt	Cl	$FeCl_3•6H_2O$	48	[14]
O	Ph	Br	Br	$FeBr_3$	66[a]	[14]
O	4-ClC_6H_4	Br	Br	$FeBr_3$	89	[14]
O	4-$O_2NC_6H_4$	OEt	Br	$FeBr_3$	54	[14]

[a] Ratio (*E*/*Z*) 10:1.

Interestingly, when the reaction is run in acetone and in the absence of acetyl chloride, iron(III) chloride catalysis results in highly efficient formation of five-membered cyclic enones **13** (Scheme 8). Moreover, the reaction also works well with longer tethers, thereby offering facile access to six-membered-ring enones **13** (e.g., Z = OCH_2) in excellent yields and even seven-membered-ring products **13** [e.g., Z = $(CH_2)_2$NTs] albeit in low to fair yields. A hydroxy-functionalized alkynyl acetal is converted into the corresponding bicyclic enone in 70% yield (Scheme 8). The reaction mechanism, as supported by DFT calculations, entails the formation of an oxete intermediate via a stepwise (2+2) cycloaddition, its subsequent cycloreversion, and dealkylation.

Scheme 8 Efficient Iron(III) Chloride Catalyzed Synthesis of Cyclic Enones[14]

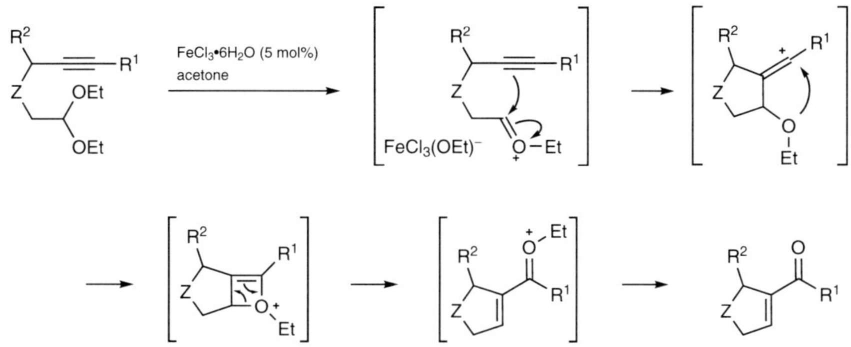

Z	R^1	R^2	Temp	Time (h)	Yield (%)	Ref
O	Ph	H	rt	2	93	[14]
O	3,5-$Me_2C_6H_3$	H	rt	0.5	99	[14]
O	Ph	Et	rt	0.5	94	[14]
O	4-HOC_6H_4	H	50 °C	0.2	97	[14]
O	4-AcC_6H_4	H	50 °C	5	89	[14]
NTs	Ph	H	50 °C	1.5	87	[14]
$C(CO_2Et)_2$	(*E*)-CH=CHPh	H	50 °C	0.2	95	[14]
OCH_2	Ph	H	50 °C	0.3	98[a]	[14]
$(CH_2)_2NTs$	Ph	H	80 °C	1.5	51[b]	[14]

[a] The product is (5,6-dihydro-2*H*-pyran-3-yl)(phenyl)methanone.
[b] The product is phenyl(1-tosyl-2,5,6,7-tetrahydro-1*H*-azepin-4-yl)methanone.

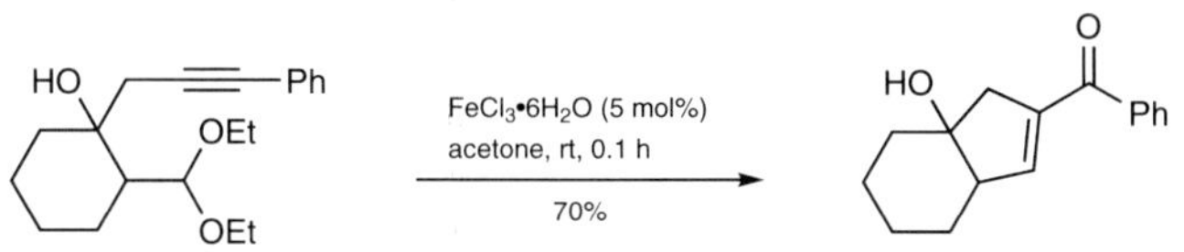

(2,5-Dihydrofuran-3-yl)(3,5-dimethylphenyl)methanone (13, R^1 = 3,5-$Me_2C_6H_3$; R^2 = H; Z = O); Typical Procedure:[14]

$FeCl_3•6H_2O$ (6.8 mg, 5 mol%) was added to a soln of 1-[3-(2,2-diethoxyethoxy)prop-1-ynyl]-3,5-dimethylbenzene (0.5 mmol) in acetone (5 mL) under air. The mixture was stirred at rt, and the reaction was monitored by TLC analysis (silica gel). After the substrate had been completely consumed (0.5 h), all volatile substances were evaporated under reduced pressure. The resulting residue was purified by flash column chromatography [silica gel, petroleum ether (60–90 °C)/Et_2O 10:1] to afford a white solid; yield: 99%.

1.5.1.1.3 Carbonyl-Ene Reactions

Most intramolecular carbonyl-ene reactions[2] are promoted by stoichiometric Lewis acids, although it is foreseeable that the corresponding catalytic versions could be developed. One of the early catalytic examples is the preparation of L-isopulegol from D-citronellal en route to the manufacture of L-menthol.[15,16] Recent developments have been centered on enantioselective carbonyl-ene reactions catalyzed by chiral Lewis acids.[17–19]

1.5.1.1.3.1 Chiral Copper(II)-Catalyzed Enantioselective Intramolecular Carbonyl-Ene Reactions of Unsaturated α-Oxo Esters

α-Oxo esters can be employed as bidentate substrates for copper(II)-catalyzed intramolecular carbonyl-ene reactions to give cyclic carboxylates **15** (Scheme 9).[17] With (4*S*,4′*S*)-2,2′-(propane-2,2-diyl)bis(4-phenyl-4,5-dihydrooxazole) (**14**) as the chiral ligand, the reactions of achiral substrates proceed with excellent to serviceable enantioselectivity (entries 1 and 2). With a chiral substrate (entries 3–5), the use of **14** as ligand (entry 4) apparently creates a matched scenario with the inherent stereoselectivity (see entry 3). Consequently, the stereochemical outcome is better than that without the ligand. Consistent with this observation, the reaction using the antipode of ligand **14** displays stereoselectivity inferior to that without ligand and hence is a mismatched case. Interestingly, the reaction outcome can be improved by a seemingly innocent switch of the methyl ester (entry 4) to a benzyl counterpart (entry 6).

Scheme 9 Copper(II) Trifluoromethanesulfonate/(4*S*,4′*S*)-2,2′-(Propane-2,2-diyl)bis(4-phenyl-4,5-dihydrooxazole) Catalyzed Asymmetric Synthesis of Cyclic Carboxylates[17]

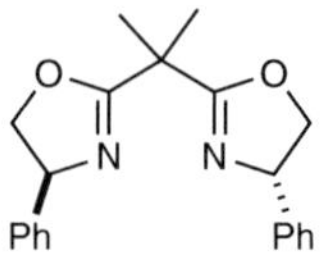

14

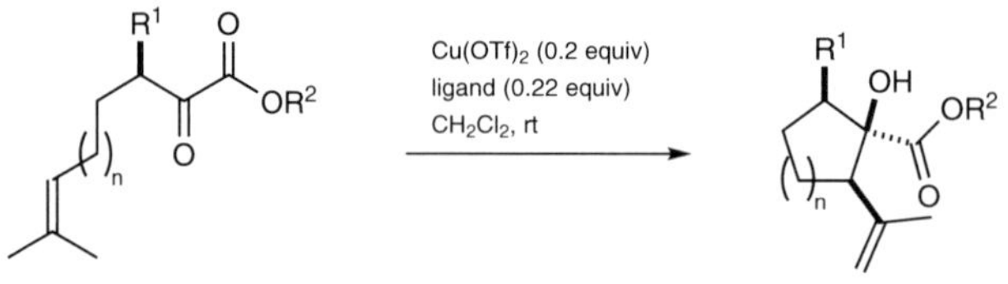

$Cu(OTf)_2$ (0.2 equiv)
ligand (0.22 equiv)
CH_2Cl_2, rt

15

Entry	R^1	R^2	n	Ligand	Yield[a] (%)	dr	ee (%)	Ref
1	H	Et	2	**14**	81[b]	>50:1	91	[17]
2	H	Et	1	**14**	78	46:1	71	[17]
3	Me	Me	1	–	76	7.3:1[c]	93[d]	[17]

Entry	R^1	R^2	n	Ligand	Yield[a] (%)	dr	ee (%)	Ref
4	Me	Me	1	**14**	91	24:1[c]	97[d]	[17]
5	Me	Me	1	*ent*-**14**	54	1.3:1[c]	87[d]	[17]
6	Me	Bn	1	**14**	94	34:1[c]	99.3[d]	[17]

[a] Determined by ^{1}H NMR spectroscopy using 1,2-diphenylprop-1-ene as an internal standard.
[b] A reaction using $Cu(OTf)_2$ (1 equiv) and **14** (1.1 equiv) gave **15** in 90% yield; dr >50:1; 87% ee.
[c] Minor isomer has only the ester group *cis* to the ring methyl group.
[d] ee of the major isomer.

Ethyl (1*R*,2*R*)-1-Hydroxy-2-(prop-1-en-2-yl)cyclohexane-1-carboxylate (15, R^1 = H; R^2 = Et; n = 2); Typical Procedure:[17]
(4*S*,4′*S*)-2,2′-(Propane-2,2-diyl)bis(4-phenyl-4,5-dihydrooxazole) (**14**; 66.2 mg, 0.20 mmol) and $Cu(OTf)_2$ (68.2 mg, 0.19 mmol) were mixed in anhyd CH_2Cl_2 (4 mL) at rt for 0.5 h. Ethyl 8-methyl-2-oxonon-7-enoate (40 mg, 0.19 mmol) was added. The mixture was stirred at rt. When the reaction was complete (TLC), the mixture was filtered through a thin pad of silica gel and then concentrated to give an analytically pure compound as a colorless oil; yield: 90%; dr >50:1; 87% ee.

1.5.1.1.3.2 Chromium(III)-Catalyzed Enantioselective Catalytic Carbonyl-Ene Cyclization Reactions

Certain enals undergo highly asymmetric carbonyl-ene reaction with chiral Schiff base ligated chromium(III) dimeric catalyst **16**.[18] Functionalized tetrahydrofurans (Table 1, entries 1–3), pyrrolidines (entry 5), methylenecyclohexanes (entries 4 and 8), and cyclopentanes (entries 6 and 7) are formed in good to excellent yields and with mostly excellent diastereo- and enantioselectivities. In this synthetically highly versatile reaction, the tether linking the aldehyde moiety and the alkene part can contain a heteroatom such as oxygen and nitrogen or be all-carbon, and several substitution patterns are allowed on the alkene including a tetrasubstituted alkene (entry 3) and a terminal alkene substituted at the 2-position (entry 4). In the latter case, the substrate undergoes 6-*endo* cyclization instead of the otherwise observed 5-*exo* cyclization. Of synthetic importance are desymmetrization reactions, where quaternary carbon centers are generated with excellent enantioselectivity (entries 6–8). An interesting double intramolecular carbonyl-ene reaction with a dialdehyde substrate forms a bridged bicyclic product in a serviceable yield and with excellent enantiomeric excess (entry 9).

Table 1 Schiff Base/Chromium(III)-Catalyzed Enantioselective Intramolecular Carbonyl-Ene Reactions[18]

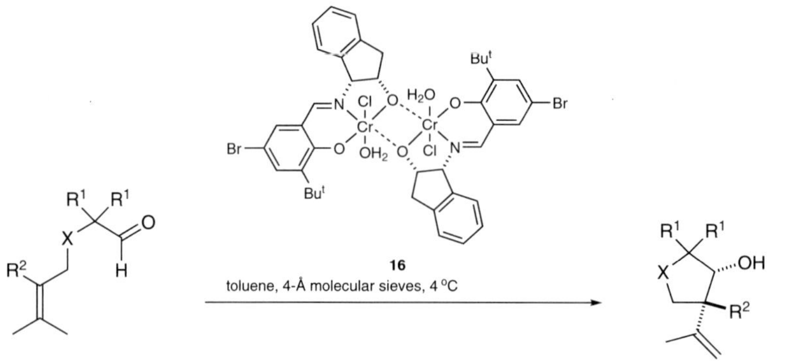

16

toluene, 4-Å molecular sieves, 4 °C

Entry	Substrate	Product	Catalyst (mol%)	Yield[a] (%)	dr	ee (%)	Ref
1	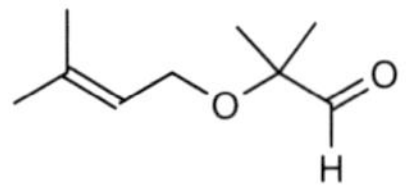	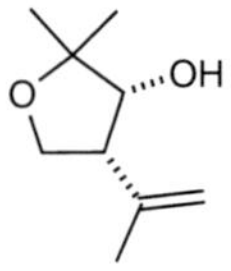	0.8	77	>30:1	93	[18]
2	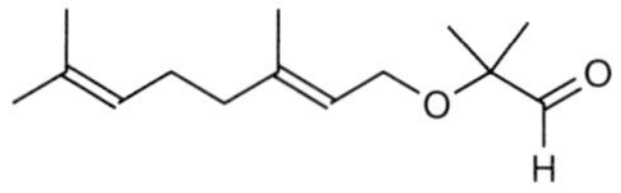	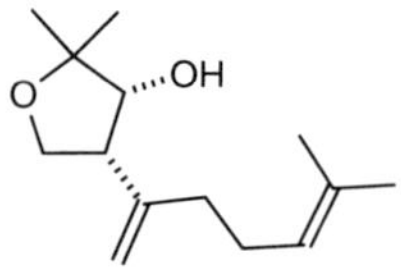	1	94	20:1	96	[18]
3	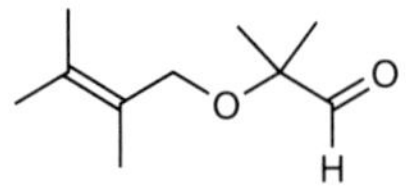	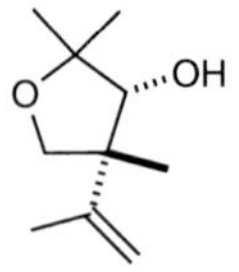	5	78	>30:1	75	[18]
4	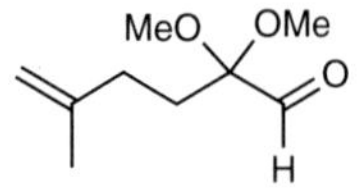	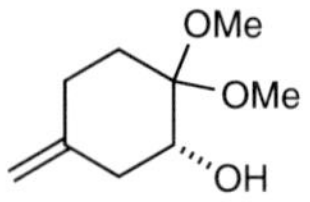	2.5	88	–	94	[18]
5	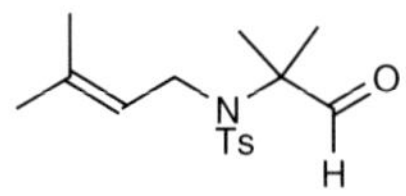	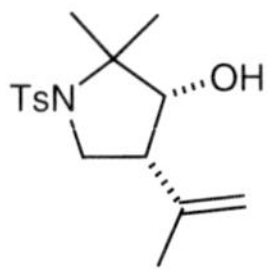	2	98	>30:1	95	[18]
6	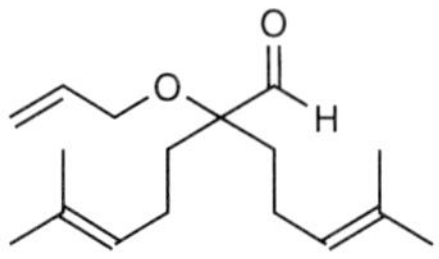	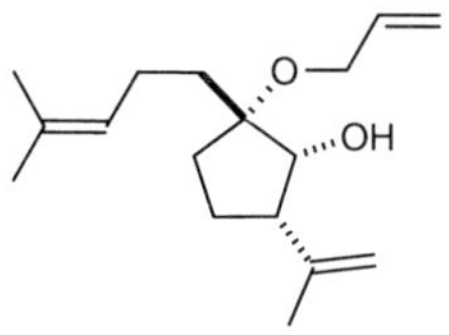	2	87	7:1	99	[18]

Table 1 (cont.)

Entry	Substrate	Product	Catalyst (mol%)	Yield[a] (%)	dr	ee (%)	Ref
7	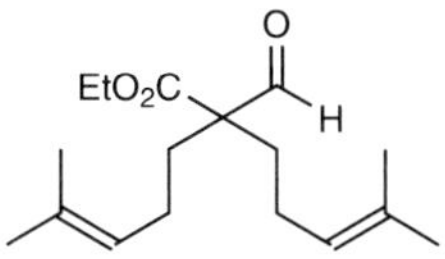	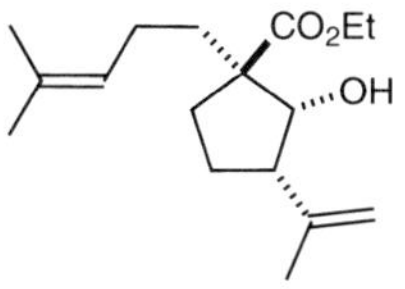	2	95	>30:1	98	[18]
8	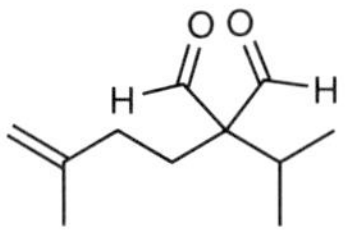	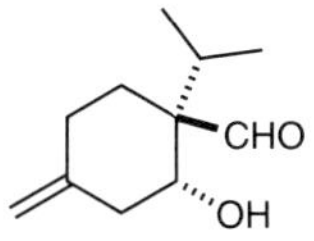	1	57	2.2:1	91	[18]
9	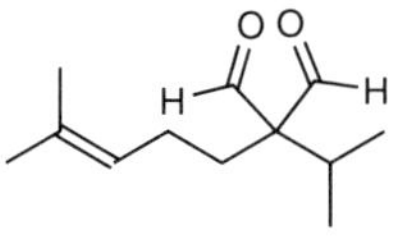	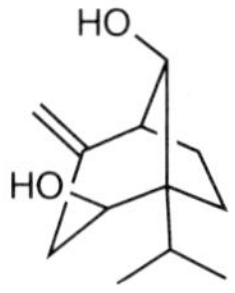	10	46	>30:1	92	[18]

[a] Isolated yield of the indicated isomer.

(3*R*,4*R*)-2,2-Dimethyl-4-(prop-1-en-2-yl)tetrahydrofuran-3-ol (Table 1, Entry 1); Typical Procedure:[18]

Toluene (25 μL) and 2-methyl-2-[(3-methylbut-2-enyl)oxy]propanal (31 mg, 0.2 mmol) were added to a cooled (0 °C), flame-dried, and N_2-protected 0.5-dram reaction vial containing 4-Å molecular sieves (40 mg) and the chiral dimeric chromium catalyst **16** (1.6 mg, 1.6 μmol). The mixture was warmed to 4 °C and stirred until conversion of the substrate was deemed complete by TLC (ca. 30 h). The mixture was diluted with Et_2O/hexanes (1:1; 0.5 mL) and loaded onto a column (silica gel). Purification by flash column chromatography (Et_2O/hexanes 1:9) afforded the product as a volatile, colorless oil; yield: 24 mg (77%); dr >30:1 (determined by ^{1}H NMR spectroscopy); 93% ee {determined by chiral GC [γ-TA, 85 °C isothermal; t_R (minor) 19.5 min; t_R (major) 27.1 min]}.

1.5.1.1.3.3 Chromium(III)-Catalyzed Enantioselective Transannular Ketone-Ene Reactions

Although ketones are typically not suitable for carbonyl-ene reactions, (*E*)-5-methylcyclodec-5-enones **17** undergo transannular ketone-ene reaction by using a similar chiral Schiff base–chromium dimer complex **18** as catalyst to give bicyclic alcohols **19** (Scheme 10).[19] The examples shown exhibit good yields, and excellent diastereoselectivities and enantiomeric excesses; however, substrates **17** (Z = O; X = CH_2) or a cyclononenone analogue do not react well, and the yields are <20%. The counteranion of the chiral catalyst has a significant impact on the reaction outcome. In the reaction of **17** (Z = X = CH_2), use of trifluoromethanesulfonate as the anion leads to 93% ee and a decent yield; in contrast, the use of hexafluoroantimonate(V), hexafluorophosphate, and bis(trifluoromethylsulfonyl)amide as counteranions all result in much lower enantiomeric excesses, albeit in excellent yields.

for references see p 314

Scheme 10 Chromium(III)-Catalyzed Enantioselective Transannular Ketone-Ene Reactions[19]

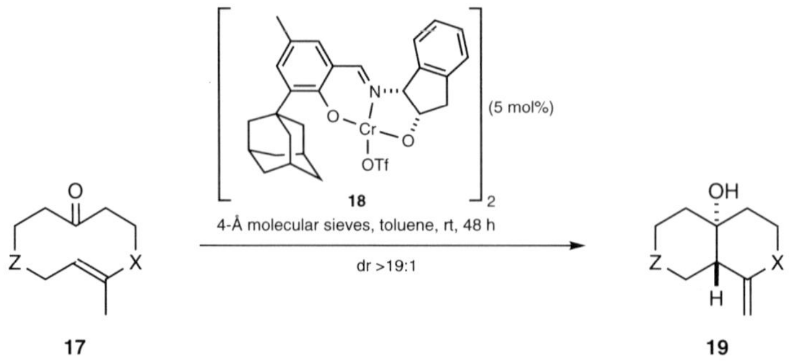

Z	X	Yield (%)	ee (%)	Ref
CH_2	CH_2	81	93	[19]
CMe_2	CH_2	96	94	[19]
$C(OCH_2)_2$	CH_2	87	96	[19]
$C{=}CH_2$	CH_2	62	94	[19]
O	CH_2	13	49	[19]
CH_2	CMe_2	84	94	[19]

Bicyclic Alcohols 19; General Procedure:[19]
An oven-dried, 0.5-dram screw-top vial charged with a stirrer bar and activated 4-Å molecular sieves (10 mg) was sealed with a cap containing a Teflon-lined septum. The molecular sieves were flame-dried under reduced pressure (1 Torr) and allowed to cool to rt under N_2. To the cooled vial was added the Cr catalyst dimer **18** (12.7 mg, 0.02 mmol, 5 mol%; 10 mol% based on Cr). The (*E*)-5-methylcyclodec-5-enone **17** (0.2 mmol) was added to the vial, followed by toluene (50 μL) by microliter syringe (or in the opposite order). The N_2 line was removed, the cap was wrapped with Parafilm, and the mixture was stirred at rt for 48 h. An aliquot (ca. 2 μL) was removed from the vial and diluted into an NMR tube with $CDCl_3$ to determine the dr of the product. The NMR sample along with the remainder of the crude mixture was directly loaded onto a column (silica gel, neutral alumina or Davisil) and chromatographed.

1.5.1.2 Iminium-Based Electrophiles in Aza-Prins Reactions

Analogous to the Prins cyclization, the aza-Prins cyclization employs an iminium ion as the electrophilic moiety to enable cyclization by an alkene or alkyne. The reaction offers access to piperidines and tetrahydropyridines in an overall (5+1)-annulation process.

1.5.1.2.1 Iron(III)-Catalyzed Aza-Prins Cyclizations

In the report on iron(III)-catalyzed Prins cyclization (see Section 1.5.1.1.1.1),[5] it was also disclosed that aza-Prins cyclization can readily occur under similar conditions. By using a halotrimethylsilane as the halide source, the reaction between an *N*-homopropargyl or *N*-homoallyl sulfonamide (e.g., **20** or **22**, respectively) and an aldehyde is catalyzed efficiently by iron(III) chloride or tris(acetylacetonato)iron(III) at room temperature to give the corresponding piperidine products **21** and **23** (Scheme 11).[5,20] In the cases of the alkenyl sulfonamides, *trans*-disubstituted piperidines are the major products; this is in stark contrast to the corresponding Prins cyclization, where *cis* products are predominantly formed. The tosyl group can be replaced by more easily removable 4-nitrophenylsulfonyl (nosyl) or mesyl groups without having much impact on efficiency, and various aldehydes including functionalized ones are allowed, reflecting the mild nature of the reaction conditions.

Scheme 11 Iron(III)-Catalyzed Synthesis of Tetrahydropyridines and Piperidines[5,20]

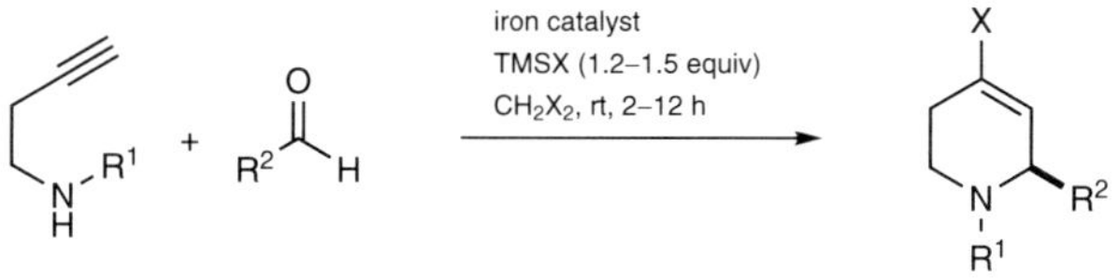

iron catalyst
TMSX (1.2–1.5 equiv)
CH_2X_2, rt, 2–12 h

20 → 21

R^1	R^2	X	Catalyst (mol%)	Yield (%)	Ref
Ts	iBu	Cl	$Fe(acac)_3$ (7.5)	80	[5]
Ts	Cy	Cl	$Fe(acac)_3$ (7.5)	79	[5]
Ts	iBu	Br	$Fe(acac)_3$ (7.5)	80	[5]
Ts	Bn	Br	$Fe(acac)_3$ (7.5)	88	[5]
$4-O_2NC_6H_4SO_2$	iBu	Br	$FeCl_3$ (15)	97	[20]
Ms	iBu	Br	$FeCl_3$ (10)	82	[20]

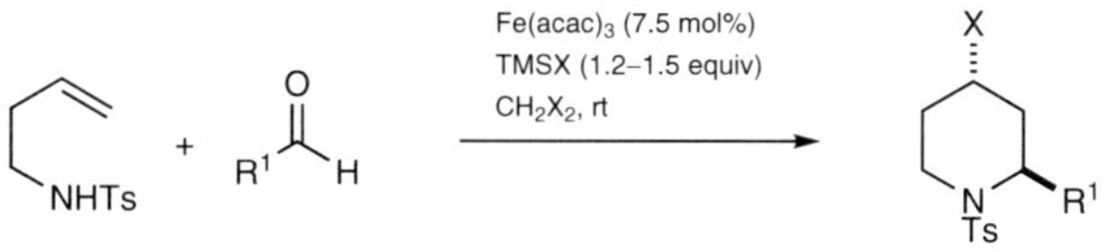

$Fe(acac)_3$ (7.5 mol%)
TMSX (1.2–1.5 equiv)
CH_2X_2, rt

22 → 23

R^1	X	Yield (%)	dr (*trans*/*cis*)	Ref
Bn	Cl	99	83:17	[5]
iBu	Br	>99	94:6	[5]
iBu	I	92[a]	95:5	[5]
$(CH_2)_2CH{=}CH_2$	Cl	85	83:17	[5]
$(CH_2)_2OBn$	Cl	85	95:5	[5]

[a] MeI was used as solvent.

***trans*-2-[2-(Benzyloxy)ethyl]-4-chloro-1-tosylpiperidine** [23, $R^1 = (CH_2)_2OBn$]; **Typical Procedure:**[5]

To a soln of *N*-(but-3-enyl)-4-toluenesulfonamide (**22**; 1.2 equiv) in anhyd CH_2Cl_2 (0.1 M) was added $Fe(acac)_3$ (7.5 mol%). The resulting soln was cooled to 0 °C using an ice bath and then TMSCl (1.2 equiv) was added. After 5 min, 3-(benzyloxy)propanal (1.5 equiv) was added dropwise. The mixture was stirred until TLC analysis showed complete formation of the product. The reaction was then quenched by addition of H_2O with stirring and the mixture was extracted with CH_2Cl_2. The combined organic layers were dried ($MgSO_4$), and the solvent was removed under reduced pressure. The residue was purified by flash column chromatography (silica gel, hexane/EtOAc); yield: 85%.

1.5.1.2.2 Scandium(III) Trifluoromethanesulfonate Catalyzed Aza-Prins Cyclizations for the Synthesis of Heterobicycles

Similar to the work discussed in Section 1.5.1.1.1.3, an aza-Prins cyclization can be coupled with an intramolecular trapping of the developing positive charge in the initially formed ring, thereby yielding bicyclic piperidines. As a consequence of the trapping, metal salts can be used as catalyst to promote this type of reaction. The reactions of *N,N′*-(hex-3-ene-1,6-diyl)bis(4-toluenesulfonamides) with various aldehydes are catalyzed by scandium(III) trifluoromethanesulfonate at 80 °C to afford bicyclic piperidines in mostly good yields (Table 2, entries 1–4).[21] The reaction is highly stereoselective with regard to the substrate double bond geometry (compare entries 2 and 3), suggesting that the bicyclic skeleton is formed concertedly. An interesting case shows that the alternative Prins cyclization outcompetes the aza-Prins cyclization and the tetrahydropyran product is formed selectively (entry 5).

Table 2 Scandium(III) Trifluoromethanesulfonate Catalyzed Synthesis of Heterobicycles[21]

TsHN, NHTs + R^1CHO (1.5 equiv) → $Sc(OTf)_3$ (10 mol%), 1,2-dichloroethane, 80 °C → NTs, H, H, R^1, N, Ts

Entry	Substrates: Amide	Substrates: Aldehyde	Time (h)	Product	Yield (%)	dr	Ref
1	TsHN, NHTs	Ph, O, H	6	NTs, H, H, Ph, N, Ts	78	95:5	[21]
2	NHTs, NHTs	O, H, Br	11	NTs, H, H, Br, N, Ts	65	–[a]	[21]

Table 2 (cont.)

Entry	Substrates		Time (h)	Product	Yield (%)	dr	Ref
	Amide	Aldehyde					
3	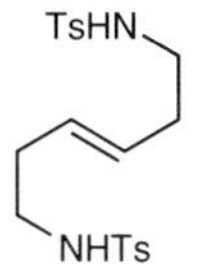	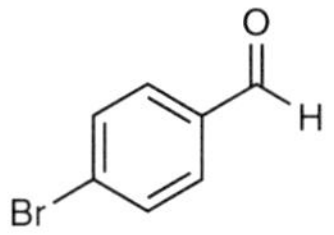	10	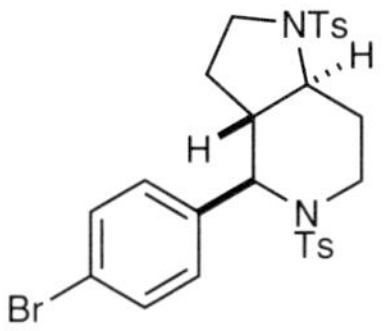	67	90:10	[21]
4	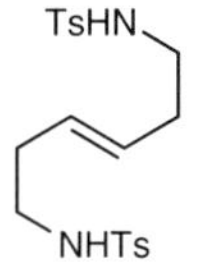	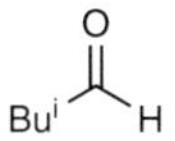	6	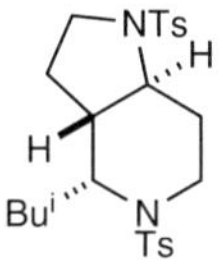	70	95:5	[21]
5	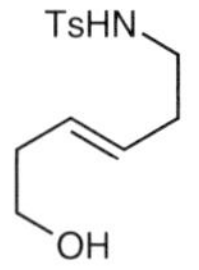	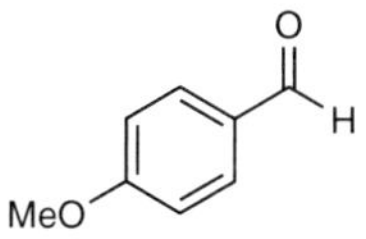	6	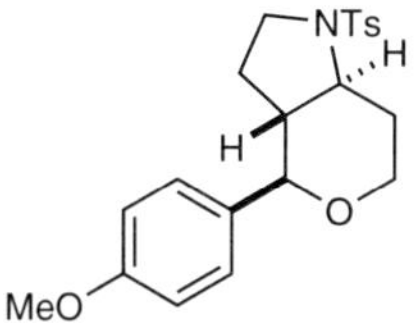	75	95:5	[21]

a Product formed exclusively; no other diastereomers were reported to be formed.

(3a*R,4*R**,7a*S**)-4-[(*E*)-Styryl]-1,5-ditosyloctahydro-1*H*-pyrrolo[3,2-*c*]pyridine (Table 2, Entry 1); Typical Procedure:**[21]
To a soln of (*E*)-*N*,*N*′-(hex-3-ene-1,6-diyl)bis(4-toluenesulfonamide) (211 mg, 0.50 mmol) and cinnamaldehyde (99 mg, 0.75 mmol) in anhyd 1,2-dichloroethane (5 mL) was added $Sc(OTf)_3$ (10 mol%) and the resulting mixture was heated at 80 °C for 6 h. After completion of the reaction as indicated by TLC analysis, the organic layer was washed with brine (3 × 2 mL), dried (Na_2SO_4), and concentrated under reduced pressure. The resulting crude product was purified by column chromatography [silica gel (100–200 mesh; Merck), EtOAc/hexane 8:92 (50 mL), EtOAc/hexane 13:87 with two drops of Et_3N (50 mL), EtOAc /hexane 18:82 with four drops of Et_3N (100 mL)]; yield: 210 mg (78%); dr 95:5.

1.5.2 Metal-Activated Alkenes and Alkynes as Electrophiles

The coordination of a π-system, be it a C=C or a C≡C bond, to a Lewis acidic metal complex, lowers the energy of the π*-orbital and thereby facilitates attack by nucleophiles in an *anti* manner. This general reaction pattern coupled with efficient regeneration of the active metal complex, as outlined in Scheme 12, constitutes a synthetically versatile metal-catalyzed addition of nucleophiles to alkenes and alkynes. In this type of reaction, electron-rich or electron-neutral alkenes and alkynes typically experience a reversal of polarity and become electrophilic. The intramolecular version, i.e. cyclization, is generally facile and can be promoted by various transition metals and especially late transition metals such as palladium, rhodium, ruthenium, copper, silver, platinum, gold, and mercury. With alkyne substrates, cationic gold(I) complexes are in general the most versatile among various metals, and have become the preferred catalysts due to generally high reaction efficiency, tolerance of moisture and air, and mild reaction conditions, despite the cost of the noble metal.

An alternative mechanism for this type of cyclization may entail a *syn* addition, where both the nucleophile and the π-bond are coordinated to the metal catalyst before the cyclization step (Scheme 12).

Because cycloisomerization reactions involving multiple unsaturated bonds are discussed in Section 1.7 and amination and C—O bond-forming cyclizations are covered in Section 1.8, this section will be limited to nucleophiles based on stabilized carbanions and arenes. Few cyclizations by heteronucleophiles other than those based on oxygen and nitrogen are synthetically versatile and these are thus not discussed.

Scheme 12 Metal-Catalyzed Additions of Nucleophiles to Alkenes or Alkynes

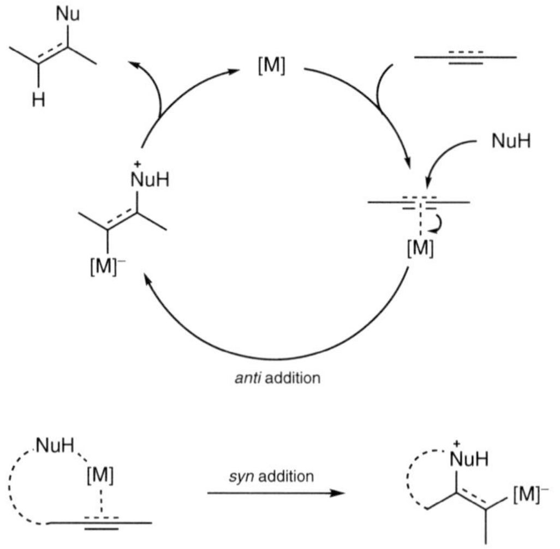

1.5.2.1 1,3-Dicarbonyl-Based Carbon Nucleophiles

Intramolecular cyclization of tethered 1,3-dicarbonyl carbon nucleophiles, such as β-oxo esters, to alkenes and alkynes belong to the Conia-ene reaction.[22] This approach provides a rather efficient synthesis of carbocycles owing to the ease of substrate preparation. Although the original reported conditions (≥250 °C) are harsh, metal catalysts can dramatically lower the reaction temperature and improve the reaction outcome.[23] As shown in Scheme 13, there are competitive *endo*- or *exo*-cyclization modes, and the catalyst used and the substrate structure affect regioselectivity.

Scheme 13 Metal-Catalyzed Conia-Ene Reactions

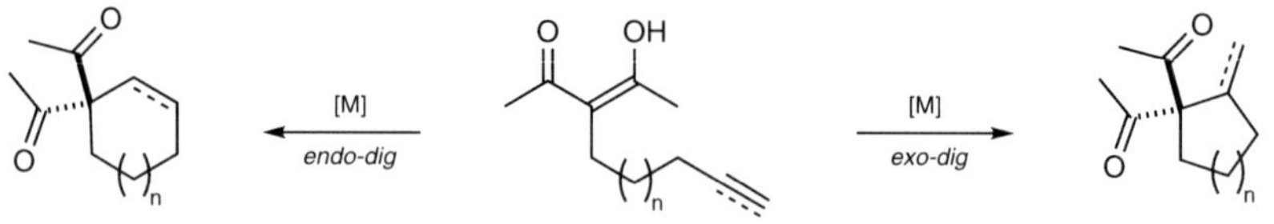

1.5.2.1.1 Cyclizations onto Alkenes

The metal-catalyzed cyclization of tethered 1,3-dicarbonyl carbon nucleophiles to non-electron-deficient alkenes constitutes a method for the intramolecular addition of a labile C—H bond across an unactivated C=C bond. This strategy offers a synthetically versatile approach to the construction of carbocycles that are otherwise challenging. Advances in this area have been recently reported using palladium[24–27] or gold catalysis.[28,29]

1.5.2.1.1.1 Palladium-Catalyzed 6-*endo-trig* Cyclization of Alkenyl 1,3-Dicarbonyls

The cyclization of alkenyl 1,3-diketones **24** is catalyzed by bis(acetonitrile)dichloropalladium(II) with surprising ease; the reaction proceeds at room temperature to give 2-acylcyclohexanones **25** (Scheme 14).[24,25] Substitution at the alkene terminus is tolerated. The proposed reaction mechanism entails acidic palladium(II) cation promoted nucleophilic attack at the C=C bond in a highly selective 6-*endo-trig* cyclization and subsequent protonation of the nascent Pd—C(sp^3) bond. Surprisingly, no β-hydride elimination product derived from the alkylpalladium intermediate is detected. When less acidic β-oxo esters are used as the tethered nucleophiles, the initial conditions lead to poor yields. However, the addition of chlorotrimethylsilane improves the reaction significantly, which can be attributed to the in situ generation of the more nucleophilic silyl enol ether derivatives of the carbon nucleophiles. In some cases, copper(II) chloride (1 equiv) is used to prevent the reduction of the palladium(II) catalyst.

Scheme 14 Palladium-Catalyzed Synthesis of 2-Acylcyclohexanones and 2-Oxocyclohexanecarboxylates[24,25]

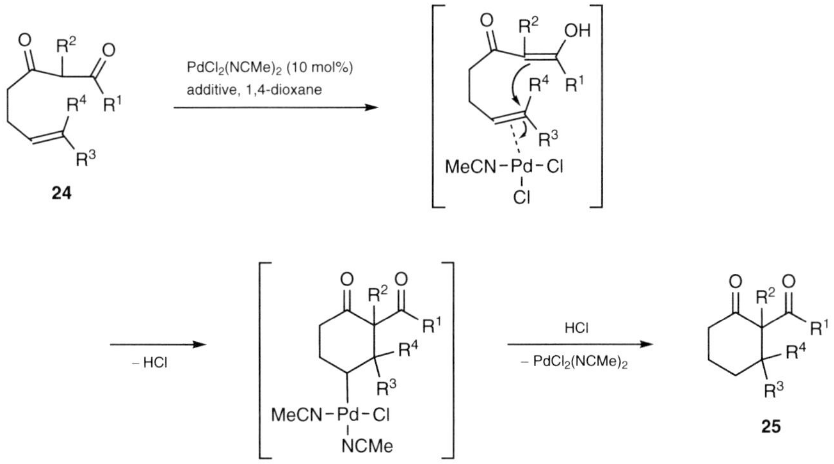

R^1	R^2	R^3	R^4	Additive	Yield (%)	Ref
Me	H	H	H	–	81	[24]
t-Bu	H	H	H	–	81	[24]
Me	Bn	H	H	–	70[a]	[24]
Me	H	Me	Me	–	71	[24]
Me	H	Bu	H	–	89[a]	[24]
OMe	H	H	H	TMSCl (2 equiv)	91	[25]
OMe	H	Et	H	TMSCl (2 equiv), $CuCl_2$ (1 equiv)	82	[25]
OMe	H	Ph	H	TMSCl (2 equiv)	93	[25]

[a] 20 mol% of catalyst was used.

Methyl 2-Ethyl-6-oxocyclohexane-1-carboxylate (25, R^1 = OMe; R^2 = R^4 = H; R^3 = Et); Typical Procedure:[25]

A suspension of $PdCl_2(NCMe)_2$ (15 mg, 0.06 mmol), $CuCl_2$•$2H_2O$ (94 mg, 0.53 mmol), methyl (*E*)-3-oxonon-6-enoate (**24**, R^1 = OMe; R^2 = R^4 = H; R^3 = Et; 100 mg, 0.55 mmol), and TMSCl (0.14 mL, 1.1 mmol) in 1,4-dioxane (15 mL) was stirred at 55 °C for 12 h. The resulting suspension was concentrated under reduced pressure, treated with 1 M aq HCl (20 mL), and

for references see p 314

extracted with Et_2O (3 × 60 mL). The combined extracts were washed with brine, dried ($MgSO_4$), and concentrated. The residue was purified by chromatography (hexanes/Et_2O 20:1 to 5:1) to give the product as a pale yellow oil; yield: 82 mg (82%); R_f 0.32 (hexanes/EtOAc 5:1).

1.5.2.1.1.2 Gold-Catalyzed 5-*exo-trig* Cyclization of Alkenyl β-Oxo Amides

In contrast to the palladium catalysis described in Section 1.5.2.1.1.1, related alkenyl β-oxo amides undergo exclusive *exo-trig* cyclization in the presence of cationic gold(I) catalyst **26** to deliver substituted γ-lactams (Table 3, entries 1–6) or piperidin-2-ones (entry 7) in mostly excellent yields.[29] The ketone moiety can be a phenone (entry 2) or a cyclic derivative (entries 3 and 4). In the latter cases, spirolactams are generated, albeit with moderate diastereoselectivities. A phenyl group at the allylic position does not affect the reaction (e.g., entry 5). However, methyl substitution at the internal position of the alkene substantially slows the reaction (entry 6), and substrates with 1,2-disubstituted alkenes do not undergo the desired cyclization.

Table 3 Efficient Gold(I)-Catalyzed Synthesis of Highly Substituted Lactams[29]

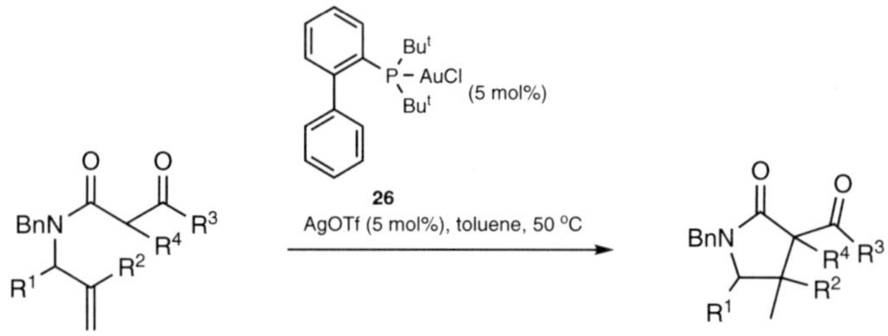

26 (5 mol%), AgOTf (5 mol%), toluene, 50 °C

Entry	Substrate	Time (h)	Product	Yield (%)	Ref
1	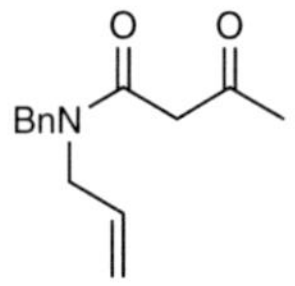	5	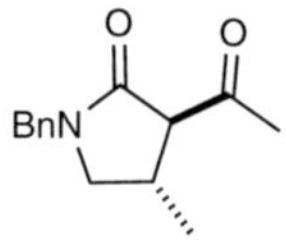	99	[29]
2	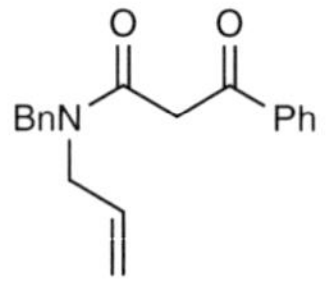	4	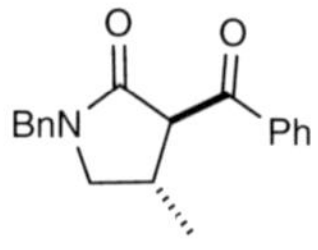	99	[29]
3	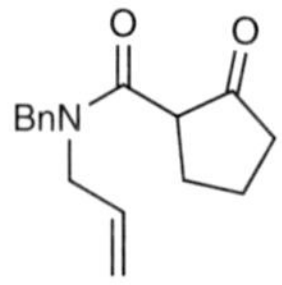	4	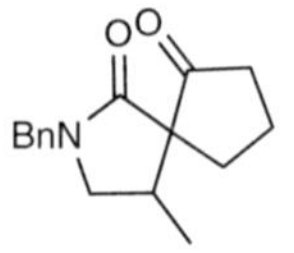	99[a]	[29]
4	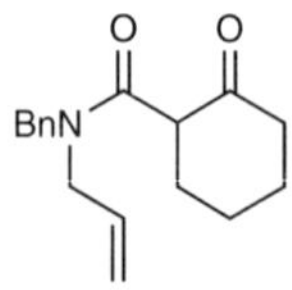	4	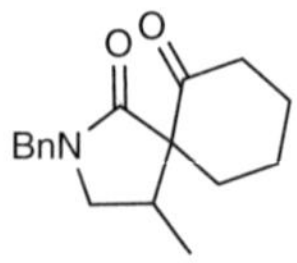	99[a]	[29]

Table 3 (cont.)

Entry	Substrate	Time (h)	Product	Yield (%)	Ref
5	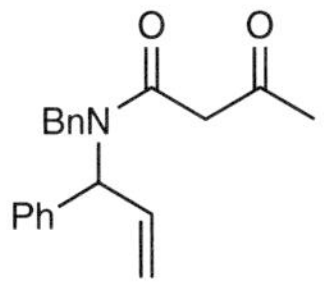	5	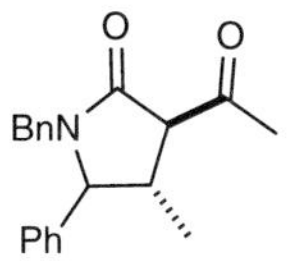	91[b]	[29]
6	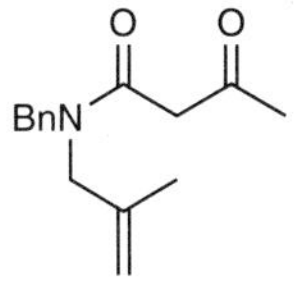	12	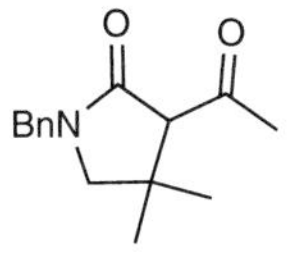	95[c]	[29]
7	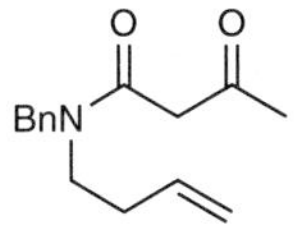	5	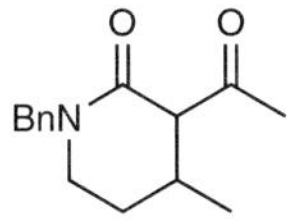	98	[29]

[a] dr 3:1.
[b] dr 1.5:1.
[c] Reaction temperature was 60 °C.

***trans*-3-Acetyl-1-benzyl-4-methylpyrrolidin-2-one (Table 3, Entry 1); Typical Procedure:**[29]
To a toluene (3 mL) soln containing gold(I) chloride catalyst **26** (0.02 mmol) and AgOTf (0.02 mmol) was added *N*-allyl-*N*-benzyl-3-oxobutanamide (0.4 mmol), and the resultant mixture was stirred at 50 °C for 5 h. The solvent was removed and the crude residue was purified by column chromatography (silica gel); yield: 99%.

1.5.2.1.1.3 Gold-Catalyzed *exo-trig* Cyclization of Alkenyl Ketones

The palladium-catalyzed method discussed in Section 1.5.2.1.1.1 can be extended by replacing the dicarbonyl-based carbon nucleophiles with less acidic dialkyl ketones and performing the reaction in the presence of hydrogen chloride (10 mol%) and copper(II) chloride (30 mol%) at 70 °C.[26] The reactions again exhibit excellent regioselectivity for 6-*endo-trig* over 5-*exo-trig* cyclization. In the same vein, gold catalysis can be extended to dialkyl and alkyl aryl ketone substrates.[28] The use of gold(I) chloride catalyst **27** offers quick access to cyclopentyl or cyclohexyl ketones, with some selected examples shown in Table 4.[28] In addition to using malonate as the linker for the two reacting partners (entries 1–7), the reduced form (entry 8) and the tosylamino group (entry 9) work equally well, although no case with a methylene linker group has been reported. Methyl substitution at the internal position of the alkene can be tolerated (entry 7), but *gem*-dimethyl groups at its terminus lead to no reaction. The broad range of suitable ketones for this reaction is noteworthy, and all the reactions again exhibit exclusive *exo* selectivity.

Table 4 Efficient Gold(I)-Catalyzed Synthesis of Cyclopentyl or Cyclohexyl Ketones[28]

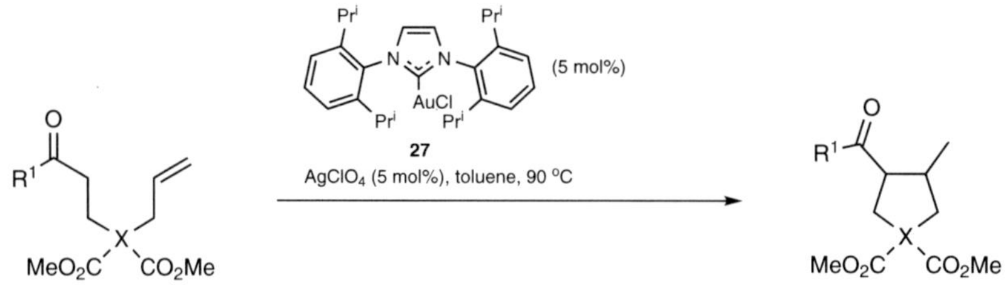

27 (5 mol%), $AgClO_4$ (5 mol%), toluene, 90 °C

Entry	Substrate	Time (h)	Product	Yield (%)	dr	Ref
1	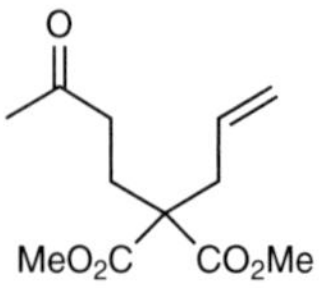	4.5	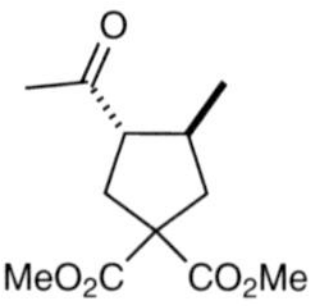	99	8.6:1	[28]
2	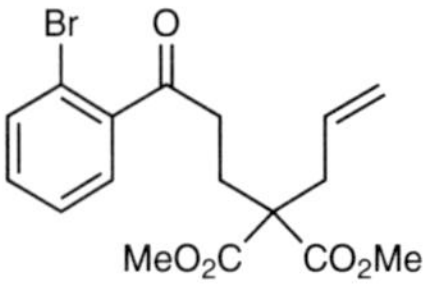	9	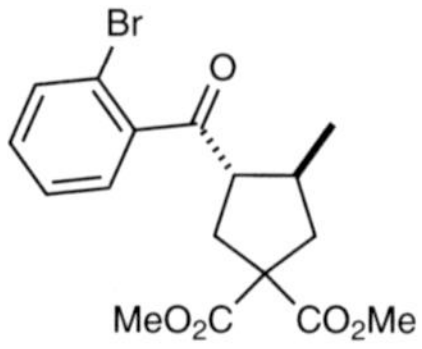	89	9.0:1	[28]
3	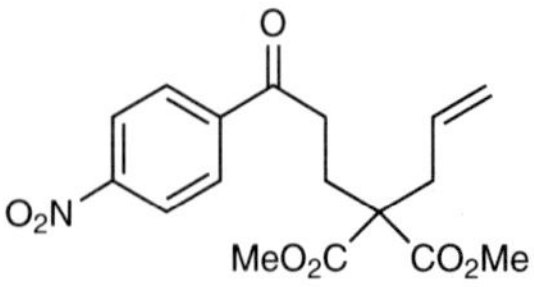	10	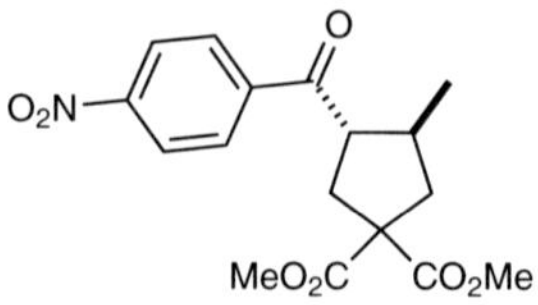	89	8.0:1	[28]
4	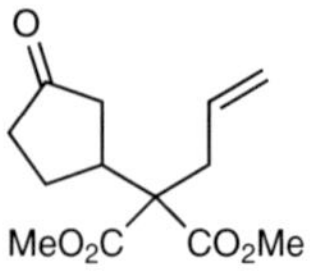	13	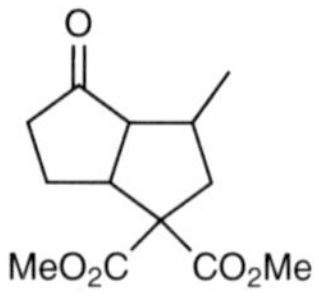	99	4.0:1.5:1	[28]
5	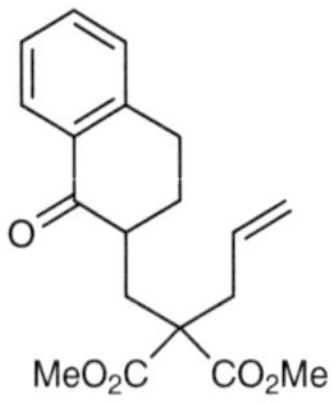	5	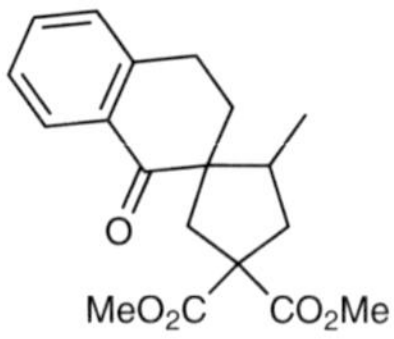	81	1.7:1	[28]
6	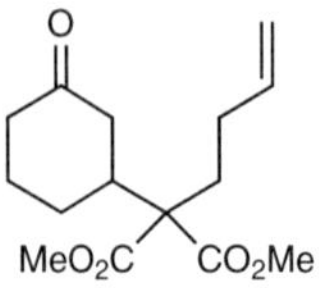	9	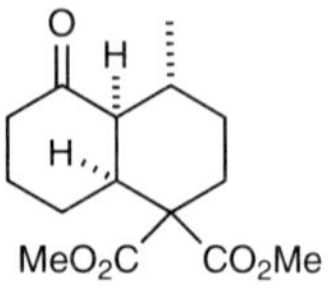	90	9.5:1	[28]

Table 4 (cont.)

Entry	Substrate	Time (h)	Product	Yield (%)	dr	Ref
7	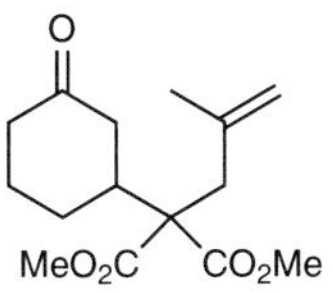 O; MeO_2C, CO_2Me	16	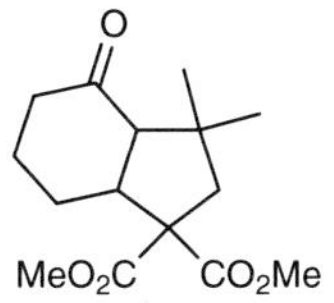 O; MeO_2C, CO_2Me	86[a]	6.0:1	[28]
8	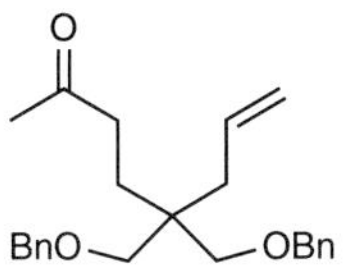 O; BnO, OBn	3	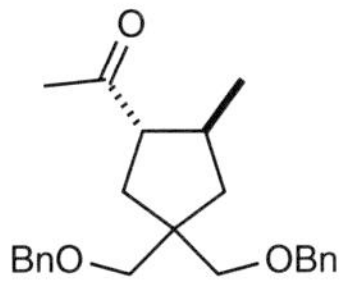 O; BnO, OBn	71	3.0:1	[28]
9	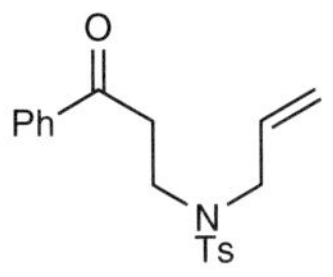 O; Ph; N, Ts	4.5	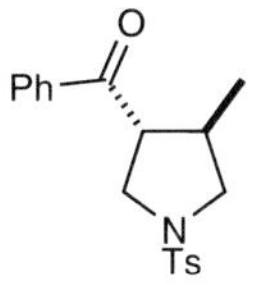 O; Ph; N, Ts	71	1.5:1	[28]

[a] 10 mol% of catalyst **27** and $AgClO_4$ was used at 110 °C.

Dimethyl *trans*-3-Acetyl-4-methylcyclopentane-1,1-dicarboxylate (Table 4, Entry 1); Typical Procedure:[28]

CAUTION: *Perchlorate salts are potentially explosive and should be handled with great caution.*

A mixture of gold(I) chloride catalyst **27** (7.8 mg, 12.5 μmol), $AgClO_4$ (2.5 mg, 12.5 μmol), and dimethyl 2-allyl-2-(3-oxobutyl)malonate (0.25 mmol) in toluene (0.5 mL) was stirred at 90 °C under argon and the reaction was monitored by TLC. Upon completion, the solvent was removed under reduced pressure, and the residue was purified by column chromatography (silica gel, EtOAc/petroleum ether 1:12 to 1:6); yield: 99%; dr 8.6:1.

1.5.2.1.2 Cyclizations onto Alkynes

Various metals including gold,[30–32] nickel,[33] zinc,[34] copper/silver,[35] and iron[36] can catalyze the Conia-ene reaction with 1,3-dicarbonyl-functionalized alkynes under mild conditions. Cationic gold(I) complexes are highly versatile and can catalyze this reaction at ambient temperature, whereas less alkynophilic Lewis acids such as zinc(II) chloride[34] and iron(III) chloride[36] require heating and typically longer reaction times and higher catalyst loadings, although the metals are inexpensive.

1.5.2.1.2.1 Cationic Gold(I) Catalyzed Conia-Ene Reactions

(Triphenylphosphine)gold(I) trifluoromethanesulfonate, generated in situ by mixing chloro(triphenylphosphine)gold(I) (1 mol%) and 1 equivalent of silver(I) trifluoromethanesulfonate, is a versatile catalyst to facilitate highly regioselective 5-*exo-dig* cyclizations of β-oxo esters to tethered terminal alkynes (Table 5).[30] The reaction permits highly efficient access to a fused bicyclic lactone (entry 3), fused bicyclic ketones (entries 2 and 4) of varying ring size, and even a bridged bicycle (entry 5) with excellent diastereoselectivities. In addition, fair to good diastereoselectivities can be achieved with substrates possessing substitution at the backbone (e.g., entry 6). The alkyne terminus cannot be substituted under the reaction conditions, although a gold catalyst based on a designed semi-hollow ligand tolerates this type of substrate.[32] One apparent omission in the reported scope is the lack of any examples using stabilized carbon nucleophiles other than β-oxo esters.

Table 5 Gold(I)-Catalyzed Conia-Ene 5-*exo-dig* Cyclization Reactions of Alkynes[30]

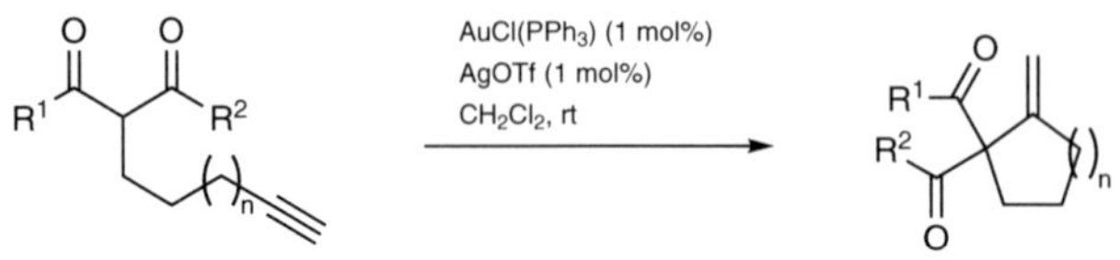

$AuCl(PPh_3)$ (1 mol%), AgOTf (1 mol%), CH_2Cl_2, rt

Entry	Substrate	Time	Product	Yield (%)	Ref
1	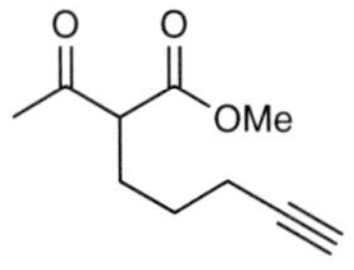	15 min	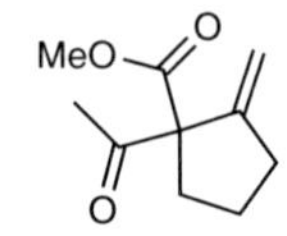	96	[30]
2	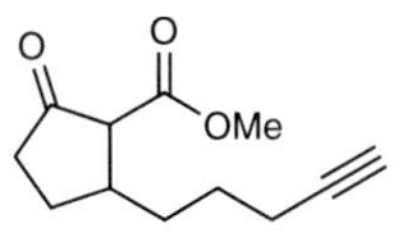	18 h	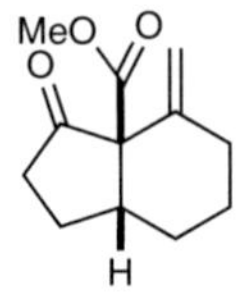	90[a]	[30]
3	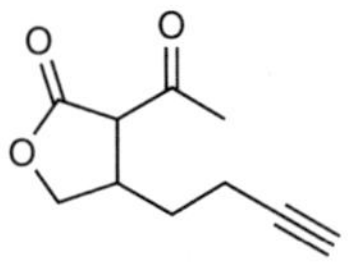	16 h	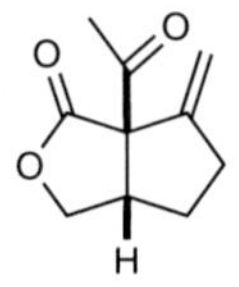	88	[30]
4	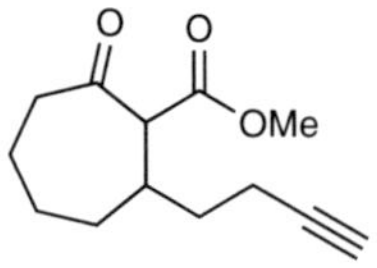	30 min	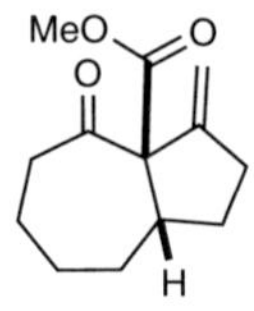	86[a]	[30]
5	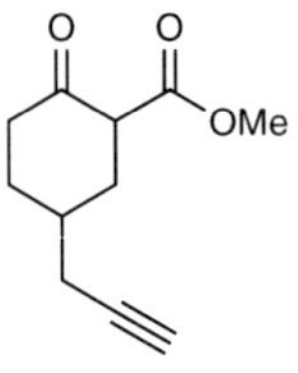	5 min	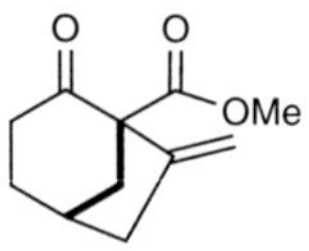	99	[30]
6	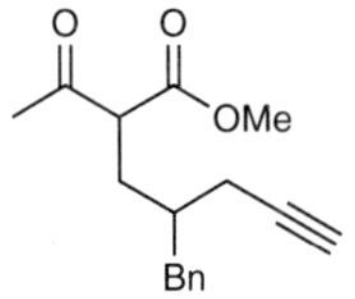	1 h	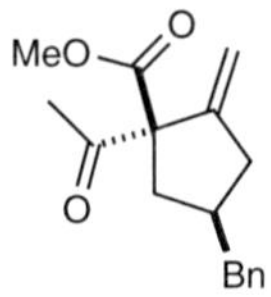	96[b]	[30]

[a] 5 mol% of $AuCl(PPh_3)$ and AgOTf was used.
[b] dr 4:1.

With the tether between the β-oxo ester moiety and the C≡C bond shortened to two carbons, the gold catalysis proceeds selectively via 5-*endo-dig* cyclization to afford cyclopentene products under similarly mild conditions (Table 6).[31] Notably, various types of substituent at the alkyne terminus including hydrogen (entry 1), phenyl, alkyl (entries 2–5) and even iodide (entry 6) are allowed. Due to the mild reaction conditions, acid-sensitive functional groups such as a tetrahydropyran-protected hydroxy group (entry 3) and a tertiary propargylic ether (entry 4) are tolerated. The reaction also enables quick access to bicyclic products (entries 1, 4, and 6). In contrast to the above *exo-dig* cyclization chemistry, this *endo* cyclization also allows the use of 1,3-diketones as carbon nucleophiles (entry 5).

Table 6 Gold(I)-Catalyzed Conia-Ene 5-*endo-dig* Cyclization Reactions of Alkynes[31]

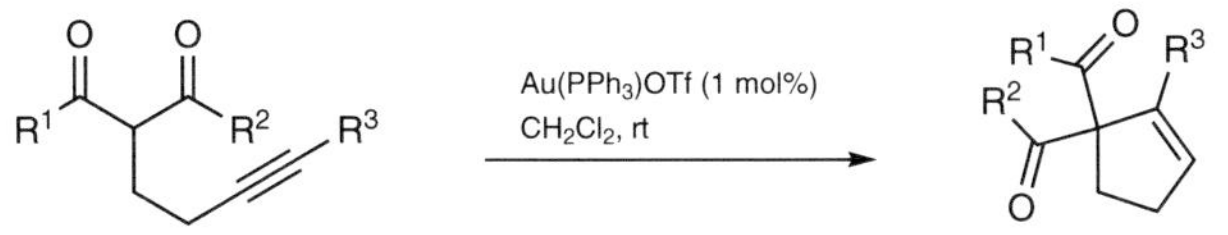

$Au(PPh_3)OTf$ (1 mol%)
CH_2Cl_2, rt

Entry	Substrate	Time	Product	Yield (%)	Ref
1	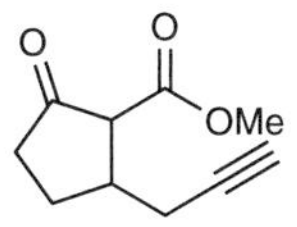	1 h	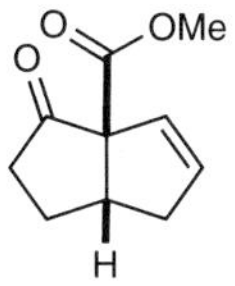	83	[31]
2	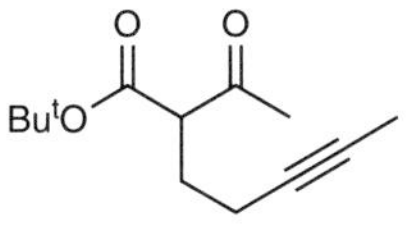	10 min	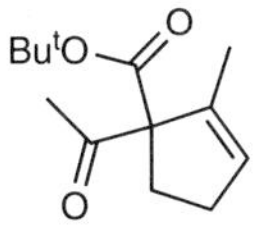	93[a]	[31]
3	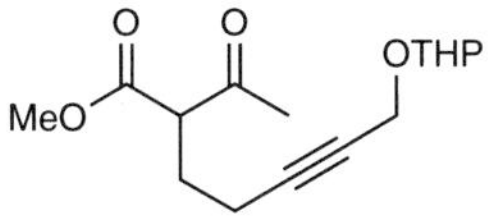	45 min	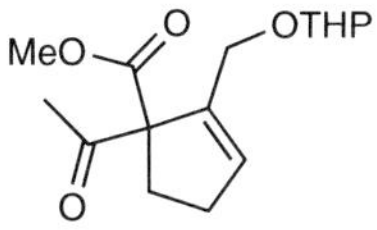	80	[31]
4	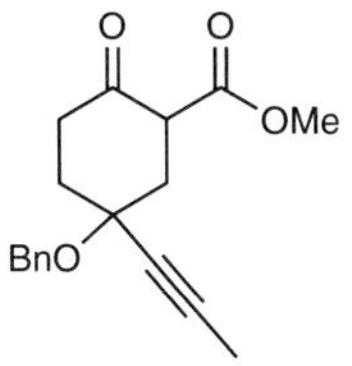	6 min	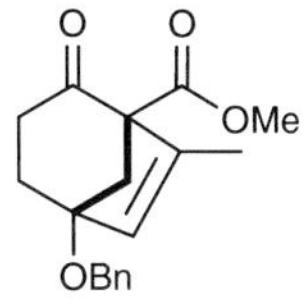	99	[31]
5	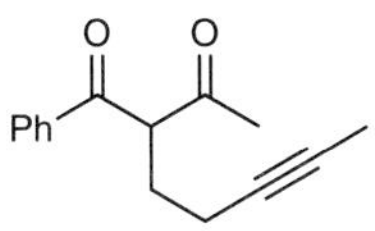	10 min	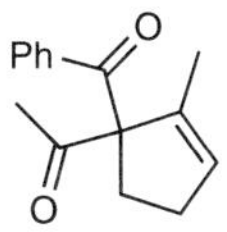	90	[31]
6	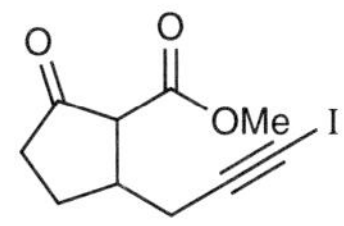	–[b]	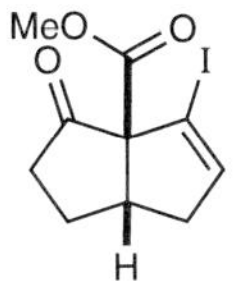	93	[31]

[a] No reaction using $(CuOTf)_2$•toluene (5 mol%) and <5% conversion using AgOTf (5 mol%) for 18 h.
[b] Reaction time not reported.

for references see p 314

Methyl 1-Acetyl-2-methylenecyclopentane-1-carboxylate (Table 5, Entry 1); Typical Procedure:[30]

To a small screw-cap vial equipped with a magnetic stirrer bar and charged with a 0.4 M soln of methyl 2-acetylhept-6-ynoate (ca. 150 mg, 1 equiv) in CH_2Cl_2 was added $AuCl(PPh_3)$ (1 mol%) followed by AgOTf (1 mol%). The cloudy white mixture was then stirred at rt and the reaction was monitored periodically by TLC. Upon completion, the mixture was loaded directly onto a column (silica gel) and chromatographed (hexanes/EtOAc); yield: 96%.

1.5.2.1.2.2 Zinc-Catalyzed Conia-Ene Reactions

By using zinc(II) chloride, cyclization of alkyne-tethered dicarbonyl compounds **28** can be performed at 100 °C in the absence of solvent (with solid substrates, 1,2-dichloroethane is used) to give the cyclic products **29** (Scheme 15).[34] The reactions are generally highly efficient, and the substrate scope is broad. In particular, previously challenging and hence little reported dicarbonyl compounds including malonates, 1,3-diketones, β-oxo amides, and malonate monoamides are all suitable as nucleophiles, and rare 4-*exo-dig* cyclization (i.e., n = 1) is possible, thereby offering rapid access to cyclobutane products. In addition, substitution at the alkyne terminus, including by iodide, is permitted. The products **29** have the thermodynamically more stable *E* configuration at the exocyclic double bond.

Scheme 15 Zinc(II) Chloride Catalyzed Conia-Ene Reaction of 1,3-Dicarbonyl Compounds with Tethered Alkynes[34]

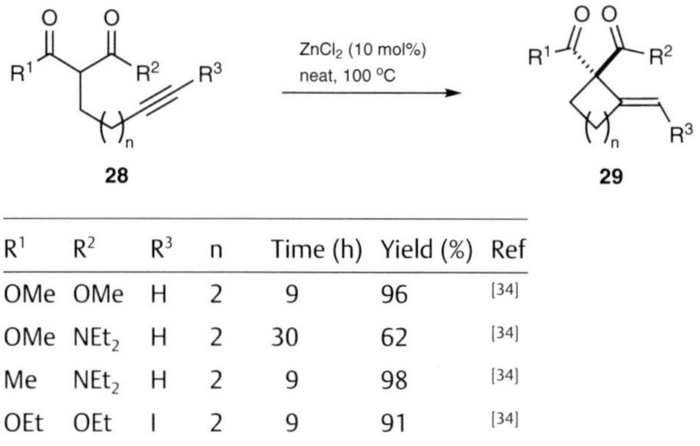

R^1	R^2	R^3	n	Time (h)	Yield (%)	Ref
OMe	OMe	H	2	9	96	[34]
OMe	NEt_2	H	2	30	62	[34]
Me	NEt_2	H	2	9	98	[34]
OEt	OEt	I	2	9	91	[34]
Ph	Me	H	2	9	94	[34]
OMe	Me	H	3	30	90[a]	[34]
Ph	Ph	H	1	10	51	[34]

[a] 40 mol% of $ZnCl_2$ was used.

1,3-Dicarbonyl Compounds 29; **General Procedure:**[34]

1,3-Dicarbonyl compound **28** (0.3 mmol) and $ZnCl_2$ (10 mol%), under neat conditions or in 1,2-dichloroethane (2 mL), were added to a 25-mL sealed glass tube. The mixture was stirred under argon at 100 °C for 9 h. The starting material was completely consumed, as monitored by TLC and GC/MS analysis. The mixture was diluted with Et_2O, washed with sat. aq NaCl, and concentrated under reduced pressure. The residue was purified by flash column chromatography (hexane/EtOAc).

1.5.2.1.2.3 Iron(III)-Catalyzed Stannyl Conia-Ene Cyclization

Iron(III) chloride can be an efficient yet inexpensive catalyst for the Conia-ene reaction of 2-alkynyl 1,3-dicarbonyl compounds. Stannyl enol ethers **30** and **32**, generated in situ from the acetates of dicarbonyl substrates, undergo a Conia-ene reaction in acceptable overall (two-step) yields to give the stannylated cyclic products **31** and **33**, respectively (Scheme 16).[36] In these cases, iron(III) trifluoromethanesulfonate is a much more efficient catalyst. Depending on the length of the tether between the stannyl enol ether moiety and the C≡C bond, the cyclization can proceed via either 5-*exo-dig* or 5-*endo-dig* cyclization. Of practical importance is that the products generated are functionalized with a tributyltin group, and can serve as organostannane substrates for the Kosugi–Migita–Stille coupling.

Scheme 16 Iron(III) Trifluoromethanesulfonate Catalyzed Synthesis of Stannylated Cyclopentanes and Cyclopentenes[36]

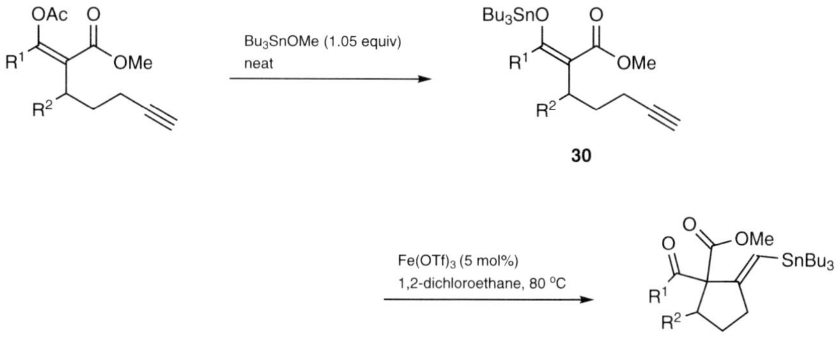

OAc O R1 OMe R2 → Bu_3SnOMe (1.05 equiv), neat → Bu_3SnO O R1 OMe R2 **30**

→ $Fe(OTf)_3$ (5 mol%), 1,2-dichloroethane, 80 °C → O O OMe $SnBu_3$ R1 R2 **31**

R^1	R^2	Yield[a] (%)	Ref
Me	H	91	[36]
$(CH_2)_2$		68	[36]

[a] Overall yield for both steps.

for references see p 314

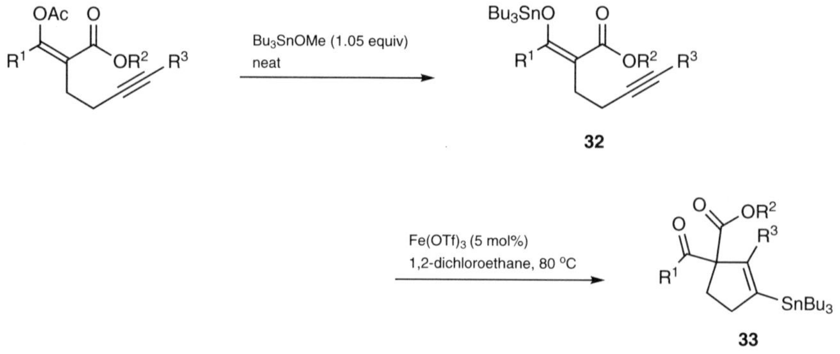

R^1	R^2	R^3	Yield[a] (%)	Ref
Me	Me	Me	63	[36]
Ph	Et	Et	66	[36]
iPr	Et	Et	65	[36]

[a] Overall yield for both steps.

(*E*)-1-Acyl-2-[(tributylstannyl)methylene]cyclopentane-1-carboxylates 31 **and 1-Acyl-2-alkyl-3-(tributylstannyl)cyclopent-2-ene-1-carboxylates** 33**; General Procedure:**[36]
Anhyd $Fe(OTf)_3$ (5 mol%) was carefully weighed and stirred in 1,2-dichloroethane (2 mL). The crude stannyl enol ether intermediate **30** or **32** (0.3 mmol, 1.0 equiv), generated by reacting the corresponding acetate with neat Bu_3SnOMe, was then added, and the mixture was heated to 80 °C. The residual crude product was concentrated under reduced pressure and purified by flash chromatography.

1.5.2.2 Arene-Based Carbon Nucleophiles (Friedel–Crafts-Type)

Arenes, especially electron-rich ones, are capable of attacking metal-activated and electrophilic alkenes and alkynes in hydroarylation reactions.[37–40] In intramolecular cases, the reaction offers access to synthetically useful carbo- and heterocycle-fused arenes. With electron-deficient alkenes, the metal activation usually occurs at the tethered electron-withdrawing groups[41,42] instead of the C=C bond to promote Michael-type addition and will not be discussed here.

1.5.2.2.1 Cyclizations onto Alkenes

Although this type of cyclization offers a quick and straightforward approach to construct C(aryl)—C bonds with newly formed chiral centers without prefunctionalization of the arene ring, it remains largely underdeveloped.[37] Several synthetically useful developments are discussed below.

1.5.2.2.1.1 Ruthenium(III)-Catalyzed Intramolecular Hydroarylation of Unactivated Alkenes

Ruthenium(III) chloride/silver(I) trifluoromethanesulfonate can be an efficient hydroarylation catalyst for alkenylarenes with diverse structural features (Table 7).[43] Substrates with electronically neutral phenyl rings (entries 1, 2, 5, and 7) undergo the reaction with good efficiency, as do those containing activated phenyl rings (entries 3, 4, and 6). Various substitution patterns on the alkene are permitted. A range of other metal catalysts have been screened for the reaction of 4-(3,5-dimethylphenoxy)but-1-ene (entry 6) and the yield with the second best catalyst [Sc(OTf)$_3$] is substantially lower (56%).

Table 7 Ruthenium(III)-Catalyzed Synthesis of Arene-Fused Carbo- and Heterocycles[43]

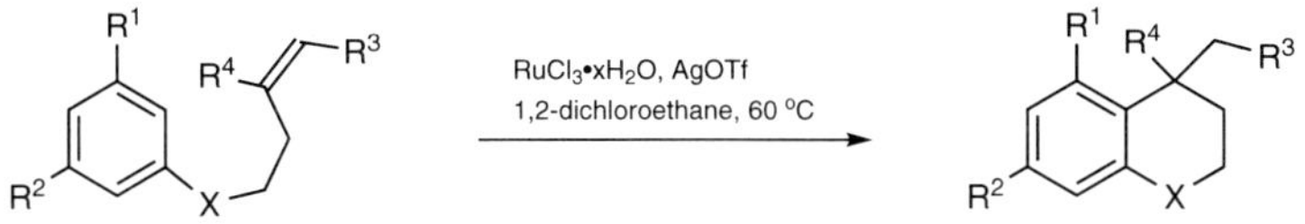

RuCl$_3$•xH$_2$O, AgOTf
1,2-dichloroethane, 60 °C

Entry	Substrate	RuCl$_3$•xH$_2$O (mol%)	AgOTf (mol%)	Product	Yield (%)	Ref
1	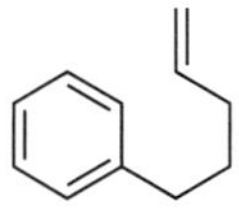	1	2	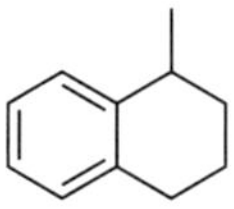	82	[43]
2	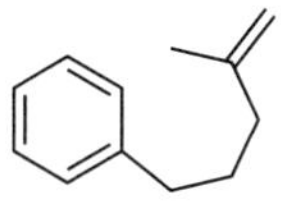	1	2	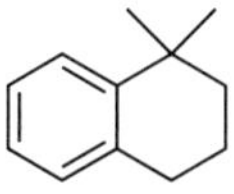	92	[43]
3	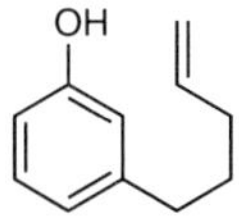 OH	2	4	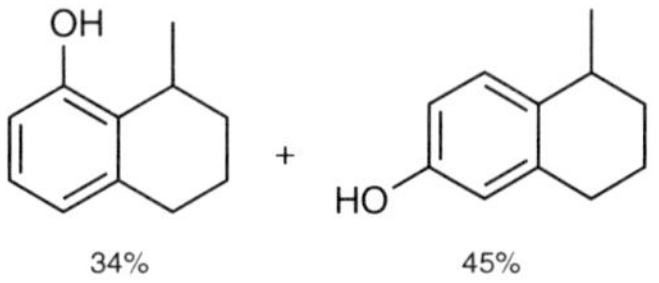 OH 34% + HO 45%	79	[43]
4	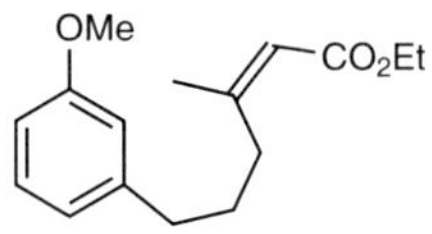 OMe, CO$_2$Et	5	10	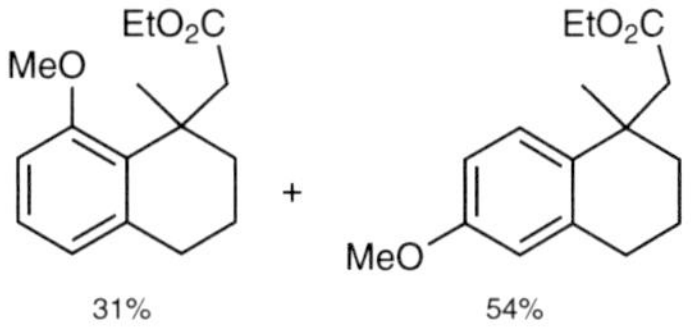 MeO, EtO$_2$C 31% + MeO, EtO$_2$C 54%	85	[43]
5	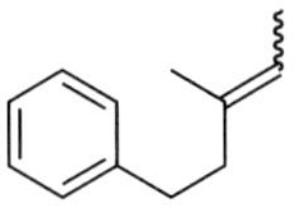	5	10	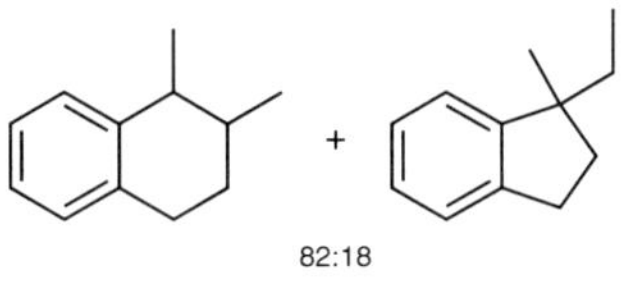 + 82:18	80	[43]
6	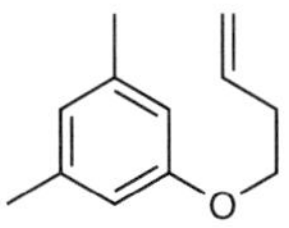 O	5	10	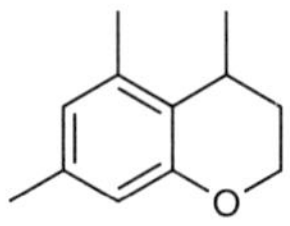 O	80	[43]
7	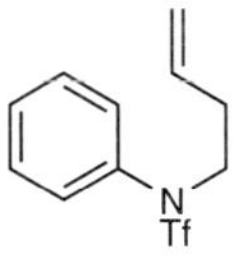 N Tf	10	20	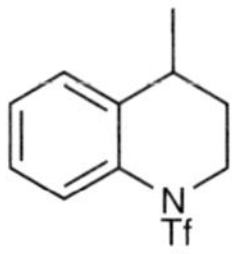 N Tf	53[a]	[43]

[a] At 80 °C.

1,1-Dimethyl-1,2,3,4-tetrahydronaphthalene (Table 7, Entry 2); Typical Procedure:[43]
A mixture of $RuCl_3$•xH_2O (1 mol%) and AgOTf (2 mol%) in 1,2-dichloroethane was stirred vigorously for 1 h. To the resulting soln was added a 0.2 M soln of 2-methyl-5-phenylpent-1-ene in 1,2-dichloroethane, and the resulting mixture was heated at 60 °C for 1 h. The solvent was then evaporated and the residue was purified by column chromatography (silica gel); yield: 92%.

1.5.2.2.1.2 Metal-Catalyzed Cyclization of Aryl Allylic Alcohols

The last step in metal-catalyzed nucleophilic additions to alkenes is protonation of the alkyl–metal bond, which is essential for catalyst turnover. However, for late transition metals this type of bond has high covalent character and hence reacts slowly with protons. A solution is to use allylic alcohols as the alkene reacting partners, because a dehydrative pathway can facilitate the fragmentation of the metal—C(sp^3) bond to afford an alkenyl product (Table 8). This reaction resembles an oxidative Heck reaction,[44,45] but its mechanism is distinctively different.

With mercury(II) trifluoromethanesulfonate as catalyst,[46] the dehydrative cyclization occurs in high efficiencies with substrates possessing electron-rich arenes (entries 1, 4, and 5). The reaction proceeds without incident even with a simple benzene substrate albeit in a lower yet serviceable yield (entry 3). In comparison to allylic alcohols, a methyl allyl ether is a suitable but less effective alkene partner (entry 2). The reaction involving a 6-*endo-trig* cyclization is also highly efficient (entry 5).

With cationic gold(I) catalysts equally capable of promoting reactions of allylic alcohols as electrophiles,[47] a chiral digold complex generated in situ from (*S*)-(+)-5,5′-bis[bis(3,5-di-*tert*-butyl-4-methoxyphenyl)phosphino]-4,4′-bi-1,3-benzodioxole [**34**, (*S*)-DTBM-SEGPHOS], gold(I) chloride, and silver(I) trifluoromethanesulfonate catalyzes the cyclization of indoles to the Z-allylic alcohol moiety in high enantioselectivity (entries 6–8);[48] the *E*-isomers do not react. The reaction tolerates indole rings with electron-donating and halide substituents and can accommodate the tethering of the allylic alcohol moiety both at the indole 3-position (entries 7 and 8) and at the 2-position (entry 6).

Table 8 Metal-Catalyzed Cyclization of Aryl Allylic Alcohols[46,48]

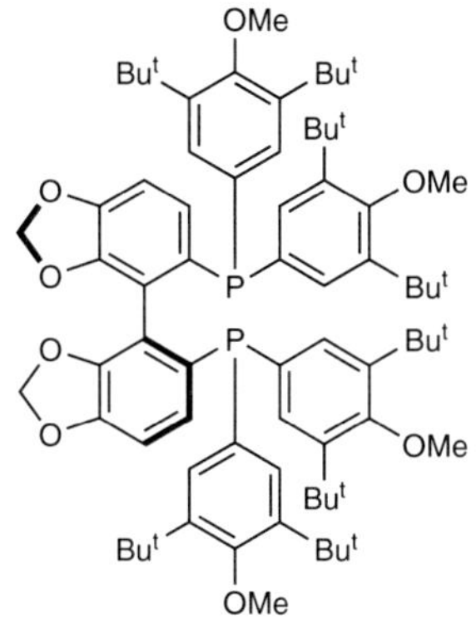

OMe Bu^t Bu^t Bu^t OMe Bu^t P P Bu^t OMe Bu^t Bu^t Bu^t OMe

34

[M] → [M]-- OH → −H$^+$ → $^-$[M] OH → H$^+$ → $^-$[M] $\overset{+}{O}H_2$ → −[M] −H_2O

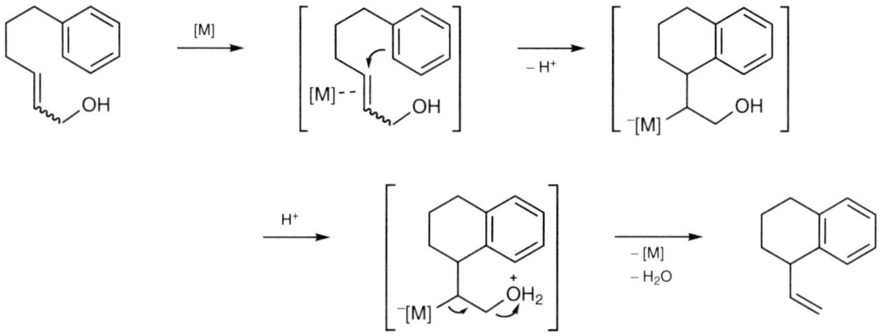

Entry	Substrate	Conditions	Product	Yield (%)	ee (%)	Ref
1	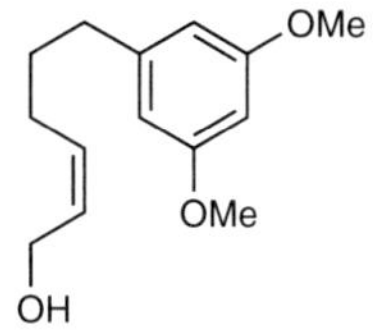	$Hg(OTf)_2$ (0.5 mol%), toluene, reflux, 5 min	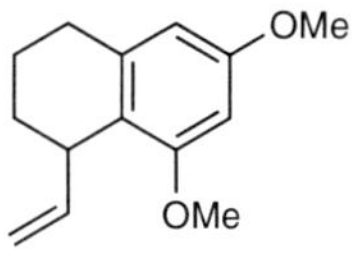	96	–	[46]
2	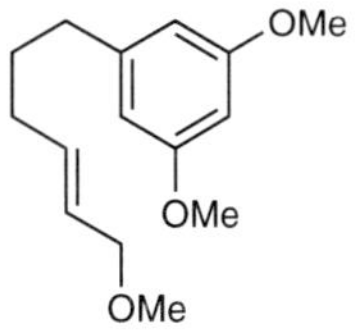	$Hg(OTf)_2$ (0.5 mol%), toluene, reflux, 1 h	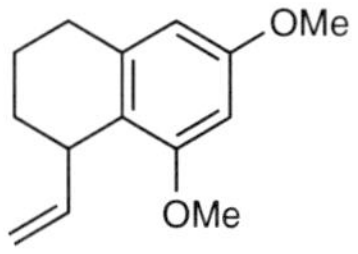	76	–	[46]
3	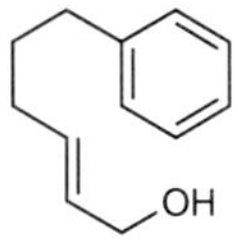	$Hg(OTf)_2$ (1 mol%), toluene, reflux, 5 min	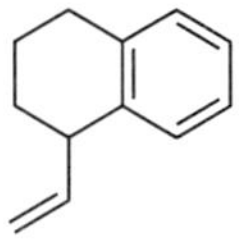	65	–	[46]

for references see p 314

Table 8 (cont.)

Entry	Substrate	Conditions	Product	Yield (%)	ee (%)	Ref
4	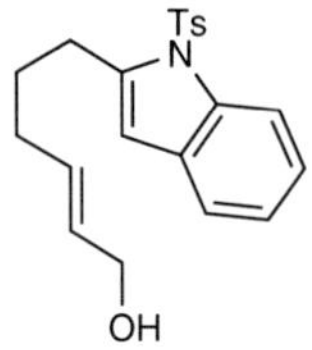	$Hg(OTf)_2$ (1 mol%), toluene, reflux, 5 min	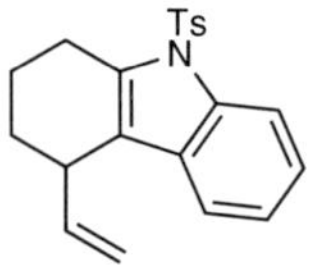	90	–	[46]
5	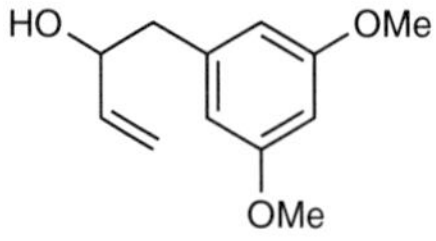	$Hg(OTf)_2$ (1 mol%), toluene, reflux, 10 h	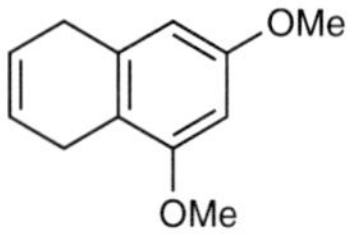	99	–	[46]
6	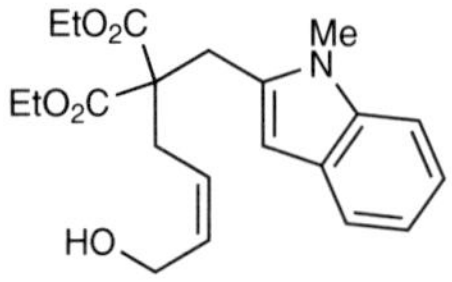	$Au_2Cl_2(L)$ (10 mol%),[a] AgOTf (20 mol%), toluene, 0°C	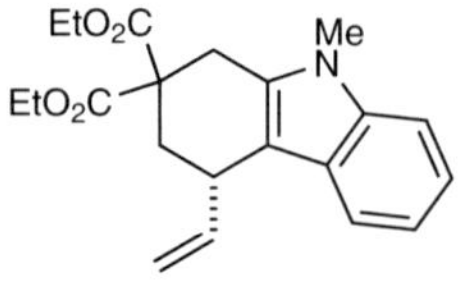	79	86[b]	[48]
7	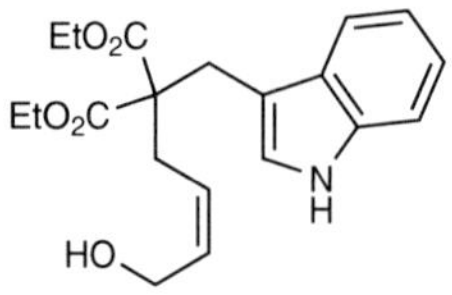	$Au_2Cl_2(L)$ (10 mol%),[a] AgOTf (20 mol%), toluene, 0°C, 48 h	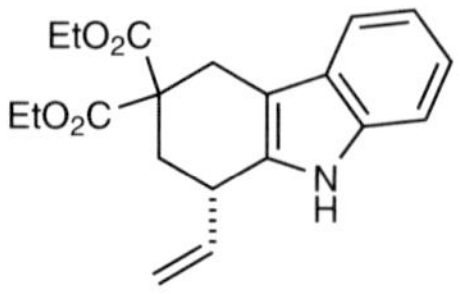	95	90[b]	[48]
8	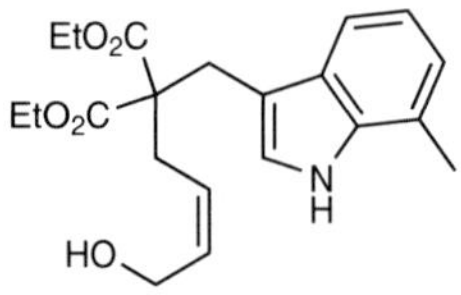	$Au_2Cl_2(L)$ (10 mol%),[a] AgOTf (20 mol%), toluene, 0°C, 48 h	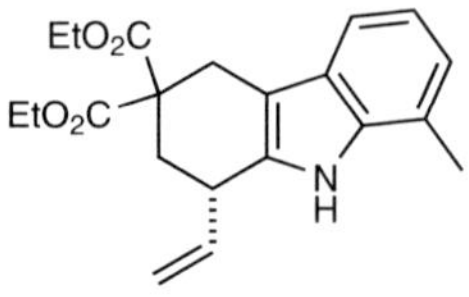	91	83[b]	[48]

[a] L = (*S*)-DTBM-SEGPHOS (**34**).
[b] The absolute stereoconfiguration was not determined.

6,8-Dimethoxy-1-vinyl-1,2,3,4-tetrahydronaphthalene (Table 8, Entry 1); Typical Procedure:[46]
A 0.1 M soln of $Hg(OTf)_2$ in MeCN (0.1 mL, 0.01 mmol) was added to a dried flask under argon, and the soln was concentrated to dryness under reduced pressure. Toluene (10.6 mL) and a soln of (*Z*)-6-(3,5-dimethoxyphenyl)hex-2-en-1-ol (500 mg, 2.1 mmol) in toluene (10.6 mL) were added, and the mixture was heated at reflux for 5 min. After addition of Et_3N (14 μL), inorganic material was removed by filtration through silica gel. The concentrated filtrate was subjected to column chromatography (silica gel, hexane/EtOAc 20:1) to give the desired product as a colorless syrup; yield: 441.6 mg (96%).

1.5.2.2.2 Cyclizations onto Alkynes

The metal-catalyzed cyclization of arenes to alkynes represents a versatile approach to various arene-fused carbo- and heterocycles. The ease of protodemetalation of alkenylmetal intermediates that are generated upon cyclization, or access to alternative mechanistic pathways, make this type of cyclization more facile and synthetically much more versatile than the corresponding alkene counterpart.[39]

1.5.2.2.2.1 Formation of Dihydronaphthalenes

Metal-catalyzed cyclization of 4-arylbut-1-ynes offers direct access to dihydronaphthalenes. The catalysts can be cationic mercury(II),[49] cationic gold(I),[50] and even iron(III) salts.[51,52] When using mercury(II) trifluoromethanesulfonate–tetramethylurea (TMU) complex as catalyst,[49] the reaction is only applicable to electron-rich arenes (Table 9, entries 1 and 2). With nontoxic chloro(triphenylphosphine)gold(I) as the precatalyst (e.g., entry 3),[50] the cyclization occurs smoothly with a phenyl group. With inexpensive and nontoxic iron(III) chloride hexahydrate as catalyst, the cyclization works with substrates possessing more reactive aryl-terminated alkynes, even at room temperature or under mild heating (entries 4–6).[51] Similarly, electronically activated alkynyl sulfides or ynamides are amenable to iron(III) catalysis, although a more cationic and thereby more acidic iron(III) catalyst is generated by the combination of iron(III) chloride and the halide scavenger silver(I) trifluoromethanesulfonate (entries 7–9).[52] In both iron-catalyzed methods, the nucleophilic phenyl ring does not need to be activated by electron-donating groups.

Table 9 Metal-Catalyzed Synthesis of Dihydronaphthalenes[49–52]

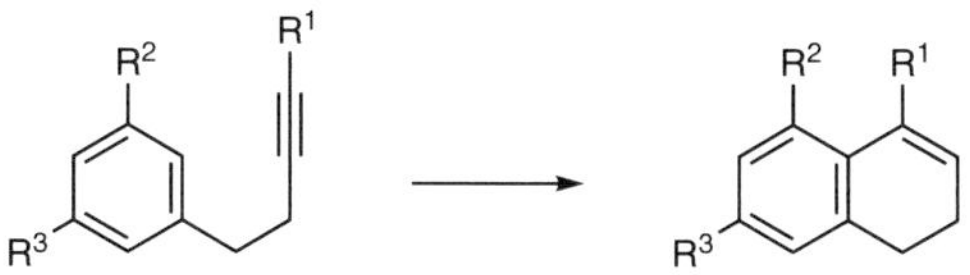

Entry	Substrate	Conditions	Product	Yield (%)	Ref
1	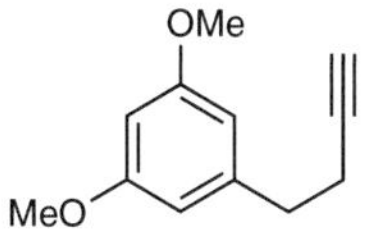	$Hg(OTf)_2(TMU)_3$ (2 mol%),[a] rt, MeCN, 0.8 h	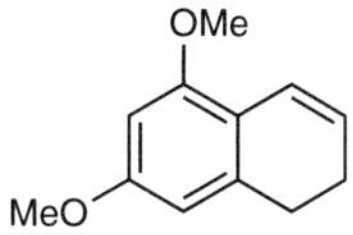	91	[49]
2	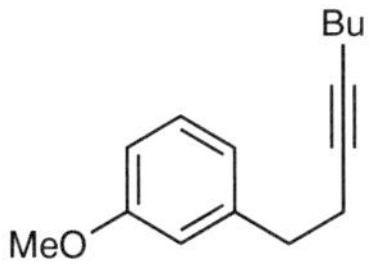	$Hg(OTf)_2(TMU)_3$ (2 mol%),[a] rt, MeCN, 6 h	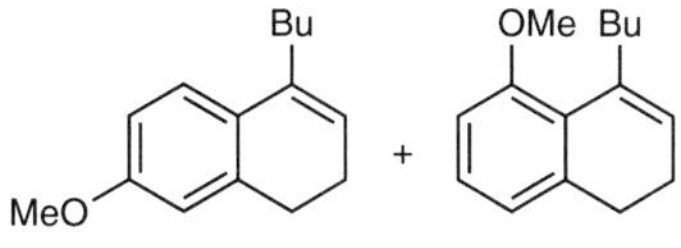 4.5:1	100	[49]
3	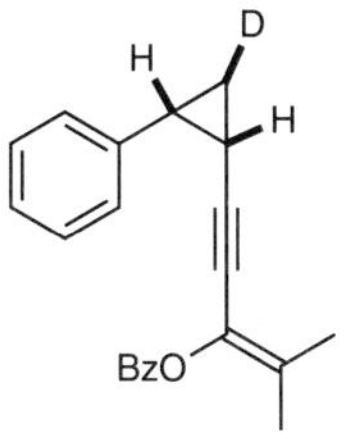	$AuCl(PPh_3)/AgSbF_6$ (5 mol%), rt	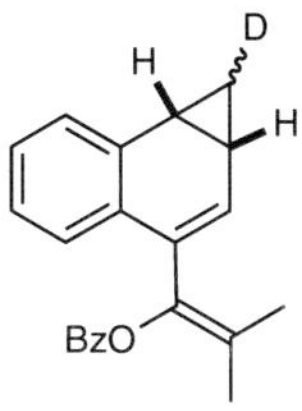 dr 1.5:1	77	[50]
4	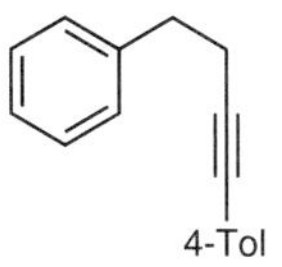	$FeCl_3$•$6H_2O$ (10 mol%), 1,2-dichloroethane, rt, 1 h	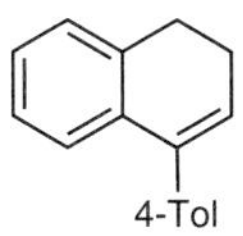	84	[51]

Table 9 (cont.)

Entry	Substrate	Conditions	Product	Yield (%)	Ref
5	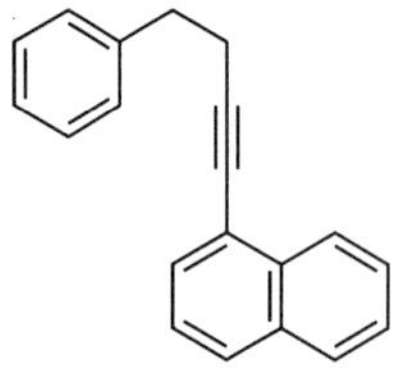	$FeCl_3$•$6H_2O$ (10 mol%), 1,2-dichloroethane, rt, 5 h	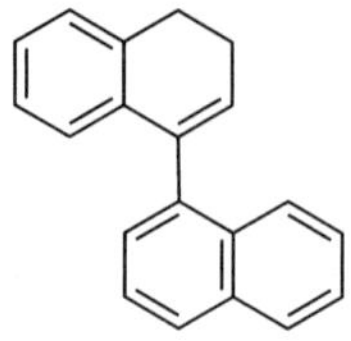	87	[51]
6	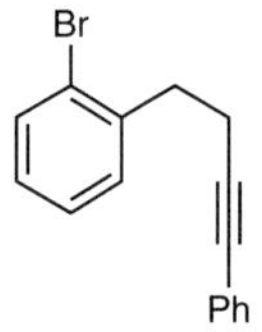 Br, Ph	$FeCl_3$•$6H_2O$ (10 mol%), 1,2-dichloroethane, 40 °C, 2 h	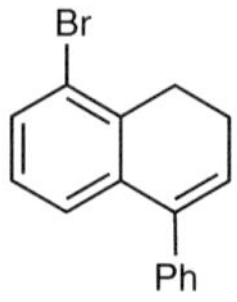 Br, Ph	68	[51]
7	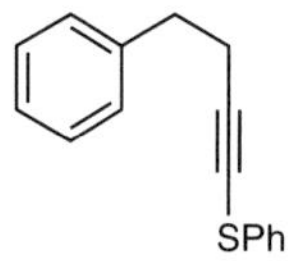 SPh	$FeCl_3$ (5 mol%), AgOTf (15 mol%), 1,2-dichloroethane, 25 °C, 20 min	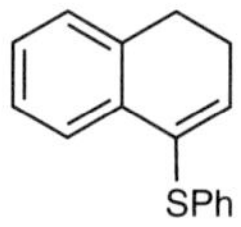 SPh	88	[52]
8	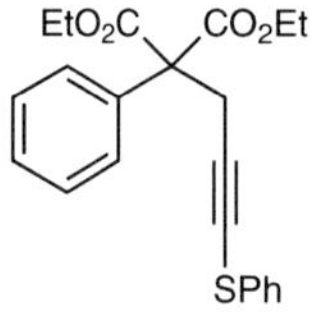 EtO_2C, CO_2Et, SPh	$FeCl_3$ (5 mol%), AgOTf (15 mol%), 1,2-dichloroethane, 25 °C, 60 min	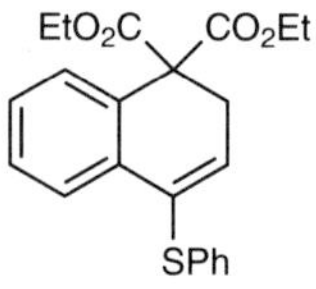 EtO_2C, CO_2Et, SPh	79	[52]
9	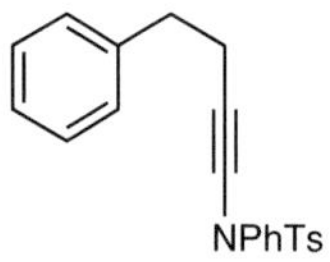 NPhTs	$FeCl_3$ (5 mol%), AgOTf (15 mol%), 1,2-dichloroethane, 25 °C, 20 min	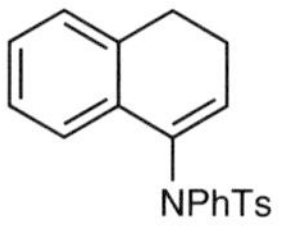 NPhTs	85	[52]

[a] TMU = tetramethylurea.

1-(Phenylsulfanyl)-3,4-dihydronaphthalene (Table 9, Entry 7); Typical Procedure:[52]
A suspension of $FeCl_3$ (2.4 mg, 0.015 mmol, 5 mol%) and AgOTf (11.6 mg, 0.045 mmol, 15 mol%) in 1,2-dichloroethane (0.8 mL) was stirred at 25 °C for 5 min. A soln of 4-phenyl-1-(phenylsulfanyl)but-1-yne (71.4 mg, 0.3 mmol) in 1,2-dichloroethane (0.7 mL) was added under N_2. After the mixture had been stirred at 25 °C for 20 min, it was quenched with H_2O. The aqueous layer was extracted with CH_2Cl_2 (2 × 15 mL), and the combined organic layers were washed with H_2O and brine, filtered, and dried under reduced pressure. The residue was purified by column chromatography (silica gel, EtOAc/hexane 1:15); yield: 63 mg (88%).

1.5.2.2.2.2 Formation of Phenanthrenes and Other Related Arenes

The metal-catalyzed cyclization of 2-alkynyl-1,1′-biphenyls and related biaryls is a straightforward route for the synthesis of phenanthrenes and other fused arenes.[38] Platinum(II) chloride is a generally effective catalyst for the cyclization (Table 10, entries 2–4).[53] In certain cases, gold(III) chloride (entry 1) or indium(III) chloride (entry 5) are better catalysts, suggesting the need for catalyst screening to achieve optimal outcomes in this type of cyclization. The cyclization of alkynyl selenides can be realized uneventfully with indium(III) trifluoromethanesulfonate as the catalyst (entry 6); however, when [1,3-bis(2,6-diisopropylphenyl)imidazol-2-ylidene]gold(I) hexafluoroantimonate is the catalyst, the phenylselanyl group undergoes 1,2-migration to afford the isomeric product exclusively (entry 7).[54] This outcome is likely due to the intermediacy of an electrophilic gold vinylidene species, generated via cationic gold(I) catalyst enabled isomerization of the phenylselanyl-terminated alkyne to a (phenylselanyl)vinylidene.

A notable advance in this type of cyclization is the use of polycationic-ligand-coordinated cationic gold catalyst **35**.[55] By using this highly electron-deficient gold(I) catalyst, highly sterically congested and hence challenging phenanthrenes (entries 8–10) can be prepared, and some can serve as key intermediates to natural products (e.g., entry 10 for coeloginin). For biphenyl substrates with an electronically deactivated nucleophilic phenyl ring, iron(III) trifluoromethanesulfonate is an effective catalyst, although the alkyne terminus must be substituted with a phenyl group (entries 11 and 12).[56] Even a nitrobenzene ring can cyclize to the C≡C bond under the iron catalysis.

Ruthenium-catalyzed isomerization of terminal alkynes into ruthenium vinylidene intermediates is a versatile and reliable approach to convert the alkyne terminus into an electrophilic carbene center.[57,58] The benzannulation strategy based on ruthenium vinylidene intermediates derived using ruthenium catalyst **36** provides expeditious and efficient access to various anthracene and coronene derivatives (entries 13–15).[59] In the case of 9,10-di(penta-1,4-diyn-3-ylidene)-9,10-dihydroanthracene (entry 14), the reaction efficiency is highly dependent on the substrate concentration; higher yields are obtained under more dilute conditions. This phenomenon is ascribed to π–π stacking of substrate molecules facilitating unwanted intermolecular reactions.

for references see p 314

Table 10 Metal-Catalyzed Synthesis of Phenanthrenes and Other Arenes[53–56,59]

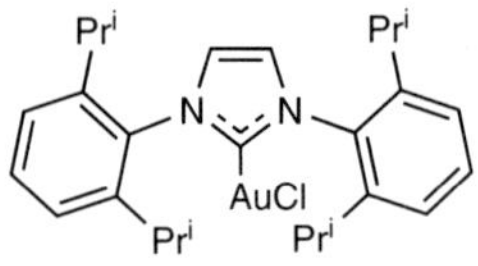

27 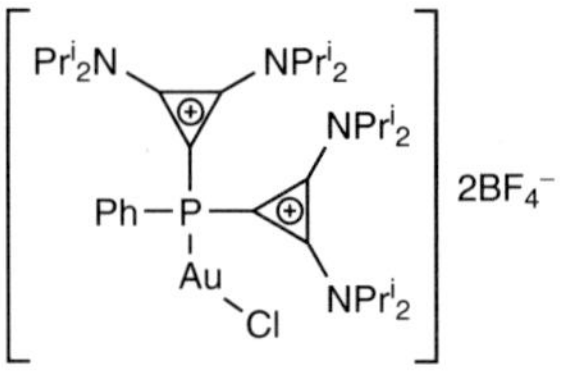 **35** 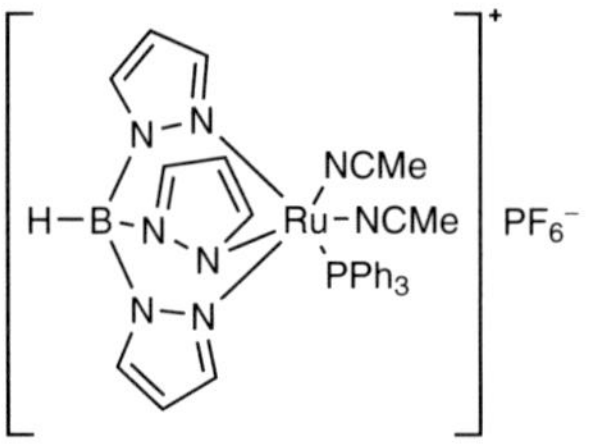 **36**

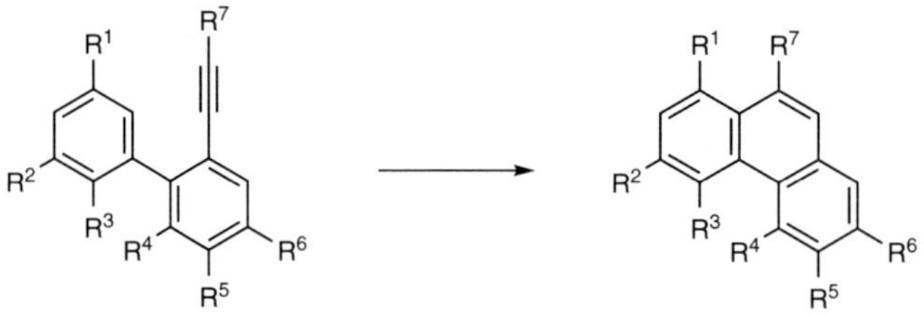

Entry	Substrate	Conditions	Product	Yield (%)	Ref
1	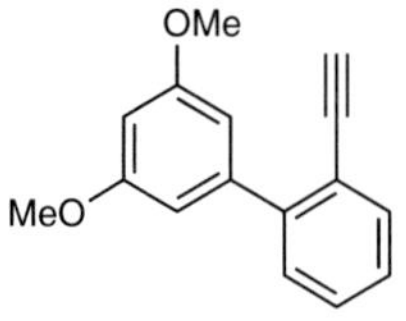	$AuCl_3$ (5 mol%), toluene, 80 °C, 22 h	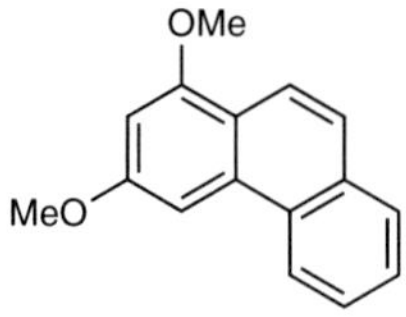	95[a]	[53]
2	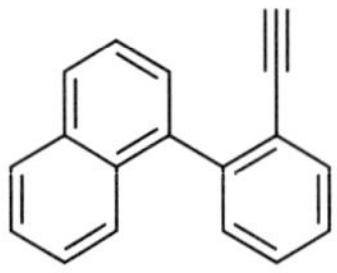	$PtCl_2$ (5 mol%), toluene, 80 °C, 20 h	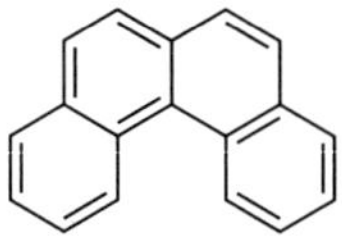	65	[53]
3	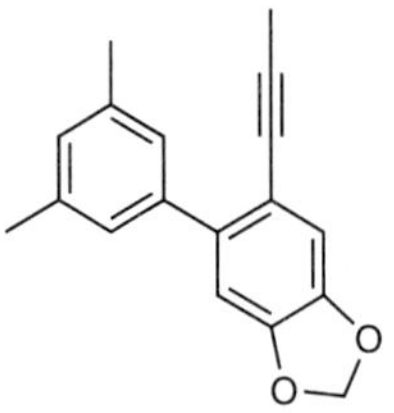	$PtCl_2$ (5 mol%), toluene, 80 °C, 24 h	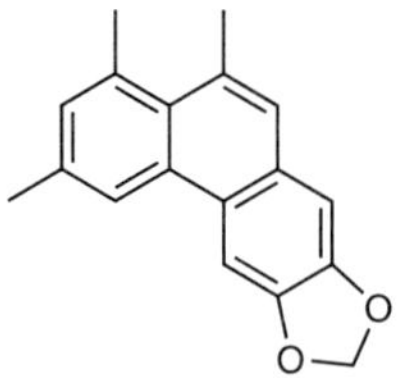	94	[53]

Table 10 (cont.)

Entry	Substrate	Conditions	Product	Yield (%)	Ref
4	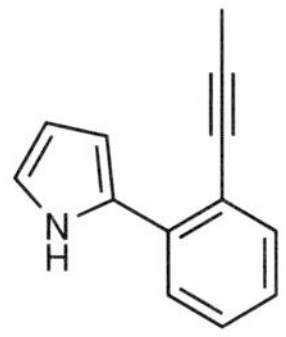	$PtCl_2$ (5 mol%), toluene, 80 °C, 20 h	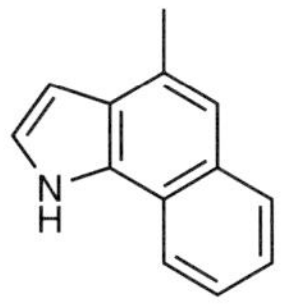	76	[53]
5	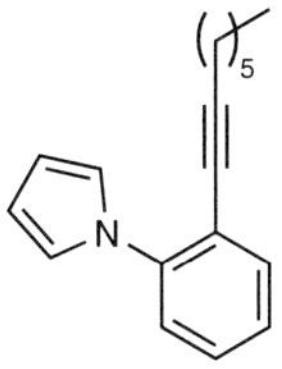	$InCl_3$ (5 mol%), toluene, 100 °C, 20 h	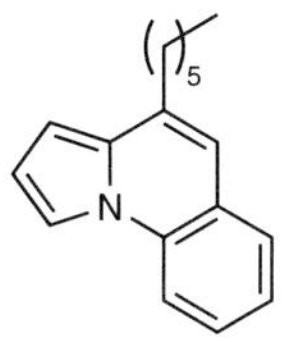	91[b]	[53]
6	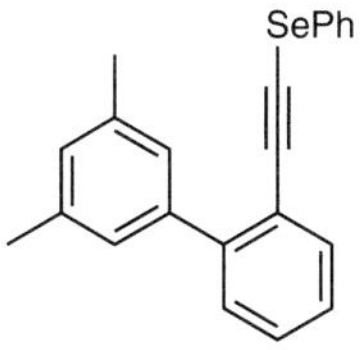 SePh	$In(OTf)_3$ (5 mol%), toluene, 80 °C, 4 h	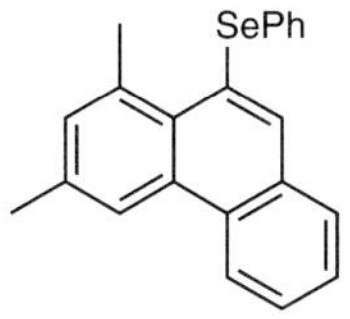 SePh	93	[54]
7	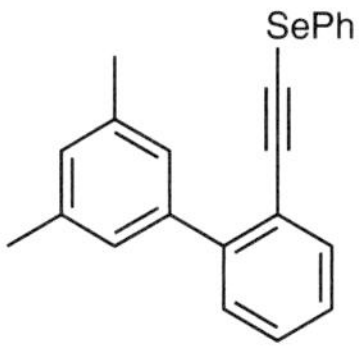 SePh	**27** (5 mol%), $AgSbF_6$ (5 mol%), CH_2Cl_2, rt, 4 h	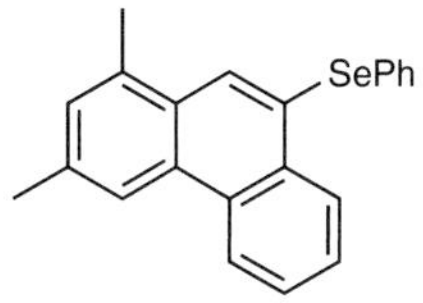 SePh	97	[54]
8	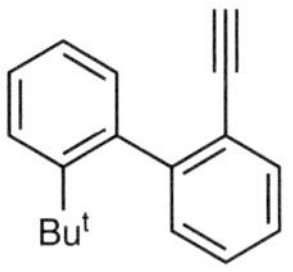 Bu^t	**35** (2 mol%), $AgSbF_6$ (2 mol%), CH_2Cl_2, rt	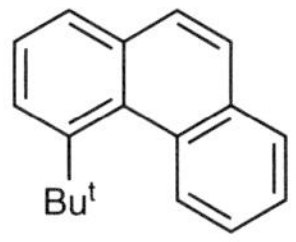 Bu^t	95	[55]
9	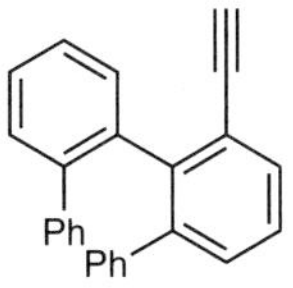 Ph, Ph	**35** (2 mol%), $AgSbF_6$ (2 mol%), CH_2Cl_2, rt	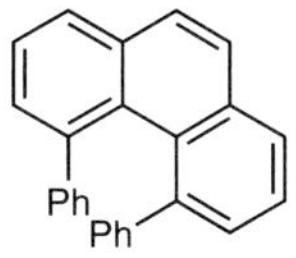 Ph, Ph	90	[55]
10	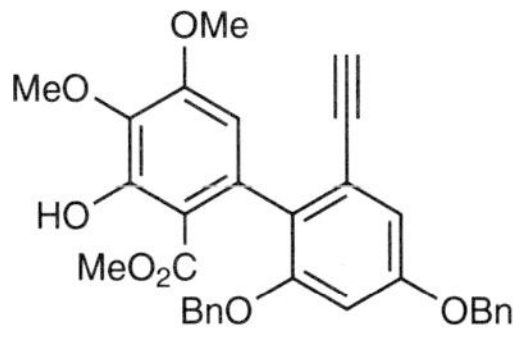 OMe, MeO, HO, MeO_2C, BnO, OBn	**35** (5 mol%), $AgSbF_6$ (5 mol%), 1,2-dichloroethane, rt	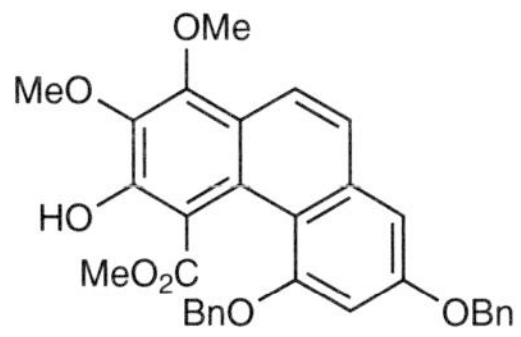 OMe, MeO, HO, MeO_2C, BnO, OBn	75	[55]

Table 10 (cont.)

Entry	Substrate	Conditions	Product	Yield (%)	Ref
11	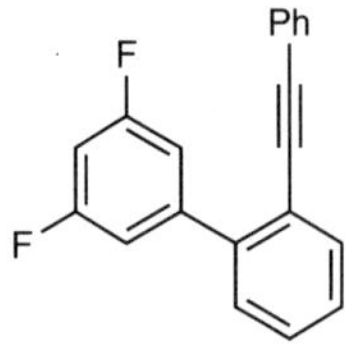	$Fe(OTf)_3$ (10 mol%), 1,2-dichloroethane, 80 °C, 2 h	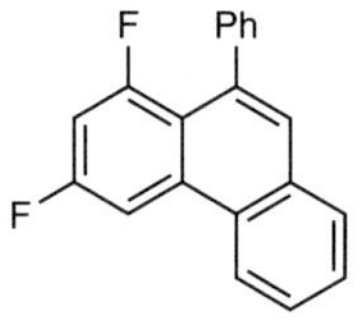	91	[56]
12	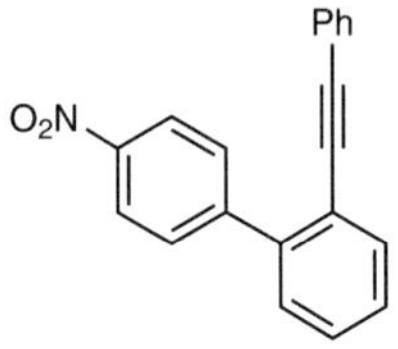	$Fe(OTf)_3$ (10 mol%), 1,2-dichloroethane, 80 °C, 6 h	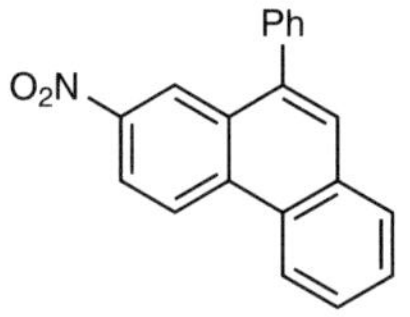	79	[56]
13	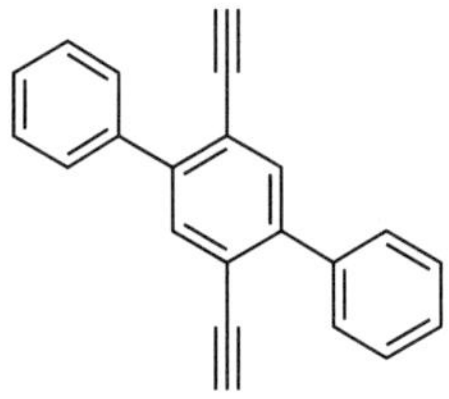	**36** (10 mol%), 1,2-dichloroethane (0.05 M), 80 °C, 24 h	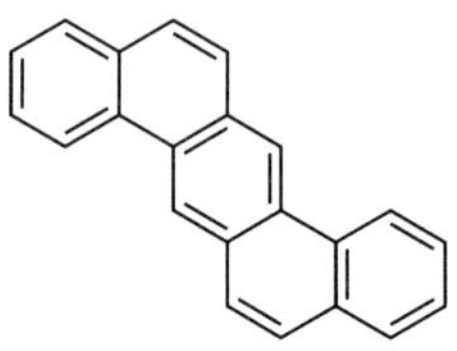	79	[59]
14	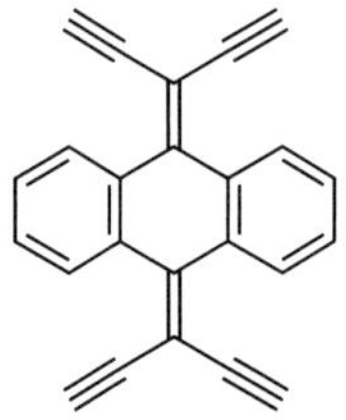	**36** (5 mol%), 1,2-dichloroethane (0.1 mM), 80 °C, 24 h	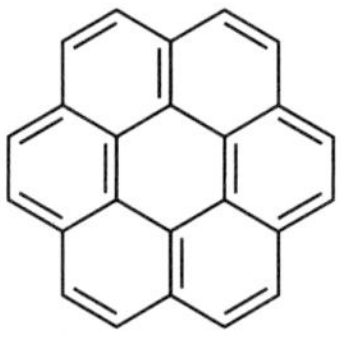	86	[59]
15	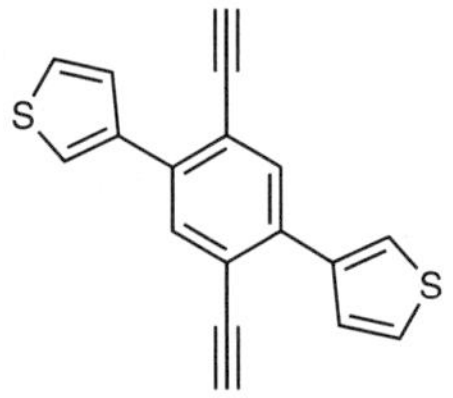	**36** (5 mol%), 1,2-dichloroethane (0.05 M), 80 °C, 24 h	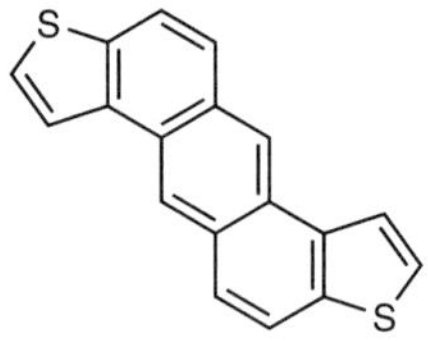	74	[59]

[a] 76% yield using $PtCl_2$ as catalyst.
[b] <40% yield using $PtCl_2$ or $AuCl_3$ as catalyst.

2,4,5-Trimethylphenanthro[2,3-*d*][1,3]dioxole (Table 10, Entry 3); Typical Procedure:[53]
A soln of 5-(3,5-dimethylphenyl)-6-(prop-1-ynyl)benzo[*d*][1,3]dioxole (528 mg, 2.0 mmol) and $PtCl_2$ (26.6 mg, 0.1 mmol) in toluene (10 mL) was stirred for 24 h at 80 °C under argon until GC analysis showed complete conversion of the substrate. The solvent was evaporated, and the residue was purified by flash chromatography (silica gel, hexanes) to give the desired product as a white solid; yield: 495 mg (94%).

1.5.2.2.2.3 Dehydrative Formation of Carbazoles

With 1-(indol-2-yl)alk-3-yn-1-ols **37** as substrates, the gold(III) chloride catalyzed intramolecular hydroarylation of the C≡C bond and subsequent dehydrative aromatization proceeds at ambient temperature in toluene (Scheme 17).[60] The substrates are readily accessible from indole-1-carbaldehydes and nonterminal secondary propargyl bromides, and thus this method offers expedient access to substituted carbazoles (e.g., **38**), which are of importance in medicinal chemistry. The reaction generally works well for cases where R^2 and R^3 are either an alkyl or a phenyl group and with a variety of substituents on the indole ring. It should be noted that steric hindrance imposed at the C4-position (R^4 = Me), electron donation at the C5-position (R^5 = OMe), or a slightly electron-withdrawing substituent at the indole ring nitrogen (R^1 = Bn, PMB), as well examples where R^2 = Ph, lead to decreased yields.

Scheme 17 Gold-Catalyzed Synthesis of Substituted Carbazoles[60]

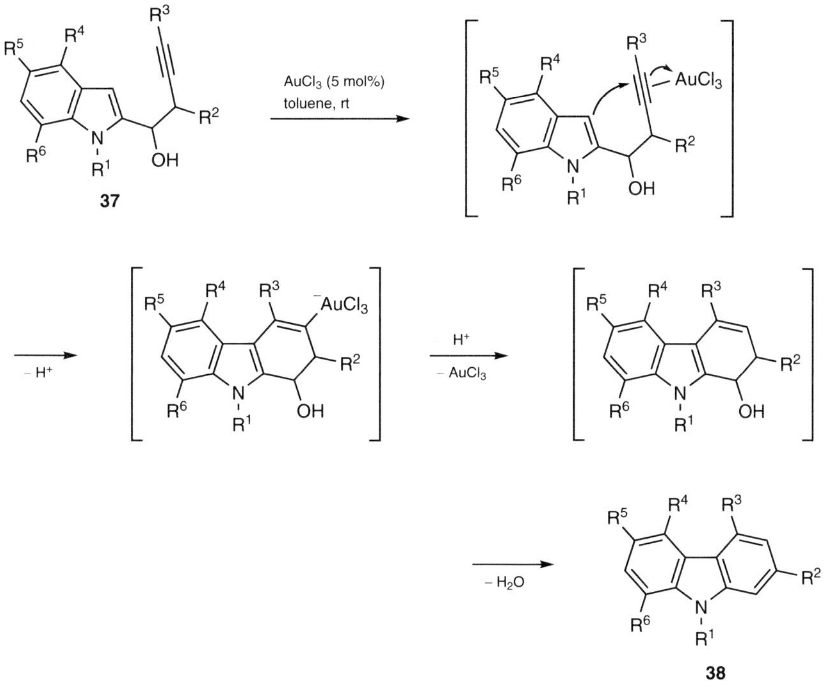

R^1	R^2	R^3	R^4	R^5	R^6	Yield (%)	Ref
Et	Me	Ph	H	H	H	90	[60]
Et	Me	Et	H	H	H	81	[60]
Et	Me	Bu	H	Me	H	72	[60]
Et	Et	Ph	H	Me	H	79	[60]
Et	Ph	Bu	H	Me	H	64	[60]
Et	Me	Bu	H	OMe	H	60	[60]
Bn	Et	Ph	H	Br	H	72	[60]
Et	Me	Bu	Me	H	H	65	[60]

R^1	R^2	R^3	R^4	R^5	R^6	Yield (%)	Ref
Et	Me	Ph	H	H	Me	90	[60]
PMB	Me	Bu	H	H	H	58	[60]
Bn	Et	Ph	H	H	H	62	[60]

9-Ethyl-2-methyl-4-phenyl-9*H*-carbazole (38, R^1 = Et; R^2 = Me; R^3 = Ph; R^4 = R^5 = R^6 = H); Typical Procedure:[60]

To a dry Schlenk tube were added sequentially $AuCl_3$ (3.1 mg, 0.01 mmol), 1-(1-ethyl-1*H*-indol-2-yl)-2-methyl-4-phenylbut-3-yn-1-ol (**37**, R^1 = Et; R^2 = Me; R^3 = Ph; R^4 = R^5 = R^6 = H; 60.6 mg, 0.2 mmol), and toluene (1.0 mL) under N_2. The mixture was stirred at rt and the reaction was complete after 3 h (monitored by TLC). The mixture was filtered and the solvent was evaporated. The residue was purified by column chromatography (silica gel, petroleum ether/EtOAc 40:1) to afford the product as a solid; yield: 51.3 mg (90%).

1.5.2.2.2.4 Formation of 2*H*-1-Benzopyran-2-ones and Quinolin-2(1*H*)-ones

Cyclization of aryl alkynoates or *N*-arylalkynamides in the presence of transition-metal catalysts represents one of the most facile accesses to 2*H*-1-benzopyran-2-ones (coumarins) and quinolin-2(1*H*)-ones under mild conditions (Table 11). Palladium(II) acetate in the presence of trifluoroacetic acid as a cosolvent is a versatile catalyst[61] for the cyclization of both types of substrates, offering facile access to both 2*H*-1-benzopyran-2-ones and quinolin-2(1*H*)-ones.[62] The cyclization of alkynoates can accommodate various substituents at the alkyne terminus (e.g., entries 1 and 2) and tolerates steric hindrance (entries 4 and 5). For the transformation of 4-*tert*-butylphenyl propynoate, catalysis by hafnium(IV) trifluoromethanesulfonate instead of palladium(II) acetate requires a higher catalyst loading (10 mol%), a longer reaction time (10 h), and results in a lower yield (72%).[63]

The reactions of alkynoates can also be effectively catalyzed by a combination of gold(III) chloride (5 mol%) and silver(I) trifluoromethanesulfonate (15 mol%), the latter salt acting as halide scavenger.[64] Even though the catalyst system is more costly, the reaction with 4-*tert*-butylphenyl propynoate is more efficient (entry 6) than that using palladium(II) acetate (entry 3). This phenomenon is again observed with the reaction of 2-naphthyl propynoate (entry 4; 99% instead of 65% yield); however, with 4-*tert*-butylphenyl 3-phenylpropynoate (entry 1), the yields are comparable.

The conditions of this gold catalysis, however, are most likely harsh because of the highly acidic nature of the catalyst system (the formation of naked Au^{3+} is inferred but unlikely). A much milder version using gold catalyst **39**[65] displays efficiency comparable to the palladium(II) acetate catalyzed method for a substrate that possesses alkyl substitution on the benzene ring (entry 7 vs. entry 3);[66] however, only two examples were reported. This protocol has been successfully employed as the last step of the synthesis of pimpinellin (entry 8)[67] and columbianetin (entry 9).[68] Similarly, alkynamides can be converted into quinolin-2(1*H*)-ones efficiently in the presence of catalytic palladium(II) acetate (e.g., entry 10).[62]

Table 11 Metal-Catalyzed Synthesis of 2*H*-1-Benzopyran-2-ones and Quinolin-2(1*H*)-ones[62,64,66–68]

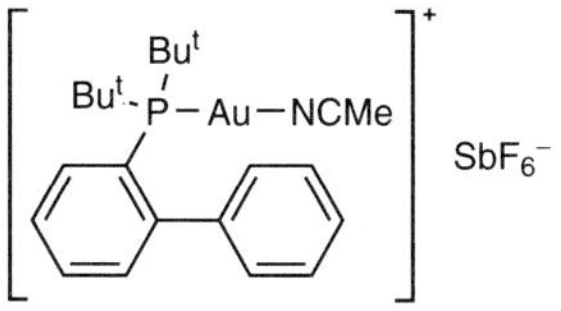

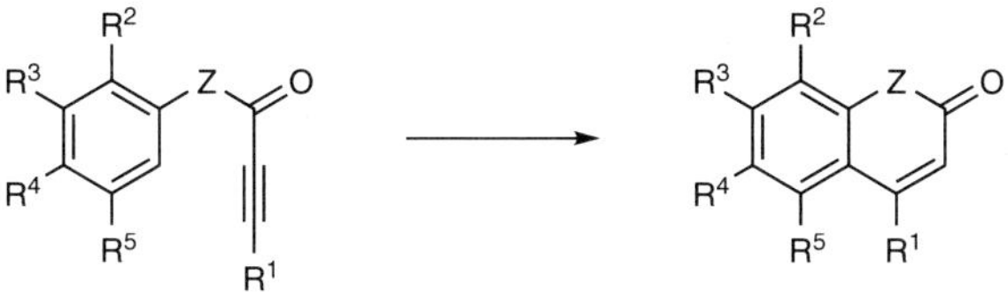

Entry	Substrate	Conditions	Product	Yield (%)	Ref
1	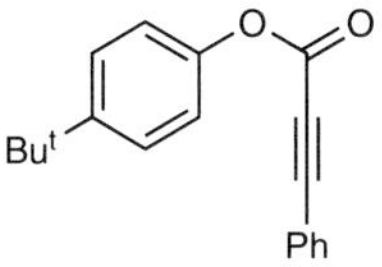	$Pd(OAc)_2$ (1 mol%), TFA, CH_2Cl_2, rt, 0.5 h	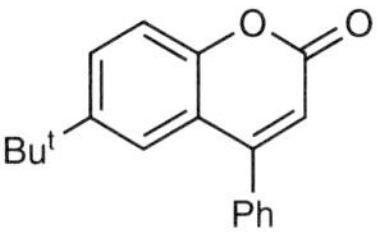	90[a]	[62]
2	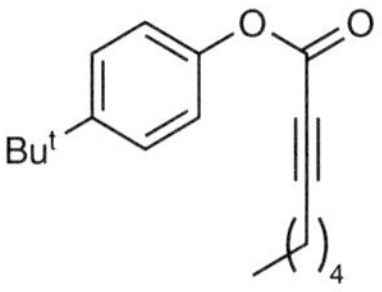	$Pd(OAc)_2$ (1 mol%), TFA, CH_2Cl_2, rt, 0.5 h	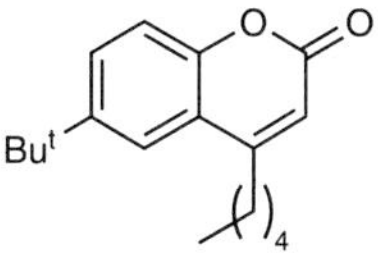	71	[62]
3	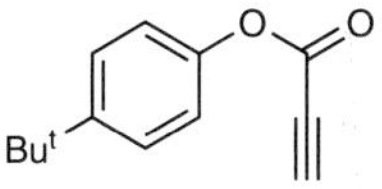	$Pd(OAc)_2$ (1 mol%), TFA, CH_2Cl_2, rt, 0.5 h	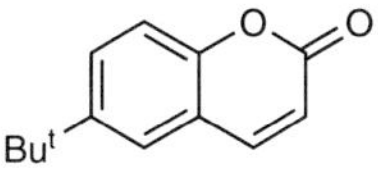	60	[62]
4	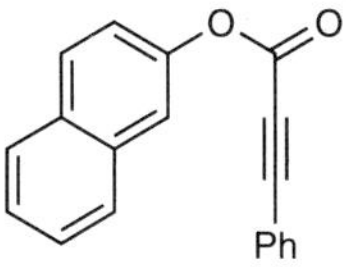	$Pd(OAc)_2$ (1 mol%), TFA, CH_2Cl_2, rt, 0.5 h	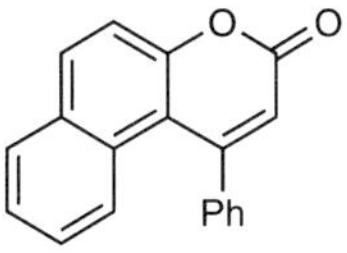	65[b]	[62]
5	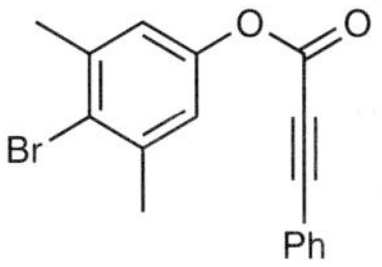	$Pd(OAc)_2$ (1 mol%), TFA, CH_2Cl_2, rt, 0.5 h	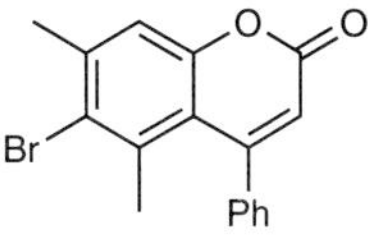	85	[62]
6	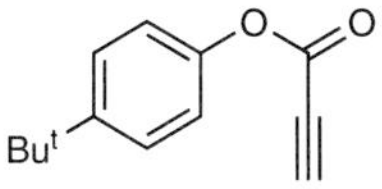	$AuCl_3$ (5 mol%), $Ag(OTf)_3$ (15 mol%), 1,2-dichloroethane, 50 °C	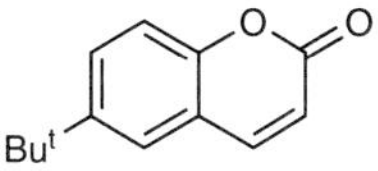	99	[64]
7	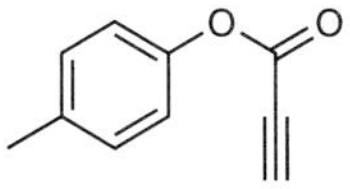	**39** (5 mol%), CH_2Cl_2, 18 °C, 12 h	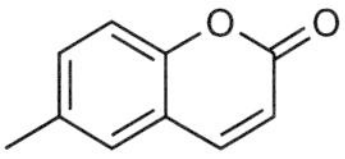	60	[66]

for references see p 314

Table 11 (cont.)

Entry	Substrate	Conditions	Product	Yield (%)	Ref
8	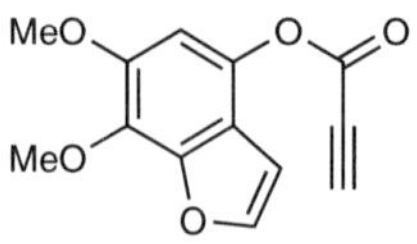	**39** (5 mol%), CH_2Cl_2, 25 °C, 0.5 h	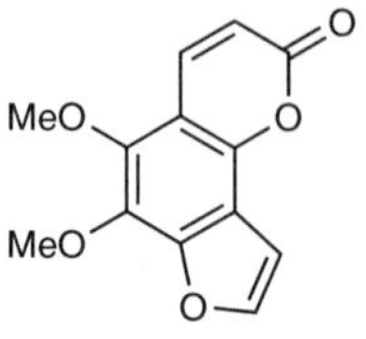 pimpinellin	72	[67]
9	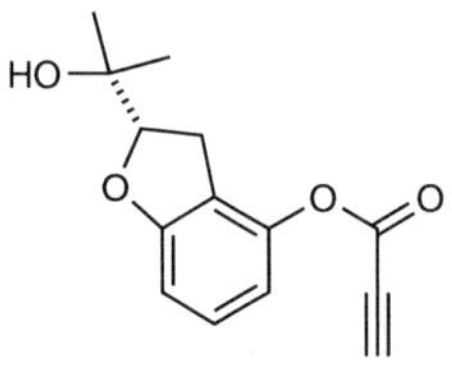	**39** (5 mol%), CH_2Cl_2, 25 °C, 0.5 h	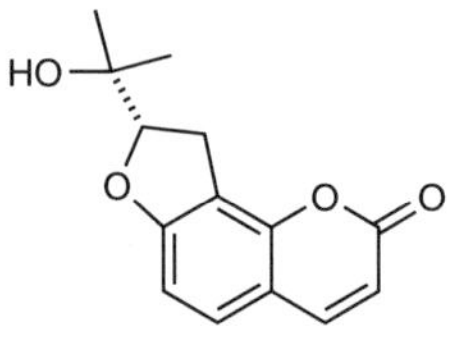 columbianetin	75	[68]
10	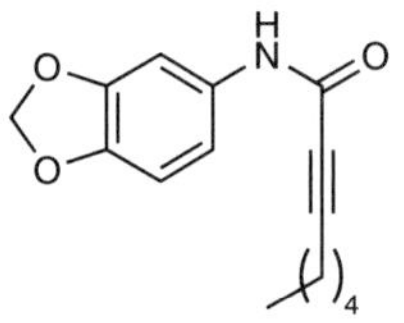	$Pd(OAc)_2$ (2 mol%), TFA, CH_2Cl_2, rt, 1 h	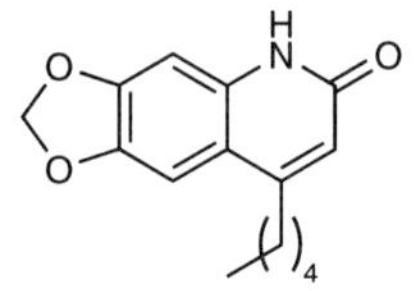	91	[62]

[a] 92% yield using $AuCl_3$ (5 mol%) and $Ag(OTf)_3$ (15 mol%) in 1,2-dichloroethane at 50 °C.
[b] 99% yield using $AuCl_3$ (5 mol%) and $Ag(OTf)_3$ (15 mol%) in 1,2-dichloroethane at 50 °C.

2*H*-1-Benzopyran-2-ones (Table 11); General Procedure Using Palladium(II) Acetate:[62]
An aryl alkynoate (1 mmol), $Pd(OAc)_2$ (1–2 mol%), TFA (1.5 mL), and CH_2Cl_2 (0.5 mL) were mixed in a 25-mL dry Pyrex tube and the mixture was stirred at rt until disappearance of the starting material as monitored by GC, TLC, or ^1H NMR spectroscopy. The mixture was poured into sat. aq NaCl and extracted with Et_2O. The ethereal layer was washed with sat. aq NaCl, neutralized with aq $NaHCO_3$, and dried (Na_2SO_4). The solvent was removed under reduced pressure, and the products were purified by flash column chromatography (silica gel) and recrystallization (EtOAc/hexanes mixtures).

In related preparations of quinolin-2(1*H*)-ones from *N*-arylalkynamides, the products were precipitated by pouring the mixture into sat. aq NaCl. The precipitate was washed with H_2O, aq $NaHCO_3$, and hexane, and then crystallized (hexane/$CHCl_3$ or hexane/EtOH).

1.5.2.2.2.5 Formation of 2*H*-1-Benzopyrans and Dihydroquinolines

6-*endo-dig* cyclization of aryl propargyl ethers or the corresponding *N*-propargylanilines provides one-step access to 2*H*-1-benzopyrans and 1,2-dihydroquinolines. Late transition metals such as platinum and gold are the most versatile in promoting these reactions regioselectively over the competing 5-*exo-dig* cyclization.[69] As shown in Table 12, bis(acetonitrile)dichloroplatinum(II) (0.09 mol%) is an excellent catalyst for the cyclization of the electron-rich substrate 5-(prop-2-ynyloxy)benzo[*d*][1,3]dioxole (entry 1), and gives a better yield than (triphenylphosphine)gold(I) hexafluoroantimonate.[70] Platinum(IV) chloride is in general a highly efficient catalyst for a broad range of substrates (entries 2–5).[71] It is notable there is no erosion of stereopurity (entry 3), which suggests that the Claisen-type rearrangement that is frequently observed in gold catalysis is not an issue.[68] The presence of a (*tert*-butoxycarbonyl)amino group (entry 4) is tolerated, suggesting the mild nature of the reaction conditions, and an electron-withdrawing acetyl group (entry 5) on the aryl ring is also allowed. In a three-step synthesis of deguelin,[72] the cyclization to construct the key 2*H*-1-benzopyran ring (entry 6), however, is best catalyzed by platinum(II) chloride even though in other cases involving ynoate and ynone substrates platinum(IV) chloride is better.

In the synthesis of dihydroquinolines, the aniline is typically employed as a sulfonamide (entries 7–9), and gold catalyst **39**[65] is effective in catalyzing the cyclization of these compounds with varying substitution on the benzene ring.[66] Carbamate derivatives are, in general, not suitable substrates for this method due to cyclization of the carbonyl group to the alkyne; however, platinum(IV) chloride is an effective catalyst for such transformations (entry 10).[71] In addition, terminally iodinated anilides (e.g., entry 11) can undergo gold-catalyzed cyclization, although iodide migration occurs preferentially.[73] The reaction likely involves a gold vinylidene intermediate.[74]

As in the case of arylalkynes (see Table 9, Section 1.5.2.2.2.1), propargyl ethers and anilines possessing electronically activated alkynes in the form of ynyl sulfides (Table 12, entries 12 and 13) or ynamides (entry 14) readily undergo cyclization under iron(III) catalysis, affording C4-functionalized 2*H*-1-benzopyrans and dihydroquinolines.[52]

Table 12 Metal-Catalyzed Synthesis of 2*H*-1-Benzopyrans and Dihydroquinolines[52,66,70–73]

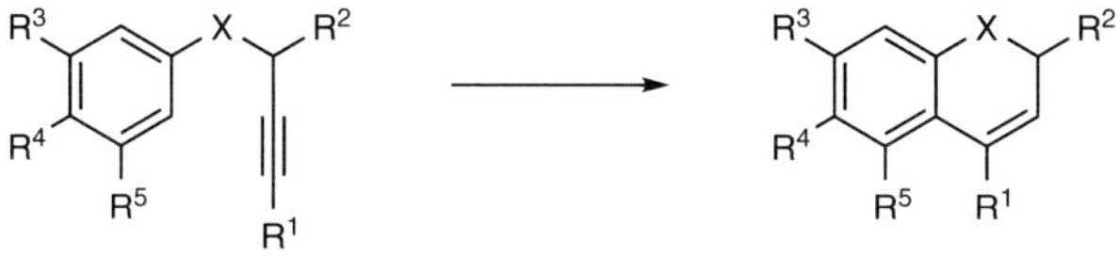

Entry	Substrate	Conditions	Product	Yield (%)	Ref
1	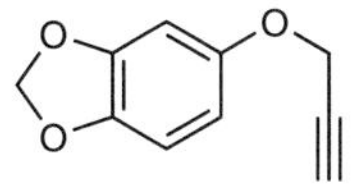	$PtCl_2(NCMe)_2$ (0.09 mol%), acetone, reflux, 16 h	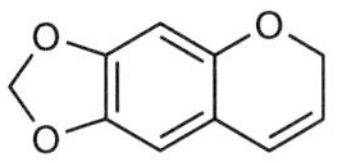	95[a]	[70]
2	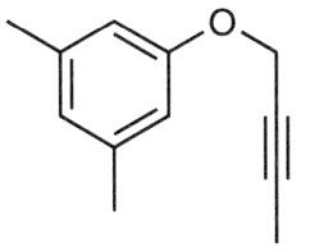	$PtCl_4$ (1 mol%), dioxane, rt, 1 h	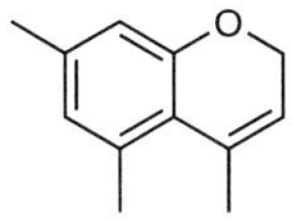	86 (9)[b]	[71]
3	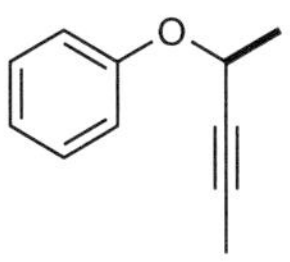 89% ee	$PtCl_4$ (5 mol%), dioxane, rt, 1 h	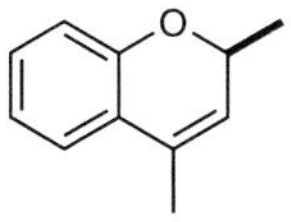 89% ee	82	[71]
4	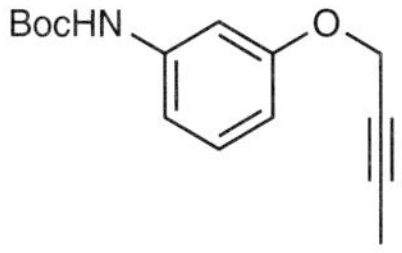 BocHN	$PtCl_4$ (2 mol%), 1,2-dichloroethane, rt, 6 h	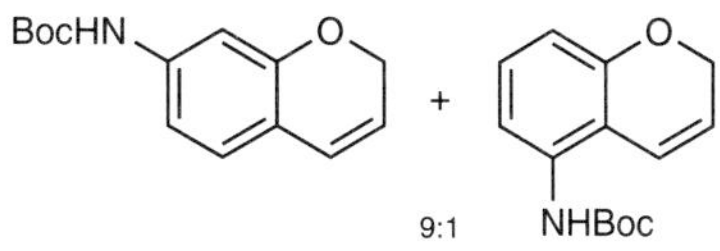 BocHN + NHBoc 9:1	67	[71]

Table 12 (cont.)

Entry	Substrate	Conditions	Product	Yield (%)	Ref
5	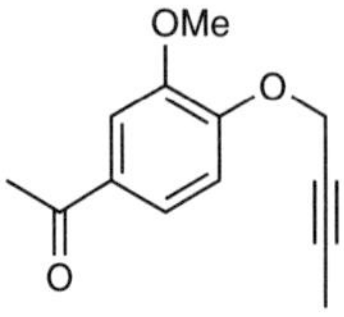	$PtCl_4$ (2 mol%), dioxane, rt, 3 h	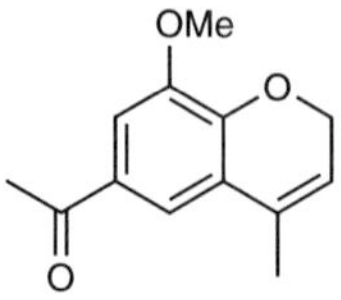	92	[71]
6	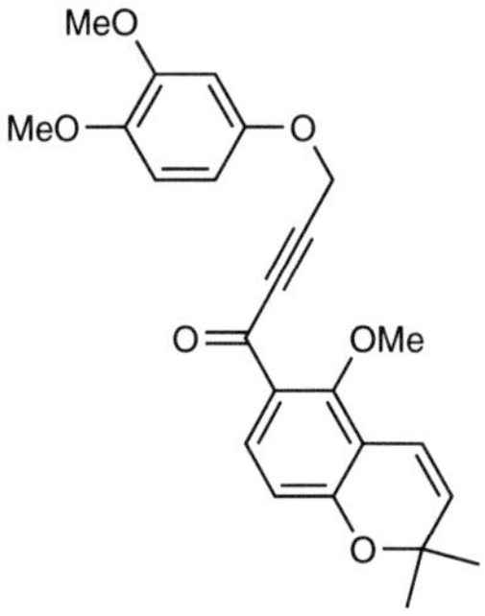	$PtCl_2$ (5 mol%), toluene, 55 °C	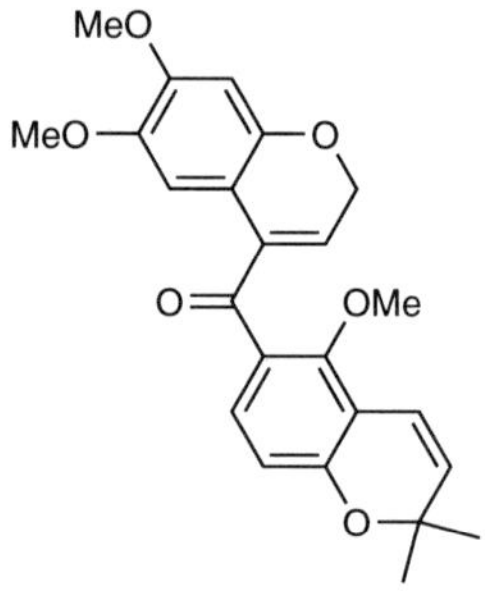	91[c]	[72]
7	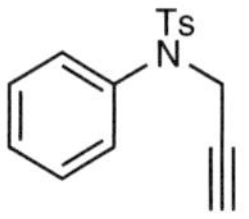	**39** (5 mol%), CH_2Cl_2, 40 °C, 2 h	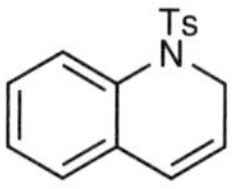	81	[66]
8	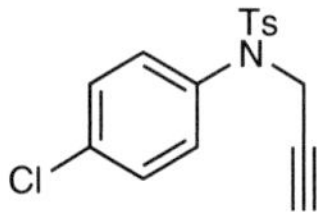	**39** (5 mol%), CH_2Cl_2, 40 °C, 48 h	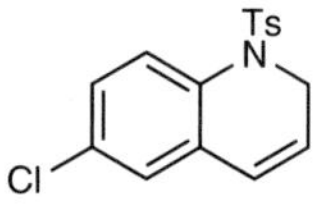	52	[66]
9	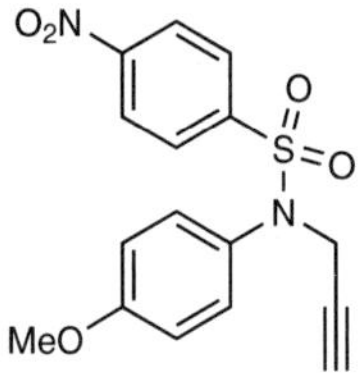	**39** (5 mol%), CH_2Cl_2, 40 °C, 24 h	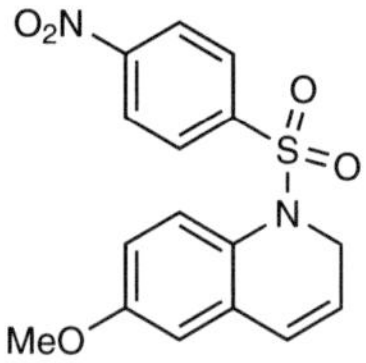	75	[66]
10	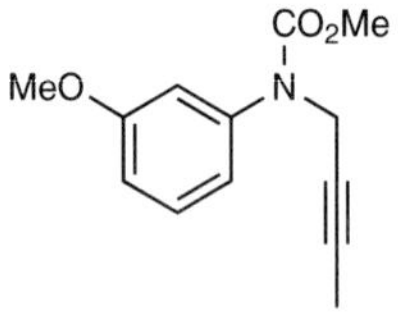	$PtCl_4$ (5 mol%), 1,2-dichloroethane, 70 °C, 2 h	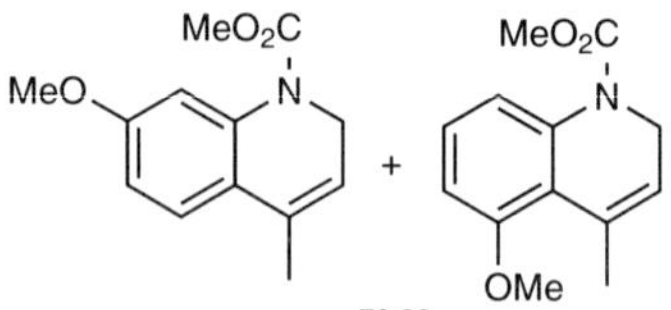 78:22	82	[71]
11	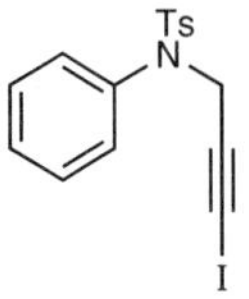	Au(IPr)NTf_2,[d] 1,2-dichloroethane, rt, 24 h	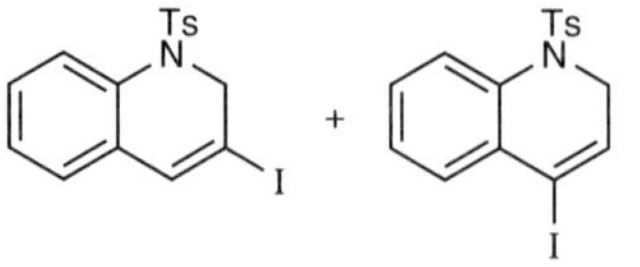 14:1	85	[73]

Table 12 (cont.)

Entry	Substrate	Conditions	Product	Yield (%)	Ref
12	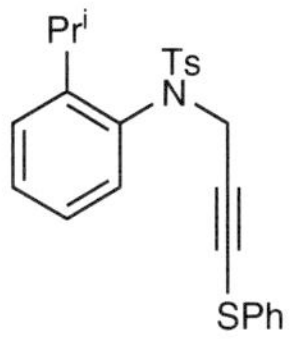	$FeCl_3$ (5 mol%), AgOTf (15 mol%), 1,2-dichloroethane, 25 °C, 10 min	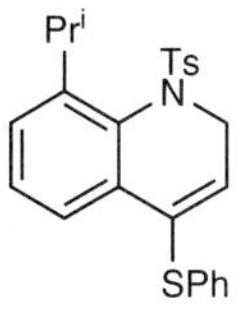	95	[52]
13	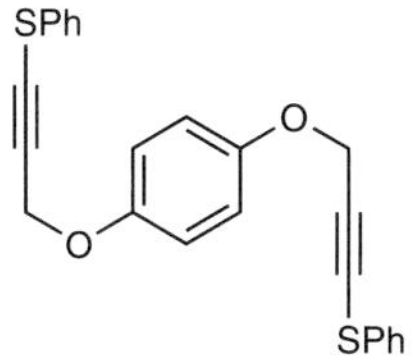	$FeCl_3$ (5 mol%), AgOTf (15 mol%), 1,2-dichloroethane, 25 °C, 1 h	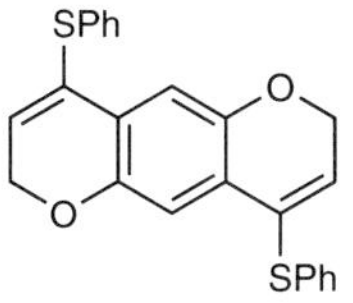	72	[52]
14	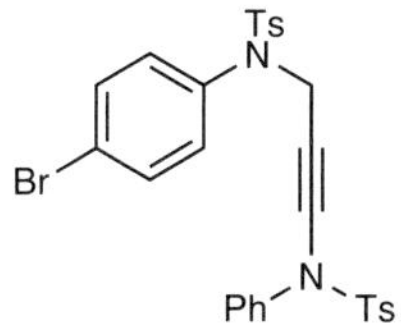	$FeCl_3$ (5 mol%), AgOTf (15 mol%), 1,2-dichloroethane, 25 °C, 1 h	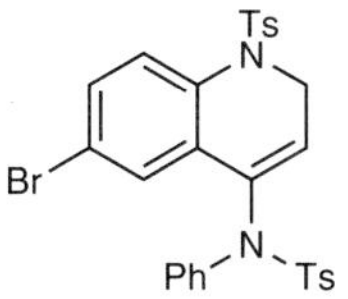	93	[52]

[a] 70% yield using $AuCl(PPh_3)$ (3 mol%) and HBF_4 (6 mol%) in CH_2Cl_2 at 23 °C.
[b] Yield of the 4*H*-1-benzopyran isomer in parentheses.
[c] 40% yield using $PtCl_4$ (5 mol%) in dioxane at 65 °C.
[d] IPr = 1,3-bis(2,6-diisopropylphenyl)imidazol-2-ylidene.

(*S*)-2,4-Dimethyl-2*H*-1-benzopyran (Table 12, Entry 3); Typical Procedure:[71]
A 0.2 M soln of (*S*)-(pent-3-yn-2-yloxy)benzene (89% ee) in dioxane and $PtCl_4$ (5 mol%) was stirred under air until the substrate had been completely consumed, as monitored by TLC (1 h). The solvent was evaporated, and the resulting residue was purified by column chromatography (silica gel); yield: 82%; 89% ee.

1.5.2.2.2.6 Formation of Seven-Membered-Ring Annulated Indoles

The method for iron-catalyzed cyclizations outlined in Table 9 (entries 7–9) and Table 12 (entries 12–14) is also applicable to the construction of 6,7-dihydro-5*H*-benzo[7]annulene and 2,3-dihydro-1*H*-benzo[*b*]azepine frameworks.[52] However, the most notable developments in the formation of seven-membered-ring annulated arenes concern electron-rich indoles, and often involve gold-catalyzed regioselective formal 7-*endo-dig* cyclization. Pioneering studies on the cyclization of indole–yne substrates in the presence of gold or platinum catalysts have revealed competing modes of cyclization and rearrangement.[75,76]

1.5.2.2.2.6.1 Annulation with Nitroenynes

In a sequential organocatalysis/gold catalysis process (Scheme 18),[77] alkyne substrates **41** for the gold-catalyzed step are generated in high enantiomeric excess via organocatalyzed Friedel–Crafts-type alkylation at the 1*H*-indole 3-position by a nitroenyne. In a one-pot process, the subsequent gold catalysis is realized readily using the prototypical gold catalyst {bis[(trifluoromethyl)sulfonyl]imidate}(triphenylphosphine)gold(I) [$Au(PPh_3)NTf_2$] in the presence of 4-toluenesulfonic acid as additive, which serves to protonate the basic organocatalyst **40** used in the previous step and hence prevents deactivation of the gold catalyst. The reaction is generally highly enantioselective, and the overall yields of the tetracyclic indole products **42** are mostly good to excellent. The reported reaction scope includes electron-donating substituents on the indole benzene ring and minor modifications of the nitroenyne reacting partner.

More variations with regard to the substituents on the benzene ring of 1*H*-indole substrates **43** and the R^4 group of nitroenynes **44** have been reported in a racemic version of the method singly catalyzed by gold complex **39** to give 5,12-dihydrobenzo[4,5]cyclohepta[1,2-*b*]indoles **45** (Scheme 18).[78] Notably, the reactions are performed in aqueous media and display mostly good efficiencies, and alkyl groups can be tolerated at the C6-position of the 1*H*-indole substrates (R^3). An alkyne of type **41** is likely the intermediate in this process.

Scheme 18 Metal-Catalyzed 7-*endo-dig* Cyclization as the Key Step in Bimolecular Annulations between 1*H*-Indoles and Nitroenynes[77,78]

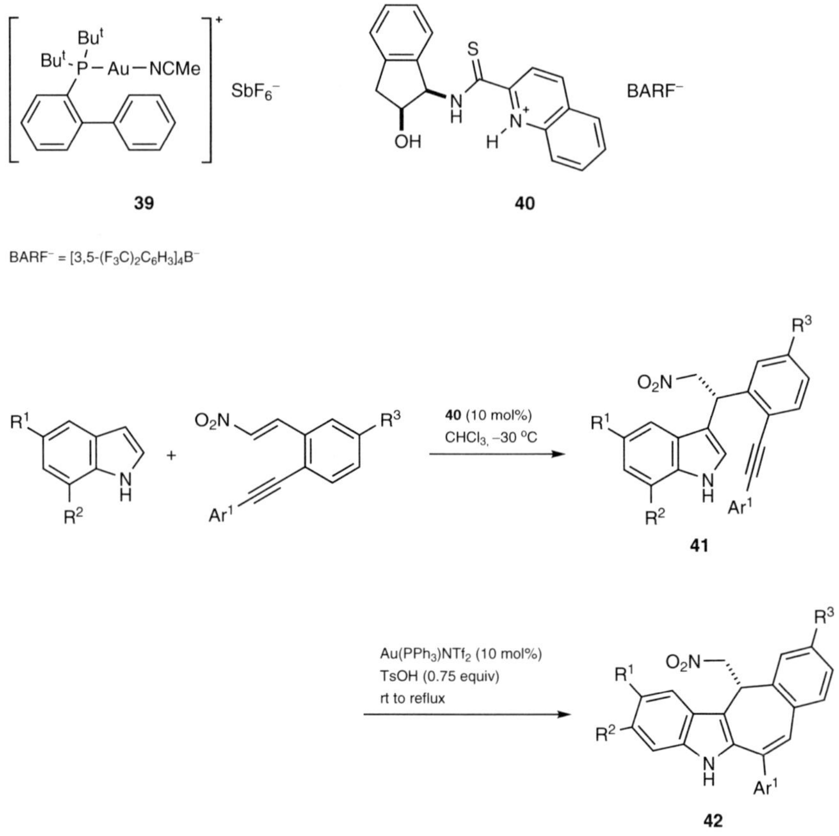

$[Bu^t_2P(biphenyl)Au{-}NCMe]^+$ SbF_6^- **39**

40 BARF⁻

$BARF^- = [3,5\text{-}(F_3C)_2C_6H_3]_4B^-$

40 (10 mol%), $CHCl_3$, −30 °C → **41**

$Au(PPh_3)NTf_2$ (10 mol%), TsOH (0.75 equiv), rt to reflux → **42**

R^1	R^2	R^3	Ar^1	Time (h)		Yield (%)	ee (%)	Ref
				Step 1	Step 2			
OMe	H	H	Ph	24	19	78	97	[77]
H	Me	H	Ph	37	31	70	98	[77]
H	H	H	3-Tol	25	67	79	97	[77]
Me	H	F	Ph	16	29	96	95	[77]

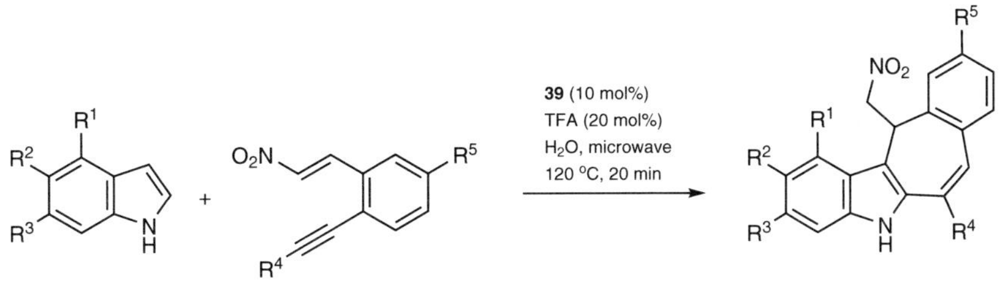

R^1	R^2	R^3	R^4	R^5	Yield (%)	Ref
H	H	H	Ph	H	86	[78]
H	Cl	H	Ph	H	62	[78]
H	OMe	OMe	Ph	H	78	[78]
Me	H	H	4-Tol	Me	66	[78]
H	H	H	Pr	H	53	[78]
H	H	H	cyclopentyl	H	76	[78]
H	F	H	Ph	F	71	[78]

12-(Nitromethyl)-6-phenyl-5,12-dihydrobenzo[4,5]cyclohepta[1,2-*b*]indoles 45 **(R^4 = Ph; R^5 = H); General Procedure:**[78]
To a mixture of 1-[(*E*)-2-nitrovinyl]-2-(phenylethynyl)benzene (**44**, R^4 = Ph; R^5 = H; 0.2 mmol) and an indole **43** (0.3 mmol) in H_2O (3 mL) were added gold catalyst **39** (0.02 mmol) and TFA (0.04 mmol). The vial was sealed and the mixture was then irradiated (microwave) at 120 °C for 20 min. After the mixture had cooled to ambient temperature, it was concentrated under reduced pressure, and the resulting residue was purified by flash chromatography (petroleum ether/EtOAc 10:1).

1.5.2.2.2.6.2 Annulation with Enynones

A sodium tetrachloroaurate(III) catalyzed bimolecular annulation between 1*H*-indoles **47** and enynones **46** offers rapid access to medically important [6,5,7]-tricyclic 1*H*-indoles **50** (Scheme 19).[79] The first step of this tandem process is a Michael addition to the enone moiety by the electron-rich 1*H*-indole, forming the observable ynone intermediate **48**. Subsequent cyclization, promoted by the same gold catalyst, affords the product. Mechanistically, the high selectivity toward 7-*endo-dig* cyclization can be rationalized by an initial, kinetically more favorable 6-*endo-dig* cyclization to form the spiro intermediate **49** followed by selective ring enlargement. This reaction pathway is supported by the observation of spirobicyclic intermediates related to **49** and is consistent with Echavarren's proposal,[75,76] and can also rationalize the selective formation of seven-membered rings in reactions using nitroenynes described in Section 1.5.2.2.2.6.1.

The reaction exhibits a broad scope; both electron-withdrawing substituents (e.g., CO_2Me, Cl, and even 4,4,5,5-tetramethyl-1,3,2-dioxaborolan-2-yl) and electron-donating substituents (e.g., OMe and free OH) can be tolerated on the indole ring, and sterically encumbering groups can be accommodated at the C4- and C7-positions. The enynones **46** can have various substituted phenyl groups and alkyl groups at the terminal alkyne position (R^3), as well as different combinations of R^1 and R^2; however, in some cases higher catalyst loadings (5 mol%) and heating are required. Similarly, a one-pot synthesis of indole-fused scaffolds is achieved via gold-catalyzed tandem annulation reactions of 1,2-dialkynyl-2-en-1-ones with indoles.[80]

for references see p 314

Scheme 19 Synthesis of 9,10-Dihydrocyclohepta[*b*]indol-8(5*H*)-ones[79]

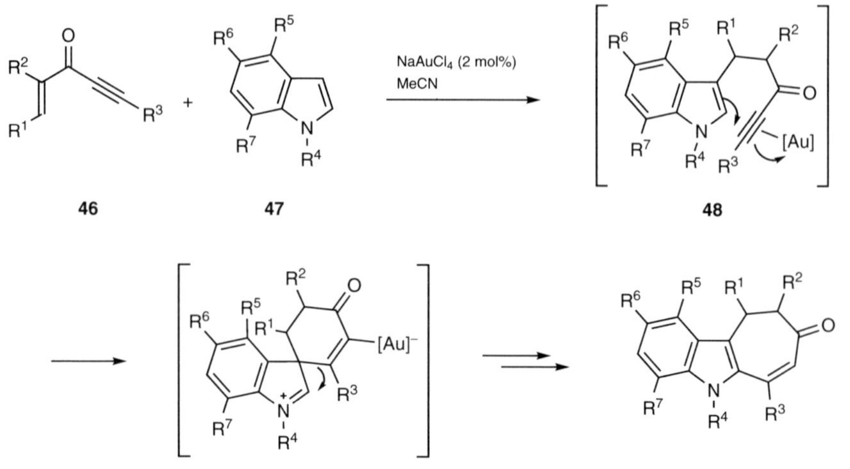

NaAuCl4 (2 mol%)
MeCN

46 47 48 49 50

R^1	R^2	R^3	R^4	R^5	R^6	R^7	Time (h)	Temp	Yield[a] (%)	Ref
Me	H	Ph	H	H	H	H	4	rt	100[b]	[79]
Me	H	Ph	H	H	OMe	H	4	rt	90	[79]
Me	H	Ph	H	H	CO_2Me	H	4	rt	71	[79]
Me	H	Ph	H	Me	H	H	4	rt	93	[79]
Me	H	Ph	H	Cl	H	H	4	rt	80	[79]
Me	H	Ph	H	H	H	Me	4	rt	95	[79]
Me	H	Ph	H	H	OH	H	4	rt	92	[79]
Me	H	Ph	H	H	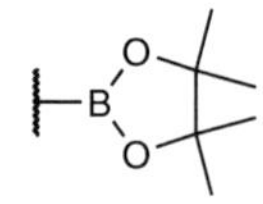 Bpin (4,4,5,5-tetramethyl-1,3,2-dioxaborolan-2-yl)	H	4	rt	88	[79]
Me	H	4-Tol	H	H	H	H	6	rt	98[b]	[79]
Me	H	4-$MeOC_6H_4$	H	H	H	H	24	rt	82[b]	[79]
Me	H	4-$F_3CC_6H_4$	H	H	H	H	16	rt	73[b]	[79]
Ph	H	Ph	H	H	H	H	3	82 °C	100[b,c]	[79]
Ph	H	Bu	H	H	H	H	18	82 °C	69[b,c,d]	[79]
Me	H	Bu	H	H	H	H	4	82 °C	82[b,c]	[79]
H	H	Bu	H	H	H	H	96	82 °C	60[b]	[79]
Me	Me	Ph	H	H	H	H	18	rt	73[b,e]	[79]
Me	H	Ph	Me	H	H	H	20	rt	71[b]	[79]

[a] Isolated yield.
[b] 5 mol% of $NaAuCl_4$ was used.
[c] 2 equiv of H_2O was added.
[d] Based on recovered starting material.
[e] Ratio (*syn*/*anti*) 4:1, based on ^1H NMR analysis of the crude mixture.

10-Methyl-6-phenyl-9,10-dihydrocyclohepta[*b*]indol-8(5*H*)-one (50, R^1 = Me; R^2 = R^4 = R^5 = R^6 = R^7 = H; R^3 = Ph;); **Typical Procedure:**[79]

To 1*H*-indole (**47**, R^4 = R^5 = R^6 = R^7 = H; 24.6 mg, 0.21 mmol, 1.05 equiv) in a round-bottomed flask was added a soln of (*E*)-1-phenylhex-4-en-1-yn-3-one (**46**, R^1 = Me; R^2 = H; R^3 = Ph;

34.0 mg, 0.20 mmol, 1 equiv) in MeCN (1 mL) followed by a soln of $NaAuCl_4 \cdot H_2O$ (0.05 equiv) in MeCN (1 mL). The mixture was stirred at rt for 4 h, filtered through Celite, and concentrated under reduced pressure. The residue was purified by flash column chromatography (petroleum ether/EtOAc 8:1), affording a yellow solid; yield: 57 mg (quant).

1.5.3 Applications in the Synthesis of Natural Products and Relevant Structures

The array of versatile synthetic methods developed based on metal-catalyzed cyclization of alkynes and alkenes offer powerful tools for the construction of complex structures, including natural products and related core structures. Several illustrative examples are discussed in the following sections.

1.5.3.1 Based on the Prins Cyclization

The Prins cyclization[81] and the pinacol-terminated Prins cyclization[82] offer expedient and often stereoselective access to substituted tetrahydropyrans, tetrahydrofurans, and relevant carbocycles, and have hence been applied as key steps in the total synthesis of various natural products. Most of these cases are promoted by stoichiometric Lewis or Brønsted acids or catalyzed by Brønsted acids; however, metal-catalyzed Prins cyclizations have been successfully applied in several examples.

Perhaps one of the most elegant applications of the Prins cyclization is in the enantioselective total synthesis of briarellins E and F (Scheme 20).[83] The condensation of the 1,2-diol **51** and the enal **52** at low temperature leads to the formation of acetals **53** as a mixture of four inconsequential diastereomers. This crude mixture is subjected to tin(IV) chloride catalyzed Prins cyclization, which is terminated by a pinacol-type rearrangement. This two-step annulation reaction offers rapid access to the highly substituted tetrahydrofuran product **54** in excellent overall yield. This intermediate possesses the requisite functionalized bicyclic structure to allow the implementation of the first total synthesis of the targeted natural products.

for references see p 314

Scheme 20 Enantioselective Total Synthesis of Briarellins E and F via a Key Tin(IV) Chloride Catalyzed Pinacol-Terminated Prins Cyclization[83]

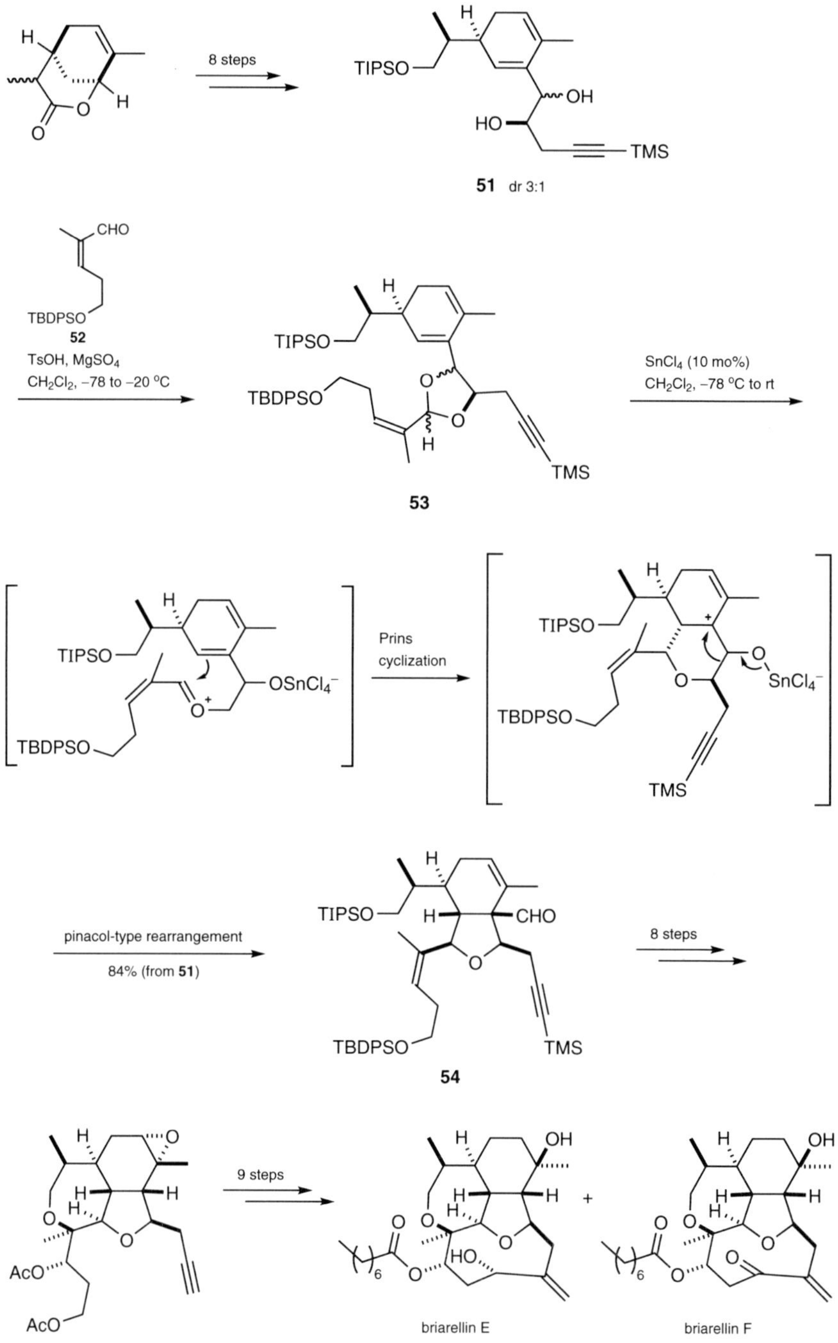

The synthetic utility of the indium(III) bromide catalyzed synthesis of 3-oxaterpenoids (see Scheme 5, Section 1.5.1.1.1.4) is demonstrated by its application as the key transformation in the total synthesis of moluccanic acid methyl ester (Scheme 21).[9] Notably in this case, an acetal, 2-chloro-1,1-dimethoxyethane, was used instead of an aldehyde in the scope studies, due to the difficulty in accessing nonhydrated chloroacetaldehyde.

Scheme 21 Total Synthesis of Moluccanic Acid Methyl Ester[9]

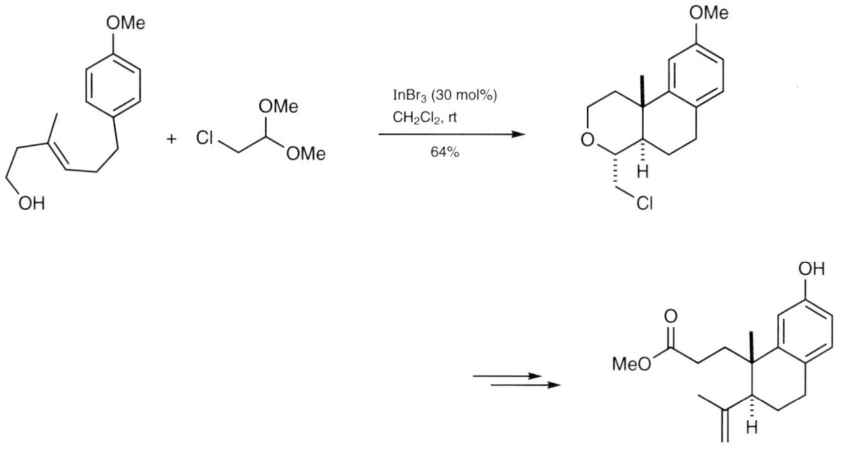

OMe OH + Cl OMe OMe → $InBr_3$ (30 mol%), CH_2Cl_2, rt, 64% → OMe O H Cl

→ → MeO O OH H

moluccanic acid methyl ester

1.5.3.2 Based on Nucleophilic Addition to Metal-Activated Alkenes and Alkynes

Metal-catalyzed cyclizations of carbon nucleophiles to π-systems are of great synthetic utility, and provide versatile access to key synthetic intermediates suitable for the total synthesis of natural products or construction of their core structures. For example, methyl 5,7-bis(benzyloxy)-3-hydroxy-1,2-dimethoxyphenanthrene-4-carboxylate (see Table 10, entry 10, Section 1.5.2.2.2.2) can be elaborated to the phenanthrenoid coeloginin (Scheme 22).[55]

Scheme 22 Coeloginin, a Phenanthrenoid[55]

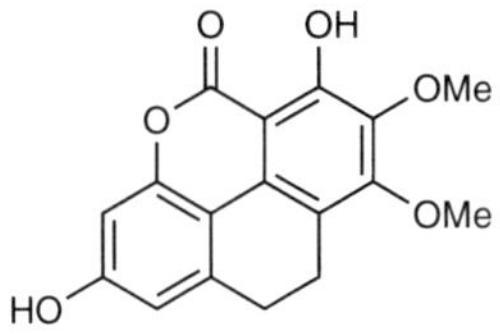

O OH O OMe OMe HO

The gold-catalyzed cycloisomerization of an *N*-propargylindole-2-carboxamide enables access to analogues of the antitumor antibiotic lavendamycin (Scheme 23).[84] The indole **55** undergoes a gold(III) chloride catalyzed cyclization to afford the carbolinone **56** in 60% yield. It is noteworthy that with secondary amides (e.g., Cbz is replaced by H) cyclization of the amide group to the C≡C bond becomes overwhelming, resulting in the exclusive formations of oxazole products. Carbolinone **56** is converted into the chlorinated β-carboline **57** (80% yield over two steps), which is readily transformed to the pyridine-substituted β-carboline **58** (Z = N) via an uneventful palladium-catalyzed Migita–Kosugi–Stille coupling. A related Suzuki–Miyaura coupling leads to its phenyl counterpart **58** (Z = CH). Compound **58** (Z = N) possesses most of the key structural features of lavendamycin, except the benzoquinone moiety.

Scheme 23 Access to Analogues of the Antitumor Antibiotic Lavendamycin[84]

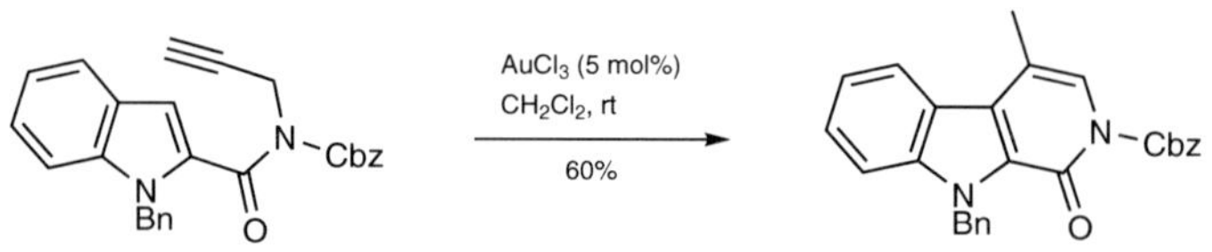

AuCl3 (5 mol%)
CH2Cl2, rt
60%

55 56

1. TFA
2. POCl3
80%

PdCl2(PPh3)

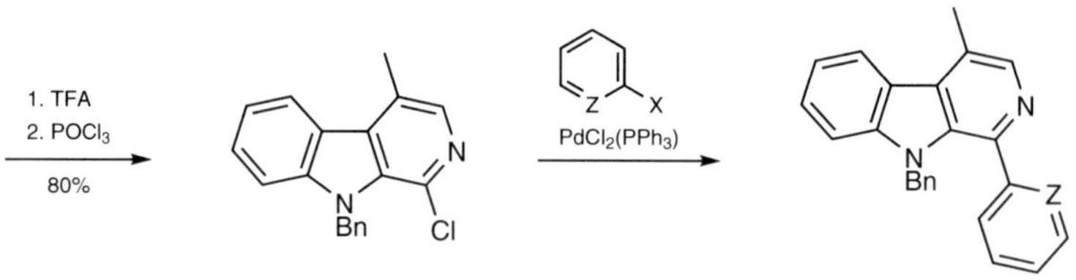

57 58

X	Z	Yield (%)	Ref
$SnBu_3$	N	79	[84]
$B(OH)_2$	CH	77	[84]

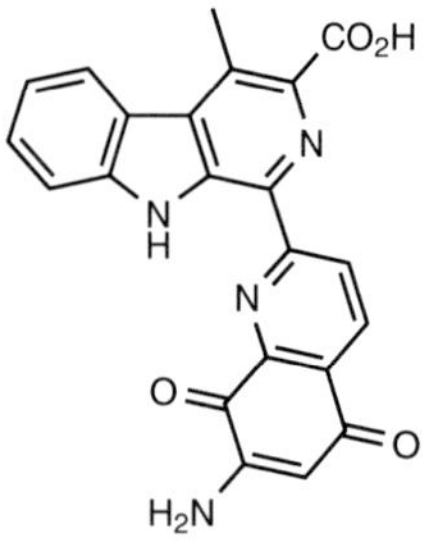

lavendamycin

A gold(III) chloride catalyzed 8-*endo-dig* cyclization of an unprotected indole ring onto a terminal alkyne provides, despite the moderate yield, an elegant and expedient access to the tetracyclic skeleton of lundurines and especially lundurine A (Scheme 24).[85] Notably, under the shown conditions, as well as with the alternative use of gold(I) chloride as catalyst, the side product derived from the typically competing 7-*exo-dig* cyclization is not detected. In contrast, when cationic gold(I) catalysts such as acetonitrile(triphenylphosphine)gold(I) hexafluoroantimonate {$[Au(NCMe)(PPh_3)]^+ SbF_6^-$} are used, the 7-*exo-dig* cyclization is indeed competitive.

Scheme 24 Access to the Tetracyclic Skeleton of Lundurine A[85]

2 steps

MeO_2C(NH$_3^+$ Cl$^-$)$CH_2CH_2CO_2Me$, $NaBH(OAc)_3$, Et_3N, CH_2Cl_2, 8 h, rt

4 steps

$AuCl_3$ (5 mol%), CH_2Cl_2, rt, 24 h; 55%

lundurine A

1.5.4 Conclusions and Future Perspectives

Though the author strives to cover the most synthetically versatile transformations based on metal-catalyzed cyclizations of alkenes and alkynes, the reactions discussed in this chapter belong to a subset of what this general strategy could offer, and there must be worthy reactions that are left out.

A large array of cyclic structural motifs are accessible, many in stereoselective manners, via metal-catalyzed cyclization of alkenes and alkynes. They include tetrahydrofuran, tetrahydropyran, cycloalkenes, dihydronaphthalene, carbazole, coumarin, quinolinone, chromene, dihydroquinoline, phenanthrene, etc., and are essential structural components in various bioactive compounds including natural products. Applications of these methods in the syntheses of natural products and their relevant structures have been documented.

Despite the myriad developments in metal-catalyzed cyclizations of alkenes and alkynes, there are undoubtedly still many opportunities for new discovery. With the demand for more sustainable and "greener" transformations, further effort would likely be directed toward catalysis by base metals and/or with ultra-low loadings of expensive metal catalysts.

References

[1] Pastor, I. M.; Yus, M., *Curr. Org. Chem.*, (2007) **11**, 925.
[2] Clarke, M. L.; France, M. B., *Tetrahedron*, (2008) **64**, 9003.
[3] Snider, B. B., In *Comprehensive Organic Synthesis*, Trost, B. M.; Fleming, I., Eds.; Pergamon: Oxford (1991); Vol. 2, p 527.
[4] Zhang, W.-C.; Viswanathan, G. S.; Li, C.-J., *Chem. Commun. (Cambridge)*, (1999), 291.
[5] Miranda, P. O.; Carballo, R. M.; Martín, V. S.; Padrón, J. I., *Org. Lett.*, (2009) **11**, 357.
[6] Tadpetch, K.; Rychnovsky, S. D., *Org. Lett.*, (2008) **10**, 4839.
[7] Reddy, B. V. S.; Jalal, S.; Kumar Singarapu, K., *RSC Adv.*, (2014) **4**, 16 739.
[8] Lalli, C.; van de Weghe, P., *Chem. Commun. (Cambridge)*, (2014) **50**, 7495.
[9] Li, B.; Lai, Y.-C.; Zhao, Y.; Wong, Y.-H.; Shen, Z.-L.; Loh, T.-P., *Angew. Chem. Int. Ed.*, (2012) **51**, 10 619.
[10] Nakamura, M.; Niiyama, K.; Yamakawa, T., *Tetrahedron Lett.*, (2009) **50**, 6462.
[11] Zheng, K.; Shi, J.; Liu, X.; Feng, X., *J. Am. Chem. Soc.*, (2008) **130**, 15 770.
[12] Zheng, K.; Liu, X.; Qin, S.; Xie, M.; Lin, L.; Hu, C.; Feng, X., *J. Am. Chem. Soc.*, (2012) **134**, 17 564.
[13] Miranda, P. O.; Ramírez, M. A.; Martín, V. S.; Padrón, J. I., *Chem.–Eur. J.*, (2008) **14**, 6260.
[14] Xu, T.; Yang, Q.; Li, D.; Dong, J.; Yu, Z.; Li, Y., *Chem.–Eur. J.*, (2010) **16**, 9264.
[15] Nakatani, Y.; Kawashima, K., *Synthesis*, (1978), 147.
[16] Kočovský, P.; Ahmed, G.; Šrogl, J.; Malkov, A. V.; Steele, J., *J. Org. Chem.*, (1999) **64**, 2765.
[17] Yang, D.; Yang, M.; Zhu, N.-Y., *Org. Lett.*, (2003) **5**, 3749.
[18] Grachan, M. L.; Tudge, M. T.; Jacobsen, E. N., *Angew. Chem. Int. Ed.*, (2008) **47**, 1469.
[19] Rajapaksa, N. S.; Jacobsen, E. N., *Org. Lett.*, (2013) **15**, 4238.
[20] Carballo, R. M.; Valdomir, G.; Purino, M.; Martín, V. S.; Padrón, J. I., *Eur. J. Org. Chem.*, (2010), 2304.
[21] Reddy, B. V. S.; Borkar, P.; Chakravarthy, P. P.; Yadav, J. S.; Gree, R., *Tetrahedron Lett.*, (2010) **51**, 3412.
[22] Conia, J. M.; Le Perchec, P., *Synthesis*, (1975), 1.
[23] Patil, N. T.; Kavthe, R. D.; Shinde, V. S., *Tetrahedron*, (2012) **68**, 8079.
[24] Pei, T.; Widenhoefer, R. A., *J. Am. Chem. Soc.*, (2001) **123**, 11 290.
[25] Pei, T.; Widenhoefer, R. A., *Chem. Commun. (Cambridge)*, (2002), 650.
[26] Wang, X.; Pei, T.; Han, X.; Widenhoefer, R. A., *Org. Lett.*, (2003) **5**, 2699.
[27] Yang, D.; Li, J.-H.; Gao, Q.; Yan, Y.-L., *Org. Lett.*, (2003) **5**, 2869.
[28] Xiao, Y.-P.; Liu, X.-Y.; Che, C.-M., *Angew. Chem. Int. Ed.*, (2011) **50**, 4937.
[29] Zhou, C.-Y.; Che, C.-M., *J. Am. Chem. Soc.*, (2007) **129**, 5828.
[30] Kennedy-Smith, J. J.; Staben, S. T.; Toste, F. D., *J. Am. Chem. Soc.*, (2004) **126**, 4526.
[31] Staben, S. T.; Kennedy-Smith, J. J.; Toste, F. D., *Angew. Chem. Int. Ed.*, (2004) **43**, 5350.
[32] Ito, H.; Makida, Y.; Ochida, A.; Ohmiya, H.; Sawamura, M., *Org. Lett.*, (2008) **10**, 5051.
[33] Gao, Q.; Zheng, B.-F.; Li, J.-H.; Yang, D., *Org. Lett.*, (2005) **7**, 2185.
[34] Deng, C.-L.; Song, R.-J.; Liu, Y.-L.; Li, J.-H., *Adv. Synth. Catal.*, (2009) **351**, 3096.
[35] Deng, C.-L.; Zou, T.; Wang, Z.-Q.; Song, R.-J.; Li, J.-H., *J. Org. Chem.*, (2009) **74**, 412.
[36] Chan, L. Y.; Kim, S.; Park, Y.; Lee, P. H., *J. Org. Chem.*, (2012) **77**, 5239.
[37] Bandini, M.; Emer, E.; Tommasi, S.; Umani-Ronchi, A., *Eur. J. Org. Chem.*, (2006), 3527.
[38] de Mendoza, P.; Echavarren, A. M., *Pure Appl. Chem.*, (2010) **82**, 801.
[39] Kitamura, T., *Eur. J. Org. Chem.*, (2009), 1111.
[40] Yamamoto, Y., *Chem. Soc. Rev.*, (2014) **43**, 1575.
[41] Evans, D. A.; Fandrick, K. R.; Song, H.-J., *J. Am. Chem. Soc.*, (2005) **127**, 8942.
[42] Agnusdei, M.; Bandini, M.; Melloni, A.; Umani-Ronchi, A., *J. Org. Chem.*, (2003) **68**, 7126.
[43] Youn, S. W.; Pastine, S. J.; Sames, D., *Org. Lett.*, (2004) **6**, 581.
[44] Zhang, H.; Ferreira, E. M.; Stoltz, B. M., *Angew. Chem. Int. Ed.*, (2004) **43**, 6144.
[45] Ferreira, E. M.; Stoltz, B. M., *J. Am. Chem. Soc.*, (2003) **125**, 9578.
[46] Namba, K.; Yamamoto, H.; Sasaki, I.; Mori, K.; Imagawa, H.; Nishizawa, M., *Org. Lett.*, (2008) **10**, 1767.
[47] Biannic, B.; Aponick, A., *Eur. J. Org. Chem.*, (2011), 6605.
[48] Bandini, M.; Eichholzer, A., *Angew. Chem. Int. Ed.*, (2009) **48**, 9533.
[49] Nishizawa, M.; Takao, H.; Yadav, V. K.; Imagawa, H.; Sugihara, T., *Org. Lett.*, (2003) **5**, 4563.
[50] Gorin, D. J.; Dubé, P.; Toste, F. D., *J. Am. Chem. Soc.*, (2006) **128**, 14 480.
[51] Dal Zotto, C.; Wehbe, J.; Virieux, D.; Campagne, J.-M., *Synlett*, (2008), 2033.
[52] Eom, D.; Mo, J.; Lee, P. H.; Gao, Z.; Kim, S., *Eur. J. Org. Chem.*, (2013), 533.

[53] Fürstner, A.; Mamane, V., *J. Org. Chem.*, (2002) **67**, 6264.
[54] Lim, W.; Rhee, Y. H., *Eur. J. Org. Chem.*, (2013), 460.
[55] Carreras, J.; Gopakumar, G.; Gu, L.; Gimeno, A.; Linowski, P.; Petuskova, J.; Thiel, W.; Alcarazo, M., *J. Am. Chem. Soc.*, (2013) **135**, 18 815.
[56] Komeyama, K.; Igawa, R.; Takaki, K., *Chem. Commun. (Cambridge)*, (2010) **46**, 1748.
[57] Bruneau, C.; Dixneuf, P. H., *Acc. Chem. Res.*, (1999) **32**, 311.
[58] Bruneau, C.; Dixneuf, P. H., *Metal Vinylidenes and Allenylidenes in Catalysis: From Reactivity to Applications in Synthesis*, Wiley-VCH: Weinheim, Germany, (2008).
[59] Shen, H.-C.; Tang, J.-M.; Chang, H.-K.; Yang, C.-W.; Liu, R.-S., *J. Org. Chem.*, (2005) **70**, 10 113.
[60] Qiu, Y.; Kong, W.; Fu, C.; Ma, S., *Org. Lett.*, (2012) **14**, 6198.
[61] Jia, C.; Piao, D.; Oyamada, J.; Lu, W.; Kitamura, T.; Fujiwara, Y., *Science (Washington, D. C.)*, (2000) **287**, 1992.
[62] Jia, C.; Piao, D.; Kitamura, T.; Fujiwara, Y., *J. Org. Chem.*, (2000) **65**, 7516.
[63] Yoon, M. Y.; Kim, J. H.; Choi, D. S.; Shin, U. S.; Lee, J. Y.; Song, C. E., *Adv. Synth. Catal.*, (2007) **349**, 1725.
[64] Shi, Z.; He, C., *J. Org. Chem.*, (2004) **69**, 3669.
[65] Herrero-Gómez, E.; Nieto-Oberhuber, C.; López, S.; Benet-Buchholz, J.; Echavarren, A. M., *Angew. Chem. Int. Ed.*, (2006) **45**, 5455.
[66] Menon, R. S.; Findlay, A. D.; Bissember, A. C.; Banwell, M. G., *J. Org. Chem.*, (2009) **74**, 8901.
[67] Cervi, A.; Aillard, P.; Hazeri, N.; Petit, L.; Chai, C. L. L.; Willis, A. C.; Banwell, M. G., *J. Org. Chem.*, (2013) **78**, 9876.
[68] Harris, E. B. J.; Banwell, M. G.; Willis, A. C., *Tetrahedron Lett.*, (2011) **52**, 6887.
[69] Nevado, C.; Echavarren, A. M., *Chem.–Eur. J.*, (2005) **11**, 3155.
[70] Martín-Matute, B.; Nevado, C.; Cárdenas, D. J.; Echavarren, A. M., *J. Am. Chem. Soc.*, (2003) **125**, 5757.
[71] Pastine, S. J.; Youn, S. W.; Sames, D., *Tetrahedron*, (2003) **59**, 8859.
[72] Pastine, S. J.; Sames, D., *Org. Lett.*, (2003) **5**, 4053.
[73] Morán-Poladura, P.; Suárez-Pantiga, S.; Piedrafita, M.; Rubio, E.; González, J. M., *J. Organomet. Chem.*, (2011) **696**, 12.
[74] Ye, L.; Wang, Y.; Aue, D. H.; Zhang, L., *J. Am. Chem. Soc.*, (2012) **134**, 31.
[75] Ferrer, C.; Amijs, C. H. M.; Echavarren, A. M., *Chem.–Eur. J.*, (2007) **13**, 1358.
[76] Ferrer, C.; Echavarren, A. M., *Angew. Chem. Int. Ed.*, (2006) **45**, 1105.
[77] Loh, C. C. J.; Badorrek, J.; Raabe, G.; Enders, D., *Chem.–Eur. J.*, (2011) **17**, 13 409.
[78] Xu, S.; Zhou, Y.; Xu, J.; Jiang, H.; Liu, H., *Green Chem.*, (2013) **15**, 718.
[79] Heffernan, S. J.; Tellam, J. P.; Queru, M. E.; Silvanus, A. C.; Benito, D.; Mahon, M. F.; Hennessy, A. J.; Andrews, B. I.; Carbery, D. R., *Adv. Synth. Catal.*, (2013) **355**, 1149.
[80] Xie, X.; Du, X.; Chen, Y.; Liu, Y., *J. Org. Chem.*, (2011) **76**, 9175.
[81] Han, X.; Peh, G.; Floreancig, P. E., *Eur. J. Org. Chem.*, (2013), 1193.
[82] Overman, L. E.; Pennington, L. D., *J. Org. Chem.*, (2003) **68**, 7143.
[83] Corminboeuf, O.; Overman, L. E.; Pennington, L. D., *J. Am. Chem. Soc.*, (2003) **125**, 6650.
[84] England, D. B.; Padwa, A., *Org. Lett.*, (2008) **10**, 3631.
[85] Ferrer, C.; Escribano-Cuesta, A.; Echavarren, A. M., *Tetrahedron*, (2009) **65**, 9015.

1.6 Metal-Catalyzed Cyclization Reactions of Allenes

A. M. Phelps, J. M. Alderson, and J. M. Schomaker

General Introduction

Allenes have long fascinated organic chemists, but the early lack of general syntheses and their relative unfamiliarity compared to alkenes and alkynes largely relegated them to laboratory curiosities, as opposed to useful building blocks. Fortunately, the last decade has seen a rapid increase in new methods for the preparation of both racemic and enantioenriched allenes from simple precursors.[1] This, in turn, has led to an explosion in the development of powerful transformations of allenes that reveal their extensive potential in synthesis for the construction of complex molecules.

Several unique features of allenes distinguish their chemistry from that of other common unsaturated compounds. First, the cumulated double bonds in allenes are quite strained; the C=C π-bond is approximately 10 kcal•mol^{-1} less stable than the corresponding bond in a simple alkene.[2] The relief of strain drives addition reactions not readily achieved with alkenes, rendering allenes popular substrates for a variety of transition-metal-catalyzed reactions. Second, the presence of three unsaturated carbons, as opposed to the two sp^2 or sp carbons contained in alkenes and alkynes, respectively, means that a greater degree of complexity is obtained in the products of allene functionalizations. Finally, a highly useful feature of allenes is their ability to exhibit axial chirality. The axial chirality can often be transferred to point or central chirality in the products, or otherwise influence the diastereoselectivity of allene functionalization.

One popular strategy to transform allenes into useful and highly substituted carbocyclic and heterocyclic products is to employ a transition-metal-catalyzed cyclization reaction. A host of metals have been utilized for this purpose, including silver, gold, palladium, and rhodium. The transition-metal catalyst can play a variety of roles, depending on the structure of the allene and the mechanism of the cyclization. The role of carbophilic, soft Lewis acids, such as silver and gold, is to coordinate to the allene and activate it toward nucleophilic attack. In other cases involving metals such as palladium and rhodium, the catalyst plays a dual role in activating the allene and promoting tandem reactions subsequent to cyclization to increase the complexity of the products. Examples of both mechanistic pathways are described below in Schemes 1 and 2.

Simple intramolecular allene cyclizations typically employ silver(I) and gold(I) complexes containing phosphine ligands, as well as non-coordinating counteranions. These cyclizations generally occur with high transfer of axial-to-point chirality. The reactions are highly atom-economical, as the majority of the atoms in the starting material are transferred to the product. Reactions generally involve a substrate containing a pendant nucleophilic group X (Scheme 1).[3–5] These groups can consist of alcohols, deactivated amines, enols, enolates, or enamines, although sulfides, nucleophilic arenes, and other electron-rich alkenes also participate in cyclization. The precursor is treated with a Lewis acidic, carbophilic transition metal that binds to one of the allene double bonds. The nature of the metal and the substitution pattern of the allene controls which of the double bonds prefers to coordinate to the catalyst. For example, metals with relatively large ionic radii, such as gold(I) supported by bulky ligands, may prefer to interact with the terminal or less sterically encumbered bond. In contrast, if the metal is a "harder" Lewis acid, such as palladium(II), coordination to the more electron-rich double bond is favored over a more sterically accessible alkene. Depending on the binding site that is favored, a subsequent *endo* or *exo* cyclization occurs to form the ring and produce a vinylmetal intermediate. Protodemetalation yields the final product and regenerates the metal catalyst.

Scheme 1 Cyclization of Allenes Using Soft Lewis Acids[3–5]

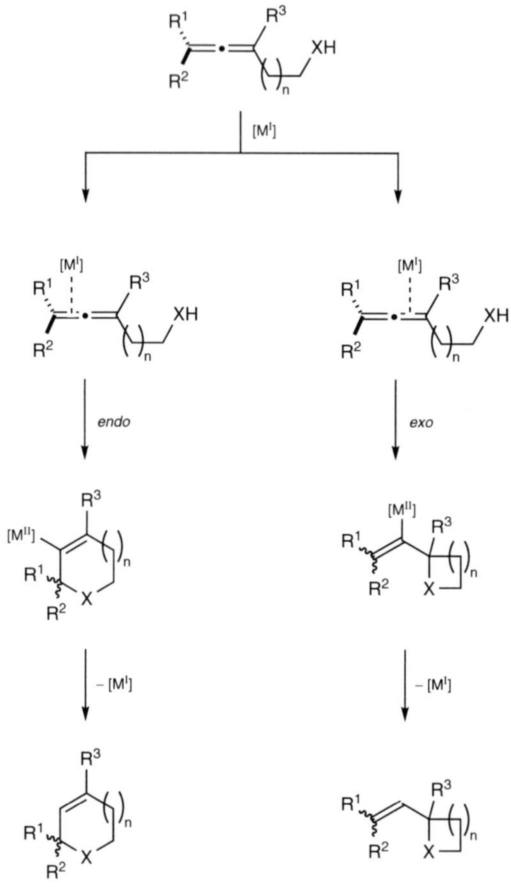

M = metal catalyst

Another commonly invoked reaction pathway for metal-catalyzed cyclizations of allenes is typically promoted by palladium catalysts. An advantage of this strategy is that the vinylmetal intermediate arising from cyclization can be trapped by a variety of other reactants used in palladium-catalyzed cross-coupling and π-allyl chemistry. However, predicting the product distribution and the stereochemical outcome can be difficult if coordination of the palladium is not selective for either the distal or the proximal double bond of the allene, as illustrated in Scheme 2 for a pendant amine nucleophile.[6] In addition to issues with regioselectivity in the coordination of the metal to the double bond, there are two competing pathways that lead to cyclization: aminopalladation or carbopalladation.

The two different intermediates **1** and **3** can be formed by aminopalladation, each resulting from activation of a different double bond of the allene (Scheme 2). If the arylpalladium halide reacts with the distal double bond, reaction occurs from the less hindered side to yield product **2A** in a stereospecific fashion, following reductive elimination. In contrast, reaction at the proximal double bond in an *exo* fashion yields the regioisomeric product **4**.

Considering the carbopalladation pathway, activation of the distal bond from the less hindered side, followed by *anti* cyclization by attack of the amine nucleophile on the η^3-allylpalladium, gives the *endo*-product **2A**. In contrast, if this same reaction occurs at the proximal double bond instead, *endo* cyclization to **2B** results, which has the opposite configuration as compared to **2A**. Cyclization in *exo* mode results in yet another pathway to **4**. Clearly, these different pathways will need to be modulated to access a desired product in high yield when using palladium catalysis.

for references see p 347

Scheme 2 Dual Roles in Catalysis of Allene Cyclization[6]

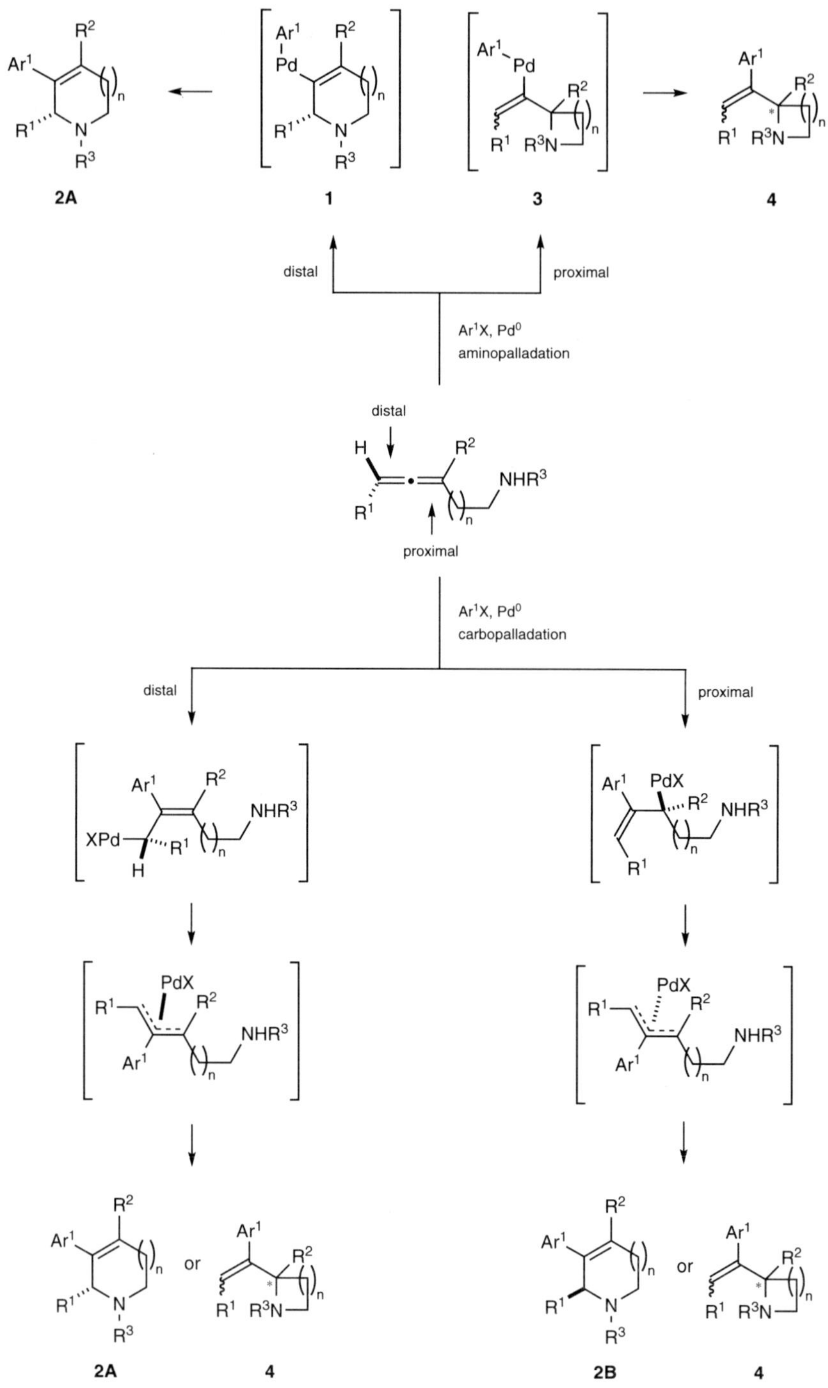

Overall, these unique features of allenes and the variety of activation modes have resulted in many powerful transformations to yield diverse heterocycles and carbocycles. The transformations discussed in this chapter will show the wide scope of cyclic products that can be accessed when utilizing each of these mechanistic paradigms.

1.6.1 Intramolecular C–O Bond Formation

1.6.1.1 Palladium-Catalyzed Synthesis of 2,3-Dihydrofurans from Allenic β-Oxo Esters

The coupling and subsequent cyclization of allenic β-oxo esters **5** with allyl bromides **6** using bis(dibenzylideneacetone)palladium(0) as a catalyst yields 2,3-dihydrofurans **7** via a 5-*exo* cyclization of the oxygen atom onto the allene (Scheme 3).[3] The reaction is easily performed at room temperature using potassium carbonate as the base; other bases fail to achieve the desired reaction. The beauty of this chemistry is that a variety of terminal and 1,2-disubstituted allyl halides can be employed as coupling partners to form an additional C—C bond following the metal-catalyzed cyclization event. Mechanistically, the palladium catalyst reacts with the allyl halide to yield a π-allyl complex that complexes to the more substituted and electron-rich double bond of the allene. Attack of the enolate oxygen at the proximal allene carbon in an *exo* fashion forms the heterocycle and a vinylpalladium species. Reductive elimination forms the final C—C bond and regenerates the active catalyst.

Scheme 3 Palladium-Catalyzed Synthesis of 2,3-Dihydrofurans from Allenic β-Oxo Esters and Allyl Bromides[3]

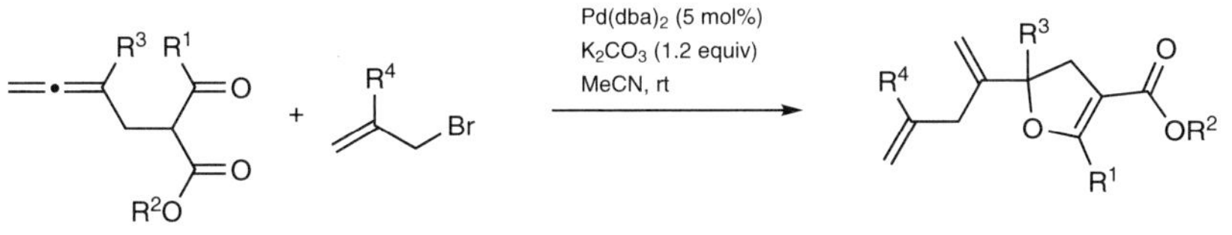

R1	R2	R3	R4	Yield (%)	Ref
Bn	Me	H	H	70	[3]
Me	Et	Bn	H	61	[3]
Et	Me	H	Ph	59	[3]
iPr	Me	H	Ph	51	[3]
iPr	Me	H	Bu	66	[3]
Me	*t*-Bu	H	Ph	52	[3]
Pr	Me	H	Ph	53	[3]

2,3-Dihydrofurans 7; General Procedure:[3]
A mixture of an allene **5** (0.25 mmol), $Pd(dba)_2$ (7.0 mg, 12.2 μmol), K_2CO_3 (42 mg, 0.30 mmol), and an allyl bromide **6** (0.50 mmol) in MeCN (2 mL) was stirred in a flame-dried Schlenk tube at rt for 13 h. After the reaction was complete (monitored by TLC), the solvent was removed by rotary evaporation and the residue was purified by flash chromatography (silica gel) to afford the product as a liquid.

1.6.1.2 Palladium-Catalyzed Synthesis of 2,5-Dihydrofurans from α-Hydroxyallenes

The use of palladium catalysis also permits tandem reactions that convert α-hydroxyallenes **8** and **9** into 2,5-dihydrofurans **10** via a heterodimeric coupling–cyclization reaction (Scheme 4).[7] In this transformation, coordination of palladium(II) iodide to the terminal, less hindered allene double bond of a 1,1′-disubstituted allene **8** is followed by a 5-*endo* attack of the hydroxy group at the terminal allene carbon. The intermediate vinylpalladi-

for references see p 347

um species produced in the cyclization reacts with a monosubstituted α-hydroxyallene **9** to give a π-allylpalladium species that undergoes *trans*-β-hydroxide elimination mediated by a Lewis acid to afford a 2,5-dihydrofuran **10** and hydroxypalladium(II) iodide [PdI(OH)]. Hydroxypalladium(II) iodide is converted back into the catalytically active palladium(II) iodide by reaction with the hydrogen iodide generated in the first step. When secondary α-hydroxyallenes **9** are used, high stereoselectivity for *E*-configured 2,5-dihydrofurans **10** is observed. The use of boron trifluoride–diethyl ether complex allows the reaction to proceed with only a slight excess of the α-hydroxyallene substrate **9** instead of multiple equivalents relative to the α-hydroxyallene substrate **8**.

Scheme 4 Palladium-Catalyzed Synthesis of 2,5-Dihydrofurans from α-Hydroxyallenes[7]

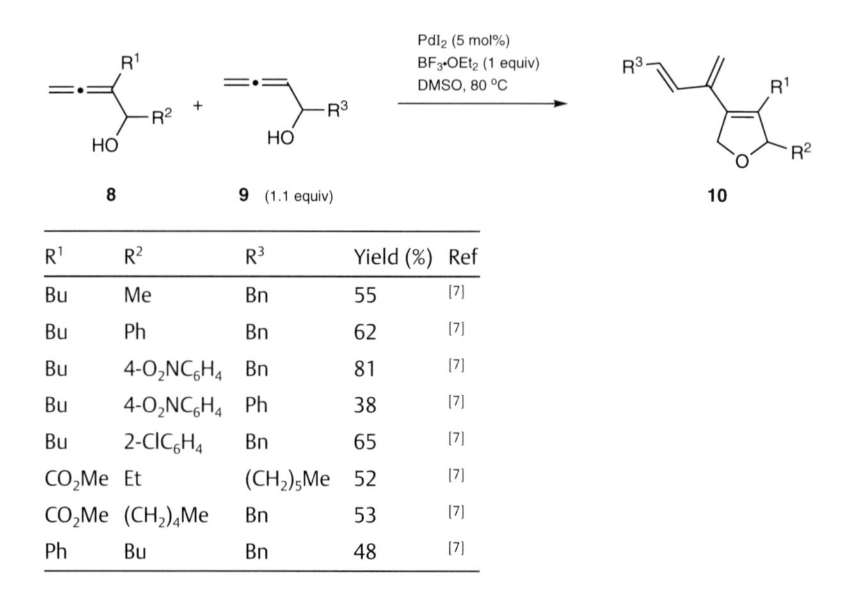

R^1	R^2	R^3	Yield (%)	Ref
Bu	Me	Bn	55	[7]
Bu	Ph	Bn	62	[7]
Bu	$4-O_2NC_6H_4$	Bn	81	[7]
Bu	$4-O_2NC_6H_4$	Ph	38	[7]
Bu	$2-ClC_6H_4$	Bn	65	[7]
CO_2Me	Et	$(CH_2)_5Me$	52	[7]
CO_2Me	$(CH_2)_4Me$	Bn	53	[7]
Ph	Bu	Bn	48	[7]

2,5-Dihydrofurans 10; **General Procedure:**[7]

BF_3•OEt_2 (127 μL, 1.0 mmol), an α-hydroxyallene substrate **8** (1.0 mmol), and DMSO (2.5 mL) were added sequentially to a mixture of PdI_2 (18.5 mg, 0.051 mmol, 5 mol%) and an α-hydroxyallene substrate **9** (1.11 mmol) in DMSO (2.5 mL), and the mixture was stirred at 80 °C for 1.5 h. After the reaction had gone to completion, as determined by TLC, the mixture was cooled to rt and the reaction was quenched with H_2O (10 mL). The mixture was extracted with Et_2O (3 × 25 mL). The combined organic layers were washed with sat. aq $Na_2S_2O_3$ and brine, dried (Na_2SO_4), and concentrated. The residue was purified by column chromatography (silica gel) to afford the product as an oil.

1.6.1.3 Gold-Catalyzed Cyclization of Allenoates To Form Functionalized Butenolides

Heterocycles can also be formed using soft Lewis acids, most notably gold, to promote allene cyclization. For example, cyclization of *tert*-butyl allenoates **11** to give substituted butenolides [furan-2(5*H*)-ones] **12** is observed using a low catalyst loading of gold(III) chloride (Scheme 5),[8] comparing favorably to previous methods that require stoichiometric amounts of electrophilic reagents and hydrolysis of the allene ester to the acid.[9,10] These mild reaction conditions have the advantage of preventing problems with isomerization of allenes to alkynyl acetic acid derivatives and racemization of axial chirality under hydrolytic conditions. The most likely mechanistic scenario involves the use of gold(III) to promote cleavage of the *tert*-butyl group to generate an allenic acid derivative, which immediately cyclizes to the butenolide through a pathway similar to that described in Scheme 1.

Scheme 5 Gold-Catalyzed Synthesis of Functionalized Butenolides from Allenoates[8]

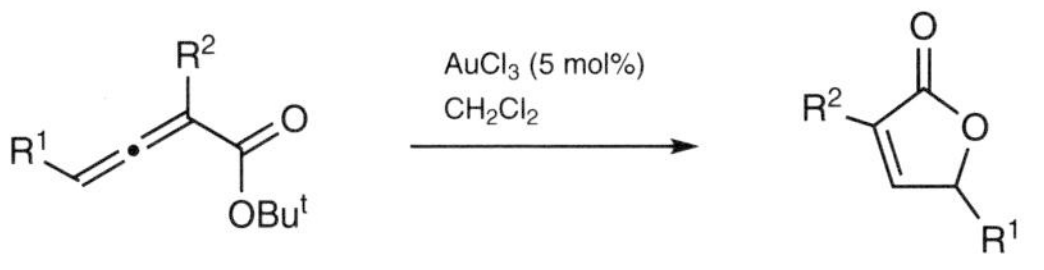

$AuCl_3$ (5 mol%), CH_2Cl_2; **11** → **12**

R^1	R^2	Temp (°C)	Time (h)	Yield (%)	Ref
Bn	H	80	2	88	[8]
Bn	Me	80	5	77	[8]
Bn	Bn	80	0.17	72	[8]
Bn	Bn	rt	1.5	96	[8]
Ph	Bn	80	20	32	[8]
$CH_2CH{=}CH_2$	H	80	2	57	[8]
$CH_2CH{=}CH_2$	Bn	80	0.5	81	[8]
$(CH_2)_8Me$	H	80	1.5	65	[8]

Furan-2(5*H*)-ones 12; General Procedure:[8]
To a soln of a *tert*-butyl allenoate **11** (0.20 mmol) in anhyd CH_2Cl_2 (1 mL) was added $AuCl_3$ (3.0 mg, 10 μmol) and the mixture was stirred at 80 °C in a sealed tube. Upon completion of the reaction, as indicated by TLC, the solvent was removed under reduced pressure. The resulting oil was purified by column chromatography (silica gel, EtOAc/hexane) to provide the product as a colorless oil.

1.6.1.4 Gold-Catalyzed Cyclization of α-Hydroxyallenes to 2,5-Dihydrofurans

Densely functionalized tri- and tetrasubstituted 2,5-dihydrofurans **14** can be obtained in good yield from α-hydroxyallenes **13** using gold(III) chloride as a catalyst (Scheme 6).[11] This electrophilic cyclization method is effective for both alkyl and alkenyl substrates, as well as sterically hindered allenes, with little difference noted between the performance of tri- and tetrasubstituted substrates in the reaction. Compared to previous silver-catalyzed methods, this strategy improves the rate of the reaction and significantly expands the substrate scope to include those containing difficult alcohol and silyl ether functionalities.[12] These gold-catalyzed conditions are also milder than reported cyclizations that employ acidic resins. Importantly, these reactions all proceed with perfect stereocontrol and the dihydrofuran products are obtained with >90% diastereomeric excess.

Scheme 6 Gold-Catalyzed Synthesis of 2,5-Dihydrofurans from α-Hydroxyallenes[11]

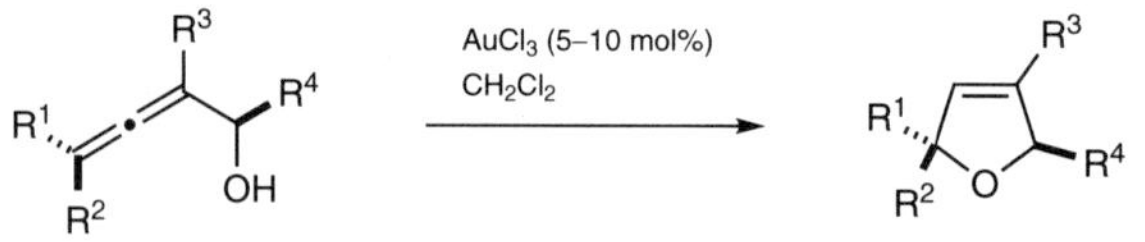

$AuCl_3$ (5–10 mol%)
CH_2Cl_2

13 → **14**

R^1	R^2	R^3	R^4	$AuCl_3$ (mol%)	Yield (%)	Ref
t-Bu	Me	H	CO_2Et	5	74	[11]
t-Bu	Me	Me	CO_2Et	10	94	[11]
t-Bu	H	Me	CO_2Me	5	78	[11]
t-Bu	Me	H	CH_2OH	5	24	[11]
t-Bu	H	Me	$CH_2OTBDMS$	7	95	[11]
H	Me	Me	$CH_2OTBDMS$	5	77	[11]
t-Bu	Me	Me	CH_2OMe	10	90	[11]
$(CH_2)_2CH{=}CH_2$	Me	Me	CH_2OMe	10	86	[11]

2,5-Dihydrofurans 14; **General Procedure:**[11]
Note: $AuCl_3$ is hygroscopic; the reaction proceeds slowly with material that has been exposed to moisture. To a soln of an α-hydroxyallene **13** (2.0 mmol) in anhyd CH_2Cl_2 (5 mL) was added $AuCl_3$ (5–10 mol%) under argon. The mixture was stirred at rt and monitored by TLC. After completion of the reaction, the solvent was evaporated under reduced pressure and the residue was purified by flash chromatography over a short column (silica gel, cyclohexane/Et_2O 10:1).

1.6.2 Intramolecular C—N Bond Formation

1.6.2.1 Rhodium-Catalyzed Synthesis of Nitrogen-Containing Stereotriads from Allenes

Nitrogen-containing stereotriads with the general structural motif C—X/C—N/C—Z (X, Z = heteroatomic groups based on sulfur, nitrogen, halogens, or oxygen) can be readily synthesized from allenic sulfamates in a two-step sequence (Scheme 7).[13] The allenic sulfamates (e.g., **15**) are rapidly cyclized to bicyclic methyleneaziridines (e.g., **16**) using dinuclear rhodium(II) complexes [e.g., rhodium(II) triphenylacetate dimer] with excellent regioselectivity for the proximal allene double bond. Only the *E*-configured bicyclic methylene aziridine product is observed; subsequent in situ aziridine ring-opening reaction with a diverse array of nucleophiles delivers exclusively *E*-enesulfamates. Trapping with an electrophile yields 1,2,3-oxathiazepane 2,2-dioxides **17** as nitrogen-containing stereotriads in a single reaction vessel. A wide variety of functional groups can be employed as the nucleophile and electrophile, providing a general and modular method for the oxidative transformation of simple allenes to densely functionalized amine stereotriads with excellent levels of chemo-, regio-, and stereoselectivity. Transfer of the axial chirality of the substrate **15** to point chirality in the product **17** with >98% fidelity provides access to enantioenriched amines.[13]

Scheme 7 Rhodium-Catalyzed Synthesis of Nitrogen-Containing Stereotriads from Allenic Sulfamates[13]

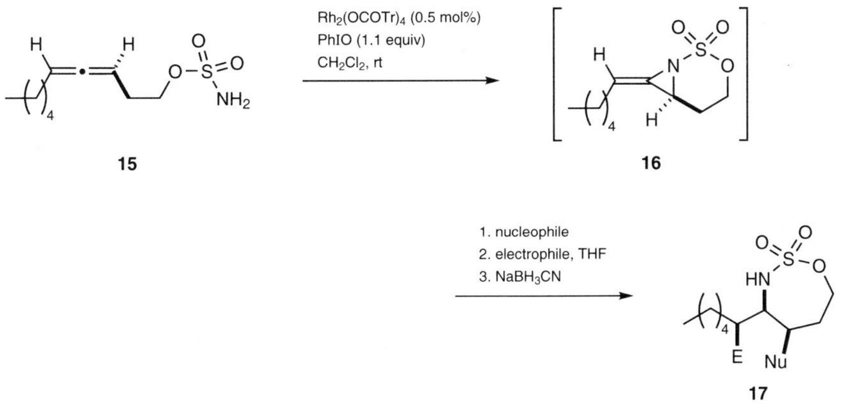

$Rh_2(OCOTr)_4$ (0.5 mol%)
PhIO (1.1 equiv)
CH_2Cl_2, rt

15 → 16

1. nucleophile
2. electrophile, THF
3. $NaBH_3CN$

→ 17

Nucleophile	Nu	Electrophile	E	Time (min)	Temp (°C)	Yield (%)	dr	Ref
AcOH	OAc	NBS	Br	120	rt	60	5:1	[13]
AcOH	OAc	NBS	Br	15	0	61	20:1	[13]
MeOH	OMe	NBS	Br	45	0	60	1.7:1	[13]
MeOH	OMe	NBS	Br	10	−10	58	2.6:1	[13]
MeOH	OMe	DIAD[a]	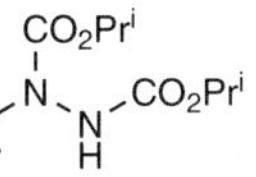 $N(CO_2Pr^i)NHCO_2Pr^i$	120	70	64	4.6:1	[13]
MeOH	OMe	PhSCl	SPh	30	rt	74	2.6:1	[13]

[a] DIAD = diisopropyl azodicarboxylate; Celite filtration before addition of DIAD.

1,2,3-Oxathiazepane 2,2-Dioxides 17; General Procedure:[13]
The allenic sulfamate **15** (0.441 mmol) and rhodium(II) triphenylacetate dimer (3.0 mg, 2.15 μmol, 0.5 mol%) were added to a dry 25-mL round-bottomed flask. The mixture was kept under N_2 and CH_2Cl_2 (4.4 mL) was added. The resulting blue-green mixture was stirred for 5 min at rt, and then powdered 4-Å molecular sieves (100 mg) were added in one portion. The mixture was stirred for 5 min at rt, and PhIO (113 mg, 0.515 mmol, 1.2 equiv) was added in one portion. After a total reaction time of 90 min, TLC of the mixture indicated consumption of the sulfamate **15**. The nucleophile (2.15 mmol, 5.0 equiv) was added in one portion at rt, and the mixture was heated at reflux for 3.5 h. The soln was concentrated by rotary evaporation. THF (6 mL) was added, and the mixture was cooled to 0 °C under N_2. The electrophile (0.644 mmol, 1.5 equiv) was added in one portion, and the mixture was stirred at 0 °C. After 15 min, $NaBH_3CN$ (81 mg, 1.29 mmol, 3.0 equiv) dissolved in anhyd MeOH (3 mL) was added, and the mixture was warmed to rt, stirred for 1 h, and concentrated by rotary evaporation. EtOAc (15 mL) was added to the residue and the mixture was washed with sat. aq $NaHCO_3$ (2 × 10 mL) and brine (2 × 10 mL). The organic layer was dried (Na_2SO_4) and concentrated by rotary evaporation. The residue was purified by chromatography (silica gel) to give the product as a white solid.

1.6.2.2 Gold-Catalyzed Enantioselective Intramolecular Hydroamination of Allenes with Ureas

Enantioselective hydroamination has traditionally been a challenging transformation, but gold-catalyzed approaches are uniquely suited to this task. A 1:2 mixture of the dimeric enantioenriched gold catalyst **18** and silver(I) tetrafluoroborate catalyzes the cyclization of allene-containing ureas **19** to give pyrrolidines **20** with up to 93% enantiomeric excess (Scheme 8).[14] The reaction is sensitive to the nature of the solvent, with diethyl ether giving the highest yields and enantiomeric excesses. The insolubility of the substrates in diethyl ether suggests that low concentration is important to the success of the reaction. This expands the scope from previous hydroamination reactions, which are generally performed with carbamates as the nucleophile. However, attempts to utilize gold(I)-catalyzed dynamic kinetic enantioselective hydroamination with trisubstituted allenes resulted in low diastereoselectivities due to the slower rate of allene isomerization as compared to the rate of cyclization, indicating that hydroamination occurs mainly through a static, catalyst-controlled pathway.

Scheme 8 Gold-Catalyzed Enantioselective Synthesis of Pyrrolidines from Allenyl Ureas[14]

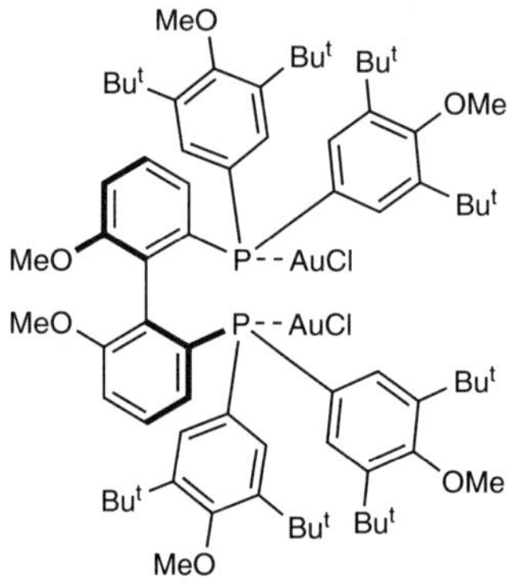

MeO, Bu^t, Bu^t Bu^t, OMe, Bu^t, MeO, P- -AuCl, MeO, P- -AuCl, Bu^t, OMe, Bu^t, Bu^t Bu^t, MeO

18

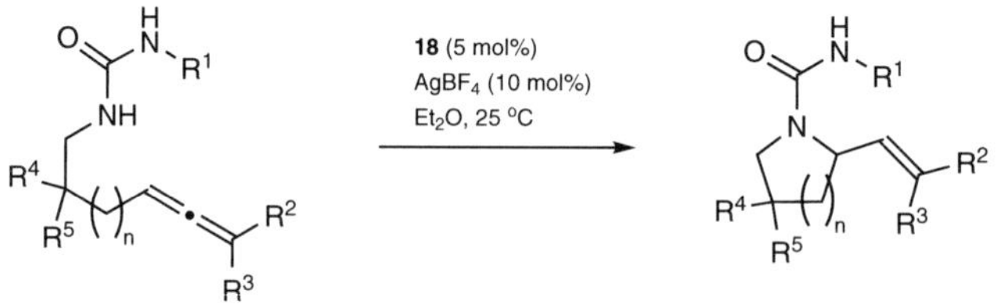

19 → **18** (5 mol%), $AgBF_4$ (10 mol%), Et_2O, 25 °C → **20**

R^1	R^2	R^3	R^4	R^5	n	Time (h)	Yield (%)	ee[a] (%)	Ref
$4\text{-}O_2NC_6H_4$	H	H	Ph	Ph	1	48	90	93	[14]
Ph	H	H	Ph	Ph	1	10	98	82	[14]
$4\text{-}MeOC_6H_4$	H	H	Ph	Ph	1	16	90	72	[14]
Bu	H	H	Ph	Ph	1	20	90	50	[14]
$4\text{-}O_2NC_6H_4$	Me	Me	Ph	Ph	1	48	91	56	[14]
$4\text{-}O_2NC_6H_4$	H	H	Ph	Ph	2	48	89	53	[14]
$4\text{-}O_2NC_6H_4$	H	H	H	H	1	48	82	7	[14]

[a] Absolute stereochemistry unknown.

1-Amido-2-vinylpyrrolidines 20; General Procedure:[14]
A suspension of a urea **19** (0.05 mmol), chiral gold catalyst **18** (4.0 mg, 2.5 μmol), and $AgBF_4$ (1.0 mg, 5 μmol) in Et_2O (0.5 mL) was stirred at rt for the indicated time. The crude mixture was loaded directly onto a column (silica gel) and chromatographed.

1.6.2.3 Gold-Catalyzed Enantioselective Intramolecular Hydroamination of Allenes with Sulfonamides

The enantioselective hydroamination of unactivated allenes is a challenging goal in transition-metal catalysis, further compounded by the preferred linear geometry of many of the gold(I) catalysts used in this chemistry. A solution to this problem is achieved by replacing the typical chloride, tetrafluoroborate, and benzoate counteranions in chiral dinuclear phosphinegold(I) complexes with 4-nitrobenzoate. This change imparts enantioselectivities of up to 99% in the hydroamination of allenes **22** with the chiral gold(I) 4-nitrobenzoate catalyst **21** via this remarkable counterion effect (Scheme 9).[15] Simple amine substrates are cyclized in high yields to pyrrolidines **23**, with particularly flexible substitution at the allene terminus. Extending the tether length between the amine and the allene by one carbon allows for the stereocontrolled synthesis of piperidines. An expanded substrate scope can be achieved with ligand reoptimization for each group of substrates.

Scheme 9 Gold-Catalyzed Synthesis of Pyrrolidines from Simple Allene-Containing Amines[15]

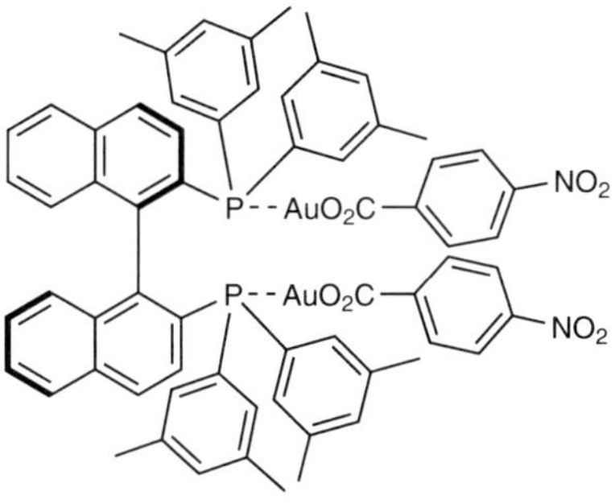

P--AuO2C, NO2, P--AuO2C, NO2, 21

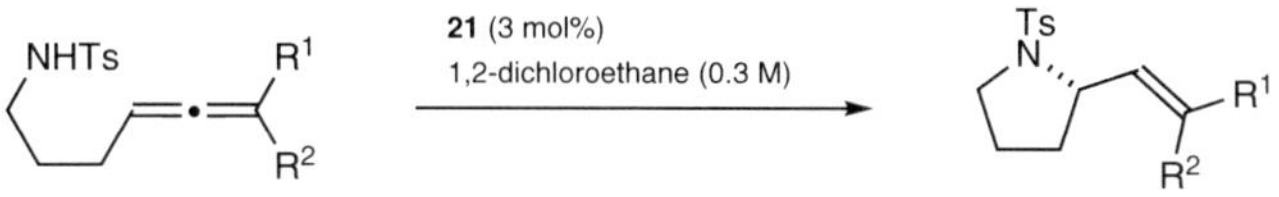

NHTs, R1, R2, 22; 21 (3 mol%), 1,2-dichloroethane (0.3 M); Ts, N, R1, R2, 23

R^1	R^2	Temp	Time (h)	Yield (%)	ee (%)	Ref
Me	Me	rt	15	98	99	[15]
$(CH_2)_4$		rt	15	75	83	[15]
$(CH_2)_5$		rt	17	88	98	[15]
$(CH_2)_6$		rt	15	88	98	[15]
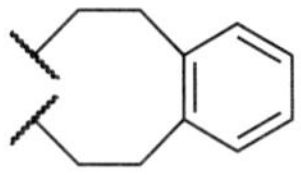 [structure: benzo-fused ring]		rt	25	80	98	[15]
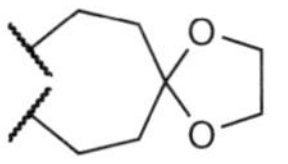 [structure: dioxolane spiro ring, O, O]		50 °C	25	79	98	[15]

Pyrrolidines 23; General Procedure:[15]
To a soln of a tosylamine **22** (1 equiv) in 1,2-dichloroethane (0.30 M) was added the chiral gold catalyst **21** (3 mol%). The resulting homogeneous mixture was protected from ambient light and stirred at rt for the indicated time. Upon completion of the reaction, as indicated by TLC, the soln was loaded directly onto a column (silica gel). Purification by flash column chromatography afforded the product.

1.6.2.4 Silver-Catalyzed Synthesis of 4-Vinyloxazolidin-2-ones from Allenyl Carbamates

Early studies in metal-catalyzed cyclizations of allenes focused heavily on silver catalysis. Carbamates were typically used in these reactions but require an electron-withdrawing group, such as a toluenesulfonyl or acyl group, on the nitrogen. For example, a series of 4-vinyloxazolidin-2-ones **25** are synthesized from buta-2,3-dienyl carbamates **24** using either silver(I) trifluoromethanesulfonate or silver(I) isocyanate as the catalyst (Scheme 10).[4] The presence of triethylamine is essential and promotes the addition of nitrogen to the proximal allene carbon in good yield under very simple conditions. Both *trans*- and *cis*-oxazolidin-2-ones **25** can be produced from this reaction, with a higher diastereomeric ratio observed when more sterically bulky substituents are installed at the carbamate-bearing carbon. This can be mechanistically explained through the avoidance of gauche interactions in the transition state of the cyclization and gives a *trans*/*cis* ratio of 1.8:1 when R^1 = Me, and up to >30:1 when R^1 = *t*-Bu.

Scheme 10 Silver-Catalyzed Synthesis of 4-Vinyloxazolidin-2-ones from Allenyl Carbamates[4]

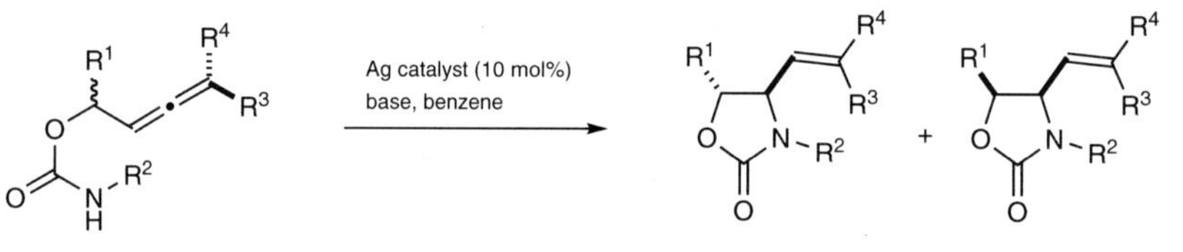

Ag catalyst (10 mol%)
base, benzene

24 → **25A** + **25B**

R^1	R^2	R^3	R^4	Base	Catalyst	Temp (°C)	Time (h)	Yield (%)	Ratio (**25A**/**25B**)	Ref
H	Ts	H	H	Et_3N	AgNCO	50	6	78	–	[4]
Et	Ts	H	H	Et_3N	AgOTf	50	6	74	1.5:1	[4]
iPr	Ts	H	H	Et_3N	AgOTf	50	43	74	7.1:1	[4]
t-Bu	Ts	H	H	Et_3N	AgOTf	50, then reflux	41	53	>30:1	[4]
Et	Ts	Me	Me	Et_3N	AgOTf	50	32	57	4.9:1	[4]
Me	Ts	Me	H	Et_3N	AgNCO	50	5	91	1.8:1	[4]
Et	Ac	H	H	*t*-BuOK	AgNCO	reflux	49	34	1.8:1	[4]

4-Vinyloxazolidin-2-ones 25; General Procedure:[4]
A flask containing a buta-2,3-dienyl carbamate **24** (1 mmol) and AgNCO (0.1 mmol) was purged with argon. Benzene (6 mL) (**CAUTION:** *carcinogen*) and Et_3N (0.1 mmol) were added. The heterogeneous mixture was stirred at 50 °C or at reflux for 6–49 h. After workup, a mixture of diastereomers was isolated by column chromatography (silica gel, benzene). The pure *trans*-product **25A** was obtained by recrystallization (benzene/hexane).

1.6.2.5 Silver-Catalyzed Dynamic Kinetic Enantioselective Intramolecular Hydroamination of Allenes

Another advantage of employing allenes in metal-catalyzed cyclizations is the potential for carrying out dynamic kinetic asymmetric transformations. If the axial chirality of the allene substrate undergoes racemization at a rate faster than that of cyclization, there is a good chance that a racemic allene substrate might be converted into an enantioenriched product in the presence of an asymmetric catalyst. For example, the reaction of trisubstituted γ-allenylcarbamates **26** (Scheme 11) in the presence of the chiral gold(I) catalyst **18** results in a dynamic kinetic enantioselective hydroamination.[5] This reaction produces chiral, nonracemic pyrrolidines **27**; the observed enantioselectivity develops from matched and mismatched pairings of the substrate isomers with the catalyst. This dynamic kinetic asymmetric transformation is unique in that it involves both C—X bond formation and addition across a C=C bond, two features that are rare in typical dynamic kinetic asymmetric transformations. It is hypothesized that the trisubstituted allene helps to retard the rate of C—N bond formation relative to racemization, resulting in this enantioselective reaction, which proceeds in moderate to high yields for a variety of trisubstituted allenes with enantiomeric excesses reaching up to 96%.

Scheme 11 Silver-Catalyzed Dynamic Kinetic Enantioselective Hydroamination of Allenes To Give Pyrrolidines[5]

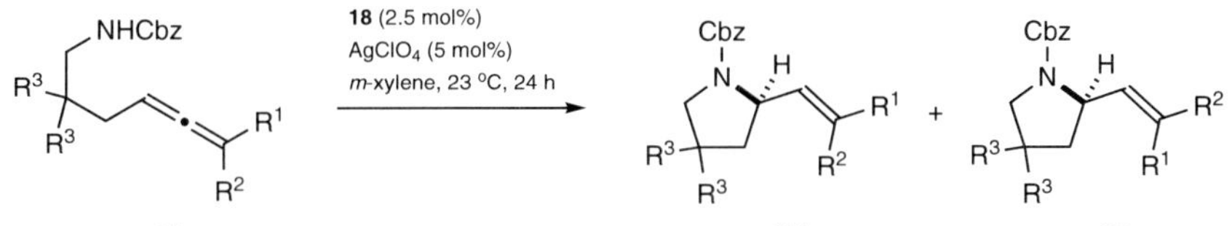

NHCbz; R^3; R^3; R^1; R^2; *rac*-**26**

18 (2.5 mol%)
$AgClO_4$ (5 mol%)
m-xylene, 23 °C, 24 h

Cbz; N; H; R^1; R^2; R^3; R^3; (*Z*)-**27** + Cbz; N; H; R^2; R^1; R^3; R^3; (*E*)-**27**

R^1	R^2	R^3	Yield[a] (%)	Ratio (*Z*/*E*)	ee (%)		Ref
					(*Z*)-**27**	(*E*)-**27**	
Me	Et	Ph	94	3.1:1	96	76	[5]
Me	$(CH_2)_5Me$	Ph	99	10.1:1	91	9	[5]
Me	iBu	Ph	99[b]	2.6:1	87	54	[5]
Me	iPr	Ph	94	2.0:1	95	67	[5]
Me	*t*-Bu	Ph	52[c]	≤1:25	2	–	[5]
Et	$(CH_2)_5Me$	Ph	86	4.3:1	84	47	[5]
Me	$(CH_2)_5Me$	H	87	2.4:1	75	45	[5]

[a] Yield of isolated material of >95% purity.
[b] Reaction run at 0 °C for 24 h followed by 23 °C for 24 h.
[c] Reaction run at 60 °C for 212 h followed by 100 °C for 48 h.

1-(Benzyloxycarbonyl)-2-vinylpyrrolidines 27**; General Procedure:**[5]
A mixture of chiral gold catalyst **18** (3.8 μmol) and $AgClO_4$ (7.5 μmol) in *m*-xylene (0.2 mL) was stirred at rt for 5 min, and then treated with a soln of a carbamate **26** (0.15 mmol) in *m*-xylene (0.3 mL). The mixture was stirred at rt for 24 h, and purified by column chromatography. The enantiomeric purities were determined by chiral HPLC analysis. The stereochemistry of all pyrrolidines was verified by NOE analysis.

1.6.2.6 Gold-Catalyzed Cycloisomerization of α-Aminoallenes to 2,5-Dihydropyrroles

Analogous to the chemistry described in Section 1.6.1.4, the cycloisomerization of α-aminoallenes **28** can be catalyzed by a gold(III) complex to form 2,5-dihydro-1*H*-pyrroles **29** (Scheme 12).[10] This system utilizes low catalyst loadings in the order of 2 mol% gold(III) chloride to produce the cyclized products in good to excellent yields. Interestingly, the identity of the N-protecting group (R^3) affects not only the reactivity, but also the chirality transfer of the reaction. Amines containing either no protecting group or a strongly electron-withdrawing group on the nitrogen give excellent chirality transfer, but the use of an acetyl or *tert*-butoxycarbonyl group on the nitrogen significantly decreases the diastereomeric ratio of the 2,5-dihydro-1*H*-pyrrole products. Although both protected and unprotected amine functionalities are tolerated in the cyclization, unprotected aminoallenes require a significantly longer reaction time (5 days vs. 30 minutes for a tosyl-protected amine). Mechanistically, the gold(III) chloride catalyzed cycloisomerization is believed to occur via coordination of the carbophilic gold catalyst to the distal double bond of the allene, followed by formation of a metallacyclopropane. The increased electrophilicity of the allene favors cyclization via an S_N2-type transition state; subsequent proton transfer produces the 2,5-dihydro-1*H*-pyrrole with good axial-to-center chirality transfer. However, when the nitrogen is protected with an oxygen-containing group, stabilization of a zwitterionic intermediate can lead to partial isomerization which lowers the diastereoselectivity.

Scheme 12 Gold(III)-Catalyzed Cycloisomerization of α-Aminoallenes to 2,5-Dihydro-1*H*-pyrroles[10]

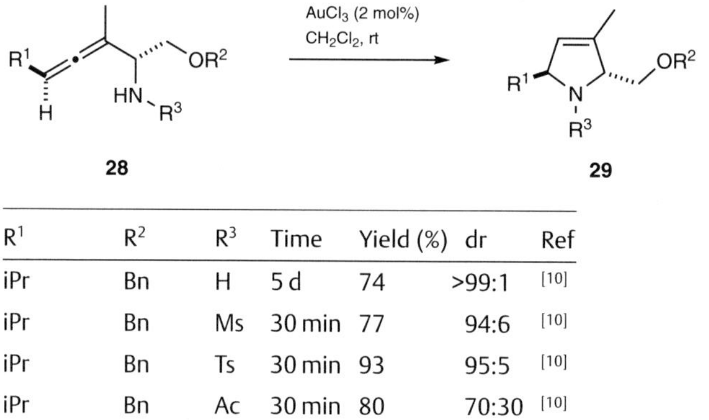

$AuCl_3$ (2 mol%)
CH_2Cl_2, rt

28 → **29**

R^1	R^2	R^3	Time	Yield (%)	dr	Ref
iPr	Bn	H	5 d	74	>99:1	[10]
iPr	Bn	Ms	30 min	77	94:6	[10]
iPr	Bn	Ts	30 min	93	95:5	[10]
iPr	Bn	Ac	30 min	80	70:30	[10]
iPr	Bn	Boc	30 min	69	46:54	[10]
Me	Bn	H	5 d	71	90:10	[10]
$(CH_2)_5Me$	TBDMS	H	5 d	82	85:15	[10]
Ph	TBDMS	H	5 d	79	>99:1	[10]

2,5-Dihydro-1*H*-pyrroles 29; General Procedure:[10]
To a soln of an α-aminoallene **28** (0.17 mmol) in anhyd CH_2Cl_2 (5 mL) was added a 0.165 M soln of $AuCl_3$ in MeCN (20 μL, 3.3 μmol, 2 mol%) at rt under argon. The reaction was monitored by TLC and, upon completion, the solvent was removed under reduced pressure. The crude product was purified by column chromatography ($MeOH/CH_2Cl_2$ 1:10).

1.6.3 C–C Bond Formation

1.6.3.1 Palladium-Catalyzed Carbocyclization of Allenes

Allenes are excellent substrates for achieving tandem reactions that convert simple precursors into products of significantly increased complexity. Cyclization reactions of allenes involving carbopalladation represent a powerful strategy to construct complex carbocycles. The diastereoselective carbocyclization reactions of allenes **30** tethered to aldehyde or ketone functionalities are achieved via cooperative catalysis of pyrrolidine and palladium(II) acetate (Scheme 13).[16] This method represents a logical extension of alkyne carbocyclization chemistry and is based on the activation of the tethered carbonyl group by the amine, while the allene is activated by palladium.[16] Mechanistically, reaction of the amine with the carbonyl forms an enamine, while the palladium catalyst binds to the proximal double bond of the allene. Nucleophilic attack of the enamine carbon on the allene in a 5-*exo* fashion yields cyclopentanes or pyrrolidines **31** with high diastereoselectivity via catalyst control. The scope of the reaction is broad and includes some substrates with a nitrogen atom in the chain; however, the best results are achieved with allenes that contain all-carbon chains and geminal substitution, presumably due to a Thorpe–Ingold effect. As expected, the *trans*-product is preferred in ratios typically greater than 10:1.

Scheme 13 Palladium-Catalyzed Carbocyclization of Allenyl Aldehydes to Cyclopentanes or Pyrrolidines[16]

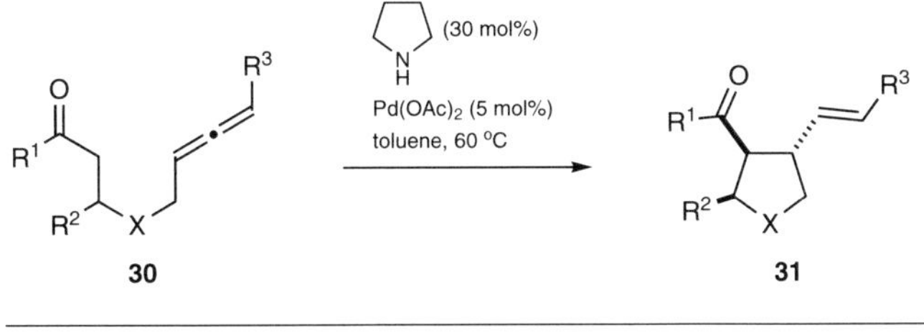

R^1	R^2	R^3	X	Time (h)	Yield[a] (%)	dr[b]	Ref
H	H	H	$C(CO_2Me)_2$	12	68	13:1	[16]
H	H	Me	$C(CO_2Me)_2$	24	60	12:1	[16]
H	H	H	$C(CH_2OMe)_2$	16	65	14:1	[16]
H	H	H	$C(CO_2Me)COEt$	12	71	22:16:1.5:1	[16]
H	H	H	NTs	16	56	10:1	[16]
$(CH_2)_2$		H	$C(CO_2Me)_2$	14	75	14:1	[16]
$(CH_2)_3$		H	$C(CO_2Me)_2$	20	78	2:1[c]	[16]

[a] Isolated yield of diastereomers.
[b] Determined by ^1H NMR spectroscopy after flash column chromatography.
[c] Separable isomers.

When a chiral diarylprolinol-based organocatalyst (e.g., **32**) is used instead of the pyrrolidine catalyst, the reaction is rendered enantioselective, affording cyclopentane and pyrrolidine products **34** with moderate enantiomeric excesses for a variety of substituted allenes **33** (Scheme 14).[16] The steric bulk of the silyl group is an important factor for improved enantioselectivity.

for references see p 347

Scheme 14 Palladium-Catalyzed Enantioselective Carbocyclization of Allenyl Aldehydes to Cyclopentanes or Pyrrolidines[16]

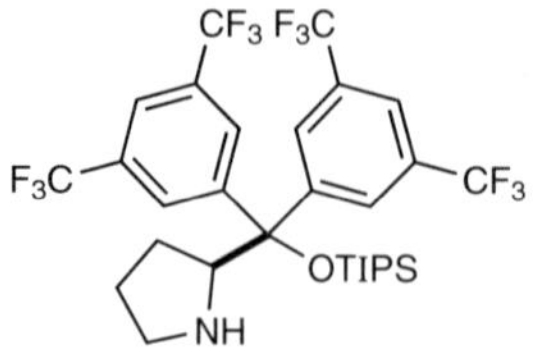

CF_3 F_3C F_3C CF_3 OTIPS NH **32**

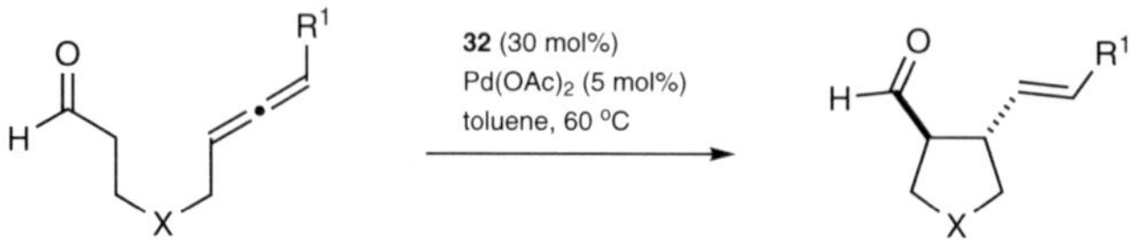

33 → **32** (30 mol%), $Pd(OAc)_2$ (5 mol%), toluene, 60 °C → **34**

R^1	X	Yield[a] (%)	dr[b]	ee[c] (%)	Ref
H	$C(CO_2Me)_2$	72	13:1[d]	82	[16]
Me	$C(CO_2Me)_2$	65	20:1	51	[16]
H	NTs	48	16:1	63	[16]
H	NBoc	51	18:1	79	[16]

[a] Isolated yield of two diastereomers.
[b] Determined by ^{1}H NMR spectroscopy after flash column chromatography.
[c] Determined by chiral HPLC analysis of the purified benzoate ester derivatives.
[d] The absolute stereochemistry of (3*R*,4*S*) was determined by single-crystal X-ray analysis of a derivative.

Racemic Carbocyclization Products 31**; General Procedure:**[16]
In a sealed vial, an allene substrate **30** (0.20 mmol) and pyrrolidine (5.0 μL, 0.06 mmol) were added to a soln of $Pd(OAc)_2$ (0.01 mmol) in toluene (2 mL). The mixture was stirred at 60 °C and monitored by TLC and, upon completion of the reaction, the mixture was concentrated under reduced pressure. The crude material was purified by flash column chromatography (silica gel).

Nonracemic Carbocyclization Products 34**; General Procedure:**[16]
In a sealed vial, an allene substrate **33** (0.20 mmol) and chiral organocatalyst **32** (39.9 mg, 0.06 mmol) were added to a soln of $Pd(OAc)_2$ (0.01 mmol) in toluene (2 mL). The mixture was stirred at 60 °C and monitored by TLC and, upon completion of the reaction, the mixture was concentrated under reduced pressure. The crude material was purified by flash column chromatography (silica gel).

For the preparation of benzoate ester derivatives for chiral HPLC analysis, $NaBH_4$ (11.8 mg, 0.310 mmol) was added to a soln of a carbocyclization product **34** (0.125 mmol) in MeOH (1.3 mL) at 0 °C. The mixture was stirred at 0 °C for 30 min and diluted with Et_2O, and sat. aq NH_4Cl was slowly added. The organic layer was washed with H_2O, dried (Na_2SO_4), filtered, and concentrated under reduced pressure. The residue was dissolved in CH_2Cl_2 (1.3 mL), and Et_3N (52 μL, 0.375 mmol), BzCl (29.0 μL, 0.250 mmol), and DMAP (1.5 mg, 0.013 mmol) were added. The mixture was warmed to rt and stirred for 16 h. H_2O was added and the organic layer was separated. The aqueous layer was extracted with CH_2Cl_2, and the combined organic layers were washed with brine, dried (Na_2SO_4), filtered, and concentrated under reduced pressure. The residue was purified by flash column chromatography (silica gel) to afford the benzoate ester for HPLC analysis.

1.6.3.2 Rhodium- or Copper-Catalyzed Divergent Carbene Reactivity with Allenic Diazo Esters

Divergent chemoselective transformations of allenic diazo esters to give either C—H insertion products or methylenecyclopropanes can be achieved by judicious choice of metal. Dinuclear rhodium catalysts, such as dirhodium(II) acetate or 3,3′-(1,4-phenylene)bis(2,2-dimethylpropanoate) complex **35** [$Rh_2(esp)_2$], facilitate chemoselective C—H insertion of allenic diazo esters **36** to form allene derivatives **37** bearing a pendant lactone (X = O) or cyclopentanone (X = CH_2) moiety with excellent *trans* diastereoselectivity (Scheme 15).[17]

Scheme 15 Rhodium-Catalyzed C—H Insertion of Allenic Diazo Esters To Give Allenyl Furan-2-ones[17]

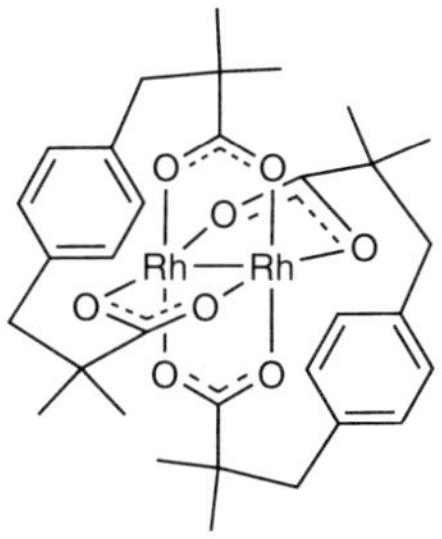

35 $Rh_2(esp)_2$

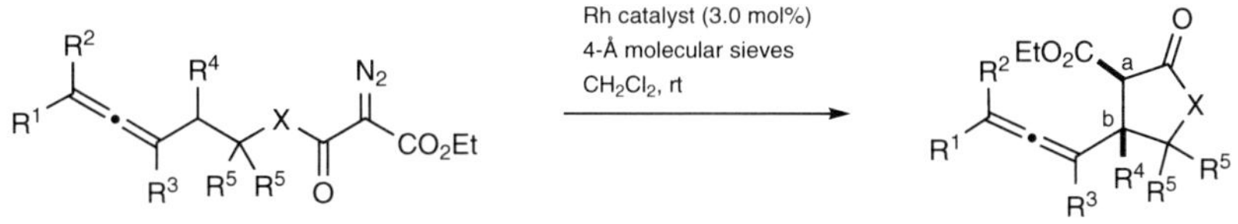

36 → **37**: Rh catalyst (3.0 mol%), 4-Å molecular sieves, CH_2Cl_2, rt

R^1	R^2	R^3	R^4	R^5	X	Catalyst	Yield (%)	dr[a]	Ref
$(CH_2)_2Ph$	H	H	H	H	O	$Rh_2(OAc)_4$	70	10:1	[17]
$(CH_2)_2Ph$	H	H	H	H	O	$Rh_2(esp)_2$	80	10:1	[17]
$(CH_2)_4Me$	H	H	H	H	O	$Rh_2(OAc)_4$	73	10:1	[17]
$(CH_2)_2OTBDMS$	H	H	H	H	O	$Rh_2(OAc)_4$	76	6:1	[17]
Pr	Pr	H	H	H	O	$Rh_2(esp)_2$	65	>10:1	[17]
Pr	Pr	H	H	H	O	$Rh_2(OAc)_4$	88	>10:1	[17]
$(CH_2)_4Me$	H	Me	H	H	O	$Rh_2(OAc)_4$	74	>10:1	[17]
Ph	Me	H	H	H	O	$Rh_2(OAc)_4$	34	4.4:1	[17]
$(CH_2)_4Me$	H	H	H	Me	O	$Rh_2(OAc)_4$	70	>10:1	[17]
$(CH_2)_4Me$	H	H	H	Me	O	$Rh_2(esp)_2$	97	>10:1	[17]
$(CH_2)_4Me$	H	H	H	H	CH_2	$Rh_2(OAc)_4$	92	>14:1	[17]

[a] The dr of the chiral centers at a and b.

Conversely, a copper catalyst yields exclusive cyclopropanation of allenic diazo derivatives **38** to give bicyclic methylenecyclopropanes **39** with moderate to good *E*/*Z* ratios (Scheme 16).[17] The scope is limited to acceptor–acceptor carbenes, but a variety of substitution patterns on both the allene and the tether are well-tolerated with less sterically demanding substrates yielding better results.

Scheme 16 Copper-Catalyzed Cyclopropanation of Allenic Diazo Esters[17]

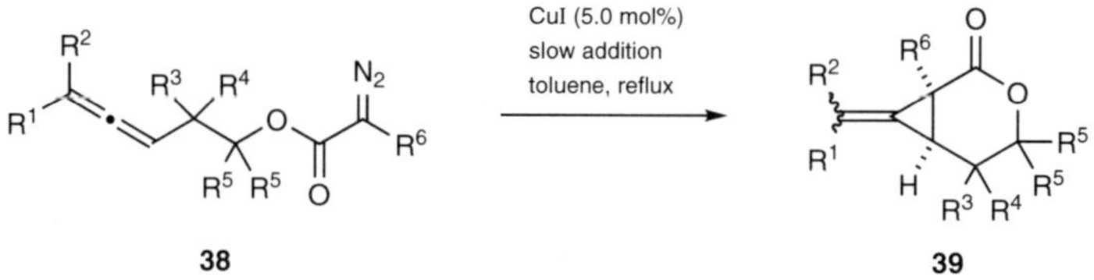

R^1	R^2	R^3	R^4	R^5	R^6	Yield (%)	Ratio (*E*/*Z*)	Ref
$(CH_2)_2Ph$	H	H	H	H	CO_2Et	77	7.9:1	[17]
$(CH_2)_4Me$	H	H	H	H	CO_2Et	67	6.7:1	[17]
Pr	Pr	H	H	H	CO_2Et	23	–	[17]
$(CH_2)_4Me$	H	Me	H	H	CO_2Et	68	>20:1	[17]
$(CH_2)_4Me$	H	H	H	Me	CO_2Et	18	19:1	[17]
$(CH_2)_4Me$	H	Me	Me	H	CO_2Et	33	24:1	[17]
$(CH_2)_4Me$	H	H	H	H	H	68	2.3:1	[17]
$(CH_2)_4Me$	H	H	H	H	CO_2Me	30	9.3:1	[17]

Ethyl 4-Allenyl-2-oxotetrahydrofuran-3-carboxylates 37 **(X = O); General Procedure:**[17]
To a vigorously stirred mixture of $Rh_2(OAc)_4$ or $Rh_2(esp)_2$ (3.0 mol%) and activated 4-Å molecular sieves (100 mg/0.1 mmol) in CH_2Cl_2 (0.1 M) was added a 0.1 M soln of a diazomalonate **36** (X = O) in CH_2Cl_2, using a syringe pump, at a rate of 1.0 mmol/h under N_2. The reaction was monitored by TLC and upon completion, the mixture was filtered through Celite and concentrated under reduced pressure. The crude material was purified by column chromatography to yield the cyclized product as a mixture of diastereomers.

Methylenecyclopropanes 39**; General Procedure:**[17]
To a vigorously stirred mixture of CuI (5.0 mol%) in CH_2Cl_2 (0.1 M) was added a 0.1 M soln of a diazo allene **38** in toluene under N_2 over a period of 12–24 h. When the reaction was complete, as indicated by TLC, the mixture was concentrated under reduced pressure and the residue was purified by column chromatography to yield the products as inseparable mixtures of *E*- and *Z*-isomers.

1.6.3.3 Gold-Catalyzed Intramolecular Hydroarylation of 2-Allenic Indoles

In addition to the typical nitrogen, oxygen, and carbon nucleophiles, the hydrofunctionalization of allenes can be extended to include 2-allenic indoles **41**, where reaction with gold(I) catalyst **40** and silver(I) trifluoromethanesulfonate induces intramolecular *exo* hydroarylation to yield 4-vinyltetrahydrocarbazoles **42** (n = 1) (Scheme 17).[2] In comparison to other *exo* hydrofunctionalizations, which often suffer from limited substrate scopes and low reactivity, this high-yielding reaction proceeds for 2-allenic indoles substituted with either electron-donating or electron-withdrawing groups and tolerates substitution at the terminal or internal allenyl carbon. The axial chirality of the 2-allenic indole can be completely transferred to the newly formed stereocenter. Mechanistically, the high selectivity in chirality transfer, as well as the high diastereomeric ratio of the gold-catalyzed hydroarylation of 2-allenic indoles, presents strong evidence that the chemistry occurs by outer-sphere attack of the indole on a *cis* gold–allene complex.

Scheme 17 Intramolecular Hydroarylation of 2-Allenic Indoles[2]

R^1	R^2	R^3	R^4	R^5	R^6	n	Yield[a] (%)	Ref
CO_2Me	CO_2Me	H	H	H	H	1	87	[2]
CO_2Me	CO_2Me	H	H	H	OMe	1	89	[2]
CO_2Me	CO_2Me	H	H	H	F	1	91	[2]
CO_2Me	CO_2Me	Me	H	H	H	1	71	[2]
CO_2Me	CO_2Me	H	$(CH_2)_4Me$	H	H	1	82[b]	[2]
CO_2Me	CO_2Me	H	Me	Me	H	1	92	[2]
CO_2Me	H	H	H	H	H	1	94[c]	[2]
CH_2OH	CH_2OH	H	H	H	H	1	82	[2]
CO_2Me	CO_2Me	H	H	H	H	2	70[d]	[2]

[a] Isolated material of >95% purity.
[b] 52% ee of starting material and product.
[c] dr 5:1.
[d] Reaction run for 22 h.

Dimethyl 9-Methyl-4-vinyl-1,3,4,9-tetrahydro-2*H*-carbazole-2,2-dicarboxylate (42, $R^1 = R^2 = CO_2Me$; $R^3 = R^4 = R^5 = R^6 = H$; n = 1); Typical Procedure:[2]
A mixture of gold catalyst **40** (6.6 mg, 13 μmol) and AgOTf (3.2 mg, 13 μmol) in dioxane (0.1 mL) was stirred for 10 min at rt. A soln of dimethyl 2-(buta-2,3-dienyl)-2-[(1-methyl-1*H*-indol-2-yl)methyl]malonate (**41**, $R^1 = R^2 = CO_2Me$; $R^3 = R^4 = R^5 = R^6 = H$; n = 1; 82 mg,

0.25 mmol) in dioxane (0.4 mL) was added and the resulting mixture was stirred for 30 min. The crude material was purified by column chromatography (silica gel, hexanes/EtOAc 10:1 to 5:1) to give the product as a pale yellow oil; yield: 71 mg (87%).

1.6.3.4 Gold-Catalyzed Hydroarylation of Allenic Anilines and Phenols

The hydroarylation of allenic substrates has been extended to include allenic anilines and phenols to offer a convenient route to dihydroquinoline and benzopyran derivatives under mild conditions.[18] When substrates such as allenic aniline derivatives **43** are reacted with a catalytic mixture of gold(I) catalyst **40** and silver(I) trifluoromethanesulfonate, cyclization via a 6-*endo* hydroarylation occurs to produce methyl quinoline-1(4*H*)-carboxylates **44**, which are usually unstable and therefore hydrogenated to the corresponding methyl 3,4-dihydroquinoline-1(2*H*)-carboxylates **45** (Scheme 18).[18] Various substitution patterns on the allene are tolerated, including unactivated allenes; however, unsubstituted aryl groups exhibit lower reactivity. The addition of electron-donating groups on the aromatic ring results in good to high yields of the cyclized product. The mechanism is proposed to occur through activation of the allene by coordination of the cationic gold catalyst to the distal double bond of the substrate. Electrophilic aromatic substitution with the electron-rich arene occurs to yield a vinylgold intermediate that undergoes deprotonation to a neutral vinylgold species. Cleavage of the gold—carbon bond by the proton generated in the previous step affords the product and regenerates the cationic gold catalyst.

Scheme 18 Gold-Catalyzed Cyclization of *N*-Allenylamines to Dihydroquinolines[18]

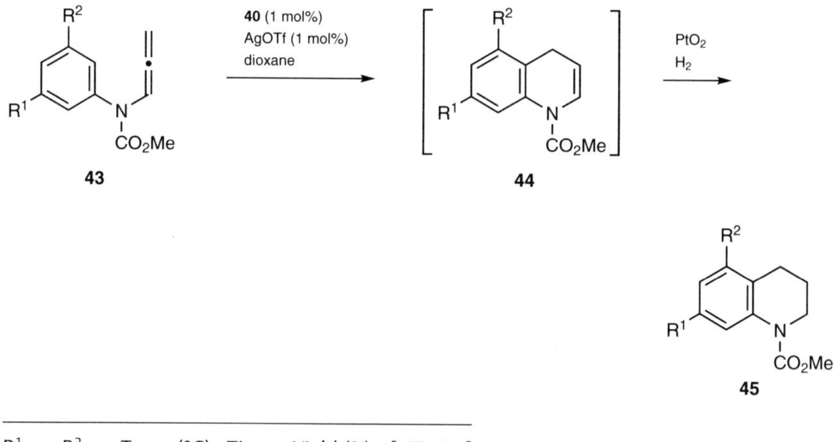

R^2 R^1 N CO_2Me **43** → **40** (1 mol%), AgOTf (1 mol%), dioxane → [R^2 R^1 N CO_2Me **44**] → PtO_2, H_2 → R^2 R^1 N CO_2Me **45**

R^1	R^2	Temp (°C)	Time	Yield (%) of **45**	Ref
OMe	OMe	25	5 min	92	[18]
OCH_2O		60	1 h	88	[18]
Me	Me	60	1 h	88	[18]
OMe	H	100	1 h	90	[18]
Me	H	reflux	3 h	44[a]	[18]
H	H	reflux	3 h	40[b]	[18]

[a] 28% of the regioisomer **45** (R^1 = H; R^2 = Me) was also isolated.
[b] 3 mol% of gold catalyst **40** and AgOTf was used.

With a slight alteration in the conditions and the addition of an extra carbon between the heteroatom and the allene, the protocol can be further applied to other aniline derivatives **46** (X = NCO_2Me) and their phenol analogues **46** (X = O) (Scheme 19). In these cases, the cyclization generally occurs via a 6-*exo* mode to give the corresponding products **47**, as opposed to the *endo* mode described above, although the *endo* product **48** is sometimes obtained in significant quantities when phenol derivatives **46** (X = O) are used.

Scheme 19 Gold-Catalyzed Cyclization of Other Aniline Derivatives and Phenol Analogues[18]

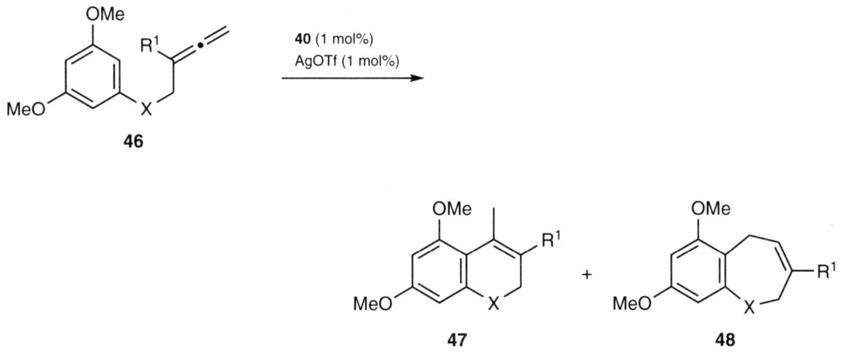

40 (1 mol%)
AgOTf (1 mol%)

46 → **47** + **48**

X	R^1	Solvent	mol% of Catalyst	Temp (°C)	Time (h)	Yield[a] (%) of **47**	Ref
NCO_2Me	H	dioxane	3	60	1.5	63	[18]
NCO_2Me	H	AcOH	3	25	1	82	[18]
NCO_2Me	H	AcOH	1	60	1	85	[18]
NCO_2Me	Me	AcOH	3	25	4	75	[18]
NCO_2Me	Me	AcOH	1	60	1	74	[18]
O	H	dioxane	1	60	1	98[b]	[18]
O	Me	dioxane	1	60	4	68[c]	[18]
O	Me	dioxane/AcOH (4:1)	1	60	3.5	99[d]	[18]

[a] Isolated yield.
[b] Isolated yield after hydrogenation.
[c] Ratio (**47**/**48**) 59:41, as determined by ^1H NMR spectroscopy.
[d] Ratio (**47**/**48**) 48:52, as determined by ^1H NMR spectroscopy.

Methyl 5,7-Dimethoxy-3,4-dihydroquinoline-1(2*H*)-carboxylate (45, $R^1 = R^2 =$ OMe); Typical Procedure:[18]

A flask was charged with methyl (3,5-dimethoxyphenyl)(propa-1,2-dienyl)carbamate (**43**, $R^1 = R^2 =$ OMe; 99.7 mg, 0.40 mmol), gold(I) catalyst **40** (2.12 mg, 1 mol%) and AgOTf (1.03 mg, 1 mol%) under argon. Dioxane (4.0 mL) was added and the mixture was stirred at rt for 5 min. PtO_2 (27.2 mg, 0.3 equiv) was added to the crude mixture and a hydrogen balloon was connected to the reaction flask. The mixture was stirred for 4 h at rt, filtered through Celite, and concentrated under reduced pressure. The residue was purified by column chromatography (hexane/EtOAc 15:1); yield: 90.5 mg (92%).

1.6.3.5 Gold-Catalyzed Intramolecular Hydroarylation of Allenes

Electron-rich aromatic compounds can also participate in nucleophilic attack on allenes activated by transition-metal catalysts. For example, vinyl-substituted tetrahydronaphthalenes **50** are produced via an *exo* cyclization when allenic arenes **49** react with a highly electrophilic phosphite-supported gold(I) complex (Scheme 20).[19] This bench-stable catalyst offers many advantages, including the ability to use unpurified commercial solvents

for references see p 347

and low loadings. The scope is limited to electron-rich arenes, but the conditions are tolerant of a variety of functional groups including ethers, esters, and pyrroles, producing the products in good to excellent yields. Coordinating aromatics, such as triazoles and oxazoles, are ineffective under the optimized reaction conditions.

Scheme 20 Gold-Catalyzed Cyclohydroarylation of Allenes To Give Vinyltetrahydronaphthalenes and an 8-Vinyl-5,6,7,8-tetrahydroindolizine[19]

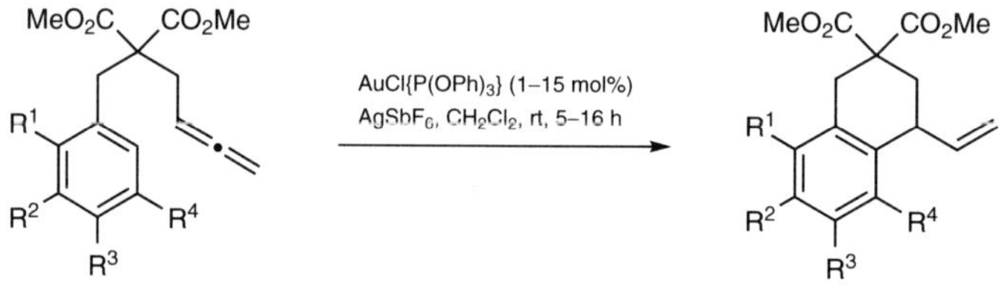

49 → 50: $AuCl\{P(OPh)_3\}$ (1–15 mol%), $AgSbF_6$, CH_2Cl_2, rt, 5–16 h

R^1	R^2	R^3	R^4	Au (mol%)	Time (h)	Yield (%)	Ref
H	OMe	H	OMe	3	6	85	[19]
(CH=CH)$_2$		H	H	10	16	87	[19]
H	OCH$_2$O		H	3	16	93	[19]
OMe	OMe	OMe	H	5	16	59	[19]
H	OMe	OMe	OMe	5	16	75	[19]
H	OMe	H	H	10	5	60	[19]

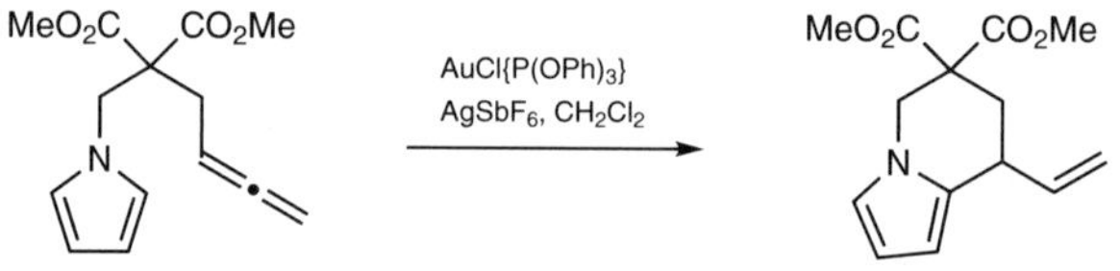

$AuCl\{P(OPh)_3\}$, $AgSbF_6$, CH_2Cl_2

Au (mol%)	Time (h)	Yield (%)	Ref
3	6	79	[19]
1	24	73	[19]

Dimethyl 4-Vinyl-3,4-dihydronaphthalene-2,2(1*H*)-dicarboxylates 50; **General Procedure:**[19]

A 5-mL vial was charged with gold catalyst $AuCl\{P(OPh)_3\}$ (27.2 mg, 0.05 mmol, 0.1 equiv) and $AgSbF_6$ (24.0 mg, 0.07 mmol, 0.14 equiv). CH_2Cl_2 (1.0 mL) was added and a white-gray suspension formed. An allenic arene **49** (0.5 mmol, 10 equiv) was added by pipet after 2 min. Within 20 min, the mixture turned deep green. After 4–24 h, the mixture was loaded directly onto a column (silica gel) and subjected to chromatography (hexanes/EtOAc).

1.6.3.6 Palladium-Catalyzed Synthesis of Cyclopropanes via Coupling/Cyclization of Allenic Malonates with Organic Halides

The palladium-catalyzed coupling/cyclization of allenic malonates **51** with organic halides provides a convenient route to polysubstituted cyclopropanes **52** with high regio- and stereoselectivity (Scheme 21).[20] This reaction is a nice addition to the synthetic lexicon, because other methods that show high selectivity for less-substituted cyclopropane products fail when additional substitution is introduced on the allene. The ability of this chemistry to tolerate substitution on both the allene and the tether between the allene and the active methylene nucleophile leads to fully substituted cyclopropanes. A variety of differentially substituted allenes and halides can react to form cyclized products in good to excellent yields with the *trans*-isomer as the predominant product. In a mechanistic sense, the reaction proceeds through an initial palladium-catalyzed coupling of the organic halide with the allene to form an intermediate π-allylpalladium species. The carbanion formed by deprotonation of the acidic proton flanked by two electron-withdrawing groups attacks the π-allylpalladium to form the cyclopropane product and regenerate the catalyst.

Scheme 21 Palladium-Catalyzed Cyclization of Allenic Malonates To Give Cyclopropanes[20]

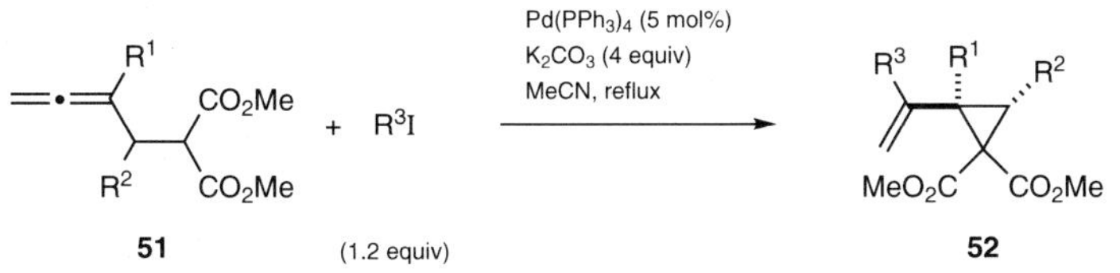

$Pd(PPh_3)_4$ (5 mol%)
K_2CO_3 (4 equiv)
MeCN, reflux

51 + R^3I (1.2 equiv) → 52

R^1	R^2	R^3	Additive (mol%)	Time (h)	Yield (%)	Ratio (*cis*/*trans*)	Ref
$(CH_2)_5Me$	Me	Ph	–	10	93	6:94	[20]
$(CH_2)_5Me$	Me	Ph	TBAB (10)	17	86	5:95	[20]
$(CH_2)_5Me$	Ph	Ph	–	13	91	4:96	[20]
$(CH_2)_5Me$	Ph	Ph	TBAB (10)	36	86	4:96	[20]
Me	Me	Ph	–	25	72	5:95	[20]
Me	Me	Ph	TBAB (10)	21	93	6:94	[20]
$(CH_2)_5Me$	Me	$4-MeO_2CC_6H_4$	–	35	49	6:94	[20]
$(CH_2)_5Me$	Me	$4-MeO_2CC_6H_4$	TBAB (10)	9	77	6:94	[20]
$(CH_2)_5Me$	Me	(*E*)-CH=CHPh	–	13	72	7:93	[20]
$(CH_2)_5Me$	Me	(*E*)-CH=CHPh	TBAB (10)	20	98	11:89	[20]
$(CH_2)_5Me$	Me	2-thienyl	–	13	61	7:93	[20]
$(CH_2)_5Me$	Me	2-thienyl	TBAB (10)	21	83	8:92	[20]
$(CH_2)_5Me$	Ph	$4-MeO_2CC_6H_4$	–	72	60	7:93	[20]
$(CH_2)_5Me$	Ph	$4-MeO_2CC_6H_4$	TBAB (10)	99	76	5:95	[20]
$(CH_2)_5Me$	Ph	(*E*)-CH=CHPh	–	13	88	6:94	[20]
$(CH_2)_5Me$	Ph	(*E*)-CH=CHPh	TBAB (10)	20	86	6:94	[20]

Dimethyl 2-Vinylcyclopropane-1,1-dicarboxylates 52; General Procedure:[20]
A flask was charged with a mixture of K_2CO_3 (112 mg, 0.8 mmol, 4 equiv), TBAB (6.4 mg, 10 mol%), and $Pd(PPh_3)_4$ (12 mg, 5 mol%) in MeCN (2 mL) under N_2. An allenic dimethyl malonate **51** (0.2 mmol) and PhI (49 mg, 1.2 equiv, 0.24 mmol) were subsequently added, and the mixture was stirred for 24 h and monitored by TLC. Once the reaction was complete, the mixture was filtered, the collected material was washed with Et_2O, and the filtrate was concentrated. The crude material was purified by flash chromatography (silica gel, petroleum ether/EtOAc).

1.6.3.7 Palladium-Catalyzed Synthesis of Eight- to Ten-Membered Lactones from Allenic 3-Oxoalkanoates and Organic Halides

Medium-sized lactones are common motifs in a variety of natural products with potentially useful bioactivities, but can be challenging to synthesize. Palladium-catalyzed allene coupling/cyclization represents a useful strategy to achieve this goal through formation of a π-allylpalladium intermediate, which can be trapped by a nucleophile to form carbocycles of varying ring sizes. This strategy is particularly useful for the synthesis of medium-sized lactones in a highly chemo-, regio-, and stereoselective manner from allenic 3-oxoalkanoates (e.g., **53**) and organic halides (Scheme 22). Interestingly, this strategy favors the formation of the unexpected eight-membered ring over the more stable six-membered ring. A possible rationale to explain the high regio- and stereoselectivity hinges on the presence of the phenyl group in the intermediate resulting from initial coupling of the allene with the aryl halide. The presence of the halide strongly favors the formation of an *anti*-π-allylic palladium intermediate to avoid steric congestion between the phenyl group and the chain bearing the nucleophilic functionality that would arise in the *syn*-intermediate. This sets up a chemoselective nucleophilic attack of the carbon nucleophile on the π-allyl species at the less-substituted terminal carbon to yield the eight-membered lactone. This chemistry can also be employed for the preparation of nine- and ten-membered rings simply by increasing the tether length between the allene and ester. A variety of substitutions on the chain and halide result in good yields of the cyclized products **54**; however, only unsubstituted allenes work under these conditions, presumably because more highly substituted allenes would override the preference for the *anti*-π-allylic palladium intermediate leading to the macrocycle. Despite this limitation in scope, this method represents a new and highly selective way to synthesize medium-sized lactones that are traditionally challenging to prepare.[21]

Scheme 22 Synthesis of Lactones by Palladium-Catalyzed Coupling/Cyclization of Allenic 3-Oxoalkanoates and Organic Iodides[21]

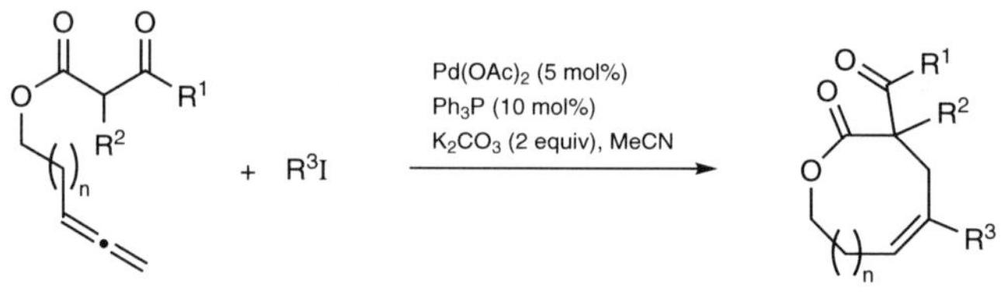

53 + R^3I (1.5 equiv) → $Pd(OAc)_2$ (5 mol%), Ph_3P (10 mol%), K_2CO_3 (2 equiv), MeCN → **54**

n	R^1	R^2	R^3	Temp (°C)	Time (h)	Yield (%)	Ref
1	Me	H	Ph	85	4.5	73	[21]
1	Ph	H	Ph	85	1.5	79	[21]
1	Me	Pr	Ph	100	82	67[a]	[21]
1	Me	$CH_2CH=CH_2$	Ph	100	29	69	[21]
1	Ph	H	$4\text{-}MeOC_6H_4$	85	5	69	[21]
1	Ph	H	$4\text{-}EtO_2CC_6H_4$	85	3	58	[21]
1	Ph	H	(*E*)-CH=CHBu	85	2.5	59	[21]
2	Me	H	Ph	85	11	82	[21]
3	Me	H	$4\text{-}BrC_6H_4$	85	6	69	[21]

[a] 2.5 equiv of PhI was used.

(*E*)-3-Acetyl-5-phenyl-3,4,7,8-tetrahydro-2*H*-oxocin-2-one (54, R^1 = Me; R^2 = H; R^3 = Ph; n = 1); Typical Procedure:[21]

To a flame-dried Schlenk flask were added sequentially $Pd(OAc)_2$ (5.7 mg, 0.025 mmol), Ph_3P (13.3 mg, 0.050 mmol), K_2CO_3 (1.0 mmol, 2 equiv), MeCN (2 mL), PhI (153.6 mg, 0.75 mmol), MeCN (1 mL), penta-3,4-dienyl 3-oxobutanoate (**53**, R^1 = Me; R^2 = H; n = 1; 83.2 mg, 0.50 mmol), and MeCN (1 mL) under argon. The mixture was stirred at 85 °C in a preheated oil bath and the reaction was monitored by TLC. Upon completion of the reaction, the resulting mixture was filtered through a short column (silica gel) and concentrated. The residue was purified by column chromatography (petroleum ether/Et_2O 10:1); yield: 88.6 mg (73%).

1.6.3.8 Palladium-Catalyzed Synthesis of Nine- to Twelve-Membered Cyclic Compounds from Allenes Containing a Nucleophilic Functionality and Organic Halides

In a similar vein to the chemistry described in Section 1.6.3.7, the palladium-catalyzed cyclization of allenes has been extended to include the formation of nine- to twelve-membered rings (e.g., **56**) from tandem coupling/cyclization between functionalized allenes (e.g., **55**) and organic halides. These products are formed in a highly regioselective manner in moderate to high yields (Scheme 23). The *E* stereoselectivity is believed to be a result of avoiding steric hindrance between the R^3 group and the nucleophilic substituent, resulting in nucleophilic attack on the less-substituted terminal end of the allylpalladium intermediate. A variety of substitution patterns on the carbon chain and groups on the iodide are tolerated, and use of an aminoallene provides an efficient route to azacycles. Overall, this method utilizes mild conditions without the need for very dilute conditions, representing a new and efficient route to these cyclized products.[22]

Scheme 23 Synthesis of Nine- to Twelve-Membered Rings by Palladium-Catalyzed Coupling/Cyclization of Allenes with Organic Iodides[22]

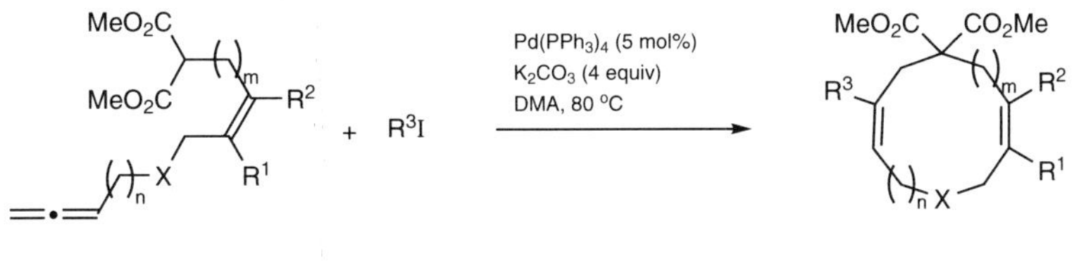

MeO2C, MeO2C, R2, X, n, m, R1 — **55** + R^3I (2 equiv) → Pd(PPh3)4 (5 mol%), K2CO3 (4 equiv), DMA, 80 °C → MeO2C, CO2Me, R3, R2, R1, X, n, m — **56**

X	n	m	R^1	R^2	R^3	Time (h)	Yield (%)	Ref
$C(CO_2Me)_2$	1	1	$(CH{=}CH)_2$		Ph	11	91	[22]
$C(CO_2Me)_2$	1	1	$(CH{=}CH)_2$		$4-ClC_6H_4$	17	85	[22]
$C(CO_2Me)_2$	1	1	$(CH{=}CH)_2$		4-pyridyl	14	84	[22]
$C(CO_2Me)_2$	1	1	$(CH{=}CH)_2$		(*E*)-CH=CHBu	72	60[a]	[22]
$C(CO_2Me)_2$	1	1	H	H	Ph	16.7	89	[22]
NTs	1	1	$(CH{=}CH)_2$		$4-ClC_6H_4$	8	82	[22]
$C(CO_2Me)_2$	2	1	$(CH{=}CH)_2$		Ph	11	73	[22]
$C(CO_2Me)_2$	2	2	$(CH{=}CH)_2$		Ph	13	73	[22]

[a] 3 equiv of (*E*)-1-iodohex-1-ene was used.

for references see p 347

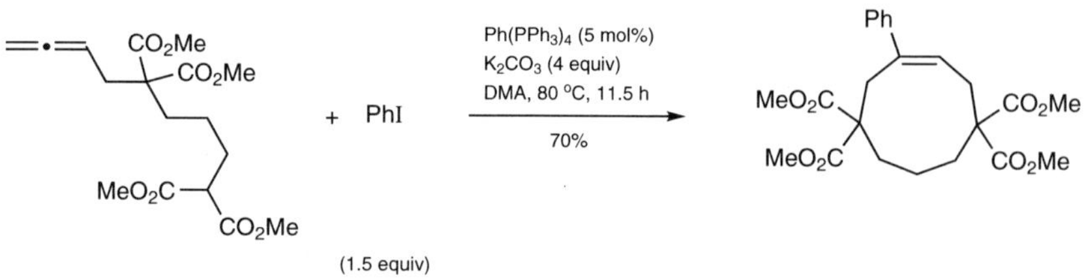

Ten- to Twelve-Membered Rings 56; General Procedure:[22]
$Pd(PPh_3)_4$ (5 μmol), a functionalized allene **55** (0.1 mmol), PhI (0.2 mmol), and DMA (1.5 mL) were added sequentially to a Schlenk tube containing K_2CO_3 (0.4 mmol) under N_2. The Schlenk tube was cooled using a dry ice/acetone bath and back-filled with N_2 (3 ×). The mixture was stirred at 80 °C until completion of the reaction, as monitored by TLC (petroleum ether/EtOAc 5:1). H_2O (10 mL) was added and the mixture was extracted with Et_2O (3 × 30 mL). The combined extracts were washed with H_2O and brine, dried (Na_2SO_4), filtered, and concentrated. The residue was purified by chromatography (silica gel, petroleum ether/EtOAc/CH_2Cl_2 5:1:1).

1.6.4 C—S Bond Formation

1.6.4.1 Cycloisomerization of Sulfanylallenes to 2,5-Dihydrothiophenes

Although most of the nucleophiles employed in metal-catalyzed cyclizations of allenes are based on nitrogen, oxygen, or carbon, there are scattered examples of more unusual nucleophiles. For example, the gold-catalyzed cycloisomerization of sulfanylallenes **57** efficiently forms 2,5-dihydrothiophenes **58** in moderate yields with excellent diastereoselectivities (Scheme 24).[23] Due to the tendency of organosulfur compounds to strongly coordinate to transition metals, the success of this chemistry is surprising. Although silver and copper catalysts are not successful, several gold precatalysts promote the cyclization, provided that the gold catalyst is sufficiently Lewis acidic. The proposed mechanism is similar to the corresponding reactions of hydroxy- and aminoallenes, where coordination of the metal to the distal allene double bond is followed by an S_N2-like attack of the thiol on the allene. The intermediate zwitterionic species undergo protodemetalation to yield the thiophene with excellent fidelity in the transfer of axial-to-center chirality. This methodology represents a unique advance in the field and serves as the first example of a gold-catalyzed C—S bond formation. The protocol tolerates differentially substituted allenes and requires either a gold(I) or gold(III) catalyst for the reaction to proceed.

Scheme 24 Gold-Catalyzed Cycloisomerization of Sulfanylallenes to 2,5-Dihydrothiophenes[23]

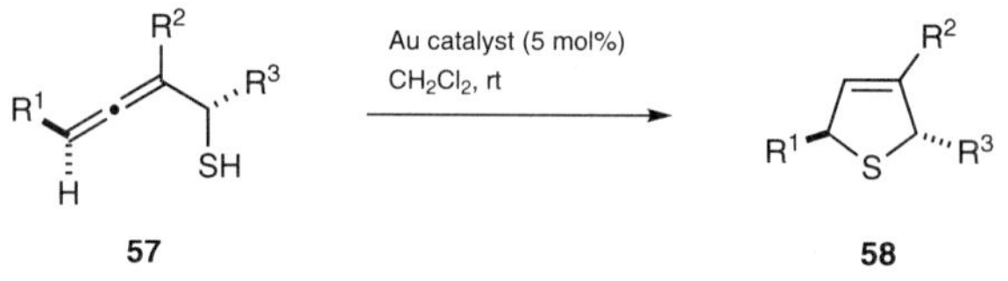

R^1	R^2	R^3	dr of **57**	Catalyst	Time	Yield (%)	dr of **58**	Ref
iPr	Me	CH_2OMe	>99:1	AuCl	1.5 h	88	>99:1	[23]
iPr	Me	CH_2OBn	95:5	AuCl	1.5 h	86	95:5	[23]
iPr	Me	CH_2OBn	95:5	AuI	5 min	87	95:5	[23]

R^1	R^2	R^3	dr of **57**	Catalyst	Time	Yield (%)	dr of **58**	Ref
iPr	Me	4-$F_3CC_6H_4OCH_2$	>99:1	AuCl	4 h	67	>99:1	[23]
$(CH_2)_5Me$	Me	CH_2OBn	>99:1	AuCl	2 h	82	>99:1	[23]
$(CH_2)_7CH{=}CH_2$	H	H	–	AuCl	1.5 h	43	–	[23]

5-Isopropyl-2-(methoxymethyl)-3-methyl-2,5-dihydrothiophene (58, **R^1 = iPr; R^2 = Me; R^3 = CH_2OMe); Typical Procedure:**[23]

To a stirred soln of 1-methoxy-3,6-dimethylhepta-3,4-diene-2-thiol (**57**, R^1 = iPr; R^2 = Me; R^3 = CH_2OMe; 50 mg, 0.27 mmol; dr >99:1) in CH_2Cl_2 (5 mL) was added AuCl (3 mg, 13 μmol) at rt under argon. The mixture was stirred for 90 min and the reaction was monitored by TLC. After completion of the reaction, the mixture was concentrated under reduced pressure and the residue was purified by column chromatography (silica gel, cyclohexane/EtOAc 30:1); yield: 44 mg (88%); dr >99:1.

1.6.5 Total Synthesis Using Allene Cyclization

1.6.5.1 Gold-Catalyzed Cycloisomerization of Dihydroxyallenes to 2,5-Dihydrofurans

The ability of metal-catalyzed allene cyclizations to convert simple precursors into complex and densely functionalized products makes this chemistry ideal for the stereoselective synthesis of natural products and related analogues. For example, a gold(III)-catalyzed cycloisomerization of dihydroxyallene **59** comprises a key step in a synthesis of (+)-linalool oxide and shows excellent transfer of axial to central chirality in the product **60** using a low loading of the gold catalyst (0.05–0.1 mol%) (Scheme 25).[24] Similar strategies have been used to access more complex natural products that also contain a 2,5-dihydrofuran motif, including (−)-isocyclocapitelline and (−)-isochrysotricine. Because the absolute stereochemistry of these compounds was not known previously, gold-catalyzed allene cycloisomerization allowed for the identification of the absolute configuration of both of these natural products.

Scheme 25 Gold-Catalyzed Synthesis of (+)-Linalool Oxide[24]

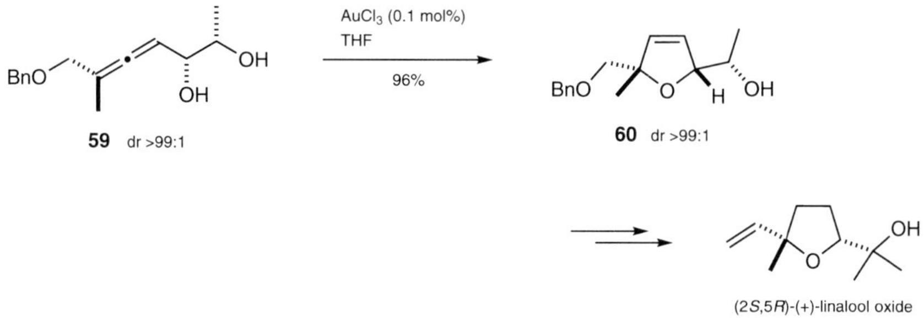

(S)-1-{(2R,5S)-5-[(Benzyloxy)methyl]-5-methyl-2,5-dihydrofuran-2-yl}ethan-1-ol (60):[24]

To a soln of (2S,3R,5R)-7-(benzyloxy)-6-methylhepta-4,5-diene-2,3-diol (**59**; 2.31 g, 9.3 mmol) in anhyd THF (93 mL) was added a 0.166 M soln of $AuCl_3$ in MeCN (56 μL, 9.3 μmol) at rt. After 30 min, the solvent was removed under reduced pressure and the crude product was purified by column chromatography (silica gel, cyclohexane/EtOAc 8:2 to 7:3) to afford the product as a colorless oil; yield: 2.22 g (96%); dr >99:1.

for references see p 347

1.6.5.2 Silver-Catalyzed Cyclization of Allenic Hydroxylamines

The tricyclic piperidine alkaloid porantheridine has been synthesized using either silver- or gold-catalyzed cyclizations of an allenic hydroxylamine **61** as a key step. The use of silver(I) tetrafluoroborate (10 mol%) leads to the isoxazolidine **62** in good yield and with high stereoselectivity (Scheme 26); a higher catalyst loading was shown to decrease the selectivity.[25] Silver-catalyzed reactions also show superior diastereoselectivities as compared to both gold(I) and gold(III) catalysts[26] and other silver catalysts with more coordinating anions as compared to tetrafluoroborate.[27] The isoxazolidine resulting from this chemistry was carried on to deliver porantheridine, but the power of allene cyclization to construct amino alcohol moieties could prove useful in many other contexts.

Scheme 26 Silver-Catalyzed Synthesis of an Isoxazolidine from an Allenic Hydroxylamine[25]

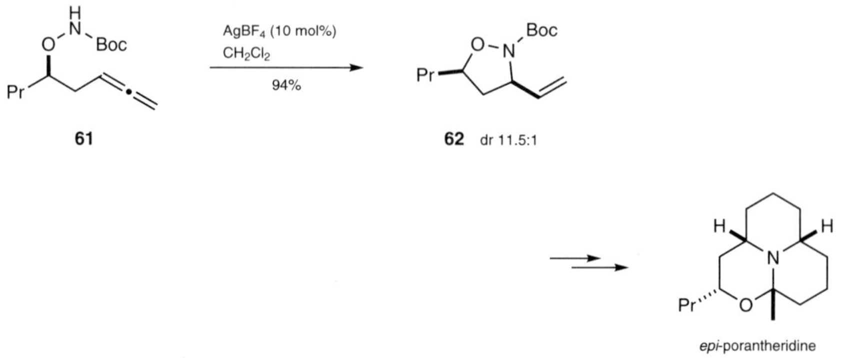

(3*R*,5*R*)-2-(*tert*-Butoxycarbonyl)-5-propyl-3-vinylisoxazolidine (62):[25]
$AgBF_4$ (16.3 mg 0.083 mmol) was added to a soln of *N-tert*-butoxycarbonyl-*O*-[(*R*)-(octa-6,7-dien-4-yl]hydroxylamine (**61**; 200 mg, 0.83 mmol) in anhyd CH_2Cl_2 (6 mL) at rt. The mixture was stirred at rt in the absence of light for 8 h and then filtered through Celite. The filtrate was washed with sat. aq $NaHCO_3$ and brine, dried (Na_2SO_4), filtered, and concentrated under reduced pressure. The residue was purified by flash chromatography (silica gel, EtOAc/hexane 1:9) to afford an inseparable mixture of diastereomers; yield: 188 mg (94%); dr (*cis*/*trans*) 11.5:1.

1.6.5.3 Synthesis of Flinderoles B and C through Gold-Catalyzed Allene Hydroarylation

Flinderoles B and C are members of a new class of antimalarial bisindole alkaloids isolated from plants of the *Flindersia* genus. These compounds have demonstrated antimalarial activity by selective growth inhibition against Dd2 (choroquine-resistant) *P. falciparum* malaria strain with IC_{50} values between 0.15–1.42 mM. A gold-catalyzed allene hydroarylation was proposed as a key step in the synthesis. The indole **63** was subjected to a variety of gold(I) catalysts.[28] Initial attempts using 5 mol% (triphenylphosphine)gold(I) as the metal catalyst failed to promote any cyclization of the allene. However, a more electropositive N-heterocyclic carbene catalyst, [1,3-bis(2,6-diisopropylphenyl)imidazol-2-ylidene]-gold(I) hexafluoroantimonate(V), promote the cyclization to the desired pyrrolidine derivative **64** as a single diastereomer in 88% yield (Scheme 27). This cyclization installs two key functional groups and proceeds under mild reaction conditions. Indole hydroarylation reactions of this type have potential to be useful across a wide variety of biologically active bisindole natural products.

Scheme 27 Synthesis of a Flinderole Intermediate through Gold-Catalyzed Hydroarylation of an Indole[28]

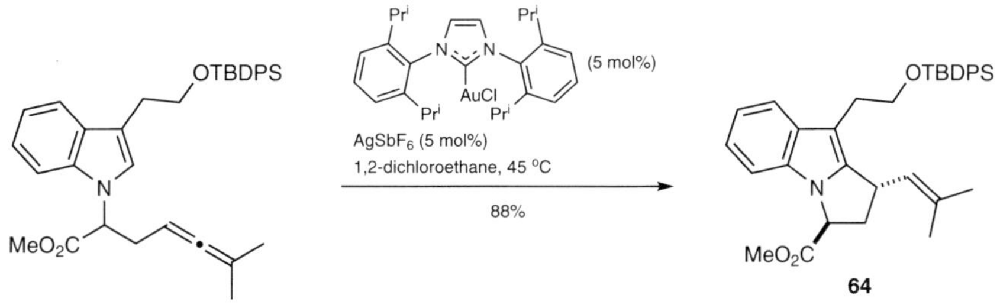

Methyl 9-[2-(*tert*-Butyldiphenylsiloxy)ethyl]-1-(2-methylprop-1-enyl)-2,3-dihydro-1*H*-pyrrolo[1,2-*a*]indole-3-carboxylate (64):[28]
Chloro[1,3-bis(2,6-diisopropylphenyl)imidazol-2-ylidene]gold(I) (28 mg, 0.044 mmol, 5 mol%) and $AgSbF_6$ (15 mg, 0.044 mmol, 5 mol%) in 1,2-dichloroethane (2 mL) were mixed in the dark in a sealed vial. After 5 min, the catalyst mixture was passed through a glass wool filter into a soln of methyl 2-{3-[2-(*tert*-butyldiphenylsiloxy)ethyl]-1*H*-indolyl}-6-methylhepta-4,5-dienoate (**63**; 489 mg, 0.886 mmol, 1.0 equiv) in 1,2-dichloroethane (5 mL). The filter was washed with 1,2-dichloroethane (2 × 0.5 mL) and then the reaction vessel was sealed and submerged in a 45 °C oil bath for 4 h. The mixture was then filtered over a silica gel plug and the filtrate was concentrated under reduced pressure. The product was isolated as a single diastereomer by flash chromatography (EtOAc/hexanes 5:95) as a white solid; yield: 432 mg (88%).

1.6.5.4 Palladium-Catalyzed Domino Cyclization of Allenes Bearing Amino and Bromoindolyl Groups

Palladium-catalyzed domino cyclizations of allenes bearing amino and bromoindolyl nucleophiles can be employed for the synthesis of members of the ergot alkaloid family, including (+)-lysergic acid, (+)-lysergol, and (+)-isolysergol. Domino cyclization of the allenic amide **65** allows for an efficient synthesis of a key C/D ring system in this class of natural products, which has been synthesized in a linear fashion in previous syntheses (Scheme 28).[6] A diastereomeric ratio of 94:6 in the substrate **65** translates to a diastereomeric ratio of 92:8 in the product **66**. This control of stereochemistry in the key step comes from good transfer of the axial chirality of the allene into central chirality in the product; in addition, this transformation also shows high regiocontrol in comparison to previous palladium-catalyzed cyclizations of allenes containing amino groups.

for references see p 347

Scheme 28 Palladium-Catalyzed Synthesis of the C/D Ring System of Ergot Alkaloids from an Allene Bearing Amino and Bromoindolyl Groups[6]

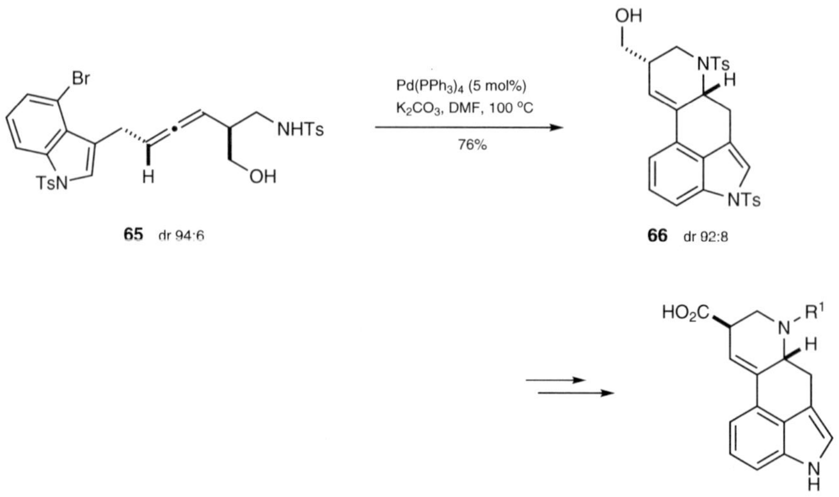

$Pd(PPh_3)_4$ (5 mol%)
K_2CO_3, DMF, 100 °C
76%

65 dr 94:6

66 dr 92:8

[(6a*R*,9*S*)-4,7-Ditosyl-4,6,6a,7,8,9-hexahydroindolo[4,3-*fg*]quinolin-9-yl]methanol (66):[6]
To a stirred mixture of *N*-[(2*S*,4*R*)-6-(4-bromo-1-tosyl-1*H*-indol-3-yl]-2-(hydroxymethyl)hexa-3,4-dien-1-yl)-4-toluenesulfonamide (**65**; dr 94:6; 248 mg, 0.39 mmol) in DMF (8.0 mL) were added $Pd(PPh_3)_4$ (22.8 mg, 0.020 mmol) and K_2CO_3 (162 mg, 1.17 mmol) at rt under argon, and the mixture was stirred for 2.5 h at 100 °C. Concentration under reduced pressure gave an oily residue. The residue was dissolved in EtOAc, and the soln was washed with sat. aq NH_4Cl, H_2O, and brine, dried ($MgSO_4$), filtered, and concentrated under reduced pressure to give a brown oil, which was purified by flash chromatography (silica gel, hexane/EtOAc 1:1), followed by another flash chromatography (Chromatorex, hexane/EtOAc 1:1 to 1:2) to give the product as a pale brown amorphous solid; yield: 162 mg (76%); dr 92:8. The pure diastereomer was isolated by preparative TLC.

1.6.6 Conclusions

In summary, this chapter has demonstrated the synthetic utility of transition-metal-catalyzed cyclizations of allenes as a means to access substituted heterocycles and carbon scaffolds. Many of the desired scaffolds are formed in high diastereo- and enantioselectivities using simple transition-metal catalysts, and leveraging of the axial chirality of the allene into point chirality is a reaction feature wholly unique to allene chemistry. These high stereoselectivities, in particular, are one of the main reasons that allene cyclizations are useful in achieving total syntheses of biologically relevant compounds. While these cyclization reactions have opened up new routes to a large amount of chemical space, there are sure to be new and creative uses of the allene scaffold moving forward.

References

[1] Yu, S.; Ma, S., *Chem. Commun. (Cambridge)*, (2011) **47**, 5384.
[2] Zhang, Z.; Liu, C.; Kinder, R. E.; Han, X.; Qian, H.; Widenhoefer, R. A., *J. Am. Chem. Soc.*, (2006) **128**, 9066.
[3] Jiang, X.; Ma, X.; Zheng, Z.; Ma, S., *Chem.–Eur. J.*, (2008) **14**, 8572.
[4] Kimura, M.; Fugami, K.; Tanaka, S.; Tamaru, Y., *Tetrahedron Lett.*, (1991) **32**, 6359.
[5] Zhang, Z.; Bender, C. F.; Widenhoefer, R. A., *J. Am. Chem. Soc.*, (2007) **129**, 14148.
[6] Inuki, S.; Iwata, A.; Oishi, S.; Fujii, N.; Ohno, H., *J. Org. Chem.*, (2011) **76**, 2072.
[7] Deng, Y.; Li, J.; Ma, S., *Chem.–Eur. J.*, (2008) **14**, 4263.
[8] Kang, J.-E.; Lee, E.-S.; Park, S.-I.; Shin, S., *Tetrahedron Lett.*, (2005) **46**, 7431.
[9] Marshall, J. A.; Wolf, M. A.; Wallace, E. M., *J. Org. Chem.*, (1997) **62**, 367.
[10] Krause, N.; Morita, N., *Org. Lett.*, (2004) **6**, 4121.
[11] Hoffmann-Röder, A.; Krause, N., *Org. Lett.*, (2001) **3**, 2537.
[12] Marshall, J. A.; Bartley, G. S., *J. Org. Chem.*, (1994) **59**, 7169.
[13] Adams, C. S.; Boralsky, L. B.; Guzei, I. A.; Schomaker, J. M., *J. Am. Chem. Soc.*, (2012) **134**, 10807.
[14] Li, H.; Lee, S. D.; Widenhoefer, R. A., *J. Organomet. Chem.*, (2011) **696**, 316.
[15] LaLonde, R. L.; Sherry, B. D.; Kang, E. J.; Toste, F. D., *J. Am. Chem. Soc.*, (2007) **129**, 2452.
[16] Li, M.; Datta, S.; Barber, D. M.; Dixon, D. J., *Org. Lett.*, (2012) **14**, 6350.
[17] Phelps, A. M.; Dolan, N. S.; Connell, N. T.; Schomaker, J. M., *Tetrahedron*, (2013) **69**, 5614.
[18] Watanabe, T.; Oishi, S.; Fujii, N.; Ohno, H., *Org. Lett.*, (2007) **9**, 4821.
[19] Tarselli, M. A.; Gagné, M. R., *J. Org. Chem.*, (2008) **73**, 2439.
[20] Ma, S.; Jiao, N.; Yang, Q.; Zheng, Z., *J. Org. Chem.*, (2004) **69**, 6463.
[21] Wan, B.; Jia, G.; Ma, S., *Adv. Synth. Catal.*, (2011) **353**, 1763.
[22] Jiang, X.; Yang, Q.; Yu, Y.; Fu, C.; Ma, S., *Chem.–Eur. J.*, (2009) **15**, 7283.
[23] Morita, N.; Krause, N., *Angew. Chem. Int. Ed.*, (2006) **45**, 1897.
[24] Volz, F.; Wadman, S. H.; Hoffmann-Röder, A.; Krause, N., *Tetrahedron*, (2009) **65**, 1902.
[25] Bates, R. W.; Lu, Y., *J. Org. Chem.*, (2009) **74**, 9460.
[26] Bates, R. W.; Dewey, M. R., *Org. Lett.*, (2009) **11**, 3706.
[27] Bates, R. W.; Nemeth, J.; Snell, R. H., *Synthesis*, (2008), 1033.
[28] Zeldin, R. M.; Toste, F. D., *Chem. Sci.*, (2011) **2**, 1706.

1.7 Cycloisomerizations of Substrates with Multiple Unsaturated Bonds

Y. Yamamoto

General Introduction

Cycloisomerizations are literally cyclization reactions that proceed with the partial reorganization of atom connectivity (isomerization). Therefore, all atoms contained in a starting material will be retained in the final product. Because of this intrinsic atom-economical nature,[1] cycloisomerizations have received considerable attention as green transformations that produce valuable cyclic frameworks from simpler acyclic starting materials and, naturally, efficient transition-metal catalysts have been identified to realize mild and selective protocols.[2–5] Although intramolecular thermal Alder–ene reactions of oct-6-en-1-yne and octa-1,6-diene can proceed without a catalyst, these reactions require severe pyrolysis conditions at 400 and 457 °C and the yields are not high (65 and 35%, respectively).[6,7] In contrast, cycloisomerizations of similar 1,6-enynes and 1,6-dienes catalyzed by transition-metal complexes proceed under much milder conditions at temperatures below 100 °C.[2–5] However, the catalyzed protocols often suffer from difficulties in selectively obtaining the desired isomer (isomer selectivity). Therefore, judicious choice of a catalytic system that is suitable for a specific substrate type is the key to obtaining successful results. In the following sections, synthetically useful examples of catalyzed cycloisomerizations of substrates bearing multiple unsaturated carbon–carbon bonds will be discussed on the basis of the classification of the substrate types and the transition-metal catalysts used. The scope of this chapter is limited to cycloisomerizations involving the formation of one carbon–carbon bond with concomitant hydrogen-atom transfer. Thus, cycloadditions and metathesis-type reactions (including skeletal reorganizations)[8] are not covered (see Sections 2.3–2.8 and Section 2.10), although they also fall into the category of formal cycloisomerization.

1.7.1 Cycloisomerizations of α,ω-Enynes

Transition-metal catalysts enable the selective cycloisomerization of α,ω-enynes under reaction conditions that are milder than those used in uncatalyzed reactions, and diverse cyclic products are obtained using different catalysts.[9] The palladium-catalyzed cycloisomerization of enynes such as **1** normally produces Alder–ene products **2** or exocyclic 1,3-dienes **3** (Scheme 1). Although other transition-metal catalysts deliver similar products, other product types such as **4** may be observed, depending on the structures of the substrates and the catalytic systems employed.[9]

Scheme 1 Transition-Metal-Catalyzed Cycloisomerization of 1,6-Enynes[9]

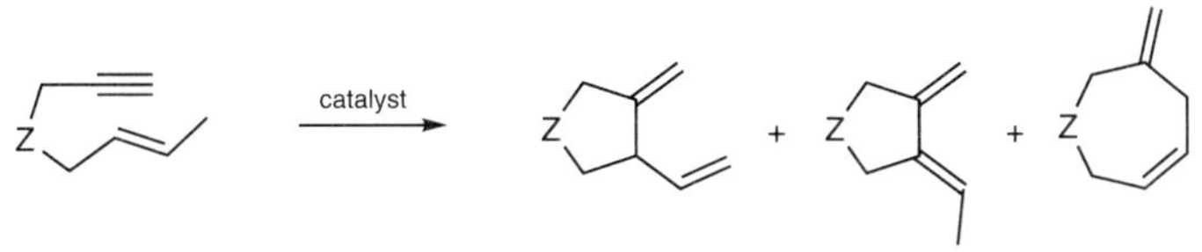

Z
catalyst
Z + Z + Z

1 2 3 4

As for the mechanism, several pathways have been proposed depending on the types of catalysts and substrates.[9] Scheme 2 depicts two major routes for the cycloisomerization of enynes **5** into exocyclic dienes **6**. One route follows the so-called hydride mechanism, in which metal hydrides are involved as the catalytically active species. Hydrometalation of the alkyne moiety and subsequent intramolecular carbometalation of the pendant alkene lead to an alkylmetal intermediate that then undergoes β-hydride elimination to afford **6**. The second route follows a metallacycle mechanism, in which oxidative cyclization of **5** onto a low-valent metal center affords a bicyclic metal complex as the key intermediate. This intermediate undergoes β-hydride elimination and subsequent reductive elimination to afford **6**.

Scheme 2 Representative Mechanisms for the Transition-Metal-Catalyzed Cycloisomerization of 1,6-Enynes[9]

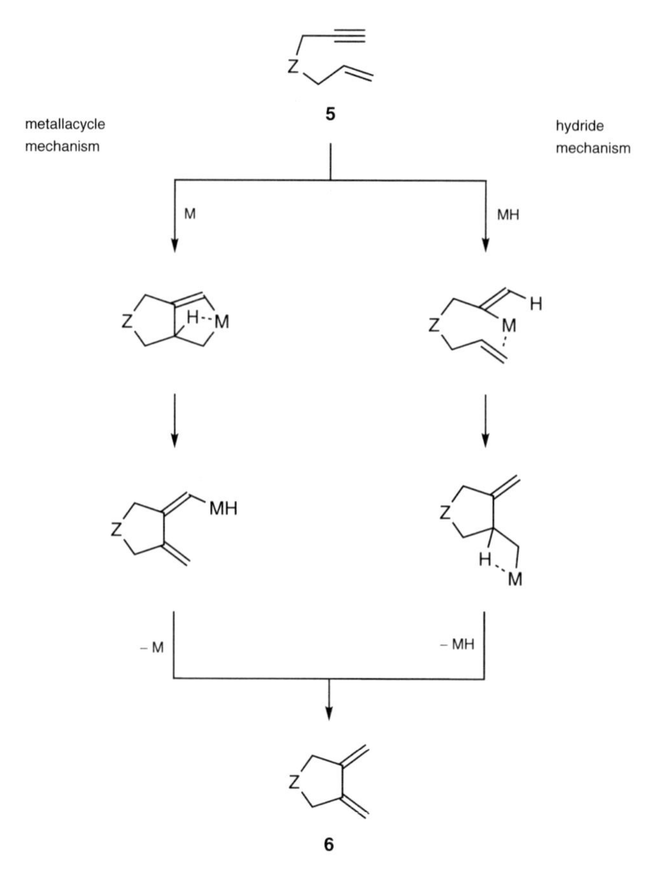

On the other hand, mechanistically different modes of enyne cycloisomerization have also been developed using gold and platinum complexes as soft, alkynophilic catalysts.[10,11] In these novel methods, the gold and platinum catalysts activate the alkyne moiety toward electrophilic attack by the pendant alkene and lead to diverse reactions such as skeletal rearrangement, cyclopropanation, and cyclobutene formation.

Representative selective protocols for enyne cycloisomerizations involving intramolecular hydrogen-atom transfer are discussed in Sections 1.7.1.1–1.7.1.5.

1.7.1.1 Palladium-Catalyzed Cycloisomerization of α,ω-Enynes

Typical examples of the palladium-catalyzed cycloisomerization of 1,6- and 1,7-enynes are summarized in Table 1.[12,13] The reactions of 1,6-enynes **7** possessing an internal alkene moiety proceed efficiently at 60 °C in the presence of palladium(II) acetate and a phosphine ligand to selectively afford the corresponding Alder–ene products **8** in yields of 67–77% (entries 1–4).[12] Triphenylphosphine is the standard ligand; however, tri-2-tolylphosphine is also used as the ligand for the cyclization of 4-(*tert*-butoxydimethylsiloxy)-2,4-dimethyloct-2-en-7-yne (entry 3), as the former is less effective in this case. Moderate to high diastereoselectivities can be observed for reactions of substrates possessing a chiral center adjacent to the alkene moiety (entries 2 and 3). The diastereoselective synthesis of bicyclic compounds is also possible if enynes possessing a cycloalkene moiety are employed (entry 4). On the other hand, the palladium-catalyzed reactions of 1,6-enynes **7** possessing a terminal alkene moiety afford the corresponding exocyclic 1,3-dienes **9** (entries 5 and 6).[13] In these examples, *N*,*N*-bis(benzylidene)ethylenediamine is a better ligand than triphenylphosphine. The formation of a six-membered-ring compound from a 1,7-enyne is also possible, but in this case no ligand is necessary (entry 7). As described above, enynes **7** possessing an *E*-1,2-disubstituted alkene moiety (e.g., entry 1) tend to afford Alder–ene products **8**. However, a 1,6-enyne that has an alkene terminal substituent with branching at the allylic carbon atom selectively produces the exocyclic 1,3-diene of type **9** using triphenylphosphine as the ligand (entry 8).

Table 1 Palladium-Catalyzed Cycloisomerization of α,ω-Enynes[12,13]

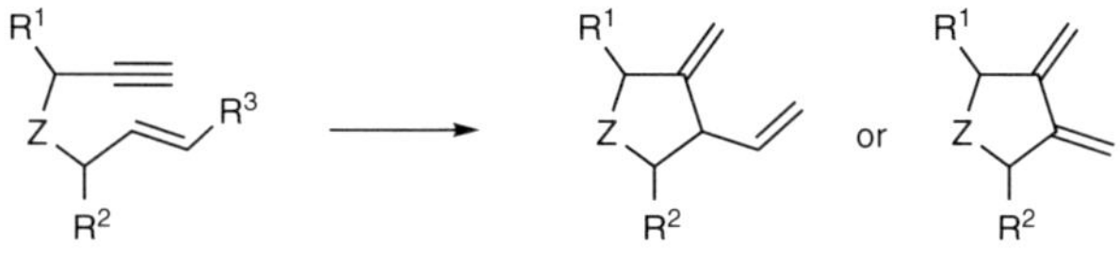

Entry	Substrate	Conditions	Product	Yield (%)	Ref
1	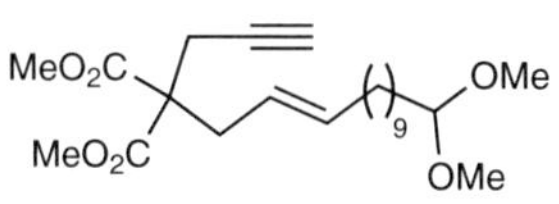	$Pd(OAc)_2 \bullet (Ph_3P)_2$ (5 mol%), benzene-d_6, 60 °C, 1 h	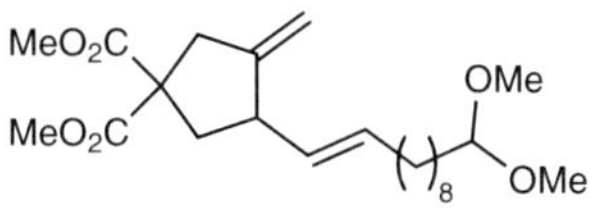	71	[12]
2	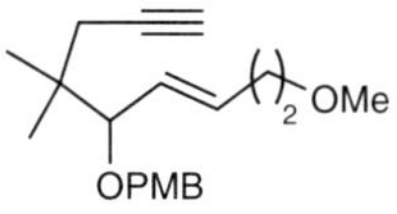	$Pd(OAc)_2 \bullet (Ph_3P)_2$ (5 mol%), benzene, 60 °C, 0.5 h	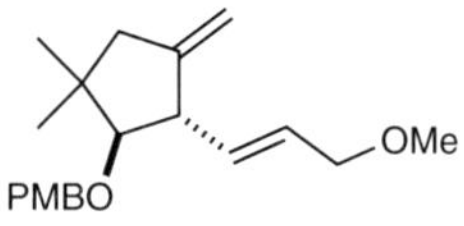	77[a]	[12]
3	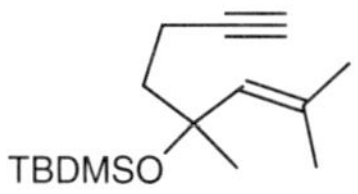	$Pd(OAc)_2 \bullet (2\text{-}Tol_3P)_2$ (5 mol%), benzene, 60 °C, 0.5 h	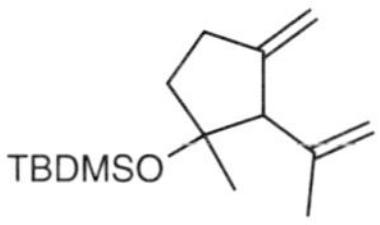	67[b]	[12]
4	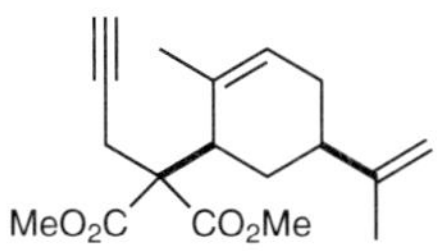	$Pd(OAc)_2 \bullet (Ph_3P)_2$ (5 mol%), benzene, 60 °C, 1 h	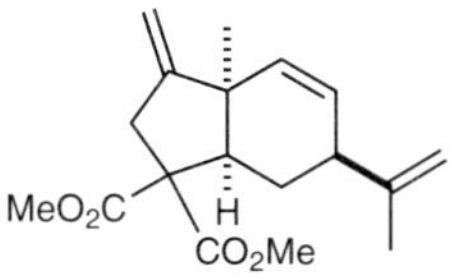	73	[12]

for references see p 383

Table 1 (cont.)

Entry	Substrate	Conditions	Product	Yield (%)	Ref
5	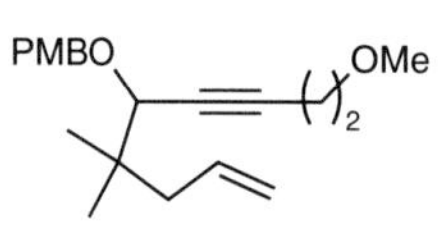	$Pd(OAc)_2$ (10 mol%), BBDEA (10 mol%),[c] benzene-d_6, 60 °C, 14.75 h	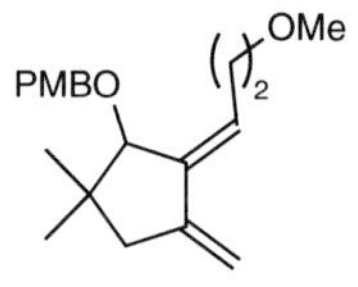	93[d]	[13]
6	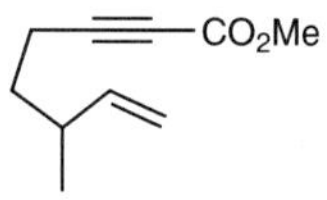	$Pd(OAc)_2$ (5 mol%), BBDEA (5 mol%),[c] benzene, 45–50 °C, 1 h	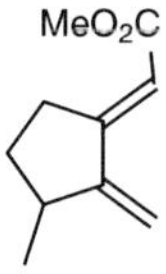	83	[13]
7	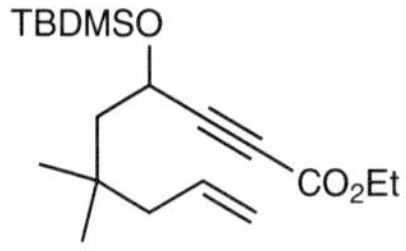	$Pd(OAc)_2$ (5 mol%), $CDCl_3$, 45 °C, 7.25 h	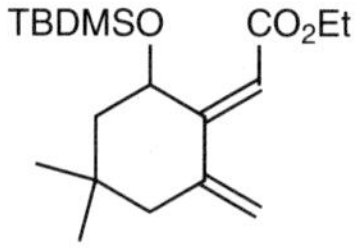	60	[13]
8	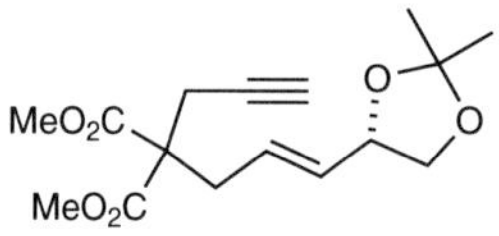	$Pd(OAc)_2$•$(Ph_3P)_2$ (4.4 mol%), Ph_3P (4.4 mol%), benzene-d_6, reflux, 1.75 h	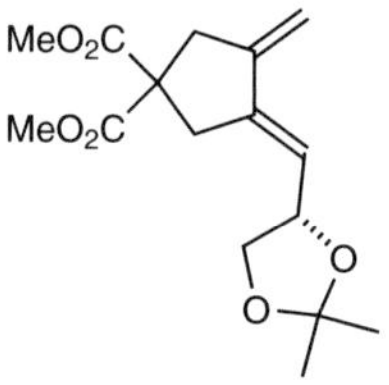	71	[13]

[a] dr >99:1.
[b] dr 2.3:1.
[c] BBDEA = *N*,*N*-bis(benzylidene)ethylenediamine.
[d] Ratio (*Z*/*E*) >11:1.

2-[(4-Methoxybenzyl)oxy]-3-[(*E*)-3-methoxyprop-1-enyl]-1,1-dimethyl-4-methylenecyclopentane (Table 1, Entry 2); Typical Procedure:[12]

A soln of 1-methoxy-4-({[(6*E*)-9-methoxy-4,4-dimethylnon-6-en-1-yn-5-yl]oxy}methyl)benzene (155 mg, 0.490 mmol) and $Pd(OAc)_2$•$(Ph_3P)_2$ (18.4 mg, 0.025 mmol) in dry benzene (5 mL) (**CAUTION:** *carcinogen*) was heated at 60 °C for 0.5 h. Direct flash column chromatography (silica gel, hexane/Et_2O 19:1 to 9:1) gave the title compound; yield: 120 mg (77%); dr >99:1.

2-[(4-Methoxybenzyl)oxy]-3-[(*Z*)-3-methoxypropylidene]-1,1-dimethyl-4-methylenecyclopentane (Table 1, Entry 5); Typical Procedure:[13]

A soln of 1-methoxy-4-{[(9-methoxy-4,4-dimethylnon-1-en-6-yn-5-yl)oxy]methyl}benzene (57.9 mg, 0.183 mmol), $Pd(OAc)_2$ (4.3 mg, 0.019 mmol), and *N*,*N*-bis(benzylidene)ethylenediamine (4.5 mg, 0.019 mmol) in dry benzene-d_6 (0.6 mL) (**CAUTION:** *carcinogen*) was heated at 60 °C for 14.75 h. Direct flash column chromatography (silica gel, hexane/Et_2O 10:1 to 4:1) gave the title compound; yield: 53.6 mg (93%); ratio (*Z*/*E*) >11:1.

1.7.1.2 Titanium-Catalyzed Cycloisomerization of α,ω-Enynes

In the presence of dicarbonylbis(η^5-cyclopentadienyl)titanium(II) (5–20 mol%), 1,6-enynes **10** possessing both an internal alkyne and an *E*-1,2-disubstituted alkene undergo cycloisomerization in toluene at 95–105 °C to afford Alder–ene products **11** (Table 2).[14] As an example, the titanium-catalyzed reaction of diethyl (*E*)-but-2-enyl(but-2-ynyl)malonate stereoselectively affords the Alder–ene product in 97% yield (entry 1). This method is compatible with amine-tethered enynes, which afford pyrrolidine derivatives (entry 2). Enynes possessing a propyl group at the alkene terminus afford *Z*/*E* mixtures (entry 3). Furthermore, if the propargylic position is unsubstituted, a single diastereomeric product is obtained, but if the propargylic position bears a methyl group, a mixture of diastereomers is formed (compare entries 5 and 6).

Table 2 Titanium-Catalyzed Cycloisomerization of α,ω-Enynes[14]

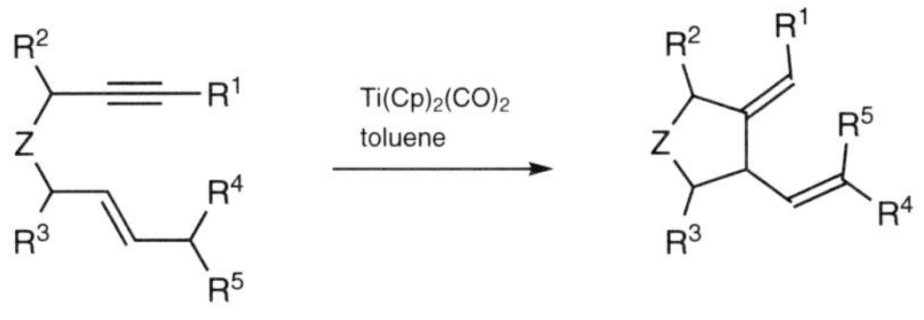

10 → **11** ($Ti(Cp)_2(CO)_2$, toluene)

Entry	Substrate	Catalyst (mol%)	Conditions	Product	Yield (%)	Ref
1	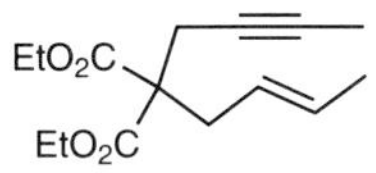	10	95 °C, 24 h	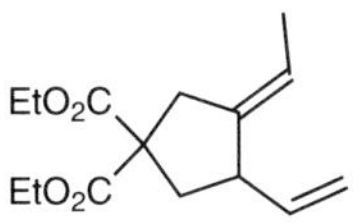	97	[14]
2	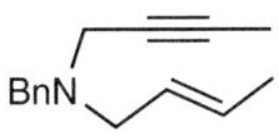	5	95 °C, 4 h	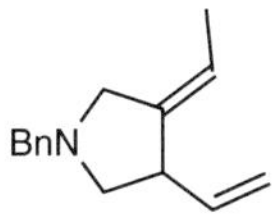	85	[14]
3	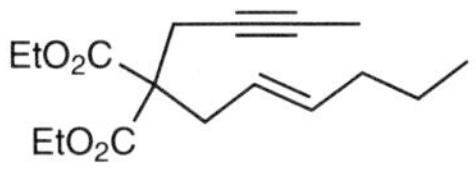	15	95 °C, 27 h	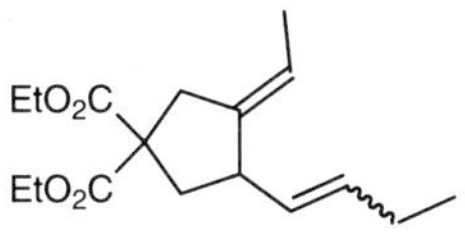	88[a]	[14]
4	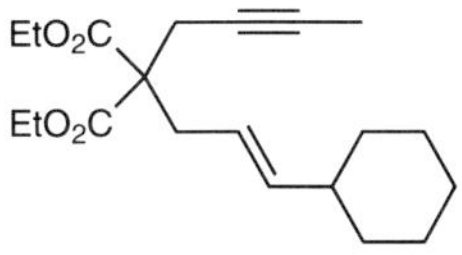	20	105 °C, 24 h	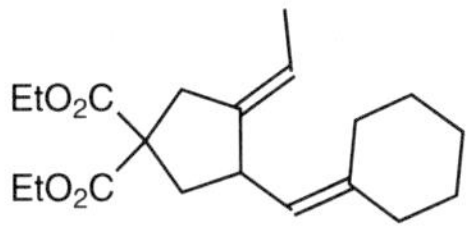	87	[14]
5	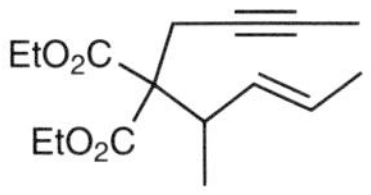	10	105 °C, 38 h	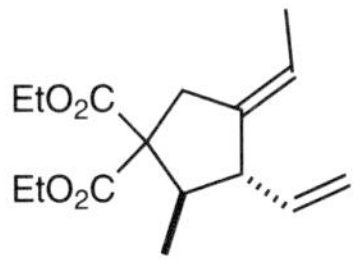	92	[14]
6	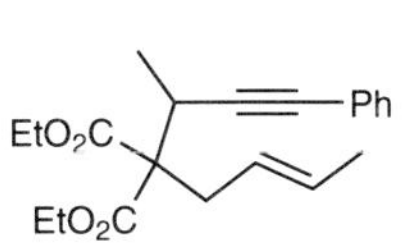	20	95 °C, 22 h	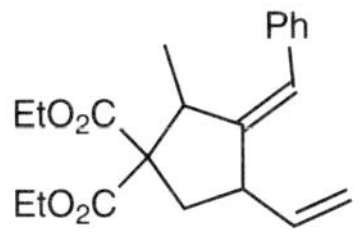	80[b]	[14]

[a] Isomeric ratio 1:3.5.
[b] Isomeric ratio 1:1.3.

Diethyl (*E*)-3-Ethylidene-4-vinylcyclopentane-1,1-dicarboxylate (Table 2, Entry 1); Typical Procedure:[14]

A soln of diethyl (*E*)-but-2-enyl(but-2-ynyl)malonate (67 mg, 0.25 mmol) and $Ti(Cp)_2(CO)_2$ (6 mg, 0.026 mmol) in dry toluene (2 mL) was heated at 95 °C for 24 h. The crude mixture was concentrated and purified by flash column chromatography (silica gel, hexane/Et_2O 50:1) to give the title compound as a clear oil; yield: 65 mg (97%).

1.7.1.3 Ruthenium-Catalyzed Cycloisomerization of α,ω-Enynes

In the presence of tris(acetonitrile)(cyclopentadienyl)ruthenium(II) hexafluorophosphate (10 mol%), the Alder–ene-type cycloisomerization of enynes **12** effectively proceeds in dimethylformamide or acetone as solvent at room temperature to afford cyclic products **13** (Table 3).[15] As an example, the reaction of dimethyl [(*E*)-3-cyclohexylprop-2-enyl](prop-2-ynyl)malonate, possessing a terminal alkyne and an *E*-1,2-disubstituted alkene moiety, affords the product in 82% yield (entry 1). Trisubstituted alkene and sulfone moieties are also compatible with this method, although lower yields are obtained (entry 2). Tetrahydrofuran and piperidine derivatives can be obtained from ether- and tosylamide-tethered enynes (entries 3 and 4). On the other hand, enynes possessing an electron-deficient internal alkyne afford only modest yields (entry 5). In striking contrast, an unusual cycloisomerization involving allylic C—H activation occurs if similar enynes possessing a quaternary propargylic carbon center together with an electron-deficient alkyne are used as substrates. As a result, cycloheptene derivatives **14** are obtained in yields of 53–83% (entries 6–8).[16] The reaction of an enyne possessing two different substituents on the terminus of the alkene (entry 8) selectively proceeds at the *cis*-methyl substituent at elevated temperatures.

Table 3 Ruthenium-Catalyzed Cycloisomerization of α,ω-Enynes[15,16]

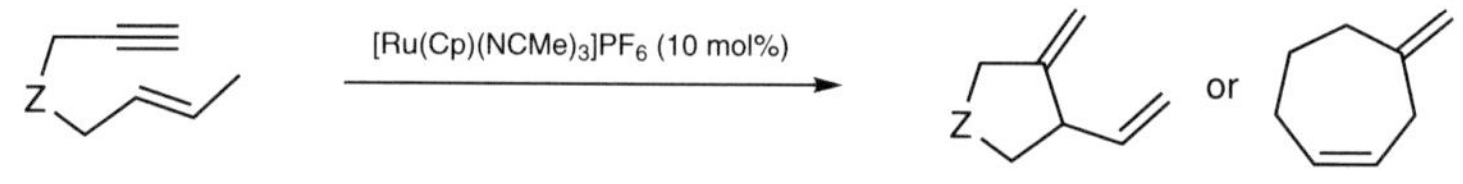

12 → [Ru(Cp)(NCMe)$_3$]PF$_6$ (10 mol%) → 13 or 14

Entry	Substrate	Conditions	Product	Yield (%)	Ref
1	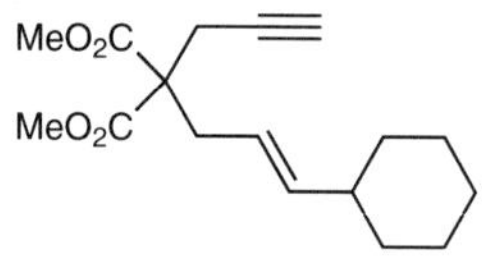 MeO2C, MeO2C	DMF, rt, 1 h	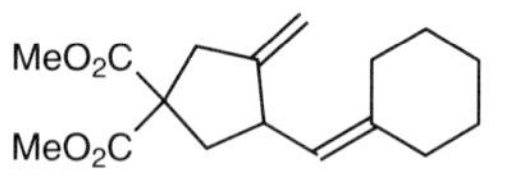 MeO2C, MeO2C	82	[15]
2	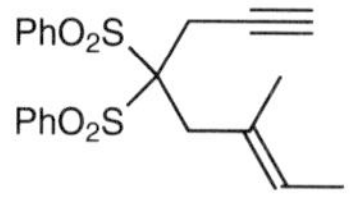 PhO2S, PhO2S	DMF, rt, 4 h	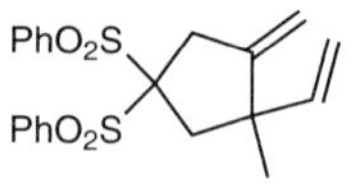 PhO2S, PhO2S	69	[15]
3	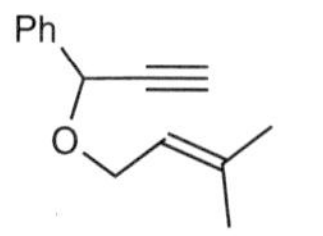 Ph, O	acetone, rt, 2 h	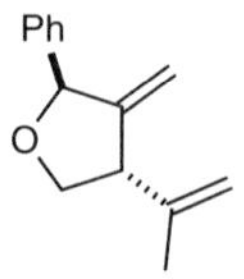 Ph, O	62[a]	[15]

Table 3 (cont.)

Entry	Substrate	Conditions	Product	Yield (%)	Ref
4	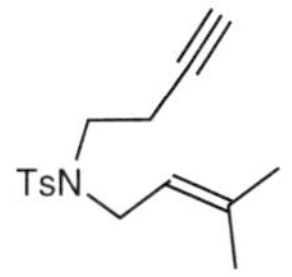 TsN	acetone, rt, 2 h	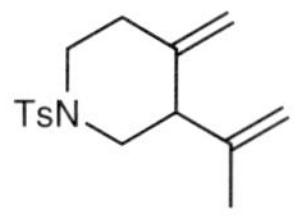 TsN	75	[15]
5	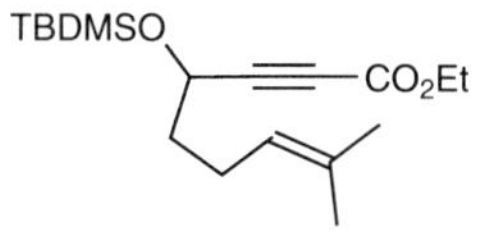 TBDMSO, CO_2Et	DMF, rt, 6 h	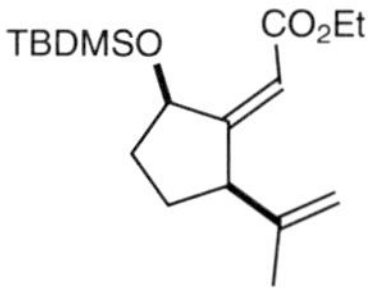 TBDMSO, CO_2Et	54[b]	[15]
6	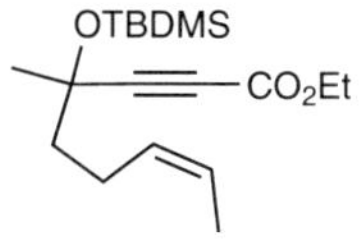 OTBDMS, CO_2Et	DMF, rt, 4 h	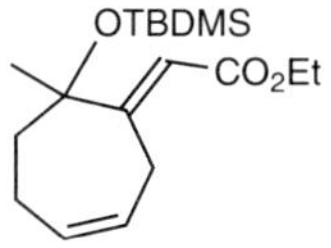 OTBDMS, CO_2Et	83	[16]
7	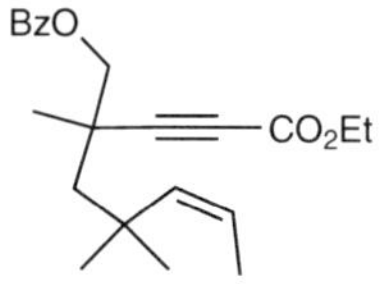 BzO, CO_2Et	DMF, rt, 6 h	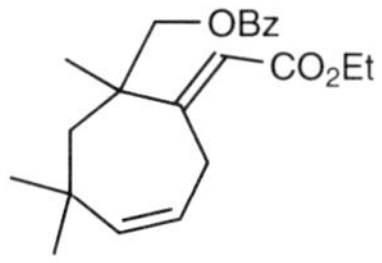 OBz, CO_2Et	65	[16]
8	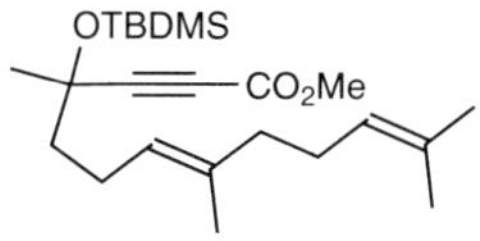 OTBDMS, CO_2Me	DMF, 60 °C, 6 h	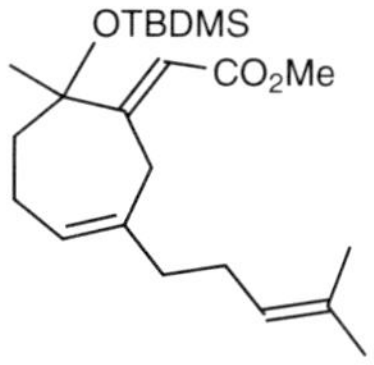 OTBDMS, CO_2Me	53	[16]

[a] dr 8:1.
[b] dr 1.1:1.

Dimethyl 3-(Cyclohexylidenemethyl)-4-methylenecyclopentane-1,1-dicarboxylate (Table 3, Entry 1); Typical Procedure:[15]
A soln of dimethyl [(*E*)-3-cyclohexylprop-2-enyl](prop-2-ynyl)malonate (50 mg, 0.171 mmol) and $[Ru(Cp)(NCMe)_3]PF_6$ (8 mg, 0.018 mmol) in degassed DMF (0.8 mL) was stirred for 1 h at rt under an atmosphere of argon. The soln was diluted with Et_2O (15 mL), washed with H_2O (2 × 20 mL), dried ($MgSO_4$), and concentrated. Flash column chromatography (silica gel, petroleum ether/Et_2O 10:1) gave the title compound as a colorless liquid; yield: 41 mg (82%).

1.7.1.4 Iron-Catalyzed Cycloisomerization of α,ω-Enynes

In the presence of an iron ferrate complex {$[Li(TMEDA)][Fe(Cp)(H_2C{=}CH_2)_2]$; 5 mol%}, the Alder–ene-type cycloisomerization of enynes **15** possessing a cycloalkene moiety proceeds in toluene at 80–90 °C to afford bicyclic products **16** in yields of 61–96% (Table 4).[17] A cyclohexene derivative affords a *cis*-fused product (entry 1), whereas *trans*-fused products are predominant in the reactions of enynes bearing larger cycloalkene moieties (entries 2–5). Although the geometries of smaller endocyclic alkenes (entries 1–3) are *Z*, in larger cycles (entries 4 and 5) the endocyclic *E*-alkene moieties are selectively formed, even though their precursors are *Z*/*E* mixtures. The aryl substituent at the terminus of the alkyne is not essential; diethyl (*Z*)-cyclooct-2-enyl(prop-2-ynyl)malonate (entry 3) affords a 5,8-fused product in high yield. Although a simpler acyclic enyne fails to undergo cycloisomerization with this iron catalyst, the use of a different iron catalyst [Li(DME)][Fe(Cp)(cod)] enables the formation of the desired product in 83% yield (entry 6).[18]

Table 4 Iron-Catalyzed Cycloisomerization of α,ω-Enynes[17,18]

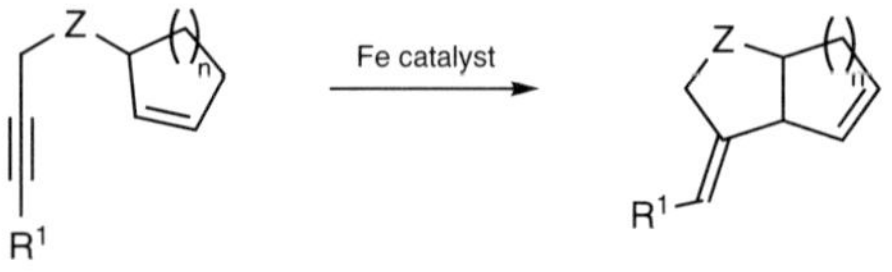

Entry	Substrate	Conditions	Product	Yield (%)	Ref
1	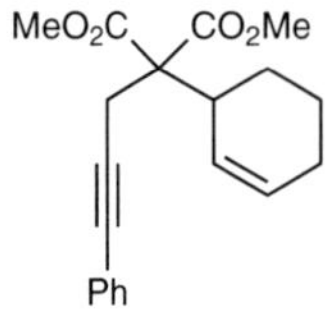	$[Li(TMEDA)][Fe(Cp)(H_2C{=}CH_2)_2]$ (5 mol%), toluene, 80–90 °C, 3–6 h	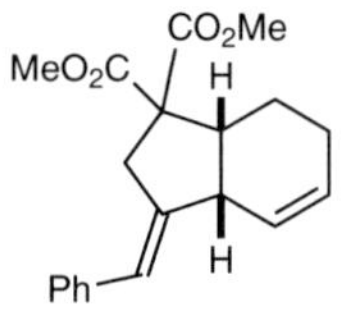	86	[17]
2	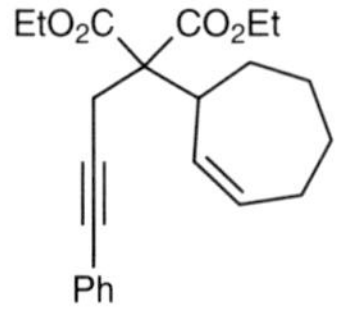	$[Li(TMEDA)][Fe(Cp)(H_2C{=}CH_2)_2]$ (5 mol%), toluene, 80–90 °C, 3–6 h	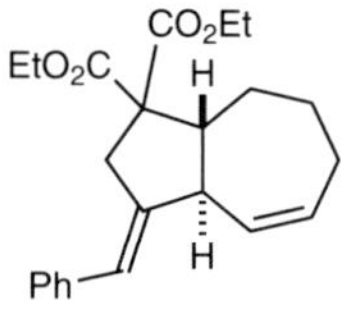	63	[17]
3	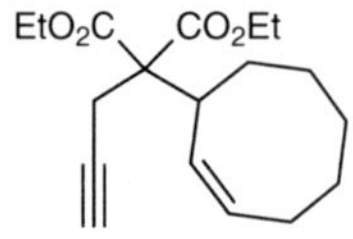	$[Li(TMEDA)][Fe(Cp)(H_2C{=}CH_2)_2]$ (5 mol%), toluene, 80–90 °C, 6 h	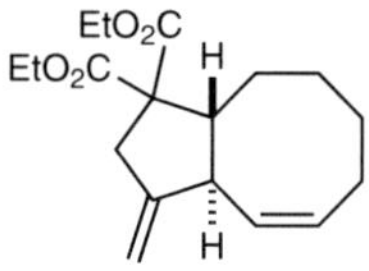	83	[17]

Table 4 (cont.)

Entry	Substrate	Conditions	Product	Yield (%)	Ref
4		$[Li(TMEDA)][Fe(Cp)(H_2C{=}CH_2)_2]$ (5 mol%), toluene, 80–90 °C, 3–6 h		89	[17]
5		$[Li(TMEDA)][Fe(Cp)(H_2C{=}CH_2)_2]$ (5 mol%), toluene, 80–90 °C, 3–6 h		96	[17]
6		[Li(DME)][Fe(Cp)(cod)] (5 mol%), toluene, reflux, 20 h		83	[18]

Diethyl (*Z*)-3-Methylene-3,3a,7,8,9,9a-hexahydro-1*H*-cyclopenta[8]annulene-1,1(2*H*,6*H*)-dicarboxylate (Table 4, Entry 3); Typical Procedure:[17]

A soln of diethyl (*Z*)-cyclooct-2-enyl(prop-2-ynyl)malonate (123 mg, 0.40 mmol) in dry toluene (2 mL) was added to a soln of $[Li(TMEDA)][Fe(Cp)(H_2C{=}CH_2)_2]$ (6 mg, 0.020 mmol) in dry toluene (2 mL), and the resulting mixture was stirred for 6 h at 80–90 °C under an atmosphere of argon. The soln was concentrated, and the residue was purified by flash column chromatography (silica gel, hexane/EtOAc) to give the title compound as a colorless oil; yield: 102 mg (83%).

1.7.1.5 Enantioselective Cycloisomerization of α,ω-Enynes

In the presence of silver(I) hexafluoroantimonate, a rhodium(I) complex with a 1,4-bis(diphenylphosphino)butane ligand efficiently catalyzes the cycloisomerization of α,ω-enynes **17** possessing a *Z*-1,2-disubstituted alkene moiety to afford Alder–ene products **18**.[19] An enantioselective version of this method is also possible by combining a rhodium catalyst with a chiral BINAP ligand (Table 5).[20–22] In the presence of 10 mol% of the cationic rhodium catalyst derived in situ from chloro(cycloocta-1,5-diene)rhodium(I) dimer, BINAP, and silver(I) hexafluoroantimonate, ether-tethered enynes undergo cycloisomerization in 1,2-dichloroethane within 5 minutes at room temperature to afford tetrahydrofuran derivatives in almost quantitative yields with enantiomeric excess values over 99% (entries 1 and 2).[20] In the same manner, cycloisomerization of amide-tethered enynes affords γ-lactams in excellent yields with >99% ee (entries 3 and 4).[21] Furthermore, γ-lactones (entries 5 and 6) are also obtained from ester-tethered enynes with similar efficiencies.[22] Notably, allylic acetate and allylic alcohol moieties on these substrates are tolerated, and the latter is ultimately transformed into a formyl group via an intermediary enol.

Table 5 Rhodium-Catalyzed Enantioselective Cycloisomerization of α,ω-Enynes[20–22]

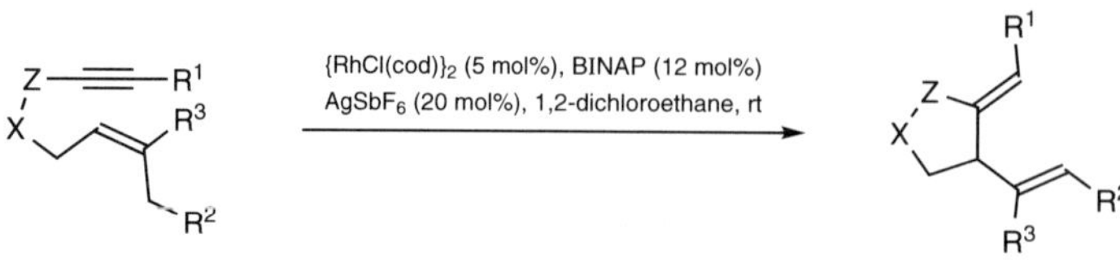

17 → 18: $\{RhCl(cod)\}_2$ (5 mol%), BINAP (12 mol%), $AgSbF_6$ (20 mol%), 1,2-dichloroethane, rt

Entry	Substrate	Config of BINAP	Time	Product[a]	ee (%)	Yield (%)	Ref
1	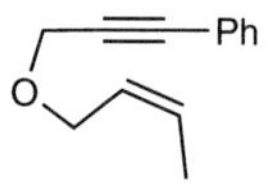	*R*	<5 min	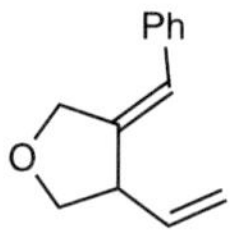	>99.9	96	[20]
2	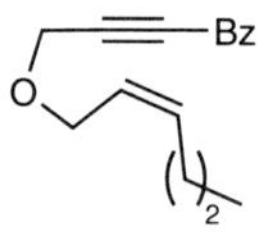	*S*	<5 min	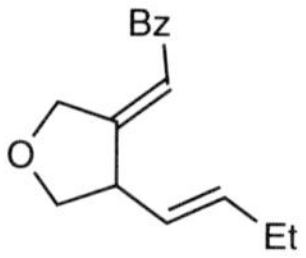	>99.9	99	[20]
3	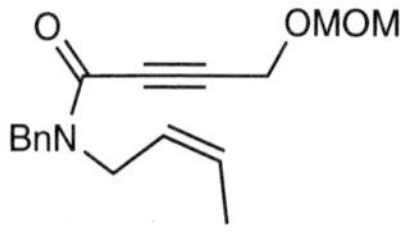	*S*	<5 min	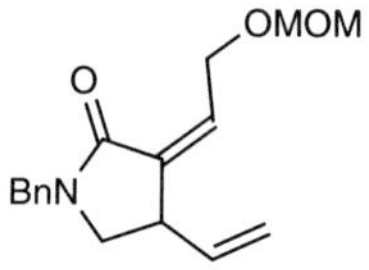	>99	87	[21]
4	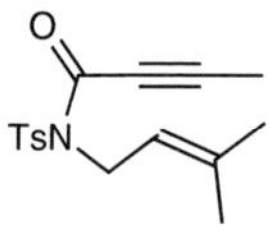	*S*	24 h	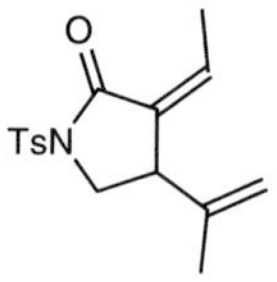	>99	91	[21]
5	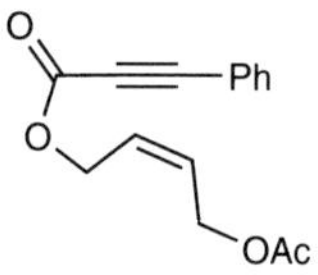	*R*	<10 min	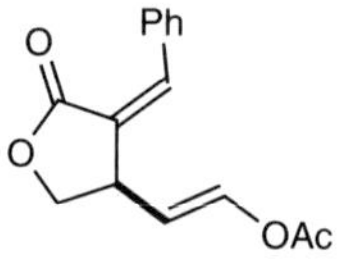	>99	96	[22]
6	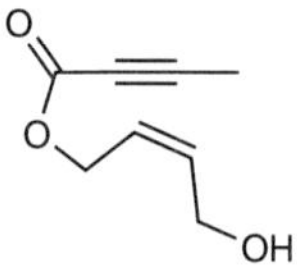	*R*	<10 min	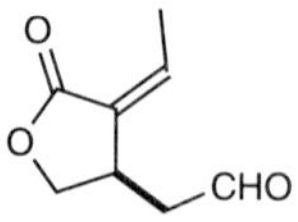	>99	99	[22]

[a] The absolute stereochemistry of the products given in entries 1–4 is not shown in the original report.

In the presence of formic acid (1 equiv), the enantioselective Alder–ene-type cycloisomerization of aniline-tethered enynes **19** is promoted by 5 mol% of a catalyst derived from the cationic palladium complex, tetrakis(acetonitrile)palladium(II) bis(tetrafluoroborate), and (*S*)-BINAP and quantitatively affords quinoline derivatives **20** possessing a quaternary carbon stereocenter (Table 6). An ester-substituted alkyne gives the product as a single enantiomer (entry 1),[23] but the ester group is not essential, as a terminal alkyne (entry 2) gives the corresponding product with similar efficiency. An enyne possessing a cyclohexene moiety (entry 3) affords an interesting spirocyclic quinoline derivative in good yield, although the enantioselectivity is significantly lower because of the migration of the resultant endocyclic alkene moiety. Similar enynes with a dihydropyran moiety (entries 4 and 5) can be quantitatively transformed into the corresponding spirocyclic products as single enantiomers, as alkene migration is not possible in this case. Similarly, an enyne with a 15-membered cycloalkene (entry 6) still affords the product with good enantioselectivity, albeit in modest yield, even though migration of the endocyclic alkene occurs.

Table 6 Palladium-Catalyzed Enantioselective Cycloisomerization of α,ω-Enynes[23]

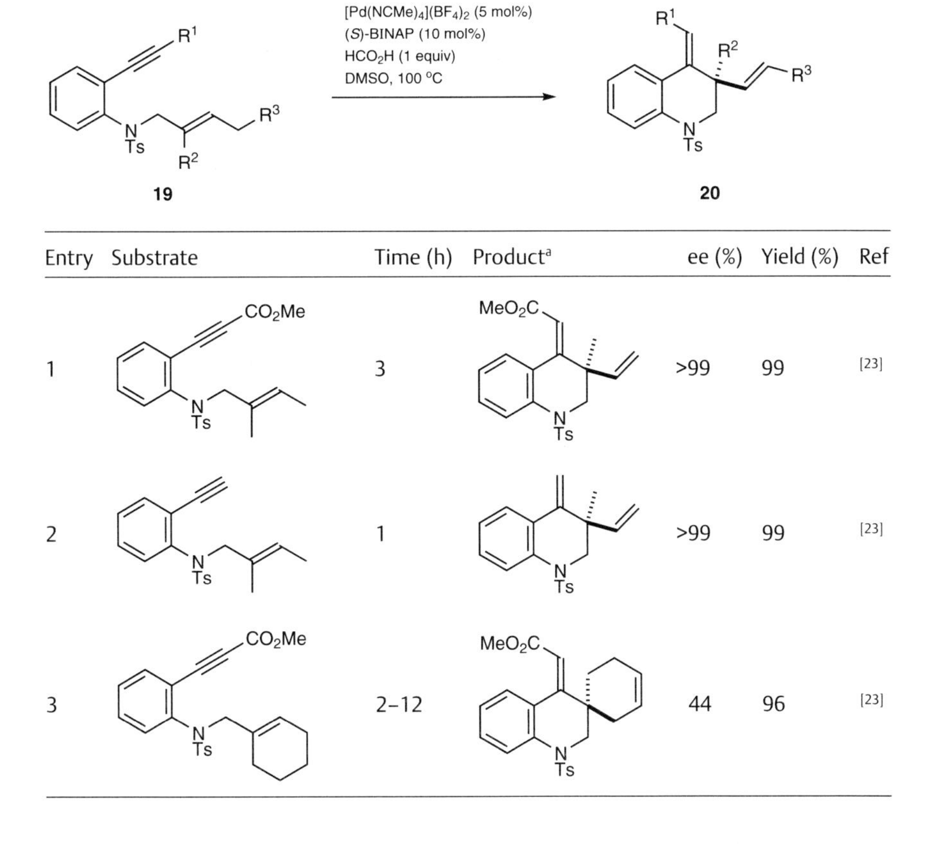

Entry	Substrate	Time (h)	Product[a]	ee (%)	Yield (%)	Ref
1		3		>99	99	[23]
2		1		>99	99	[23]
3		2–12		44	96	[23]

for references see p 383

Table 6 (cont.)

Entry	Substrate	Time (h)	Product[a]	ee (%)	Yield (%)	Ref
4	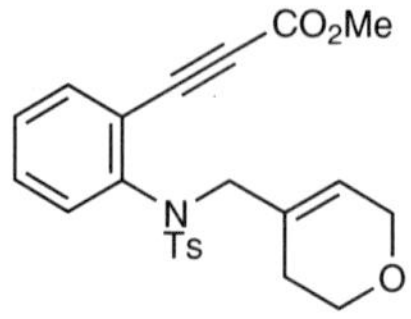	1	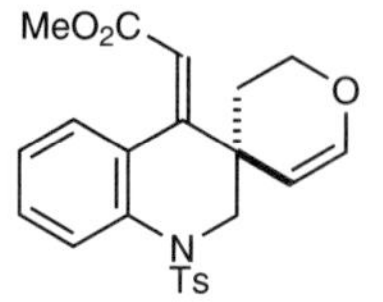	>99	>99	[23]
5	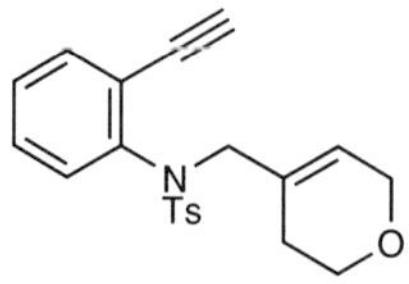	1	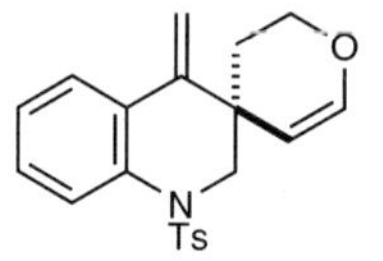	98	>99	[23]
6	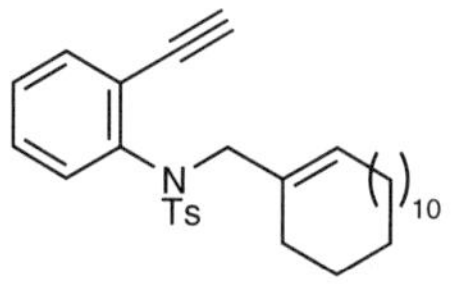	1	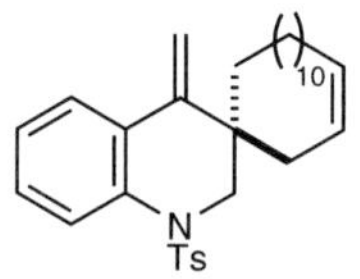	86	53	[23]

[a] Structures of the major products are shown.

(−)-(Z)-3-Benzylidene-4-vinyltetrahydrofuran (Table 5, Entry 1); Typical Procedure:[20]
Freshly prepared {3-[(Z)-but-2-enyloxy]prop-1-ynyl}benzene (37.2 mg, 0.20 mmol) was added to a soln of $\{RhCl(cod)\}_2$ (4.9 mg, 0.01 mmol) and (S)-BINAP (13.8 mg, 0.022 mmol) in freshly distilled 1,2-dichloroethane (2 mL) at rt under an atmosphere of N_2. After stirring for 1 min, $AgSbF_6$ (13.7 mg, 0.040 mmol) was added to the mixture, and the reaction was complete within 5 min. The mixture was directly subjected to flash column chromatography to give the title compound; yield: 35.8 mg (96%); >99.9% ee.

4-Methylene-1-[(4-methylphenyl)sulfonyl]-3-vinyl-1,2,3,4-tetrahydroquinolines 20; **General Procedure:**[23]
Degassed DMSO (2.0 mL) was added to a mixture of $[Pd(NCMe)_4](BF_4)_2$ (2.2 mg, 0.005 mmol) and (S)-BINAP (6.2 mg, 0.010 mmol) under an atmosphere of argon, and the mixture was stirred for 5 min at rt. The enyne **19** (0.10 mmol) and HCO_2H (3.7 μL, 0.10 mmol) were added, and the mixture was stirred for 1–3 h at 100 °C. The mixture was washed with brine and extracted with Et_2O. The combined organic layer was concentrated under reduced pressure, and the residue was purified by short column chromatography (silica gel, pentane/Et_2O 2:1) to give the title compound.

1.7.2 Cycloisomerizations of α,ω-Dienes

The cycloisomerization of α,ω-dienes involving intramolecular hydrogen-atom transfer using various transition-metal catalysts also proceeds under conditions milder than those used in the corresponding thermal process.[24] However, the cycloisomerization of 1,6-dienes such as **21** generally produces several isomeric products such as **22–24** depending on the catalyst and the substrate type (Scheme 3); therefore, identification of an isomer-selective diene cycloisomerization reaction with wide generality has remained a difficult challenge. Moreover, the control of regiochemistry in the cycloisomerization of dienes is a formidable task, especially if unsymmetrical dienes containing alkene termini with similar reactivities are used.[25] This situation is in striking contrast to the cycloisomerization of enynes described in Section 1.7.1, for which the alkyne and alkene termini have markedly different reactivities toward transition-metal catalysts. In Sections

1.7.2.1–1.7.2.4 representative protocols for diene cycloisomerizations with practical levels of isomer selectivity are presented.

Scheme 3 Transition-Metal-Catalyzed Cycloisomerization of 1,6-Dienes[24]

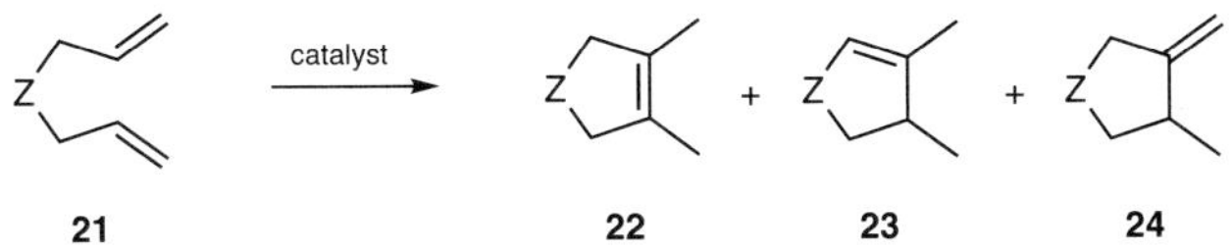

catalyst

21 22 23 24

1.7.2.1 Palladium-Catalyzed Cycloisomerization of α,ω-Dienes

In the presence of triethylsilane, the cationic palladium complex generated in situ from η^3-allyl(chloro)(tricyclohexylphosphine)palladium(II) and sodium tetrakis[3,5-bis(trifluoromethyl)phenyl]borate (NaBARF) efficiently catalyzes the cycloisomerization of 1,6-dienes **25** possessing a quaternary carbon atom on the tether at room temperature over 20 minutes to afford cyclopentenes **26**, possessing a thermodynamically favorable tetrasubstituted alkene moiety, in yields of 82–99% with an isomeric ratio of ~50:1 (Table 7, entries 1–4).[26] A substituent at one of the allylic methylene carbon atoms (entry 4) is also tolerated, as this diene is transformed into the product with similar efficiency. On the other hand, cyclopentene isomers **27** with a thermodynamically less-favorable trisubstituted alkene moiety are obtained from cycloisomerizations catalyzed by bis(pivalonitrile)dichloropalladium(II) without an additive (Table 8).[27] As an example, dimethyl diallylmalonate is subjected to cycloisomerization in the presence of the catalyst (5 mol%) at 40 °C for 1.5 hours to afford the cyclopentene product in 96% yield with >95% selectivity (entry 1). A spirocyclic product is formed in 90% yield from a derivative of Meldrum's acid (entry 2). Reaction of a monoprotected diol derivative affords the cyclopentene product in good yield, albeit with low diastereoselectivity (17% de; entry 3). In addition, an *N*-(trifluoroacetyl)-2,3-dihydropyrrole derivative can be obtained from an *N*,*N*-diallylamine substrate, albeit in diminished yield (entry 4).

for references see p 383

Table 7 Palladium-Catalyzed Cycloisomerization of α,ω-Dienes To Give Tetrasubstituted Alkenes[26]

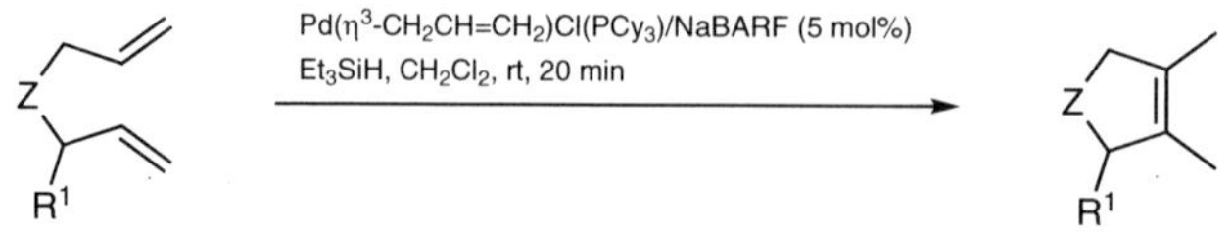

Pd(η^3-$CH_2CH{=}CH_2$)Cl(PCy_3)/NaBARF (5 mol%)
Et_3SiH, CH_2Cl_2, rt, 20 min

25 → **26**

Entry	Substrate	Et_3SiH (Equiv)	Product	Isomeric Ratio	Yield (%)	Ref
1	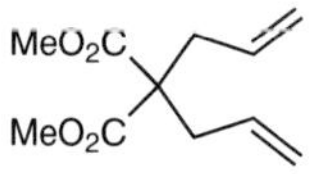 MeO2C, MeO2C	1.5	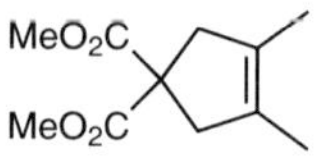 MeO2C, MeO2C	49:1	89	[26]
2	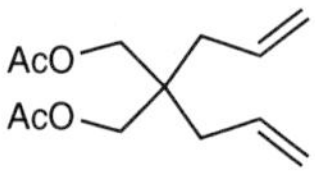 AcO, AcO	1.2	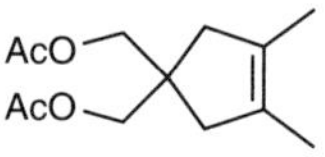 AcO, AcO	52:1	99	[26]
3	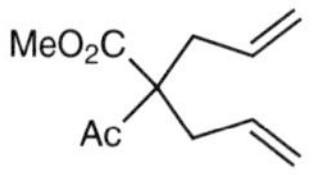 MeO2C, Ac	1.2	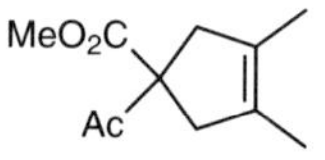 MeO2C, Ac	50:1	89	[26]
4	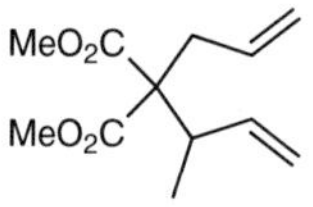 MeO2C, MeO2C	1.2	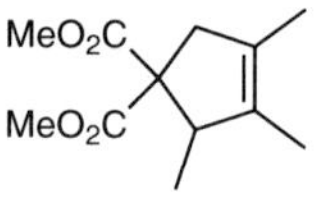 MeO2C, MeO2C	52:1	82	[26]

Table 8 Palladium-Catalyzed Cycloisomerization of α,ω-Dienes To Give Trisubstituted Alkenes[27]

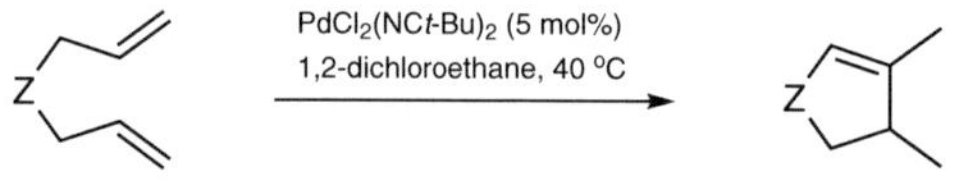

$PdCl_2$(NC*t*-Bu)$_2$ (5 mol%)
1,2-dichloroethane, 40 °C

→ **27**

Entry	Substrate	Time (h)	Product	Yield[a] (%)	Ref
1	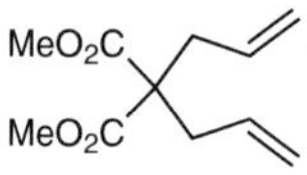 MeO2C, MeO2C	1.5	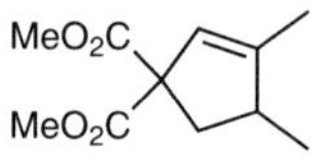 MeO2C, MeO2C	96	[27]
2	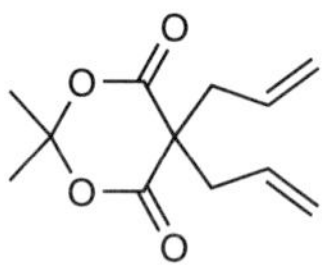 O, O, O, O	4	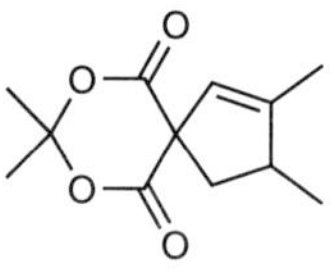 O, O, O, O	90	[27]
3	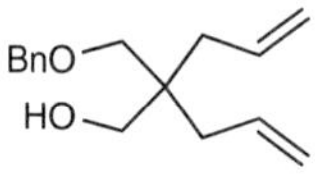 BnO, HO	1	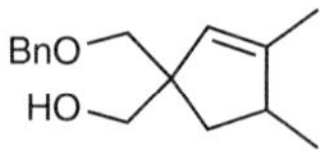 BnO, HO	82[b]	[27]
4	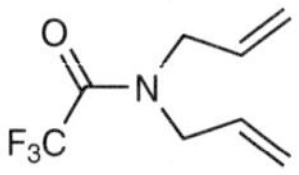 O, N, F3C	2	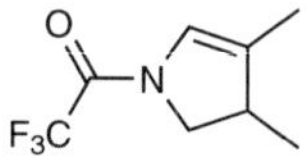 O, N, F3C	50	[27]

[a] Regioselectivity was >95% in all cases.
[b] 17% de.

Dimethyl 3,4-Dimethylcyclopent-3-ene-1,1-dicarboxylate (Table 7, Entry 1); Typical Procedure:[26]

Dimethyl diallylmalonate (100 mg, 0.47 mmol) and Et_3SiH (80 mg, 0.69 mmol) were added sequentially to a soln of $Pd(\eta^3\text{-}CH_2CH{=}CH_2)Cl(PCy_3)$ (10 mg, 0.022 mmol) and NaBARF (24 mg, 0.027 mmol) in dry CH_2Cl_2 (8 mL) at 0 °C under an atmosphere of N_2. The resultant yellow soln was stirred for 20 min at rt. The solvent and silane were evaporated under reduced pressure, and the residue was purified by flash column chromatography (silica gel, hexane/EtOAc 12:1) to give the title compound as a colorless oil; yield: 89 mg (89%); isomeric ratio 49:1.

2,3,8,8-Tetramethyl-7,9-dioxaspiro[4.5]dec-1-ene-6,10-dione (Table 8, Entry 2); Typical Procedure:[27]

A soln of 5,5-diallyl-2,2-dimethyl-1,3-dioxane-4,6-dione (80.2 mg, 0.36 mmol) in dry 1,2-dichloroethane (1 mL) was added to a stirred soln of $PdCl_2(NC\textit{t}\text{-}Bu)_2$ (6.1 mg, 0.018 mmol) in dry 1,2-dichloroethane (1 mL). The mixture was stirred for 4 h at 40 °C and then filtered through a pad of silica gel (2 cm × 1.5 cm, CH_2Cl_2, 20 mL). The filtrate was concentrated under reduced pressure, and the residue was purified by flash column chromatography (silica gel, hexane/EtOAc 10:1) to give the title compound as a clear oil; yield: 72.1 mg (90%).

1.7.2.2 Titanium-Catalyzed Cycloisomerization of α,ω-Dienes

In contrast to the palladium-catalyzed reactions outlined in Section 1.7.2.1, thermodynamically much less favorable *exo*-methylenecycloalkanes **29** are obtained by the cycloisomerization of 1,6-dienes **28** using a titanium(IV) phenolate as the precatalyst (Table 9).[28] Typically, 1,6-dienes (entries 1 and 2) are treated with the titanium(IV) phenolate (10 mol%) and cyclohexylmagnesium chloride (25 mol%) at 0 °C to room temperature to afford spirocyclic products in yields greater than 80%, although the corresponding saturated byproducts are formed in small amounts. A silyl ether (entry 3) and a silane-tethered diene (entry 4) are also compatible with this protocol and good yields are obtained. Similarly, a cyclohexane derivative (entry 5) is formed with high diastereoselectivity from a 1,7-diene. Instead of the titanium(IV) phenolate, commercially available dichlorobis(η^5-cyclopentadienyl)titanium(IV) (titanocene dichloride) can also be used as the precatalyst along with butylmagnesium bromide as the activator.[29] Thus, 1,6-dienes can be transformed in high yields upon treatment with titanocene dichloride (5 mol%) and butylmagnesium bromide (15 mol%) at room temperature (entry 6). The corresponding nitrogen heterocycle (entry 7) is also accessible from an *N*,*N*-diallylamine derivative, although an increased catalyst loading is necessary.

for references see p 383

Table 9 Titanium-Catalyzed Cycloisomerization of α,ω-Dienes[28,29]

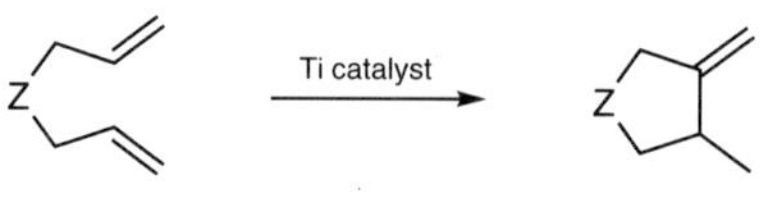

28 → (Ti catalyst) → 29

Entry	Substrate	Conditions	Product	Yield (%)	Ref
1	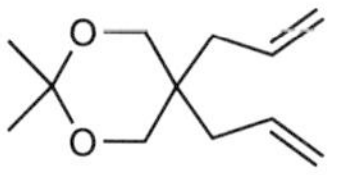	$Ti(2,6-Me_2C_6H_3O)_4$ (10 mol%), CyMgCl (25 mol%), THF, 0°C to rt, 24 h	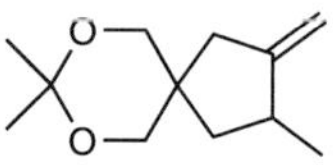	85[a]	[28]
2	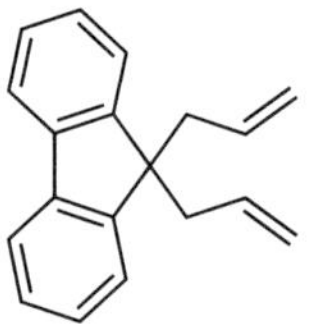	$Ti(2,6-Me_2C_6H_3O)_4$ (10 mol%), CyMgCl (25 mol%), THF, 0°C to rt, 24 h	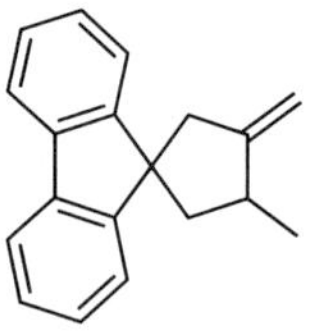	88[a]	[28]
3	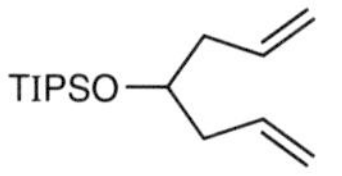 TIPSO	$Ti(2,6-Me_2C_6H_3O)_4$ (10 mol%), CyMgCl (25 mol%), THF, 0°C to rt, 24 h	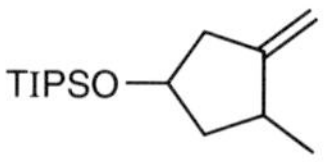 TIPSO	85[a,b]	[28]
4	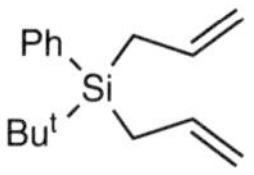 Ph, Bu^t, Si	$Ti(2,6-Me_2C_6H_3O)_4$ (10 mol%), CyMgCl (25 mol%), THF, 0°C to rt, 24 h	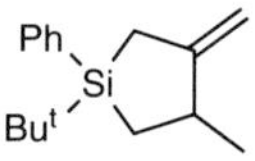 Ph, Bu^t, Si	73[a,c]	[28]
5	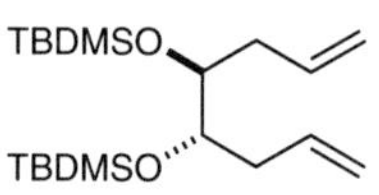 TBDMSO, TBDMSO	$Ti(2,6-Me_2C_6H_3O)_4$ (10 mol%), CyMgCl (25 mol%), THF, 0°C to rt, 24 h	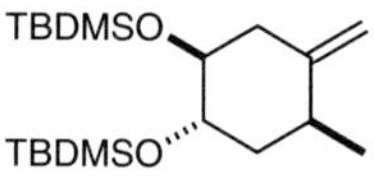 TBDMSO, TBDMSO	71[a,d]	[28]
6	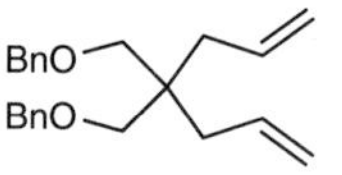 BnO, BnO	$Ti(Cp)_2Cl_2$ (5 mol%), BuMgBr (15 mol%), Et_2O, rt, 24–48 h	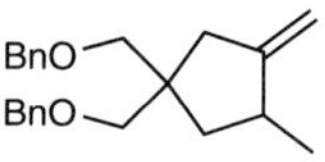 BnO, BnO	92	[29]
7	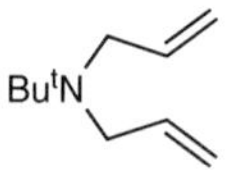 Bu^{t}N	$Ti(Cp)_2Cl_2$ (10 mol%), BuMgBr (30 mol%), Et_2O, rt, 24–48 h	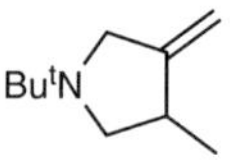 Bu^{t}N	87[e]	[29]

[a] Saturated compounds are formed as byproducts in yields of 5–10%.
[b] dr 52:48.
[c] dr 60:40.
[d] dr 96:4.
[e] Yield determined by NMR spectroscopy.

1-Methyl-2-methylenecyclopentanes 29; General Procedure:[28]
A 0.96 M soln of CyMgCl in Et_2O (130 μL, 0.125 mmol) was added dropwise to a mixture of α,ω-diene **28** (0.5 mmol) and $Ti(2,6-Me_2C_6H_3O)_4$ (0.05 mmol) in dry degassed THF (2 mL) at 0°C under an atmosphere of argon. The resultant mixture was warmed to rt over ~10 h and was stirred for a total of 24 h. Sat. aq $NaHCO_3$ (~120 μL), hexanes (3 mL), Celite (0.8 g), and $MgSO_4$ (0.8 g) were then sequentially added. The resultant suspension was filtered using hexanes as the eluant, and the filtrate was concentrated under reduced pressure. The obtained residue was purified by column chromatography (silica gel, hexanes/Et_2O) to give the title compound.

1.7.2.3 Ruthenium-Catalyzed Cycloisomerization of α,ω-Dienes

Although the titanium-catalyzed cycloisomerization of α,ω-dienes (Section 1.7.2.2) selectively affords *exo*-methylenecycloalkanes under mild reaction conditions, the scope of this method is limited to relatively less functionalized α,ω-dienes.[28,29] This problem can be overcome in the ruthenium-catalyzed cycloisomerization of α,ω-dienes **30** to afford *exo*-methylenecycloalkanes **31**. The cycloisomerization is performed with the use of dichloro(cycloocta-1,5-diene)ruthenium(II) polymer as the catalyst and has wide functional group compatibility; thus, various functionalized 1,6-dienes can be used (Table 10).[30–32] As an example, in the presence of the ruthenium(II) catalyst (1 mol%), dimethyl diallylmalonate undergoes cycloisomerization at 90 °C in propan-2-ol as the solvent over 24 hours to afford the *exo*-methylenecyclopentane product (entry 1) in 92% yield with 95% isomeric purity.[30] Remarkably, this protocol is also applicable to unsymmetrical 1,6-dienes possessing an internal alkene moiety (entries 4 and 7), with the *exo*-methylenecyclopentane derivatives obtained in high yields and with high isomer selectivity. If a butenolide-derived diene (entry 5) is subjected to the ruthenium-catalyzed cycloisomerization conditions with the use of ethanol as the solvent for 12 hours, the spirocyclic product is delivered with high diastereoselectivity.[31] Furthermore, the reaction time can be considerably shortened under microwave irradiation conditions.[32] Accordingly, the cycloisomerization reactions of *N*,*N*-diallylbenzamide (entry 6) and a 1,6-diene possessing a cyclohexene moiety (entry 7) are complete within 30 minutes and yield a pyrrolidine derivative and a bicyclic compound, respectively, with improved yields and isomeric purities. Moreover, the use of a different ruthenium catalyst system (10 mol% second-generation Grubbs catalyst and 1 equiv of vinyl trimethylsilyl ether) in refluxing xylene or toluene enables the tandem alkene transposition/cycloisomerization of *N*-allyl-*N*-(2-vinylphenyl)tosylamides and allyl 2-vinylphenyl ethers, to afford *exo*-methylenedihydroindole derivatives and *exo*-methylene-2,3-dihydrobenzofurans in good yields.[33]

Table 10 Ruthenium-Catalyzed Cycloisomerization of α,ω-Dienes[30–32]

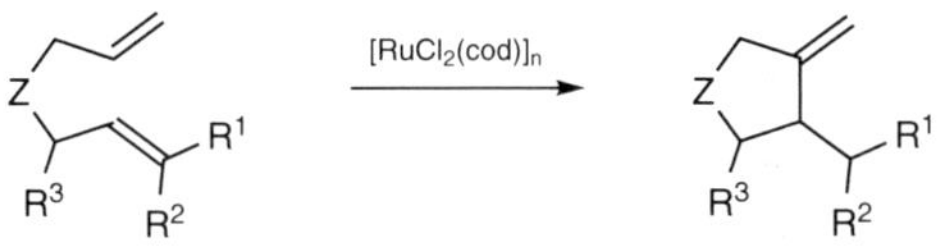

Z, R^1, R^2, R^3 **30** → ($[RuCl_2(cod)]_n$) → Z, R^1, R^2, R^3 **31**

Entry	Substrate	Catalyst (mol%)	Conditions	Product	Isomeric Purity[a] (%)	Yield (%)	Ref
1	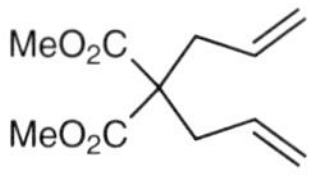 MeO_2C, MeO_2C	1	iPrOH, 90 °C, 24 h	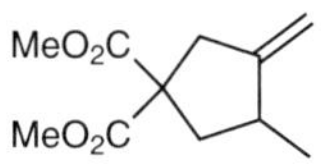 MeO_2C, MeO_2C	95	92	[30]
2	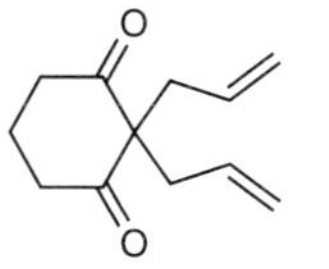 O, O	5	iPrOH, 90 °C, 24 h	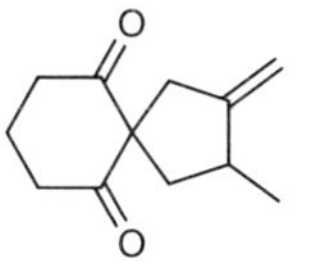 O, O	98	83	[30]
3	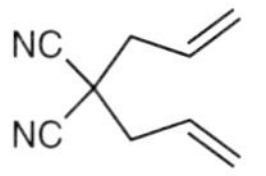 NC, NC	5	iPrOH, 90 °C, 24 h	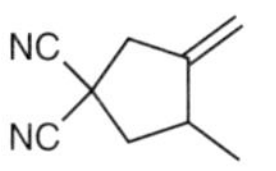 NC, NC	99	91	[30]

for references see p 383

Table 10 (cont.)

Entry	Substrate	Catalyst (mol%)	Conditions	Product	Isomeric Purity[a] (%)	Yield (%)	Ref
4	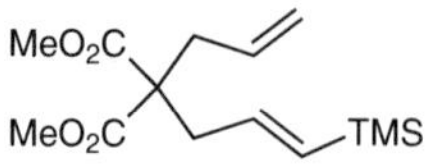 MeO2C, MeO2C, TMS	2.5	iPrOH, 90 °C, 24 h	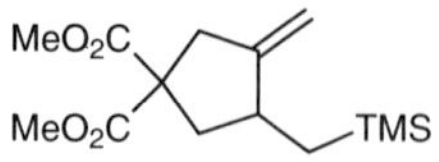 MeO2C, MeO2C, TMS	93	94	[30]
5	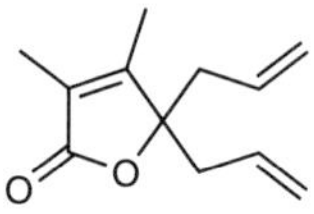 O, O	3	EtOH, 80 °C, 12 h	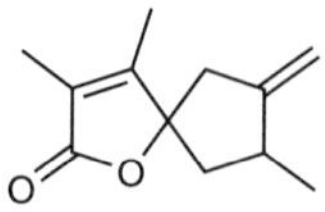 O, O	–[b]	83	[31]
6	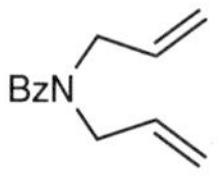 BzN	5	iPrOH, microwave (300 W), 90 °C, 0.5 h	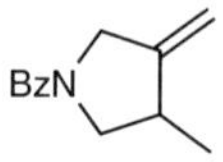 BzN	92	97	[32]
7	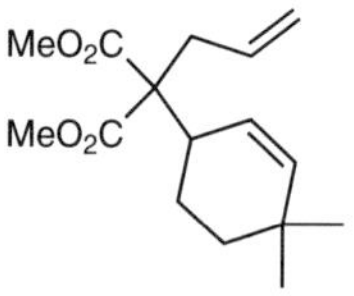 MeO2C, MeO2C	10	iPrOH, microwave (300 W), 90 °C, 0.25 h	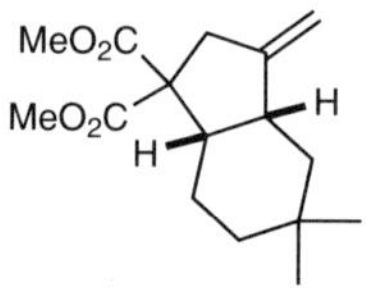 MeO2C, MeO2C, H, H	92	96	[32]

[a] Percentage of the major product among other inseparable constitutional isomers.
[b] dr 95:5.

Dimethyl 3-Methyl-4-methylenecyclopentane-1,1-dicarboxylate (Table 10, Entry 1); Typical Procedure:[30]

$[RuCl_2(cod)]_n$ (2.8 mg, 0.010 mmol) was added to a soln of dimethyl diallylmalonate (212 mg, 1.0 mmol) in degassed iPrOH (4 mL) at rt. The mixture was stirred for 24 h at 90 °C under an atmosphere of argon. The solvent was evaporated under reduced pressure, and the residue was purified by flash column chromatography (silica gel, hexane/EtOAc 20:1) to give the title compound as a colorless oil; yield: 195 mg (92%); 95% isomeric purity (GC).

1.7.2.4 Enantioselective Cycloisomerization of α,ω-Dienes

The enantioselective cycloisomerization of 1,6-dienes **33** to afford 1-methyl-2-methylenecyclopentanes **34** is achieved using a cationic nickel precatalyst with chiral azaphospholene ligand **32** (Table 11).[34] In the presence of the nickel catalyst (0.5 mol%) and **32** (0.25 mol%), dimethyl diallylmalonate undergoes cycloisomerization at 20 °C to afford the *exo*-methylenecyclopentane product (entry 1) with good conversion and selectivity. However, the enantioselectivity is modest (67% ee). The enantioselectivity is improved to 72 and 88% ee for bulkier ethyl ester (entry 2) and *tert*-butyl ester (entry 3) groups, respectively. The reaction of a diol (entry 4) is sluggish, even at a catalyst loading of 5 mol%. In contrast, the corresponding protected derivatives (entries 5 and 6) undergo cycloisomerization with good conversions and selectivity at the same catalyst loading (5 mol%), and the expected products are obtained with 71 and 86% ee, respectively.

Table 11 Nickel-Catalyzed Enantioselective Cycloisomerization of α,ω-Dienes[34]

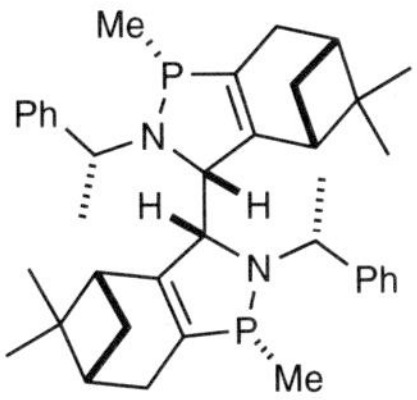

32

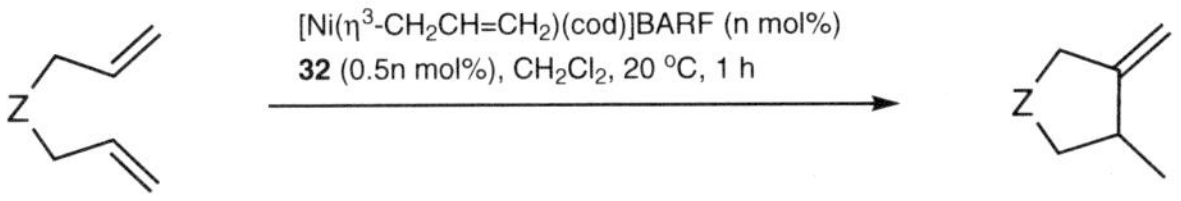

[Ni(η^3-$CH_2CH{=}CH_2$)(cod)]BARF (n mol%)
32 (0.5n mol%), CH_2Cl_2, 20 °C, 1 h

33 → 34

BARF = tetrakis[3,5-bis(trifluoromethyl)phenyl]borate

Entry	Diene	n	Product	Isomeric Selectivity[a] (%)	ee[b] (%)	Conversion (%)	Ref
1	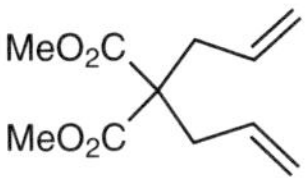 MeO_2C, MeO_2C	0.5	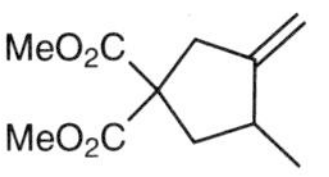 MeO_2C, MeO_2C	97	67	91	[34]
2	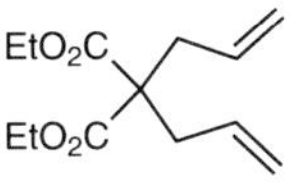 EtO_2C, EtO_2C	0.5	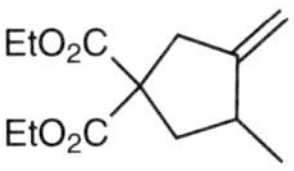 EtO_2C, EtO_2C	91	72	72	[34]
3	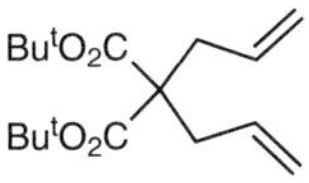 Bu^tO_2C, Bu^tO_2C	0.7	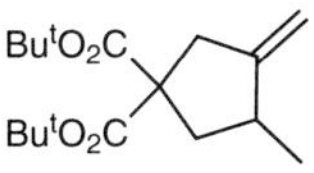 Bu^tO_2C, Bu^tO_2C	89	88	55	[34]
4	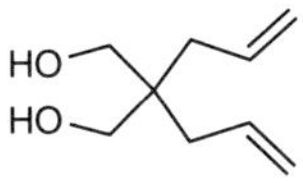 HO, HO	5	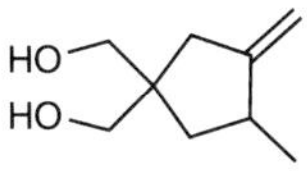 HO, HO	>99	91	39	[34]
5	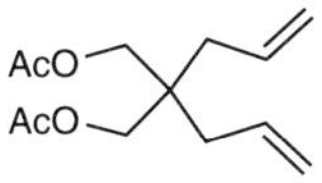 AcO, AcO	5	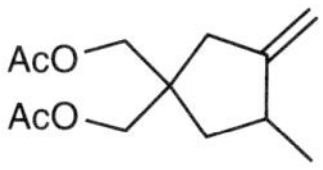 AcO, AcO	83	71	>99	[34]
6	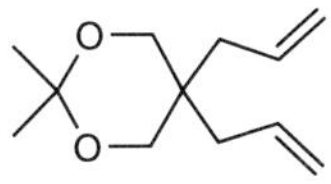 O, O	5	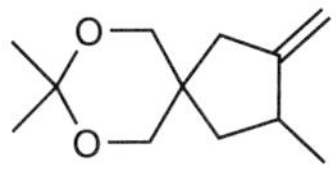 O, O	87	86	>99	[34]

[a] Percentage of the major product among other inseparable constitutional isomers.
[b] The (+)-enantiomers were predominant.

1-Methyl-2-methylenecyclopentanes 34; **General Procedure:**[34]

A soln of the azaphospholene ligand **32** (0.03 mmol) in CH_2Cl_2 (5 mL) was added to a soln of [Ni(η^3-$CH_2CH{=}CH_2$)(cod)][BARF] (0.06 mmol) in CH_2Cl_2 (5 mL) at rt under an atmosphere of argon. After 15 min, the 1,6-diene **33** was added to the resultant yellow catalyst soln, and the mixture was stirred at 20 °C for 1 h. The reaction was quenched by adding aq NH_3 (1 mL). The organic phase was washed with H_2O (3 × 2 mL), dried (Na_2SO_4), and analyzed by GC and GC/MS.

1.7.3 Cycloisomerizations Involving 1,2- and 1,3-Dienes

1,2-Dienes (allenes) and 1,3-dienes are generally more reactive than nonactivated alkenes. Therefore, replacement of the alkyne or alkene terminus of α,ω-enynes and α,ω-dienes with these groups provides intriguing substrates for transition-metal-catalyzed cycloisomerizations.[35,36] In the following sections, selected examples of practical cycloisomerizations involving allenes and 1,3-dienes are discussed.

1.7.3.1 Cycloisomerizations of Allenynes

In the presence of a rhodium complex possessing carbonyl ligands, the Alder–ene-type cycloisomerization of allenynes **35** selectively proceeds at the distal double bond of the allene moiety to afford cross-conjugated trienes **36** (Table 12).[37–41] As an example, diethyl deca-2,3-dienyl(prop-2-ynyl)malonate is treated with the rhodium complex (2 mol%) at 90 °C in toluene for 1 hour to afford the *exo*-methylenecyclohexene product in 80% yield as a mixture of *E*/*Z*-isomers (entry 1).[37] Allenynes with a heteroatom tether (entries 2–4) are transformed into heterocyclic products in good yields. The formation of a seven-membered ring is also possible if an *N*-acylpyrrolidine-derived allenyne is used (entry 5); the pyrroloazepinone derivative is obtained in 62% yield.[39] Moreover, an activated alkyne (entry 6) and an allenamide (entry 7) are compatible with this rhodium-catalyzed Alder–ene-type cycloisomerization protocol.

Table 12 Rhodium-Catalyzed Cycloisomerization of Allenynes[37–41]

35 (Z, R^1, R^2, R^3, R^4, R^5) $\xrightarrow{\{RhCl(CO)_2\}_2}$ **36**

Entry	Substrate	Catalyst (mol%)	Conditions	Product	Yield (%)	Ref
1	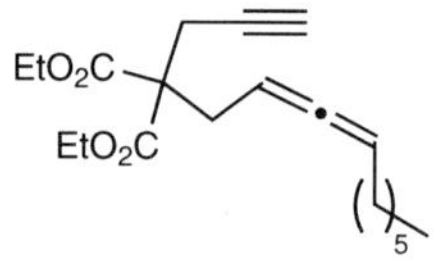 EtO_2C, EtO_2C	2	toluene, 90 °C, 1 h	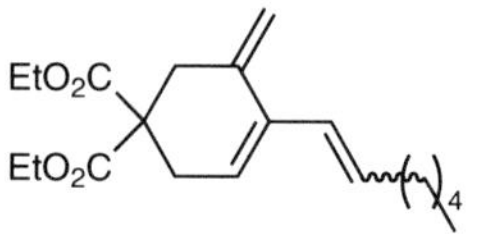 EtO_2C, EtO_2C	80[a]	[37]
2	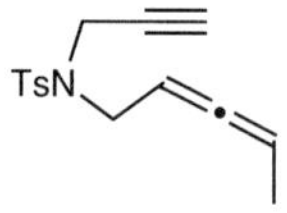 TsN	2	toluene, rt, 3 h	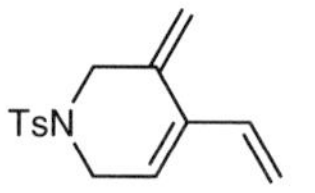 TsN	93	[37]
3	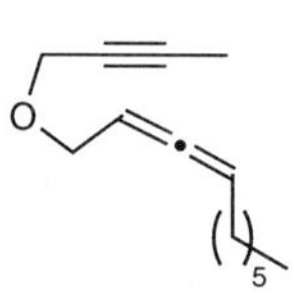 O	2	toluene, 90 °C, 1 h	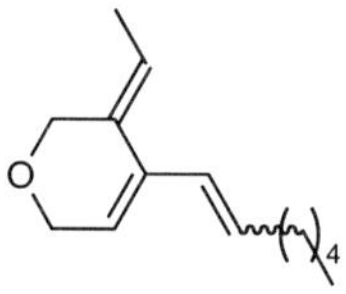 O	74[b]	[37]

Table 12 (cont.)

Entry	Substrate	Catalyst (mol%)	Conditions	Product	Yield (%)	Ref
4	CbzN, Ph, MeO2C	5	toluene, rt, 10 min	Ph, CbzN, MeO2C	95	[38]
5	N, O	10	toluene, 50 °C, 1.5 h	N, O	62	[39]
6	O, TMS	3	toluene, 0 °C, 5 min	TMS, O	95	[40]
7	O, N, O	10	toluene, rt, 3.5 h	O, N, O, O	72	[41]

[a] Ratio (*E*/*Z*) 3:1.
[b] Ratio (*E*/*Z*) 6:1.

A tris(phosphinegold)oxonium complex is effective in similar Alder–ene-type cycloisomerizations of allenynes **37** that result in the formation of cross-conjugated trienes **38** (Table 13).[42] As an example, 7-methylocta-5,6-dien-1-yne undergoes cycloisomerization in the presence of the gold catalyst (2 mol%) in $CHCl_3$ at 60 °C to afford the cross-conjugated triene product in 84% yield (entry 1). The product yield is improved if a quaternary carbon atom is involved in the tether (entry 2). In contrast to the above-described rhodium-based catalytic system, substrates are confined to those possessing a terminal alkyne and two alkyl substituents on the allene terminal carbon atom, and relatively unfunctionalized allenynes (entries 3 and 4) are used. A fused bicyclic product can be obtained from a *cis*-disubstituted cyclohexane derivative (entry 5). Formation of a six-membered ring is also possible using a malonate-tethered allenyne, although the reaction proceeds with lower efficiency (entry 6).

Table 13 Gold-Catalyzed Cycloisomerization of Allenynes[42]

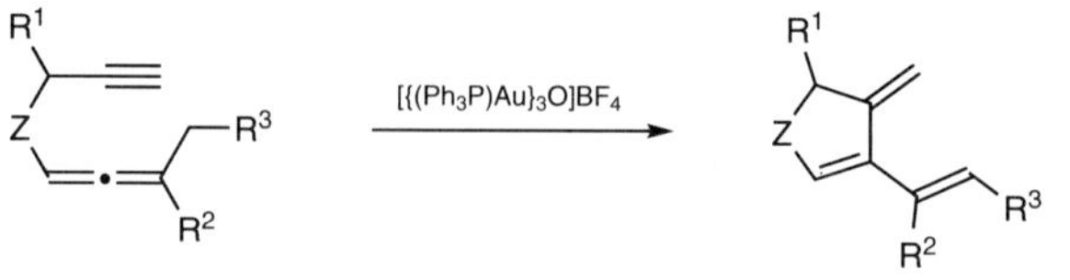

37 $\xrightarrow{[\{(Ph_3P)Au\}_3O]BF_4}$ **38**

Entry	Substrate	Catalyst (mol%)	Conditions	Product	Yield (%)	Ref
1	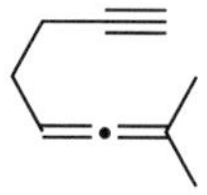	2	$CHCl_3$, 60 °C, 48 h	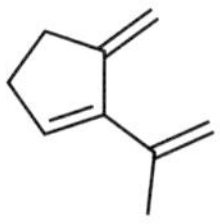	84	[42]
2	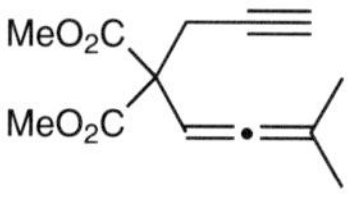 MeO_2C, MeO_2C	2	CH_2Cl_2, 40 °C, 18 h	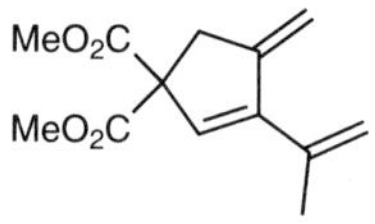 MeO_2C, MeO_2C	99	[42]
3	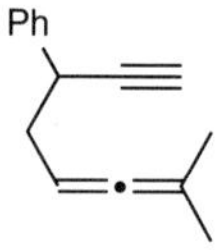 Ph	2	CH_2Cl_2, 40 °C, 18 h	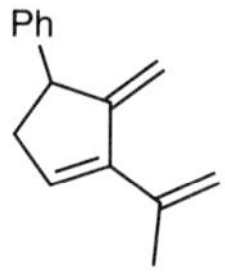 Ph	89	[42]
4	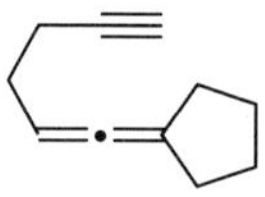	3	$CHCl_3$, 60 °C, 48 h	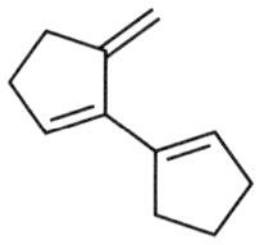	70	[42]
5	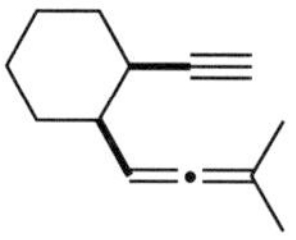	1	$CHCl_3$, 60 °C, 6 h	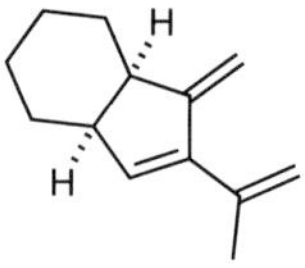 H, H	78	[42]
6	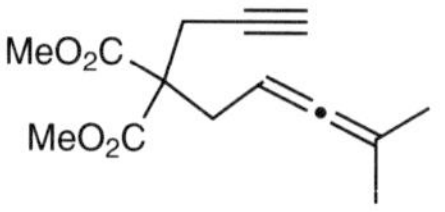 MeO_2C, MeO_2C	5	$CHCl_3$, 60 °C, 48 h	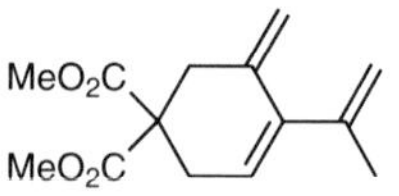 MeO_2C, MeO_2C	40	[42]

Diethyl 4-[(*E*)-Hept-1-enyl]-5-methylenecyclohex-3-ene-1,1-dicarboxylate (Table 12, Entry 1); Typical Procedure:[37]

$\{RhCl(CO)_2\}_2$ (1 mg, 0.003 mmol) was added to a soln of diethyl deca-2,3-dienyl(prop-2-ynyl)malonate (53 mg, 0.16 mmol) in degassed toluene (0.8 mL) under an atmosphere of N_2. The mixture was stirred at 90 °C for 1 h under an atmosphere of N_2. The solvent was evaporated under reduced pressure, and the residue was purified by flash column chromatography (silica gel, hexanes/EtOAc 9:1) to give the title compound as a colorless oil; yield: 42.9 mg (80%); ratio (*E*/*Z*) 3:1 (as determined by ^1H NMR spectroscopy).

5-Methylene-1-(prop-1-en-2-yl)cyclopentenes 38**; General Procedure:**[42]

$[\{(Ph_3P)Au\}_3O]BF_4$ (1–5 mol%) was added to a soln of the allenyne **37** in $CHCl_3$ or CH_2Cl_2, and the mixture was stirred at 40 or 60 °C until TLC analysis indicated consumption of the starting material. The solvent was removed under reduced pressure and the residue was purified by flash column chromatography (silica gel).

1.7.3.2 Cycloisomerizations of Allenenes and Bisallenes

Similar to allenynes (Section 1.7.3.1), allenenes **39** undergo cycloisomerization catalyzed by chloro(cycloocta-1,5-diene)rhodium(I) dimer with tri-2-tolyl phosphite as the ligand in refluxing 1,4-dioxane (Table 14).[43] The carbon–carbon bond is formed at the proximal double bond of the allene terminus, which leads to cycloalkanes **40** possessing *exo*-methylene and alkenyl substituents. Cycloisomerization products are obtained in high yields with good selectivity from malonate- and tosylamide-derived allenenes (entries 1 and 2). If an allenene with an *E*-1,2-disubstituted alkene terminus (entry 3) is used as the substrate, the corresponding product with an *E*-alkylidene moiety is stereoselectively formed. The reaction of an allenene possessing a single terminal substituent on the allene moiety (entry 4) results in the formation of an *E*/*Z* mixture. In this case, the use of tri([1,1′-biphenyl]-2-yl) phosphite instead of tri-2-tolyl phosphite improves the stereoselectivity. A cyclohexane derivative (entry 5) can also be obtained through this process, although a higher catalyst loading is required. The use of dicarbonyl(chloro)rhodium(I) dimer as the catalyst completely alters the reaction pathway and leads to medium-sized-ring compounds **41** (Table 15).[43–46] In the presence of 5 mol% of this rhodium catalyst, allenenes are converted into tetrahydroazepine derivatives or tetrahydrooxepin derivatives (entries 1 and 2).[44] An allenene possessing a 1,1-disubstituted alkene is also converted into a cycloheptene derivative in high yield by performing the reaction in refluxing dioxane under a carbon monoxide atmosphere (entry 3).[43] The formation of an eight-membered ring from an allenene possessing a cyclopropylidene unit (entry 4) proceeds in a mixed solvent (toluene/acetonitrile) at 100 °C to afford a hexahydroazocine derivative in 77% yield.[45] Furthermore, a nonracemic unsymmetrical bisallene (entry 5) also undergoes a similar cycloisomerization in refluxing acetonitrile to regioselectively afford the product without losing optical purity.[46]

Table 14 Rhodium-Catalyzed Cycloisomerization of Allenenes To Give 1-Methylene-2-vinylcycloalkanes[43]

$\{RhCl(cod)\}_2$

39 → **40**

Entry	Substrate	Catalyst (mol%)	Conditions	Product	Isomeric Purity[a] (%)	Yield (%)	Ref
1		2.5	$(2\text{-TolO})_3P$ (10 mol%), 1,4-dioxane, reflux, 18 h		99	92	[43]
2		2.5	$(2\text{-TolO})_3P$ (10 mol%), 1,4-dioxane, reflux, 3 h		86	93	[43]
3		5	$(2\text{-TolO})_3P$ (20 mol%), 1,4-dioxane, reflux, 24 h		92	94	[43]
4		7.5	$(2\text{-PhC}_6H_4O)_3P$ (30 mol%), 1,4-dioxane, reflux, 24 h		99	91[b]	[43]
5		10	$(2\text{-TolO})_3P$ (40 mol%), 1,4-dioxane, reflux, 24 h		93	83	[43]

[a] Percentage of the major product among other inseparable constitutional isomers.
[b] Ratio (*E*/*Z*) 8.3:1.

Table 15 Rhodium-Catalyzed Cycloisomerization of Allenenes To Give Methylenecycloalkenes[43–46]

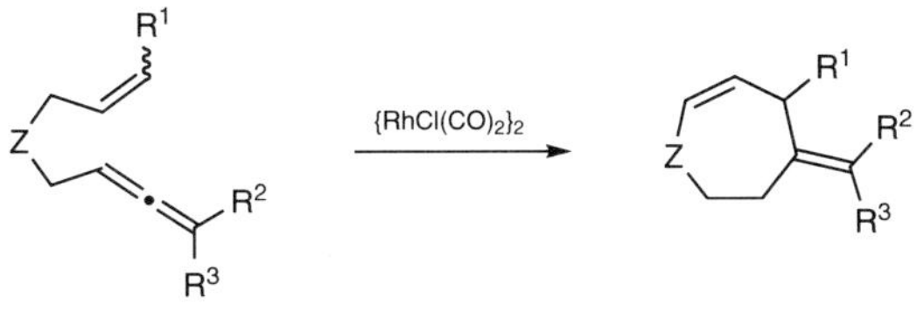

$\{RhCl(CO)_2\}_2$

41

Entry	Substrate	Catalyst (mol%)	Conditions	Product	Yield (%)	Ref
1	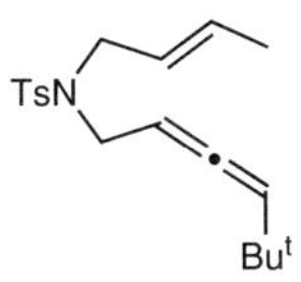	5	1,2-dichloroethane, 90 °C, 1.5 h	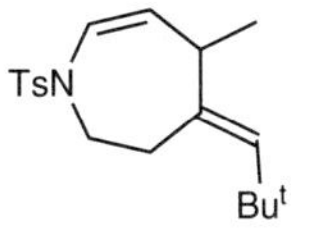	95	[44]
2	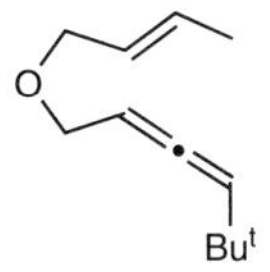	5	1,2-dichloroethane, 90 °C, 1 h	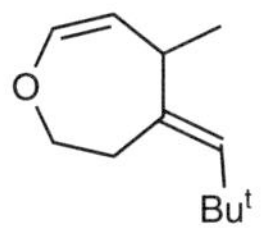	55	[44]
3	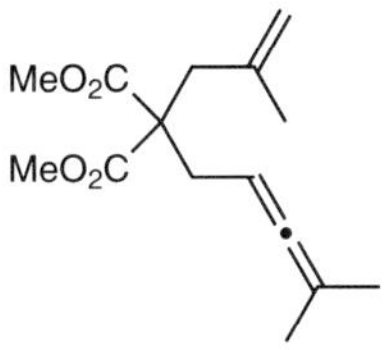	2.5	CO (1 atm), 1,4-dioxane, reflux, 1 h	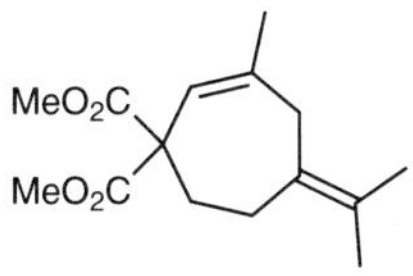	91	[43]
4	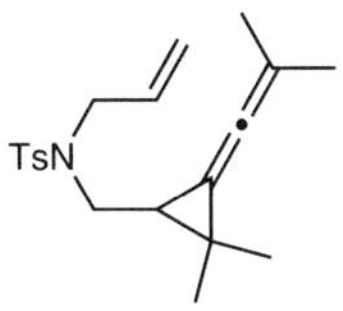	5	toluene/MeCN (2:1), 100 °C, 2.5 h	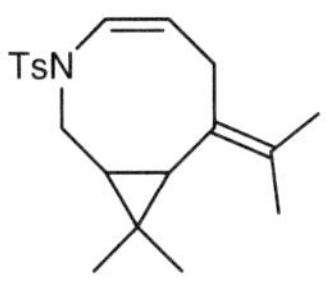	78	[45]
5	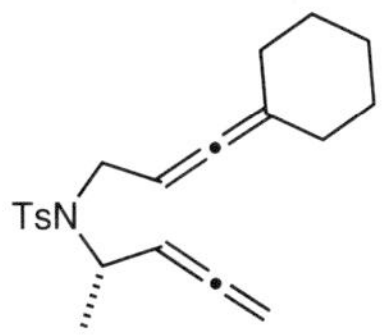	2	MeCN, reflux, 24 h	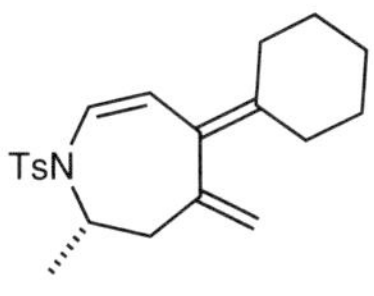	67[a]	[46]

[a] 84% ee (S).

Dimethyl 3-Methylene-4-(2-methylprop-1-enyl)cyclopentane-1,1-dicarboxylate (Table 14, Entry 1); Typical Procedure:[43]

A soln of dimethyl allyl(4-methylpenta-2,3-dienyl)malonate (62.1 mg, 0.25 mmol) in 1,4-dioxane (1.5 mL) was added to a soln of $\{RhCl(cod)\}_2$ (3.1 mg, 0.0063 mmol) and tri-2-tolyl phosphite (9.4 mg, 0.027 mmol) in 1,4-dioxane (1 mL). The mixture was stirred for 18 h at 110 °C under an atmosphere of argon. The solvent was evaporated under reduced pressure, and the residue was purified by flash column chromatography (silica gel, hexanes/EtOAc 20:1) to give the title compound as a colorless oil; yield: 56.9 mg (92%); 99% purity (GC).

(*E*)-4-(2,2-Dimethylpropylidene)-5-methyl-1-tosyl-2,3,4,5-tetrahydro-1*H*-azepine (Table 15, Entry 1); Typical Procedure:[44]
$\{RhCl(CO)_2\}_2$ (1.7 mg, 0.0045 mmol) was added to a degassed soln of *N*-[(*E*)-but-2-enyl]-*N*-(5,5-dimethylhexa-2,3-dienyl)-4-toluenesulfonamide (29.9 mg, 0.09 mmol) in 1,2-dichloroethane (0.5 mL). The mixture was stirred for 1.5 h at 90 °C under an atmosphere of argon and was monitored by GC until the starting material was consumed. The solvent was evaporated under reduced pressure, and the residue was purified by flash column chromatography (silica gel, hexane/EtOAc 9:1) to give the title compound as a colorless oil; yield: 28.3 mg (95%).

1.7.3.3 Cycloisomerizations of (1,3-Diene)enes

Upon treatment with an active iron catalyst derived from tris(acetylacetonato)iron(III), 2,2′-bipyridine, and triethylaluminum, trienes **42** possessing an allylic benzyl ether terminus undergo cycloisomerization to produce dialkenylcycloalkanes **43** as an *E*/*Z* mixture (Table 16).[47] Acetalization of the crude product then affords single products **44**. Treatment of a triene possessing an *E* allylic benzyl ether moiety under the iron-catalyzed cycloisomerization conditions affords the *cis*-disubstituted cyclopentane product in 81% yield after acetalization (entry 1).[48] In contrast, a *trans*-disubstituted cyclohexane is obtained, albeit in lower yield, if a substrate possessing a *Z* allylic ether terminus with a longer tether is used as the substrate (entry 2).[49] *trans*-Disubstituted piperidine derivatives (entries 3 and 4) can be obtained in good yields from the reactions of amide-tethered trienes.[49] The formation of tetrahydropyran derivatives can also be achieved under mild conditions upon using ether-tethered trienes (entries 5 and 6).[50] In particular, the trisubstituted product shown in entry 6 is obtained diastereoselectively, albeit in a lower yield than the disubstituted product shown in entry 5.

Table 16 Iron-Catalyzed Cycloisomerization of (1,3-Diene)enes[47–50]

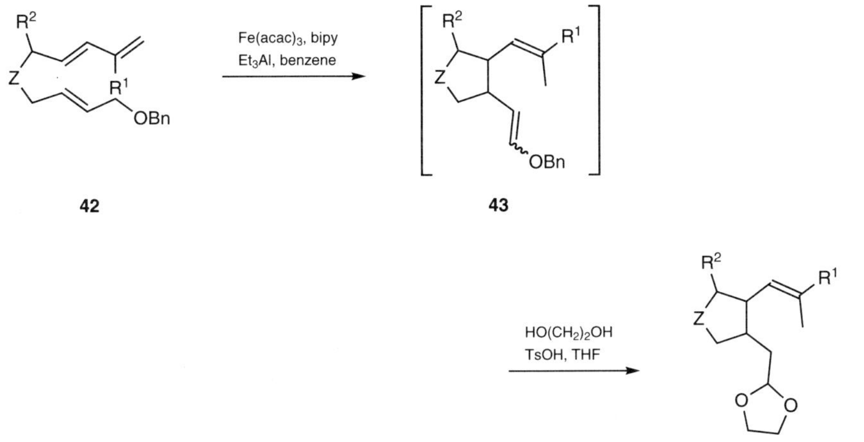

Fe(acac)$_3$, bipy
Et$_3$Al, benzene

42 → **43**

HO(CH$_2$)$_2$OH
TsOH, THF

→ **44**

Entry	Substrate	Catalyst (mol%)	Conditions[a]	Product	Yield (%)	Ref
1	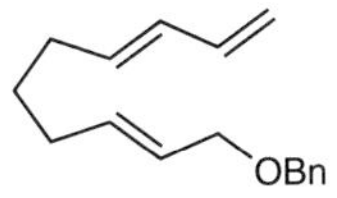 OBn	32	50°C, 15 h	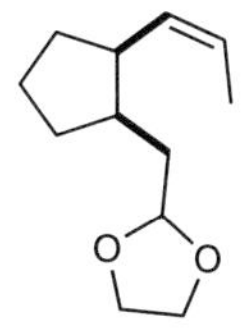	81	[48]
2	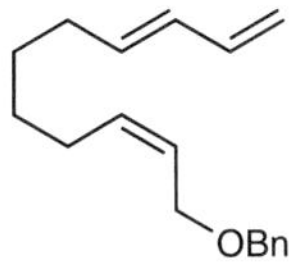 OBn	15	45°C, 8 h	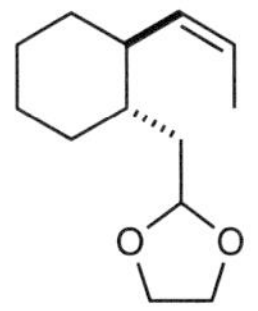	68	[49]
3	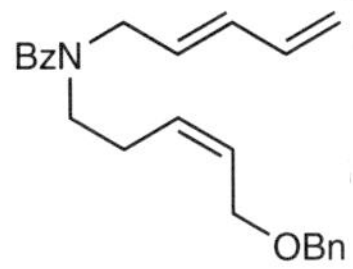 BzN OBn	15	45°C, 8 h	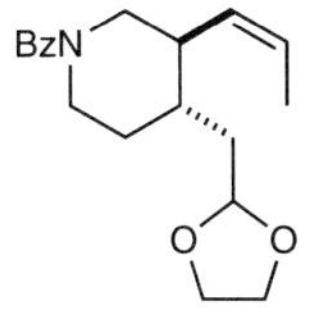 BzN	71	[49]

for references see p 383

Table 16 (cont.)

Entry	Substrate	Catalyst (mol%)	Conditions[a]	Product	Yield (%)	Ref
4	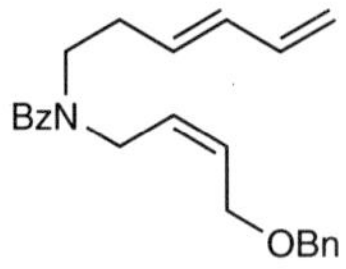 BzN, OBn	15	25 °C, 0.5 h	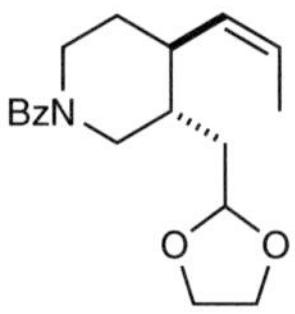 BzN	85	[49]
5	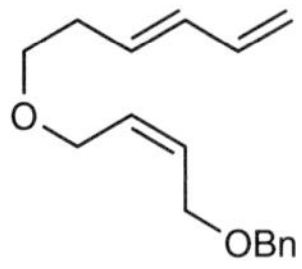 O, OBn	10	25 °C, 7 h	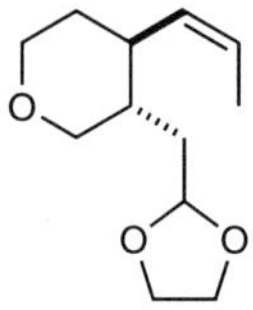 O	84	[50]
6	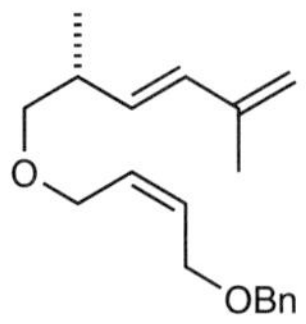 O, OBn	10	25 °C, 8 h	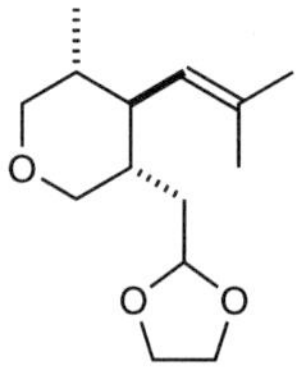 O	68	[50]

[a] For cycloisomerization step.

Enantioselective cycloisomerization of (1,3-diene)enes **46** is achieved using a cationic rhodium catalyst system derived from the chlorobis(cyclooctene)rhodium(I) dimer [{RhCl(coe)$_2$}$_2$; coe = cyclooctene], chiral phosphoramidite ligand **45**, and silver(I) trifluoromethanesulfonate to afford trisubstituted five-membered ring products **47** with moderate to high enantioselectivity and >19:1 diastereoselectivity (Table 17).[51] The formation of two consecutive stereocenters, including one quaternary carbon atom, is effectively controlled. This protocol also tolerates tosylamides, malonates, and ether tethers.

Table 17 Rhodium-Catalyzed Enantioselective Cycloisomerization of (1,3-Diene)enes[51]

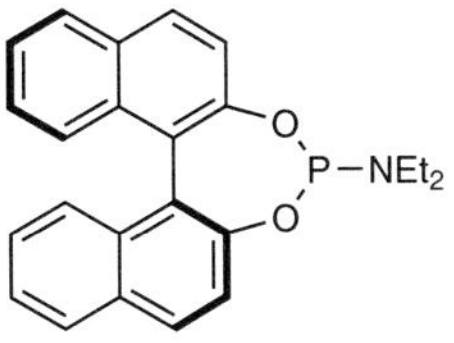

P–NEt2

45

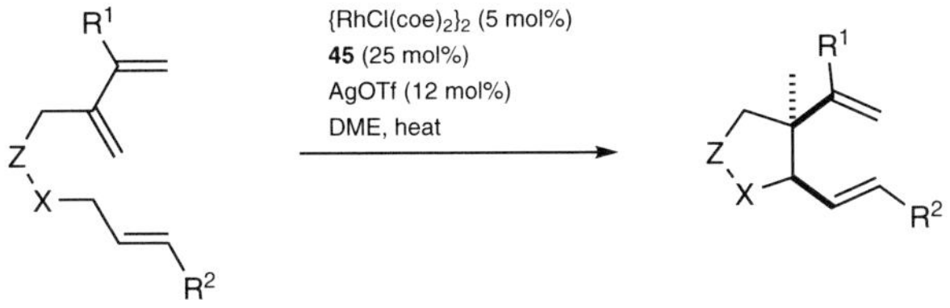

R1

Z

X

R2

46

{RhCl(coe)2}2 (5 mol%)
45 (25 mol%)
AgOTf (12 mol%)
DME, heat

R1

Z

X

R2

47

Starting Material	Conditions	Product	ee (%)	Yield (%)	Ref
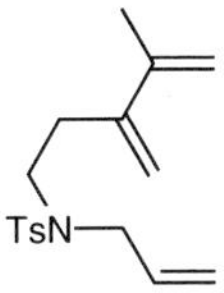 TsN	70 °C, 8 h	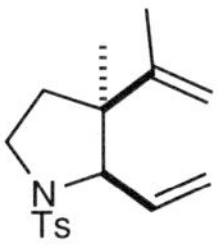 N Ts	90	90	[51]
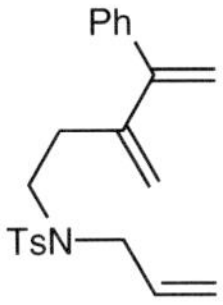 Ph TsN	80 °C, 11 h	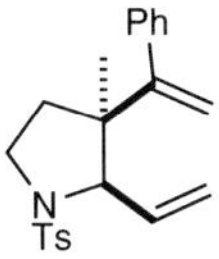 Ph N Ts	77	75[a]	[51]
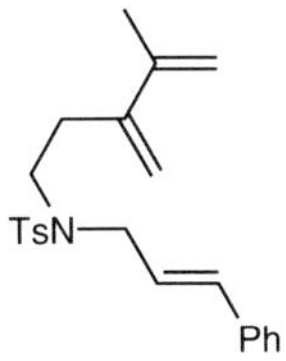 TsN Ph	75 °C, 11 h	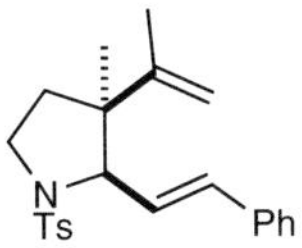 N Ts Ph	85	59[b]	[51]

Table 17 (cont.)

Starting Material	Conditions	Product	ee (%)	Yield (%)	Ref
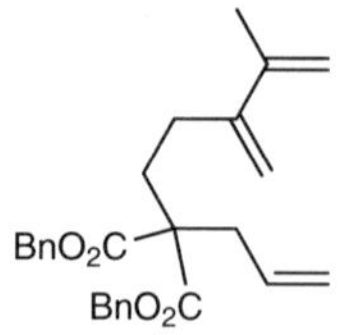 BnO_2C, BnO_2C	75 °C, 11 h	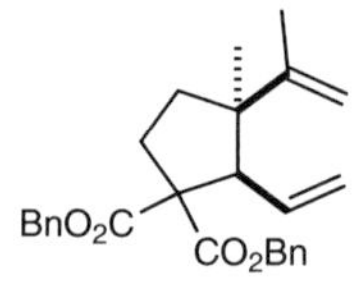 BnO_2C, CO_2Bn	81	74	[51]
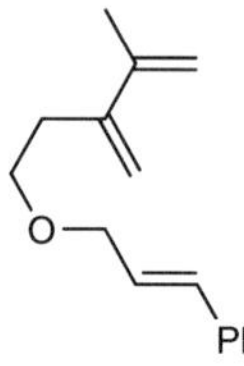 O, Ph	60 °C, 11 h	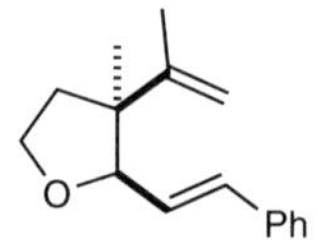 O, Ph	64	64[c]	[51]
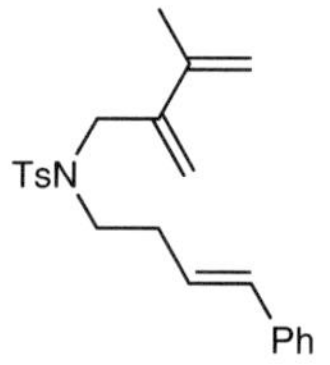 TsN, Ph	80 °C, 11 h	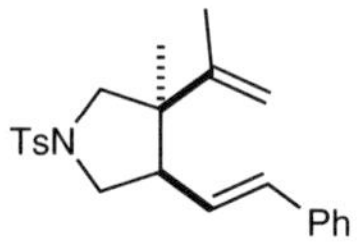 TsN, Ph	84	85	[51]

[a] 91% conversion.
[b] 80% conversion.
[c] 75% conversion.

2-({(1*R,2*R**)-2-[(*Z*)-Prop-1-enyl]cyclopentyl}methyl)-1,3-dioxolane (Table 16, Entry 1); Typical Procedure:**[48]
A 1.9 M soln of Et_3Al in toluene (0.60 mL, 1.14 mmol) was added dropwise to a stirred soln (0–5 °C) of $Fe(acac)_3$ (126 mg, 0.36 mmol), bipy (63 mg, 0.40 mmol), and furan (0.5 mL) in dry degassed benzene (30 mL) (**CAUTION:** *carcinogen*). The resulting black soln was stirred for 5 to 15 min warming from 0 to 25 °C. A soln of benzyl (2*E*,7*E*)-deca-2,7,9-trienyl ether (304.6 mg, 1.26 mmol) in dry degassed benzene (5 mL) was added dropwise to the catalyst soln, and the mixture was stirred for 15 h at 50 °C under an atmosphere of N_2. The soln was filtered through a plug of silica gel (60–200 mesh, hexane/EtOAc 1:1) and was concentrated to give the crude enol ether (303.7 mg). Acetalization of the crude product was performed [$HO(CH_2)_2OH$ (ca. 1 mL), TsOH (ca. 20 mg), THF (ca. 5 mL)]. Upon complete conversion of the enol ethers, as judged by TLC analysis, the mixture was partitioned between Et_2O (ca. 50 mL) and sat. aq $NaHCO_3$ (ca. 50 mL). The organic layer was washed with brine (ca. 50 mL), dried ($MgSO_4$), and concentrated. The residue was purified by flash column chromatography (silica gel, hexane/EtOAc 95:5) to give the title compound as a clear oil; yield: 200 mg (81%).

(2*R*,3*S*)-3-Methyl-3-(prop-1-en-2-yl)-1-tosyl-2-vinylpyrrolidine (Table 17, Entry 1); Typical Procedure:[51]
Anhyd DME (1.0 mL) was added to a mixture of $\{RhCl(coe)_2\}_2$ (2.5 mg, 0.0035 mmol) and AgOTf (2.1 mg, 0.0082 mmol) under an atmosphere of argon. The mixture was stirred for 15 min at rt. This mixture was added to ligand **45** (6.8 mg, 0.0176 mmol) under an atmosphere of argon, and the mixture was stirred for another 15 min at rt. A soln of *N*-allyl-*N*-(4-methyl-3-methylenepent-4-enyl)-4-toluenesulfonamide (21.3 mg, 0.070 mmol) in dry DME (1.3 mL) was added dropwise to the resultant yellow suspension, and the mixture was stirred for 8 h at 70 °C under an atmosphere of argon. The solvent was removed

under reduced pressure. The residue was purified by flash column chromatography (silica gel, hexanes/EtOAc 100:1 to 30:1) to give the title compound as a colorless oil; yield: 19.2 mg (90%); 90% ee (HPLC).

1.7.3.4 Cycloisomerizations of Bis(1,3-dienes)

In the presence of palladium(II) acetate and triphenylphosphine, tetraenes **48** undergo cycloisomerization to afford *trans*-disubstituted cycloalkane derivatives **49** (Table 18). In the case of diethyl (2*E*,4*E*)-hexa-2,4-dienyl[(*E*)-penta-2,4-dienyl]malonate, the product is obtained in 95% yield with an isomeric purity over 95% (entry 1).[52] The cycloisomerization reactions of tetraenes can also be performed using a different catalytic system comprising tetrakis(acetonitrile)palladium(II) bis(tetrafluoroborate) and triphenylphosphine (entries 2 and 3), which predominantly affords *trans*-disubstituted products in high yields. Six-membered-ring formation is also successful using the standard catalyst, and this affords a *trans*-disubstituted cyclohexane derivative in 90% yield with high isomeric purity (>95%) (entry 4). The corresponding *trans*-disubstituted piperidine derivatives (entries 5 and 6) are obtained in good yields if tosylamide-tethered tetraenes are used as the substrates.

Table 18 Palladium-Catalyzed Cycloisomerization of Bis(1,3-dienes)[52]

48 49

Entry	Substrate	Conditions	Product	Yield (%)	Ref
1	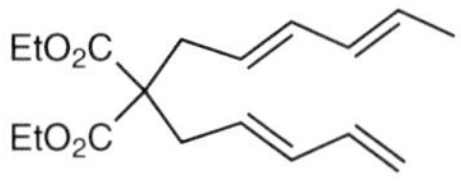	$Pd(OAc)_2$ (5 mol%), Ph_3P (10 mol%), THF, reflux, 14 h	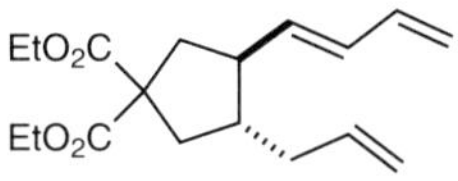	95	[52]
2	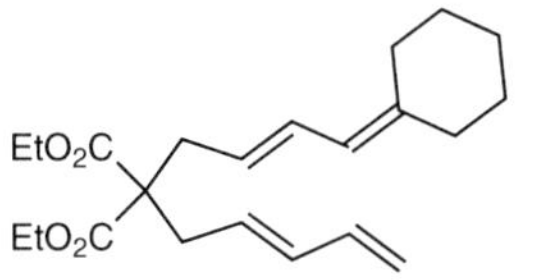	$[Pd(NCMe)_4](BF_4)_2$ (5 mol%), Ph_3P (15 mol%), Et_3N (5 equiv), THF, reflux, 24 h	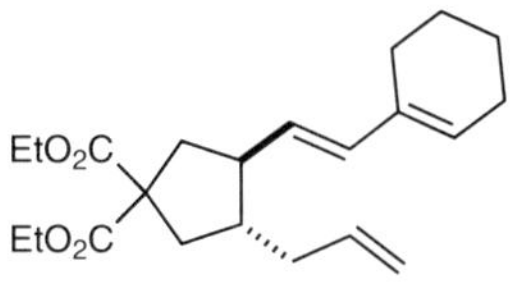	90	[52]
3	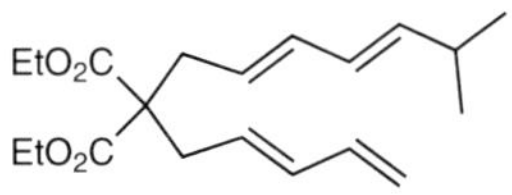	$[Pd(NCMe)_4](BF_4)_2$ (5 mol%), Ph_3P (15 mol%), Et_3N (5 equiv), THF, reflux, 20 h	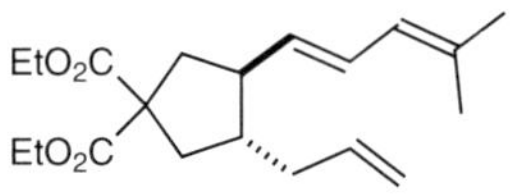	96	[52]
4	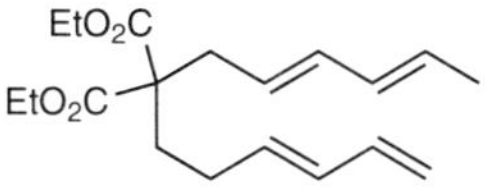	$Pd(OAc)_2$ (6 mol%), Ph_3P (10 mol%), Et_3N (3 equiv), THF, reflux, 16 h	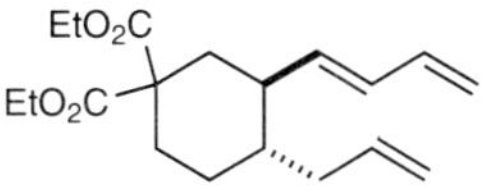	90	[52]
5	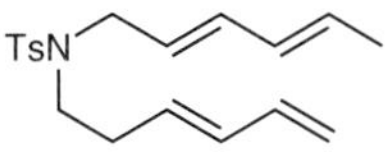	$Pd(OAc)_2$ (6.6 mol%), Ph_3P (13.2 mol%), Et_3N (3.2 equiv), THF, reflux, 40 h	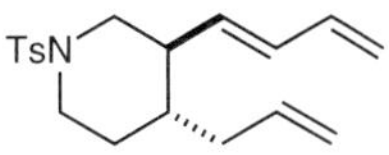	72	[52]
6	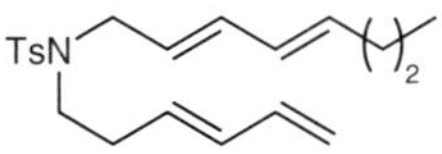	$Pd(OAc)_2$ (7 mol%), Ph_3P (14 mol%), Et_3N (3.6 equiv), THF, reflux, 20 h	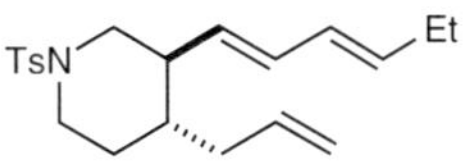	84	[52]

for references see p 383

Diethyl *trans*-3-Allyl-4-[(*E*)-buta-1,3-dienyl]cyclopentane-1,1-dicarboxylate (Table 18, Entry 1); Typical Procedure:[52]

A soln of diethyl (2*E*,4*E*)-hexa-2,4-dienyl[(*E*)-penta-2,4-dienyl]malonate (306.0 mg, 1.0 mmol), $Pd(OAc)_2$ (12.2 mg, 0.054 mmol), and Ph_3P (26–39 mg, 0.10–0.15 mmol) in THF (5 mL) was heated to reflux for 14 h and then cooled and concentrated under reduced pressure. The residue was purified by column chromatography (silica gel, hexane/EtOAc 95:5) to give the title compound; yield: 290.0 mg (95%); >95% isomeric purity.

1.7.4 Applications of Cycloisomerizations to Natural Product Synthesis

As discussed above, various efficient and selective protocols for transition-metal-catalyzed cycloisomerizations have been developed and some of these atom-economical cyclizations have been applied in the synthesis of natural products that display important biological activities. In this section, significant examples of natural product synthesis using transition-metal-catalyzed cycloisomerization as the key step will be discussed.

The palladium-catalyzed Alder–ene-type cycloisomerization of α,ω-enynes is a powerful method to construct valuable cyclic frameworks equipped with methylene and alkenyl groups, which are useful synthetic handles for further transformations (see Section 1.7.1.1). Thus, under various modified conditions, this method has been applied to the total synthesis of many natural products such as chokol C,[53] (+)-cassiol,[54] (+)-saponaceolide B,[55] and 7-*O*-methyldehydropinguisenol.[56] In particular, the Alder–ene-type cycloisomerization of nonracemic cyclohexene derivative **52** has been successfully implemented in the total synthesis of several picrotoxane sesquiterpenes including picrotoxinin and corianin through the formation of tricycle **53** as an intermediate (Scheme 4).[57,58] Moreover, the alternate mode of cycloisomerization, leading to exocyclic 1,3-dienes, has also been exploited along with a subsequent Diels–Alder reaction in the total synthesis of natural products such as sterepolide[59] and echinopines A and B.[60] Ruthenium-catalyzed Alder–ene-type cycloisomerization (Section 1.7.1.3) has also been applied in the enantioselective total synthesis of (+)-allocyathin B_2.[61]

Scheme 4 Total Synthesis of Picrotoxane Sesquiterpenes by Palladium-Catalyzed Alder–Ene-Type Cycloisomerization of a 1,6-Enyne[58]

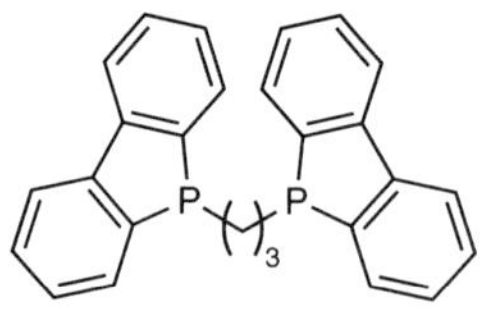

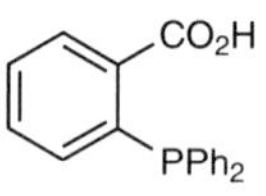

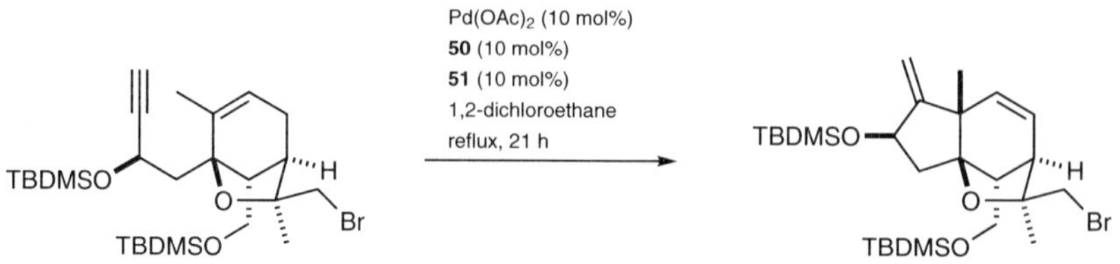

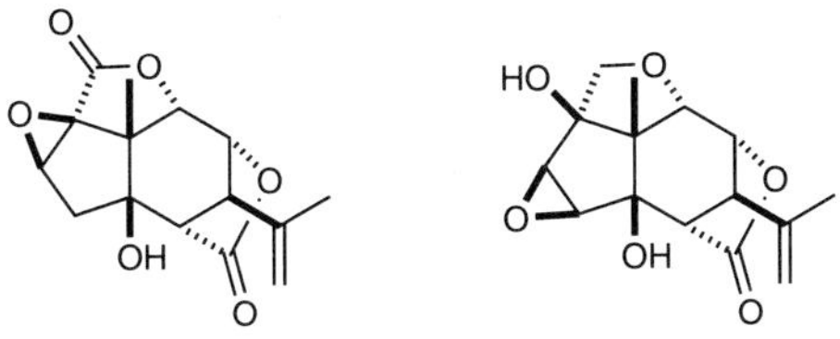

picrotoxinin

corianin

As discussed in Section 1.7.1.5, the enantioselective cycloisomerization of α,ω-enynes is a straightforward method to obtain optically active building blocks for the asymmetric total synthesis of natural products. In fact, this rhodium-catalyzed enantioselective enyne cycloisomerization has been applied to the total synthesis of 15-deoxy-$\Delta^{12,14}$-prostaglandin J_2.[62] Moreover, in the enantioselective total synthesis of platensimycin, a modified protocol involving the use of a preformed chiral rhodium catalyst enables the efficient transformation of cyclohexadienone-derived enyne **54** into chiral spirocyclic aldehyde (−)-**55** in 86% yield with almost perfect enantioselectivity (Scheme 5).[63]

Scheme 5 Total Synthesis of Platensimycin by Rhodium-Catalyzed Enantioselective Cycloisomerization of a 1,6-Enyne[63]

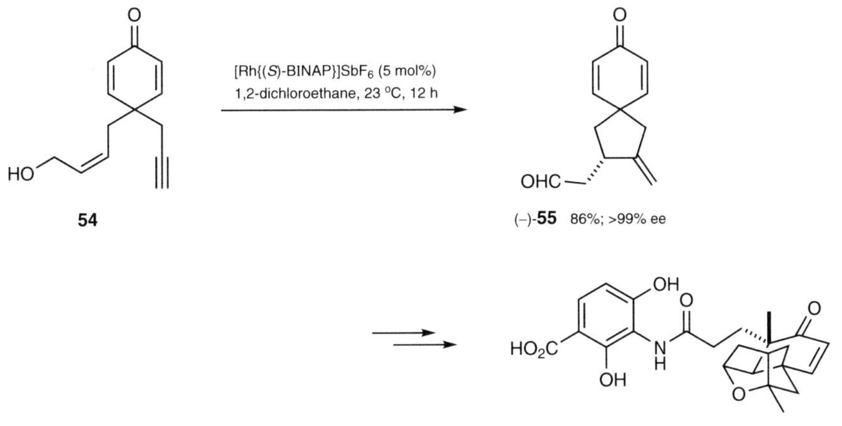

[Rh{(*S*)-BINAP}]SbF_6 (5 mol%)
1,2-dichloroethane, 23 °C, 12 h

54

(−)-**55** 86%; >99% ee

platensimycin

In addition to the above enyne cycloisomerizations, the iron-catalyzed cycloisomerization of (1,3-diene)enes has also been applied to the enantioselective synthesis of natural products such as (−)-mitsugashiwalactone,[64] (+)-isoiridomyrmecin,[64] (−)-protoemetinol,[65] and (−)-gibboside.[66]

(1*R*,3*R*,5*S*,8*R*,9*S*,11*R*)-9-(Bromomethyl)-3-(*tert*-butyldimethylsiloxy)-11-[(*tert*-butyldimethylsiloxy)methyl]-5,9-dimethyl-4-methylene-10-oxotricyclo[6.2.1.0^{1,5}]undec-6-ene (53); Typical Procedure:[58]

$Pd(OAc)_2$ (297 mg, 1.32 mmol), phosphinobenzoic acid **51** (407.7 mg, 1.32 mmol), and diphosphine **50** (541.8 mg, 1.32 mmol) were added to a soln of enyne **52** (7.40 g, 13.26 mmol) in 1,2-dichloroethane (175 mL) at rt. The mixture was stirred for 20 min at rt and was then gently heated at reflux for 21 h. The solvent was removed under reduced pressure, and the obtained residue was chromatographed (silica gel, hexane/CH_2Cl_2 4:1) to give the title compound; yield: 5.17 g (70%); $[\alpha]_D$ −71.5 (*c* 2.28, CH_2Cl_2).

for references see p 383

(*S*)-2-(3-Methylene-8-oxospiro[4.5]deca-6,9-dien-2-yl)acetaldehyde [(−)-55]; Typical Procedure:[63]

A 0.04 M soln of [Rh{(*S*)-BINAP}]SbF_6 in 1,2-dichloroethane (7.8 mL) was quickly added to a flame-dried flask containing enyne **54** (1.27 g, 6.28 mmol) under an atmosphere of argon. The mixture was stirred for 12 h at 23 °C under an atmosphere of argon. The solvent was removed under reduced pressure, and the obtained residue was purified by flash column chromatography (silica gel, hexanes/EtOAc 2:1) to give the title compound as a pale yellow oil; yield: 1.09 g (86%); >99% ee; $[\alpha]_D$ −68.4 (20 °C, *c* 0.75, $CHCl_3$).

1.7.5 Conclusions and Future Perspectives

The development of cutting-edge technology that enables the efficient construction of complex cyclic scaffolds with high environmental compatibility is continuously sought in modern organic synthesis. In line with this requirement, recent advances in transition-metal-catalyzed cycloisomerizations have provided highly feasible and reliable protocols that transform readily accessible α,ω-enynes, α,ω-dienes, and their allenic or 1,3-dienyl congeners into diverse carbocyclic and heterocyclic molecules.[2–5] In particular, highly enantioselective methods have also been realized, and they provide efficient routes to valuable chiral building blocks. As a consequence, streamlined total syntheses of various complex natural products have been achieved using transition-metal-catalyzed cycloisomerizations. Although this chapter focuses on cycloisomerizations through the formation of one carbon–carbon bond with concomitant hydrogen-atom transfer, this research area has been rapidly expanded to more elaborate protocols. For example, the advent of alkynophilic transition metal–Lewis acid catalysts as a novel class of catalysts has opened the door to an unprecedented mode of cycloisomerization involving the formation of multiple bonds and/or their fission.[10,11] The application of underused transition-metal complexes, along with newly designed ligands, will lead to significant advances in this research area in the future.

References

[1] Trost, B. M., *Science (Washington, D. C.)*, (1991) **254**, 1471.
[2] Trost, B. M.; Krische, M. J., *Synlett*, (1998), 1.
[3] Belmont, P.; Parker, E., *Eur. J. Org. Chem.*, (2009), 6075.
[4] Watson, I. D. G.; Toste, F. D., *Chem. Sci.*, (2012) **3**, 2899.
[5] Marinetti, A.; Jullien, H.; Voituriez, A., *Chem. Soc. Rev.*, (2012) **41**, 4884.
[6] Huntsman, W. D.; Hall, R. P., *J. Org. Chem.*, (1962) **27**, 1988.
[7] Huntsman, W. D.; Solomon, V. C.; Eros, D., *J. Am. Chem. Soc.*, (1958) **80**, 5455.
[8] Lee, S. I.; Chatani, N., *Chem. Commun. (Cambridge)*, (2009), 371.
[9] Michelet, V.; Toullec, P. Y.; Genêt, J.-P., *Angew. Chem.*, (2008) **120**, 4338; *Angew. Chem. Int. Ed.*, (2008) **47**, 4268.
[10] Zhang, L.; Sun, J.; Kozmin, S. A., *Adv. Synth. Catal.*, (2006) **348**, 2271.
[11] Jiménez-Núñez, E.; Echavarren, A. M., *Chem. Rev.*, (2008) **108**, 3326.
[12] Trost, B. M.; Lautens, M.; Chan, C.; Jebaratnam, D. J.; Mueller, T., *J. Am. Chem. Soc.*, (1991) **113**, 636.
[13] Trost, B. M.; Tanoury, G. J.; Lautens, M.; Chan, C.; MacPherson, D. T., *J. Am. Chem. Soc.*, (1994) **116**, 4255.
[14] Sturla, S. J.; Kablaoui, N. M.; Buchwald, S. L., *J. Am. Chem. Soc.*, (1999) **121**, 1976.
[15] Trost, B. M.; Toste, F. D., *J. Am. Chem. Soc.*, (2000) **122**, 714.
[16] Trost, B. M.; Toste, F. D., *J. Am. Chem. Soc.*, (1999) **121**, 9728.
[17] Fürstner, A.; Martin, R.; Majima, K., *J. Am. Chem. Soc.*, (2005) **127**, 12 236.
[18] Fürstner, A.; Majima, K.; Martin, R.; Krause, H.; Kattnig, E.; Goddard, R.; Lehmann, C. W., *J. Am. Chem. Soc.*, (2008) **130**, 1992.
[19] Cao, P.; Wang, B.; Zhang, X., *J. Am. Chem. Soc.*, (2000) **122**, 6490.
[20] Lei, A.; He, M.; Wu, S.; Zhang, X., *Angew. Chem.*, (2002) **114**, 3607; *Angew. Chem. Int. Ed.*, (2002) **41**, 3457.
[21] Lei, A.; Waldkirch, J. P.; He, M.; Zhang, X., *Angew. Chem.*, (2002) **114**, 4708; *Angew. Chem. Int. Ed.*, (2002) **41**, 4526.
[22] Lei, A.; He, M.; Zhang, X., *J. Am. Chem. Soc.*, (2002) **124**, 8198.
[23] Hatano, M.; Mikami, K., *J. Am. Chem. Soc.*, (2003) **125**, 4704.
[24] Yamamoto, Y., *Chem. Rev.*, (2012) **112**, 4736.
[25] Grigg, R.; Malone, J. F.; Mitchell, T. R. B.; Ramasubbu, A.; Scott, R. M., *J. Chem. Soc., Perkin Trans. 1*, (1984), 1745.
[26] Kisanga, P.; Widenhoefer, R. A., *J. Am. Chem. Soc.*, (2000) **122**, 10 017.
[27] Bray, K. L.; Fairlamb, I. J. S.; Kaiser, J.-P.; Lloyd-Jones, G. C.; Slatford, P. A., *Top. Catal.*, (2002) **19**, 49.
[28] Okamoto, S.; Livinghouse, T., *J. Am. Chem. Soc.*, (2000) **122**, 1223.
[29] Okamoto, S.; Livinghouse, T., *Organometallics*, (2000) **19**, 1449.
[30] Yamamoto, Y.; Nakagai, Y.; Ohkoshi, N.; Itoh, K., *J. Am. Chem. Soc.*, (2001) **123**, 6372.
[31] Michaut, M.; Santelli, M.; Parrain, J.-L., *Tetrahedron Lett.*, (2003) **44**, 2157.
[32] Fairlamb, I. J. S.; McGlacken, G. P.; Weissberger, F., *Chem. Commun. (Cambridge)*, (2006), 988.
[33] Arisawa, M.; Terada, Y.; Takahashi, K.; Nakagawa, M.; Nishida, A., *J. Org. Chem.*, (2006) **71**, 4255.
[34] Böing, C.; Hahne, J.; Franciò, G.; Leitner, W., *Adv. Synth. Catal.*, (2008) **350**, 1073.
[35] Aubert, C.; Fensterbank, L.; Garcia, P.; Malacria, M.; Simonneau, A., *Chem. Rev.*, (2011) **111**, 1954.
[36] Takacs, J. M.; Boito, S. C.; Myoung, Y.-C., *Curr. Org. Chem.*, (1998) **2**, 233.
[37] Brummond, K. M.; Chen, H.; Sill, P.; You, L., *J. Am. Chem. Soc.*, (2002) **124**, 15 186.
[38] Brummond, K. M.; Mitasev, B., *Org. Lett.*, (2004) **6**, 2245.
[39] Brummond, K. M.; Painter, T. O.; Probst, D. A.; Mitasev, B., *Org. Lett.*, (2007) **9**, 347.
[40] Brummond, K. M.; Chen, D.; Painter, T. O.; Mao, S.; Seifried, D. D., *Synlett*, (2008), 759.
[41] Brummond, K. M.; Yan, B., *Synlett*, (2008), 2303.
[42] Cheong, P. H.-Y.; Morganelli, P.; Luzung, M. R.; Houk, K. N.; Toste, F. D., *J. Am. Chem. Soc.*, (2008) **130**, 4517.
[43] Makino, T.; Itoh, K., *J. Org. Chem.*, (2004) **69**, 395.
[44] Brummond, K. M.; Chen, H.; Mitasev, B.; Casarez, A. D., *Org. Lett.*, (2004) **6**, 2161.
[45] Lu, B.-L.; Shi, M., *Angew. Chem.*, (2011) **123**, 12 233; *Angew. Chem. Int. Ed.*, (2011) **50**, 12 027.
[46] Lu, P.; Ma, S., *Org. Lett.*, (2007) **9**, 2095.
[47] Takacs, J. M.; Anderson, L. G., *J. Am. Chem. Soc.*, (1987) **109**, 2200.
[48] Takacs, J. M.; Myoung, Y.-C.; Anderson, L. G., *J. Org. Chem.*, (1994) **59**, 6928.
[49] Takacs, B. E.; Takacs, J. M., *Tetrahedron Lett.*, (1990) **31**, 2865.

[50] Takacs, J. M.; Anderson, L. G.; Creswell, M. W.; Takacs, B. E., *Tetrahedron Lett.*, (1987) **28**, 5627.
[51] Li, Q.; Yu, Z.-X., *Angew. Chem.*, (2011) **123**, 2192; *Angew. Chem. Int. Ed.*, (2011) **50**, 2144.
[52] Takacs, J. M.; Clement, F.; Zhu, J.; Chandramouli, S. V.; Gong, X., *J. Am. Chem. Soc.*, (1997) **119**, 5804.
[53] Trost, B. M.; Phan, L. T., *Tetrahedron Lett.*, (1993) **34**, 4735.
[54] Trost, B. M.; Li, Y., *J. Am. Chem. Soc.*, (1996) **118**, 6625.
[55] Trost, B. M.; Corte, J. R.; Gudiksen, M. S., *Angew. Chem.*, (1999) **111**, 3945; *Angew. Chem. Int. Ed.*, (1999) **38**, 3662.
[56] Harada, K.; Tonoi, Y.; Kato, H.; Fukuyama, Y., *Tetrahedron Lett.*, (2002) **43**, 3829.
[57] Trost, B.; Krische, M. J., *J. Am. Chem. Soc.*, (1999) **121**, 6131.
[58] Trost, B. M.; Haffner, C. D.; Jebaratnam, D. J.; Krische, M. J.; Thomas, A. P., *J. Am. Chem. Soc.*, (1999) **121**, 6183.
[59] Trost, B. M.; Chung, J. Y. L., *J. Am. Chem. Soc.*, (1985) **107**, 4586.
[60] Peixoto, P. A.; Richard, J.-A.; Severin, R.; Chen, D. Y.-K., *Org. Lett.*, (2011) **13**, 5724.
[61] Trost, B. M.; Dong, L.; Schroeder, G. M., *J. Am. Chem. Soc.*, (2005) **127**, 10 259.
[62] Kim, N.-J.; Moon, H.; Park, T.; Yun, H.; Jung, J.-W.; Chang, D.-J.; Kim, D.-D.; Suh, Y.-G., *J. Org. Chem.*, (2010) **75**, 7458.
[63] Nicolaou, K. C.; Li, A.; Edmonds, D. J.; Tria, G. S.; Ellery, S. P., *J. Am. Chem. Soc.*, (2009) **131**, 16 905.
[64] Takacs, J. M.; Myoung, Y. C., *Tetrahedron Lett.*, (1992) **33**, 317.
[65] Takacs, J. M.; Boito, S. C., *Tetrahedron Lett.*, (1995) **36**, 2941.
[66] Takacs, J. M.; Vayalakkada, S.; Mehrman, S. J.; Kingsbury, C. L., *Tetrahedron Lett.*, (2002) **43**, 8417.

1.8 Metal-Catalyzed Intramolecular C—N and C—O Bond Formation

E. M. Beccalli, A. Bonetti, and A. Mazza

General Introduction

The formation of C—N and C—O bonds continues to stimulate considerable interest due to the wide presence of nitrogen- and oxygen-containing heterocyclic structures in natural products and medicinally important compounds. The utilization of unsaturated carbon–carbon systems containing tethered nucleophiles represents one of the most versatile and efficient methods for the preparation of heterocycles. Different types of C—N and C—O functionalities may be formed by addition of nitrogen and oxygen nucleophiles to unsaturated bonds. The reaction types considered in this chapter are: aminations, oxidative aminations, hydroaminations, carboaminations, carbamoylations, alkoxylations, oxidative alkoxylations, hydroalkoxylations, hydroacyloxylations, carboalkoxylations, and alkoxycarbonylations of alkenes and alkynes (allenes are treated in Section 1.6). In comparison to C—N addition reactions, C—O bond formation proceeds under milder conditions, giving high yields of the products with good or excellent regio- and stereoselectivities.

The nature of the transition-metal complex plays a key role in all of these reactions and the main problem is to find a suitable catalyst, which should not only be efficient but also nontoxic, relatively inexpensive, and stable. In this field, palladium catalysts dominate by far. Recently, catalysis by coinage metals and bimetallic catalytic systems have emerged as interesting tools for the formation of various carbon—heteroatom bonds as the key step toward the synthesis of heterocycles.

1.8.1 Reaction Mechanisms of Transition-Metal-Catalyzed C—N and C—O Bond Formation

The reaction mechanisms can be quite different depending on the unsaturated substrates and the transition metals involved. Two potential mechanisms are most commonly accepted. Scheme 1 depicts the coordination of a carbon–carbon multiple bond to an electrophilic metal center, thus activating the unsaturated system toward outer-sphere attack by a protic nucleophile to give a β-substituted alkylmetal species **1**. The newly formed M—C bond is then cleaved by protonolysis to regenerate the catalyst.[1]

Scheme 1 Outer-Sphere Mechanism for Cyclization of Heteroatom-Substituted Alkenes or Alkynes[1]

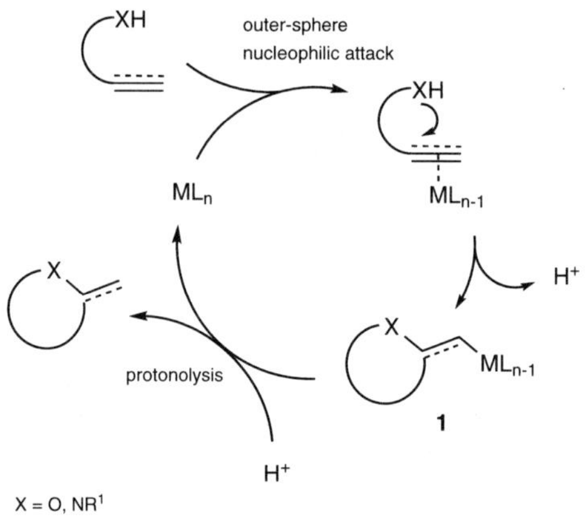

Scheme 2 shows an alternative inner-sphere mechanism, first involving the initial addition of the metal to the nucleophile followed by alkene/alkyne insertion into the M—X bond of intermediate **2**. The resulting M—C bond is cleaved by a C—H reductive elimination or by protonolysis. Mechanistic studies, both experimental and theoretical, have demonstrated that both mechanisms can in fact be operative. While the latter is generally preferred for more electron-rich metals such as rhodium and iridium, several studies suggest that platinum- and palladium-catalyzed addition of N—H or O—H nucleophiles is more likely to run by the outer-sphere, electrophilic activation mechanism.[1] In some cases, starting from alkynes, the formation of an allene intermediate is proposed, followed by the intermediacy of a π-allylmetal complex which undergoes attack of the nucleophile to give allyl derivatives.[2]

Scheme 2 Inner-Sphere Mechanism for Cyclization of Heteroatom-Substituted Alkenes or Alkynes[1]

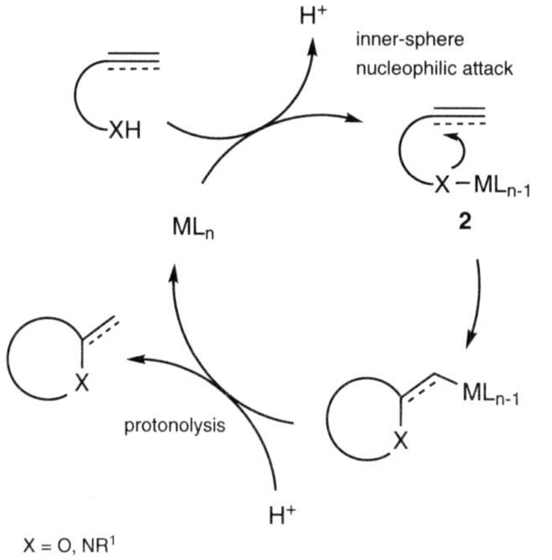

Both palladium and platinum catalysts are quite efficient in the promotion of nucleophilic addition to a coordinated alkene, and their distinct properties often lead to complementary M—C bond cleavage required to obtain a product. Specifically, palladium complexes are reactive toward ligand substitution, thus facilitating β-hydride elimination. In contrast, platinum complexes are relatively inert toward ligand substitution. This difference facilitates the development of alternative pathways for M—C bond cleavage, such as protonolysis, and reduces the problems caused by competing alkene isomerization reactions.

The impact of the transition-metal catalyst on the regioselectivity deserves a particular mention. For example, alkynylamines and alkynols show a particular regioselectivity in the presence of different transition-metal catalysts depending on the substrate and reaction conditions. Hydroalkoxylation with iridium gives a 6-*endo-dig* cyclization process, whereas palladium shows 5-*exo-dig* regioselectivity, and with platinum and rhodium 5-*endo-dig* cyclization is observed (Scheme 3).[3]

Scheme 3 Regioselective Cyclization of Heteroatom-Substituted Alkynes[3]

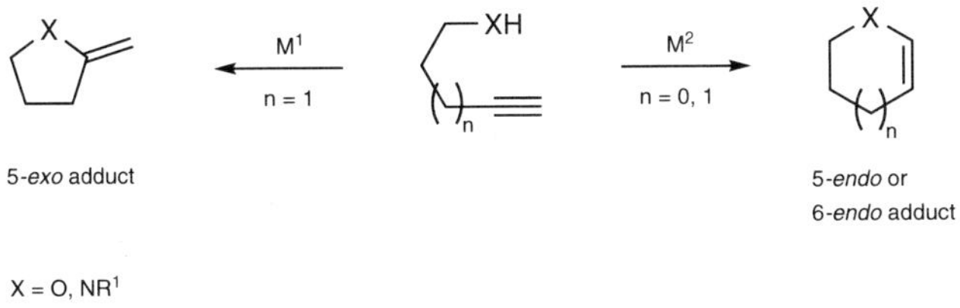

XH
M^1
n = 1
M^2
n = 0, 1
5-*exo* adduct
5-*endo* or 6-*endo* adduct
X = O, NR^1

In 1995, Buchwald and Hartwig independently discovered an important amination/alkoxylation process based on the reaction between an aryl halide or pseudohalide (e.g., trifluoromethanesulfonate) and NH or OH functional groups able to react as nucleophiles.[4] This process is realized under palladium catalysis and requires the presence of ligands and a stoichiometric amount of base, the choice of which has great influence on product formation, as do the electronic and steric properties of the nucleophiles combined with the substrate. The mechanism involves oxidative addition of the palladium to the aryl halide followed by the reaction with the nucleophile (Scheme 4).[4] The intramolecular version of the Buchwald–Hartwig reaction affords heterocyclic systems. Similar coupling reactions, named by some authors as intramolecular Ullmann condensations, are also realized under copper catalysis starting from aryl halides and nucleophilic functional groups.[5,6]

Scheme 4 Buchwald–Hartwig-Type Reaction Mechanism[4]

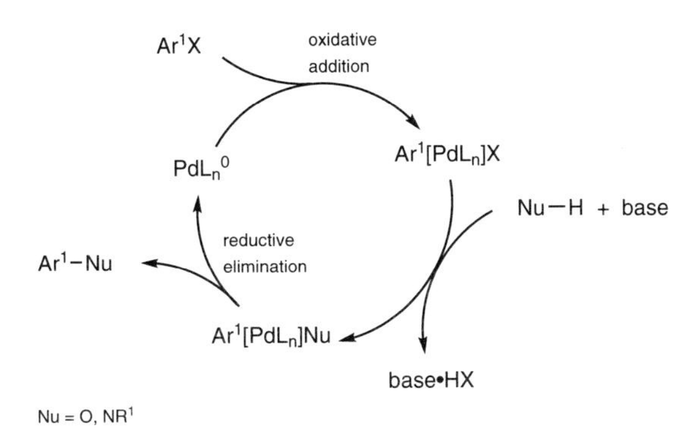

A particular case of transition-metal-catalyzed cyclization is represented by reactions occurring in the presence of carbon monoxide (aminocarbonylations and alkoxycarbonylations). The incorporation of carbon monoxide can result in a cyclization product having an endocyclic or exocyclic carbonyl group. In path A, the reductive elimination of the metal affords the product, in path B, the displacement of the metal requires the intervention of a second nucleophile, such as an alcohol (Scheme 5).[7]

for references see p 499

Scheme 5 Carbonylative Cyclization Reaction Mechanisms[7]

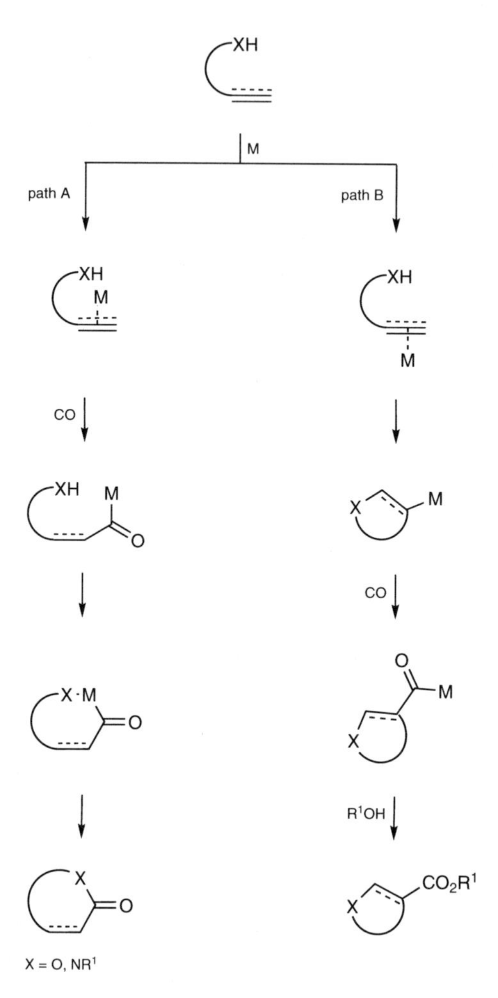

During the last decades, several reviews on transition-metal-catalyzed reactions have summarized the preparation of heterocyclic systems through inter- or intramolecular synthetic pathways.[8–13] The present chapter highlights the most significant metal-catalyzed intramolecular reactions aimed at the synthesis of heterocycles, exploiting the formation of C—N and C—O bonds. The chapter subdivision arises from the bond formed and the type of coupling reaction applied. The subchapters are divided according to the different transition-metal catalysts employed.

1.8.2 C—N Bond Formation

The reactions considered are aminations, oxidative aminations, hydroaminations, aminocarbonylations, and carboaminations of alkynes and alkenes under the most used transition-metal catalysis systems. The tethered nucleophilic nitrogen can represent amines, amides, carbamates, sulfonamides, imines, and amidines, but also nitro groups and azides.

1.8.2.1 Hydroamination of Alkynes

The hydroamination reaction is an atom-efficient pathway to add a nucleophilic nitrogen to a C≡C bond. Various kinds of metal complexes are used; in general, metals with high Lewis acidity can perform the activation of the triple bond toward nucleophilic attack. Good results have been obtained with palladium complexes,[14,15] and studies to identify the most active transition-metal catalysts and optimize the reaction conditions have been reported.[1,16,17]

The intramolecular hydroamination of alkynes proceeds in a 5-*exo-dig* or 6-*endo-dig* manner, affording five- or six-membered nitrogen heterocycles such as dihydropyrroles and tetrahydropyridines, which may also be fused with aromatic rings. When the substrates are primary amines, the initially formed enamines often isomerize to the thermodynamically more stable cyclic imines.

1.8.2.1.1 Copper Catalysis

The particular advantages of copper are its low cost and low toxicity, but, on the other hand, quite harsh reaction conditions are required. Both copper(I) and copper(II) catalysts are used for intramolecular hydroamination of alkynes and no ligands are required.

Starting from 2-alkynylanilines **3**, the indole scaffolds **4** may be obtained under copper(II) catalysis (Scheme 6).[18] It is likely that the reactivity of the substrates depends on the acidity of the proton on the nitrogen atom; in fact, the best results are obtained with sulfonamides, but primary anilines also cyclize using different copper(II) salts, such as copper(II) trifluoroacetate (20 mol%), whereas copper(II) acetate is the best catalyst for sulfonamides. The reaction, performed in 1,2-dichloroethane at reflux, tolerates both electron-withdrawing and electron-donating groups on the aromatic ring and C2-position of the indole products. However, the yields depend on the bulkiness of the substituents on the triple bond.

Scheme 6 Copper-Catalyzed Cyclization of 2-Alkynylanilines to Indoles[18]

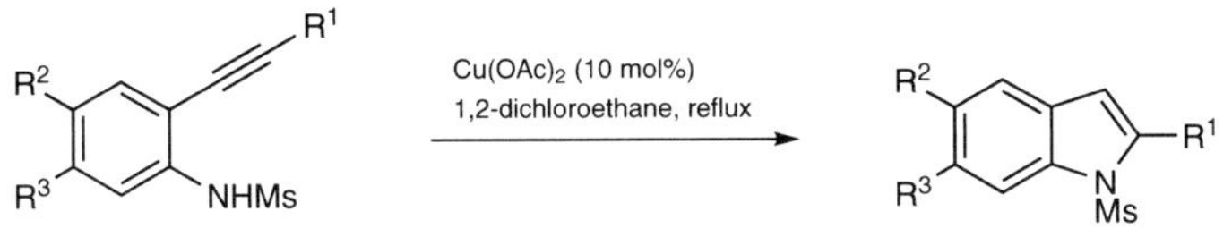

$Cu(OAc)_2$ (10 mol%)
1,2-dichloroethane, reflux

3 → **4**

R^1	R^2	R^3	Time (h)	Yield (%)	Ref
Ph	H	H	18	94	[18]
Bu	H	H	20	91	[18]
H	H	H	1.5	87	[18]
CO_2Me	H	H	24	79	[18]
Ph	Br	H	7	76	[18]
Ph	CN	H	50	74	[18]
Ph	H	OMe	38	95	[18]
Ph	Me	H	6	97	[18]
t-Bu	H	H	72	22	[18]

Pyrroles **6** can be prepared via copper-catalyzed cyclization of alkynylimines **5** (Scheme 7).[19] The reaction proceeds via isomerization to an allenylimine intermediate followed by nucleophilic attack of the nitrogen atom of the imine. This approach is also applicable to the synthesis of various types of fused nitrogen-containing heteroaromatic compounds such as indolizines.

Scheme 7 Copper-Catalyzed Cyclization of Alkynylimines to Pyrroles[19]

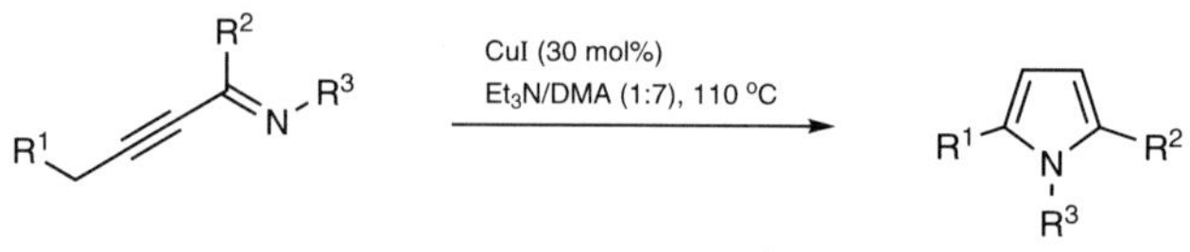

CuI (30 mol%)
Et_3N/DMA (1:7), 110 °C

5 → 6

R^1	R^2	R^3	Time (h)	Yield (%)	Ref
Bu	H	Bu	2	50	[19]
Bu	H	*t*-Bu	6	86	[19]
Bu	H	CPh_3	10	91	[19]
$(CH_2)_3CH{=}CH_2$	H	*t*-Bu	6	83	[19]
H	Ph	Ph	4	86	[19]
Bu	H	$CH(Me)CH_2CO_2Et$	2	93	[19]
$(CH_2)_2CN$	H	$CH(Me)CH_2CO_2Et$	1	51	[19]
$CH_2OTBDMS$	H	$CH(Me)CH_2CO_2Et$	1.5	52	[19]
OTBDMS	H	$CH(Me)CH_2CO_2Et$	0.5	79	[19]
H	Pr	$CH(Me)CH_2CO_2Et$	1.5	71	[19]
Pr	Me	$CH(Me)CH_2CO_2Et$	2.5	87	[19]

Substituted pyrroles **8** are obtained from enynamines **7** with copper(II) chloride in dimethylacetamide at 100 °C. This methodology is quite general; the amino group can be either primary or secondary and substituted with bulky groups, and the triple bond can be terminal or internal (Scheme 8).[20]

Scheme 8 Copper-Catalyzed Cyclization of Enynamines to Pyrroles[20]

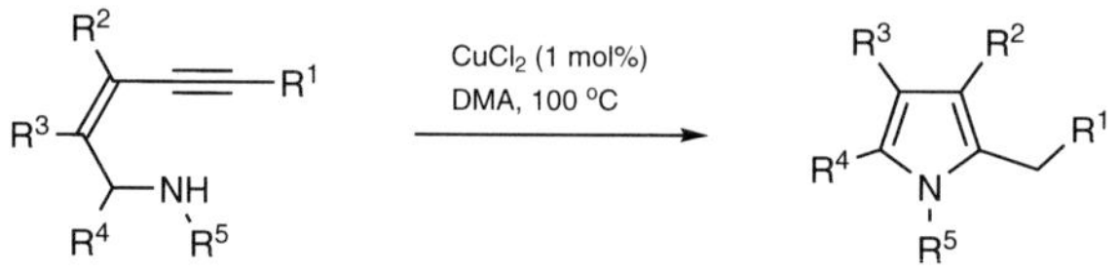

$CuCl_2$ (1 mol%)
DMA, 100 °C

7 → 8

R^1	R^2	R^3	R^4	R^5	Time (h)	Yield (%)	Ref
Bu	Me	H	H	Bn	6	91	[20]
Bu	Me	H	H	Bu	3	80	[20]
Bu	Bu	H	H	Bu	3	75	[20]
Bu	Me	H	H	*t*-Bu	26	63[a]	[20]
Bu	Me	H	H	H	2	88	[20]

R^1	R^2	R^3	R^4	R^5	Time (h)	Yield (%)	Ref
Bu	Me	H	Bu	Bn	12	90	[20]
Bu	Ph	Et	H	Bu	1.5	90	[20]
CH=CMe$_2$	Me	H	H	Bn	5	83	[20]

[a] Reaction carried out with 5 mol% $CuCl_2$.

1-(Methylsulfonyl)-1*H*-indoles 4; General Procedure:[18]
A soln of a 2-alkynylaniline **3** (1 mmol) in anhyd 1,2-dichloroethane (5 mL) was added to a suspension of $Cu(OAc)_2$ (10 mol%) in anhyd toluene or 1,2-dichloroethane (1 mL). The mixture was heated at reflux for the indicated time. H_2O (6 mL) was added and the resulting mixture was extracted with EtOAc (3 × 10 mL). The combined organic soln was washed with sat. aq NaCl, dried ($MgSO_4$), and concentrated. The residue was purified by chromatography (silica gel) to afford the product as a solid.

Pyrroles 6; General Procedure:[19]
CuI (30 mol%), anhyd DMA (1.80 mL), and imine **5** (0.4 mmol) were successively added to a 2.5-mL Wheaton microreactor under argon. After the CuI had dissolved, anhyd Et_3N (0.25 mL) was added, and the microreactor was placed in a heating block at 110 °C. The mixture was stirred with protection from light. Completeness of the reaction was monitored by GC/MS. The mixture was cooled to rt and poured into H_2O (15 mL). The aqueous phase was extracted with hexane (3 × 2 mL). The combined extracts were filtered (Na_2SO_4), and the solvent was evaporated under reduced pressure. The residue was purified by chromatography over a short column (Florisil, hexane).

Pyrroles 8; General Procedure:[20]
Anhyd $CuCl_2$ (1 mol%) was added to a soln of enynamine **7** (5 mmol) in anhyd DMA (2.5 mL) in a Schlenk flask. The resulting mixture was stirred at 100 °C for the time indicated. The solvent was evaporated and the crude product was purified by column chromatography (silica gel, hexane/EtOAc) to give the product as a yellow oil.

1.8.2.1.2 Gold Catalysis

Gold catalysts are generally more reactive than silver salts and can be used in lower loadings. The hydroamination process with alkynes bearing a primary amino group proceeds under gold catalysis in acetonitrile at reflux within 1–2 hours, but also at room temperature within 12 hours under neutral conditions, and provides 3,4-dihydro-2*H*-pyrroles **9** and **10** and 2,3,4,5-tetrahydropyridines **11** (Scheme 9).[21,22] In the case of 5-*exo-dig* cyclization, the cationic gold catalyst {bis[(2-(diphenylphosphino)ethyl]phenylphosphine}chlorogold(III) nitrate {[AuCl(triphos)](NO_3)$_2$} is more effective.[16]

Scheme 9 Gold-Catalyzed Cyclization of Alkynylamines to 3,4-Dihydro-2*H*-pyrroles and 2,3,4,5-Tetrahydropyridines[21]

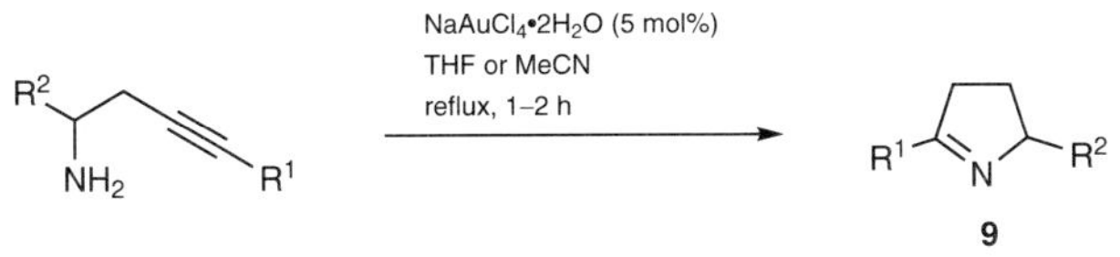

R^1	R^2	Yield (%)	Ref
$(CH_2)_7Me$	H	99	[21]
$(CH_2)_6Me$	Me	99	[21]

for references see p 499

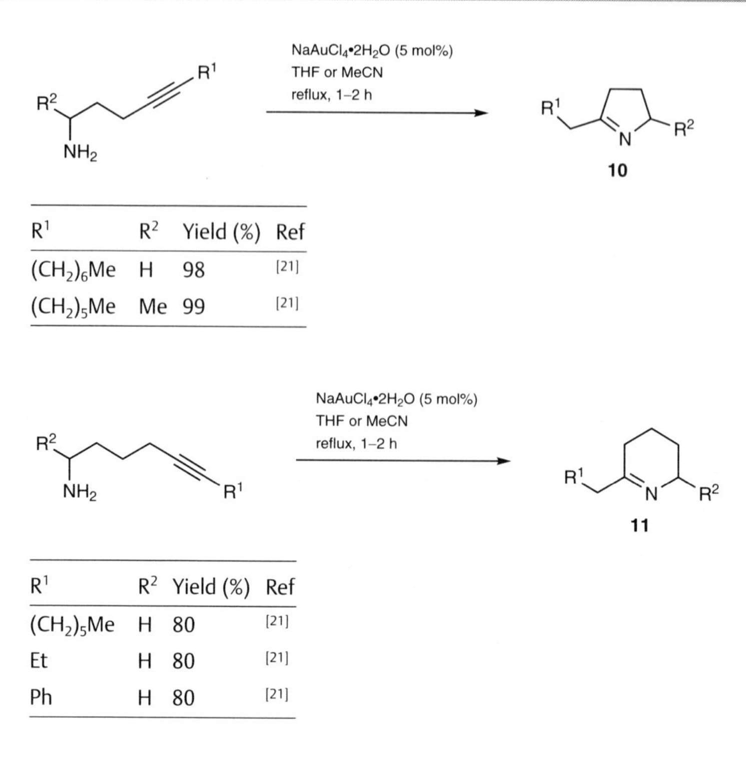

R^1	R^2	Yield (%)	Ref
$(CH_2)_6Me$	H	98	[21]
$(CH_2)_5Me$	Me	99	[21]

R^1	R^2	Yield (%)	Ref
$(CH_2)_5Me$	H	80	[21]
Et	H	80	[21]
Ph	H	80	[21]

Analogously, hydroamination is an important process in the synthesis of indoles **13** starting from 2-alkynylanilines **12**. Different transition-metal catalysts may be used, but the advantages of gold catalysis are the very mild reaction conditions and short reaction times compared to copper and palladium catalysis. The reaction is performed with sodium tetrachloroaurate(III) in a mixture of ethanol and water as solvent at room temperature (Scheme 10).[23]

Scheme 10 Gold-Catalyzed Cyclization of 2-Alkynylanilines to Indoles[23]

$NaAuCl_4 \cdot 2H_2O$ (4 mol%), EtOH/H_2O; **12** → **13**

R^1	R^2	R^3	Ratio (EtOH/H_2O)	Temp	Time (h)	Yield (%)	Ref
4-Tol	H	H	100:0	rt	2.5	80	[23]
Bu	H	H	95:5	rt	4	74	[23]
2-thienyl	H	H	50:50	rt	2.5	90	[23]
$4-ClC_6H_4$	H	H	100:0	rt	6	92	[23]
$4-MeOC_6H_4$	Cl	Cl	50:50	rt	5.5	75	[23]
(*E*)-CH=CHPh	NO_2	Cl	100:0	70 °C	5	91	[23]
H	Cl	Cl	100:0	rt	7	79	[23]
H	NO_2	Cl	100:0	rt	16	66	[23]

3,4-Dihydro-2*H*-pyrroles 9 and 10 and 2,3,4,5-Tetrahydropyridines 11; General Procedure:[21]

$NaAuCl_4$•$2H_2O$ (5 mol%) was added to a soln of the appropriate alkynylamine (1 mmol) in MeCN or THF (40 mL). The mixture was heated at reflux for 1–2 h. After evaporation of the solvent, the residue was diluted with Et_2O (30 mL) and washed with a mixture of aq NH_3 and brine (1:1; 30 mL). The organic layer was dried and concentrated to give the crude product, which was purified by distillation.

5,7-Dichloro-1*H*-indole (13, R^1 = H; R^2 = R^3 = Cl); Typical Procedure:[23]

$NaAuCl_4$•$2H_2O$ (4 mol%) was added to a soln of the 2-alkynylaniline **12** (R^1 = H; R^2 = R^3 = Cl; 0.75 mmol) in EtOH (6 mL). The mixture was stirred at rt for 7 h under N_2. The solvent was removed under reduced pressure and the crude product was purified by chromatography (silica gel, hexane/EtOAc) to give the product; yield: 79%.

1.8.2.1.3 Palladium Catalysis

One of the first works on the hydroamination of alkynes provided a wide variety of substituted indoles starting from amides, carbamates, ureas, and sulfonamides of 2-alkynylanilines.[24] In particular, cyclization of *N*-acetyl-2-alkynylanilines **14** provides 2-substituted indoles **15** using a palladium(II) catalyst without the need for an oxidizing reagent (Scheme 11). Alkynes bearing aliphatic substituents on the terminus react most efficiently.

Scheme 11 Palladium-Catalyzed Cyclization of *N*-Acetyl-2-alkynylanilines to Indoles[24]

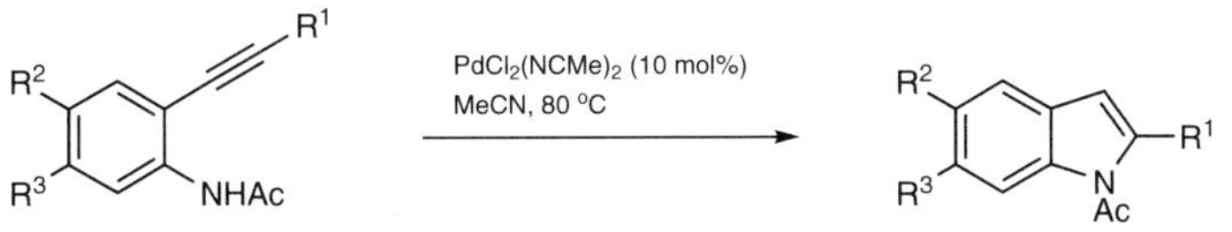

R^1	R^2	R^3	Time (h)	Yield (%)	Ref
iPr	H	CO_2Me	1.5	82	[24]
Bu	H	CO_2Me	2	71	[24]
Ph	H	CO_2Me	1.5	76	[24]
Pr	H	OMe	0.5	75	[24]
iPr	H	OMe	1	78	[24]
Bu	H	OMe	1.5	66	[24]
Pr	Me	H	1	81	[24]
Bu	Me	H	0.5	77	[24]
Ph	Me	H	1	80	[24]
Pr	Cl	H	3	76	[24]
iPr	Cl	H	1.5	80	[24]
Bu	Cl	H	2	82	[24]
Bu	OTf	H	2.5	64	[24]

Aliphatic alkynes tethered to an amino group are suitable substrates to obtain five- and six-membered nitrogen heterocycles. This kind of hydroamination requires the use of both tetrakis(triphenylphosphine)palladium(0) and triphenylphosphine under neutral

for references see p 499

conditions in benzene at 100 °C.[25] When the reactions of aliphatic alkynes **19** are performed in the presence of chiral bisphosphine ligands such as norphos derivative **16**, (*R*,*R*)-renorphos (**17**), or renorphos derivative **18**, the chiral heterocycles **20** produced are obtained with high enantioselectivity (Scheme 12).[26,27] The addition of 10 mol% benzoic acid enhances the reaction rate through the formation of hydridopalladium species. Working with this catalytic system, the best and most stable chiral ligand for hydroamination is the tolyl renorphos derivative **18**.[27]

Scheme 12 Palladium-Catalyzed Cyclization of Alkynylamines to Pyrrolidines and Piperidines[27]

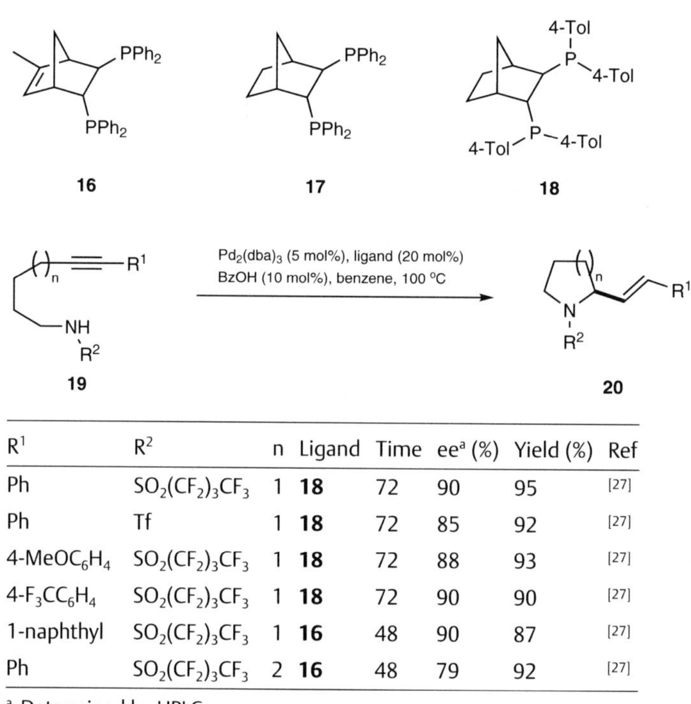

R^1	R^2	n	Ligand	Time	ee[a] (%)	Yield (%)	Ref
Ph	$SO_2(CF_2)_3CF_3$	1	**18**	72	90	95	[27]
Ph	Tf	1	**18**	72	85	92	[27]
4-$MeOC_6H_4$	$SO_2(CF_2)_3CF_3$	1	**18**	72	88	93	[27]
4-$F_3CC_6H_4$	$SO_2(CF_2)_3CF_3$	1	**18**	72	90	90	[27]
1-naphthyl	$SO_2(CF_2)_3CF_3$	1	**16**	48	90	87	[27]
Ph	$SO_2(CF_2)_3CF_3$	2	**16**	48	79	92	[27]

[a] Determined by HPLC.

N-Acetylindoles 15**; General Procedure:**[24]
To a soln of an *N*-acetyl-2-alkynylaniline **14** (0.4 mmol) in anhyd MeCN (4 mL), $PdCl_2(NCMe)_2$ (10 mol%) was added and the mixture was heated at 80 °C. After the indicated time, the solvent was removed under reduced pressure and the product was purified by column chromatography (silica gel, hexane/EtOAc).

Pyrrolidines 20 **(n = 1) and Piperidines** 20 **(n = 2); General Procedure:**[27]
An alkynylamine **19** (0.125 mmol) in benzene (1.5 mL) (**CAUTION:** *carcinogen*) was added to a soln of $Pd_2(dba)_3$•$CHCl_3$ (5 mol%), BzOH (10 mol%), and norphos derivative **16** (20 mol%) or renorphos derivative **18** (20 mol%) in benzene (2 mL) in a screw-capped vial under argon. After the mixture had been heated at 100 °C for the indicated time, it was filtered through a short column (silica gel, Et_2O). The solvent was evaporated, and the residue was purified by column chromatography (silica gel, hexane/EtOAc).

1.8.2.1.4 **Rhodium Catalysis**

A variety of 2-alkynylanilines **21** are converted into the corresponding indoles **22** through hydroamination under rhodium catalysis in the presence of triphenylphosphine or tris(4-fluorophenyl)phosphine as ligand in dimethylformamide at 85 °C (Scheme 13). The catalytic system is compatible with a wide range of functional groups on the aryl ring but the triple bond must be terminal, suggesting a mechanism that involves the coordination of the metal to the triple bond resulting in a vinylidene intermediate.[28]

Scheme 13 Rhodium-Catalyzed Cyclization of 2-Alkynylanilines to Indoles[28]

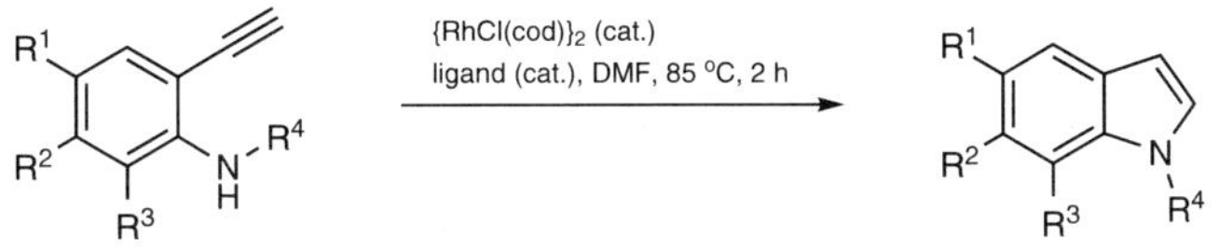

{RhCl(cod)}$_2$ (cat.)
ligand (cat.), DMF, 85 °C, 2 h

21 → **22**

R^1	R^2	R^3	R^4	mol% of [RhCl(cod)]$_2$	Ligand (mol%)	Yield (%)	Ref
Me	H	H	H	1	Ph_3P (4)	52	[28]
Me	H	H	H	5	$(4\text{-}FC_6H_4)_3P$ (60)	79	[28]
Me	H	H	H	1	$(4\text{-}FC_6H_4)_3P$ (4)	72	[28]
Cl	H	H	H	1	Ph_3P (4)	77	[28]
Cl	H	H	H	5	$(4\text{-}FC_6H_4)_3P$ (60)	84	[28]
Cl	H	H	H	1	$(4\text{-}FC_6H_4)_3P$ (4)	88	[28]
CO_2Et	H	H	H	1	Ph_3P (4)	84	[28]
CO_2Et	H	H	H	5	$(4\text{-}FC_6H_4)_3P$ (60)	98	[28]
CO_2Et	H	H	H	1	$(4\text{-}FC_6H_4)_3P$ (4)	89	[28]
NC	H	H	H	1	Ph_3P (4)	90	[28]
NC	H	H	H	5	$(4\text{-}FC_6H_4)_3P$ (60)	98	[28]
NC	H	H	H	1	$(4\text{-}FC_6H_4)_3P$ (4)	89	[28]
Ac	H	H	H	1	Ph_3P (4)	77	[28]
Ac	H	H	H	5	$(4\text{-}FC_6H_4)_3P$ (60)	93	[28]
Ac	H	H	H	1	$(4\text{-}FC_6H_4)_3P$ (4)	85	[28]
NO_2	H	H	H	1	Ph_3P (4)	81	[28]
NO_2	H	H	H	5	$(4\text{-}FC_6H_4)_3P$ (60)	79	[28]
NO_2	H	H	H	1	$(4\text{-}FC_6H_4)_3P$ (4)	80	[28]
Cl	H	H	Bn	1	Ph_3P (4)	61	[28]
Cl	H	H	Bn	5	$(4\text{-}FC_6H_4)_3P$ (60)	88	[28]
Cl	H	H	Bn	1	$(4\text{-}FC_6H_4)_3P$ (4)	73	[28]
Cl	H	H	PMB	1	Ph_3P (4)	52	[28]
Cl	H	H	PMB	5	$(4\text{-}FC_6H_4)_3P$ (60)	94	[28]
Cl	H	H	PMB	1	$(4\text{-}FC_6H_4)_3P$ (4)	74	[28]
Cl	$(CH_2)_4$		H	1	Ph_3P (4)	59	[28]
Cl	$(CH_2)_4$		H	5	$(4\text{-}FC_6H_4)_3P$ (60)	83	[28]
Cl	$(CH_2)_4$		H	1	$(4\text{-}FC_6H_4)_3P$ (4)	73	[28]

Indoles 22; General Procedure:[28]
DMF (2.50 mL) was degassed (argon balloon) for 20 min and then added via cannula to a mixture of 2-alkynylaniline **21** (0.50 mmol), $\{RhCl(cod)\}_2$ (2.5 mg, 5 μmol), and Ph_3P (5.2 mg, 20 μmol). The soln was stirred for 2 h at 85 °C under argon, and cooled to 25 °C. Sat. aq $NaHCO_3$ was added and the mixture was extracted with Et_2O or EtOAc. The combined organic extracts were washed with sat. aq $NaHCO_3$, dried ($MgSO_4$), filtered, and concentrated under reduced pressure. The residue was purified by flash column chromatography (silica gel).

1.8.2.1.5 Ruthenium Catalysis

Alkynylamines of the type used in the palladium-catalyzed hydroamination reaction (Section 1.8.2.1.3), but differing in length and substitution (e.g., **23**) are able to give intramolecular hydroamination products using the simple triruthenium dodecacarbonyl catalyst with the advantage of avoiding the use of ligands or additives. The cyclization rates depend on the ring size, following the trend five- > six- >> seven-membered nitrogen heterocycles. The reaction is applicable to primary and secondary amines and gives 3,4-dihydro-2*H*-pyrroles (e.g., **24**), 3,4,5,6-tetrahydropyridines, or 3,4,5,6-tetrahydro-2*H*-azepines, and 2-vinylidenepyrrolidines (e.g., **25**), respectively (Scheme 14).[29] 2-Alkynylanilines afford indoles.

Scheme 14 Ruthenium-Catalyzed Cyclization of Alkynylamines[29]

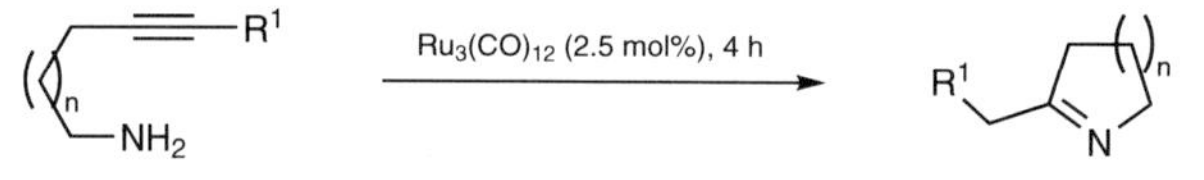

$Ru_3(CO)_{12}$ (2.5 mol%), 4 h

23 → **24**

R^1	n	Solvent	Temp (°C)	Yield (%)	Ref
Ph	1	diglyme	110	84[a]	[29]
Me	1	$(BuOCH_2CH_2)_2O$	110	60[b]	[29]
H	1	$(BuOCH_2CH_2)_2O$	110	78[b]	[29]
Ph	2	diglyme	120	77[b,c]	[29]
Ph	3	diglyme	140	21[b]	[29]

[a] Isolated yield.
[b] Yield calculated by GLC.
[c] Reaction time was 6 h.

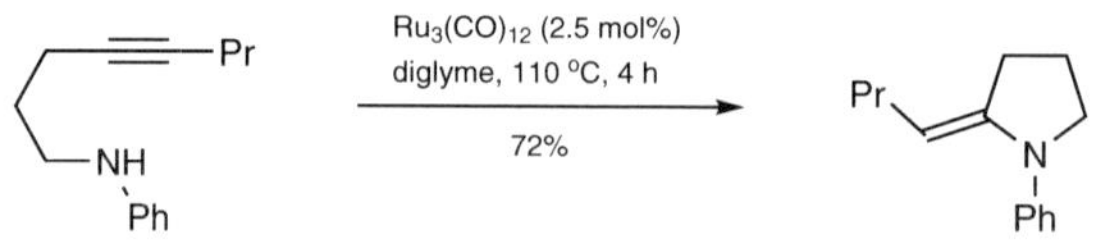

$Ru_3(CO)_{12}$ (2.5 mol%)
diglyme, 110 °C, 4 h
72%

25

Nitrogen Heterocycles 24 and 25; General Procedure:[29]
A mixture of an aminoalkyne (2.5 mmol), $Ru_3(CO)_{12}$ (2.5 mol%), and diglyme (4.0 mL) was placed in a two-necked Pyrex flask equipped with a magnetic stirrer bar and a reflux condenser under argon. The mixture was stirred at 110–140 °C for 4–6 h, cooled, and analyzed by GLC. The product was isolated by Kugelrohr distillation and/or recycling preparative HPLC.

1.8.2.1.6 Silver Catalysis

The low cost is the main advantage of silver catalysts, but their use is limited to certain reaction types, such as hydroamination of alkynes and hydroalkoxylation of alkenes. Silver salts are used more frequently as additives with other transition-metal complexes to generate the more reactive cationic species (see Section 1.8.2.1.7). Frequently, silver-catalyzed reactions involve allenes rather than alkynes as substrates.[30] The utility of allenes is due to their potential axial chirality with the advantage of allowing enantio- and diastereoselective reactions.

Primary aminoalkynes **26** undergo intramolecular hydroamination under the catalysis of silver–2,9-dimethyl-1,10-phenanthroline complex **27** to form 3,4-dihydro-2*H*-pyrroles or 2,3,4,5-tetrahydropyridines **28**. The advantages of this silver complex are its stability to air and moisture and easy recyclability (Scheme 15).[31]

Scheme 15 Silver-Catalyzed Cyclization of Primary Alkynylamines to 3,4-Dihydro-2*H*-pyrroles and 3,4,5,6-Tetrahydropyridines[31]

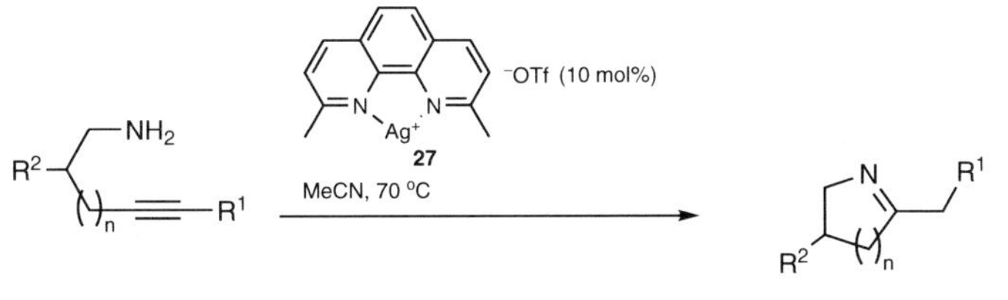

$^{-}$OTf (10 mol%)
27
MeCN, 70 °C

26 → **28**

R^1	R^2	n	Time (h)	Yield[a] (%)	Ref
Ph	H	1	4	95	[31]
Ph	H	2	6	77[b]	[31]
H	H	1	5	95	[31]
H	H	2	5	78	[31]
Me	H	1	10	90	[31]
Ph	$CH_2C{\equiv}CPh$	1	4	72[b]	[31]

[a] Yields were calculated from ^{1}H NMR spectra using mesitylene as an internal standard.
[b] 1,10-Phenanthroline was used as the ligand instead of 2,9-dimethyl-1,10 phenanthroline.

3,4-Dihydro-2*H*-pyrroles 28 **(n = 1) and 2,3,4,5-Tetrahydropyridines** 28 **(n = 2); General Procedure:**[31]
Silver catalyst **27** (10 mol%) was added to a soln of aminoalkyne **26** (0.6 mmol) in anhyd MeCN (4 mL). The mixture was heated at 70 °C for 4–10 h in the absence of light, cooled to rt, and passed through a short column (basic alumina, CH_2Cl_2). For volatile compounds, purification was achieved by distillation. Solvent evaporation gave the products as clear oils.

1.8.2.1.7 Silver–Gold Heterobimetallic Catalysis

Different heterobimetallic complexes are exploited to afford 1,2-dihydroisoquinolines **30** starting from (2-alkynylbenzyl)carbamates and -sulfonamides **29**. The reaction is performed in 1,2-dichloroethane at room temperature. The presence of ethanol accelerates the reaction, enabling the catalyst loading to be reduced to 1 mol%. Electron-withdrawing groups are required on the nitrogen atom and the substituents on the alkynyl group significantly affect the outcome of the reaction by changing the electron density on the triple bond; electron-donating groups furnish the desired 6-*endo* adducts **30** in good yields (Scheme 16).[32]

Scheme 16 Silver–Gold Catalyzed Cyclization of (2-Alkynylbenzyl)carbamates and -sulfonamides to 1,2-Dihydroisoquinolines[32]

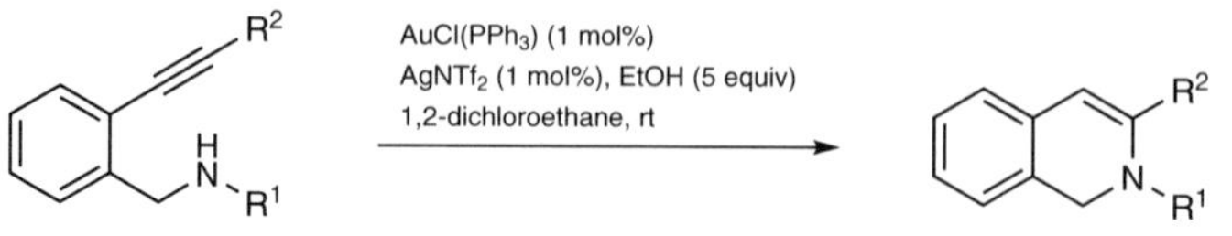

AuCl(PPh_3) (1 mol%)
$AgNTf_2$ (1 mol%), EtOH (5 equiv)
1,2-dichloroethane, rt

29 → **30**

R^1	R^2	Time (h)	Yield (%)	Ref
Boc	Ph	2	83	[32]
Cbz	Ph	4	83	[32]
Ms	Ph	7	81	[32]
Boc	3,5-$Me_2C_6H_3$	2	87	[32]
Boc	4-ClC_6H_4	2	71	[32]
Boc	4-$HOCH_2C_6H_4$	3	79	[32]
Boc	4-$MeOC_6H_4$	2	87[a]	[32]
Boc	Pr	0.2	98[a]	[32]
Boc	Ph	0.5	91[a]	[32]

[a] Reaction carried out without EtOH and using 1 mol% AuCl(**31**) instead of AuCl(PPh_3).

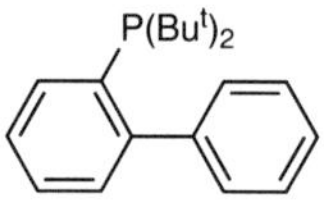

P(Bu^t)$_2$

31

1,2-Dihydroisoquinolines 30; General Procedure:[32]
To a soln of a (2-alkynylbenzyl)carbamate **29** (1.0 mmol) in 1,2-dichloroethane (2 mL) was added a mixture of AuCl(PPh_3) (1 mol%), $AgNTf_2$ (1 mol%), EtOH (5 equiv), and 1,2-dichloroethane (1 mL) at rt. After the mixture had been stirred for the required time, sat. aq $NaHCO_3$ was added and the resultant mixture was extracted with $CHCl_3$. The combined organic extracts were washed with brine, dried (Na_2SO_4), filtered, and concentrated under reduced pressure. The residue was purified by column chromatography (silica gel, hexane/EtOAc) to afford the product as colorless crystals.

1.8.2.1.8 Zinc Catalysis

Compared to other transition-metal catalysts, the advantages of zinc compounds are high stability toward moisture and air, low cost, nontoxicity, and compatibility with various polar functional groups (Scheme 17).[33] Nonactivated alkynes (e.g., **32**) and alkenes, bearing various functional groups, give access to five-, six- (e.g., **34**), and seven-membered rings in benzene at 120 °C using zinc catalyst **33**, which is readily available from *N*-isopropyl-2-(isopropylamino)troponimine and dimethylzinc. The catalyst loading may be lowered from 10 to 1 mol% by the addition of an equimolar amount of dimethyl(phenyl)ammonium tetrakis(pentafluorophenyl)borate {[$PhNMe_2H$][$B(C_6F_5)_4$]} as cocatalyst. In the case of alkenes, which are less reactive than alkynes, the presence of geminal substituents β to the amino group is required to obtain good yields, in accordance with the Thorpe–Ingold effect.

Scheme 17 Zinc-Catalyzed Cyclization of Functionalized Aminoalkynes[33]

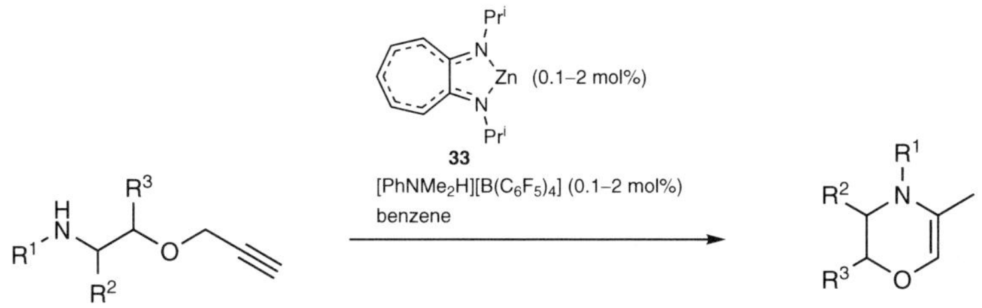

R^1	R^2	R^3	mol% of Catalyst	mol% of Activator	Temp (°C)	Time (h)	Conversion[a] (%)	Ref
PMB	H	H	1	1	60	39	>99	[33]
Bn	Et	H	0.1	0.1	120	8	>99	[33]
Bn	iPr	H	1	1	120	144	>99	[33]
H	H	Ph	2	2	120	14	>99	[33]
$(CH_2)_3$		H	0.1	0.1	120	8	>99	[33]

[a] Determined by ^{1}H NMR spectroscopy.

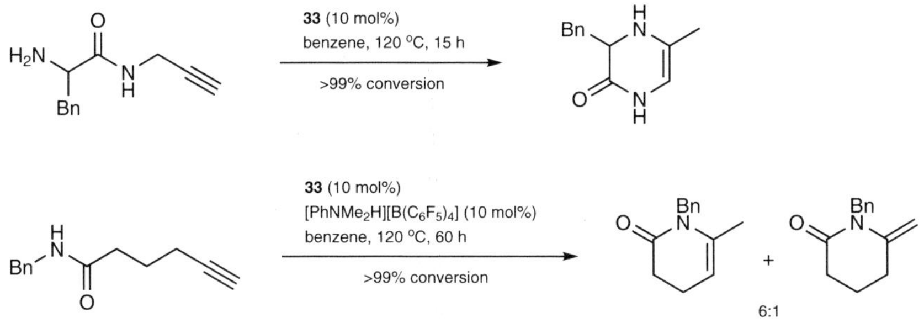

2-Alkynyl-*N*-tosylanilines **35** react under zinc catalysis in toluene at reflux to afford 2-substituted *N*-tosylindoles **36** (Scheme 18). Performing the reaction in the presence of an acyl chloride gives C3-substituted indoles in high yields via a tandem process of cyclization and electrophilic addition.[34]

Scheme 18 Zinc-Catalyzed Cyclization of 2-Alkynyl-*N*-tosylanilines to *N*-Tosylindoles[34]

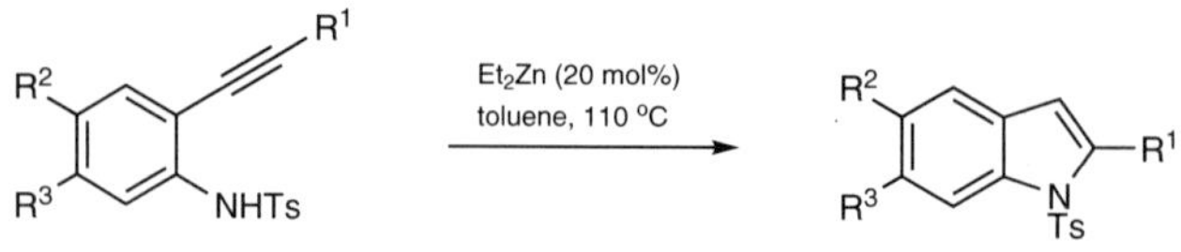

R^1	R^2	R^3	Time (h)	Yield (%)	Ref
Ph	Me	H	4	90	[34]
Ph	H	H	3	98	[34]
Ph	F	H	2.5	93	[34]
Ph	Cl	H	2.5	92	[34]
Ph	Br	H	2.5	92	[34]
Ph	NO_2	H	2	94	[34]
Ph	Cl	NO_2	2	93	[34]
Bu	H	H	3	95	[34]
H	H	H	3	97	[34]
CH_2OH	H	H	3	98	[34]

The sequential process of amination and cyclization is an elegant method to obtain 2-arylindoles **39** starting from (2-bromophenyl)acetylenes **37**. For the substitution with aqueous ammonia, copper catalysis with a suitable ligand is required. The cyclization step is then performed in the presence of zinc bromide as catalyst. Both electron-rich and electron-deficient substituents are tolerated on the substrates **38** (Scheme 19).[35]

Scheme 19 Zinc-Catalyzed Cyclization of 2-Alkynylanilines to 2-Arylindoles[35]

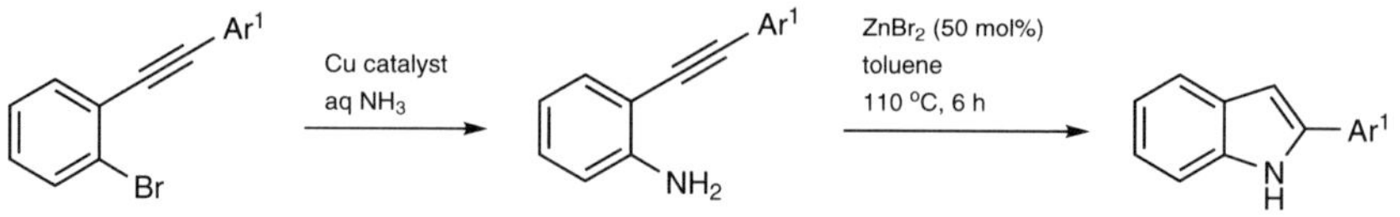

Ar^1	Yield[a] (%)	Ref
Ph	87	[35]
4-Tol	85	[35]
$4\text{-}ClC_6H_4$	84	[35]
$4\text{-}MeOC_6H_4$	87	[35]
$3,5\text{-}(MeO)_2C_6H_3$	83	[35]
1-naphthyl	83	[35]
$4\text{-}AcC_6H_4$	68[b]	[35]

Ar^1	Yield[a] (%)	Ref
$4\text{-}FC_6H_4$	88[b]	[35]
$4\text{-}F_3CC_6H_4$	83[b]	[35]
2-thienyl	57	[35]

[a] Isolated overall yield of amination and cyclization.
[b] Reaction time was 15 h.

5-Methyl-3,4-dihydro-2*H*-1,4-oxazines 34; **General Procedure:**[33]
All NMR tube scale reactions were performed in a N_2-filled glovebox. An aminoalkyne **32** (430 μmol) was dissolved in benzene-d_6 (0.5 mL) (**CAUTION:** *carcinogen*) and then added to the zinc catalyst **33** (0.1–2 mol%) and activator $[PhNMe_2H][B(C_6F_5)_4]$ (0.1–2 mol%). An NMR tube was then charged with the resulting yellow soln, removed from the glovebox, and flame-sealed under reduced pressure. The mixture was heated at 120 °C for the indicated time. NMR yields were determined by comparing the integration of a well-resolved signal for the starting material with a well-resolved signal for the heterocyclic product.

***N*-Tosylindoles** 36; **General Procedure:**[34]
Et_2Zn (20 mol%) was added to a soln of a 2-alkynyl-*N*-tosylaniline **35** (0.3 mmol) in anhyd toluene (6 mL). The mixture was stirred at reflux. At the end of the reaction, the mixture was cooled to rt and sat. aq NH_4Cl (2 mL) was added. The mixture was extracted with Et_2O (3 × 25 mL), washed with brine, dried (Na_2SO_4), and concentrated. Purification of the crude product by column chromatography (silica gel, hexane/EtOAc) afforded the product as a solid.

2-Arylindoles 39; **General Procedure:**[35]
A 2-alkynylaniline **38** arising from amination of the corresponding (2-bromophenyl)acetylene **37** was directly treated with $ZnBr_2$ (50 mol%) and toluene (4 mL). The mixture was heated under reflux at 110 °C for 6 or 15 h. After completion of the reaction, as detected by HPLC, toluene was removed under reduced pressure. The residue was purified by column chromatography (silica gel, petroleum ether/EtOAc).

1.8.2.2 Hydroamination of Alkenes

Hydroamination of unactivated alkenes bearing a tethered amino group is one of the simplest methods to obtain nitrogen heterocycles and for this reason it remains a topic under study. Many efforts have been made toward the exploitation of this methodology in the field of natural product synthesis. Various metal catalysts may be used depending on the substrates. All the metals give good results with terminal alkenes, but for internal alkenes only iron and rhodium catalysts afford the cyclization products. Regarding the nucleophilic amino group, with unsubstituted primary amines only copper, iridium, and rhodium are used. By contrast, gold and iron need the substrate to contain amides or sulfonamides. Another important point with copper and iridium is the need for β-substituents on the linker chain to bias the substrate toward the cyclization. In terms of costs, iron is the least expensive catalyst. Compared to other metals, gold and iridium do not need ligands. Various solvents are used and the temperature always ranges from 60 °C to 100 °C.

1.8.2.2.1 Copper Catalysis

The hydroamination of unactivated terminal alkenes **40** bearing an aminoalkyl substituent or unsubstituted amino group gives pyrrolidine and piperidine derivatives **42** (n = 1 and 2, respectively) in excellent yields and in an economical way (Scheme 20).[36] The

for references see p 499

substrates must bear geminal substituents on the linker chain to bias the residue toward cyclization in accordance with the Thorpe–Ingold effect. The reaction performed with copper salts and ligands [e.g., copper(I) *tert*-butoxide and 4,5-bis(diphenylphosphino)-9,9-dimethylxanthene (**41**, Xantphos)] in the absence of a base and heated in an alcoholic solvent system is applicable to both primary and secondary amines and tolerates a variety of functional groups.

Scheme 20 Copper-Catalyzed Cyclization of Aminoalkenes to Pyrrolidines and Piperidines[36]

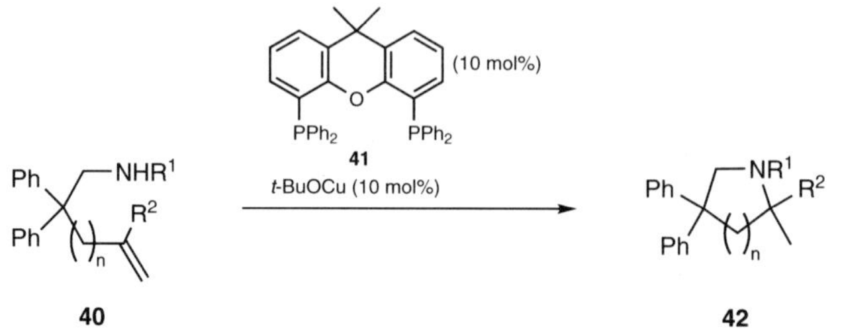

R^1	R^2	n	Solvent	Temp (°C)	Time (h)	Yield (%)	Ref
Et	H	1	MeOH/toluene (1:1)	100	48	87	[36]
Pr	H	1	MeOH/toluene (1:1)	100	48	93	[36]
iBu	H	1	MeOH/toluene (1:1)	100	48	88	[36]
$(CH_2)_2Ph$	H	1	MeOH/toluene (1:1)	100	48	87	[36]
CO_2Me	H	1	MeOH/toluene (1:1)	100	48	96	[36]
Bz	H	1	MeOH/*o*-xylene (1:1)	140	48	94	[36]
Ph	H	1	MeOH/toluene (1:1)	100	48	99	[36]
$4-MeOC_6H_4$	H	1	MeOH/toluene (1:1)	100	48	94[a]	[36]
$4-FC_6H_4$	H	1	MeOH/toluene (1:1)	100	48	98[a]	[36]
$4-NCC_6H_4$	H	1	MeOH/toluene (1:1)	100	48	86[a]	[36]
$4-MeO_2CC_6H_4$	H	1	MeOH/toluene (1:1)	100	48	86[a]	[36]
Me	Me	1	MeOH/*o*-xylene (1:1)	140	48	89[a]	[36]
Me	H	2	MeOH/toluene (1:1)	100	24	92	[36]
H	H	2	MeOH/toluene (1:1)	100	72	87	[36]

[a] 15 mol% of catalyst components was used.

2-Methylpyrrolidines 42 **(n = 1) and 2-Methylpiperidines** 42 **(n = 2); General Procedure:**[36]
In a glovebox, *t*-BuOCu (10–15 mol%) and Xantphos (**41**; 10–15 mol%) were placed in a screw-capped vial. Anhyd degassed MeOH/toluene or MeOH/*o*-xylene (1:1; 0.5 mL) was added and the mixture was stirred at rt for 15 min to give a pale yellow soln. A soln of an aminoalkene **40** (0.5 mmol) in the mixed solvent (0.5 mL) was added. The vial was sealed with a screw cap and removed from the glovebox. The mixture was stirred at 100 °C (MeOH/toluene) or 140 °C (MeOH/*o*-xylene) for the required time. After the mixture had been cooled to rt, it was purified by column chromatography (hexane/EtOAc).

1.8.2.2.2 Gold Catalysis

Nonactivated alkenes (e.g., **43**) tethered with tosylamides, sulfonamides, or benzamides provide cyclized products (e.g., **44**) through hydroamination in the presence of the catalytic system comprising chloro(triphenylphosphine)gold(I)/silver(I) trifluoromethanesulfonate (Scheme 21).[37] The use of microwave heating shortens the reaction time remarkably. Alkenylureas also furnish the cyclized products under similar conditions in dioxane as solvent and with thermal heating.[38]

Scheme 21 Gold-Catalyzed Cyclization of Alkene- and Arenesulfonamides to 1,2-Thiazinane 1,1-Dioxides[37]

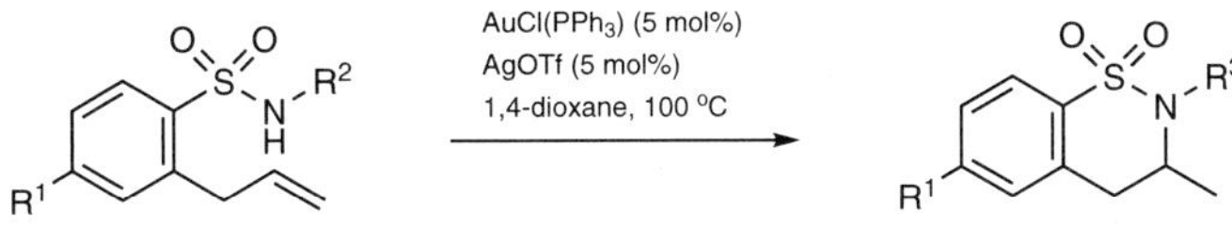

AuCl(PPh_3) (5 mol%), AgOTf (5 mol%), 1,4-dioxane, 100 °C

43 → **44**

R^1	R^2	R^3	Time (h)	Yield[a] (%)	Ref
Me	Et	Et	24	95	[37]
Me	*t*-Bu	H	48	99	[37]
Me	H	H	48	99	[37]
H	H	H	48	95	[37]
OMe	H	H	48	99	[37]

[a] Yield determined by ^1H NMR spectroscopy.

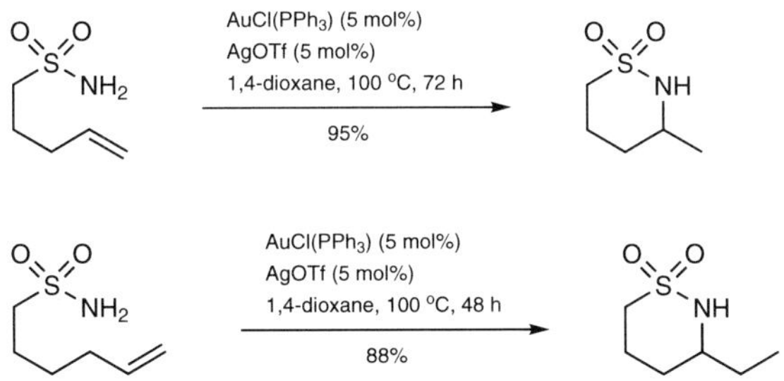

AuCl(PPh_3) (5 mol%), AgOTf (5 mol%), 1,4-dioxane, 100 °C, 72 h; 95%

AuCl(PPh_3) (5 mol%), AgOTf (5 mol%), 1,4-dioxane, 100 °C, 48 h; 88%

Cyclic Sulfonamides, e.g. 44; General Procedure:[37]

1,4-Dioxane (0.50 mL) was added to a mixture of an arenesulfonamide (e.g., **43**; 0.25 mmol), AuCl(PPh_3) (5 mol%), and AgOTf (5 mol%), and the resulting suspension was stirred at 100 °C for the indicated time. The resulting mixture was concentrated and the residue was purified by chromatography (Et_2O/CH_2Cl_2) to give the product as a white solid.

1.8.2.2.3 Iridium Catalysis

Intramolecular hydroamination of unactivated terminal alkenes **45** bearing primary as well as secondary alkyl- and arylamines employs a commercially available single-component iridium precatalyst at relatively low loadings and without the need for added ligands or cocatalysts. The presence of substituents β to the amino group is required to bias the substrates toward cyclization to the heterocyclic products **46**. The reaction is conducted at 110 °C without phosphine coligands. In the case of primary amines, the use of triethylammonium chloride as cocatalyst is required (Scheme 22).[39,40] *N*-*tert*-Butoxycarbonylamines are unreactive under these conditions.

Scheme 22 Iridium-Catalyzed Cyclization of Alkenylamines to Pyrrolidines and Piperidines[40]

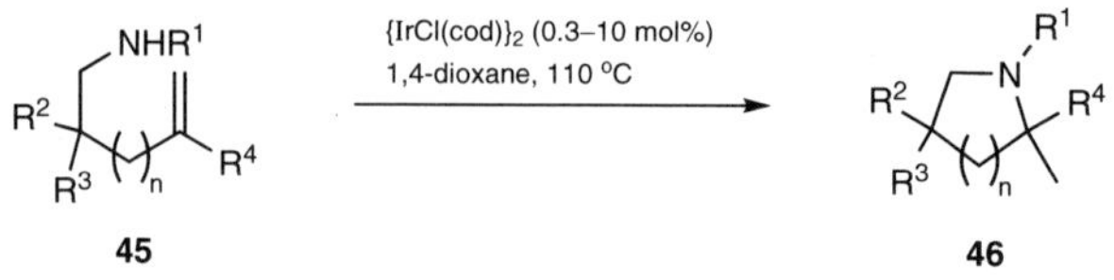

R^1	R^2	R^3	R^4	n	mol% of Catalyst	Time (h)	Yield (%)	Ref
Bn	Ph	Ph	H	1	0.3	3	88	[40]
$4\text{-ClC}_6H_4CH_2$	Ph	Ph	H	1	1.0	3	85	[40]
$4\text{-MeO}_2CC_6H_4CH_2$	Ph	Ph	H	1	1.0	3	87	[40]
$4\text{-MeOC}_6H_4CH_2$	Ph	Ph	H	1	2.5	3	89	[40]
CH_2Cy	Ph	Ph	H	1	1.0	7	88	[40]
Bn	Me	Me	H	1	5.0	24	84	[40]
H	Ph	Ph	H	2	5.0	24	72[a]	[40]
Bn	Ph	Ph	Me	1	2.5	16	88	[40]
H	Ph	Ph	Me	1	10	48	87[a]	[40]
Bn	$(CH_2)_5$		H	1	2.5	7	86	[40]
Bn	$(CH_2)_5$		Me	1	5.0	48	74	[40]
Ph	$(CH_2)_5$		H	1	5.0	16	95	[40]
H	$(CH_2)_5$		H	1	5.0	24	75[a]	[40]
Ph	Ph	Ph	H	1	2.5	7	95	[40]
4-Tol	Ph	Ph	H	1	1.0	7	96	[40]
4-MeOC_6H_4	Ph	Ph	H	1	0.5	7	94	[40]
4-FC_6H_4	Ph	Ph	H	1	2.5	7	95	[40]
H	Ph	Ph	H	1	5.0	24	89[a]	[40]

[a] Triethylammonium chloride [2 equiv with respect to $\{IrCl(cod)\}_2$] was added.

2-Methylpyrrolidines 46 **(n = 1) and 2-Methylpiperidines** 46 **(n = 2); General Procedure:**[40]
To a screw-capped vial containing an aminoalkene **45** (0.25 mmol) and a stirrer bar were added a stock soln of $\{IrCl(cod)\}_2$ in 1,4-dioxane (0.3–10 mol%) and 1,4-dioxane (0.375 mL; total reaction volume = 0.5 mL). The vial was sealed under N_2 with a cap containing a PTFE septum and, once all the material had dissolved, it was removed from the glovebox and placed in a temperature-controlled aluminum heating block set at 110 °C. After magnetic stirring for the appropriate time, the vial was cooled to rt. The mixture was diluted with CH_2Cl_2 (2 mL) and washed with brine (2 × 5 mL). The organic extracts were combined, dried (Na_2SO_4), filtered, and concentrated under reduced pressure. The resulting residue was purified by flash column chromatography (silica gel, hexane/EtOAc 20:1).

1.8.2.2.4 Iron Catalysis

The use of inexpensive iron salts as catalysts is the key point of an intramolecular hydroamination of unactivated (*N*-tosylamino)alkenes **47** resulting in the formation of pyrrolidines **48**. Although 2,2-disubstituted aminoalkenes show high reactivity, the reaction also runs with unsubstituted alkenes. The reaction proceeds under air and employs iron(III) chloride in 1,2-dichloroethane as solvent. No cyclization product is obtained with other iron salts such as iron(III) nitrate, iron(III) sulfate, or iron(III) acetylacetonate. The yields depend on the solvent, and 1,2-dichloroethane is the best choice (Scheme 23).[41] The method is also applicable to the synthesis of six-membered azacycles.

Scheme 23 Iron-Catalyzed Cyclization of Alkenyl-*N*-tosylamines to *N*-Tosylpyrrolidines[41]

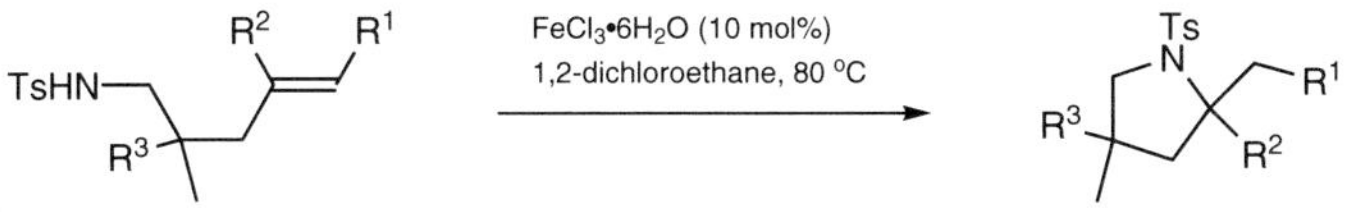

$FeCl_3 \cdot 6H_2O$ (10 mol%), 1,2-dichloroethane, 80 °C; **47** → **48**

R^1	R^2	R^3	Time (h)	Yield (%)	dr[a]	Ref
H	H	Me	2	97	–	[41]
Me	H	Me	3	95	–	[41]
Et	H	Me	3	95	–	[41]
H	Me	Me	2	99	–	[41]
H	H	CH_2OMe	9	96	1.1:1	[41]
H	H	CH_2OTs	21	81	1.3:1	[41]
H	H	I	4	93	2.1:1	[41]

[a] Determined by ^1H or ^{13}C NMR spectroscopy.

2,4,4-Trisubstituted *N*-Tosylpyrrolidines 48**; General Procedure:**[41]
A mixture of an alkenyl-*N*-tosylamine **47** (1 mmol) and $FeCl_3 \cdot 6H_2O$ (10 mol%) in 1,2-dichloroethane (10 mL) was heated at 80 °C for the indicated time and then cooled. H_2O (10 mL) was added and the aqueous phase was extracted with Et_2O (20 mL). The combined organic layers were washed with brine (10 mL), dried ($MgSO_4$), filtered, and concentrated. The crude product was purified by column chromatography (silica gel, hexane/EtOAc) to provide the product as a white solid.

1.8.2.2.5 Platinum Catalysis

Hydrofunctionalization of unactivated terminal alkenes with secondary amines, amides, and other nucleophiles, in diglyme at 60–80 °C using a platinum(II) chloride/ligand catalyst system, gives heterocycles in an atom-economical approach.[42] The catalytic system depends on sterically hindered monophosphine ligands (e.g., **49** and **50**) and cyclizes alkenylamines **51** to the corresponding heterocycles **52**. Substrates **51** ($R^1 = R^2 = H$) without substituents on the linker chain also display good activity (Scheme 24).[43]

Scheme 24 Platinum-Catalyzed Cyclization of Alkenyl(benzyl)amines to N-Benzylpyrrolidines and -piperidines[43]

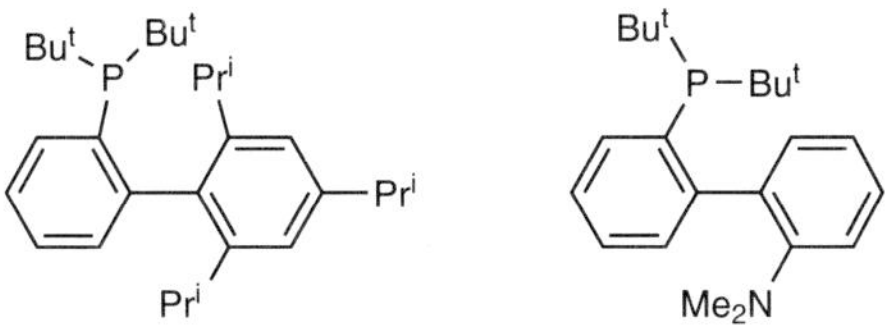

49 **50**

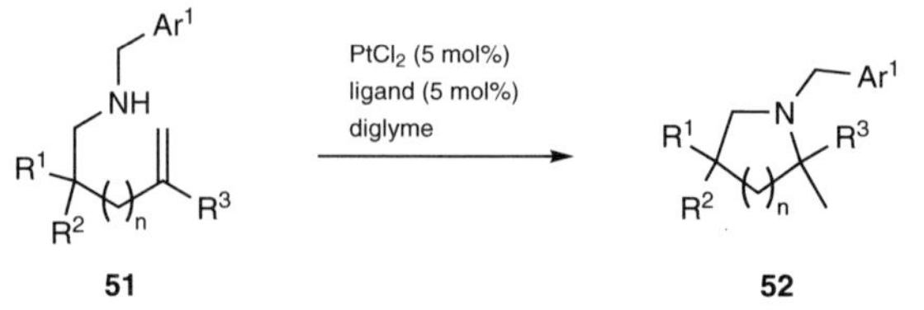

51 → 52: $PtCl_2$ (5 mol%), ligand (5 mol%), diglyme

Ar^1	R^1	R^2	R^3	n	Ligand	Temp (°C)	Time (h)	Yield (%)	Ref
Ph	Ph	Ph	H	1	**50**	60	14	86	[43]
Ph	$(CH_2)_5$		H	1	**49**	60	10	86	[43]
4-$MeOC_6H_4$	$(CH_2)_5$		H	1	**49**	60	10	86	[43]
4-BrC_6H_4	$(CH_2)_5$		H	1	**50**	60	9	77	[43]
4-$O_2NC_6H_4$	$(CH_2)_5$		H	1	**50**	60	13	84	[43]
2-naphthyl	$(CH_2)_5$		H	1	**49**	60	9	81	[43]
Ph	$(CH_2)_5$		Me	1	**50**	80	64	85	[43]
Ph	$(CH_2)_5$		H	2	**49**	60	9	76	[43]
Ph	H	H	H	2	**50**	60	96	66	[43]

1-Benzyl-2-methyl-4,4-diphenylpyrrolidine (52, $Ar^1 = R^1 = R^2 = Ph$; $R^3 = H$; n = 1); Typical Procedure:[43]
A suspension of *N*-benzyl-2,2-diphenylpent-4-en-1-amine (**51**, $Ar^1 = R^1 = R^2 = Ph$; $R^3 = H$; n = 1; 0.75 mmol), $PtCl_2$ (5 mol%), and ligand **50** (5 mol%) in diglyme (0.75 mL) was stirred at 60 °C for 14 h. The resulting mixture was concentrated and the obtained oily residue was purified by bulb-to-bulb distillation (90 °C, 75 mTorr) to give the product as a colorless oil; yield: 86%.

1.8.2.2.6 Rhodium Catalysis

Bis(cycloocta-1,5-diene)rhodium(I) tetrafluoroborate with 2-(dicyclohexylphosphino)-2′-(dimethylamino)biphenyl (**54**, DavePhos) as the ligand is a suitable system to achieve hydroamination of unactivated terminal and internal alkenes **53** bearing primary and secondary amines, with or without substituents that favor cyclization, with formation of five- or six-membered rings **55** (Scheme 25).[44] The reaction is performed under mild conditions and allows the presence of common functional groups, including free hydroxy groups, esters, chlorides, and an enolizable ketone.[45]

Scheme 25 Rhodium-Catalyzed Cyclization of Aminoalkenes to Pyrrolidines and Piperidines[44]

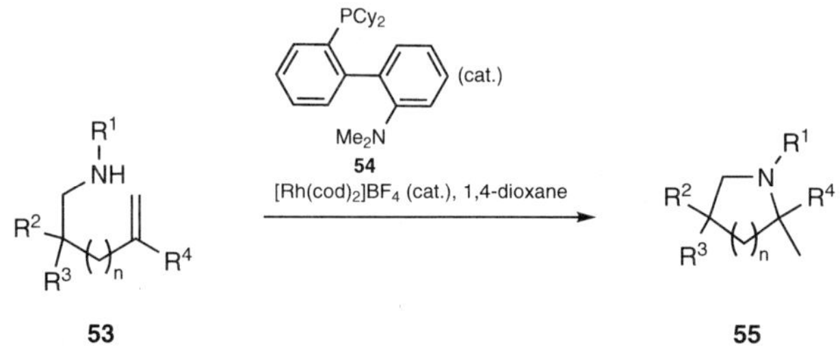

53 → 55: **54** (cat.), $[Rh(cod)_2]BF_4$ (cat.), 1,4-dioxane

R^1	R^2	R^3	R^4	n	mol% of Catalyst/Ligand	Temp (°C)	Time (h)	Yield (%)	Ref
CH_2Cy	Ph	Ph	H	1	1.5/3	70	7	91	[44]
Bn	Ph	Ph	H	1	1.5/3	70	7	91	[44]
H	Ph	Ph	H	1	5/6	100	10	83	[44]
Bn	Me	Me	H	1	1.5/3	70	7	83	[44]
Bn	$(CH_2)_5$		H	1	1.5/3	70	7	92	[44]
$4\text{-}ClC_6H_4CH_2$	H	H	H	1	1.5/3	70	7	69	[44]
$4\text{-}NCC_6H_4CH_2$	H	H	H	1	1.5/3	70	7	62	[44]
$4\text{-}MeO_2CC_6H_4CH_2$	H	H	H	1	1.5/3	70	7	72	[44]
Me	Ph	Ph	Me	1	1.5/3	70	7	96	[44]
H	Ph	Ph	Me	1	5/6	100	24	96	[44]
Me	Ph	H	Me	1	7.5/9	120	7	80	[44]
H	H	H	H	2	1.5/3	70	7	74	[44]
Bn	Ph	Ph	H	2	1.5/3	70	7	86	[44]
H	Ph	Ph	H	2	5/6	100	10	84	[44]

The first enantioselective intramolecular hydroamination of unactivated terminal alkenes **58** under rhodium catalysis gives enantiomerically enriched pyrrolidines **59**. The nature of the protecting group on the nitrogen atom has a pronounced influence on the outcome of the reaction, and the 2-methylbenzyl group gives higher enantioselectivity than the benzyl group. The cyclized products are formed in up to 91% enantiomeric excess in the presence of chiral binaphthylphosphine ligands **56** or **57** in 1,4-dioxane at 70 °C (Scheme 26).[46]

Scheme 26 Rhodium-Catalyzed Enantioselective Hydroamination of Aminoalkenes to Pyrrolidines[46]

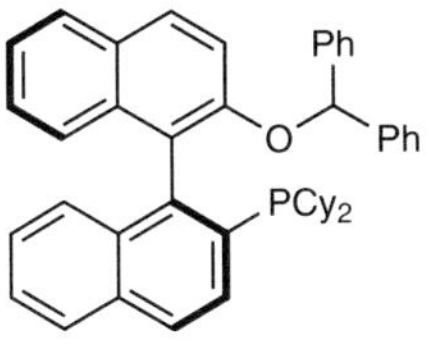

Ph
O
Ph
PCy_2
56

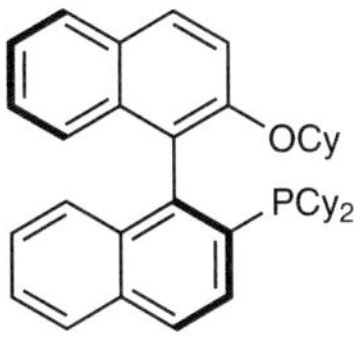

OCy
PCy_2
57

for references see p 499

[Rh(cod)$_2$]BF$_4$ (cat.)
ligand (cat.)
1,4-dioxane, 70 °C

58 → 59

R^1	R^2	R^3	Ligand	mol% of Catalyst/Ligand	Time (h)	Yield (%)	ee[a] (%)	Ref
Ph	Ph	Ph	**56**	5/6	15	91	80	[46]
2-Tol	Ph	Ph	**56**	5/6	15	92	84	[46]
2-Tol	Me	Me	**56**	5/6	20	75	62	[46]
2-Tol	$(CH_2)_5$		**56**	5/6	20	80	63	[46]
Ph	H	H	**56**	2.5/3	20	48	90	[46]
2-Tol	H	H	**57**	10/12	20	50	86	[46]
4-ClC_6H_4	H	H	**57**	10/12	20	63	85	[46]
4-$MeOC_6H_4$	H	H	**57**	10/12	20	35	85	[46]
4-$MeO_2CC_6H_4$	H	H	**57**	10/12	20	61	83	[46]

[a] ee values were determined by chiral HPLC, GC, or ^{1}H NMR spectroscopy of derivatives.

2-Methylpyrrolidines 55 **(n = 1) and 2-Methylpiperidines** 55 **(n = 2); General Procedure:**[44]
An aminoalkene **53** (0.50 mmol) was added to a screw-capped vial containing the two solids [Rh(cod)$_2$]BF$_4$ (1.5 mol%) and ligand 2-(dicyclohexylphosphino)-2′-(dimethylamino)biphenyl (**54**, 3 mol%) at rt. 1,4-Dioxane (0.5 mL) was then added. The vial was sealed with a cap containing a PTFE septum and removed from the drybox. The mixture was stirred at 70 °C for 7 h. Brine (20 mL) was added, and the resulting mixture was extracted with CH_2Cl_2 (2 × 20 mL). The combined organic fractions were dried (Na_2SO_4), filtered, and concentrated under reduced pressure. The residue was purified by flash chromatography (silica gel).

(*S*)-2-Methylpyrrolidines 59**; Typical Procedure:**[46]
An aminoalkene **58** (0.50 mmol) was added to a dry screw-capped test tube. The tube was transferred into a glovebox and [Rh(cod)$_2$]BF$_4$ (5 mol%), a chiral phosphine ligand **56** (6 mol%), and 1,4-dioxane (0.50 mL) were added. The vial was sealed with a PTFE cap and removed from the glovebox. The mixture was stirred at 70 °C for 20 h. Brine (20 mL) was added, and the mixture was extracted with CH_2Cl_2 (2 × 20 mL). The combined organic layers were dried (Na_2SO_4) and concentrated under reduced pressure. The crude product mixture was purified by flash column chromatography (silica gel pretreated with 2% Et_3N in hexane before loading the crude product mixture).

1.8.2.3 Oxidative Amination of Alkenes

The peculiarity of the oxidative amination of alkenes consists of maintaining the substrate and the product in the same oxidation state. There are several advantages to this process: (1) the unsaturated system is preserved in the product; (2) the resulting product is the same as that which would be obtained through a hydroamination process starting from alkynes, with the advantage of a less expensive and more easily accessible substrate; (3) compared to cross-coupling reactions, the substrate is an unfunctionalized molecule. Considering the mechanistic aspects, the initial activation of the substrate through coordination to the electrophilic metal is followed by addition of the nucleophilic nitrogen. The key step is the fate of the intermediate β-aminoalkylmetal complex **60** (Scheme 27).[1] In fact, the reaction leads to the oxidative amination product **62** if β-hydride elimination happens. On the contrary, if the newly formed M—C bond is cleaved by protonolysis, the hydroamination product **61** is obtained. Of course, the final step generates the reduced form of the metal [i.e., Pd(II) to Pd(0)], which must be reoxidized if the catalytic cycle is to resume. Thus, one key component of the oxidative amination process is the demand for an efficient oxidizing agent [benzo-1,4-quinone, copper(II) chloride, or, in some cases, molecular oxygen].

Scheme 27 Oxidative Amination Mechanism[1]

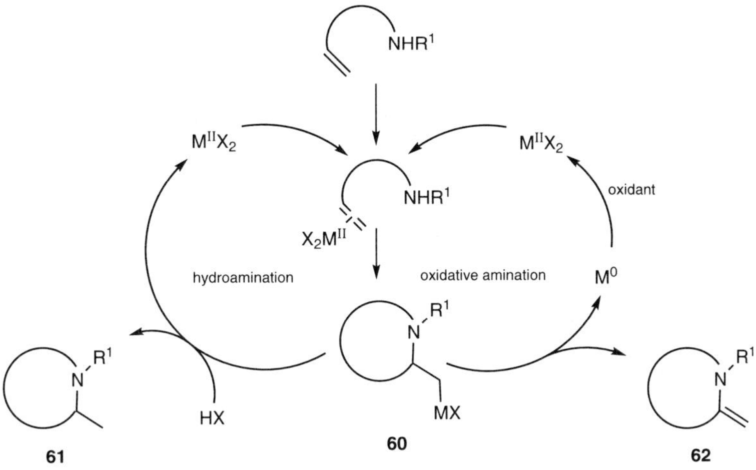

NHR^1

$M^{II}X_2$ $M^{II}X_2$ oxidant

NHR^1 X_2M^{II}

hydroamination oxidative amination M^0

R^1 N HX MX

61 **60** **62**

1.8.2.3.1 Palladium Catalysis

A landmark in the field of aminopalladation reactions is represented by the formation of indoles **64** starting from 2-allylaniline derivatives **63**. Initially, stoichiometric amounts of the palladium(II) catalyst were employed, but later catalytic amounts of palladium(II) coupled with benzo-1,4-quinone as reoxidant were used (Scheme 28).[47] Attempts to cyclize nonaromatic aminoalkenes highlight the need to increase the acidity of the NH group; alkenes tethered to a tosylamide group are sufficiently reactive to give cyclization, but the reaction is of synthetic utility only for the formation of *cis*-fused 5,5 ring systems.[48]

Scheme 28 Palladium-Catalyzed Cyclization of 2-Allylanilines to Indoles[47]

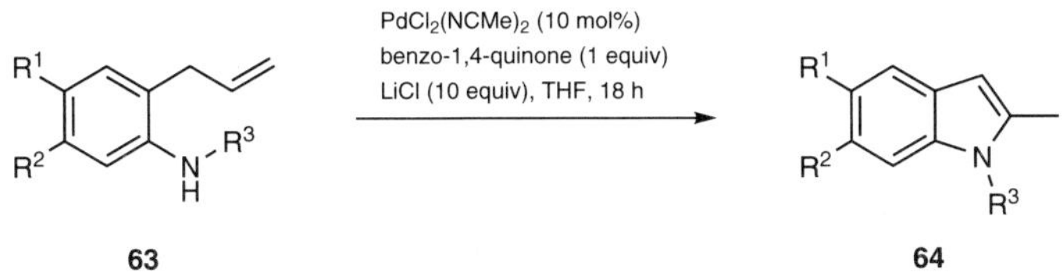

$PdCl_2(NCMe)_2$ (10 mol%)
benzo-1,4-quinone (1 equiv)
LiCl (10 equiv), THF, 18 h

63 **64**

R^1	R^2	R^3	Temp	Yield (%)	Ref
H	H	H	reflux	86	[47]
H	H	Me	reflux	89	[47]
H	H	Ac	reflux	71	[47]
OMe	OMe	H	25 °C	48	[47]
OMe	H	H	25 °C	32	[47]

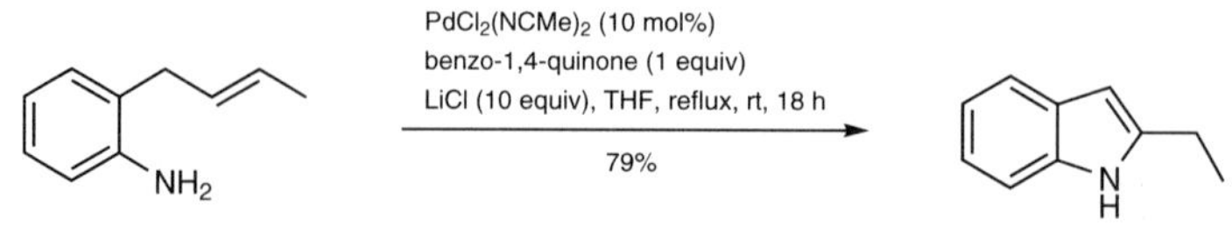

PdCl$_2$(NCMe)$_2$ (10 mol%)
benzo-1,4-quinone (1 equiv)
LiCl (10 equiv), THF, reflux, rt, 18 h
79%

Application of the amination process to the easily accessible 1-allylindole-2-carboxamides **65** leads to the formation of pyrazino[1,2-*a*]indoles **66** exploiting the amide group as nucleophile. The reaction is performed in dimethylformamide at 100 °C over 24 hours using palladium(II) acetate as catalyst and benzo-1,4-quinone as oxidant in the presence of sodium carbonate as base and tetrabutylammonium chloride as additive (Scheme 29).[49]

Scheme 29 Palladium-Catalyzed Cyclization of 1-Allylindole-2-carboxamides to Pyrazino[1,2-*a*]indol-1(2*H*)-ones[49]

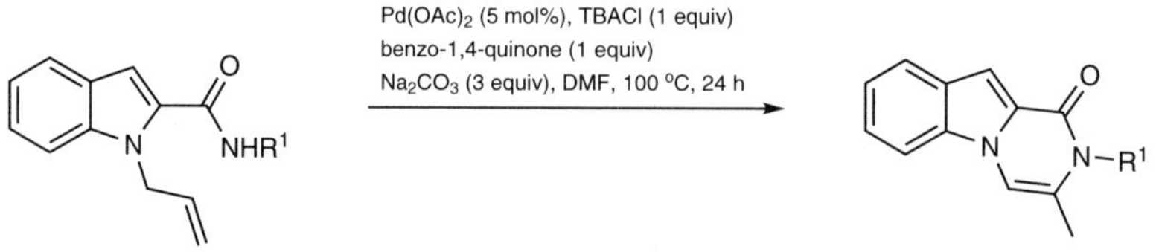

Pd(OAc)$_2$ (5 mol%), TBACl (1 equiv)
benzo-1,4-quinone (1 equiv)
Na$_2$CO$_3$ (3 equiv), DMF, 100 °C, 24 h

65 → **66**

R^1	Yield (%)	Ref
Ph	78	[49]
4-Tol	83	[49]
4-MeOC$_6$H$_4$	77	[49]
4-O$_2$NC$_6$H$_4$	75	[49]
Cy	28	[49]
Bn	36	[49]
CH$_2$CO$_2$Et	55	[49]

Tosylamides (e.g., **67** and **69**) tethered to a double bond undergo regioselective cyclization to give five- or seven-membered ring lactams (e.g., **68**, **70**, and **71**) through the formation of a π-allylpalladium intermediate. The subsequent nucleophilic attack may afford five- or seven-membered rings depending on whether the attack proceeds through an *exo* or *endo* mode (Scheme 30).[50] In the case of monosubstituted alkenes **69**, the presence of a Brønsted base can modulate the regioselectivity, promoting the selective generation of seven-membered rings (e.g., **70**). When no base is added, five-membered rings (e.g., **71**) are formed selectively.

Scheme 30 Palladium-Catalyzed Cyclization of *N*-Tosylhex-5-enamides to Lactams[50]

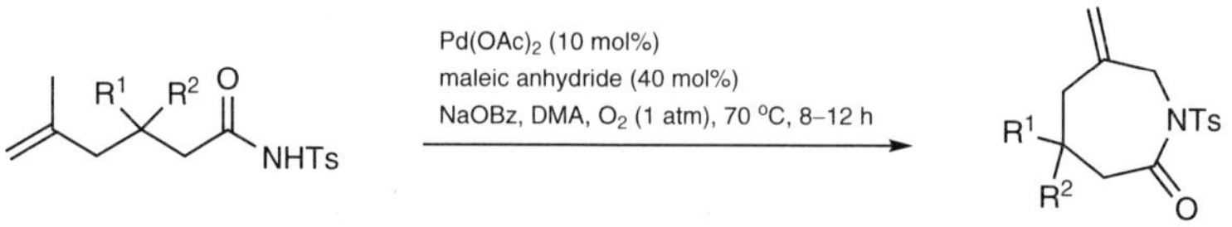

Pd(OAc)$_2$ (10 mol%)
maleic anhydride (40 mol%)
NaOBz, DMA, O$_2$ (1 atm), 70 °C, 8–12 h

67 → **68**

R^1	R^2	Equiv of NaOBz	Yield (%)	Ref
H	Me	1.5	76[a]	[50]
H	Ph	2.5	58	[50]
H	OMe	1	82	[50]
H	CO_2Me	1	70	[50]
Me	Me	1	91	[50]
Me	OMe	1	93	[50]
$(CH_2)_4$		1	82	[50]
$(CH_2)_5$		1	81	[50]

[a] Reaction carried out for 20 h.

69 → 70 + 71

$Pd(OAc)_2$ (10 mol%)
additive, base, DMA
O_2 (1 atm), 8–10 h

R^1	R^2	Additive	Base	Temp (°C)	Yield (%)	Ratio (**70**/**71**)	Ref
H	H	maleic anhydride (40 mol%)	NaOBz (1 equiv)	70	88	86:14	[50]
H	H	CrCl(salen) (10 mol%)	–	50	61	<3:97	[50]
Me	H	maleic anhydride (40 mol%)	NaOBz (1 equiv)	70	83	82:18	[50]
Me	H	CrCl(salen) (10 mol%)	–	50	66	<3:97[a]	[50]
$(CH_2)_4Me$	H	maleic anhydride (40 mol%)	NaOBz (1 equiv)	70	64	81:19	[50]
$(CH_2)_4Me$	H	CrCl(salen) (10 mol%)	–	50	71	6:94[b]	[50]
iPr	H	maleic anhydride (40 mol%)	NaOBz (1 equiv)	70	56	81:19	[50]
iPr	H	CrCl(salen) (10 mol%)	–	50	41	<3:97[c]	[50]
Me	Me	maleic anhydride (40 mol%)	NaOBz (1 equiv)	70	74	84:16	[50]
Me	Me	CrCl(salen) (10 mol%)	–	50	71	<3:97	[50]

[a] Ratio (*trans*/*cis*) of **71** = 10:1.
[b] Ratio (*trans*/*cis*) of **71** = 6:1.
[c] Ratio (*trans*/*cis*) of **71** = 9:1.

Alkenylureas (e.g., **72**) are particularly suitable substrates to give a double amination process under oxidative conditions, with bis(acetonitrile)dichloropalladium(II) as catalyst and copper(II) chloride as oxidant in the presence of a base under microwave activation, resulting in the formation of bicyclic imidazo-piperazinones **73**. Under milder reaction conditions and in the absence of base, alkenylureas **72** undergo a different aminooxygenation process leading to oxazolopiperazinone systems **74** (Scheme 31).[51]

Scheme 31 Palladium-Catalyzed Cyclization of Alkenylureas to Fused Piperazinones[51]

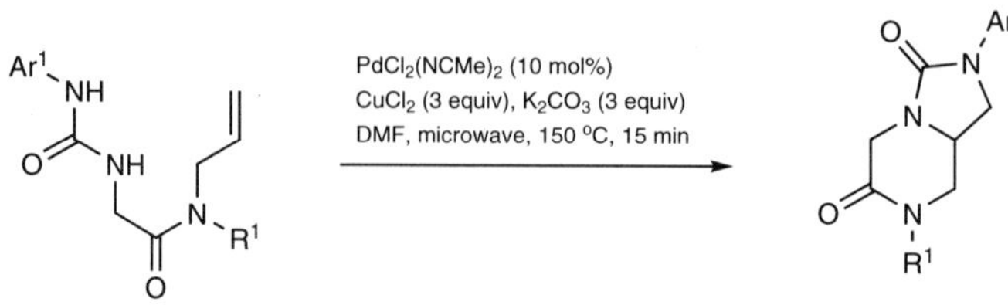

$PdCl_2(NCMe)_2$ (10 mol%)
$CuCl_2$ (3 equiv), K_2CO_3 (3 equiv)
DMF, microwave, 150 °C, 15 min

72 → **73**

R^1	Ar^1	Yield (%)	Ref
Cy	$4\text{-}ClC_6H_4$	73	[51]
Cy	$4\text{-}O_2NC_6H_4$	78	[51]
Cy	$2\text{-}O_2NC_6H_4$	80	[51]
Cy	$4\text{-}MeOC_6H_4$	79	[51]
Bn	Ph	74	[51]
Bn	$4\text{-}ClC_6H_4$	72	[51]
Bn	$4\text{-}O_2NC_6H_4$	76	[51]
Bn	$2\text{-}O_2NC_6H_4$	78	[51]
Bn	$4\text{-}MeOC_6H_4$	73	[51]

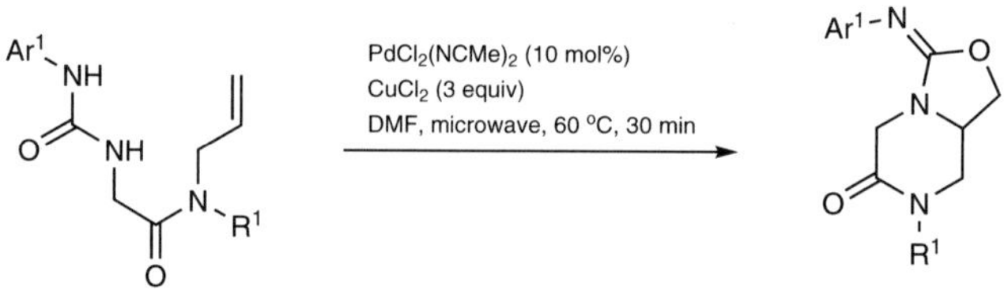

$PdCl_2(NCMe)_2$ (10 mol%)
$CuCl_2$ (3 equiv)
DMF, microwave, 60 °C, 30 min

72 → **74**

R^1	Ar^1	Yield (%)	Ref
Cy	$4\text{-}ClC_6H_4$	75	[51]
Cy	$4\text{-}O_2NC_6H_4$	71	[51]
Cy	$2\text{-}O_2NC_6H_4$	77	[51]
Cy	$4\text{-}MeOC_6H_4$	78	[51]
Bn	Ph	79	[51]
Bn	$4\text{-}ClC_6H_4$	76	[51]
Bn	$4\text{-}O_2NC_6H_4$	72	[51]
Bn	$2\text{-}O_2NC_6H_4$	74	[51]
Bn	$4\text{-}MeOC_6H_4$	77	[51]

2-Methylindoles 64; General Procedure:[47]
In a one-necked flask were placed $PdCl_2(NCMe)_2$ (10 mol%), benzo-1,4-quinone (1 equiv), and LiCl (10 equiv). THF (14 mL/mmol substrate) was then added, and the mixture was stirred for 3–5 min. A 2-allylaniline **63** in THF (3–4 mL/mmol substrate) was added to the flask via syringe and the soln was heated at reflux for 18 h. THF was removed on a rotary evaporator, and the residue was taken up in Et_2O (20 mL) and stirred for 20 min with a

small amount of decolorizing charcoal. The soln was filtered and the residue was purified by preparative layer chromatography.

3-Methylpyrazino[1,2-*a*]indol-1(2*H*)-ones 66; General Procedure:[49]
A suspension of a 1-allylindole-2-carboxamide **65** (1.0 mmol), $Pd(OAc)_2$ (5 mol%), Na_2CO_3(3.0 equiv), TBACl (1.0 equiv), and benzo-1,4-quinone (1 equiv) in DMF (10 mL) was stirred for 24 h at 100 °C. Brine (25 mL) was added and the mixture was extracted with Et_2O (2 × 25 mL). The organic layer was dried (Na_2SO_4), filtered, and concentrated to dryness under reduced pressure. The residue was purified by chromatography (silica gel, light petroleum ether/Et_2O 12:1).

6-Methylene-1-tosylazepan-2-ones 68; General Procedure:[50]
In a glass tube, a tosylamide **67** (0.2 mmol), $Pd(OAc)_2$ (10 mol%), maleic anhydride (0.4 equiv), NaOBz (1 equiv), and 4-Å molecular sieves (30 mg) were combined in DMA (2.0 mL). The reaction tube was placed into a custom 9-well parallel reactor mounted in a 300-mL Parr bomb and sealed. The whole system was purged with molecular O_2 (ca. 10 ×) and placed under vacuum. Then, the oxygen pressure was increased to 1.0 atm and the reactor was warmed at 70 °C. The mixture was stirred for 8–12 h and then concentrated under reduced pressure. The crude mixture was purified by column chromatography.

2-Aryltetrahydroimidazo[1,5-*a*]pyrazine-3,6(2*H*,5*H*)-diones 73; General Procedure:[51]
A mixture of $PdCl_2(NCMe)_2$ (10 mol%), $CuCl_2$ (3 equiv), K_2CO_3 (3 equiv), and an alkenylurea **72** (1 mmol) in DMF (10 mL) was heated at 150 °C for 15 min under microwave irradiation. Brine (15 mL) was added and the mixture was extracted with CH_2Cl_2 (3 × 15 mL). The organic phase was dried (Na_2SO_4) and the solvent was removed under reduced pressure. The crude mixture was purified by column chromatography (silica gel, light petroleum ether/EtOAc 3:7).

3-(Arylimino)tetrahydro-3*H*-oxazolo[3,4-*a*]pyrazine-6(5*H*)-ones 74; General Procedure:[51]
A mixture of $PdCl_2(NCMe)_2$ (10 mol%), $CuCl_2$ (3 equiv), and an alkenylurea **72** (1 mmol) in DMF (10 mL) was heated at 60 °C for 30 min under microwave irradiation. Brine (15 mL) was added and the mixture was extracted with CH_2Cl_2 (3 × 15 mL). The organic phase was dried (Na_2SO_4) and the solvent was removed under reduced pressure. The crude mixture was purified by column chromatography (silica gel, light petroleum ether/EtOAc 3:7).

1.8.2.3.2 Ruthenium Catalysis

Amination of aminoalkenes **75** performed under ruthenium catalysis in the presence of ligands and base provides unsaturated nitrogen heterocycles as 3,4-dihydro-2*H*-pyrroles (e.g., **76**, n = 1) and 2,3,4,5-tetrahydropyridines (e.g., **76**, n = 2) without the use of expensive aminoalkynes. The choice of the appropriate solvent and ligand is critical for the success of the reaction, with *N*-methylpiperidine being the best solvent and 1,3-bis(diphenylphosphino)propane (dppp) the most effective ligand. The presence of potassium carbonate and allyl acetate is indispensable to avoid alkene isomerization and the formation of the corresponding acetamide. Moreover, the oxidative amination involves elimination of a ruthenium dihydride. The hydrogenolysis of allyl acetate regenerates the catalytically active ruthenium species via removal of the hydride. Good yields are reported for aminoalkenes with *gem*-disubstitution at the 2-position due to the Thorpe–Ingold effect. The product yield depends on the sterically controlled ring-forming transition state and follows the order five- > six- >> seven-membered rings. This process can be applied to the synthesis of 2-methylindole from 2-allylaniline (Scheme 32).[52]

for references see p 499

Scheme 32 Ruthenium-Catalyzed Cyclization of Aminoalkenes to 3,4-Dihydro-2*H*-pyrroles and 2,3,4,5-Tetrahydropyridines[52]

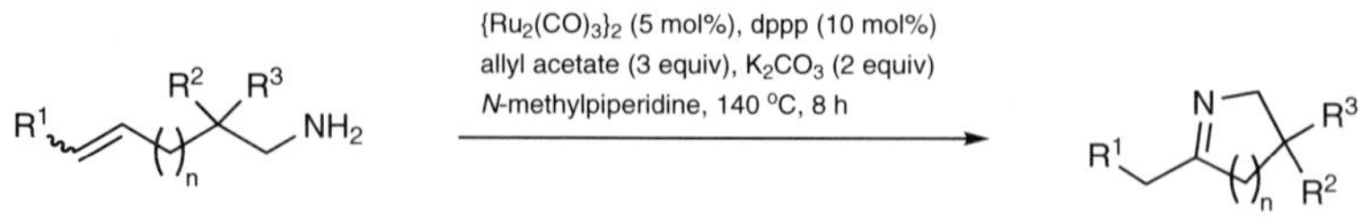

{$Ru_2(CO)_3$}$_2$ (5 mol%), dppp (10 mol%)
allyl acetate (3 equiv), K_2CO_3 (2 equiv)
N-methylpiperidine, 140 °C, 8 h

75 → **76**

R^1	R^2	R^3	n	Yield[a] (%)	Ref
H	Ph	Ph	1	87	[52]
H	Ph	Me	1	81	[52]
H	Ph	H	1	73	[52]
Me	Ph	Ph	1	87	[52]
Et	Ph	Ph	1	38	[52]
Ph	Ph	H	1	40	[52]
Ph	Me	Me	1	76[b,c]	[52]
H	Ph	Ph	2	87	[52]

[a] Isolated yield.
[b] Yield determined by ^1H NMR spectroscopy.
[c] Reaction was performed at 120 °C for 22 h.

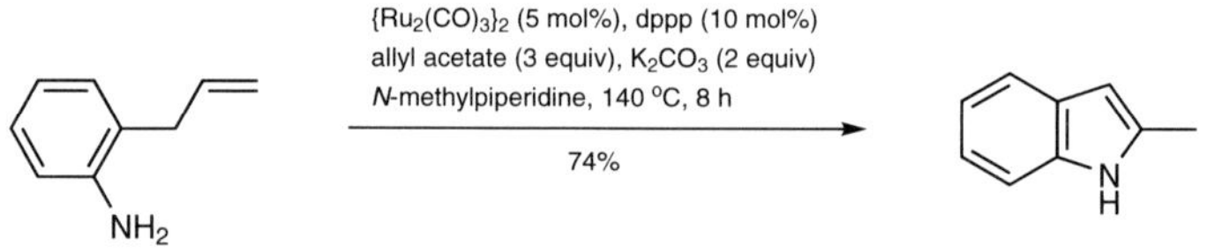

{$Ru_2(CO)_3$}$_2$ (5 mol%), dppp (10 mol%)
allyl acetate (3 equiv), K_2CO_3 (2 equiv)
N-methylpiperidine, 140 °C, 8 h

74%

3,4-Dihydro-2*H*-pyrroles 76 **(n = 1); General Procedure:**[52]

A mixture of an aminoalkene **75** (n = 1; 2.5 mmol), {$Ru_2(CO)_3$}$_2$ (0.05 mol%), dppp (0.10 mmol as a P atom), K_2CO_3 (2 equiv), allyl acetate (3 equiv), and *N*-methylpiperidine (4.0 mL) was placed in a 50-mL stainless steel autoclave equipped with a glass liner and a magnetic stirrer bar under argon. The mixture was then magnetically stirred at 120–140 °C for 8 h. The mixture was cooled and analyzed by GLC. The product was isolated by Kugelrohr distillation and/or recycling preparative HPLC.

1.8.2.4 Oxidative Amination of (Het)Arenes

Exploiting C—H bond activation, oxidative amination of (het)arenes involves the addition of a nitrogen nucleophile to a (het)arene while maintaining the substrate and the product in the same oxidation state, as in the case of oxidative amination of alkenes (Section 1.8.2.3). Compared to the cross coupling of amines and aryl halides, this process is a more effective method to access heterocycles in terms of atom economy due to the use of unfunctionalized substrates.

The mechanism involves electrophilic substitution of an aryl hydrogen by insertion of the metal to give the intermediate **77** and subsequent formation of a σ-bonded arylmetal complex **78** (Scheme 33). Nucleophilic attack of nitrogen and subsequent metal elimination as the final step close the catalytic cycle. Nevertheless, a mechanism involving initial coordination of the metal to the nitrogen nucleophile cannot be ruled out. The presence of an oxidizing agent to return the metal to the original oxidation state is essential to complete the catalytic cycle. In the case of copper catalysis, the formed copper(I) is reoxidized to copper(II).[53]

Scheme 33 Mechanism of Oxidative Amination of Arenes[53]

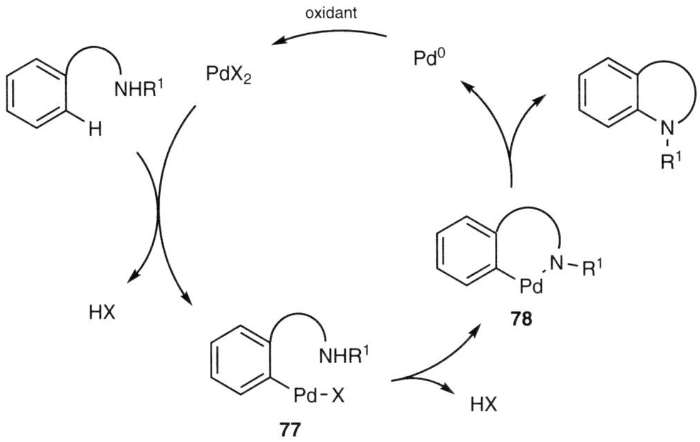

77

78

1.8.2.4.1 Copper Catalysis

Starting from arylamidines **79**, 2-arylbenzimidazoles **80** can be accessed through a sequence of C—H functionalization/C—N bond formation (Scheme 34). Limitations of this method are the requirement for an *ortho* substituent on the R^1 aryl group of the substrate and, in the case of possible regioisomers, mixtures of products are formed favoring the less sterically hindered form.[54] The reaction is performed with copper(II) acetate and oxygen as oxidant in dimethyl sulfoxide at 100 °C. The method can be extended to the preparation of 2-aryl-*N*-methylbenzimidazoles and substituted 2-*tert*-butylbenzimidazoles.

Scheme 34 Copper-Catalyzed Cyclization of Arylamidines to Benzimidazoles[54]

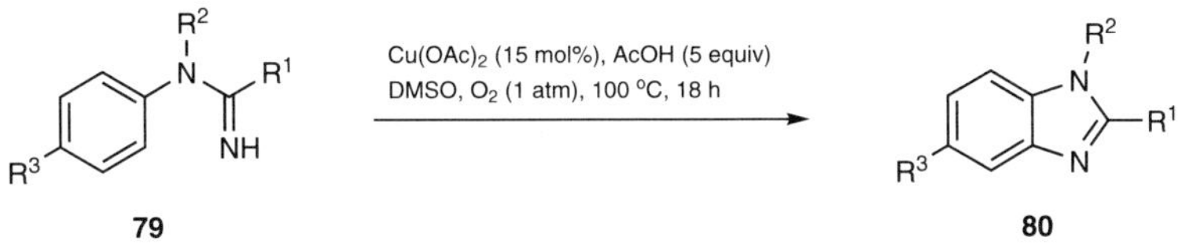

$Cu(OAc)_2$ (15 mol%), AcOH (5 equiv)
DMSO, O_2 (1 atm), 100 °C, 18 h

79 → 80

R^1	R^2	R^3	Yield (%)	Ref
2-Tol	H	H	89	[54]
2-Tol	H	OMe	70	[54]
2-Tol	H	F	86	[54]
2-Tol	H	Cl	89	[54]
2-Tol	H	Br	89	[54]
2-Tol	Me	H	68	[54]
2-$F_3CC_6H_4$	H	H	87	[54]
2-$F_3CC_6H_4$	H	OMe	74	[54]
2-$F_3CC_6H_4$	H	CO_2t-Bu	88[a]	[54]
2-$F_3CC_6H_4$	H	I	89	[54]
2-$MeOC_6H_4$	H	H	68	[54]

R^1	R^2	R^3	Yield (%)	Ref
2-$MeOC_6H_4$	H	Cl	75	[54]
2-$TBDMSC_6H_4$	H	H	80	[54]
4-Cl-2-MeC_6H_3	H	H	85	[54]
4-Cl-2-MeC_6H_3	H	Br	88	[54]
5-F-2-MeC_6H_3	H	H	89	[54]
2-ClC_6H_4	H	H	81	[54]
2-ClC_6H_4	Me	H	69	[54]
2-ClC_6H_4	Me	Br	84	[54]
t-Bu	H	H	86	[54]
t-Bu	H	Me	86	[54]
t-Bu	H	OMe	83	[54]
t-Bu	H	Br	89	[54]

[a] Reaction time was 36 h.

Aromatic C—H amination of 2-(arylamino)pyridines **81** affords pyrido[1,2-*a*]benzimidazoles **82** using the inexpensive couple of copper(II) acetate and iron(III) nitrate as cocatalyst. In this process, the pyridyl nitrogen acts as both directing group and nucleophile, and the iron helps the formation of a more electrophilic copper(III) species which facilitates the subsequent S_EAr process. The addition of pivalic acid as proton source is also required. Electron-withdrawing substituents at the *meta*-position of the aniline and at any position of the pyridine are unfavorable (Scheme 35).[55]

Scheme 35 Copper/Iron Catalyzed Cyclization of 2-(Arylamino)pyridines to Pyrido[1,2-*a*]benzimidazoles[55]

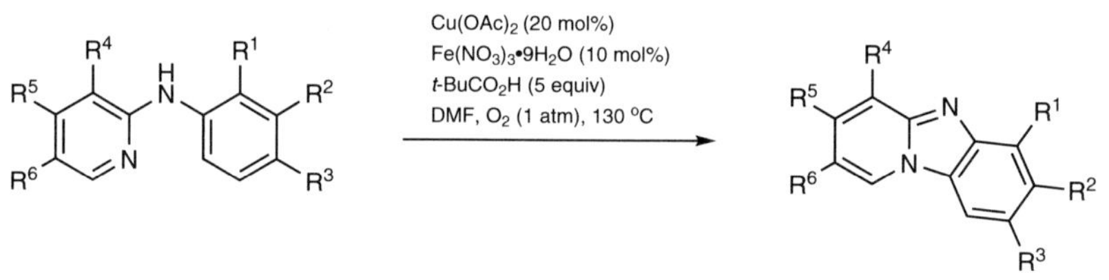

$Cu(OAc)_2$ (20 mol%)
$Fe(NO_3)_3$•$9H_2O$ (10 mol%)
t-$BuCO_2H$ (5 equiv)
DMF, O_2 (1 atm), 130 °C

81 → **82**

R^1	R^2	R^3	R^4	R^5	R^6	Time (h)	Yield (%)	Ref
Me	H	H	H	H	H	30	76	[55]
OMe	H	H	H	H	H	26	69	[55]
H	Me	H	H	H	H	28	74	[55]
H	OMe	H	H	H	H	36	62	[55]
H	H	Me	H	H	H	30	81	[55]
H	H	OMe	H	H	H	23	70	[55]
H	H	*t*-Bu	H	H	H	36	84	[55]
H	H	F	H	H	H	66	71	[55]
H	H	Cl	H	H	H	62	70	[55]
H	H	Br	H	H	H	62	68	[55]
H	H	H	Me	H	H	23	85	[55]

R^1	R^2	R^3	R^4	R^5	R^6	Time (h)	Yield (%)	Ref
H	H	H	H	Me	H	5	80	[55]
H	H	H	H	H	Me	30	82	[55]
H	H	H	$(CH{=}CH)_2$		H	23	96	[55]

Benzimidazoles 80**; General Procedure:**[54]
Outside a glovebox, a 1-dram vial equipped with a magnetic stirrer bar was charged with an arylamidine **79** (1 mmol), AcOH (5 equiv), $Cu(OAc)_2$ (15 mol%), and DMSO (5 mL). The vial was flushed with O_2 (1 atm) and capped. The contents of the sealed vial were stirred at rt for 5 min and the vial was then placed in a preheated oil bath at 100 °C. After completion of the reaction, the mixture was cooled to rt and diluted with EtOAc (50 mL). The resulting soln was washed with brine (5 × 15 mL), dried ($MgSO_4$), and concentrated under reduced pressure to a volume of about 1 mL. This mixture was subjected to chromatography (silica gel) to yield the pure product.

Pyrido[1,2-*a*]benzimidazoles 82**; General Procedure:**[55]
A mixture of a 2-(arylamino)pyridine **81** (0.5 mmol), $Cu(OAc)_2$ (20 mol%), $Fe(NO_3)_3$•$9H_2O$ (10 mol%), and *t*-$BuCO_2H$ (5 equiv) in DMF (1.0 mL) was stirred at 130 °C under O_2 (balloon pressure). The mixture was cooled to rt after complete consumption of the starting material (as monitored by TLC). H_2O (10 mL), Et_3N (1.0 mL), and EtOAc (10 mL) were added to the mixture successively. The organic phase was separated, and the aqueous phase was further extracted with EtOAc (3 × 10 mL). The combined organic layers were dried (Na_2SO_4) and concentrated. The residue was purified by flash chromatography.

1.8.2.4.2 Palladium Catalysis

Benzamidines **83** react under palladium(II) catalysis in the presence of copper(II) acetate and oxygen as final oxidant, affording 2-arylbenzimidazoles **84**. The presence of tetramethylthiourea improves the coupling, probably through a stabilizing effect on the palladium(II) species. An advantage of this method over that presented in Scheme 34 is that an *ortho* substituent on the Ar^1 group of the benzamide is not required (Scheme 36).[56]

Scheme 36 Palladium-Catalyzed Cyclization of Arylamidines to 2-Arylbenzimidazoles[56]

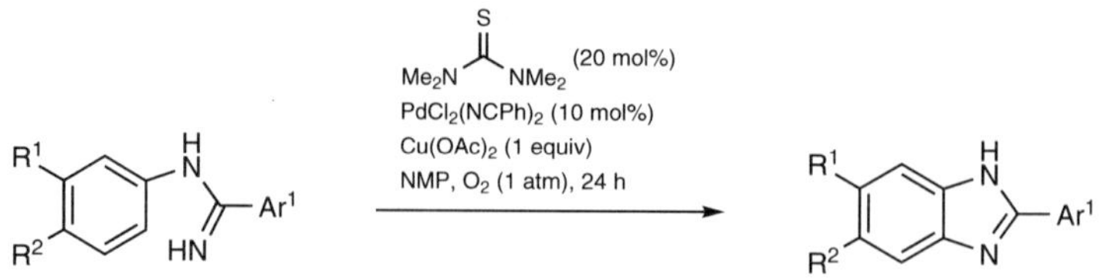

Ar[1]	R[1]	R[2]	Temp (°C)	Yield (%)	Ref
Ph	H	H	100	60	[56]
2-Tol	H	H	100	71	[56]
3-Tol	H	H	100	59	[56]
4-Tol	H	H	80	64	[56]
4-$MeOC_6H_4$	H	H	80	75	[56]
3,4-$Me_2C_6H_3$	H	H	80	63	[56]
2,3-$Me_2C_6H_3$	H	H	80	72	[56]
4-ClC_6H_4	H	H	100	40[a]	[56]
Ph	H	Me	100	67	[56]
Ph	H	OMe	80	60	[56]
Ph	OMe	OMe	80	65	[56]
Ph	$O(CH_2)_2O$		80	53	[56]

[a] 20 mol% of catalyst was used.

Beside the work on benzimidazoles, the oxidative amination of arenes has been expanded to include the preparation of different nitrogen heterocycles such as indoles,[57] dihydroindoles,[58] oxindoles,[59,60] carbazoles,[61–63] and indazoles.[64]

Starting from (2-arylethyl)amines **85**, the amination method permits access to substituted dihydroindoles **87** (Scheme 37).[58] Several alternative oxidants to copper(II) or oxygen have been explored for this transformation. Good results are obtained by using either a one-electron oxidant such as cerium(IV) sulfate or a two-electron oxidant such as *N*-fluoropyridinium reagent **86** (F^+ source). In both cases, the presence of almost one equivalent of dimethylformamide is required as a possible ligand. The reaction is of broad applicability, including substrates bearing strongly electron-withdrawing groups. Chiral dihydroindoles are also formed from chiral amino acid derived substrates.

Scheme 37 Palladium-Catalyzed Cyclization of Phenethylamines to Dihydroindoles[58]

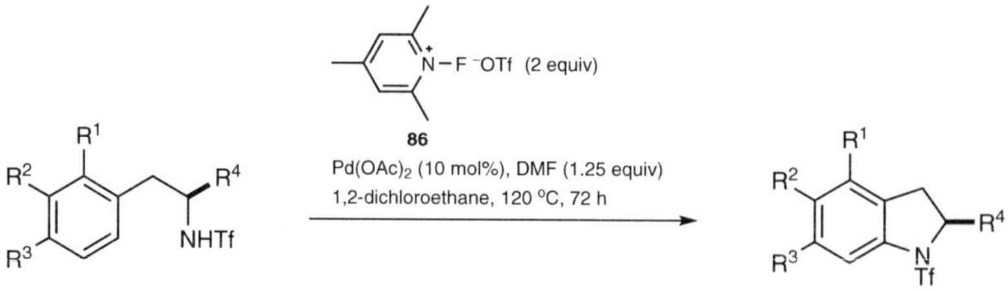

R^1	R^2	R^3	R^4	Yield (%)	Ref
H	H	H	H	75	[58]
H	H	Me	H	73	[58]
H	H	Cl	H	58	[58]
H	H	Br	H	53	[58]
H	H	Ac	H	72[a]	[58]
H	H	CO_2Me	H	82[a]	[58]
H	H	CN	H	62[a]	[58]
H	H	NO_2	H	56[a]	[58]
H	Me	H	H	74	[58]
H	OMe	H	H	70	[58]
H	Cl	H	H	68	[58]
H	Br	H	H	56	[58]
Me	H	H	H	80	[58]
OMe	H	H	H	53	[58]
F	H	H	H	55	[58]
Cl	H	H	H	62	[58]
Br	H	H	H	72	[58]
CF_3	H	H	H	65	[58]
H	H	H	CO_2Me	91 (98% ee)[b]	[58]

[a] Reaction carried out with 15 mol% $Pd(OAc)_2$.
[b] ee determined by HPLC.

C—H amination of *N*-alkoxyamides is also exploited for the formation of β-, γ-, and δ-lactams. In particular, starting from 2-aryl-*N*-methoxyacetamides **88** ($R^1 = Me$), the corresponding oxindoles **89** ($R^1 = Me$) are obtained in excellent yields. Palladium(II) and copper(II) catalysis takes place in the presence of silver(I) acetate, which as oxidant converts copper(I) into copper(II) and as ligand palladium(II) chloride into palladium(II) acetate (Scheme 38).[60] The electronic density of the arene does not affect the yield of the reaction, but the presence of hydrogens instead of alkyl groups adjacent to the amide greatly retards the C—H activation. To overcome this limitation, the α-position must be fully substituted or an α,β double bond must be present.

Scheme 38 Palladium-Catalyzed Cyclization of *N*-Alkoxyamides to 1,3-Dihydro-2*H*-indol-2-ones[60]

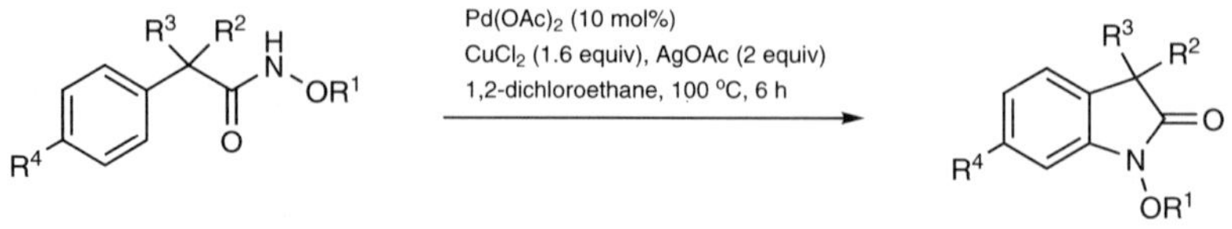

$Pd(OAc)_2$ (10 mol%)
$CuCl_2$ (1.6 equiv), AgOAc (2 equiv)
1,2-dichloroethane, 100 °C, 6 h

88 → 89

R^1	R^2	R^3	R^4	Yield (%)	Ref
Me	Me	Me	H	94	[60]
Me	$(CH_2)_2$		H	88	[60]
Me	$(CH_2)_4$		H	96	[60]
Me	$(CH_2)_4$		Cl	88	[60]
Me	$(CH_2)_4$		F	86	[60]
Me	$(CH_2)_4$		OMe	90	[60]
Bn	Me	Me	H	95	[60]
Bn	$(CH_2)_4$		Cl	84	[60]

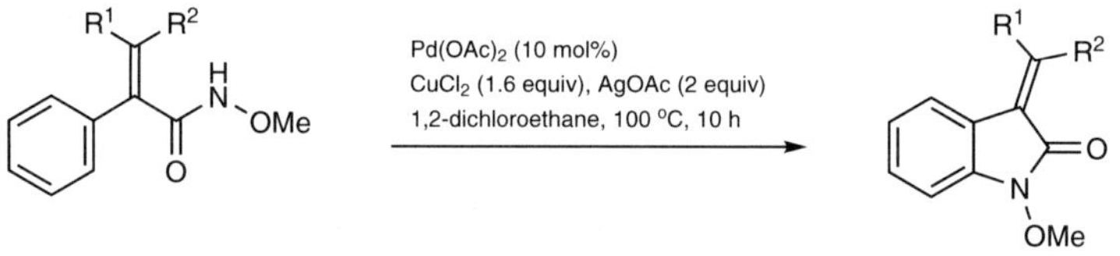

$Pd(OAc)_2$ (10 mol%)
$CuCl_2$ (1.6 equiv), AgOAc (2 equiv)
1,2-dichloroethane, 100 °C, 10 h

R^1	R^2	Yield (%)	Ref
$(CH_2)_4$		78	[60]
Ph	H	58	[60]

Carbazoles **91** can be obtained from cyclization of 2-(acetylamino)biphenyls **90** using a combination of palladium(II) acetate and copper(II) acetate as reoxidant in toluene at 120 °C under air or oxygen (Scheme 39).[61,62] The results strongly depend on the acidity of the N—H bond and thus on the protecting group present on nitrogen; the acetyl group is best, sulfones are tolerated, and carbamates and benzoyl groups give low yields. The reaction supports electron-withdrawing and electron-donating groups on both aryl rings. In the case of steric hindrance, the arene reacts at the less congested C—H bond.

Scheme 39 Palladium-Catalyzed Cyclization of 2-(Acetylamino)biphenyls to *N*-Acetylcarbazoles[61]

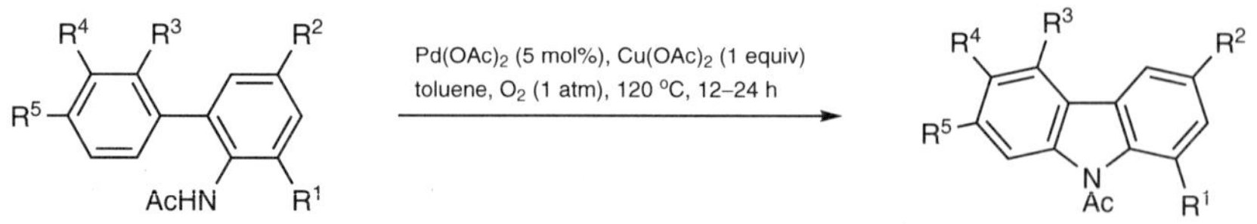

$Pd(OAc)_2$ (5 mol%), $Cu(OAc)_2$ (1 equiv)
toluene, O_2 (1 atm), 120 °C, 12–24 h

90 → 91

R^1	R^2	R^3	R^4	R^5	Yield (%)	Ref
H	H	H	H	H	94	[61]
H	Me	H	H	H	92	[61]
Me	Me	H	H	H	93	[61]
H	F	H	H	H	94	[61]
F	F	H	H	H	90	[61]
H	CF_3	H	H	H	88	[61]
H	H	H	H	Me	98	[61]
H	H	H	H	*t*-Bu	96	[61]
H	H	H	H	OMe	97	[61]
H	H	H	H	CF_3	94	[61]
H	H	H	H	F	87	[61]
H	H	OMe	H	H	81	[61]
H	H	CF_3	H	H	81	[61]
H	H	F	H	H	78	[61]
H	Me	H	H	OMe	88	[61]
H	H	OMe	H	OMe	85	[61]
H	H	H	OMe	H	41 (16:1)[a]	[61]
H	H	H	Me	H	82 (18:1)[a]	[61]

[a] Ratio of regioisomers as determined by GC; yield corresponds to the major regioisomer.

A significant improvement in the synthesis of carbazoles (e.g., **93**) is obtained by performing the oxidative amination reaction on biarylamines (e.g., **92**) without acidic hydrogens using the combination of palladium(II) acetate and (diacetoxyiodo)benzene as oxidant (Scheme 40).[63] Simple alkyl, benzyl, and allyl groups are tolerated on the nitrogen. Moreover, the reaction is also amenable to a range of substituents, with different electronic and steric properties, on the aromatic ring.

Scheme 40 Palladium-Catalyzed Cyclization of Biarylamines to Carbazoles[63]

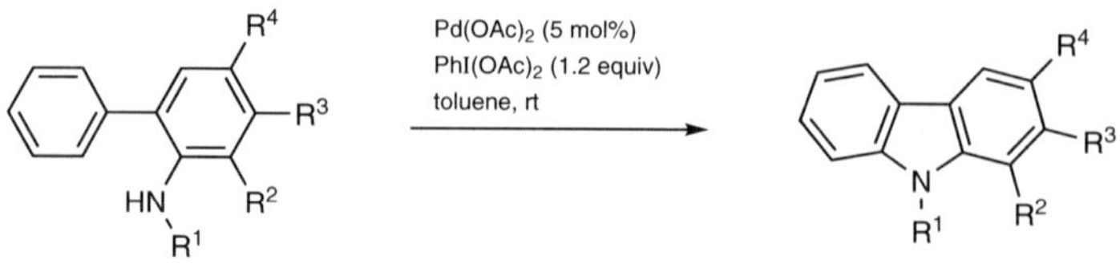

$Pd(OAc)_2$ (5 mol%)
$PhI(OAc)_2$ (1.2 equiv)
toluene, rt

92 → **93**

R^1	R^2	R^3	R^4	Time (h)	Yield (%)	Ref
Bn	H	H	H	1	96	[63]
iPr	H	H	H	4	96	[63]
Me	H	H	H	2	80	[63]
$CH_2CH=CH_2$	H	H	H	2	79[a]	[63]
t-Bu	H	H	H	1.5	80	[63]
Bn	Me	H	H	5	81	[63]
Bn	H	OMe	H	3	89	[63]
Bn	H	CF_3	H	1.5	85[a,b]	[63]
Bn	H	H	Cl	2	64	[63]
Bn	H	H	Me	1	89[b]	[63]
Bn	H	H	F	3	70	[63]
Bn	H	H	CO_2Me	24	60	[63]
iPr	H	H	I	12	71[a]	[63]

[a] Reaction carried out with 1 mol% $Pd(OAc)_2$.
[b] Reaction carried out with 1 equiv AcOH as additive.

The same catalytic system, but with added copper(II) acetate, is used to obtain phenanthridines **95** from biaryl derivatives **94** after heating the mixture at 120 °C for 24 hours. The substrates **94** are synthesized by C—H arylation of benzylamines using aryl iodides under palladium catalysis in the presence of picolinamide as an *ortho* directing group. The oxidative conditions for the cyclization of **94** result in oxidation of the benzylic position of the initially formed dihydrophenanthridine and removal of the picolinamide group, leading to isolation of the substituted phenanthridines **95**. In general, while electron-rich arenes give high yields, electron-poor substrates provide lower yields and a substrate with a *para*-nitro group fails to give any cyclized product (Scheme 41).[65]

Scheme 41 Palladium-Catalyzed Cyclization of Benzylpicolinamides to Phenanthridines[65]

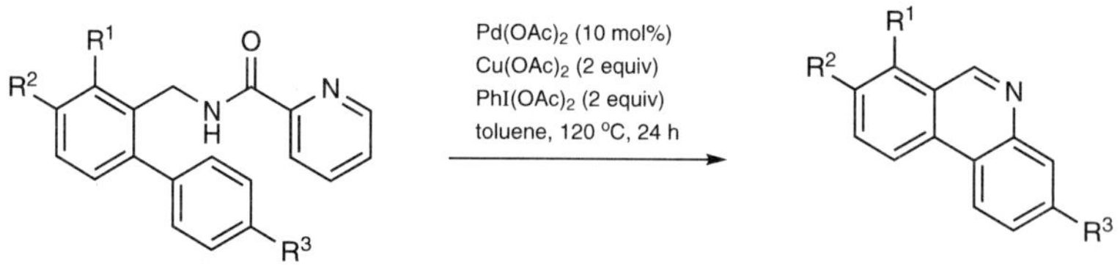

$Pd(OAc)_2$ (10 mol%)
$Cu(OAc)_2$ (2 equiv)
$PhI(OAc)_2$ (2 equiv)
toluene, 120 °C, 24 h

94 → **95**

R^1	R^2	R^3	Yield (%)	Ref
OMe	H	OMe	58	[65]
OMe	H	Br	38	[65]
OMe	H	CO_2Me	43	[65]
F	H	OMe	49	[65]
CF_3	H	OMe	51	[65]
Cl	H	OMe	48	[65]
H	Me	OMe	53	[65]

An efficient process to obtain quinolin-2-ones **97** starts from 3,3-diarylacrylamides **96** in the presence palladium(II) chloride and copper(II) acetate as cocatalyst, under an oxygen atmosphere as reoxidant in dimethyl sulfoxide at 120 °C (Scheme 42).[66] The substituent on the nitrogen atom of the amide moiety influences the yields; the tosyl group is the best choice, whereas the use of the free amide results in poor yields. In all cases, the selected reaction conditions lead to deprotection of the nitrogen atom. The reaction tolerates various substituents, including halogens. In the case of unsymmetrical substrates, the product depends on the stereochemistry of the starting double bond, with the *E*-isomer in general being the more reactive one.

Scheme 42 Palladium/Copper Catalyzed Cyclization of 3,3-Diarylacrylamides to 4-Arylquinolin-2-ones[66]

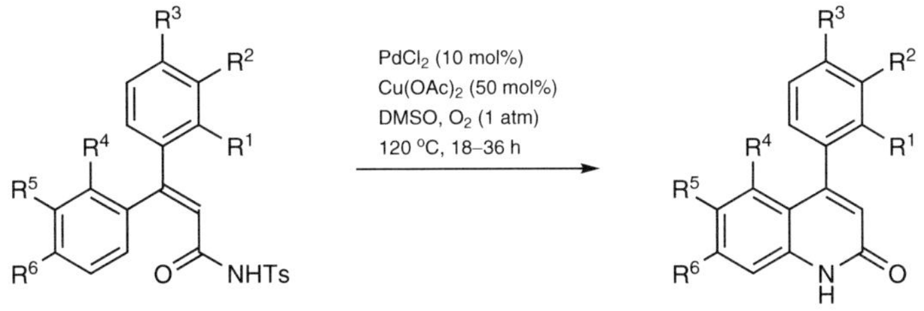

$PdCl_2$ (10 mol%)
$Cu(OAc)_2$ (50 mol%)
DMSO, O_2 (1 atm)
120 °C, 18–36 h

96 → **97**

R^1	R^2	R^3	R^4	R^5	R^6	Yield (%)	Ref
H	H	F	H	H	F	79	[66]
H	H	Br	H	H	Br	87	[66]
H	H	Cl	H	H	Cl	61	[66]
H	H	CN	H	H	CN	38	[66]
H	H	OMe	H	H	OMe	85	[66]
H	OMe	H	H	OMe	H	86	[66]
OMe	H	H	OMe	H	H	98	[66]
H	H	H	H	H	NC	74	[66]
H	H	CN	H	H	OMe	86	[66]
H	H	H	H	NC	H	>99[a]	[66]

[a] Reaction carried out starting from the *E*-isomer.

In the formation of indazoles **99** and **100** from monosubstituted arylhydrazones **98**, the regioselectivity in the cyclization step depends on the nature of the substituents. In the case of electron-donating groups, the cyclization proceeds exclusively on the ring bearing this substituent, whereas in the presence of electron-withdrawing groups, the cyclization also involves the unsubstituted benzene ring. Isomerization between *Z*- and *E*-isomers of hydrazones is observed under the conditions used, resulting in similar conversions and yields. The effective C—H activation occurs in dimethyl sulfoxide at 50 °C with the catalytic system palladium(II)/copper(II) in the presence of silver(I) trifluoroacetate as additive (Scheme 43).[64]

Scheme 43 Palladium-Catalyzed Cyclization of Arylhydrazones to Arylindazoles[64]

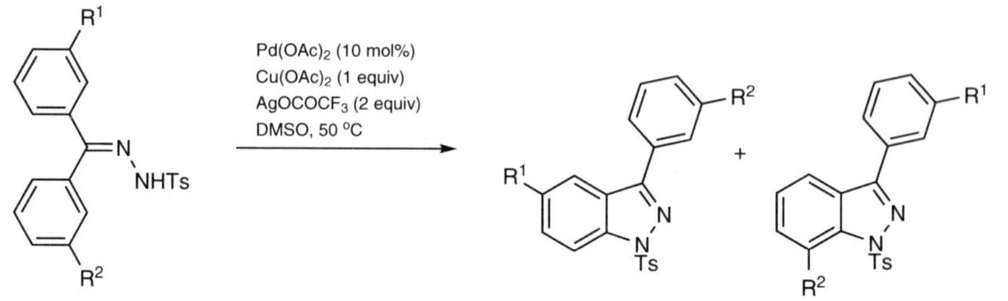

$Pd(OAc)_2$ (10 mol%)
$Cu(OAc)_2$ (1 equiv)
$AgOCOCF_3$ (2 equiv)
DMSO, 50 °C

98 → **99** + **100**

R^1	R^2	Time (h)	Yield[a] (%) **99**	Yield[a] (%) **100**	Ref
H	H	12	90	–	[64]
OMe	OMe	16	96	0	[64]
OMe	NO_2	19	99	0	[64]
OMe	CN	10	98	0	[64]
OH	H	12	75	0	[64]
NH_2	H	18[b]	66	0	[64]
OMe	H	21	54	27	[64]
Cl	H	20	19	65	[64]
Br	H	17	34	45	[64]
NO_2	H	24	0	32	[64]

[a] Isolated yield; when it was not possible to separate the regioisomers, the ratio was determined by ^{1}H NMR spectroscopy.
[b] Reaction carried out at 80 °C.

A nitro group can also be a useful nucleophile for C—N bond formation in the oxidative amination process. 2,2-Diaryl-1-nitroethenes **101** are transformed into 3-arylindoles **102** under carbon monoxide (1 atm) through cyclization/rearrangement of a putative nitrosoalkene intermediate. The carbon monoxide is used as an inexpensive stoichiometric reductant. The presence of 1,10-phenanthroline as ligand is required. The reaction conditions tolerate both electron-rich and electron-deficient aromatic substrates (Scheme 44).[67]

Scheme 44 Palladium-Catalyzed Cyclization of 2,2-Diaryl-1-nitroethenes to 3-Arylindoles[67]

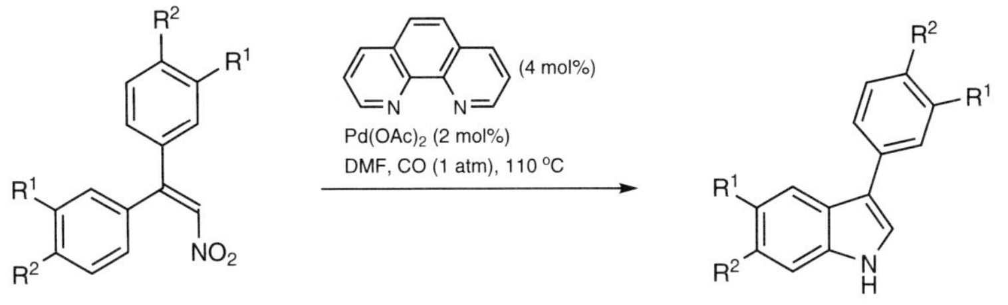

R^1	R^2	Time (h)	Yield (%)	Ref
H	H	3	97	[67]
H	Me	3	87	[67]
H	*t*-Bu	3	92	[67]
H	OMe	3	93	[67]
OMe	H	6	91 (53:47)[a]	[67]
H	Cl	6	98	[67]
Cl	H	6	91 (42:58)[a]	[67]
CF_3	H	8	86 (51:49)[a]	[67]
H	CF_3	16	58	[67]

[a] Ratio of 5-substituted 3-arylindole/7-substituted 3-arylindole, determined by ^{1}H NMR spectroscopy.

2-Arylbenzimidazoles 84**; General Procedure:**[56]
To a 25-mL Schlenk tube were added an arylamidine **83** (0.5 mmol), $PdCl_2(NCPh)_2$ (10 mol%), anhyd $Cu(OAc)_2$ (1 equiv), and tetramethylthiourea (10 mol%) (as reported). Anhyd NMP (2.0 mL) was added via syringe. The tube was sealed with a Teflon-lined cap, and the mixture was degassed and refilled with O_2 (3 ×) and then heated at the required temperature in an oil bath for 24 h. After the reaction was complete, sat. aq Na_2CO_3 (5 mL) and EtOAc (10 mL) were added. The dark solid was removed by filtration through Celite and the Celite bed was washed with EtOAc (4 × 20 mL). In most cases, the combined filtrate was washed with H_2O (4 × 8 mL) and brine (10 mL). The organic phase was dried (Na_2SO_4) and concentrated. The residue was purified by chromatography (silica gel).

N-(Trifluoromethanesulfonyl)-2,3-dihydroindoles 87**; General Procedure:**[58]
In a 20-mL sealed tube, a (2-arylethyl)amine **85** (0.2 mmol), $Pd(OAc)_2$ (10 mol%), *N*-fluoro-2,4,6-trimethylpyridinium trifluoromethanesulfonate (**86**; 2 equiv), and DMF (1.25 equiv) were dissolved in anhyd 1,2-dichloroethane (2 mL) under air. The tube was sealed with a Teflon-lined cap, and the mixture was stirred at 120 °C for 72 h. After the mixture had been cooled to rt, it was concentrated under reduced pressure and the residue was purified by column chromatography (silica gel, hexane/EtOAc gradient).

N-Alkoxy-1,3-dihydro-2*H*-indol-2-ones 89**; General Procedure:**[60]
In a 50-mL Schlenk tube, an *N*-alkoxy-2-arylacetamide **88** (0.5 mmol), $Pd(OAc)_2$ (10 mol%), AgOAc (2 equiv), and $CuCl_2$ (1.6 equiv) were dissolved in 1,2-dichloroethane (10 mL). The tube was sealed and the mixture was heated at 100 °C with vigorous stirring under N_2 for 6 h. The residue was purified by column chromatography (silica gel, hexane/Et_2O).

N-Acetylcarbazoles 91; **General Procedure:**[61]
An oven-dried 15-mL Schlenk tube was cooled under reduced pressure and a 2-(acetylamino)biphenyl **90** (0.2 mmol), $Pd(OAc)_2$ (5 mol%), and powdered activated 3-Å molecular sieves (40 mg) were added under air. The tube was evacuated and refilled with argon. Anhyd $Cu(OAc)_2$ (1 equiv) was loaded into the Schlenk tube in a N_2 glovebox. Under a positive argon pressure, toluene (2 mL) was added via syringe. The mixture was sonicated and degassed under a weak vacuum and refilled with O_2 from the double manifold (the procedure was carried out three times). The sealed Schlenk tube was lowered into an oil bath at 120 °C and stirred for the required time. After the mixture had been cooled, distilled H_2O (1 mL), 30% aq NH_3 (3 mL), and EtOAc (2 mL) were added. The organic phase was separated, dried (Na_2SO_4), and filtered through Celite. The filtrate was concentrated, and the residue was purified by chromatography (silica gel, hexane/EtOAc), affording a white powder.

N-Substituted Carbazoles 93; **General Procedure:**[63]
A biarylamine **92** (1 mmol) and $Pd(OAc)_2$ (5 mol%) were stirred in toluene (0.05 M) at rt for 1 h. $PhI(OAc)_2$ (1.2 equiv) was then added and the mixture was stirred at rt. The reaction was monitored by LC/MS and TLC until completion. The solvent was removed under reduced pressure and the residue was purified by flash column chromatography.

Phenanthridines 95; **General Procedure:**[65]
A mixture of a picolinamide **94** (0.2 mmol), $Pd(OAc)_2$ (10 mol%), $PhI(OAc)_2$ (2 equiv), and $Cu(OAc)_2$ (2 equiv) in anhyd toluene (4 mL) in a 10-mL glass vial (purged with N_2, sealed with a PTFE cap) was heated at 120 °C for 24 h. The resulting mixture was filtered through a short pad of Celite and concentrated under reduced pressure. The residue was purified by flash chromatography (silica gel, hexane/EtOAc 4:1) to give the product as a pale white solid.

4-Arylquinolin-2-ones 97; **General Procedure:**[66]
A 3,3-diarylacrylamide **96** (0.13 mmol), $PdCl_2$ (10 mol%), and $Cu(OAc)_2$ (50 mol%) were added to a pear-shaped flask. The flask was evacuated and refilled with argon. Under a positive argon pressure, DMSO (2.6 mL) was added via syringe. The flask was again evacuated and refilled with O_2. The mixture was then stirred at 120 °C for 18–36 h under O_2 (balloon). After the mixture had been cooled to rt, it was extracted with EtOAc (3 × 10 mL), and the combined organic layers were washed with brine (10 mL) and dried ($MgSO_4$). The solvent was evaporated and the residue was purified by column chromatography (silica gel, hexane/EtOAc) to give the product as a colorless solid.

3-Arylindazoles 99 **and** 100; **General Procedure:**[64]
A mixture of a hydrazone **98** (1 mmol), $Pd(OAc)_2$ (10 mol%), $Cu(OAc)_2$ (1 equiv), and $AgOCOCF_3$ (2 equiv) in DMSO (0.05 M) was stirred at 50 °C for the required time. The resulting mixture was extracted with EtOAc (3 × 5 mL), and the combined organic layers were washed with brine (10 mL) and dried ($MgSO_4$). The solvent was evaporated and the residue was purified by column chromatography (silica gel).

3-Arylindoles 102; **General Procedure:**[67]

CAUTION: *Carbon monoxide is extremely flammable and exposure to higher concentrations can quickly lead to a coma.*

An Endeavor glass liner was charged with a nitroalkene **101** (1 mmol) and a soln of $Pd(OAc)_2$ (2 mol%) and 1,10-phenanthroline (4 mol%) in DMF (5 mL). After the liner was inserted into the Endeavor pressure reactor, the reactor was sealed and purged three times with CO. The reactor was pressurized with CO (1 atm) and heated at 110 °C for 3–16 h. The red mixture was cooled to rt and concentrated. The residue was purified by chromatography (silica gel, hexane/EtOAc gradient 99:1 to 80:20).

1.8.2.5 Amination of Aryl Halides

The reaction of amines with aryl halides is one of the methodologies most widely used to achieve C—N bond formation due to the versatility and reliability of the coupling process. In particular, the most applied reactions are the palladium-catalyzed Buchwald–Hartwig and the copper-catalyzed Ullmann-type reactions. Four factors need to be considered for the efficiency of the reaction: catalytic system, ligand, base, and solvent. The ligand in general raises the electron density at the metal to facilitate the oxidative addition step and to this end a plethora of ligands with different application profiles have been developed over the years. The base is important in the deprotonation of the amine substrate and its solubility depends on the solvent, which has a prominent role compared to other transition-metal catalyzed reactions. The two types of catalysis are complementary; comparing the advantages and the disadvantages of the two methods, the lower cost of copper and the ligands used is an indubitable advantage over the more expensive palladium/ligand system. In most cases, the palladium-catalyzed reactions must be conducted under an inert atmosphere. On the contrary, the copper-catalyzed reactions tolerate atmospheric oxygen, but the reaction temperatures and reaction times of both reactions are comparable. Nowadays, the applications are abundant, also on large scale, and here only selected examples are reported. The coupling between an aryl halide and a nucleophile has also been extended to alcohol derivatives to obtain C—O bond formation through an alkoxylation process, as reported elsewhere. Considering the formation of C—N and C—O bonds, about 80% of the relevant literature is focused on C—N coupling, which is probably due to the lower nucleophilicity of oxygen compared to nitrogen.

1.8.2.5.1 Copper Catalysis (Intramolecular Ullmann Condensation)

The copper-mediated N-arylation (Ullmann condensation) is a well-known method for the formation of C—N bonds that was discovered more than a century ago. The method has been employed for several industrial applications aimed at the synthesis of intermediates, but more recently the applications have been limited due mainly to the harsh reaction conditions (polar solvents with high boiling point such as *N*-methylpyrrolidin-2-one), the necessity for stoichiometric amounts of copper salts, and the moderate yields obtained. Around the year 2000, new studies reported the fundamental role of ligands in copper-catalyzed reactions and led to a breakthrough in this coupling reaction in terms of copper loading, mild reaction conditions, substrate tolerance, and yields obtained, leading to a renewed interest in Ullmann-type reactions.[5,6] The progress has been so remarkable that the use of copper is now often an alternative to palladium-catalyzed procedures.

Preliminary intramolecular amination reactions of 2-(2-halophenyl)ethylamines and 3-(2-halophenyl)propylamines for the formation of dihydroindoles and tetrahydroquinolines, respectively, required 2 equivalents of copper(I) iodide.[68] Subsequently, some examples of the cyclization of 2-(2-halophenyl)ethylamines with a catalytic amount of copper(I) acetate have been reported.[69]

An interesting cyclization of aryl halides **103** tethered to a pendant guanidine affords 2-aminobenzimidazoles **104** in the presence of a catalytic amount of copper(I) iodide, 1,10-phenanthroline, and cesium carbonate as base in dimethoxyethane as solvent at 80 °C (Scheme 45).[70] A comparison between copper and palladium catalysts generally shows better efficiency for the former in terms of yield and selectivity.

for references see p 499

Scheme 45 Copper-Catalyzed Cyclization of Arylguanidines to 2-Aminobenzimidazoles[70]

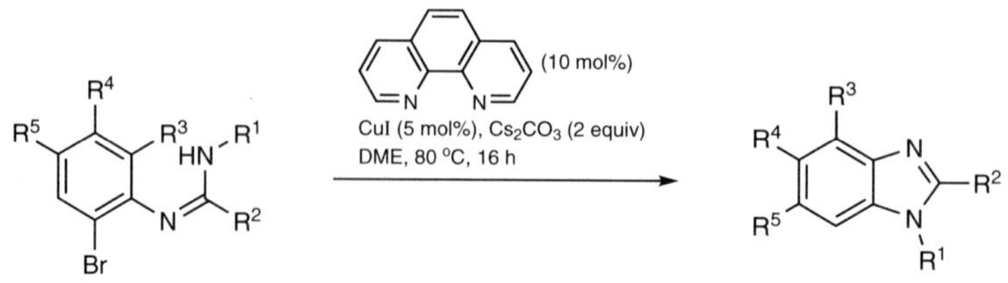

(10 mol%)
CuI (5 mol%), Cs_2CO_3 (2 equiv)
DME, 80 °C, 16 h

103 → 104

R^1	R^2	R^3	R^4	R^5	Yield[a] (%)	Ref
Bn	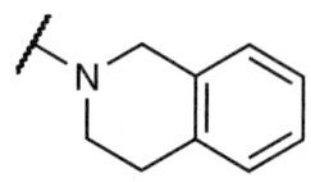	H	H	H	83 (88)	[70]
Ph	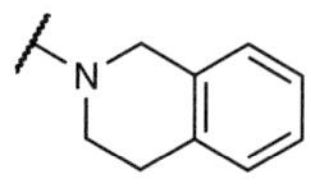	H	H	H	58 (84)	[70]
4-$MeOC_6H_4CH_2$	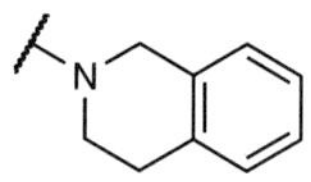	H	H	H	66 (63)	[70]
Bn	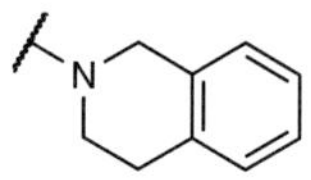	H	H	Me	90 (66)	[70]
Bn	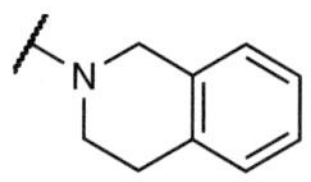	Br	H	Me	98 (76)[b]	[70]
Bn	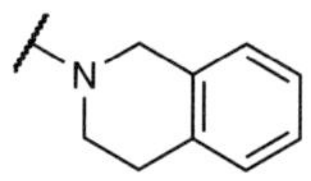	H	H	Cl	88 (81)	[70]
Bn	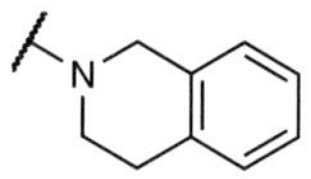	H	CF_3	H	88 (87)	[70]
4-$MeOC_6H_4CH_2$	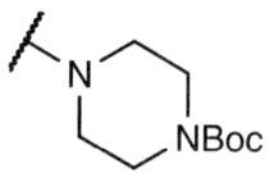 NBoc	H	H	H	97 (93)	[70]
Bn	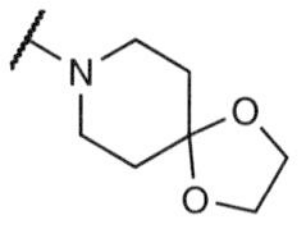	H	H	H	96 (93)	[70]
Bn	$N(Me)(CH_2)_2CN$	H	H	H	87 (31)	[70]
Ph	NHBn	H	H	H	83	[70]

[a] Values in parentheses are yields for the reaction using $Pd(PPh_3)_4$ (10 mol%) instead of CuI/1,10-phenanthroline.

[b] Product was obtained as a 1:1 mixture of monobromo and debrominated products.

An analogous cyclization process has recently been applied to electron-rich heterocycles such as the *N*′-(halopyrazolyl)amidines **105** to afford imidazo[4,5-*c*]pyrazoles **106**. In this case, the reaction is performed under conventional or microwave heating by using 5 mol% copper(I) iodide and 10 mol% *N*,*N*′-dimethylethylenediamine as ligand in the presence of potassium carbonate and dimethyl formamide as solvent (Scheme 46).[71] Both bromo and iodo derivatives readily cyclize. In the case of iodine, acetonitrile is used as solvent.

Scheme 46 Copper-Catalyzed Cyclization of *N*′-(Halopyrazolyl)amidines to Imidazo[4,5-*c*]pyrazoles[71]

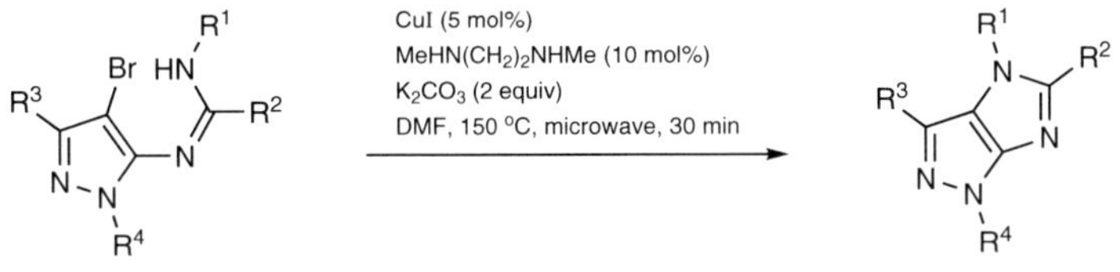

CuI (5 mol%)
$MeHN(CH_2)_2NHMe$ (10 mol%)
K_2CO_3 (2 equiv)
DMF, 150 °C, microwave, 30 min

105 → **106**

R^1	R^2	R^3	R^4	Yield (%)	Ref
Me	Ph	Me	Ph	96	[71]
Et	Ph	Me	Ph	93	[71]
Me	4-$MeOC_6H_4$	Me	Ph	88	[71]
Me	3-ClC_6H_4	Me	Ph	94	[71]
Me	Ph	Me	Me	89	[71]
Me	4-Tol	Me	Me	91	[71]
Me	Ph	H	Ph	95	[71]
Ph	Me	Me	Ph	90	[71]

Copper-catalyzed intramolecular N-arylation of diamides **107** leads to seven-membered 1,4-benzodiazepine derivatives **108** (Scheme 47).[72] The positive result depends on the choice of base; potassium carbonate is effective, as are cesium acetate and potassium phosphate, but cesium carbonate and cesium hydroxide inhibit the cyclization completely. The solvent also plays an important role; dimethyl sulfoxide and dimethylformamide are the best choices, but no reaction takes place in less polar solvents such as dioxane. Similar results are obtained under palladium catalysis. In that case, doubly functionalized substrates bearing an additional haloaryl group undergo a domino intramolecular N-arylation/C—H activation process providing dihydroazacycloheptaphenanthrene derivatives.[72]

Scheme 47 Copper-Catalyzed Cyclization of *N*-(Amidomethyl)-2-iodobenzamides to 1,4-Benzodiazepine-2,5-diones[72]

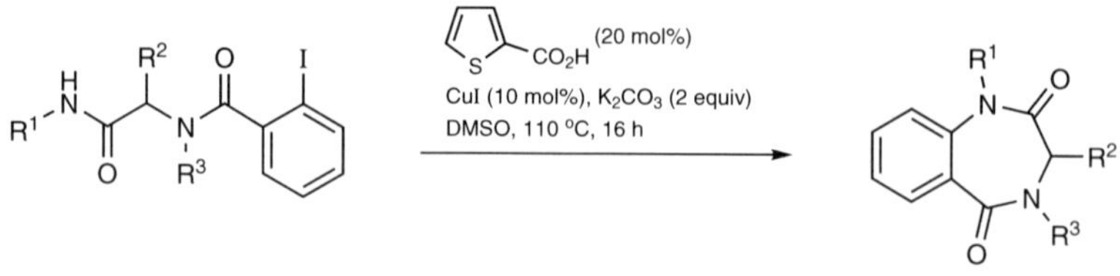

R^1	R^2	R^3	Yield (%)	Ref
CH_2CO_2Me	$(CH_2)_5Me$	Bn	86	[72]
CH_2CO_2Me	$(CH_2)_4OAc$	Bn	74	[72]
CH_2CO_2Me	$(CH_2)_5NHCbz$	Bn	88	[72]
CH_2CO_2Me	iPr	Bu	88	[72]
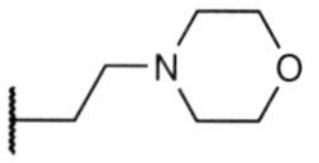 2-morpholinoethyl	iPr	Bn	99	[72]
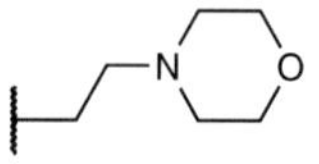 2-morpholinoethyl	iPr	$2,4-(MeO)_2C_6H_3CH_2$	97	[72]
$2,4-(MeO)_2C_6H_3CH_2$	iPr	Bn	95	[72]

1,2-Disubstituted Benzimidazoles 104; **General Procedure:**[70]
To a mixture of an arylguanidine **103** (0.20 mmol), CuI (5 mol%), 1,10-phenanthroline (10 mol%), and Cs_2CO_3 (2 equiv) was added reagent-grade DME (4 mL). The mixture was heated at 80 °C for 16 h under N_2, diluted with Et_2O (25 mL), and washed with H_2O (2 × 25 mL) and sat. aq NaCl (25 mL). The organic layer was dried ($MgSO_4$) and the solvent was removed under reduced pressure. The crude product was purified by column chromatography (silica gel).

Imidazo[4,5-*c*]pyrazoles 106; **General Procedure:**[71]
A vial was charged with a *N'*-(halopyrazolyl)amidine **105** (0.85 mmol), K_2CO_3 (2 equiv), *N,N'*-dimethylethylenediamine (10 mol%), CuI (5 mol%), and anhyd DMF (3 mL) under argon. The stirred mixture was heated for 30 min at 150 °C under microwave irradiation, concentrated, diluted with CH_2Cl_2 (100 mL), and filtered. The organic layer was washed with H_2O (2 × 50 mL), dried (Na_2SO_4), and concentrated. The crude product was purified by flash chromatography (EtOAc/hexane).

1,4-Benzodiazepine-2,5-diones 108; **General Procedure:**[72]
Freshly distilled DMSO (2.8 mL) was added to a flask containing K_2CO_3 (2 equiv), CuI (10 mol%), thiophene-2-carboxylic acid (20 mol%), and a diamide **107** (0.057 mmol). The mixture was degassed and stirred at 110 °C under argon for 16 h, cooled to rt, and extracted with EtOAc. The combined organic extracts were washed with sat. aq NH_4Cl, dried (Na_2SO_4), and concentrated. The crude material was purified by flash chromatography.

1.8.2.5.2 Palladium Catalysis (Buchwald–Hartwig Reaction)

Due to the impressive applications of this methodology, several possible alternative reaction conditions are reported.[73] Just to mention the principal requirements, electron-neutral or electron-poor aryl bromides are the most suitable substrates for the coupling with primary and secondary amines, whereas electron-rich aryl bromides give poor results. The aminations are performed in general in the presence of weak bases such as cesium carbonate with substrates bearing base-labile functional groups or sodium *tert*-butoxide with more resistant substrates. Sodium methoxide or sodium ethoxide give low conversions. The best catalyst is palladium(II) acetate, due to its low cost and easy handling, in the presence of chelating phosphines 2,2′-bis(diphenylphosphino)-1,1′-binaphthyl or 1,1′-bis(diphenylphosphino)ferrocene as ligands. Toluene is the preferred solvent.

The intramolecular reaction of aryl bromides with secondary amides or carbamates **111**, using various ligands such as 2-(diphenylphosphino)-2′-methoxy-1,1′-binaphthyl (**109**, MOP), bis[2-(diphenylphosphino)phenyl] ether (**110**, DPEphos), 4,5-bis(diphenylphosphino)-9,9-dimethylxanthene (**41**, Xantphos), or 2,2′-bis(diphenylphosphino)-1,1′-binaphthyl, affords fused five-, six-, and seven-membered heterocycles **112** (Scheme 48).[74]

Scheme 48 Palladium-Catalyzed Cyclization of Bromoaryl-Substituted Secondary Amides and Carbamates to Dihydroindoles, Tetrahydroquinolines, and Tetrahydrobenzo[*b*]azepines[74]

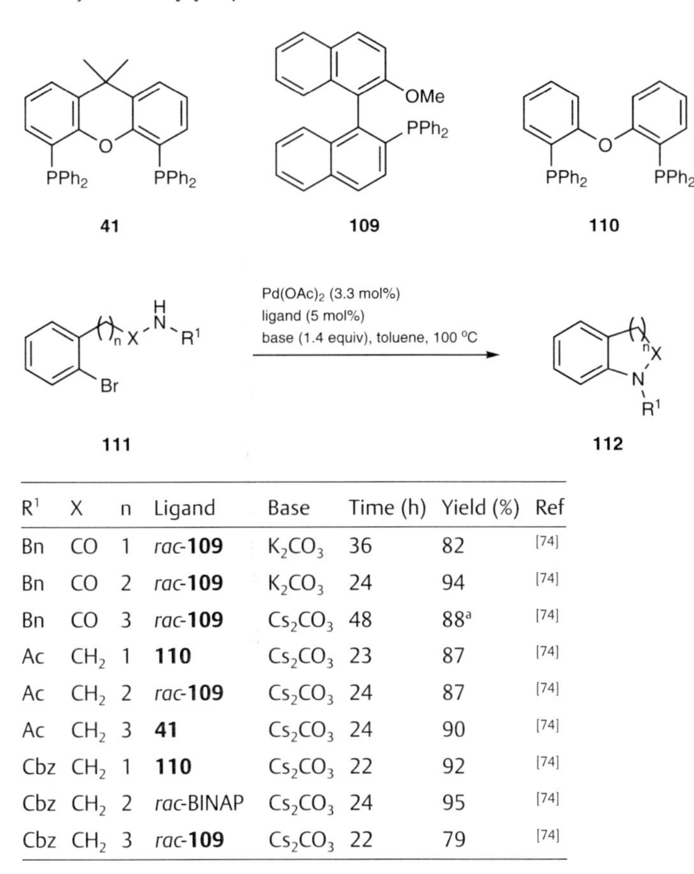

R^1	X	n	Ligand	Base	Time (h)	Yield (%)	Ref
Bn	CO	1	*rac*-**109**	K_2CO_3	36	82	[74]
Bn	CO	2	*rac*-**109**	K_2CO_3	24	94	[74]
Bn	CO	3	*rac*-**109**	Cs_2CO_3	48	88[a]	[74]
Ac	CH_2	1	**110**	Cs_2CO_3	23	87	[74]
Ac	CH_2	2	*rac*-**109**	Cs_2CO_3	24	87	[74]
Ac	CH_2	3	**41**	Cs_2CO_3	24	90	[74]
Cbz	CH_2	1	**110**	Cs_2CO_3	22	92	[74]
Cbz	CH_2	2	*rac*-BINAP	Cs_2CO_3	24	95	[74]
Cbz	CH_2	3	*rac*-**109**	Cs_2CO_3	22	79	[74]

R^1	X	n	Ligand	Base	Time (h)	Yield (%)	Ref
Boc	CH_2	1	**110**	Cs_2CO_3	36	82	[74]
Boc	CH_2	2	*rac*-BINAP	Cs_2CO_3	24	81	[74]
Boc	CH_2	3	*rac*-**109**	Cs_2CO_3	20	85	[74]

[a] Reaction carried out with 5.0 mol% $Pd(OAc)_2$ and 7.5 mol% ligand.

Analogously, N-substituted indole-2-carboxylates can be prepared in high yields via amination starting from didehydrophenylalanine derivatives, obtained from phosphonylglycinates and 2-iodobenzaldehydes, using palladium(II) chloride and 1,1′-bis(diphenylphosphino)ferrocene as ligand in dimethylformamide at 90 °C and in the presence of potassium acetate as base.[75]

Effective intramolecular N-arylation of pyrrole-2-carboxamides and indole-2-carboxamides **113** bearing a suitable *o*-halo-substituted aryl or hetaryl group at the amide nitrogen gives pyrrolo- and indoloquinoxalines **114** in the presence of palladium(0)/2,2′-bis(diphenylphosphino)-1,1′-binaphthyl and cesium carbonate in toluene at 110 °C over 24 hours (Scheme 49).[76]

Scheme 49 Palladium-Catalyzed Cyclization of *N*-(Haloaryl)pyrrole-2-carboxamides and -indole-2-carboxamides to Pyrrolo[1,2-*a*]- and Indolo[1,2-*a*]quinoxalines[76]

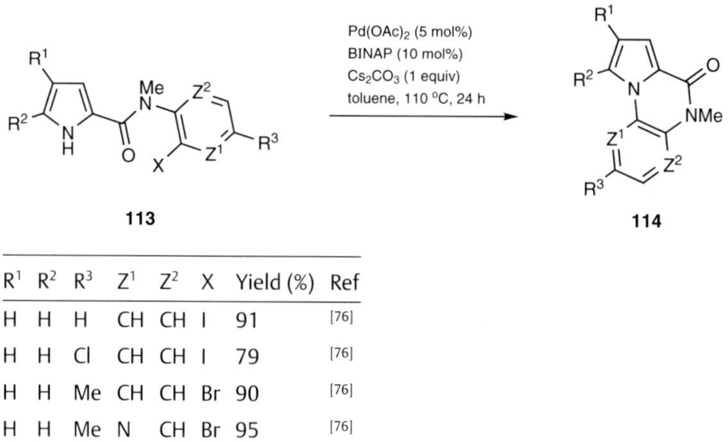

$Pd(OAc)_2$ (5 mol%), BINAP (10 mol%), Cs_2CO_3 (1 equiv), toluene, 110 °C, 24 h

113 → **114**

R^1	R^2	R^3	Z^1	Z^2	X	Yield (%)	Ref
H	H	H	CH	CH	I	91	[76]
H	H	Cl	CH	CH	I	79	[76]
H	H	Me	CH	CH	Br	90	[76]
H	H	Me	N	CH	Br	95	[76]
H	H	H	CH	N	Br	74	[76]
$(CH_2)_4$		H	CH	CH	I	98	[76]
$(CH_2)_4$		Cl	CH	CH	I	77	[76]
$(CH_2)_4$		Me	CH	CH	Br	66	[76]
$(CH_2)_4$		Me	N	CH	Br	69	[76]
$(CH_2)_4$		H	CH	N	Br	74	[76]

Azaheterocycles 112**; General Procedure:**[74]
A dry 25-mL sealable Schlenk tube was charged with $Pd(OAc)_2$ (3.3 mol%) and a ligand (5 mol%). The reaction vessel was evacuated and flushed with argon. A soln of an amide or carbamate **111** (0.45 mmol) in toluene (1 mL) was added via cannula. The mixture was heated under argon at 100 °C for 2 min to dissolve the solids. The reaction vessel was removed from the oil bath, charged quickly with Cs_2CO_3 (0.60 mmol) and toluene (0.3 mL), sealed with a Teflon screw cap, and heated at 100 °C until the aryl bromide had been consumed. The mixture was cooled to rt, filtered through a short plug of silica gel, and con-

centrated. The residue was purified by flash column chromatography (EtOAc/hexane) to afford the cyclized product as a colorless oil.

5-Methylpyrrolo[1,2-*a*]quinoxalin-4(5*H*)-one (114, $R^1 = R^2 = R^3 = H$; $Z^1 = Z^2 = CH$); Typical Procedure:[76]
A mixture of *N*-(2-iodophenyl)-*N*-methyl-1*H*-pyrrole-2-carboxamide (**113**, $R^1 = R^2 = R^3 = H$; $Z^1 = Z^2 = CH$; X = I; 1 mmol), $Pd(OAc)_2$ (0.05 mmol), BINAP (0.1 mmol), and Cs_2CO_3 (1 mmol) in toluene (10 mL) was stirred at 110 °C for 24 h. The mixture was diluted with brine and extracted with Et_2O. The organic layer was dried (Na_2SO_4) and concentrated under reduced pressure. The residue was purified by chromatography (silica gel, CH_2Cl_2/Et_2O 10:1) and the product was directly crystallized (CH_2Cl_2, CH_2Cl_2/hexane, or Et_2O); yield: 91%.

1.8.2.6 Amination of C(sp^3) Centers

Compared to the more-studied C(sp^2)—H bond activation, the direct route to obtain C—N bond formation through the functionalization of an unactivated C(sp^3)—H bond has been only recently successfully achieved.

A proposed reaction mechanism proceeds through the coordination of the metal to the nitrogen atom with metal—nitrogen bond formation. The catalytic cycle is the same for different metals but the step involving the reoxidation of the metal differs depending on the oxidation state of the metal involved. The intervention of an oxidizing agent promotes the oxidation of the metal from the reduced form to the reactive species [i.e., Pd(0) to Pd(II) and Rh(I) to Rh(III)].

1.8.2.6.1 Palladium Catalysis

The cyclization of amines and amino acids onto C(sp^3) atoms at the γ- and δ-positions of the alkyl chain is a fruitful method to obtain azetidines and pyrrolidines (Table 1).[77] The C—H activation/intramolecular amination requires the presence of a directing group on the nitrogen atom and, in the present case, the picolinamide group is exploited as a suitable removable directing group. The reaction proceeds with palladium(II) acetate and the inexpensive oxidant (diacetoxyiodo)benzene. Other oxidizing reagents completely fail to promote the cyclization.

for references see p 499

Table 1 Palladium-Catalyzed Cyclization of Amines and Amino Acids to Azetidines and Pyrrolidines[77]

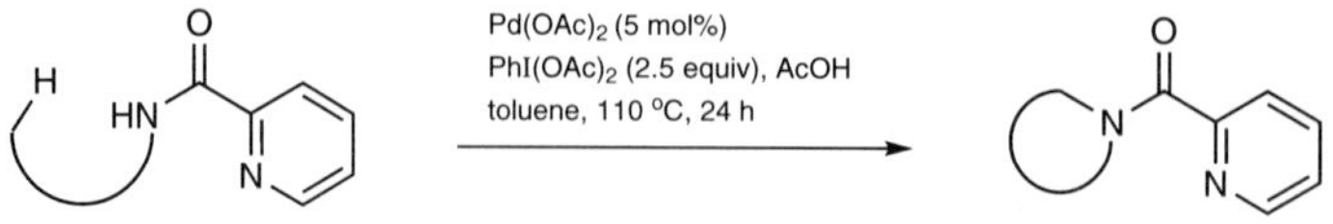

$Pd(OAc)_2$ (5 mol%)
$PhI(OAc)_2$ (2.5 equiv), AcOH
toluene, 110 °C, 24 h

Substrate	Equiv of AcOH	Product	Yield (%)	Ref
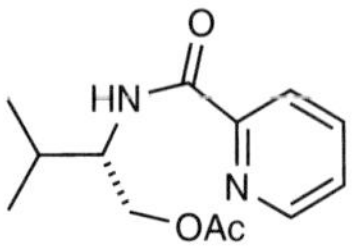	2	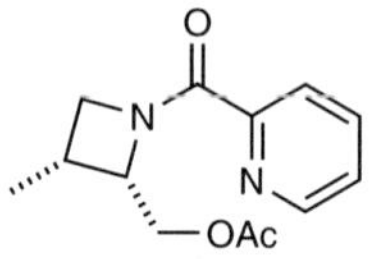	82 (dr >20:1)[a]	[77]
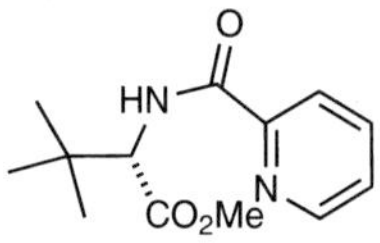	2	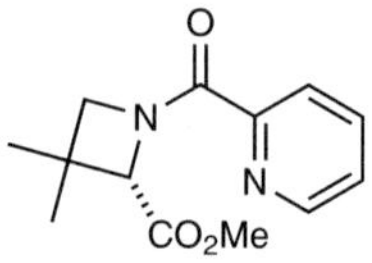	91	[77]
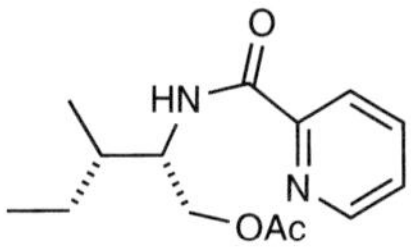	2	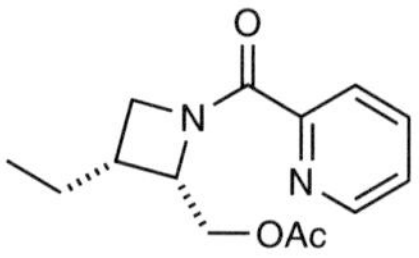	70	[77]
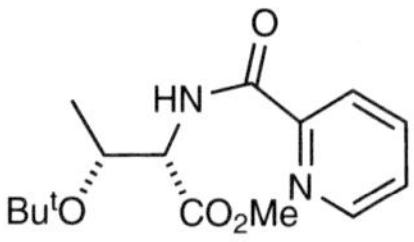	2	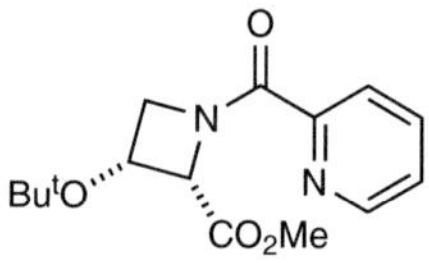	79	[77]
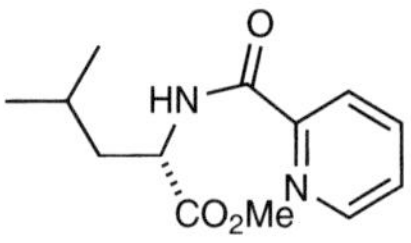	10	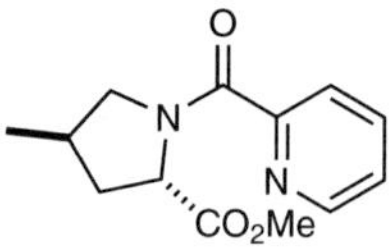	82 (dr ~7:1)[a]	[77]
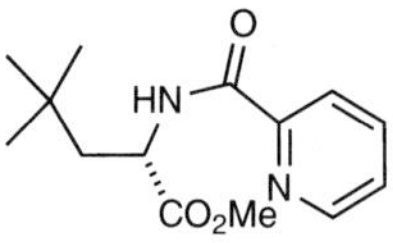	10	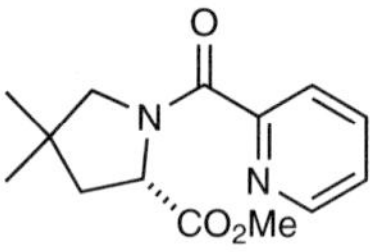	86	[77]
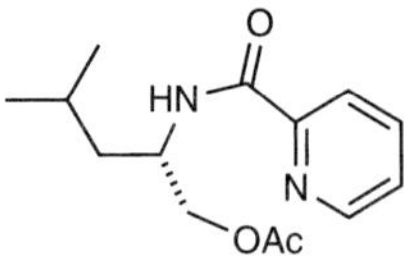	10	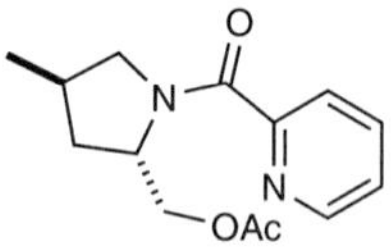	72 (dr >20:1)[a]	[77]
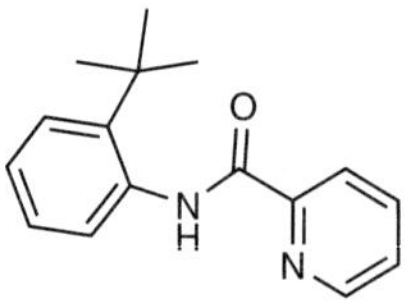	10	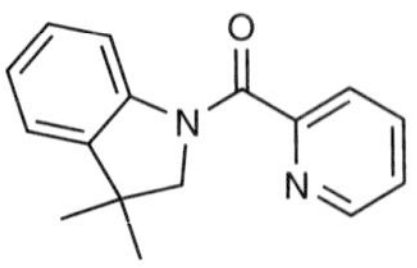	61	[77]

[a] dr determined by ^{1}H NMR spectroscopy.

The reaction mechanism involves complexation of the palladium to the acetanilide, to give intermediate **115**, forcing the metal into close proximity to the unactivated C—H bond of the alkyl group. Then, C—H bond activation can take place to give the σ-alkylpalladium(II) intermediate **116** from which reductive elimination of palladium(0) leads to the formation of the product (Scheme 50). The intervention of the oxidant promotes the oxidation of palladium(0) to palladium(II). An alternative mechanism through a more highly oxidized palladium(II)/palladium(IV) species can also be envisioned, in particular when (diacetoxyiodo)benzene is used as oxidant.[77]

Scheme 50 Amination Mechanism of C(sp^3)–H Bonds[77]

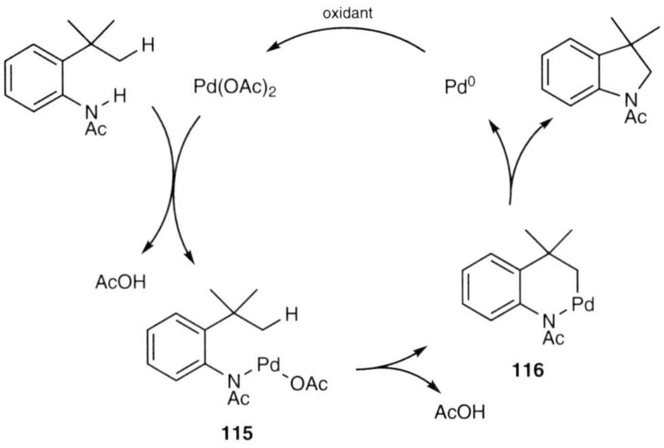

Starting from 2-*tert*-butylanilides **117**, with the aid of the amide as directing group, C(sp^3)—H activation permits the synthesis of dihydroindoles **118** through C—N bond formation using silver(I) acetate as oxidant and sodium carbonate as base in mesitylene at 140–160 °C (Scheme 51).[78]

Scheme 51 Palladium-Catalyzed Cyclization of 2-Alkylanilides to Dihydroindoles[78]

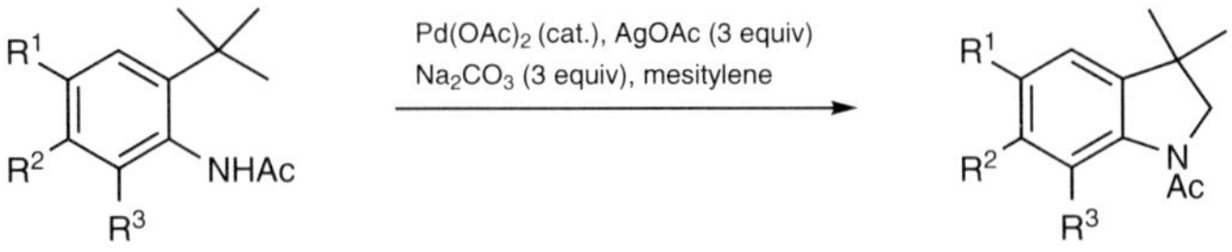

R^1	R^2	R^3	mol% of Pd(OAc)$_2$	Temp (°C)	Time (h)	Yield (%)	Ref
H	H	H	10	140	12	80	[78]
OMe	H	H	10	140	13	71	[78]
OTBDMS	H	H	10	140	12	73	[78]
OAc	H	H	10	140	16	70	[78]
TMS	H	H	10	140	12	73	[78]
Bu	H	H	10	140	13	75	[78]
Cl	H	H	10	140	13	80	[78]
Br	H	H	10	140	12	68	[78]
Ph	H	H	10	140	12	71	[78]
Ac	H	H	10	140	12	61	[78]

for references see p 499

R^1	R^2	R^3	mol% of $Pd(OAc)_2$	Temp (°C)	Time (h)	Yield (%)	Ref
CHO	H	H	10	140	13	66	[78]
CO_2Et	H	H	10	140	12	81	[78]
H	*t*-Bu	H	20	160	19	81	[78]
H	Ac	H	20	160	12	46	[78]
H	NO_2	H	30	140	12	72	[78]
H	H	Pr	10	140	36	64	[78]
H	H	Cl	10	140	36	47	[78]

***N*-(2-Pyridylcarbonyl)azetidines (Table 1); General Procedure:**[77]
A mixture of the picolinamide substrate (0.2 mmol), $Pd(OAc)_2$ (5 mol%), $PhI(OAc)_2$ (2.5 equiv), and AcOH (2.0 equiv) in anhyd toluene (2 mL) in a 10-mL glass vial was purged with argon and the vial was sealed with a PTFE cap and heated at 110 °C for 24 h. The mixture was cooled to rt and concentrated under reduced pressure. The resulting residue was purified by flash chromatography (silica gel).

***N*-Acetyl-3,3-dimethyl-2,3-dihydroindoles** 118; **General Procedure:**[78]
A sealed tube (25–30 mL) with a J. Young Teflon valve equipped with a stirrer bar was heated and cooled under reduced pressure and filled with argon. A 2-*tert*-butylanilide **117** (1 mmol), $Pd(OAc)_2$ (10 mol%), AgOAc (3.0 equiv), and Na_2CO_3 (3.0 equiv) were weighed into the reaction vessel under air and the flask was evacuated and backfilled with argon (3 ×). Anhyd mesitylene (12 mL) was added via syringe under a stream of argon. The flask was evacuated and backfilled with argon (3 ×) under efficient stirring and then closed, heated at 140 °C for the indicated time, and cooled to rt. The mixture was diluted with EtOAc (15 mL) and filtered through a short pad of Celite. The solid was washed with EtOAc (2 × 20 mL) and CH_2Cl_2 (20 mL). The combined filtrates were concentrated under reduced pressure (40–60 °C). The residue was purified by flash column chromatography (silica gel, pentane/EtOAc mixtures) to provide the pure product.

1.8.2.6.2 Rhodium Catalysis

C(sp^3)-Amination of carbamates under rhodium catalysis results in regio- and stereoselective formation of oxazolidinones under mild reaction conditions using hypervalent iodine as oxidant and magnesium oxide as a basic additive (Table 2).[79] The regioselective formation of a single product from substrates able to generate other oxazolidinones is evidence against a free nitrene intermediate.[80]

Table 2 Rhodium-Catalyzed Cyclization of Carbamates to Oxazolidin-2-ones[79]

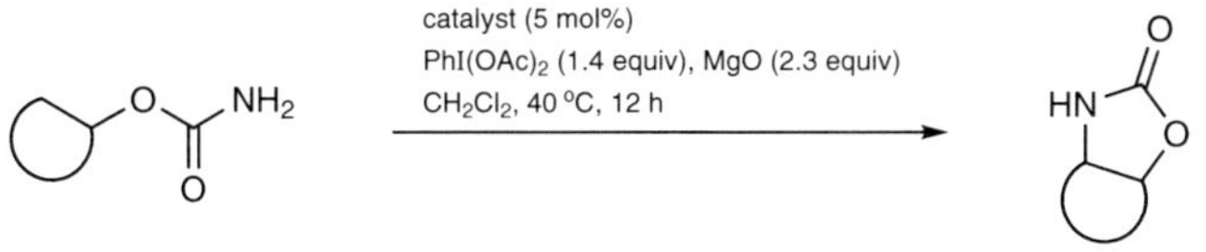

catalyst (5 mol%)
$PhI(OAc)_2$ (1.4 equiv), MgO (2.3 equiv)
CH_2Cl_2, 40 °C, 12 h

Substrate	Catalyst	Product	Yield (%)	Ref
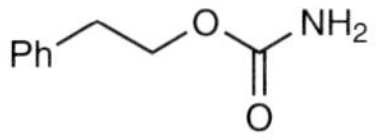	$Rh_2(O_2CCPh_3)_4$	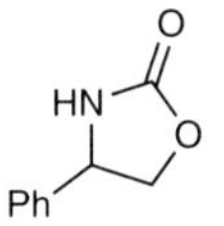	74	[79]
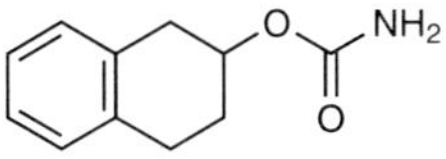	$Rh_2(O_2CCPh_3)_4$	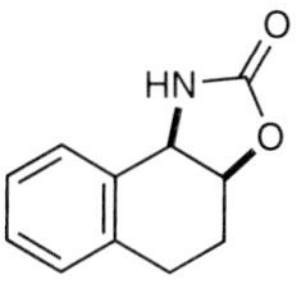	77[a]	[79]
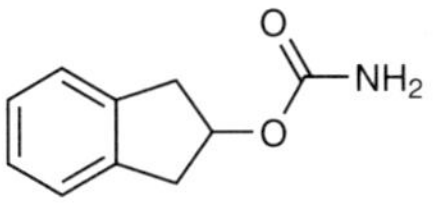	$Rh_2(OAc)_4$	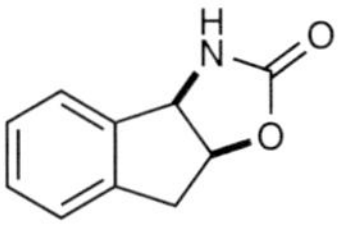	86	[79]
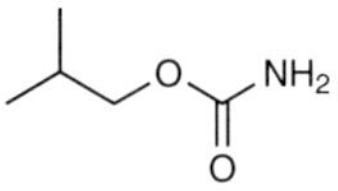	$Rh_2(OAc)_4$	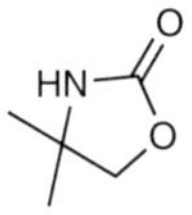	83	[79]
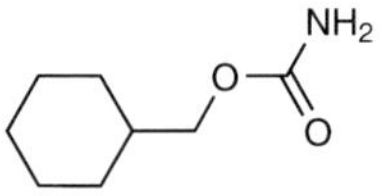	$Rh_2(O_2CCPh_3)_4$	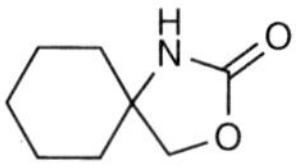	79	[79]
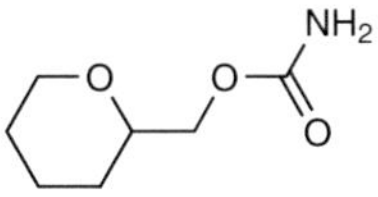	$Rh_2(O_2CCPh_3)_4$	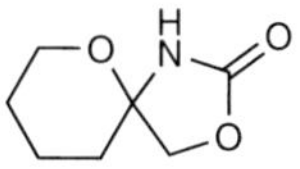	84	[79]
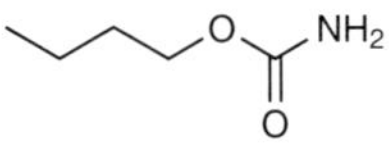	$Rh_2(O_2CCPh_3)_4$	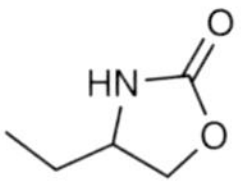	44	[79]
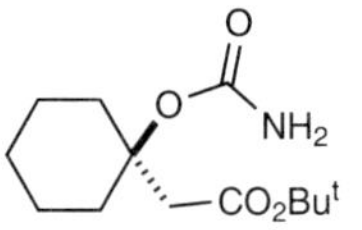	$Rh_2(OAc)_4$	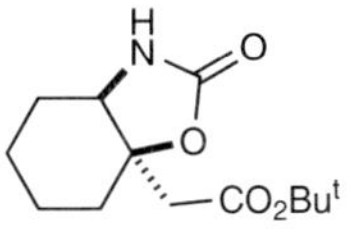	82	[79]
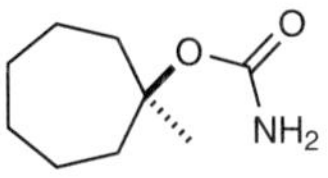	$Rh_2(OAc)_4$	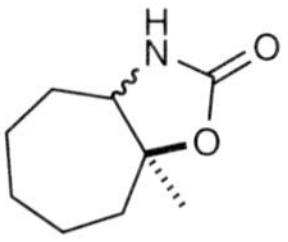	83[b]	[79]

[a] Single product as determined by ^1H NMR spectroscopy.
[b] Ratio (*cis*/*trans*) 8:1; *cis* stereochemistry determined by X-ray crystallography.

Oxazolidin-2-ones (Table 2); General Procedure:[79]
To a soln of a carbamate (1.26 mmol) in CH_2Cl_2 (8 mL) were added successively MgO (2.3 equiv), $PhI(OAc)_2$ (1.4 equiv), and a Rh(II) catalyst (5 mol%). The mixture was stirred vigorously and heated at 40 °C for 12 h, cooled to rt, diluted with CH_2Cl_2 (10 mL), and filtered through a pad of Celite (30 × 20 mm). The filter cake was rinsed with CH_2Cl_2 (2 × 10 mL). The combined filtrates were concentrated under reduced pressure and the isolated residue was purified by chromatography (silica gel).

1.8.2.7 Amination and Hydroamination Involving Azides as Nucleophiles

Alternative pathways to obtaining nitrogen-containing heterocycles involve the azide group as the nucleophile in intramolecular aminations of alkenes and arenes and hydroamination of alkynes. Both processes make use of different transition metals. Rhodium in particular gives good results and the reaction conditions are in general mild. In all cases, pentacyclic rings are formed. Pyrroles are the heterocycles obtained in most cases, but indoles and carbazoles can also be synthesized.

Compared to other nitrogen-containing groups reacting as nucleophiles in the amination process, the mechanism involving azides is not fully established and several different pathways can be envisaged (Scheme 52).[81] In all cases, the first step is the coordination of the metal to the α-nitrogen of the azide followed by loss of molecular nitrogen and formation of nitrenoid intermediate **119**, from which attack of the alkene or the aryl group produces the C—N bond. This step can occur by concerted insertion of the nitrene into an *ortho*-C—H bond via intermediate **121** or stepwise electrophilic substitution via intermediate **120**. Final metal elimination and hydride shift affords the product.

Scheme 52 Azide Amination Mechanism To Form Indoles[81]

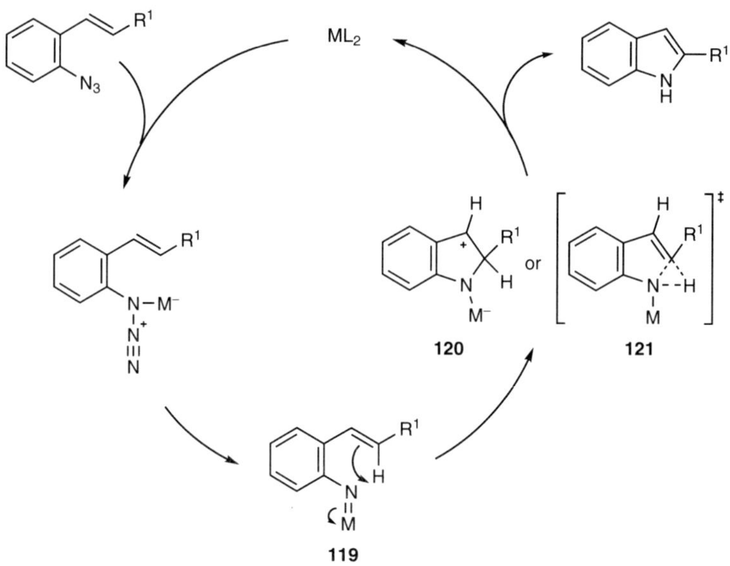

Working on alkynes as substrates, a plausible mechanism involves metal activation of the triple bond toward addition by the nucleophilic nitrogen of the azide. Subsequent loss of molecular nitrogen assisted by the metal produces cationic intermediate **122**, from which a formal 1,2-shift of hydrogen affords the heterocyclic product and regenerates the cationic catalyst (Scheme 53).[82]

Scheme 53 Azide Hydroamination Mechanism To Form Pyrroles[82]

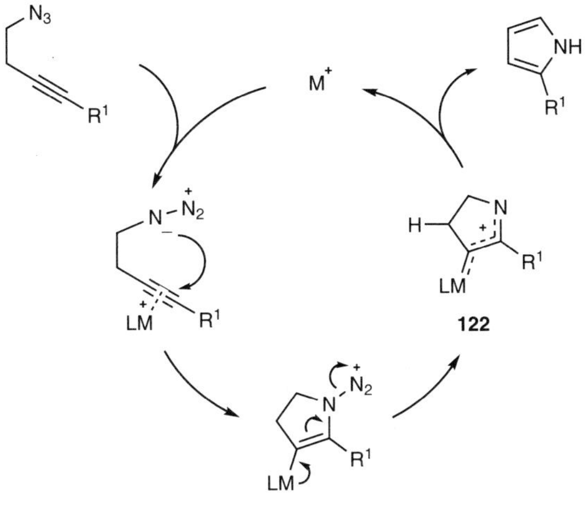

122

1.8.2.7.1 Rhodium Catalysis (Amination of Alkenes and Arenes)

2-Vinylaryl azides **123** are suitable substrates to give indoles **124** in good yields by oxidative amination (Scheme 54).[81] The reaction is performed with 5 mol% rhodium(II) perfluorobutanoate in toluene or 1,2-dichloroethane at 60 °C in the presence of molecular sieves. In the absence of water sequestering agents the yields drop precipitously. The reaction is tolerant of various functional groups but the reactivity and the yields are reduced if R^1 is an alkyl group.

Scheme 54 Rhodium-Catalyzed Cyclization of 2-Alkenylaryl Azides to Indoles[81]

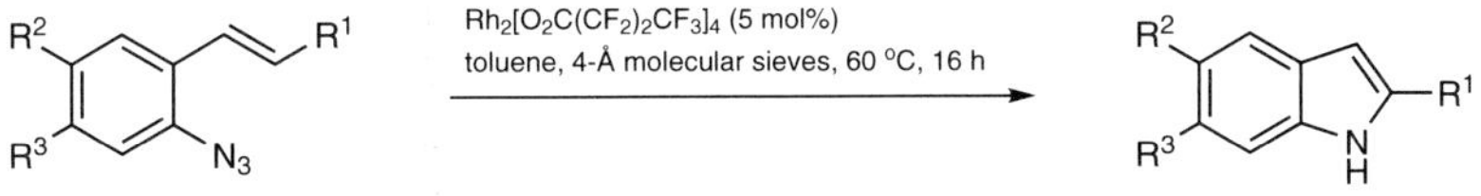

$Rh_2[O_2C(CF_2)_2CF_3]_4$ (5 mol%)
toluene, 4-Å molecular sieves, 60 °C, 16 h

123 → 124

R^1	R^2	R^3	Yield (%)	Ref
Ph	OMe	H	98	[81]
Ph	Me	H	89	[81]
Ph	H	H	94	[81]
Ph	F	H	99	[81]
Ph	OCF_3	H	95[a]	[81]
Ph	CF_3	H	82	[81]
Ph	H	OMe	88	[81]
Ph	H	Me	89	[81]
Ph	H	CF_3	89	[81]
4-Tol	H	H	96	[81]
$4\text{-}FC_6H_4$	H	H	95	[81]
$3\text{-}FC_6H_4$	H	H	91	[81]
$4\text{-}ClC_6H_4$	H	H	84	[81]
$4\text{-}F_3CC_6H_4$	H	H	95	[81]

R^1	R^2	R^3	Yield (%)	Ref
4-$F_3CC_6H_4$	OMe	H	90	[81]
4-$F_3CC_6H_4$	F	H	89	[81]
4-$F_3CC_6H_4$	OCF_3	H	91	[81]
Me	CF_3	H	61	[81]
$(CH_2)_5Me$	H	H	68	[81]

[a] 2 mol% of catalyst was used.

Substituted indoles **126** can be obtained in good yields starting from arenes **125** bearing the vinyl azide group as substituent through a formal oxidative arene amination. The reaction can be performed under rhodium (Scheme 55),[83] ruthenium,[84] and zinc[85] catalysis using different solvents under mild reaction conditions. When dichloromethane is the solvent, the reaction is performed at room temperature. The reactions are tolerant of various electron-withdrawing and electron-donating groups. The steric environment does not influence the yields.

Scheme 55 Rhodium-Catalyzed Cyclization of Vinyl Azides to Methyl Indole-2-carboxylates[83]

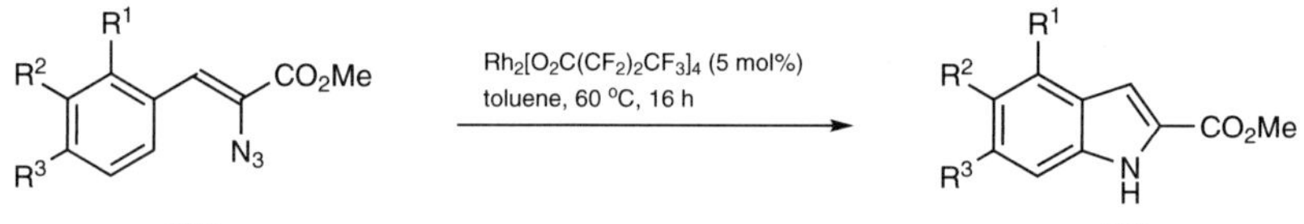

$Rh_2[O_2C(CF_2)_2CF_3]_4$ (5 mol%)
toluene, 60 °C, 16 h

125 → **126**

R^1	R^2	R^3	Yield (%)	Ref
H	H	OMe	98[a]	[83]
H	H	Me	88	[83]
H	H	iPr	91	[83]
H	H	*t*-Bu	71	[83]
H	H	H	84	[83]
H	H	Cl	85	[83]
H	H	Br	84	[83]
H	H	CF_3	88	[83]
OMe	H	H	91	[83]
Me	H	H	93	[83]
Cl	H	H	76	[83]
H	Cl	H	97 (87:13)[b]	[83]
H	Cl	OMe	92 (79:21)[a,b]	[83]
H	Br	OMe	98 (92:8)[a,b]	[83]
H	OMe	OMe	88 (95:5)[a,b]	[83]

[a] Reaction carried out at 40 °C.
[b] Yield of separable mixture of regioisomers; value in parentheses is regioisomer ratio (i.e.; 5-/5,6-substituted indole and 6/6,7-substituted indole).

The same method affords carbazoles **128** from biaryl azides **127** with similarly good results (Scheme 56).[86]

Scheme 56 Rhodium-Catalyzed Cyclization of Biaryl Azides to Carbazoles[86]

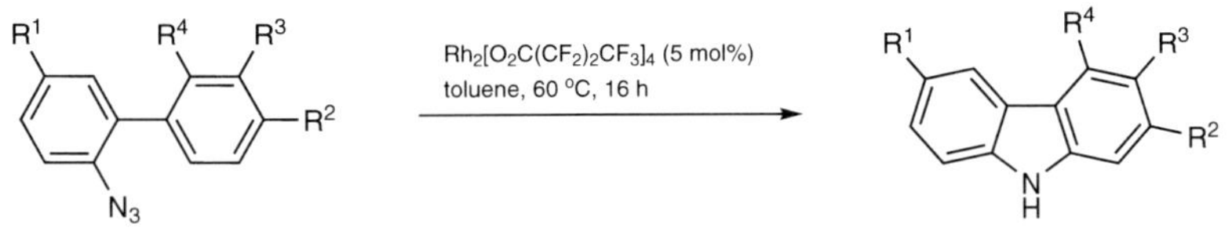

$Rh_2[O_2C(CF_2)_2CF_3]_4$ (5 mol%), toluene, 60 °C, 16 h; **127** → **128**

R^1	R^2	R^3	R^4	Yield (%)	Ref
OMe	H	H	H	71	[86]
Me	H	H	H	86	[86]
H	H	H	H	98	[86]
F	H	H	H	86	[86]
CF_3	H	H	H	83	[86]
H	H	H	Cl	93	[86]
H	H	H	F	92	[86]
H	OMe	H	H	82	[86]
H	CO_2Et	H	H	91	[86]
H	CF_3	H	H	91	[86]
H	NO_2	H	H	91[a]	[86]
H	H	Cl	H	82[b]	[86]
H	H	CF_3	H	84[b]	[86]

[a] Reaction carried out with 2 mol% $Rh_2[O_2C(CF_2)_2CF_3]_4$.
[b] Regioisomeric ratio with 1-substituted carbazole was >95:5 as determined by ^1H NMR spectroscopy.

2-Substituted Indoles 124; **General Procedure:**[81]
Toluene (0.2 mL) was added to a mixture of 2-vinylaryl azide **123** (0.3 mmol), $Rh_2[O_2C(CF_2)_2CF_3]_4$ (5 mol%), and crushed 4-Å molecular sieves to a final concentration of 1.5 M. The resulting mixture was heated to 60 °C for 16 h and cooled to rt. The heterogeneous mixture was filtered through a short pad of silica gel. The filtrate was concentrated under reduced pressure. Purification by HPLC (EtOAc/hexane 0:100 to 30:70) provided the product as a white solid.

Methyl Indole-2-carboxylates 126; **General Procedure:**[83]
Toluene (2 mL) was added to a mixture of vinyl azide **125** (0.043 mmol) and $Rh_2[O_2C(CF_2)_2CF_3]_4$ (5 mol%). The resulting mixture was heated at 60 °C for 16 h and cooled to rt. The heterogeneous mixture was filtered through silica gel. The filtrate was concentrated under reduced pressure. Purification by flash chromatography provided the product as a white solid.

Substituted Carbazoles 128; General Procedure:[86]
Toluene (0.7 mL, 0.5 M) was added to a mixture of biaryl azide **127** (0.35 mmol) and $Rh_2[O_2C(CF_2)_2CF_3]_4$ (5 mol%). The resulting mixture was stirred at 60 °C for 16 h. The heterogeneous mixture was filtered through silica gel. The filtrate was concentrated under reduced pressure. Purification by HPLC (EtOAc/hexane 0:100 to 30:70) provided the product.

1.8.2.7.2 Ruthenium Catalysis

Biaryl azides **129** and vinyl azides **131**, similar to those used above under rhodium catalysis (Section 1.8.2.7.1), also react in the presence of ruthenium complexes, affording carbazoles **130** and pyrroles **132** through C—H amination reactions (Scheme 57).[87] Compared to the rhodium-catalyzed method, the advantages of this approach are the less-expensive ruthenium(III) chloride catalyst and the wide applicability of the reaction conditions to substrates bearing substituents such as amino groups and halides.

Scheme 57 Ruthenium-Catalyzed Cyclization of Biaryl and Vinyl Azides to Carbazoles and Pyrroles[87]

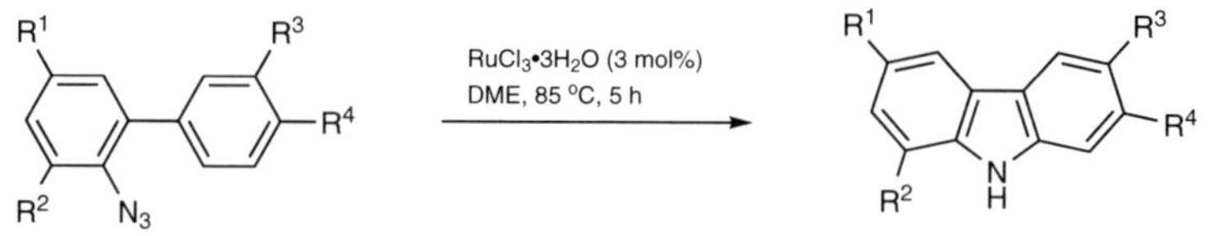

R^1	R^2	R^3	R^4	Yield (%)	Ref
H	H	H	H	90	[87]
Br	H	H	H	92	[87]
I	H	H	H	93	[87]
Br	Br	H	H	91	[87]
H	H	Ph	Ph	96	[87]

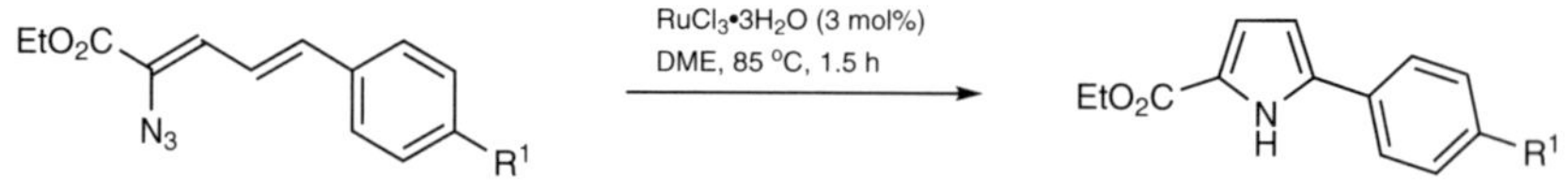

R^1	Yield (%)	Ref
H	94	[87]
OMe	92	[87]
NO_2	87	[87]
OMe	87[a]	[87]

[a] Reaction carried out at 105 °C for 3 h.

Carbazoles 130 or Ethyl 5-Arylpyrrole-2-carboxylates 132; General Procedure:[87]
A mixture of a biaryl azide **129** or vinyl azide **131** and $RuCl_3•3H_2O$ (3 mol%) in DME (3 mL) was stirred at 85 °C for 1.5–5 h to give a deep brown soln. The mixture was cooled to rt. The solvent was evaporated under reduced pressure, and the residue was purified by column chromatography (silica gel, hexane/EtOAc).

1.8.2.7.3 Zinc Catalysis

1,3-Dienyl azides **133** are aliphatic substrates that are able to undergo oxidative amination reactions. The reaction is performed under catalysis by either rhodium(II) perfluorobutanoate or zinc iodide in dichloromethane at room temperature, providing substituted pyrroles **134** (Scheme 58).[88]

Scheme 58 Zinc-Catalyzed Cyclization of 1,3-Dienyl Azides to Pyrroles[88]

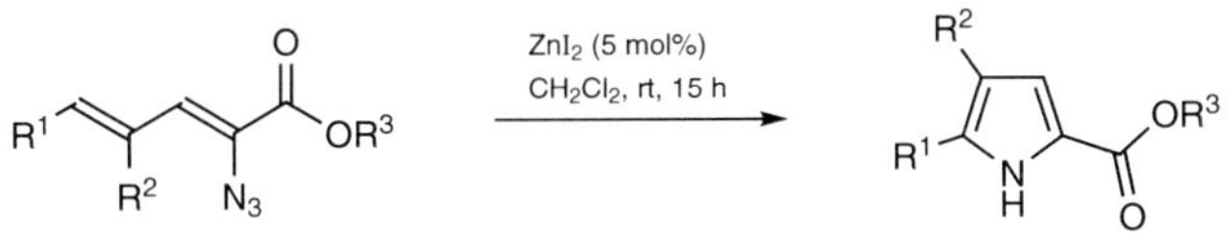

ZnI_2 (5 mol%)
CH_2Cl_2, rt, 15 h

133 → **134**

R^1	R^2	R^3	Yield (%)	Ref
Ph	H	Me	93	[88]
4-$MeOC_6H_4$	H	Me	83	[88]
2-$MeOC_6H_4$	H	Me	96	[88]
2-Tol	H	Me	90	[88]
3-Tol	H	Me	77	[88]
4-*t*-BuC_6H_4	H	Me	90	[88]
4-BrC_6H_4	H	Me	82	[88]
2-ClC_6H_4	H	Me	80	[88]
Pr	H	Me	72	[88]
$(CH_2)_4$		Me	71	[88]
Ph	Me	Me	86	[88]
Ph	Cl	Me	75	[88]
Ph	H	*t*-Bu	89	[88]

Pyrrole-2-carboxylates 134; General Procedure:[88]
CH_2Cl_2 (0.3 mL, 1.5 M) was added to a mixture of a 1,3-dienyl azide **133** (0.43 mmol) and ZnI_2 (5 mol%). The resulting mixture was stirred at rt for 15 h. The heterogeneous mixture was filtered through silica gel. The filtrate was concentrated under reduced pressure. Purification via HPLC (EtOAc/hexane 0:100 to 30:70) provided the product.

1.8.2.7.4 Platinum Catalysis (Hydroamination of Alkynes)

Starting from substituted homopropargyl azides **135**, the intramolecular hydroamination process allows the synthesis of substituted pyrroles **136** using platinum(IV) catalysis. The reaction is performed in the presence of 2,6-di-*tert*-butyl-4-methylpyridine as base in ethanol as solvent at 50 °C or reflux, under argon or aerobic conditions, with no effect on various functional groups (Scheme 59).[89]

for references see p 499

Scheme 59 Platinum-Catalyzed Cyclization of Homopropargyl Azides to Pyrroles[89]

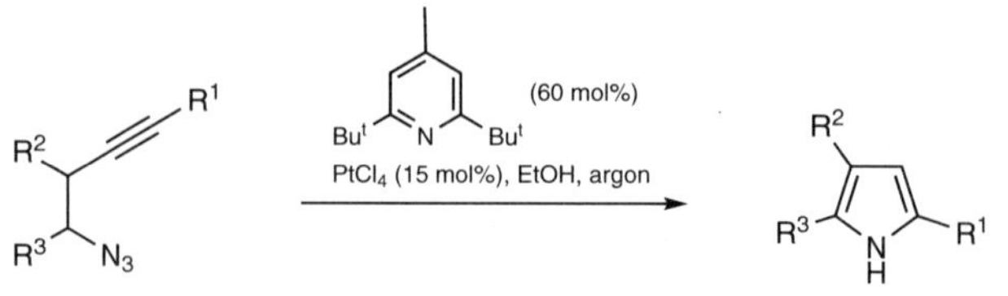

R^1	R^2	R^3	Temp	Time (h)	Yield (%)	Ref
Ph	$(CH_2)_4$		50 °C	4	88[a]	[89]
$(CH_2)_4OH$	$(CH_2)_4$		50 °C	2	72	[89]
$(CH_2)_4Me$	$(CH_2)_4$		50 °C	1.5	55	[89]
$(CH_2)_4OTBDPS$	$(CH_2)_4$		50 °C	1	65	[89]
$(CH_2)_3CO_2Et$	$(CH_2)_4$		50 °C	1	64	[89]
$(CH_2)_4N(Boc)Ts$	$(CH_2)_4$		reflux	0.5	69	[89]
Ph	H	H	reflux	6	74	[89]
$(CH_2)_{12}Me$	H	CO_2Et	reflux	14	61	[89]
Ph	H	Ph	reflux	1	57	[89]

[a] Reaction carried out with 5 mol% $PtCl_4$ and 20 mol% 2,6-di-*tert*-butyl-4-methylpyridine.

Substituted Pyrroles 136; General Procedure:[89]
2,6-Di-*tert*-butyl-4-methylpyridine (60 mol%) was added to a soln of $PtCl_4$ (15 mol%) in EtOH. The mixture was stirred at 50 °C for 1 h. A soln of a homopropargyl azide **135** in EtOH was added and the mixture was stirred at 50 °C under argon for the indicated time. The mixture was concentrated under reduced pressure and the residue was purified by column chromatography (alumina, hexane/EtOAc). Reaction under aerobic conditions gave lower yields.

1.8.2.7.5 Silver–Gold Heterobimetallic Catalysis

Using silver–gold heterobimetallic catalysis the intramolecular reaction on homopropargyl azides **137** is performed in dichloromethane as solvent under extremely mild conditions, affording substituted pyrroles **138** in good yields (Scheme 60).[82]

Scheme 60 Silver/Gold-Catalyzed Cyclization of Homopropargyl Azides to Pyrroles[82]

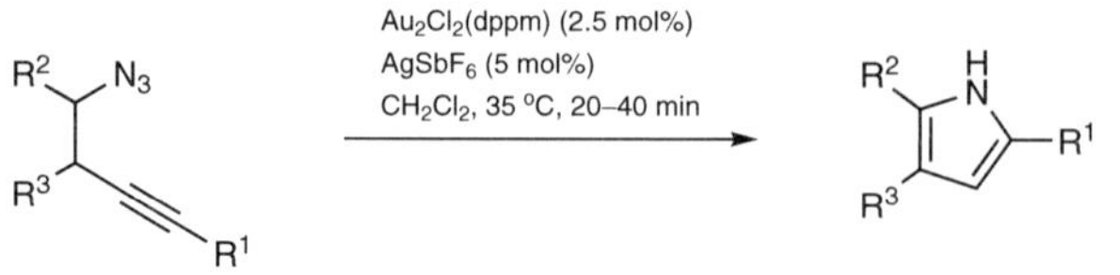

R^1	R^2	R^3	Yield (%)	Ref
Bu	Bu	H	82	[82]
$(CH_2)_5Me$	H	H	76	[82]
cyclopropyl	Bu	H	78	[82]
Ph	H	H	68	[82]
2-$MeOC_6H_4$	H	H	88	[82]
3-$F_3CC_6H_4$	H	H	93	[82]
4-IC_6H_4	H	H	87	[82]
2-furyl	H	H	61	[82]
Ph	$(CH_2)_4$		73	[82]

2-Substituted Pyrroles 138**; General Procedure:**[82]
A soln of azide **137** (1 mmol) in CH_2Cl_2 (20 mL) was added to a vial with a threaded cap containing a magnetic stirrer bar. $AgSbF_6$ (5 mol%) was added and the resulting suspension was heated at 35 °C and Au_2Cl_2(dppm) (2.5 mol%) was added. Subsequently, evolution of gas from the mixture was observed. After TLC had indicated full consumption of the starting material (20–40 min), the mixture was cooled to rt and filtered through a short pad of neutral alumina with CH_2Cl_2. The filtrate was concentrated and the residue was purified by chromatography (neutral alumina).

1.8.2.8 Carbamoylation of Arenes

The presence of carbon monoxide in a transition-metal-catalyzed reaction, in particular with palladium complexes, allows carbonylation of aromatic substrates through the insertion of carbon monoxide. This methodology has been developed for the synthesis of benzolactams, benzolactones, and biologically active heterocyclic systems.

1.8.2.8.1 Palladium Catalysis

Carbonylation of secondary ω-arylalkylamines **139** by direct aromatic metalation affords five- and six-membered benzolactams **140**. The reaction is carried out under phosphine-free catalytic conditions using palladium(II) acetate as catalyst and copper(II) acetate as oxidant under an atmosphere of carbon monoxide gas containing air (Scheme 61).[90] Five-membered rings are formed 11 times faster than six-membered rings. The electronic effects of aryl substituents have also been examined. In the case of possible regioselectivity, two regioisomers are obtained. By contrast, carbonylation of primary amines under the same conditions does not produce benzolactams, but rather ureas in good yields.

Scheme 61 Palladium-Catalyzed Cyclization of *N*-Alkyl-ω-arylalkylamines to Benzolactams[90]

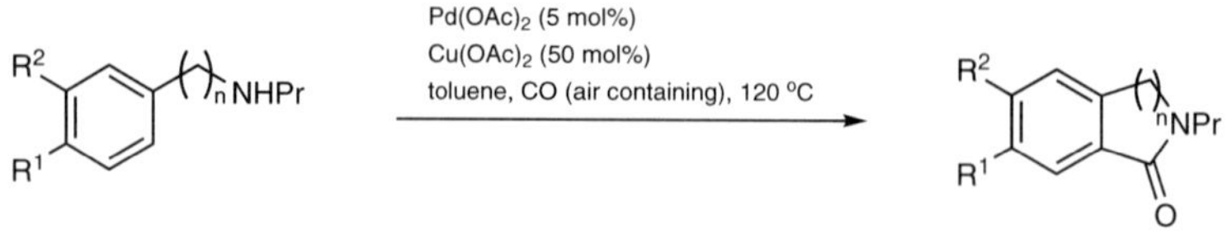

Pd(OAc)$_2$ (5 mol%)
Cu(OAc)$_2$ (50 mol%)
toluene, CO (air containing), 120 °C

139 → 140

R^1	R^2	n	Yield (%)	Ref
H	H	2	64	[90]
H	H	1	87	[90]
OMe	OMe	2	24	[90]
OCH_2O		2	86	[90]
OCH_2O		1	86 (1:4)[a]	[90]
OMe	OMe	1	88 (3:1)[a]	[90]

[a] Ratio with the 6,7-disubstituted product as determined by ^{1}H NMR spectroscopy.

***N*-Propylbenzolactams 140; General Procedure:**[90]

CAUTION: *Carbon monoxide is extremely flammable and exposure to higher concentrations can quickly lead to a coma.*

A stirred mixture of freshly prepared amine **139** (0.2 mmol), Pd(OAc)$_2$ (5 mol%), and Cu(OAc)$_2$ (0.5 equiv) in toluene (2 mL) was heated at 120 °C under CO (ca. 1~1.5 L) containing air (12 mL, corresponding to 0.1 mmol of O_2) delivered from a toy balloon for 2 h. The mixture was cooled to rt and filtered through a thin pad of powdered $MgSO_4$. Toluene was removed on a rotary evaporator. The residue was crystallized from an appropriate solvent or subjected to preparative TLC (silica gel).

1.8.2.8.2 Ruthenium Catalysis

Aryl amides **141**, bearing a pyridin-2-ylmethyl moiety on the amide nitrogen atom, react in the presence of triruthenium dodecacarbonyl as catalyst under high carbon monoxide and ethene pressure in toluene at 160 °C, affording phthalimides **142** through carbonylation of a C—H bond and C—N bond formation (Scheme 62). The conversion into the product requires the release of two hydrogen atoms and in the absence of ethene as hydrogen acceptor the product is not formed. Other alkenes known to serve as hydrogen acceptors are also ineffective.[91] Other benzamides such as *N*-benzylamides and tertiary amides are ineffective, indicating that the double N,N-coordination is key to the favorable result. This bidentate system reacts in sharp contrast to monodentate ones such as 2-arylpyridines in which the less hindered C—H bond exclusively undergoes carbonylation. A wide variety of functional groups at the *para*-position of the aromatic amide are tolerated. In the case of *meta*-substituted amides, poor regioselectivity is reported, depending mainly on steric factors.

Scheme 62 Ruthenium-Catalyzed Carbonylation of Arylamides To Give Phthalimides[91]

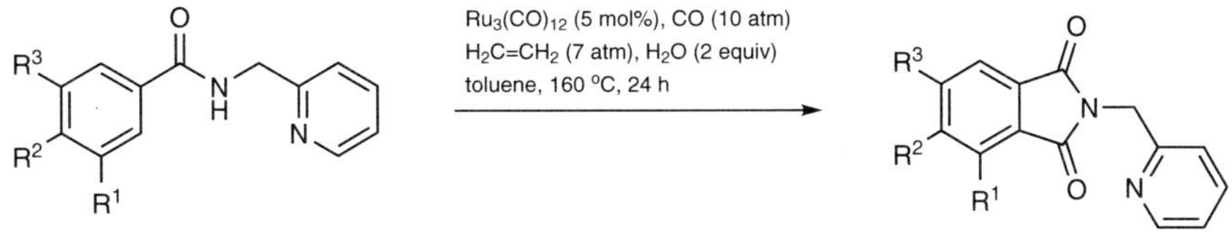

$Ru_3(CO)_{12}$ (5 mol%), CO (10 atm)
$H_2C{=}CH_2$ (7 atm), H_2O (2 equiv)
toluene, 160 °C, 24 h

141 → **142**

R^1	R^2	R^3	Yield (%)	Ref
H	H	H	77	[91]
H	Me	H	79	[91]
H	OMe	H	69	[91]
H	NMe_2	H	60	[91]
H	CO_2Me	H	82	[91]
H	Ac	H	78	[91]
H	CN	H	84	[91]
H	Cl	H	89	[91]
H	Br	H	87	[91]
H	H	Me	68 (1:1)[a]	[91]
H	H	OMe	84 (14:1)[a]	[91]
H	H	OCF_3	80 (9:1)[a]	[91]
H	H	NMe_2	86 (>20:1)[a]	[91]
H	H	Ac	82 (20:1)[a]	[91]
OMe	H	OMe	92	[91]

[a] Regioisomer ratio with the 4-substituted product, as determined by ^1H NMR spectroscopy and GC.

2-(Pyridin-2-ylmethyl)phthalimide (142, **$R^1 = R^2 = R^3 =$ H); Typical Procedure:**[91]

CAUTION: *Carbon monoxide is extremely flammable and exposure to higher concentrations can quickly lead to a coma.*

$Ru_3(CO)_{12}$ (5 mol%), *N*-(pyridin-2-ylmethyl)benzamide (**141**, $R^1 = R^2 = R^3 =$ H; 1 mmol), H_2O (2 mmol), and toluene (3 mL) were placed in a 50-mL stainless steel autoclave under N_2. After the system was flushed with 8 atm of ethene two times, it was pressurized to 7 atm and then with CO to an additional 10 atm. The autoclave was heated in an oil bath at 160 °C for 24 h, followed by cooling to rt for 1 h. CO and ethene were released. After the contents were transferred to a round-bottomed flask with $CHCl_3$, the volatiles were removed under reduced pressure, and the residue was purified by flash column chromatography (silica gel, hexane/EtOAc) to give the product as a white solid; yield: 183 mg (77%).

1.8.2.9 Carboamination of Alkynes

Carboamination reactions consist of a domino process involving the formation of both C—C and C—N bonds through the reaction between alkyne-tethered amines and aryl halides as coupling partners in the presence of a palladium(0) catalyst and base. The mechanism of the process consists of the oxidative addition of the palladium(0) to the aryl halide. Subsequent addition of the resulting palladium(II) complex to the unsaturated bond, followed by β-hydride elimination affords the product.

1.8.2.9.1 Palladium Catalysis

The milestone work on carboamination of alkynes provides a wide variety of 2,3-disubstituted indoles **144** starting from 2-alkynylaryltrifluoroacetanilides **143** and aryl-, hetaryl-, and vinyl halides or trifluoromethanesulfonates (Scheme 63).[92]

Scheme 63 Palladium-Catalyzed Carboamination of 2-Alkynylaryltrifluoroacetanilides[92]

143 → **144**

R^1X (1.5 equiv), $Pd(PPh_3)_4$ (5 mol%)
Cs_2CO_3 (1.5 equiv), MeCN, 100 °C

Ar^1	R^1	X	Time (h)	Yield (%)	Ref
Ph	Ph	Br	0.5	93	[92]
Ph	4-Tol	Br	1	98	[92]
Ph	3-Tol	Br	1	98	[92]
Ph	2-Tol	Br	1	96	[92]
Ph	3,5-$Me_2C_6H_3$	Br	0.5	98	[92]
Ph	4-$MeOC_6H_4$	Br	0.25	88	[92]
Ph	4-PhC_6H_4	Br	1	98	[92]
Ph	3-$MeOC_6H_4$	Br	0.5	86	[92]
Ph	3-FC_6H_4	Br	0.5	98	[92]
Ph	3-$F_3CC_6H_4$	Br	0.5	98	[92]
Ph	3-NCC_6H_4	Br	0.5	99	[92]
Ph	4-NCC_6H_4	Br	0.5	90	[92]
Ph	4-AcC_6H_4	Br	0.5	98	[92]
Ph	4-$O_2NC_6H_4$	Br	1	97	[92]
Ph	2-pyridyl	Br	2	94	[92]
Ph		Br	5	85	[92]
Ph		OTf	1	91[a]	[92]
Ph		OTf	1	80[a]	[92]
4-$MeOC_6H_4$	4-*t*-BuC_6H_4	Br	1	87	[92]

Ar^1	R^1	X	Time (h)	Yield (%)	Ref
4-MeOC_6H_4		Br	2	80	[92]
4-AcC_6H_4	$4\text{-}t\text{-BuC}_6H_4$	Br	1	98	[92]
$4\text{-O}_2NC_6H_4$		Br	2	40	[92]

[a] Reaction carried out with 1 equiv of R^1X.

Starting from benzaldimines **145** and aryl, allyl, and alkynyl halides the carboamination process affords isoquinolines **146** (Scheme 64).[93]

Scheme 64 Palladium-Catalyzed Carboamination of Benzaldimines To Give 3,4-Disubstituted Isoquinolines[93]

R^2X (5 equiv), $Pd(PPh_3)_4$ (5 mol%)
K_2CO_3 (5 equiv), DMF, 100 °C

145 → **146**

R^1	R^2	X	Time (h)	Yield (%)	Ref
Ph	Ph	I	12	49	[93]
Ph	$4\text{-O}_2NC_6H_4$	I	12	75	[93]
Ph	$4\text{-F}_3CC_6H_4$	I	8	65	[93]
cyclohex-1-enyl	$4\text{-O}_2NC_6H_4$	I	12	60	[93]
4-MeOC_6H_4	$4\text{-O}_2NC_6H_4$	I	10	80	[93]
cyclohex-1-enyl	$4\text{-EtO}_2CC_6H_4$	I	12	61	[93]
Ph	$CH_2CH{=}CH_2$	Br	18	65	[93]
Ph	$CH_2C(Me){=}CH_2$	Cl	24	71	[93]
4-MeOC_6H_4	$CH_2C(Me){=}CH_2$	Cl	24	88	[93]
Bu	$CH_2C(Me){=}CH_2$	Cl	48	62	[93]
Ph	$C{\equiv}C(CH_2)_7Me$	I	–	56	[93]

Substituted pyrroles **148** are obtained by the reaction of ethyl 2-[1-(2-tosylhydrazono)ethyl]pent-4-ynoate (**147**) with aryl iodides in the presence of potassium carbonate and tetrakis(triphenylphosphine)palladium(0) (Scheme 65).[94] More recently, the beneficial effect of carbon dioxide in promoting the reoxidation of palladium(0) has been reported in a study describing a new synthesis of pyrrole-2-acetic esters by palladium-catalyzed oxidative carbonylation.[95]

Scheme 65 Palladium-Catalyzed Synthesis of 1,2,3,5-Tetrasubstituted Pyrroles from Alkynyl Hydrazones and Aryl Iodides[94]

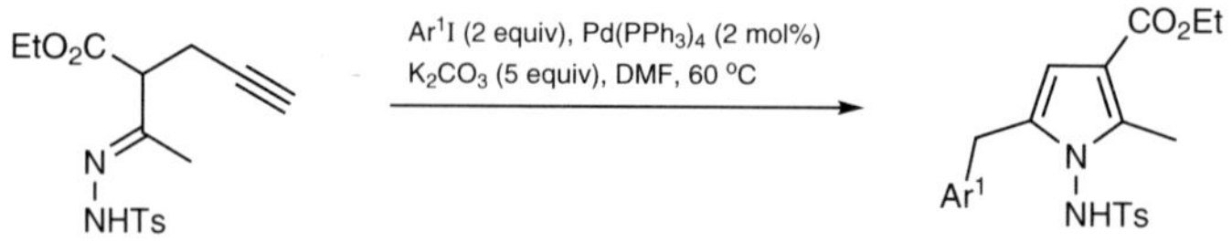

Ar^1I (2 equiv), $Pd(PPh_3)_4$ (2 mol%)
K_2CO_3 (5 equiv), DMF, 60 °C

147 → **148**

Ar^1	Time (h)	Yield (%)	Ref
$4\text{-}ClC_6H_4$	8	55	[94]
$4\text{-}MeO_2CC_6H_4$	2.5	50	[94]
$3\text{-}ClC_6H_4$	6	52	[94]
$3\text{-}F_3CC_6H_4$	6	48	[94]
$3\text{-}FC_6H_4$	7	65	[94]
$4\text{-}AcC_6H_4$	4	60	[94]
$4\text{-}O_2NC_6H_4$	4	63	[94]
Ph	5	70	[94]
3-Tol	5	62	[94]
1-naphthyl	5	61	[94]

2,3-Diphenyl-1*H*-indole (144, **$Ar^1 = R^1 =$ Ph); Typical Procedure:**[92]
MeCN (2.0 mL), 2,2,2-trifluoro-*N*-[2-(phenylethynyl)phenyl]acetamide (**143**, Ar^1 = Ph; 100 mg, 0.346 mmol), PhBr (0.519 mmol), Cs_2CO_3 (0.519 mmol), and $Pd(PPh_3)_4$ (5 mol%) were added to a 50-mL reaction tube containing a magnetic stirrer bar. The mixture was stirred for 30 min at 100 °C under argon, cooled to rt, diluted with EtOAc, and washed with H_2O. The organic phase was dried (Na_2SO_4) and concentrated under reduced pressure. The residue was purified by chromatography (hexane/EtOAc 85:15); yield: 87 mg (93%).

3,4-Diphenylisoquinoline (146, **$R^1 = R^2 =$ Ph); Typical Procedure:**[93]
A mixture of DMF (5 mL), $Pd(PPh_3)_4$ (5 mol%), K_2CO_3 (5 equiv), *N*-*tert*-butyl-2-(phenylethynyl)benzaldimine (**145**, R^1 = Ph; 65 mg, 0.25 mmol), and PhI (5 equiv) was flushed with argon at rt for 5 min and then heated to 100 °C under stirring for 12 h. The mixture was then cooled to rt, diluted with Et_2O, and washed with brine. The aqueous layer was re-extracted with Et_2O. The organic layers were combined, dried ($MgSO_4$), and filtered, and the solvent was removed under reduced pressure. The residue was purified by column chromatography (silica gel, hexane/EtOAc 10:1); yield: 34 mg (49%).

Ethyl 5-(3-Fluorobenzyl)-2-methyl-1-(tosylamino)-1*H*-pyrrole-3-carboxylate (148, **$Ar^1 = 3\text{-}FC_6H_4$); Typical Procedure:**[94]
A soln of ethyl 2-[1-(2-tosylhydrazono)ethyl]pent-4-ynoate (**147**; 229 mg, 0.68 mmol) in DMF (3 mL), 1-fluoro-3-iodobenzene (2 equiv), K_2CO_3 (5 equiv), and $Pd(PPh_3)_4$ (2 mol%) was stirred at 60 °C for 7 h under N_2, cooled, diluted with EtOAc, and washed with H_2O. The organic layer was dried (Na_2SO_4) and concentrated under reduced pressure. The residue was purified by chromatography (silica gel, hexane/EtOAc 4:1); yield: 190 mg (65%).

1.8.2.10 Carboamination of Alkenes

The domino carboamination of amines tethered to a double bond in the presence of aryl halides gives functionalized azacyclic systems.

1.8.2.10.1 Palladium Catalysis

Alkenes bearing a pendant nitrogen functionality undergo palladium-catalyzed carboamination reactions with aryl or alkenyl halides (that may be electron-rich, -neutral, or -poor) in the presence of a ligand, and base,[96] resulting in the formation of substituted nitrogen heterocycles such as pyrrolidines,[97] piperazines,[98] and imidazolidinones.[99] Functional group tolerance is improved by the use of a weak base such as cesium carbonate or potassium phosphate, allowing the use of aryl trifluoromethanesulfonates as electrophiles.

In particular, starting from *N*-acyl-, *N*-*tert*-butoxycarbonyl-, or *N*-benzyloxycarbonyl-γ-aminoalkenes **149**, the carboamination with palladium(II) acetate and bis[2-(diphenylphosphino)phenyl] ether (**110**, DPEphos) affords *trans*-2,3- or *cis*-2,5-disubstituted pyrrolidines **150**, with good diastereoselectivity in all cases (Scheme 66).[97]

Scheme 66 Palladium-Catalyzed Carboamination of γ-Aminoalkenes To Give Pyrrolidines[97]

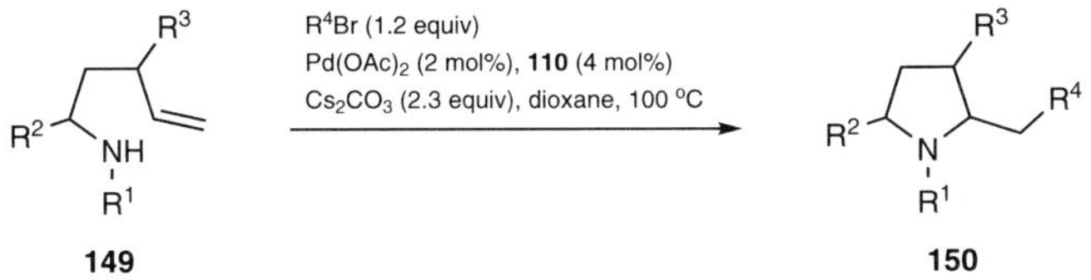

R^4Br (1.2 equiv)
$Pd(OAc)_2$ (2 mol%), **110** (4 mol%)
Cs_2CO_3 (2.3 equiv), dioxane, 100 °C

149 → **150**

R^1	R^2	R^3	R^4	Time (h)	Yield (%)	Ref
Boc	H	H	1-naphthyl	15	75	[97]
Boc	H	H	4-t-BuC_6H_4	27	82	[97]
Boc	H	H	4-OHCC_6H_4	20	78[a]	[97]
Boc	H	H	4-AcC_6H_4	18	76[a]	[97]
Boc	H	H	2-$MeO_2CC_6H_4$	28	71	[97]
Ac	H	H	1-naphthyl	18	79	[97]
Ac	H	H	4-$O_2NC_6H_4$	18	76[b]	[97]
Cbz	H	H	1-naphthyl	16	88	[97]
Cbz	H	H	2-$MeO_2CC_6H_4$	18	88[a]	[97]
Cbz	H	Me	4-$AcOCH_2C_6H_4$	20	80 (12:1)[c]	[97]
Boc	H	Me	4-$MeOC_6H_4$	16	76 (15:1)[c]	[97]
Boc	H	Me	2-$MeO_2CC_6H_4$	18	73 (14:1)[c]	[97]
Boc	Ph	H	3-$O_2NC_6H_4$	16	75 (>20:1)[d]	[97]
Cbz	Ph	H	3-pyridyl	18	74 (>20:1)[d]	[97]

[a] Reaction carried out with 4 mol% of $Pd_2(dba)_3$.
[b] Reaction carried out in DME as solvent at 85 °C.
[c] dr of *trans* product.
[d] dr of *cis* product.

Analogously, starting from chiral ethylenediamines **151**, *cis*-2,6-disubstituted piperazines **152** are obtained. The substrates can be easily prepared from amino acids and N-substituted allylamines in four steps (Scheme 67).[98]

for references see p 499

Scheme 67 Palladium-Catalyzed Carboamination of Aminoalkenes To Give Piperazines[98]

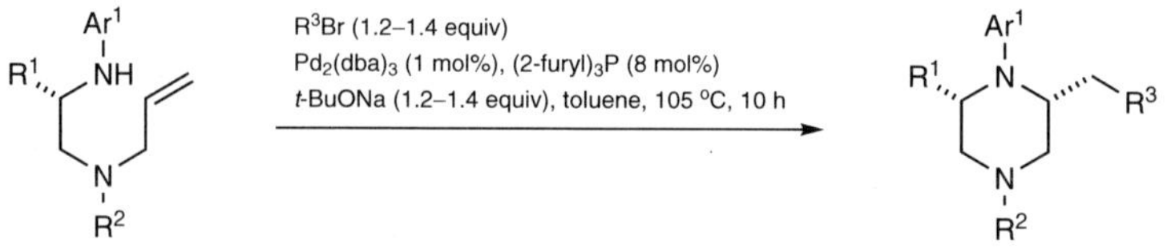

Ar^1	R^1	R^2	R^3	Yield (%)	dr	ee (%)	Ref
Ph	Bn	Bn	4-*t*-BuC_6H_4	63	>20:1	99	[98]
Ph	Bn	Bn	4-$MeOC_6H_4$	62	>20:1	99	[98]
Ph	Bn	Bn	4-*t*-$BuO_2CC_6H_4$	59	>20:1	98	[98]
Ph	Bn	Bn	(*E*)-CH=CHPh	73[a]	>20:1	95	[98]
Ph	Bn	Bn	(*E*)-CH=CHTMS	59[a]	>20:1	97	[98]
Ph	iPr	Bn	4-*t*-BuC_6H_4	51[b]	>20:1	99	[98]
Ph	iPr	Bn	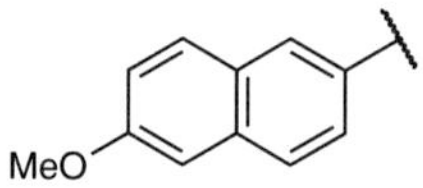 6-MeO-2-naphthyl (structure)	50	>20:1	99	[98]
4-NCC_6H_4	iPr	Bn	4-Tol	71	>14:1	97	[98]
4-NCC_6H_4	iPr	Bn	4-$F_3CC_6H_4$	69	>14:1	98	[98]
Ph	Me	$CH_2CH=CH_2$	4-$MeOC_6H_4$	53	>20:1	99	[98]
Ph	iBu	Bn	4-PhC_6H_4	57	>20:1	99	[98]

[a] Reaction carried out with 2 mol% $Pd_2(dba)_3$ and 16 mol% tri-2-furylphosphine at 90 °C.
[b] Reaction carried out at 90 °C.

Carboamination of *N*-allylureas **153** with aryl bromides affords imidazolidinones **154** with good diastereoselectivity using tris(dibenzylideneacetone)dipalladium(0) and 4,5-bis(diphenylphosphino)-9,9-dimethylxanthene (**41**, Xantphos) (Scheme 68).[99] The observed major stereoisomer arises from cyclization via the more thermodynamically stable conformation, in which the alkenyl group is in a pseudoaxial position to minimize allylic strain interactions with the urea moiety.

Scheme 68 Palladium-Catalyzed Carboamination of *N*-Allylureas To Give Imidazolidinones[99]

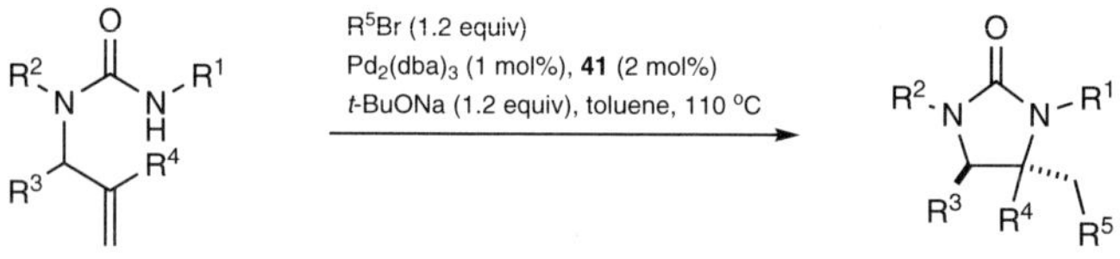

R^1	R^2	R^3	R^4	R^5	Time (h)	Yield (%)	dr	Ref
Et	Me	H	H	2-naphthyl	2	73	–	[99]
Bn	Me	H	H	4-NCC_6H_4	8	80	–	[99]
Ph	Me	H	H	2-naphthyl	1	97	–	[99]
Ph	Me	H	H	3-pyridyl	0.5	83	–	[99]
Ph	Me	H	H	4-$F_3CC_6H_4$	1	92	–	[99]

R^1	R^2	R^3	R^4	R^5	Time (h)	Yield (%)	dr	Ref
Ph	Me	H	H	4-BzC_6H_4	5	85	–	[99]
4-$MeOC_6H_4$	Bn	H	H	4-*t*-$BuO_2CC_6H_4$	8	75	–	[99]
4-$MeOC_6H_4$	Bn	H	H	2-Tol	8	71	–	[99]
4-$MeOC_6H_4$	Bn	H	H	4-*t*-BuC_6H_4	0.5	96	–	[99]
4-$MeOC_6H_4$	Bn	Me	H	4-NCC_6H_4	1	88	12:1	[99]
4-$MeOC_6H_4$	Bn	iPr	H	4-Tol	1	83	>20:1	[99]
4-$MeOC_6H_4$	$(CH_2)_4$		H	Ph	1	78	20:1	[99]
4-$MeOC_6H_4$	$(CH_2)_3$		H	3-$F_3CC_6H_4$	1	87	>20:1	[99]
Ph	Et	H	Me	2-Tol	5	89	–	[99]
4-$MeOC_6H_4$	Bn	H	Me	4-Tol	7.5	97	–	[99]
4-$MeOC_6H_4$	Bn	Me	H	(*E*)-CH=CHTMS	1	91	10:1	[99]
4-$MeOC_6H_4$	$(CH_2)_4$		H	(*E*)-CH=CHMe	40 min	87	1.6:1	[99]

Substituted Pyrrolidines 150; **General Procedure:**[97]
A flame-dried Schlenk tube equipped with a magnetic stirrer bar was cooled under a stream of N_2 and charged with an aryl bromide (1.2 equiv), $Pd(OAc)_2$ (2 mol%), bis[2-(diphenylphosphino)phenyl] ether (**110**, DPEphos; 4 mol%), and Cs_2CO_3 (2.3 equiv). The tube was purged with N_2 and a soln of N-protected amine **149** (1 equiv) in dioxane (5 mL/mmol substrate) was then added. The resulting mixture was heated at 100 °C with stirring until the starting material had been consumed, as determined by GC analysis. The mixture was cooled to rt and sat. aq NH_4Cl (1 mL) and EtOAc (1 mL) were added. The layers were separated, and the aqueous layer was extracted with EtOAc (3 × 5 mL). The combined organic layers were dried (Na_2SO_4), filtered, and concentrated under reduced pressure. The crude product was purified by flash chromatography (silica gel).

2,4,6-Trisubstituted *N*-Arylpiperazines 152; **General Procedure:**[98]
A flame-dried Schlenk tube equipped with a magnetic stirrer bar was cooled under a stream of N_2 and charged with $Pd_2(dba)_3$ (1 mol%), $(2\text{-furyl})_3P$ (8 mol%), *t*-BuONa (1.2–1.4 equiv), and an aryl bromide (1.2–1.4 equiv). The Schlenk tube was purged with N_2 and an amine **151** was added as a soln in toluene (2.5 mL/0.5 mmol substrate). The Schlenk tube was then heated at 105 °C with stirring until the starting material had been consumed, as judged by ^1H NMR analysis of a sample taken from the mixture. The mixture was then cooled to rt. Sat. aq NH_4Cl was added, and the resulting mixture was extracted with EtOAc (3 × 5 mL). The combined organic layers were dried (Na_2SO_4), filtered, and concentrated under reduced pressure. The crude product was purified by flash chromatography (silica gel).

Substituted Imidazolidin-2-ones 154; **General Procedure:**[99]
An oven- or flame-dried Schlenk tube equipped with a stirrer bar was cooled under a stream of N_2 and charged with $Pd_2(dba)_3$ (1 mol%), Xantphos (**41**; 2 mol%), *t*-BuONa (1.2 equiv), *N*-allylurea **153** (1 equiv), and an aryl bromide (1.2 equiv). The tube was purged with N_2, and undecane (0.125 equiv; internal standard) and toluene (4 mL/mmol urea substrate) were added. If the acyclic urea and/or the aryl bromide were oils they were added at the same time as toluene. The Schlenk tube was then heated at 110 °C with stirring until the starting material had been consumed, as judged by GC or ^1H NMR analysis of a sample taken from the mixture. The mixture was then cooled to rt. Sat. aq NH_4Cl was added, and the mixture was extracted with CH_2Cl_2 or EtOAc. The combined organic extracts were dried (Na_2SO_4), filtered, and concentrated under reduced pressure. The crude product was purified by flash chromatography (silica gel).

for references see p 499

1.8.2.11 Applications in the Syntheses of Natural Products and Drug Molecules

Application of the hydroamination of alkynes provides straightforward access to the nitrogen heterocyclic portion of (−)-quinocarcin, a pentacyclic tetrahydroisoquinoline alkaloid with antiproliferative activity against lymphocytic leukemia. Cationic gold complex **156** is used as catalyst to convert aminoalkyne **155** into 2a,3,4,5-tetrahydro-2*H*-furo[2,3,4-*ij*]isoquinolin-4-yl]pyrrolidine-2,4-dicarboxylate derivative **157**. Control of the regioselectivity through a 6-*endo-dig* cyclization is fundamental to the success of this strategy (Scheme 69).[100]

Scheme 69 Synthesis of a (−)-Quinocarcin Fragment by Gold-Catalyzed Hydroamination of an Aminoalkyne[100]

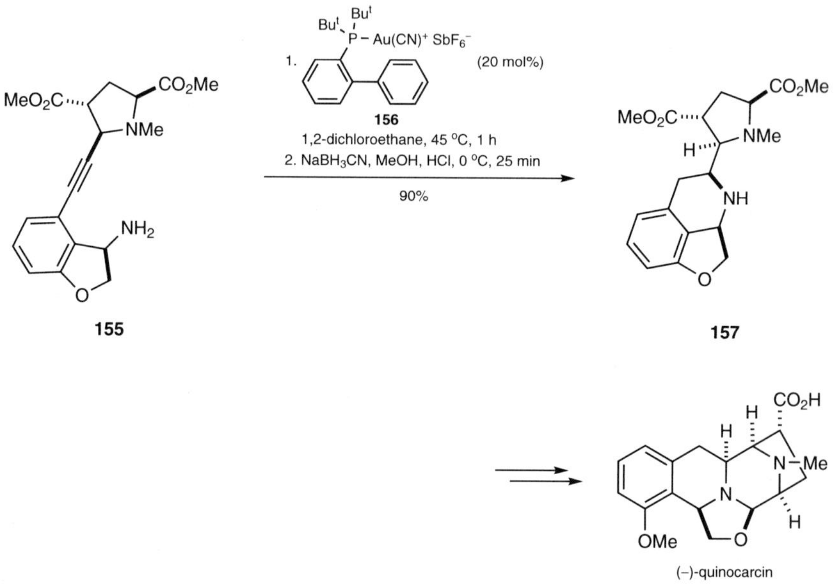

Likewise, a fine application of this pathway is devoted to the preparation of benzo[*c*]phenanthridine alkaloids such as nitidine via a tandem reaction which consists of hydroamination of alkynylbenzyl carbamates (e.g., **158**) bearing an acetal moiety followed by Michael addition of the resulting enecarbamates to give phenanthridine derivatives (e.g., **159**) (Scheme 70).[32]

Scheme 70 Synthesis of a Nitidine Precursor by Gold/Silver-Catalyzed Hydroamination of an Alkynylbenzyl Carbamate[32]

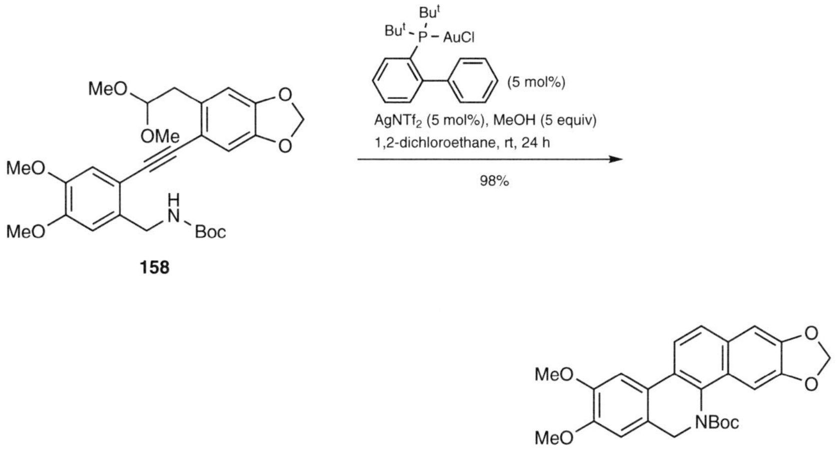

The intramolecular hydroamination between the alkyne moiety and the secondary amino group of pentacyclic ester **160**, to give the enamine intermediate **161**, is the key step in an approach to the hexacyclic core of communesin B, which belongs to a class of biologically active *Penicillium* metabolites (Scheme 71).[101]

Scheme 71 Synthesis of a Communesin B Precursor by Gold/Silver-Catalyzed Hydroamination[101]

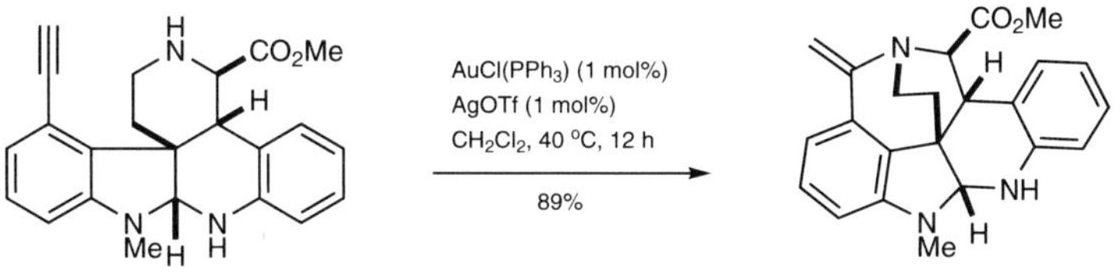

Palladium-catalyzed alkene aminations have been employed as central strategic steps in the total synthesis of a number of natural products and biologically active compounds. The tetracyclic compound (+)-agelastatin A, an alkaloid exhibiting exceptional biological activities such as nanomolar activity against a broad range of cancer cell lines, is obtained through the regio- and enantioselective formation of the precursor pyrrole–piperazinone nucleus **164** from cyclopentenyl-substituted pyrrole-2-carboxamide **162** using tris(dibenzylideneacetone)dipalladium(0) and ligand **163** (Scheme 72).[102]

for references see p 499

Scheme 72 Synthesis of an Agelastatin A Precursor by Palladium-Catalyzed Hydroamination of an Alkene[102]

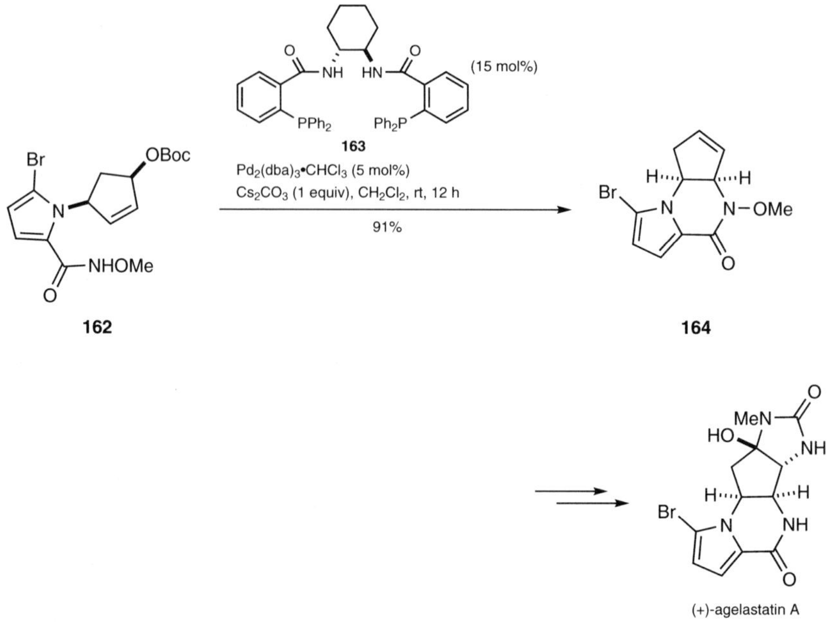

In the field of biologically active molecules, copper-catalyzed intramolecular amination of aryl halides is applied to approach β-carbolinones,[103] a class of compounds used in antitumor treatment due to their inhibition of cell proliferation, and also for the preparation of natural antitumor agents such as the duocarmycins[104] and yatakemycin A.[105] The use of copper catalysis is essential; in fact, the comparative reaction with a palladium catalyst often fails due to the presence of several reacting groups. One of the most impressive examples of the efficiency of copper-catalyzed aryl amination has been achieved for the preparation of yatakemycin A, a potent antitumor antibiotic, where no less than five C—N bonds are formed with excellent yields using ligandless conditions (Scheme 73). An analogous inexpensive and efficient catalytic method is applied for the intramolecular N-arylation of aryl bromides to prepare promazine drugs, which are used in clinical treatment for psychotropic diseases.[106]

Scheme 73 Formation of Five C—N Bonds in the Synthesis of Yatakemycin A[105]

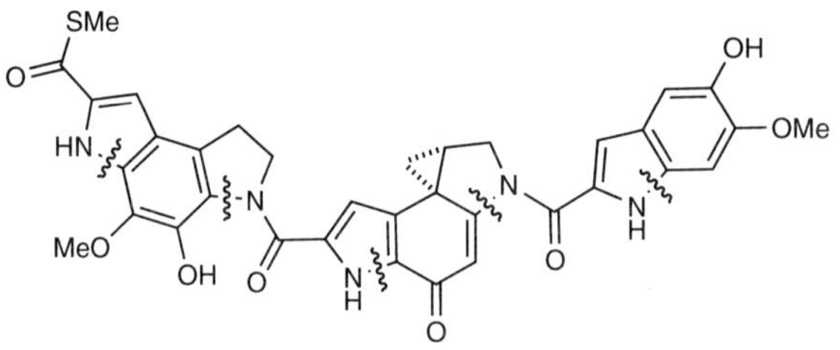

A fine example of a copper-mediated macrocyclization is reported as a late step in the syntheses of macrolactams reblastatin (**167**, R^1 = OMe)[85] and autolytimycin (**167**, R^1 = H),[84] potent inhibitors of heat shock protein 90, an important therapeutic target for cancer treatment. The corresponding drug precursors **166** are obtained by treatment of amides

165 with copper(I) iodide, *N*,*N*,*N*′,*N*′-tetramethylethylenediamine, and potassium carbonate in toluene at 100 °C (Scheme 74).

Scheme 74 Copper-Catalyzed Synthesis of Intermediates of Reblastatin and Autolytimycin by Amination of Aryl Bromides[84]

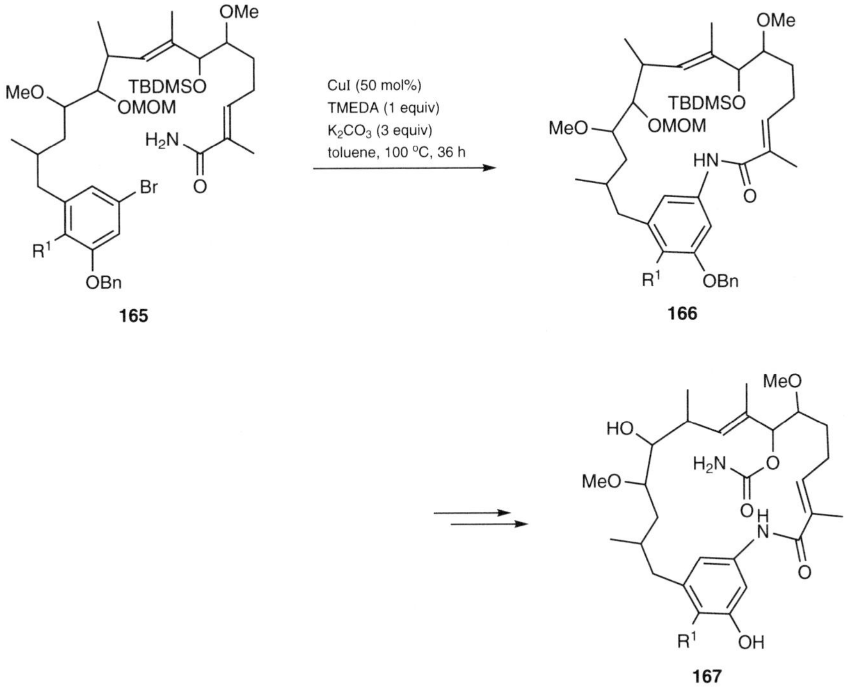

CuI (50 mol%)
TMEDA (1 equiv)
K_2CO_3 (3 equiv)
toluene, 100 °C, 36 h

R^1	Yield (%) of **166**	Ref
OMe	>80	[84]
H	80	[84]

Lotrafiban, a potent GPIIb/IIIa receptor antagonist which inhibits platelet aggregation, is efficiently prepared by exploiting the intramolecular aryl amination of aryl iodide **168** to obtain the key intermediate **169** in enantiopure form (Scheme 75).[107] Several copper-catalyzed aryl halide aminations have been applied for the synthesis of cyclopeptide alkaloids such as paliurine F.[108]

for references see p 499

Scheme 75 Copper-Catalyzed Synthesis of an Intermediate of Lotrafiban by Amination of an Aryl Iodide[107]

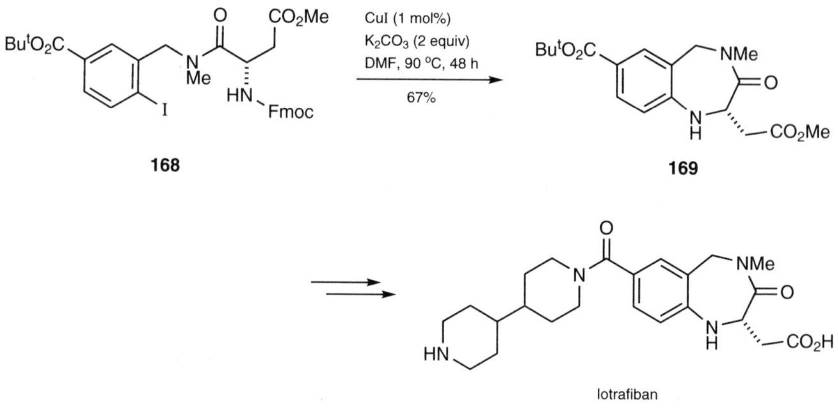

Bu^tO_2C ... 168 → CuI (1 mol%), K_2CO_3 (2 equiv), DMF, 90 °C, 48 h; 67% → 169 → lotrafiban

Palladium-catalyzed intramolecular aminations of aryl halides are key steps in the synthesis of the alkaloid tarpane[109] and the tricyclic skeleton of oxcarbazepine, one of the most prescribed drugs for the treatment of epilepsy due to the improved tolerability profile compared to carbamazepine. In this case, 1-(2-aminophenyl)-2-(2-bromophenyl)ethan-1-one (**170**) is cyclized in the presence of palladium(II) acetate and 2,2′-bis(diphenylphosphino)-1,1′-binaphthyl to give the oxcarbazepine precursor **171** (Scheme 76).[110] The reaction fails under copper-mediated Ullmann-type reaction conditions.

Scheme 76 Palladium-Catalyzed Synthesis of an Intermediate of Oxcarbazepine by Amination of an Aryl Bromide[110]

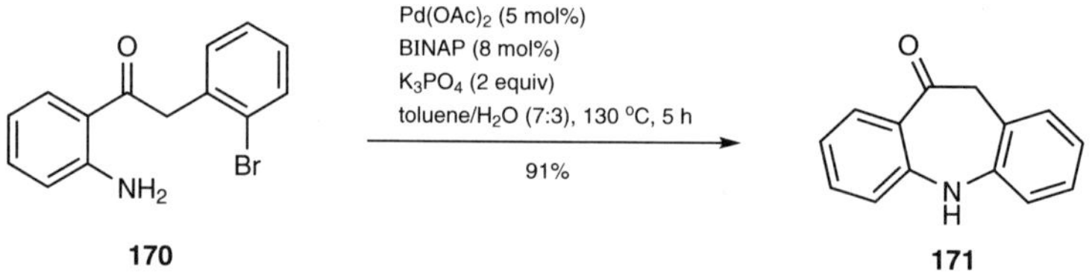

170 → $Pd(OAc)_2$ (5 mol%), BINAP (8 mol%), K_3PO_4 (2 equiv), toluene/H_2O (7:3), 130 °C, 5 h; 91% → 171

Stereoselective formation of the atropoisomeric *N*-(2-*tert*-butylphenyl)quinolinone **174**, an intermediate for norepinephrine transporter (NET) inhibitor, a useful treatment of mental disorders such as depression, is obtained by reaction of iodophenyl amide **172** in the presence of palladium(II) acetate and (*R*)-4,4′-bi-1,3-benzodioxole-5,5′-diylbis(diphenylphosphine) [**173**, (*R*)-SEGPHOS] (Scheme 77).[111]

Scheme 77 Palladium-Catalyzed Synthesis of an Intermediate of NET Inhibitor[111]

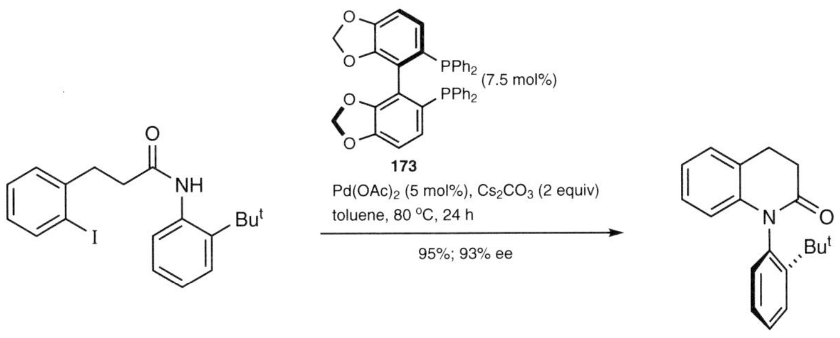

173 (7.5 mol%)
$Pd(OAc)_2$ (5 mol%), Cs_2CO_3 (2 equiv)
toluene, 80 °C, 24 h
95%; 93% ee

172 174

A useful application of the stereoselective carboamination process is in the synthesis of the alkaloid (+)-preussin,[112] known for its antifungal properties[113] and more recently indicated also for its antiviral and antitumor activity.[114] In this case, the intermediate (2*S*,3*S*,5*R*)-2-benzyl-1-(*tert*-butoxycarbonyl)-3-(*tert*-butyldimethylsiloxy)-5-nonylpyrrolidine (**176**) is formed from carbamate **175** and bromobenzene in the presence of palladium(II) acetate and bis[2-(diphenylphosphino)phenyl] ether (**110**, DPEphos) (Scheme 78).

Scheme 78 Palladium-Catalyzed Synthesis of an Intermediate of (+)-Preussin[112]

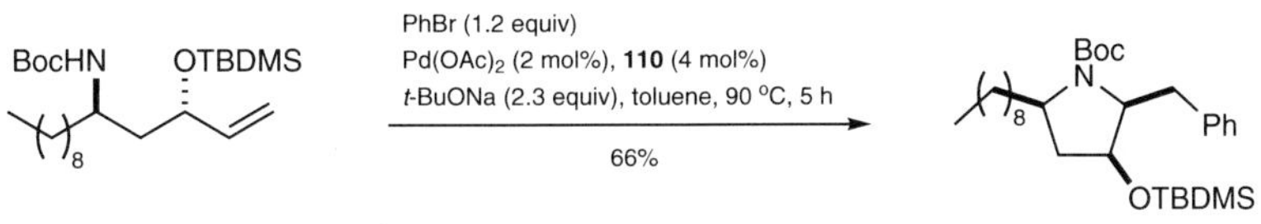

PhBr (1.2 equiv)
$Pd(OAc)_2$ (2 mol%), **110** (4 mol%)
t-BuONa (2.3 equiv), toluene, 90 °C, 5 h
66%

175 176

Dimethyl (2*S*,4*R*,5*R*)-1-Methyl-5-[(2a*R*,4*S*)-2a,3,4,5-tetrahydro-2*H*-furo[2,3,4-*ij*]isoquinolin-4-yl]pyrrolidine-2,4-dicarboxylate (157); Typical Procedure:[100]

To a stirred soln of dimethyl pyrrolidine-2,4-dicarboxylate **155** (0.17 mmol) in 1,2-dichloroethane (4 mL) under argon was added the cationic gold catalyst **156** (0.03 mmol, as reported) at rt. The mixture was warmed to 45 °C and stirred for 1 h. After the mixture had been cooled to 0 °C, $NaBH_3CN$ (0.33 mmol), MeOH (2 mL), and 1 M aq HCl (0.5 mL) were successively added to the mixture. Stirring was continued for 25 min at this temperature. The soln was made neutral with sat. aq $NaHCO_3$ at 0 °C and extracted with CH_2Cl_2. The extract was washed with brine and dried ($MgSO_4$). The filtrate was concentrated under reduced pressure to leave an oily residue, which was purified by column chromatography (N—H silica gel, hexane/EtOAc 5:1) to give the product as a pale yellow oil; yield: 90%.

***tert*-Butyl 2,3-Dimethoxy[1,3]benzodioxolo[5,6-*c*]phenanthridine-12(13*H*)-carboxylate (159); Typical Procedure:**[32]

To a soln of alkynylcarbamate **158** (0.14 mmol) in 1,2-dichloroethane (2 mL) was added a mixture of [(biphenyl-2-yl)-di-*tert*-butylphosphine]chlorogold(I) (5 mol%), $AgNTf_2$ (5 mol%), MeOH (5 equiv), and 1,2-dichloroethane (1 mL) at rt. After the mixture had been stirred for 24 h, sat. aq $NaHCO_3$ was added and the resultant mixture was extracted with $CHCl_3$. The combined organic extracts were washed with brine, dried (Na_2SO_4), filtered, and concentrated under reduced pressure. The residue was purified by column chromatography (silica gel, hexane/EtOAc 4:1) to afford the product as a white powder; yield: 98%.

Methyl (4a*R*,10*R*,10a*S*,13*S*,15*R*)-4-Methyl-14-methylene-4,4a,5,10,11,12-hexahydro-14*H*-10,13-methanoazepino[5′,4′,3′:3,4]indolo[2,3-*b*]quinoline-15-carboxylate (161); **Typical Procedure:**[101]
To a soln of methyl 1,2,3,4,9,9a,10,14b-octahydrobenzo[*c*]indolo[3,2-*j*][2,6]naphthyridine-1-carboxylate **160** (0.198 mmol) in CH_2Cl_2 (1.90 mL) was added AuCl(PPh_3) (2 μmol) and AgOTf (2 μmol). The soln was stirred overnight at 40 °C and concentrated. Purification by chromatography (silica gel, EtOAc/hexane 1:1) afforded the product as a white solid; yield: 89%.

(3a*R*,9a*S*)-8-Bromo-4-methoxy-1,3a,4,9a-tetrahydro-5*H*-cyclopenta[*e*]pyrrolo[1,2-*a*]pyrazin-5-one (164); **Typical Procedure:**[102]
A soln of $Pd_2(dba)_3$•$CHCl_3$ (0.25 μmol, as reported) and ligand **163** (0.75 μmol, as reported) in CH_2Cl_2 (1 mL), which had been stirred at rt for 10 min, was added to a mixture of *N*-methoxy-1*H*-pyrrole-2-carboxamide **162** (0.05 mmol) and Cs_2CO_3 (0.05 mmol) under argon. The mixture was stirred at rt for 12 h, then filtered through a Celite cake. The solvent was removed under reduced pressure, and the residue was purified by flash column chromatography (silica gel, light petroleum ether/Et_2O 8:1, then CH_2Cl_2/MeOH 40:1) to give the product as a white solid; yield: 91%.

Reblastatin and Autolytimycin Precursors 166; **Typical Procedure:**[84]
To a soln of amide **165** (12 μmol) in toluene (0.8 mL) in a sealed tube was added K_2CO_3 (4.96 mg, 0.0359 mmol). CuI (1.1 mg, 6.0 μmol) and TMEDA (12 μmol) were added sequentially and the tube was sealed. The green suspension was heated at 100 °C for 36 h and filtered through a plug of silica gel, washing with EtOAc. The filtrate was concentrated and the residue was purified by flash chromatography (EtOAc/hexane) to yield the product as a clear glass.

***tert*-Butyl (*R*)-2-(2-Methoxy-2-oxoethyl)-4-methyl-3-oxo-2,3,4,5-tetrahydro-1*H*-benzo[*e*]-[1,4]diazepine-7-carboxylate** (169); **Typical Procedure:**[107]
tert-Butyl 4-iodobenzoate **168** (6.06 mmol) was dissolved in anhyd DMF (20 mL). To this soln were added anhyd K_2CO_3 (13.4 mmol) and CuI (0.54 mmol). After the resultant mixture was stirred at 90 °C for 48 h under N_2, the solvent was evaporated under reduced pressure and the residue was dissolved with $CHCl_3$ (20 mL) and H_2O (10 mL). The organic layer was separated, and the aqueous layer was extracted with $CHCl_3$. The combined organic layers were washed with brine and dried (Na_2SO_4) before the solvent was evaporated. The residue was purified by chromatography (EtOAc/hexane 1:1); yield: 67%.

5,11-Dihydro-10*H*-dibenzo[*b*,*f*]azepin-10-one (171); **Typical Procedure:**[110]
A soln of 1-(2-aminophenyl)-2-(2-bromophenyl)ethan-1-one (**170**; 0.347 mmol), $Pd(OAc)_2$ (0.017 mmol), BINAP (0.027 mmol), previously ground K_3PO_4 (150 mg, 0.683 mmol), toluene (3.5 mL), and H_2O (1.5 mL) was heated at 130 °C. After 5 h, the mixture was partitioned between H_2O and CH_2Cl_2. The organic layer was dried and concentrated under reduced pressure. The crude product was purified by flash chromatography (Et_2O/CH_2Cl_2 5:95) to give the product as yellow needles; yield: 91%.

(a*R*)-1-[2-(*tert*-Butyl)phenyl]-3,4-dihydroquinolin-2(1*H*)-one (174); **Typical Procedure:**[111]
Cs_2CO_3 (2.0 mmol) was added to *N*-(2-*tert*-butylphenyl)-3-(2-iodophenyl)propanamide (**172**; 1.0 mmol) in toluene (3.0 mL) under argon. After the mixture had been stirred for 5 min at rt, the suspension of $Pd(OAc)_2$ (0.05 mmol) and (*R*)-4,4′-bi-1,3-benzodioxole-5,5′-diylbis(diphenylphosphine) (**173**; 0.075 mmol) in toluene (2.0 mL) was added. The mixture was vigorously stirred for 24 h at 80 °C, poured into 2% aq HCl, and extracted with EtOAc. The extracts were washed with brine, dried ($MgSO_4$), and concentrated to dryness. The residue was purified by column chromatography (hexane/EtOAc); yield: 95%; 93% ee.

(2S,3S,5R)-2-Benzyl-1-(*tert*-butoxycarbonyl)-3-(*tert*-butyldimethylsiloxy)-5-nonylpyrrolidine (176); Typical Procedure:[112]
A flame-dried Schlenk tube equipped with a magnetic stirrer bar was cooled under a stream of N_2 and charged with *tert*-butyl carbamate **175** (0.25 mmol), $Pd(OAc)_2$ (5 μmol), bis[2-(diphenylphosphino)phenyl] ether (**110**, DPEphos; 10 μmol), *t*-BuONa (0.575 mmol), and bromobenzene (0.3 mmol). The tube was purged with N_2 and toluene (1 mL) was added by syringe. The resulting mixture was heated at 90 °C with stirring for 5 h until the starting material had been consumed, as determined by GC analysis. The mixture was cooled to rt, and sat. aq NH_4Cl and EtOAc were added. The layers were separated, and the aqueous layer was extracted with EtOAc (3 × 5 mL). The combined organic layers were dried (Na_2SO_4), filtered, and concentrated under reduced pressure. The crude product was purified by flash chromatography (EtOAc/hexane 2.5:97.5) to afford the product as a colorless oil; yield: 66%.

1.8.3 C—O Bond Formation

Different types of C—O functionality may be formed by addition of oxygen nucleophiles to unsaturated bonds. In particular, cyclization may afford different oxygenated rings, such as furans, pyrans, and lactones, both individually and fused to aromatic rings. In all cases, the initial step is the activation of the unsaturated system by the metal catalyst, followed by nucleophilic attack of the oxygen atom. The reaction types considered are hydroalkoxylation, hydroacyloxylation, alkoxylation, oxidative alkoxylation, alkoxycarbonylation, and carboalkoxylation.

1.8.3.1 Hydroalkoxylation of Alkynes

Among the cyclization reactions involving formation of C—O bonds, the hydroalkoxylation of alkynes is the most fruitful because several different transition-metal catalysts can be used to this end. Nevertheless, differences among the metals cannot be overlooked. For example, a comparison between palladium and platinum catalysts in the activation of alkynes toward the addition of alcohols shows a much slower reaction with the homologous platinum complex, probably due to the higher stability of the Pt—C bond in the vinylplatinum complex.

1.8.3.1.1 Copper Catalysis

In comparison with silver and gold (see Section 1.8.3.1.2), only a few methods use copper salts as catalysts in intramolecular hydroalkoxylation. For example, after the pioneering work on palladium-catalyzed cyclization of β,γ-acetylenic ketones resulting in the formation of furans,[115] alkynyl ketones **177** have been used as substrates in hydroalkoxylation reactions under copper(I) catalysis, exploiting the carbonyl group as nucleophile to attack the unsaturated moiety. Subsequent aromatization of the intermediate affords the 2,5-disubstituted furans **178**. The method is compatible with both acid- and base-sensitive groups and also with 3-alkylsulfanyl substituents (Scheme 79).[116]

for references see p 499

Scheme 79 Copper-Catalyzed Cyclization of Alkynyl Ketones to Furans[116]

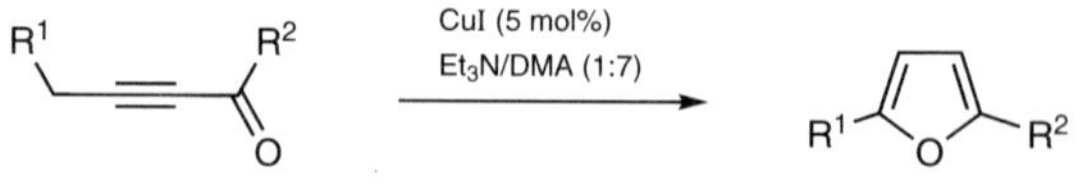

R^1	R^2	Temp (°C)	Time (h)	Yield (%)	Ref
H	Ph	80	3	85	[116]
$(CH_2)_4Me$	Me	100	27	94	[116]
Pr	Ph	100	16	92	[116]
Pr	$CH{=}CMe_2$	100	43	88[a]	[116]
Pr	$(CH_2)CO_2Me$	100	21	91	[116]
OMe	$(CH_2)_5Me$	80	21	78	[116]
CH_2OTHP	$(CH_2)_3OH$	100	9	88	[116]
OTHP	*t*-Bu	100	22	75	[116]

[a] Reaction carried out with 10 mol% of catalyst.

2,5-Disubstituted Furans 178; General Procedure:[116]

A mixture of alkynyl ketone **177** (1 mmol), CuI (0.05 mmol), anhyd DMA (2.2 mL), and Et_3N (0.3 mL) was stirred in a Wheaton microreactor (3 mL) under argon at the required temperature until the reaction was complete (monitored by GC/MS and TLC). The mixture was cooled, diluted with H_2O (15 mL), and extracted with pentane (3 × 5 mL). The combined organic extracts were filtered, dried (anhyd Na_2CO_3), and concentrated under reduced pressure. The residue was purified by chromatography (silica gel, pentane or hexane/EtOAc mixtures).

1.8.3.1.2 Gold Catalysis

Intramolecular coupling of tertiary and secondary *Z*-enynols **179** and **181** provides substituted dihydrofurans **180** and furans **182**, respectively, under gold(III) chloride or cationic gold(I) catalysis in very satisfactory yields. The reactions are carried out under neutral conditions at room temperature using 1 mol% of catalyst. In all cases, only 5-*exo-dig* cyclization is observed and, in the case of products **180**, the *Z*-5-alkylidene-2,5-dihydrofurans are obtained (Scheme 80).[117]

Scheme 80 Gold-Catalyzed Cyclization of *Z*-Enynols to Dihydrofurans or Furans[117]

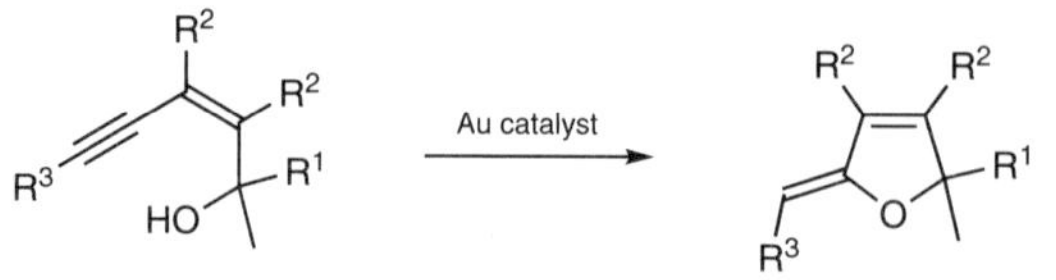

R^1	R^2	R^3	Conditions	Time (h)	Yield (%)	Ref
2-ClC_6H_4	Ph	Ph	$AuCl_3$ (1 mol%), CH_2Cl_2, rt	3	97	[117]
Ph	Ph	cyclohex-1-enyl	$AuCl_3$ (1 mol%), CH_2Cl_2, rt	1	92	[117]
4-FC_6H_4	Ph	Bu	$AuCl_3$ (1 mol%), CH_2Cl_2, rt	1	89	[117]
Ph	Ph	Ph	$AuCl_3$ (1 mol%), CH_2Cl_2, rt	3	89	[117]

R^1	R^2	R^3	Conditions	Time (h)	Yield (%)	Ref
Ph	Ph	Ph	$AuCl_3$ (1 mol%), CH_2Cl_2, rt	3	91	[117]
4-FC_6H_4	Pr	Ph	$AuCl(PPh_3)$ (1 mol%), AgOTf (1 mol%), THF, rt	3	84	[117]
2-thienyl	Ph	4-Tol	$AuCl(PPh_3)$ (1 mol%), AgOTf (1 mol%), THF, rt	3	87	[117]
2-ClC_6H_4	Ph	Bu	$AuCl_3$ (1 mol%), CH_2Cl_2, rt	1	83	[117]

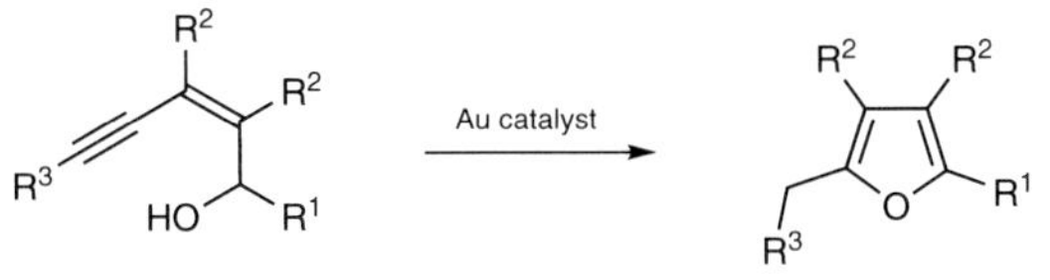

181 → 182 (Au catalyst)

R^1	R^2	R^3	Conditions	Time (h)	Yield (%)	Ref
4-Tol	Pr	Ph	$AuCl_3$ (1 mol%), CH_2Cl_2, rt	4	92	[117]
Ph	Bu	4-$MeOC_6H_4$	$AuCl(PPh_3)$ (1 mol%), AgOTf (1 mol%), THF, rt	3	85	[117]
4-$MeOC_6H_4$	Pr	Bu	$AuCl_3$ (1 mol%), CH_2Cl_2, rt	2	85	[117]
Pr	Ph	4-$MeOC_6H_4$	$AuCl_3$ (1 mol%), CH_2Cl_2, rt	3	79	[117]
4-$F_3CC_6H_4$	Et	4-ClC_6H_4	$AuCl(PPh_3)$ (1 mol%), AgOTf (1 mol%), THF, rt	2	64	[117]

Similarly, starting from alkynyl diols **183**, complex structures such as bicyclic strained ketals **184**, present in many natural systems, are accessible by exploiting a double hydroalkoxylation (Scheme 81).[118]

Scheme 81 Gold-Catalyzed Cyclization of Alkynyl Diols to Bicyclic Ketals[118]

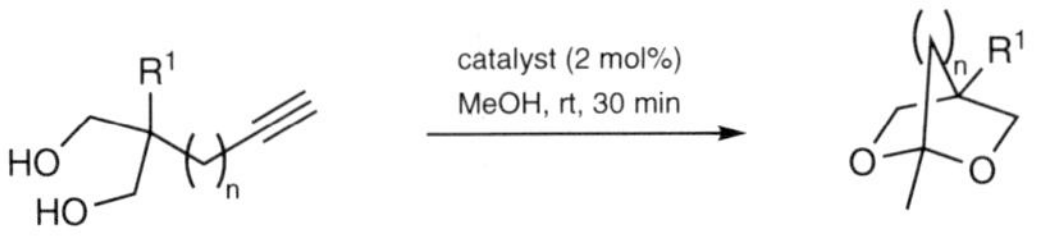

183 → 184 (catalyst (2 mol%), MeOH, rt, 30 min)

R^1	n	Catalyst	Yield (%)	Ref
(*E*)-$CH_2CH{=}CHPh$	1	AuCl	99	[118]
Bn	1	AuCl	99	[118]
Bn	1	$AuCl_3$	99	[118]
Ph	1	AuCl	99	[118]
Ph	1	$AuCl_3$	99	[118]
Bu	1	AuCl	80	[118]
(*E*)-$CH_2CH{=}CHPh$	2	AuCl	82	[118]
$CH_2CH{=}CH_2$	2	AuCl	91	[118]
$CH_2CH{=}CH_2$	1	$AuCl_3$	74[a]	[118]
cyclohex-2-enyl	1	$AuCl_3$	94	[118]

[a] Reaction time was 50 min.

Hydroalkoxylation of *N*-propargylamides **185** under mild reaction conditions results in the formation of 2,5-disubstituted oxazoles **186** through attack of the oxygen nucleophile on the triple bond, followed by isomerization of the exocyclic double bond. Various functional groups are tolerated, but the terminal alkyne is crucial for the reaction (Scheme 82).[119]

Scheme 82 Gold-Catalyzed Cyclization of *N*-Propargylamides to 2,5-Disubstituted Oxazoles[119]

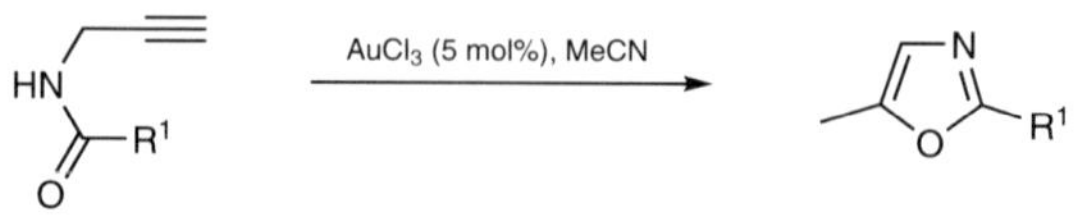

R^1	Temp (°C)	Time (h)	Yield (%)	Ref
Me	45	15	>95[a]	[119]
Ph	20	15	>95[a]	[119]
2-furyl	45	96	88[a]	[119]
1-adamantyl	45	3	98[b]	[119]
(*E*)-CH=CHPh	20	96	91[b]	[119]
CH_2CO_2Me	50	2	40[b]	[119]
$(CH_2)_2CO_2Me$	50	2	86[b]	[119]

[a] Yield determined by ^1H NMR spectroscopy.
[b] Isolated yield.

2,3,4,5-Tetrasubstituted Dihydrofurans 180**; General Procedure:**[117]
To a soln of a tertiary *Z*-enynol **179** (0.2 mmol) in THF (2 mL) was added $AuCl(PPh_3)$ (1 mol%) followed by AgOTf (1 mol%). The resulting soln was stirred at rt until the reaction was complete, as monitored by TLC. The solvent was removed under reduced pressure and the residue was purified by flash chromatography (alumina).

2,3,4,5-Tetrasubstituted Furans 182**; General Procedure:**[117]
A soln of a secondary *Z*-enynol **181** (0.2 mmol) in CH_2Cl_2 (2 mL) was treated with a 0.05 M soln of $AuCl_3$ in MeCN (40 μL, 1 mol%) under N_2. The resulting soln was stirred at rt until the reaction was complete, as monitored by TLC. The solvent was removed under reduced pressure and the residue was purified by flash chromatography (alumina).

Bicyclic Ketals 184**; General Procedure:**[118]
A mixture of alkynyl diol **183** and AuCl or $AuCl_3$ (2 mol%) in degassed MeOH (0.5 M) was stirred under argon at rt. After completion of the reaction, the mixture was filtered through a short pad of Celite (EtOAc) and the solvent was evaporated under reduced pressure to give the product.

2,5-Disubstituted Oxazoles 186**; General Procedure:**[119]
An *N*-propargylamide **185** (1 mmol) was dissolved in MeCN (500 μL) and a 10% soln of $AuCl_3$ (5 mol%) in MeCN was added. The mixture was heated at 50 °C. Conversion of the starting material was monitored via ^1H NMR spectroscopy and the crude product was purified by column chromatography (silica gel, petroleum ether/EtOAc).

1.8.3.1.3 Iridium Catalysis

Intramolecular hydroalkoxylation of *ortho*-substituted arylalkynes bearing nucleophiles gives benzopyrans **188** and benzofurans **189** through a regioselective *6-endo-dig* cyclization with a low loading (3–4 mol%) of hydridoiridium(III) catalyst complex **187** (Scheme 83).[120] The reaction is performed in chloroform or toluene at temperatures ranging from rt to reflux. Analogously, starting from 2-alkynylanilines, the hydroamination process (0.5 mol% catalyst) affords the corresponding indoles. The iridium hydride complex **187** is now commercially available or can be synthesized by C—H activation of acetophenone.[121]

Scheme 83 Iridium-Catalyzed Cyclization of 2-Substituted Arylalkynes to Benzopyrans and Benzofurans[120]

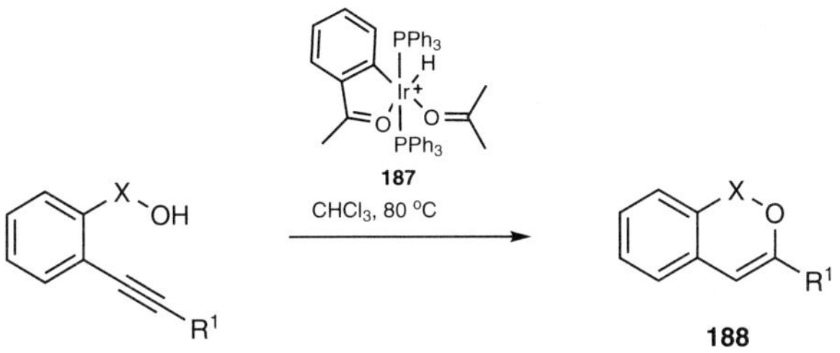

R^1	X	mol% of Catalyst	Time (h)	Yield (%)	Ref
Pr	CH_2	4	2	77	[120]
Ph	CH_2	3	2	70	[120]
Pr	C=O	3	1	89	[120]
Ph	C=O	3	1	91	[120]

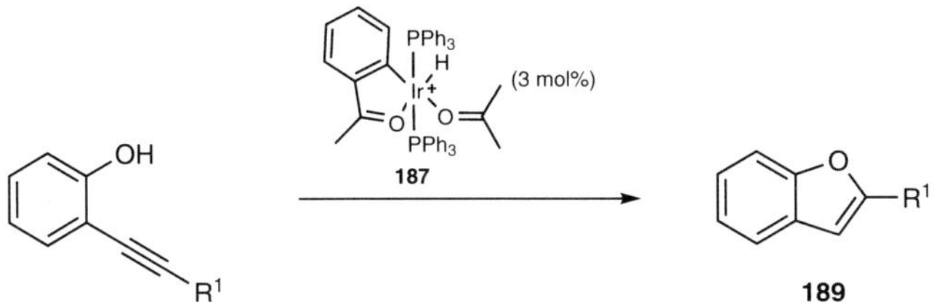

R^1	Conditions	Yield (%)	Ref
Pr	CH_2Cl_2, rt, 14 h	87	[120]
Ph	toluene, reflux, 2 h	72	[120]

3-Propyl-1*H*-2-benzopyran (188, X = CH_2; R^1 = Pr); Typical Procedure:[120]
[2-(Pent-1-ynyl)phenyl]methanol (1.32 mmol) and iridium catalyst **187** (0.052 mmol) were dissolved in $CHCl_3$ (3 mL). The soln was stirred at 80 °C for 2 h, followed by removal of all volatiles under reduced pressure. The product, which decomposed on silica gel, was purified by bulb-to-bulb vacuum distillation; yield: 77%.

1.8.3.1.4 Palladium Catalysis

Z-Alk-2-en-4-ynols **190** are used as substrates to afford furans **191** by a palladium(II)-catalyzed 5-*exo-dig* cyclization using a simple catalytic system consisting of a mixture of palladium(II) iodide and potassium iodide under neutral conditions. The use of an excess of iodide is essential for solubilizing the palladium salt (Scheme 84).[122] The methodology can be applied to obtain a variety of substituted furans, including natural products such as rosefuran (**191**, R^1 = CH=CMe$_2$; R^2 = Me; R^3 = R^4 = H), a derivative present in the oil of roses.

Scheme 84 Palladium-Catalyzed Cyclization of *Z*-Enynols to Furans[122]

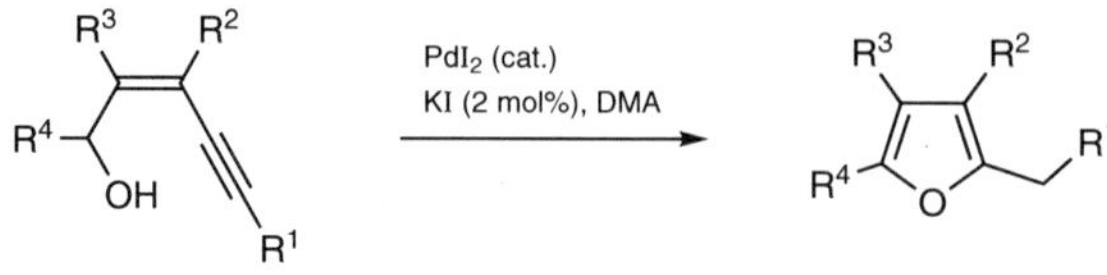

PdI_2 (cat.)
KI (2 mol%), DMA

190 → **191**

R^1	R^2	R^3	R^4	mol% of PdI_2	Temp (°C)	Time (h)	Yield (%)	Ref
H	Me	H	H	0.2	25	18	87[a]	[122]
H	Et	H	H	0.2	25	18	89[a]	[122]
H	Me	H	Et	0.2	25	20	90[a]	[122]
Bu	Me	H	H	1	100	20	81	[122]
CH=CMe$_2$	Me	H	H	1	100	24	77	[122]
Ph	Me	H	H	0.3	100	15	76	[122]
Bu	Me	H	Et	1	100	1	84	[122]
Bu	H	H	H	1	80	22	65	[122]
Bu	H	Et	H	0.3	100	15	64	[122]
Bu	H	Ph	H	0.3	100	1	68	[122]
Bu	H	Et	Et	0.5	80	3	72	[122]
Ph	Ph	Et	H	1	100	45	48	[122]

[a] Reaction carried out without solvent.

Starting from *N*-propargylamides **192**, the nucleophilic attack of the enol form on the activated triple bond permits the construction of 2-substituted oxazole-5-carbaldehydes **193** through treatment with a palladium(II)/copper(II) catalytic system and oxygen as oxidant. The formyl group arises from the nucleophilic intervention of water on the intermediate alkenylpalladium complex (Scheme 85).[123]

Scheme 85 Palladium/Copper-Catalyzed Cyclization of *N*-Propargylamides to Oxazoles[123]

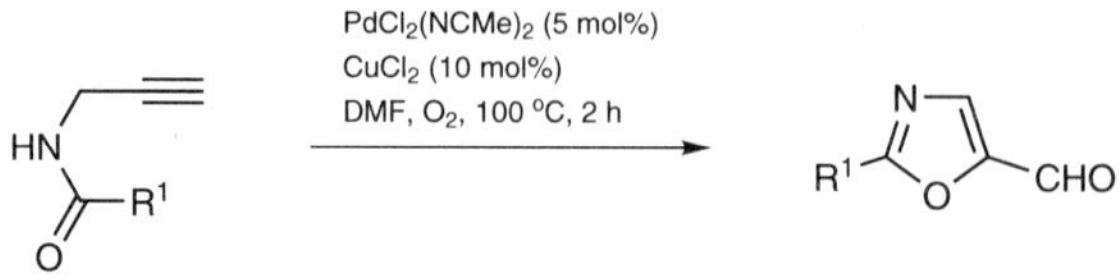

$PdCl_2(NCMe)_2$ (5 mol%)
$CuCl_2$ (10 mol%)
DMF, O_2, 100 °C, 2 h

192 → **193**

R^1	Yield (%)	Ref
4-$MeOC_6H_4$	48	[123]
4-$O_2NC_6H_4$	41[a]	[123]
Ph	54[a]	[123]
$(CH_2)_2Ph$	61	[123]
N Me	39	[123]
pyrrol-2-yl	37	[123]
2-thienyl	48	[123]
2-furyl	37	[123]
Pr^i NHBoc	42	[123]
NHBoc	39	[123]

[a] Reaction carried out with benzo-1,4-quinone (1 equiv) as oxidant instead of $CuCl_2/O_2$ in THF/DMF (5:3) at 60 °C (conventional heating) for 3 h.

Substituted Furans 191; **Typical Procedure:**[122]
Reactions were carried out on a 3–10-mmol scale. PdI_2 (1 mol%) and KI (2 mol%) were added to the pure *Z*-enynol or to a soln of the *Z*-enynol **190** in anhyd DMA in a Schlenk flask. The resulting mixture was stirred at 100 °C for the indicated time to obtain a satisfactory conversion, as shown by GLC and/or TLC analysis. The crude product derived from reaction without added solvent was purified by transfer distillation or column chromatography.

2-Substituted Oxazole-5-carbaldehydes 193; **General Procedure:**[123]
A soln of $CuCl_2$ (0.1 mmol) and $PdCl_2(NCMe)_2$ (0.05 mmol) in DMF (20 mL) was stirred at rt for 30 min. A soln of *N*-propargylamide **192** (1.0 mmol) in DMF (15 mL) was added, and the resulting mixture was warmed at 100 °C for 2 h under O_2. The soln was treated with brine and extracted with Et_2O (2 × 40 mL). The combined organic layers were dried (Na_2SO_4) and concentrated to dryness under reduced pressure. The residue was purified by column chromatography (silica gel).

1.8.3.1.5 Platinum Catalysis

The simple intramolecular 5-*endo-dig* hydroalkoxylation of 2-alkynylphenols **194** gives rise to the corresponding benzo[*b*]furans **195** on exposure to catalytic amounts of platinum(II) chloride in toluene (Scheme 86).[124] The reaction proceeds using low catalyst loadings (0.5–5 mol%) under air at room temperature, although it is significantly faster when performed at 80 °C. Compared to other metal catalysts used for similar purposes, no external base is necessary to promote the reaction. The conditions are compatible with a wide range of substrates bearing different functional groups. Almost quantitative yields of benzo[*b*]furans are obtained. The protection of the phenolic OH in ethers **196** does not preclude the cyclization, but the substituent (R^1) is transferred to the 3-position of the resulting benzo[*b*]furan **197** via an intramolecular carboalkoxylation. In this case, the rates are significantly increased when the reaction is performed under an atmosphere of carbon monoxide (Scheme 86).

Scheme 86 Platinum-Catalyzed Cyclization of 2-Alkynylphenols and 2-Alkynylphenyl Ethers to Benzo[*b*]furans[124]

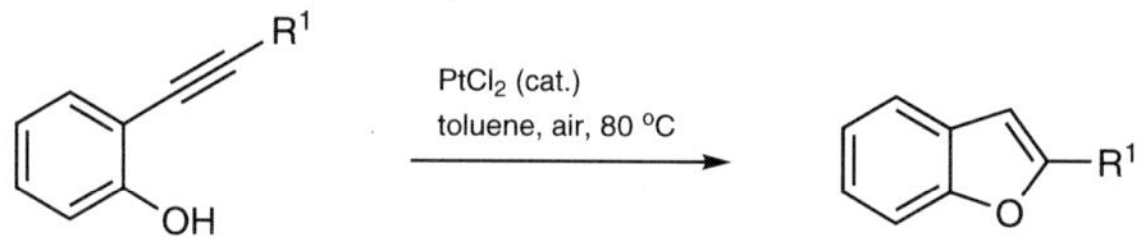

R^1	mol% of $PtCl_2$	Time (h)	Yield (%)	Ref
Pr	5	1	88	[124]
$(CH_2)_4Me$	0.5	5	98	[124]
cyclopropyl	1	1	98	[124]
$(CH_2)_2Ph$	1	5	98	[124]
Ph	5	1.5	97	[124]
4-$MeOC_6H_4$	0.5	2	95	[124]
3-$F_3CC_6H_4$	0.5	5	94	[124]

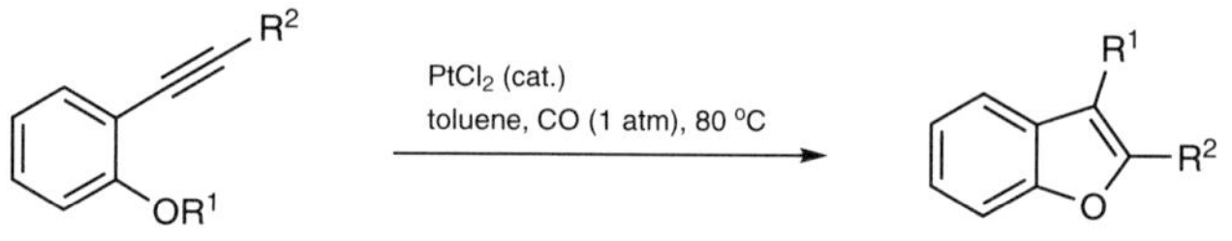

R^1	R^2	mol% of $PtCl_2$	Time (h)	Yield (%)	Ref
$CH_2CH{=}CH_2$	Pr	5	4	88	[124]
$CH_2CH{=}CH_2$	$(CH_2)_2Ph$	10	4	94	[124]
$CH_2CH{=}CH_2$	3-$MeOC_6H_4$	5	1	98	[124]
$CH_2CH{=}CH_2$	3-$F_3CC_6H_4$	5	1	94	[124]
$CH_2C(Me){=}CH_2$	$(CH_2)_4Me$	5	12	73	[124]
$CH_2C(Me){=}CH_2$	$(CH_2)_2Ph$	10	8	71	[124]
(*E*)-$CH_2CH{=}CHPh$	$(CH_2)_4Me$	10	3	68	[124]
Bn	$(CH_2)_4Me$	10	6	66	[124]
4-$MeOC_6H_4CH_2$	$(CH_2)_4Me$	10	4	76	[124]
4-$MeOC_6H_4CH_2$	3-$F_3CC_6H_4$	10	4	78	[124]
4-$MeOC_6H_4CH_2$	cyclopropyl	10	3	77	[124]

The addition of nucleophiles to cyclic and acyclic enynones followed by intramolecular hydroalkoxylation allows for the synthesis of highly substituted furans or fused bicyclic furan derivatives (Table 3).[125] Thus, nucleophilic 1,4-addition to enynones under platinum(II) catalysis generates suitable functionality in situ for the subsequent hydroalkoxylation step. The reactions are performed at temperatures ranging from rt to 60 °C and result in the simultaneous formation of a new C—O bond and a C—O, C—N, or C—C bond depending on the nucleophile.

Table 3 Platinum-Catalyzed Cyclization of Enynones to Furans[125]

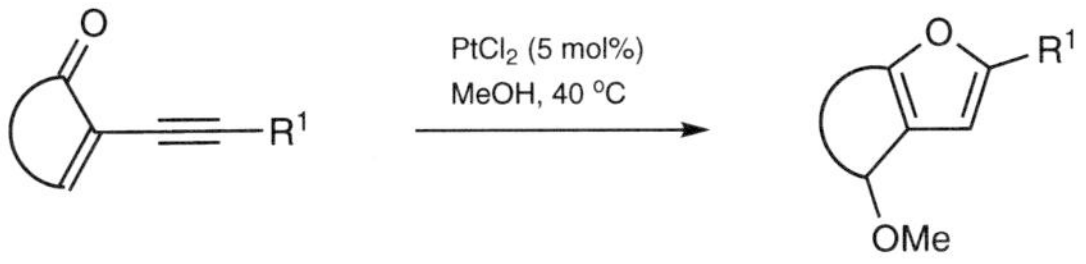

PtCl$_2$ (5 mol%)
MeOH, 40 °C

Substrate	Time (h)	Product	Yield (%)	Ref
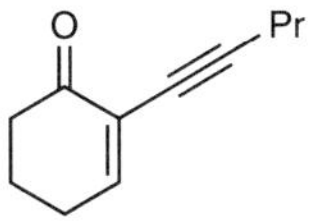	10	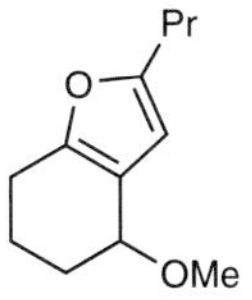	82	[125]
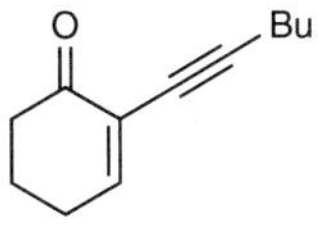	10	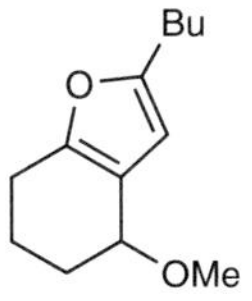	78	[125]
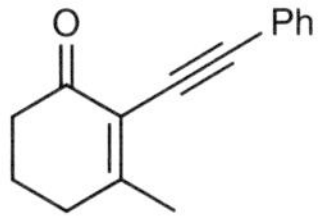	10	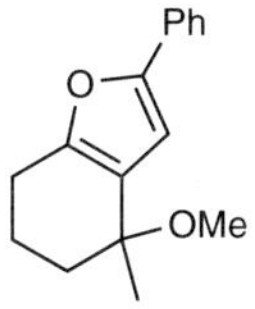	85	[125]
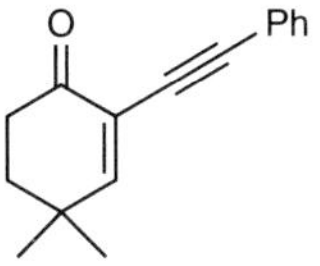	5	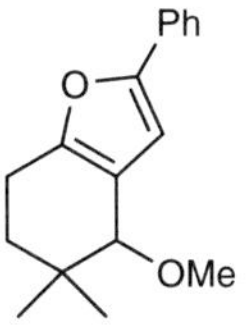	73	[125]
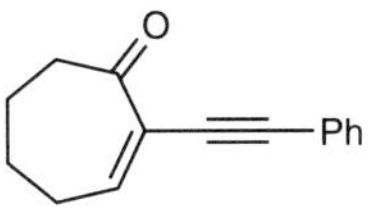	10	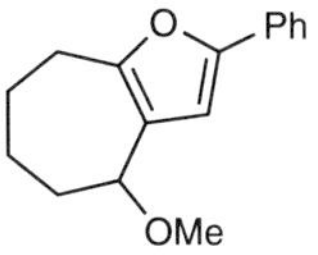	72	[125]
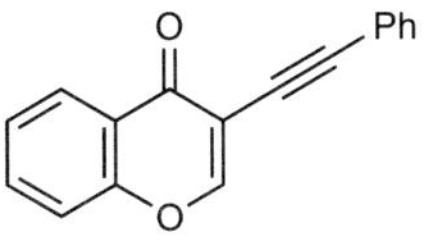	10	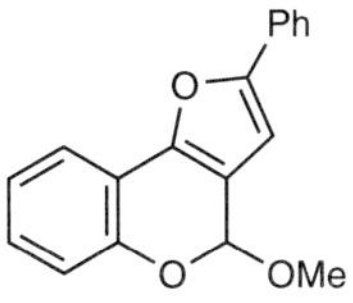	78	[125]
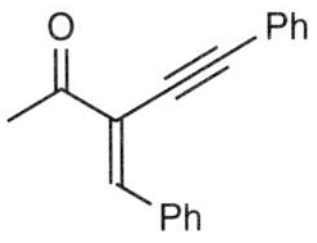	12	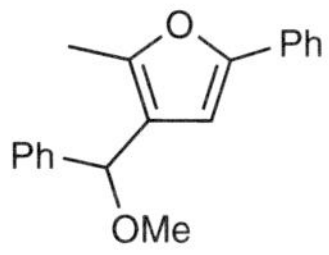	88	[125]
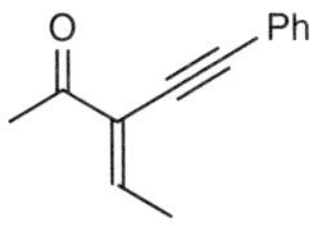	12	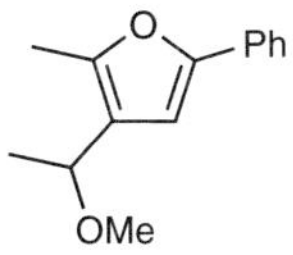	76	[125]

for references see p 499

2-Substituted Benzo[*b*]furans 195; General Procedure:[124]
A 2-alkynylphenol **194** (0.2–0.4 mmol) was weighed into a Schlenk tube and toluene was added (0.2 M). $PtCl_2$ (0.5–5 mol%) was added to the tube as a solid and the tube was sealed. The mixture was stirred at 80 °C. After the reaction was complete, the soln was cooled and directly subjected to column chromatography (*t*-BuOMe/hexane).

2,3-Disubstituted Benzo[*b*]furans 197; General Procedure:[124]

CAUTION: *Carbon monoxide is extremely flammable and exposure to higher concentrations can quickly lead to a coma.*

A 2-alkynylphenol ether **196** (0.2–0.4 mmol) was weighed into a Schlenk tube and toluene was added (0.2 M). $PtCl_2$ (5–10 mol%) was added to the tube as a solid and a CO balloon fitted to a needle was placed through a septum into the mixture. CO was bubbled through for ~30 s before the needle was removed. The mixture was stirred at 80 °C under CO. After the reaction was complete, the soln was cooled and directly subjected to column chromatography (*t*-BuOMe/hexane).

Substituted Furans (Table 3); General Procedure:[125]
$PtCl_2$ (5 mol%) was added to a mixture of an enynone (0.4 mmol) in MeOH (1.2 mL) and the resulting suspension was stirred at rt for 10 min, heated at 40 °C for the indicated time, and allowed to cool to rt. The excess solvent was removed under reduced pressure, and the residue was purified by column chromatography (silica gel).

1.8.3.1.6 Rhodium Catalysis

Based on the well-known ability of rhodium to form vinylidene complexes from terminal alkynes,[126,127] different rhodium(I)–phosphine complexes have been tested as catalysts for the cycloisomerization of ynols **198** and **200** to obtain dihydrofurans **199** and dihydropyrans **201**, respectively, through *endo-dig* cyclization (Scheme 87).[128] An excess of an electron-poor triarylphosphine, e.g. tris(3,5-difluorophenyl)phosphine, is used relative to rhodium to avoid undesirable dimerization/oligomerization processes.

Scheme 87 Rhodium-Catalyzed Cyclization of Alkynols to Dihydrofurans and Dihydropyrans[128]

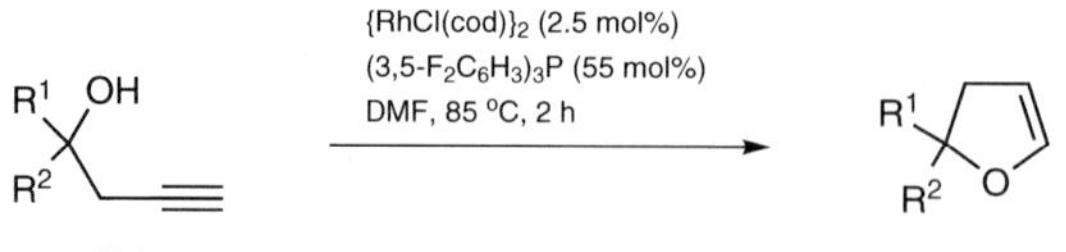

R1 OH, R2 (**198**) → {RhCl(cod)}2 (2.5 mol%), (3,5-F2C6H3)3P (55 mol%), DMF, 85 °C, 2 h → R1, R2, O (**199**)

R^1	R^2	Yield (%)	Ref
$(CH_2)_8Me$	H	69	[128]
$(CH_2)_8Me$	Me	62	[128]
CH_2CO_2t-Bu	H	71	[128]
3,4-$(MeO)_2C_6H_3$	H	74	[128]

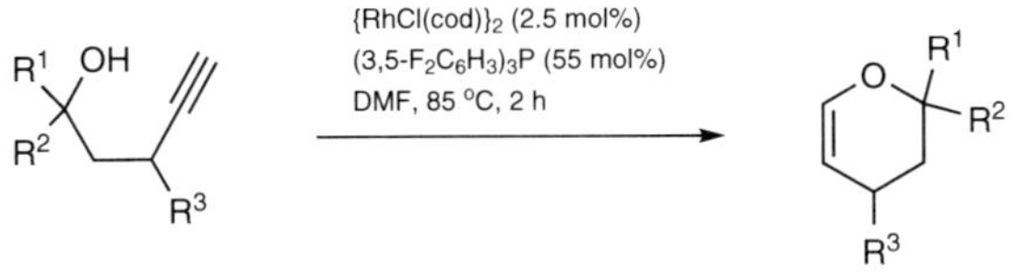

R^1	R^2	R^3	Yield (%)	Ref
$(CH_2)_6Me$	H	H	61	[128]
$(CH_2)_6Me$	Me	H	61[a]	[128]
Me	H	OPMB	58[a]	[128]
Me	H	NHTs	52	[128]

[a] Reaction carried out with 5 mol% chlorotris[(3,5-difluorophenyl)phosphine]rhodium(I) and 30 mol% $(3,5-F_2C_6H_3)_3P$ for 1 h.

Substituted Dihydrofurans 199 **or Dihydropyrans** 201**; General Procedure:**[128]
A mixture of alkynol **198** or **200** (0.509 mmol), $\{RhCl(cod)\}_2$ (0.013 mmol), and tris(3,5-difluorophenyl)phosphine (0.280 mmol) in DMF (2.6 mL) was placed in an oil bath at 85 °C and stirred for 2 h, cooled to rt, diluted with Et_2O (30 mL), and washed with H_2O (2 × 10 mL). The aqueous layers were extracted with Et_2O (2 × 20 mL). The organic phases were combined, dried ($MgSO_4$), and concentrated. The residual oil was purified by column chromatography.

1.8.3.1.7 Ruthenium Catalysis

The hydroalkoxylation reaction of terminal *Z*-enynols is a fruitful method to obtain substituted furans, with the reaction being performed under ruthenium catalysis at 60 °C over 2 hours without solvent.[129] The geometry of the double bond is an important factor as no cyclization occurs with *E*-enynols.

A particular reactivity is observed in the cyclization of alkynols in a divergent hydroalkoxylation reaction to provide dihydropyrans via *endo-dig* cyclization with ruthenium catalyst **202** (Table 4) or δ-valerolactones arising from an oxidative cyclization using ruthenium catalyst **203** (Table 5). The divergence is remarkable because it depends on the choice of phosphine ligand, requiring in the second case an electron-rich phosphine and an excess of *N*-hydroxysuccinimide as oxidant.[130]

Table 4 Ruthenium-Catalyzed Cyclization of Alkynols to Dihydropyrans[130]

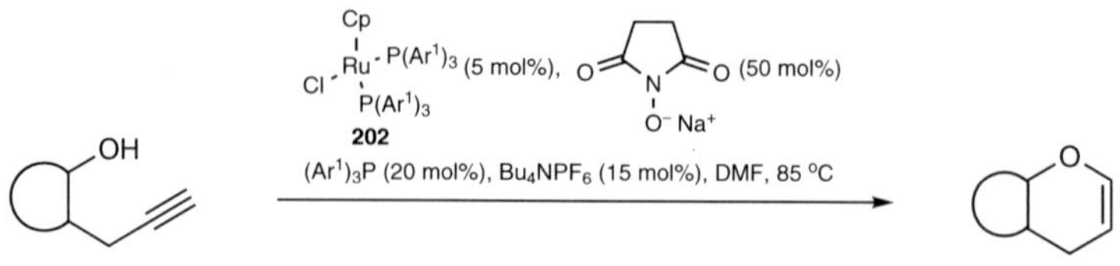

Cp(Cl)Ru[P(Ar[1])$_3$]$_2$ **202** (5 mol%), N-hydroxysuccinimide sodium salt (O$^-$ Na$^+$) (50 mol%), (Ar[1])$_3$P (20 mol%), Bu_4NPF_6 (15 mol%), DMF, 85 °C

Ar[1] = 4-FC$_6$H$_4$

Substrate	Time (h)	Product	Yield (%)	Ref
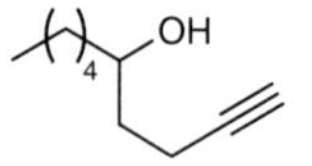	25	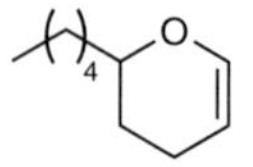	64	[130]
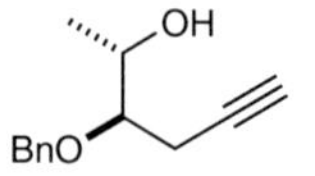	24	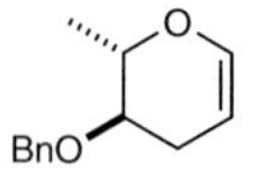	72	[130]
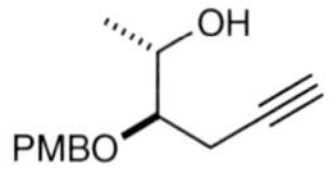	24	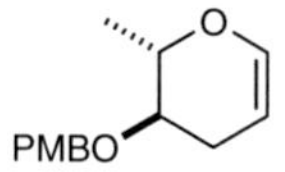	67	[130]
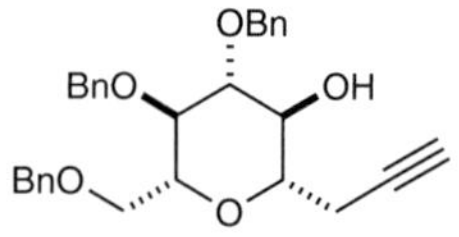	26	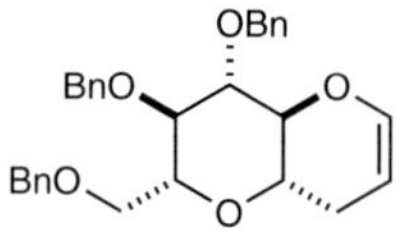	68	[130]
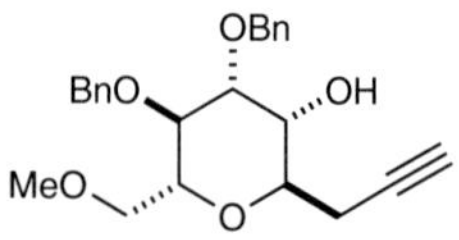	19	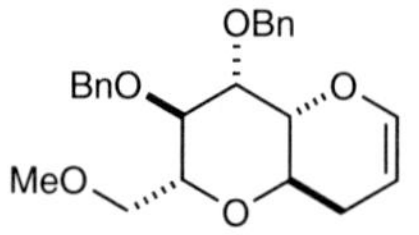	70	[130]
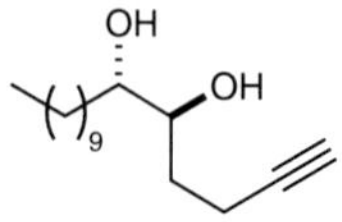	25	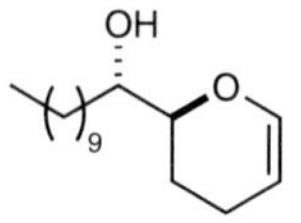	62	[130]

Table 5 Ruthenium-Catalyzed Cyclization of Alkynols to δ-Valerolactones[130]

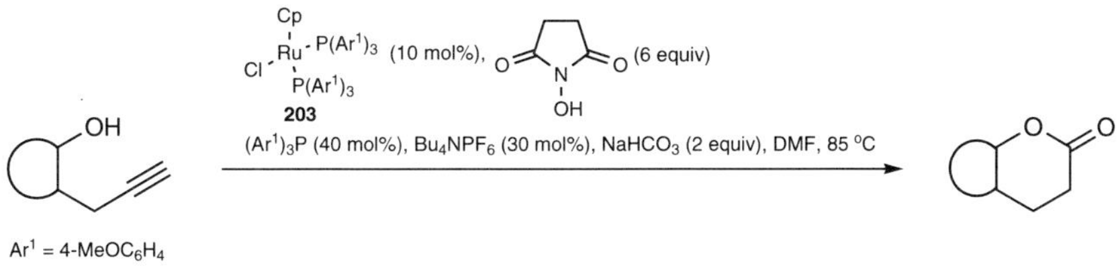

Cp
Cl–Ru–P(Ar1)3 (10 mol%), N-hydroxysuccinimide (6 equiv)
P(Ar1)3
203
(Ar1)3P (40 mol%), Bu4NPF6 (30 mol%), NaHCO3 (2 equiv), DMF, 85 °C

Ar1 = 4-MeOC6H4

Substrate	Time (h)	Product	Yield (%)	Ref
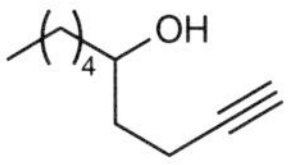	23	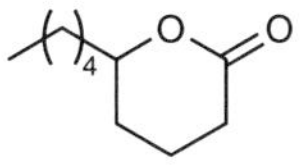	69	[130]
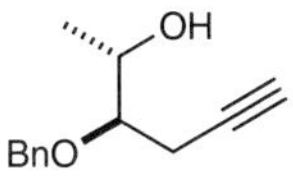	18	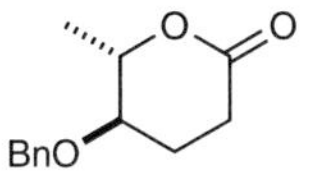	70[a]	[130]
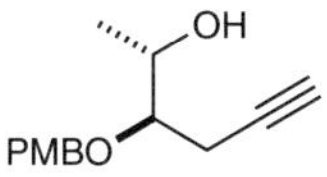	18	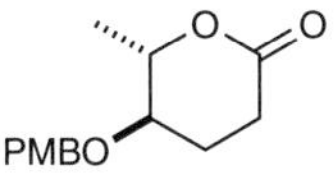	67[a]	[130]
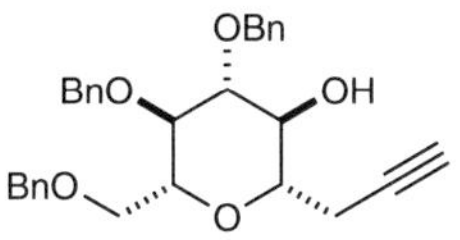	26	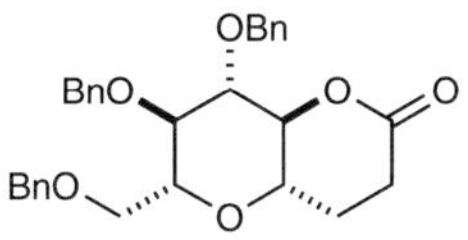	65	[130]
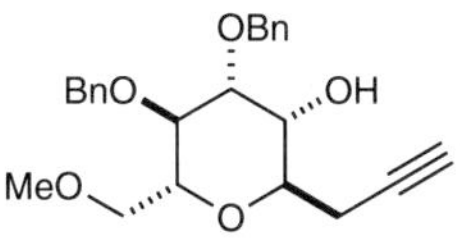	22	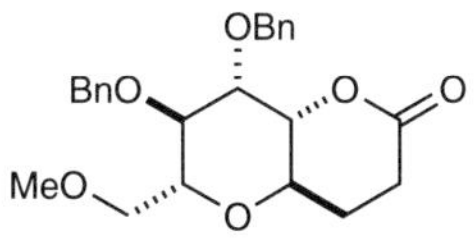	51	[130]
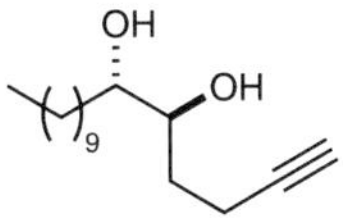	25	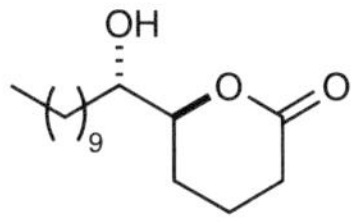	64[a]	[130]

[a] Reaction carried out with 15 mol% of catalyst, 60 mol% of ligand, and 45 mol% Bu_4NPF_6.

Dihydropyrans (Table 4); General Procedure:[130]
A mixture of an alkynol (0.768 mmol), chloro(η5-cyclopentadienyl)bis[tris(4-fluorophenyl)-phosphine]ruthenium(II) (**202**; 0.038 mmol), tris(4-fluorophenyl)phosphine (0.15 mmol), *N*-hydroxysuccinimide sodium salt (0.38 mmol), and tetrabutylammonium hexafluorophosphate (0.10 mmol) in DMF (1.9 mL, 0.4 M) was placed in a preheated oil bath at 85 °C and stirred under N_2 for the required time. The mixture was cooled to rt, diluted with Et_2O (30 mL), and washed with H_2O (2 × 10 mL). The aqueous layers were extracted with Et_2O (2 × 25 mL). The organic layers were combined, dried ($MgSO_4$), and concentrated. The residual oil was purified by column chromatography.

δ-Valerolactones (Table 5); General Procedure:[130]
A mixture of an alkynol (0.742 mmol), chloro(η5-cyclopentadienyl)bis[tris(4-methoxyphenyl)phosphine]ruthenium(II) (**203**; 0.074 mmol), tris(4-methoxyphenyl)phosphine (0.296 mmol), *N*-hydroxysuccinimide (4.44 mmol), $NaHCO_3$ (1.48 mmol), and tetrabutylammonium hexafluorophosphate (0.222 mmol) in DMF (1.9 mL, 0.4 M) was placed in a

preheated oil bath at 85 °C and stirred under N_2 for the required time. The mixture was cooled to rt, diluted with Et_2O (30 mL), and washed with H_2O (2 × 10 mL). The aqueous layers were extracted with Et_2O (2 × 25 mL). The organic layers were combined, dried ($MgSO_4$), and concentrated. The residual oil was purified by column chromatography.

1.8.3.1.8 Zinc Catalysis

Propargylic *N*-benzylhydroxylamines **204** undergo 5-*endo-dig* cyclization in the presence of zinc iodide and 4-(dimethylamino)pyridine in dichloromethane at room temperature affording 2,3,5-trisubstituted dihydroisoxazoles **205** in good yields. The simple workup of the reaction [acidic aqueous workup allows the separation of the zinc reagents and 4-(dimethylamino)pyridine from the product] enhances the usefulness of this method (Scheme 88).[131]

Scheme 88 Zinc-Catalyzed Cyclization of Propargylic *N*-Benzylhydroxylamines to 2,3,5-Trisubstituted Dihydroisoxazoles[131]

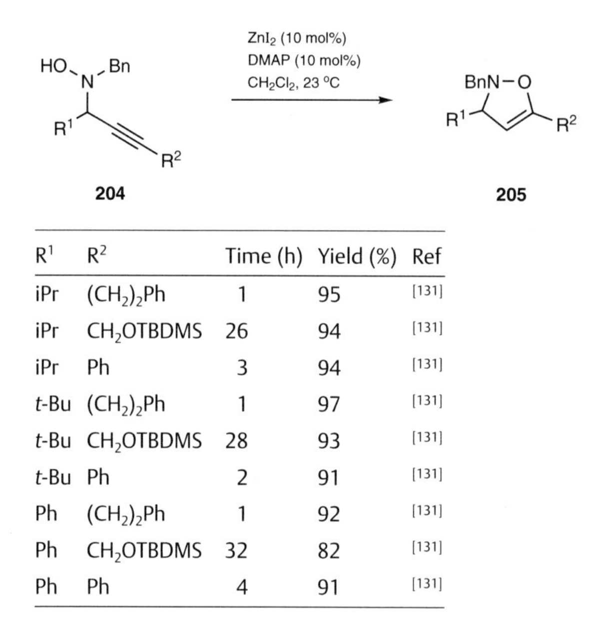

R^1	R^2	Time (h)	Yield (%)	Ref
iPr	$(CH_2)_2Ph$	1	95	[131]
iPr	$CH_2OTBDMS$	26	94	[131]
iPr	Ph	3	94	[131]
t-Bu	$(CH_2)_2Ph$	1	97	[131]
t-Bu	$CH_2OTBDMS$	28	93	[131]
t-Bu	Ph	2	91	[131]
Ph	$(CH_2)_2Ph$	1	92	[131]
Ph	$CH_2OTBDMS$	32	82	[131]
Ph	Ph	4	91	[131]

2,3,5-Trisubstituted Dihydroisoxazoles 205**; General Procedure:**[131]
A flask was charged with ZnI_2 (0.1 equiv) and purged with N_2 for 15 min. A soln of a substituted hydroxylamine **204** (1 equiv) in CH_2Cl_2 (2 mL) was added and the resulting mixture was stirred at 23 °C for 10 min. DMAP (0.1 equiv) was dissolved in CH_2Cl_2 (1 mL) and added to the flask in one portion. Upon completion, the reaction was quenched with sat. aq NH_4Cl (2 mL). The mixture was poured into a separatory funnel containing CH_2Cl_2 (10 mL). The layers were separated and the aqueous layer was extracted with CH_2Cl_2 (3 × 10 mL). The combined organic layers were washed with brine, dried ($MgSO_4$), and concentrated under reduced pressure.

1.8.3.2 Hydroacyloxylation of Alkynes

The use of the carboxylic acid group as nucleophile in the intramolecular hydroacyloxylation of C≡C bonds results in the formation of lactones.

1.8.3.2.1 Gold Catalysis

The metal-catalyzed intramolecular addition of a carboxylic acid group as nucleophile to triple bonds is generally performed in refluxing solvents and in the presence of ligands or additives. The gold-catalyzed 5-*exo-dig* cyclization of acetylenic acids **206** bearing geminal substituents in position 2 on the linker chain leads to γ-lactones **207** under very mild conditions and with short reaction times in the presence of gold(I) chloride in acetonitrile without other additives (Scheme 89).[132] Internal alkynes also cyclize, affording *Z*-*exo*-methylene-γ-lactones selectively. The total stereocontrol and the high selectivity of the reaction may be due to the Thorpe–Ingold effect as well as to the presence of substituents that could participate in intramolecular complexation with gold.

Scheme 89 Gold-Catalyzed Cyclization of Alkynoic Acids to γ-Lactones[132]

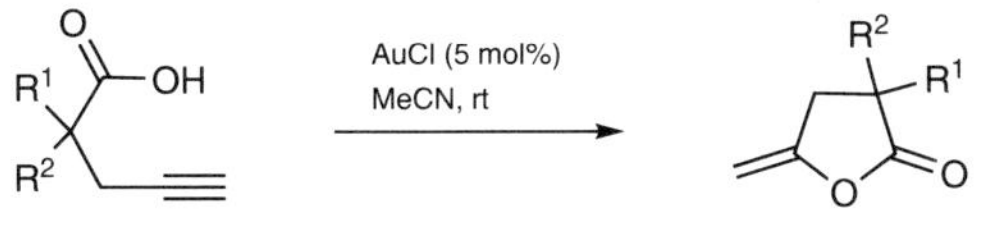

AuCl (5 mol%)
MeCN, rt

206 → 207

R^1	R^2	Time (h)	Yield (%)	Ref
CO_2Me	cyclohex-2-enyl	2	89	[132]
CO_2Me	(*E*/*Z*)-$CH_2CH{=}CHMe$	2	78	[132]
CO_2Me	$CH_2CH{=}CH_2$	2	72	[132]
CO_2Me	$(CH_2)_2CH{=}CH_2$	2	87	[132]
CO_2Me	$CH_2C{\equiv}CH$	2	97	[132]
CO_2Me	Cl	2	95	[132]
CO_2Et	Bu	2	83	[132]
CO_2Et	Bn	1	97	[132]
CH_2OBn	$CH_2C{\equiv}CH$	2	97	[132]
CO_2Me	(*E*/*Z*)-$CH_2CH{=}CHCH_2OH$	2	85	[132]
Ph	$CH_2C{\equiv}CH$	2	98	[132]
CO_2Me	(*E*)-$CH_2CH{=}CHPh$	7	90	[132]
CO_2Et	Bn	7	94	[132]

5-Methylenedihydrofuran-2(3*H*)-ones 207**; General Procedure:**[132]
A mixture of acetylenic acid **206** (1 equiv) and AuCl (5 mol%) in degassed MeCN (1.2 M) was stirred under argon at rt. After completion of the reaction, the mixture was filtered through a short pad of silica gel with EtOAc and the solvents were evaporated under reduced pressure.

1.8.3.3 Hydroalkoxylation of Alkenes

The intramolecular hydroalkoxylation reaction consists of using alkenols as substrates to obtain cyclic ethers. The reaction on aliphatic alcohols needs substrates bearing geminal substituents on the linker chain to force the cyclization in accordance with the Thorpe–Ingold effect.

1.8.3.3.1 Iron Catalysis

Hydroalkoxylation of γ-hydroxyalkenes **208** is achieved by using the cationic iron complex iron(III) trifluoromethanesulfonate under mild reaction conditions and leads to the formation of cyclic ethers **209**. The cationic iron may achieve double activation of both the hydroxy group and the alkene moiety. It is prepared in situ from iron(III) chloride and silver(I) trifluoromethanesulfonate in 1,2-dichloroethane at room temperature, but the product is also obtained in the absence of the cocatalytic silver salt (Scheme 90).[133]

Scheme 90 Iron-Catalyzed Cyclization of γ-Hydroxyalkenes to Tetrahydrofurans[133]

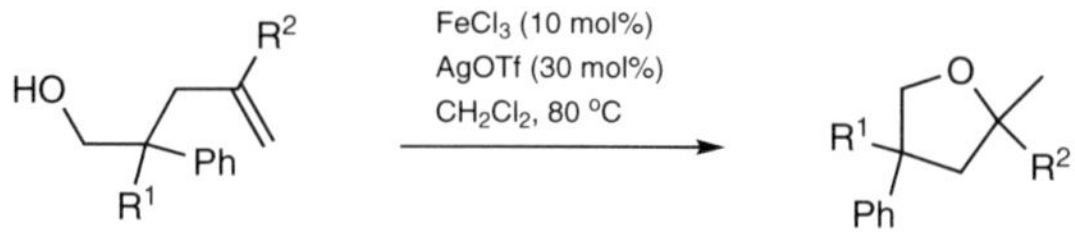

HO, R², Ph, R¹ **208** → $FeCl_3$ (10 mol%), AgOTf (30 mol%), CH_2Cl_2, 80 °C → R¹, O, Ph, R² **209**

R^1	R^2	Time (h)	Yield (%)	Ref
Ph	H	0.5	97	[133]
Ph	Me	0.5	94	[133]
CH_2OH	H	4	97	[133]
CH_2OAc	H	2	94	[133]
CH_2CO_2t-Bu	H	4	97	[133]
CH_2OBz	H	3	96	[133]

2-Methyltetrahydrofurans 209; **General Procedure:**[133]
To $Fe(OTf)_3$, which was prepared in situ from $FeCl_3$ (0.05 mmol) and AgOTf (0.15 mmol) in CH_2Cl_2 (2.5 mL) over 2 h at rt, was added a soln of hydroxyalkene **208** (0.5 mmol) in CH_2Cl_2 (2.5 mL) under N_2. The mixture was heated for the indicated time, cooled, and filtered through a short column (silica gel, hexane/EtOAc). Separation by column chromatography (hexane/EtOAc) afforded the product.

1.8.3.3.2 Platinum Catalysis

The protocol described previously for the intramolecular hydroamination of unactivated alkenes (see Section 1.8.2.2.5) can also be applied to γ- and δ-hydroxyalkenes to form cyclic ethers through outer-sphere attack of the hydroxyl group on the platinum-complexed alkenes (Table 6).[134] The process needs the presence of phosphine ligands {e.g., tris[4-(trifluoromethyl)phenyl]phosphine}. The reaction is effective on terminal and internal double bonds and tolerates substitution at the α-, β-, and γ-carbon atoms. The regioselectivity is sensitive to the substitution on the terminal alkene, giving either 6-*endo* or 5-*exo* cyclization depending on the substrate.

Table 6 Platinum-Catalyzed Cyclization of γ- and δ-Hydroxyalkenes to Cyclic Ethers[134]

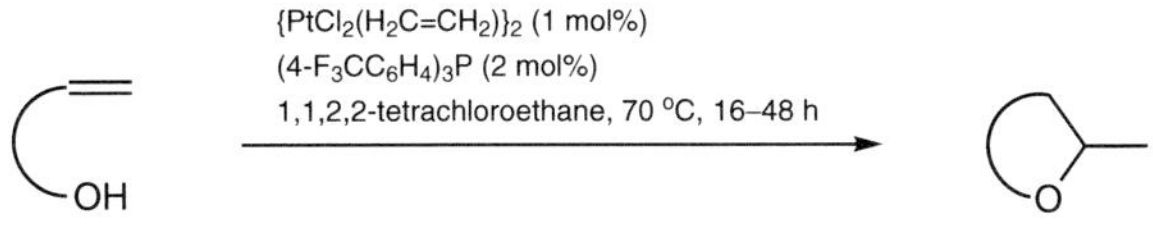

$\{PtCl_2(H_2C{=}CH_2)\}_2$ (1 mol%)
$(4\text{-}F_3CC_6H_4)_3P$ (2 mol%)
1,1,2,2-tetrachloroethane, 70 °C, 16–48 h

Entry	Substrate	Product	Yield (%)	Ref
1	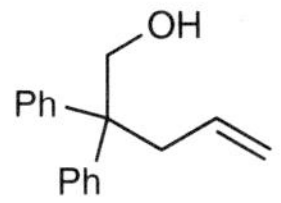	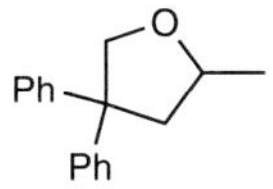	78	[134]
2	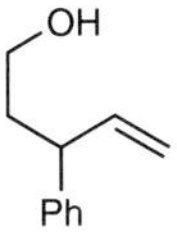	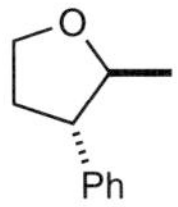	68[a]	[134]
3	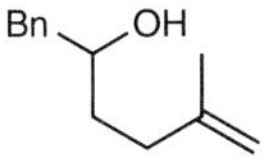	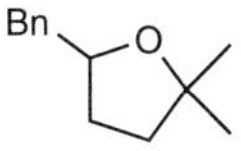	91	[134]
4	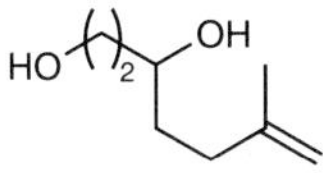	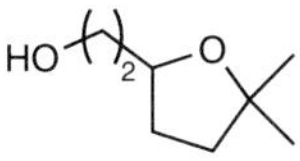	91	[134]
5	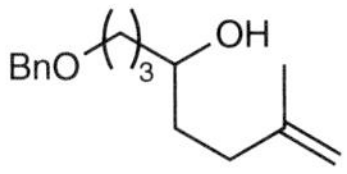	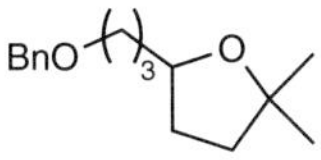	79	[134]
6	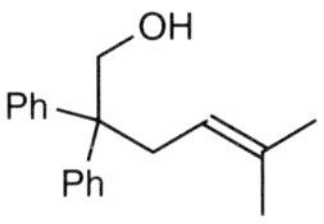	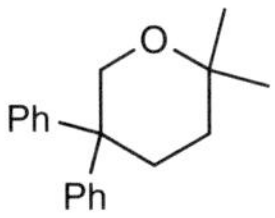	98[b]	[134]
7	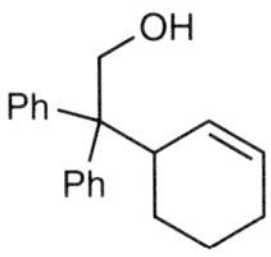	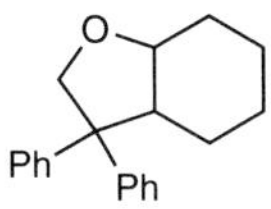	87	[134]
8	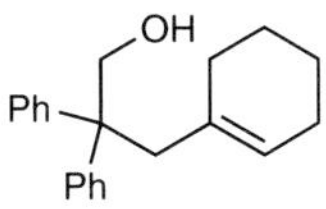	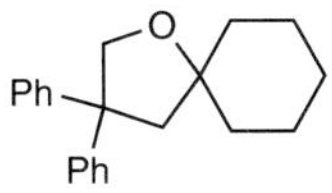	98	[134]
9	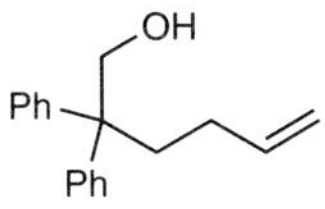	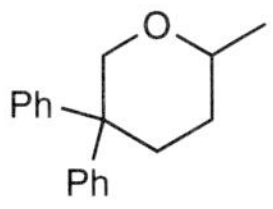	90	[134]

[a] Ratio (*trans*/*cis*) 8:1.
[b] Reaction carried out at 80 °C.

2-Methyl-4,4-diphenyltetrahydrofuran (Table 6, Entry 1); Typical Procedure:[134]
A soln of $\{PtCl_2(H_2C{=}CH_2)\}_2$ (0.010 mmol), tris[4-(trifluoromethyl)phenyl]phosphine (0.020 mmol), and 2,2-diphenylpent-4-en-1-ol (1.0 mmol) in 1,1,2,2-tetrachloroethane (0.5 mL) was stirred at 70 °C for 24 h. Column chromatography of the mixture (hexane to hexane/Et_2O 10:1) gave the product as an off-white solid; yield: 78%.

1.8.3.3.3 Ruthenium Catalysis

Hydroalkoxylation of 2-allyl- and 2-homoallylphenols **210** is performed using the combination of ruthenium(III) chloride and silver(I) trifluoromethanesulfonate in the presence of copper(I) trifluoromethanesulfonate as cocatalyst and triphenylphosphine as ligand, affording 2,3-dihydrobenzo[*b*]furans and 3,4-dihydro-2*H*-1-benzopyrans **211** (n = 1 and 2, respectively; Scheme 91).[135] This methodology can also be applied to unsaturated benzoic acids to give lactones.

Scheme 91 Ruthenium-Catalyzed Cyclization of 2-Allyl- and 2-Homoallylphenols To Give 2,3-Dihydrobenzo[*b*]furans and 3,4-Dihydro-2*H*-1-benzopyrans[135]

$RuCl_3{\bullet}nH_2O$ (10 mol%), AgOTf (30 mol%)
CuOTf (50 mol%), Ph_3P (20 mol%)
MeCN, 80 °C, 24 h

210 → **211**

R^1	R^2	R^3	n	Yield (%)	Ref
H	H	H	1	61	[135]
H	OMe	H	1	51	[135]
H	H	OMe	1	53	[135]
H	H	Me	1	65	[135]
$(CH{=}CH)_2$		H	1	58	[135]
H	H	H	2	72	[135]

2-Methyl-2,3-dihydrobenzo[*b*]furan (211, **$R^1 = R^2 = R^3 = H$; n = 1); Typical Procedure:**[135]
Into a 30-mL three-necked flask were added $RuCl_3{\bullet}nH_2O$ (0.4 mmol), MeCN (3.0 mL), and AgOTf (1.2 mmol), and the mixture was heated at 80 °C for 2 h. After the mixture had been cooled, 2-allylphenol (**210**, $R^1 = R^2 = R^3 = H$; n = 1; 4.0 mmol), Ph_3P (0.8 mmol), and $Cu(OTf)_2$ (2.0 mmol) were added, and the mixture was stirred at 80 °C for 24 h. Solid materials were removed by filtration through a Celite pad, and the solvent was evaporated. H_2O and Et_2O were added to the residue, and the organic materials were extracted with Et_2O. The solvent was removed and the product was isolated by column chromatography; yield: 61%.

1.8.3.3.4 Silver Catalysis

The advantages offered by silver catalysts compared to other metals include low toxicity and low cost. Silver(I) trifluoromethanesulfonate is the most efficient of the silver salts as catalyst for the conversion of alkenols into five- and six-membered cyclic ethers, following an *endo*- or *exo*-mode cyclization, with the prevalence in general of the latter (Table 7).[136] The reaction works with terminal and internal double bonds and tolerates various functional groups at the α-, β-, and γ-carbon atoms on the linker chain. The presence of ligands is detrimental to the yields of products.

Table 7 Silver-Catalyzed Cyclization of Alkenols to Cyclic Ethers[136]

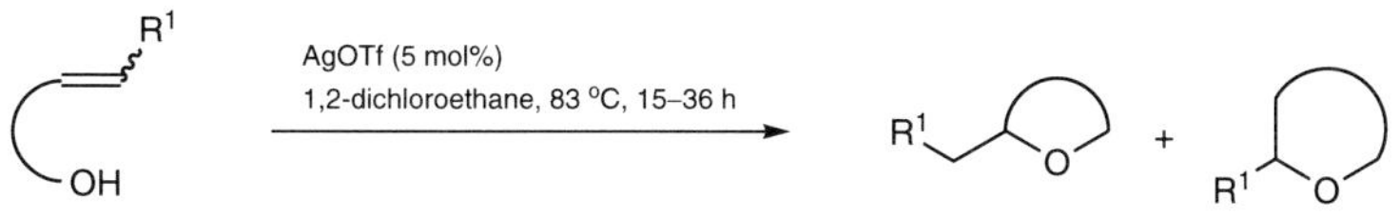

AgOTf (5 mol%), 1,2-dichloroethane, 83 °C, 15–36 h

Substrate	Product	Yield[a] (%)	Ref
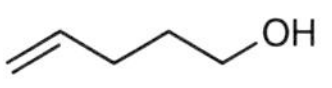	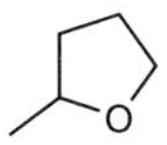	96[b]	[136]
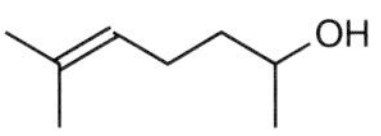	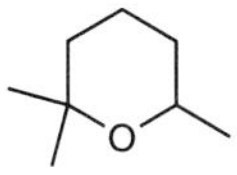	95[b]	[136]
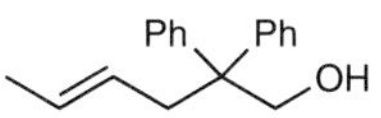	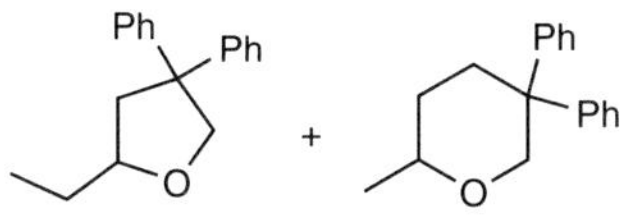	89 (10:1)[c]	[136]
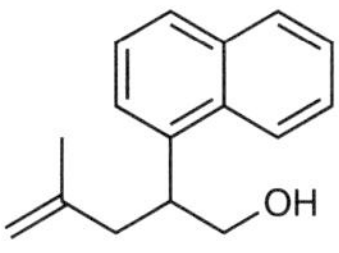	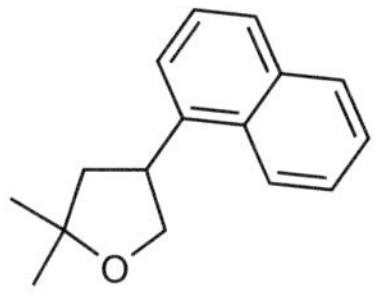	95	[136]
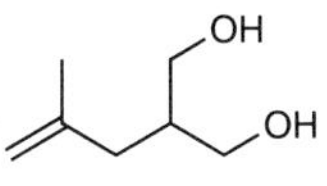	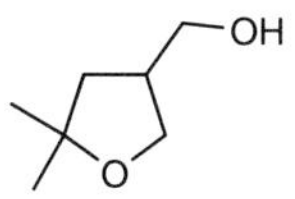	89	[136]
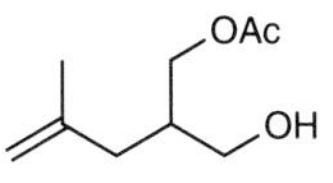	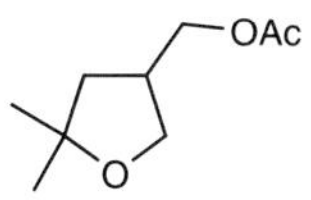	98	[136]
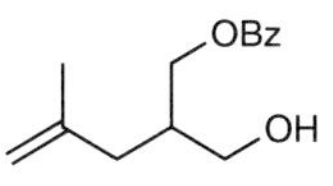	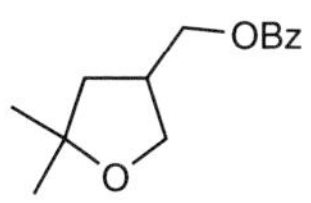	94	[136]
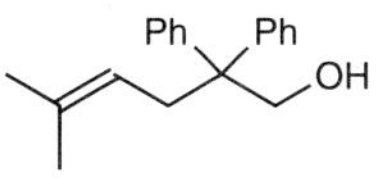	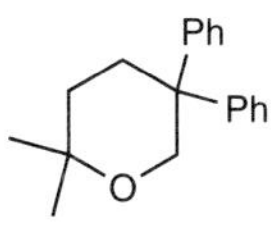	84	[136]
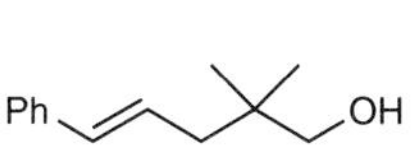	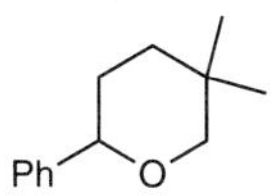	87	[136]

[a] Isolated yields.

[b] Yield determined by ^{1}H NMR spectroscopy.

[c] Ratio based on isolated yields.

Cyclic Ethers (Table 7); General Procedure:[136]
A mixture of an alkenol (1.0 mmol) and AgOTf (0.05 mmol) in anhyd 1,2-dichloroethane (2 mL) was heated at 83 °C for 15–36 h in a dry sealed tube under N_2. The mixture was cooled to rt and the catalyst was removed by filtration through a short pad of silica gel using Et_2O as eluent. The solvent was removed and the residue was purified by chromatography (silica gel).

1.8.3.4 Hydroacyloxylation of Alkenes

Analogously to the hydroacyloxylation of alkynes (see Section 1.8.3.2), when the nucleophile is a carboxyl group, the intramolecular addition to unactivated alkenes results in the formation of lactones.

1.8.3.4.1 Silver Catalysis

Silver(I) trifluoromethanesulfonate is the optimal catalyst to convert terminal alkenoic acids into lactones (Table 8).[136] As with alkenols, the mechanism suggests activation of the double bond by metal coordination, thus rendering it susceptible to attack by the oxygen nucleophile. The presence of ligands is detrimental to the yield of the products. The regioselectivity is sensitive to substitution at the terminal alkene, resulting in the predominant formation of five-membered rings.

Table 8 Silver-Catalyzed Cyclization of Alkenoic Acids to Lactones[136]

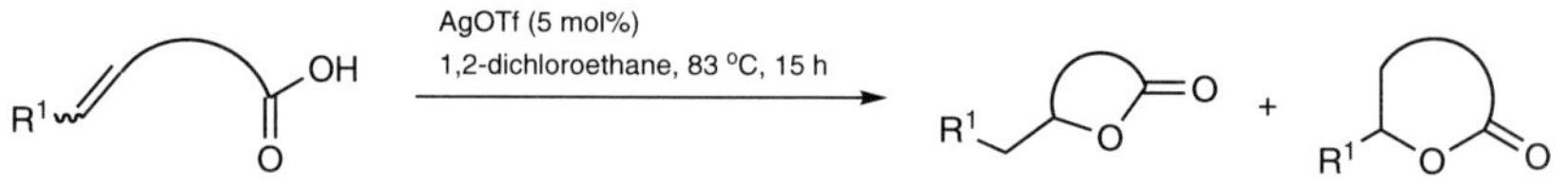

AgOTf (5 mol%), 1,2-dichloroethane, 83 °C, 15 h

Substrate	Product	Yield (%)	Ref
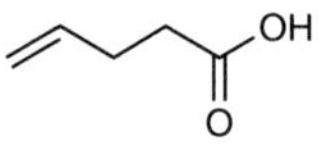	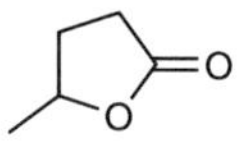	74	[136]
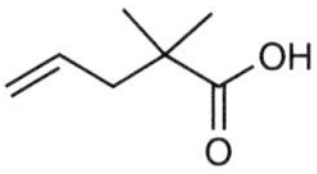	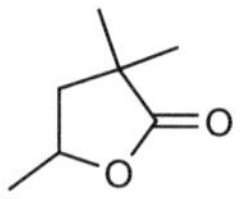	88	[136]
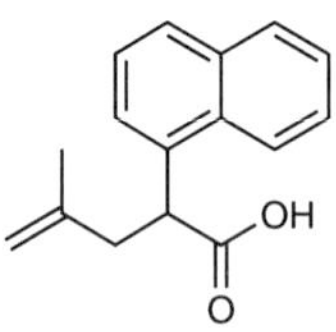	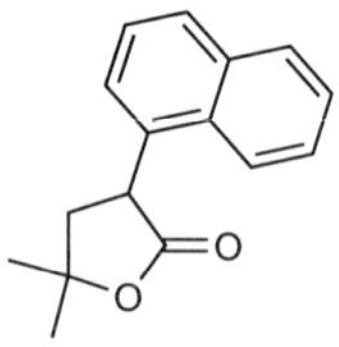	96	[136]
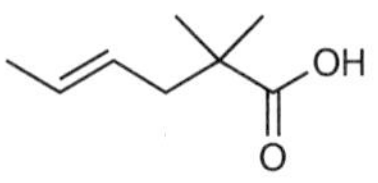	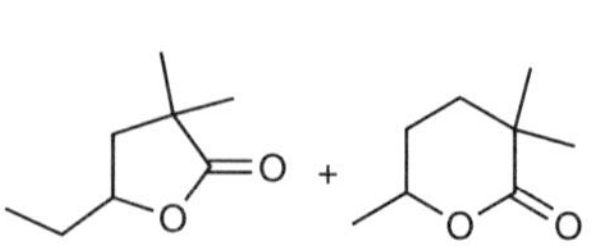	93 (3.2:1)[a]	[136]

Table 8 (cont.)

Substrate	Product	Yield (%)	Ref
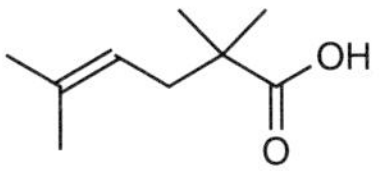	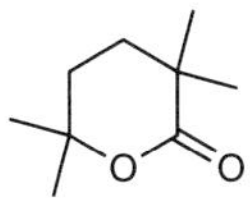	85	[136]
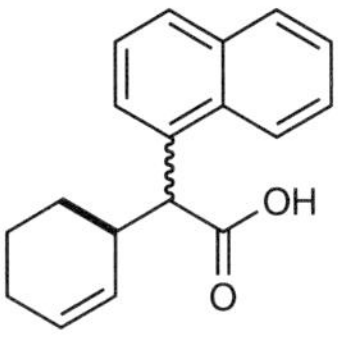	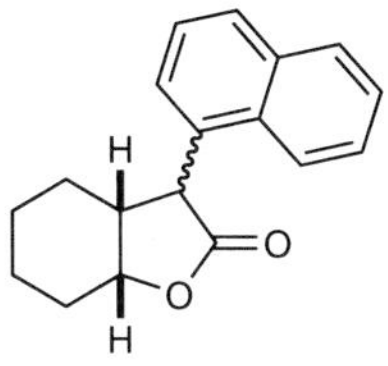	89[b]	[136]
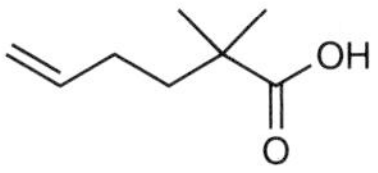	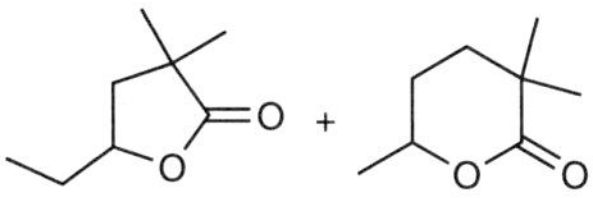	91 (1:10)[a]	[136]

[a] Ratio based on isolated yields.
[b] Reaction time was 36 h.

Substituted γ- and δ-Lactones (Table 8); General Procedure:[136]
A mixture of an alkenoic acid (1.0 mmol) and AgOTf (0.05 mmol) in anhyd 1,2-dichloroethane (2 mL) was heated at 83 °C for 15 h in a dry sealed tube under N_2. The mixture was cooled to rt and the catalyst was removed by filtration through a short pad of silica gel using Et_2O as eluent. The filtrate was concentrated and the residue was purified by chromatography (silica gel).

1.8.3.5 Oxidative Alkoxylation of Alkenes

The intramolecular nucleophilic attack on a C=C bond by the hydroxy group of an alcohol results, under oxidative conditions, in the formation of cyclic vinyl ether derivatives. The reaction is suitable for the preparation of both five- and six-membered heterocycles. Since the substrate and the product remain in the same oxidation state, the reduced form of the metal must be reoxidized by an oxidizing agent.

1.8.3.5.1 Palladium Catalysis

The cyclization of β-hydroxyenones by the combined use of palladium(II), copper(I), and oxygen, is a fruitful method for the preparation of 2,3-dihydro-4*H*-pyran-4-ones (Table 9). Enantiopure substrates are cyclized without loss of optical purity (Table 9, entries 5 and 6).[137] Under analogous conditions, the reaction has been expanded to the cyclization of α-hydroxyenones, leading to furan-3(2*H*)-one derivatives (Table 9, entries 7 and 8).[138] This class of compounds is present in many natural products such as furaneol,[139] which is often utilized as a strawberry flavoring agent, and geiparvarin,[140] which displays antiproliferative activity and has been targeted for the treatment of different types of cancer.

for references see p 499

Table 9 Palladium/Copper-Catalyzed Cyclizations of Hydroxyenones To Give 2,3-Dihydro-4*H*-pyran-4-ones and Furan-3(2*H*)-ones[137,138]

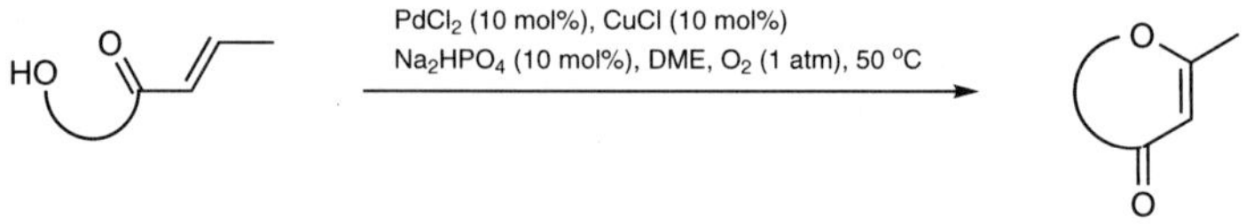

$PdCl_2$ (10 mol%), CuCl (10 mol%)
Na_2HPO_4 (10 mol%), DME, O_2 (1 atm), 50 °C

Entry	Substrate	Time (h)	Product	Yield (%)	Ref
1	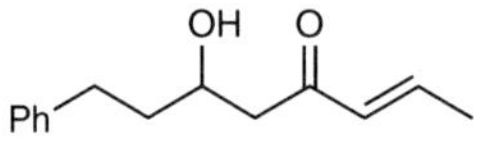	4	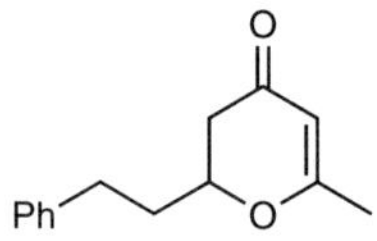	79	[137]
2	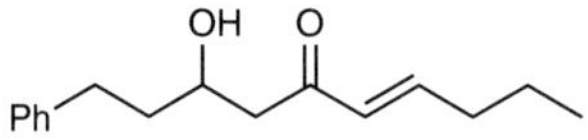	20	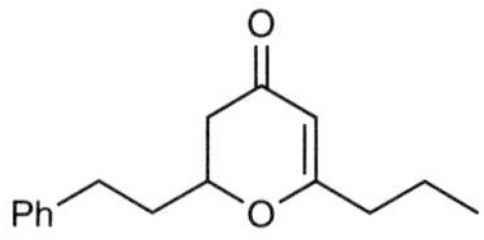	76	[137]
3	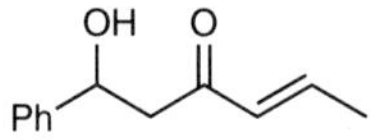	8	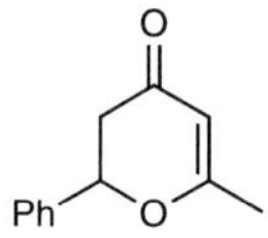	86	[137]
4	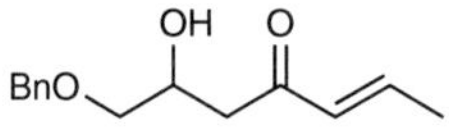	20	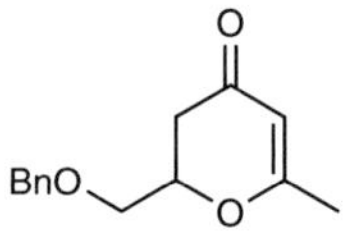	75	[137]
5	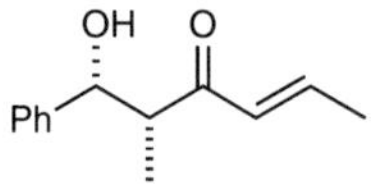 99% de; 91% ee	20	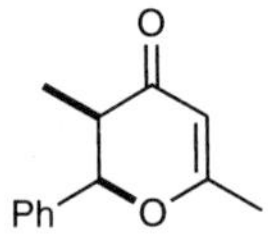 >99% de; 89% ee	74	[138]
6	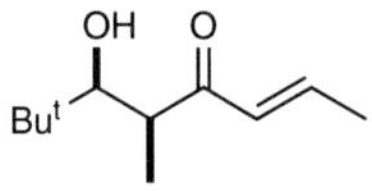 >99% de	–	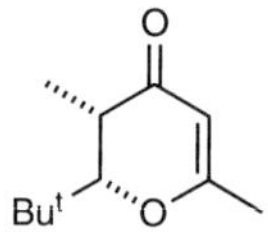 >99% de	55	[138]
7	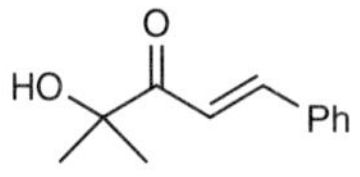	96	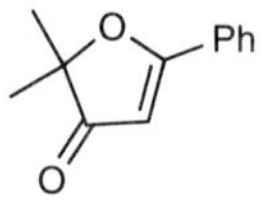	93	[138]
8	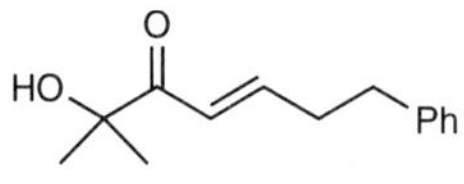	168	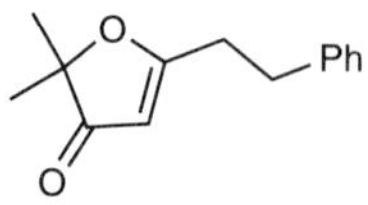	74	[138]

Treatment of α-allyl β-diketones **212** with a catalytic amount of a palladium(II) complex and a stoichiometric amount of copper(II) chloride in dioxane at 60 °C affords 2,3,5-trisubstituted furans **213** in moderate to good yields (Scheme 92).[141] The oxidative alkoxylation is possible through the enolic form of the substrate, which behaves as an alkenol.

Scheme 92 Palladium-Catalyzed Alkoxylation of α-Allyl β-Diketones to Furans[141]

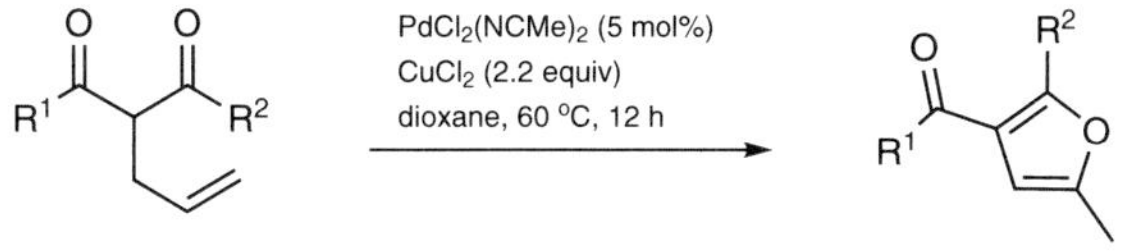

$PdCl_2(NCMe)_2$ (5 mol%)
$CuCl_2$ (2.2 equiv)
dioxane, 60 °C, 12 h

212 → **213**

R^1	R^2	Yield (%)	Ref
Et	Et	68	[141]
Ph	Ph	72	[141]
Ph	Me	69	[141]
Ph	$(CH_2)_4Me$	62	[141]
Ph	$(CH_2)_2OMe$	61	[141]
4-$MeOC_6H_4$	Me	63	[141]
4-$F_3COC_6H_4$	Me	65	[141]
2-furyl	Me	71	[141]
2-thienyl	Me	63	[141]
iPr	iPr	77	[141]

A simple alkoxylation of 3-azaalk-5-enols **214** affords dihydro-1,4-oxazines **215** under very mild reaction conditions and in an economical way using a palladium catalyst and molecular oxygen as the sole oxidant (Scheme 93).[142]

Scheme 93 Palladium-Catalyzed Cyclization of 3-Azaalk-5-enols to Dihydro-1,4-oxazines[142]

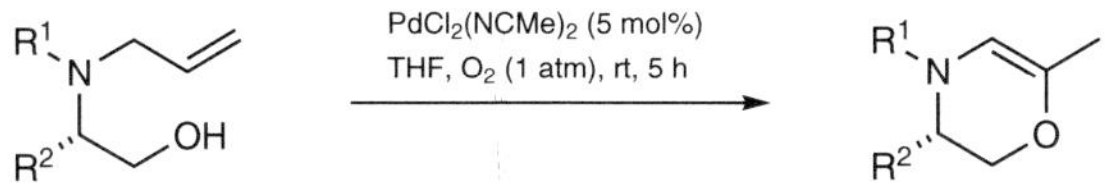

$PdCl_2(NCMe)_2$ (5 mol%)
THF, O_2 (1 atm), rt, 5 h

214 → **215**

R^1	R^2	Yield (%)	Ref
Ac	iBu	75	[142]
Ac	Et	63	[142]
Ac	Bn	63	[142]
Ts	iPr	82	[142]
Ts	Me	75	[142]
Ts	Et	74	[142]
Ts	Bn	67	[142]

A highly regioselective cyclization of 2-allylphenols to benzopyrans is achievable using bis(dibenzylideneacetone)palladium(0) or palladium(II) acetate as catalyst and air as reoxidant in dimethyl sulfoxide/water. The product is formed by the attack on the terminal alkenic carbon.[143]

Substituted 2,3-Dihydro-4*H*-pyran-4-ones and Furan-3(2*H*)-ones (Table 9); General Procedure:[137]
A 25-mL Schlenk tube was charged with $PdCl_2$ (10 mol%), CuCl (10 mol%), and Na_2HPO_4 (10 mol%) and evacuated and backfilled with O_2 (balloon) three times. The substrate (1 mmol) in anhyd DME (3 mL) was added via cannula. The resulting mixture was heated at 50 °C and stirred at this temperature until completion of the reaction. The mixture was cooled to rt, diluted with Et_2O, and filtered through a pad of silica gel. The filtrate was concentrated under reduced pressure and the residue was purified by flash column chromatography (silica gel).

2-Isopropyl-3-(2-methylpropanoyl)-5-methylfuran (213, **$R^1 = R^2 =$ iPr); Typical Procedure:**[141]
A suspension of 4-allyl-2,6-dimethylheptane-3,5-dione (**212**, $R^1 = R^2 =$ iPr; 0.51 mmol), $PdCl_2(MeCN)_2$ (0.026 mmol), and $CuCl_2$ (1.12 mmol) in dioxane (5.1 mL) was stirred at 60 °C for 12 h. The mixture was cooled to rt, filtered through a plug of silica gel (eluting with Et_2O), and concentrated under reduced pressure. The residue was purified by chromatography to give the product as a colorless oil; yield: 77%.

3,4-Dihydro-2*H*-1,4-oxazines 215; **General Procedure:**[142]
A soln of a 3-azaalk-5-enol **214** (1.0 mmol) and $PdCl_2(NCMe)_2$ (0.05 mmol) in THF (5 mL) was stirred under O_2 for 5 h at rt. The solvent was evaporated under reduced pressure. The residue was diluted with brine (10 mL) and extracted with CH_2Cl_2 (2 × 25 mL). The combined organic layers were dried (Na_2SO_4) and the solvent was removed under reduced pressure. The residue was purified by column chromatography (silica gel, pentane/EtOAc 5:1).

1.8.3.6 Oxidative Alkoxylation of Arenes

The interest in the synthesis of heterocycles starting from simple arene substrates justifies the efforts made to develop an intramolecular oxidative carbon–oxygen coupling reaction. Beside the metal, the process requires the presence of an oxidizing agent.

1.8.3.6.1 Copper Catalysis

The use of the amide oxygen as a nucleophile in an intramolecular sequence of oxidative C—H functionalization/C—O bond formation permits the synthesis of 2-arylbenzoxazoles **217** from benzanilide derivatives **216** using a catalytic copper(II)/oxygen atmosphere system in xylene at 140 °C (Scheme 94).[144] Several functional groups are tolerated including halogens and electron-donating and electron-withdrawing substituents on both of the aryl rings. The C—O bond formation takes place at the less sterically hindered position of the aryl ring. The method is not effective for the preparation of 2-alkylbenzoxazoles.

Scheme 94 Copper-Catalyzed Cyclization of *N*-Arylbenzamides to Benzoxazoles[144]

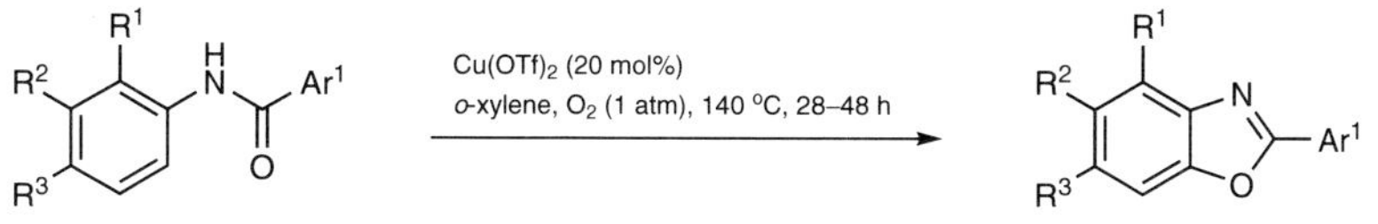

$Cu(OTf)_2$ (20 mol%)
o-xylene, O_2 (1 atm), 140 °C, 28–48 h

216 → 217

Ar^1	R^1	R^2	R^3	Yield (%)	Ref
Ph	H	H	Me	92	[144]
Ph	H	H	OEt	91	[144]
Ph	H	H	Ph	72	[144]
Ph	H	H	Bz	63	[144]
Ph	H	H	F	80	[144]
Ph	H	H	Cl	76	[144]
Ph	H	H	Br	61	[144]
Ph	H	Me	H	86	[144]
Ph	H	OMe	H	93	[144]
Ph	Me	H	H	51	[144]
Ph	OMe	H	H	55	[144]
Ph	H	$(CH{=}CH)_2$		75	[144]
Ph	$(CH{=}CH)_2$		H	70	[144]
$4-EtC_6H_4$	H	H	H	85	[144]
$4-MeOC_6H_4$	H	H	H	89	[144]
$4-MeO_2CC_6H_4$	H	H	H	83	[144]
$4-FC_6H_4$	H	H	H	80	[144]
$4-ClC_6H_4$	H	H	H	78	[144]
$4-BrC_6H_4$	H	H	H	72	[144]
$4-IC_6H_4$	H	H	H	81	[144]
2-naphthyl	H	H	H	86	[144]

2-Arylbenzoxazoles 217**; General Procedure:**[144]
To a dried Schlenk tube were added benzanilide **216** (0.25 mmol) and $Cu(OTf)_2$ (0.05 mmol). The tube and its contents were then purged with O_2 and *o*-xylene (0.5 mL) was added via syringe. The stirred mixture was heated at 140 °C for 28–48 h under O_2 (balloon). A small amount of MeOH was added to dissolve insoluble materials and the resulting mixture was purified via preparative TLC (hexane/EtOAc).

1.8.3.7 Alkoxylation of Aryl or Vinyl Halides

Among the methods available to obtain ethers, the direct intermolecular substitution of a vinyl or aryl halide with an alcohol under metal catalysis has been extensively applied. The less common but more interesting intramolecular version affords cyclic ethers or oxygenated aromatic rings under copper and palladium catalysis.

1.8.3.7.1 Copper Catalysis

Intramolecular vinylation of alcohols is a straightforward way to obtain cyclic enol ethers in good yields. In particular, starting from γ-bromo homoallylic alcohols **218**, strained 2-methyleneoxetanes **219** are efficiently and selectively obtained by exploiting an uncommon 4-*exo* ring closure, preferred over other 5-*exo*, 6-*exo*, and 6-*endo* cyclizations, under copper catalysis (Scheme 95).[145] In fact, the corresponding palladium-catalyzed process switches the selectivity to the 5-*exo* cyclization, with the formation of tetrahydrofurans. The reactivity of the alcohols follows the order aliphatic > allylic > benzylic, but primary, secondary, and tertiary alcohols react with similar rates. Substitution on the double bond discourages the cyclization and the reaction has to be conducted at higher temperatures. Beside four-membered rings, five- and six-membered heterocycles can also be prepared.

Scheme 95 Copper-Catalyzed Cyclization of 3-Bromoalk-3-enols to 2-Methyleneoxetanes[145]

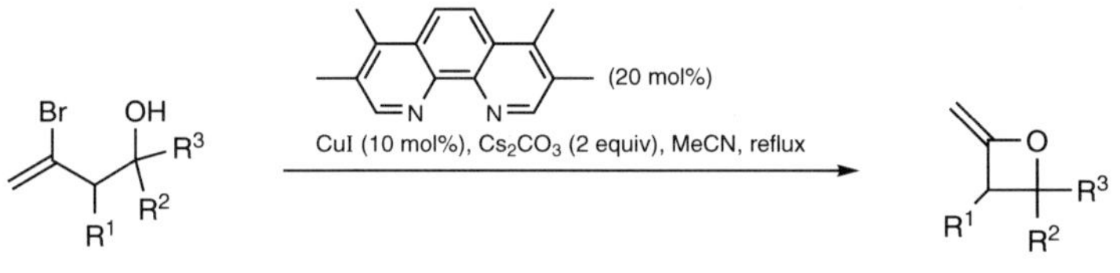

(20 mol%)
CuI (10 mol%), Cs_2CO_3 (2 equiv), MeCN, reflux

218 → **219**

R^1	R^2	R^3	Time (h)	Yield (%)	Ref
Bn	H	H	12	69	[145]
H	$(CH_2)_{10}Me$	H	8	98	[145]
H	Me	$(CH_2)_6Me$	14	74	[145]
H	Bn	$(CH_2)_6Me$	6	90	[145]
H	$(CH_2)_2Ph$	H	14	88	[145]
H	(*E*)-CH=CHPh	H	20	67	[145]
H	Ph	H	20	46	[145]
H	$(CH_2)_2OBn$	H	6	82	[145]

Carbonyl compounds such as 2-haloanilides (e.g., **222**),[146] 2-halobenzyl ketones (e.g., **220**),[147] and β-oxo esters bearing 2-bromoaryl substituents[148] can also be used as substrates for the intramolecular O-arylation leading to heterocyclization processes. In this way, a wide variety of benzo[*b*]furans **221** (Scheme 96), benzopyrans, and benzoxazoles **223** (Scheme 97) are efficiently synthesized. In these cyclizations, the use of copper catalysis proves superior to palladium catalysis. In all the cases, a variety of substituents are tolerated, both electron-withdrawing and electron-donating, as well as additional halogens. When more than one halogen is present on the aryl ring, the initial coordination of copper to the oxygen of the carbonyl group explains the selectivity toward the *ortho* halogen.

Scheme 96 Copper-Catalyzed Cyclization of 2-Halobenzyl Ketones to Benzo[*b*]furans[147]

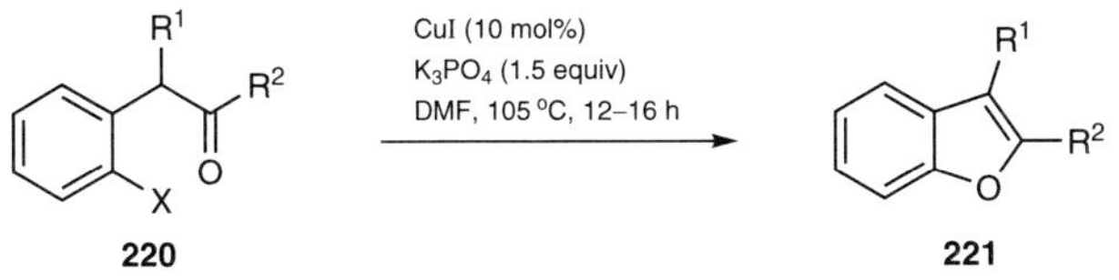

X	R^1	R^2	Yield (%)	Ref
Br	H	4-Tol	99	[147]
Br	H	$4\text{-}MeSC_6H_4$	98	[147]
I	H	$4\text{-}MeOC_6H_4$	99	[147]
Br	Me	Me	71	[147]
I	Me	$4\text{-}MeOC_6H_4$	87	[147]
I	Bn	$4\text{-}MeOC_6H_4$	76	[147]
Br	CO_2Et	Ph	88	[147]
Br	CN	Ph	91	[147]

Scheme 97 Copper-Catalyzed Cyclization of 2-Haloanilides to Benzoxazoles[146]

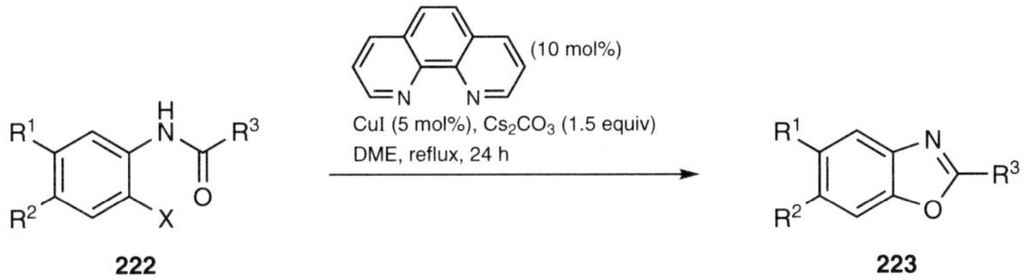

X	R^1	R^2	R^3	Yield (%)	Ref
I	H	H	Ph	>99	[146]
Br	H	H	$4\text{-}ClC_6H_4$	>99	[146]
I	H	H	$4\text{-}NCC_6H_4$	98	[146]
Br	H	H	$4\text{-}MeOC_6H_4$	>99	[146]
Br	H	Me	Ph	>99	[146]
Br	H	Cl	Ph	95	[146]
Br	H	F	Ph	99	[146]
Br	F	H	Ph	98	[146]
Br	CF_3	H	Ph	75	[146]
Br	H	H	Et	>99	[146]
Br	H	H	2-thienyl	>99	[146]

In the case of α-(2-bromobenzyl)-β-oxo esters **224** and β-diketones,[148] the O-arylation affords the corresponding substituted benzopyrans (e.g., **225**) in high yields (Scheme 98). Beside the careful choice of copper source, the base and the ligand also have a significant influence on the result.

Scheme 98 Copper-Catalyzed Cyclization of 2-(2-Bromobenzyl)-β-oxo Esters to Benzopyrans[148]

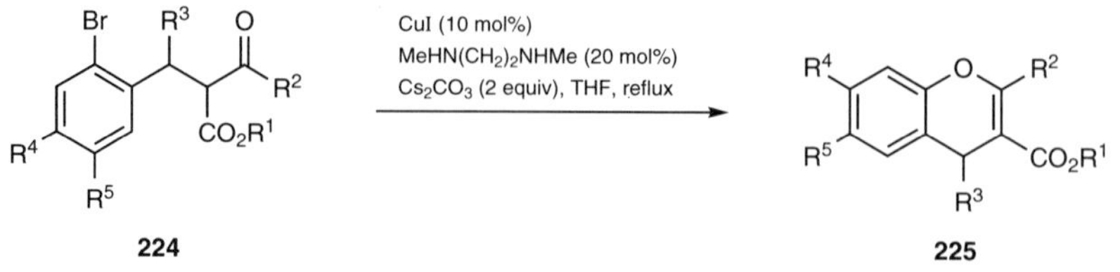

CuI (10 mol%), $MeHN(CH_2)_2NHMe$ (20 mol%), Cs_2CO_3 (2 equiv), THF, reflux

224 → 225

R^1	R^2	R^3	R^4	R^5	Time (h)	Yield (%)	Ref
Et	Me	H	H	H	1	99	[148]
Me	Et	H	H	H	3	99	[148]
Me	$(CH_2)_2CH{=}CH_2$	H	H	H	2	92	[148]
Et	Ph	H	H	H	21	80[a]	[148]
Et	Me	Ph	H	H	5	80	[148]
Me	Me	H	H	OMe	16	90	[148]
Me	Me	H	Ac	H	21	90[a]	[148]

[a] Reaction carried out in dioxane at reflux.

The O-vinylation of β-oxo esters and β-diketones tethered to vinyl bromides affords five-, six-, and seven-membered cyclic alkenyl ethers.[149]

2-Methyleneoxetanes 219; **General Procedure:**[145]
A 3-bromoalk-3-enol **218** (0.3 mmol) in anhyd MeCN (3 mL) was added to a mixture of CuI (0.03 mmol), 3,4,7,8-tetramethyl-1,10-phenanthroline (0.06 mmol), and Cs_2CO_3 (0.6 mmol) in a round-bottomed flask at rt under N_2. The mixture was stirred and heated at reflux for 8 h. TLC monitoring indicated that all the starting material had been consumed. The mixture was cooled to rt and Et_2O (20 mL) was added. The mixture was filtered and the filtrate was concentrated under reduced pressure. The crude product was purified by column chromatography (basic alumina, hexane/Et_2O/Et_3N 10:1:0.1).

2-Substituted Benzo[*b*]furans 221; **General Procedure:**[147]
A mixture of 2-bromobenzyl ketone **220** (X = Br; 2 mmol), K_3PO_4 (3 mmol), and CuI (0.2 mmol) in DMF (5 mL) was degassed via three nitrogen–vacuum cycles, heated at 105 °C for 12 h, and cooled to rt. H_2O (20 mL) was added directly to the mixture over 0.5 h to precipitate the product.

2-Substituted Benzoxazoles 223; **General Procedure:**[146]
To a mixture of amide **222** (1.0 mmol), CuI (0.05 mmol), 1,10-phenanthroline (0.10 mmol), and Cs_2CO_3 (1.5 mmol) was added DME (8 mL) at rt under N_2. The mixture was heated at reflux for 24 h and then allowed to cool to rt. H_2O (8 mL) and then CH_2Cl_2 (8 mL) were added. After the mixture had been stirred for 10 min, it was filtered through a hydrophobic membrane, with the denser DME/CH_2Cl_2 layer passing through the hydrophobic membrane and the aqueous layer remaining in the reservoir. The reservoir was washed with CH_2Cl_2 (2 × 2 mL). The combined organic layers were then evaporated under reduced pressure. The residue was passed through a short layer of silica gel to remove traces of copper salts, ligand, byproducts, or remaining starting material.

Ethyl 2-Methyl-4*H*-benzopyran-3-carboxylate (225, R^1 = Et; R^2 = Me; R^3 = R^4 = R^5 = H); Typical Procedure:[148]

N,*N*′-Dimethylethylenediamine (0.04 mmol) in THF (2 mL) was added to a mixture of CuI (3.8 mg, 0.02 mmol), ethyl 2-(2-bromobenzyl)-3-oxobutanoate (**224**, R^1 = Et; R^2 = Me; R^3 = R^4 = R^5 = H; 0.2 mmol), and Cs_2CO_3 (0.4 mmol) in a round flask under N_2. The mixture was stirred at reflux temperature for 1 h. TLC monitoring indicated that all the starting material had been consumed. The resulting mixture was cooled to rt and EtOAc (20 mL) was added. The mixture was filtered, and the filtrate was concentrated under reduced pressure. The crude product was purified by column chromatography (silica gel, hexane/EtOAc 15:1) to give the product as a colorless oil; yield: 99%.

1.8.3.7.2 Palladium Catalysis (Buchwald-Type Reaction)

A number of oxygen heterocycles (e.g., **230**) are synthesized using intramolecular etherification of aryl halides (e.g., **229**) under mild conditions in the presence of a palladium catalyst, ligand, and base. Biaryl ligands such as **226–228** display high catalytic activity, but ligand **227** is ineffective in reactions involving aryl chlorides. Although primary alcohols are more reactive, secondary alcohols are also efficiently cyclized, giving five-, six-, and seven-membered rings in good yields. Both aryl chlorides and bromides cyclize; however, the reactions of aryl bromides proceed more rapidly and cleanly. A variety of functional groups are tolerated. Optically active aryl bromides are cyclized preserving their enantiomeric purity, but aryl chlorides give a slight loss of enantiomeric excess (Scheme 99).[150]

Scheme 99 Palladium-Catalyzed Cyclization of β-(2-Bromophenyl) Alcohols to Benzopyrans, Benzodioxanes, or Benzoxazines[150]

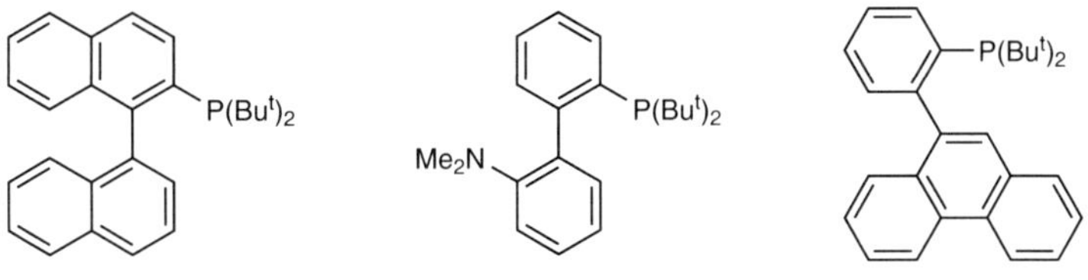

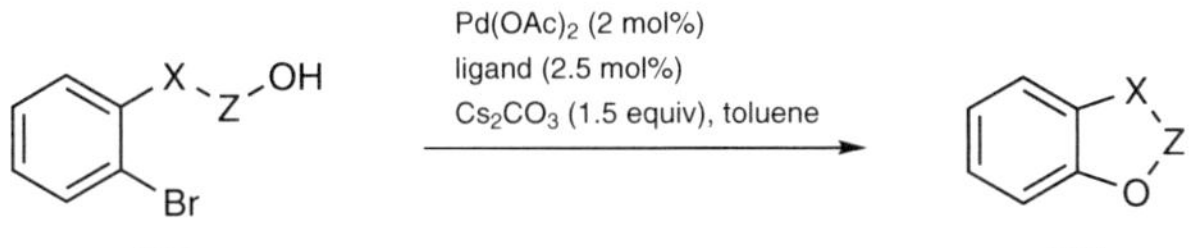

P(Bu^t)$_2$ **226** Me$_2$N P(Bu^t)$_2$ **227** P(Bu^t)$_2$ **228**

229 → Pd(OAc)$_2$ (2 mol%), ligand (2.5 mol%), Cs_2CO_3 (1.5 equiv), toluene → **230**

X	Z	Ligand	Temp (°C)	Time (h)	Yield (%)	Ref
CH_2	CH_2	**226**	50	26	85	[150]
$(CH_2)_2$	CH_2	**226**	50	21	85	[150]
$(CH_2)_2$	CHMe	**226**	65	25	79[a]	[150]
$(CH_2)_3$	CH_2	**226**	70	23	73	[150]
$(CH_2)_3$	CHMe	**226**	80	28	71[a]	[150]
O	$(CH_2)_2$	**227**	70	40	81[b]	[150]
NH	$(CH_2)_2$	**226**	70	24	84[b]	[150]
NMe	$(CH_2)_2$	**228**	70	48	74[b]	[150]

[a] Reaction carried out with 3 mol% of catalyst and 3.5 mol% of ligand.
[b] Reaction carried out with 1.2 equivalents of ligand.

for references see p 499

Heterocycles 230; General Procedure:[150]
An oven-dried 15-mL resealable Schlenk tube was charged with $Pd(OAc)_2$ (2 mol%), a ligand (2.5 mol%), and Cs_2CO_3 (1.5 equiv). The Schlenk tube was evacuated and backfilled with argon and fitted with a rubber septum. An aryl bromide **229** (0.75 mmol) and toluene (1.5 mL) were added via syringe. The resealable Schlenk tube was sealed under argon and placed in a preheated oil bath until the aryl halide had been consumed, as judged by GC analysis. The mixture was cooled to rt, diluted with pentane (2 mL), and filtered through a pad of Celite. The resulting soln was purified by flash chromatography (silica gel, hexane/EtOAc 99:1).

1.8.3.8 Alkoxycarbonylation of Alkynes

Compared to the carbamoylation process, divergent results are reported in alkoxycarbonylation reactions of alkynes. In fact, no formation of lactones is observed but the oxidative cyclization/carbonylation affords unsaturated esters as final products.

1.8.3.8.1 Palladium Catalysis

Alkoxylation of 2-alkynylphenols **231** under carbonylative conditions using methanol as solvent affords the corresponding methyl benzo[*b*]furan-3-carboxylates **232**. The reaction uses the highly effective cocatalytic palladium(II)–thiourea system and carbon tetrabromide as oxidative agent for the turnover of palladium(0) to palladium(II). The process involves the attack of the carbalkoxypalladium(II) intermediate onto the triple bond with concomitant nucleophilic attack of the phenoxy group. The cyclization gives satisfactory yields in less than half an hour and it is compatible with substrates bearing electron-donating and electron-withdrawing substituents (Scheme 100).[151]

Scheme 100 Palladium-Catalyzed Cyclization of 2-Alkynylphenols to Methyl Benzo[*b*]furan-3-carboxylates[151]

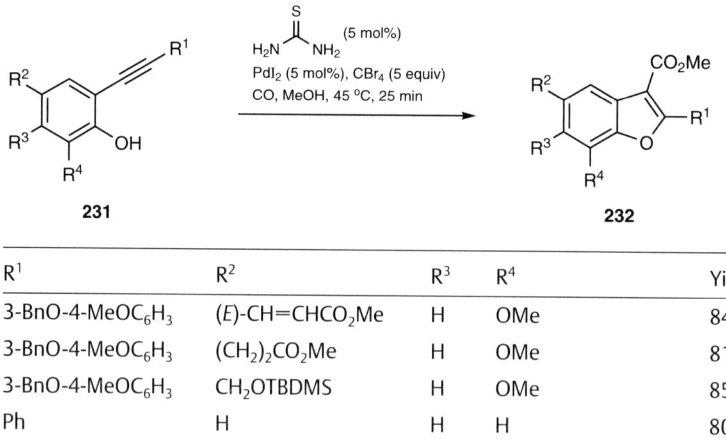

S
H2N NH2 (5 mol%)
PdI_2 (5 mol%), CBr_4 (5 equiv)
CO, MeOH, 45 °C, 25 min

R1 R2 R3 R4 OH **231**

CO2Me R1 R2 R3 R4 O **232**

R^1	R^2	R^3	R^4	Yield (%)	Ref
3-BnO-4-$MeOC_6H_3$	(*E*)-CH=$CHCO_2Me$	H	OMe	84	[151]
3-BnO-4-$MeOC_6H_3$	$(CH_2)_2CO_2Me$	H	OMe	81	[151]
3-BnO-4-$MeOC_6H_3$	$CH_2OTBDMS$	H	OMe	85	[151]
Ph	H	H	H	80	[151]
4-Tol	H	H	H	78	[151]
CH_2OMe	H	H	H	84	[151]
CH_2OMe	CO_2Me	H	OMe	80	[151]
CH_2OMe	H	OMe	(*E*)-CH=$CHCO_2Me$	79	[151]

Methyl Benzo[*b*]furan-3-carboxylates 232; General Procedure:[151]

CAUTION: *Carbon monoxide is extremely flammable and exposure to higher concentrations can quickly lead to a coma.*

A 25-mL round-bottomed flask was flame-dried under high vacuum and, on cooling, was charged with PdI_2 (0.05 mmol), thiourea (0.05 mmol), a 2-alkynylphenol **231** (1 mmol), CBr_4 (5 mmol), and MeOH (8 mL). The mixture was evacuated and backfilled with CO four times and then heated at 45 °C for 25 min. Following completion of the reaction (TLC), the mixture was cooled to rt, diluted with Et_2O (30 mL), and filtered through a pad of silica gel. The resulting soln was purified by flash chromatography (EtOAc/hexane).

1.8.3.9 Carboalkoxylation of Alkynes

The simultaneous formation of C—C and C—O bonds, known as carboalkoxylation, requires the presence of aryl or vinyl halides as intermolecular reagents to obtain the domino process. In general, the reaction gives high regio- and stereoselectivity.

1.8.3.9.1 Palladium Catalysis

The reaction of acetylenic alcohols or 2-alkynylphenols in the presence of aryl or alkyl halides under palladium catalysis and basic conditions provides 2-alkylidenetetrahydrofurans or 2-alkylidenebenzo[*b*]furans (Table 10). Beside palladium(II) acetate, triphenylphosphine and butyllithium are also essential to the reaction. Other palladium catalysts or bases, as well as different solvents instead of tetrahydrofuran, give poor results.[152]

Table 10 Palladium-Catalyzed Carboalkoxylation of Alkynyl Alcohols[152]

R^1I (1.1 equiv); $Pd(OAc)_2$ (10 mol%), Ph_3P (10 mol%), BuLi (1.1 equiv), THF, rt, 4 h

Substrate: Alkyne	Substrate: Iodide	Product	Yield (%)	Ref
	MeI		66[a]	[152]
	PhI		53	[152]
	PhI		60	[152]

for references see p 499

Table 10 (cont.)

Substrate: Alkyne	Substrate: Iodide	Product	Yield (%)	Ref
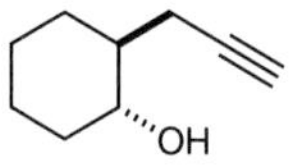	PhI	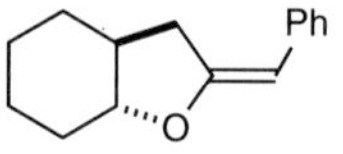	47	[152]
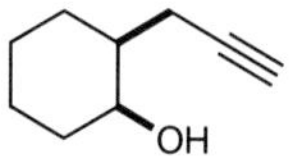	PhI	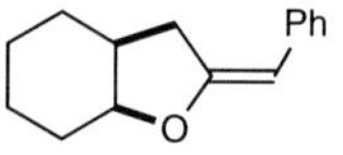	54	[152]
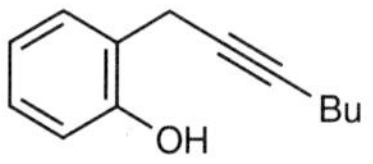	PhI	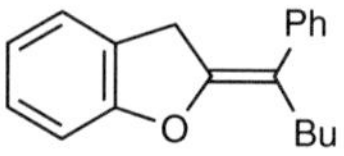	45	[152]

[a] Reaction carried out with 10 mol% $PdCl_2$.

A similar reaction is realized in acetonitrile starting from nonterminal 2-alkynylphenols **233** and aryl iodides affording a library of 2,3-diarylbenzo[*b*]furans **234** (Scheme 101).[153]

Scheme 101 Palladium-Catalyzed Cyclization of 2-Alkynylphenols to 2,3-Diarylbenzo[*b*]furans[153]

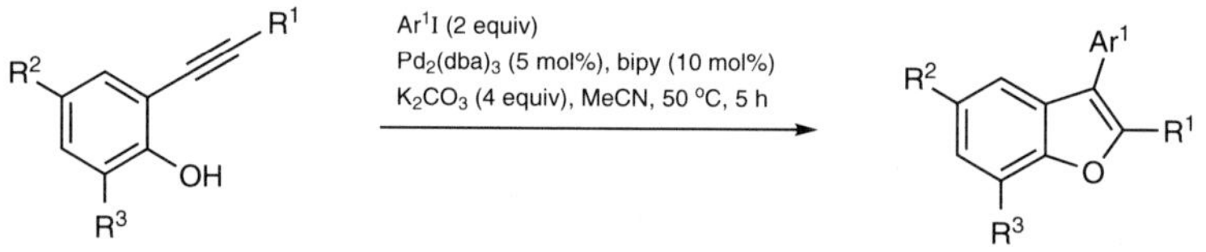

Ar^1I (2 equiv)
$Pd_2(dba)_3$ (5 mol%), bipy (10 mol%)
K_2CO_3 (4 equiv), MeCN, 50 °C, 5 h

233 → **234**

R^1	R^2	R^3	Ar^1	Yield (%)	Ref
Ph	H	H	4-Tol	64	[153]
Ph	H	H	4-$O_2NC_6H_4$	87	[153]
Ph	H	H	2-$MeOC_6H_4$	74	[153]
Ph	H	H	4-AcC_6H_4	85	[153]
Ph	H	H	Ph	52	[153]
Ph	OTBDMS	Me	4-AcC_6H_4	82	[153]
Ph	OTBDMS	Me	4-Tol	70	[153]
Ph	OTBDMS	Me	2-$MeOC_6H_4$	70	[153]
Ph	2,4-$F_2C_6H_3$	CO_2Me	4-AcC_6H_4	72	[153]
Ph	2,4-$F_2C_6H_3$	CO_2Me	4-Tol	55	[153]
CH_2OMe	CO_2Me	OMe	4-AcC_6H_4	64	[153]

The reaction of 2-ethynylphenols **235** with a variety of unsaturated halides or trifluoromethanesulfonates produces 2-aryl- or 2-vinylbenzo[*b*]furans **236** in satisfactory yields through a carboalkoxylation process (Scheme 102).[154]

Scheme 102 Palladium-Catalyzed Cyclization of 2-Ethynylphenols to 2-Substituted Benzo[b]furans[154]

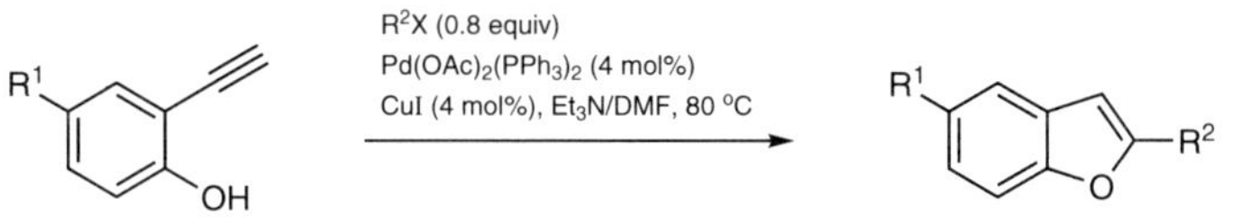

R^2X (0.8 equiv)
$Pd(OAc)_2(PPh_3)_2$ (4 mol%)
CuI (4 mol%), Et_3N/DMF, 80 °C

235 → **236**

R^1	R^2	X	Time (h)	Yield (%)	Ref
H	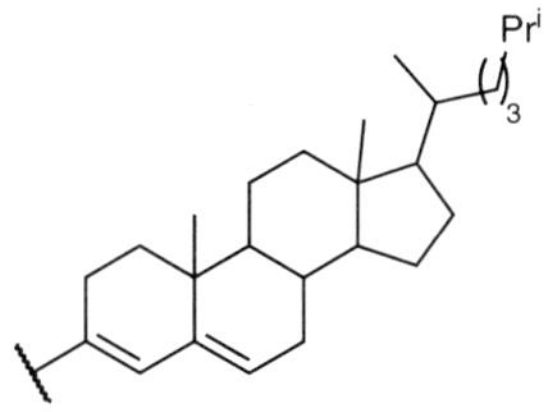	OTf	1	78	[154]
H	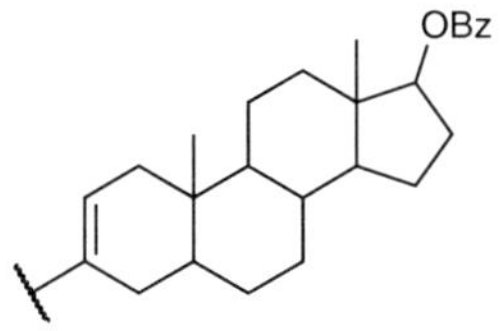	OTf	2.5	64	[154]
H	2-naphthyl	OTf	5	87	[154]
H	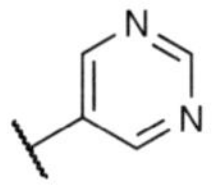	Br	4	20[a]	[154]
CN	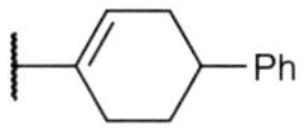	OTf	5	42	[154]
Me	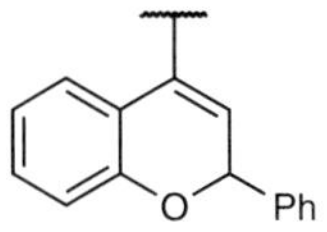	OTf	3.5	50	[154]
Me	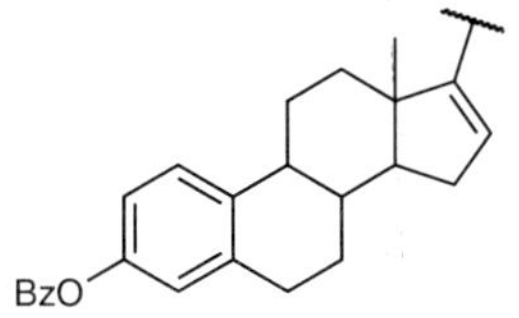	OTf	2	72	[154]

[a] The same product was also obtained in 43% yield starting from 2-[(trimethylsilyl)ethynyl]phenyl acetate.

2-Alkylidenetetrahydrofurans and 2-Alkylidenebenzo[b]furans (Table 10); General Procedure:[152]

A 1.6 M soln of BuLi in benzene (2.2 mmol) (**CAUTION:** *carcinogen*) was added dropwise to a soln of an acetylenic alcohol or a 2-alkynylphenol (2 mmol) in THF (2 mL) at 0 °C under N_2. $Pd(OAc)_2$ (45 mg, 0.2 mmol) and Ph_3P (0.2 mmol) in THF (1 mL) and then PhI (2.2 mmol) were added. The mixture was stirred at rt for 4 h and then the reaction was quenched with

H_2O (10 mL). The organic layer was separated, and the aqueous layer was extracted with Et_2O. The combined organic layers were dried ($MgSO_4$), filtered, and concentrated to give a pale yellow solid, which was purified by HPLC.

2,3-Diarylbenzo[*b*]furans 234; **General Procedure:**[153]
To a soln of an aryl iodide (2.0 mmol), bipy (0.1 mmol), and K_2CO_3 (4.0 mmol) in MeCN (3.0 mL) was added $Pd_2(dba)_3$ (0.05 mmol), and the mixture was stirred at 50 °C for 1 h. A soln of 2-alkynylphenol **233** (1.0 mmol) in MeCN (2 mL) was added, and the mixture was stirred at 50 °C for 5 h under argon. The mixture was concentrated, and the residue was filtered through a silica gel pad eluting with EtOAc. The filtrate was concentrated, and the residue was purified by flash chromatography (silica gel).

2-(Cholesta-3,5-dien-3-yl)benzo[*b*]furan (236, **R^1 = H; R^2 = Cholesta-3,5-dien-3-yl); Typical Procedure:**[154]
To a stirred soln of 2-ethynylphenol (**235**, R^1 = H; 174 mg, 1.47 mmol) and Et_3N (4 mL) in DMF (4 mL) were added cholesta-3,5-dien-3-yl trifluoromethanesulfonate (630 mg, 1.22 mmol), $Pd(OAc)_2(PPh_3)_2$ (46 mg, 0.06 mmol), and CuI (12 mg, 0.06 mmol). The mixture was stirred at 80 °C for 1 h under N_2. Et_2O and 0.1 M aq HCl were added, and the organic layer was separated, washed with H_2O, dried (Na_2SO_4), and concentrated under reduced pressure. The residue was purified by flash chromatography (silica gel, hexane/EtOAc 97:3); yield: 460 mg (78%).

1.8.3.10 Carboalkoxylation of Alkenes

Carboalkoxylation is a convenient method to afford functionalized oxygenated ring systems. The heterocyclization is performed in the presence of aryl or vinyl halides or trifluoromethanesulfonates with the insertion of a substituent onto the alkene through the concomitant formation of C—O and C—C bonds.

1.8.3.10.1 Palladium Catalysis

Starting from 2-alkenylphenols **237** and in the presence of vinyl halides or trifluoromethanesulfonates, the carboalkoxylation reaction affords substituted dihydrobenzopyrans **238** or dihydrobenzo[*b*]furans with complete regioselectivity (Scheme 103).[155]

Scheme 103 Palladium-Catalyzed Cyclization of 2-Alkenylphenols to Dihydrobenzopyrans and a Dihydrobenzo[*b*]furan[155]

R^3X (1 equiv)
$Pd(OAc)_2$ (5 mol%), Na_2CO_3 (3.5 equiv)
TBACl (1.2 equiv), DMF, 80 °C

237 (2 equiv) → **238**

R^1	R^2	R^3	X	Time (h)	Yield (%)	Ref
H	H	(*E*)-CH=CHPh	Br	24	82	[155]
H	H	(*E*)-CH=CHBu	I	36	62	[155]
H	H	CH=CMe$_2$	I	36	68	[155]
H	H	4-phenylcyclohex-1-enyl (structure) -Ph	I	6	55	[155]

R^1	R^2	R^3	X	Time (h)	Yield (%)	Ref
H	H	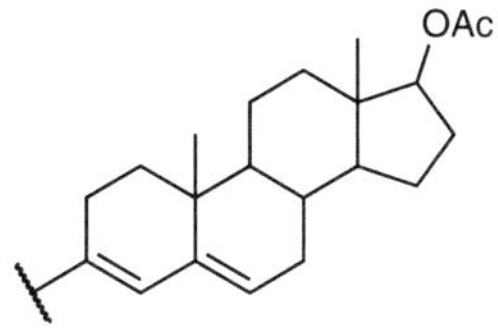 OAc	OTf	5	56	[155]
H	H	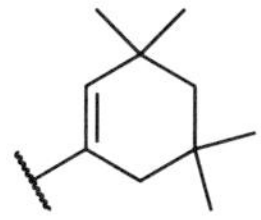	OTf	3	76	[155]
Me	H	(*E*)-CH=CHPh	Br	24	75[a]	[155]
Me	H	(*E*)-CH=CH*t*-Bu	Br	24	83[b]	[155]
H	Me	(*E*)-CH=CHPh	Br	72	56	[155]

[a] Ratio (*cis*/*trans*) 77:23.
[b] Ratio (*cis*/*trans*) 7:3.

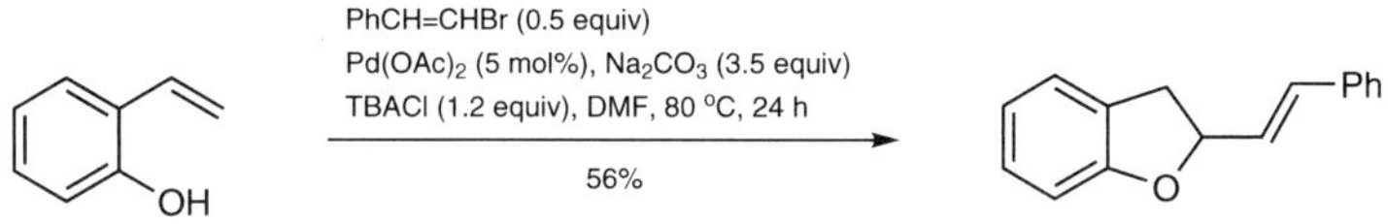

PhCH=CHBr (0.5 equiv)
$Pd(OAc)_2$ (5 mol%), Na_2CO_3 (3.5 equiv)
TBACl (1.2 equiv), DMF, 80 °C, 24 h
56%
OH
O
Ph

Dihydrobenzopyrans 238; General Procedure:[155]
A vinyl halide or trifluoromethanesulfonate (1 equiv) was added to a mixture of $Pd(OAc)_2$ (5 mol%), a 2-alkenylphenol **237** (2.0 equiv), Na_2CO_3 (3.5 equiv), and TBACl (1.2 equiv) in DMF (2 mL) in a 1-dram vial. The vial was flushed with N_2 and capped with a screw cap containing a Teflon liner. After the mixture had been heated at 80 °C for the indicated time, it was diluted with Et_2O and washed with sat. aq NH_4Cl and H_2O. The organic layer was dried ($MgSO_4$), filtered, and concentrated, and the residue was purified by flash column chromatography or preparative HPLC.

1.8.3.11 Applications in the Syntheses of Natural Products and Drug Molecules

An application of the two-fold hydroalkoxylation process (see Section 1.8.3.1.2) is exploited in the synthesis of the C19 and C34 spiroketal fragments of okadaic acid, a complex natural structure isolated from marine sponges that includes several spiroketal systems (Scheme 104).[156]

Scheme 104 Okadaic Acid[156]

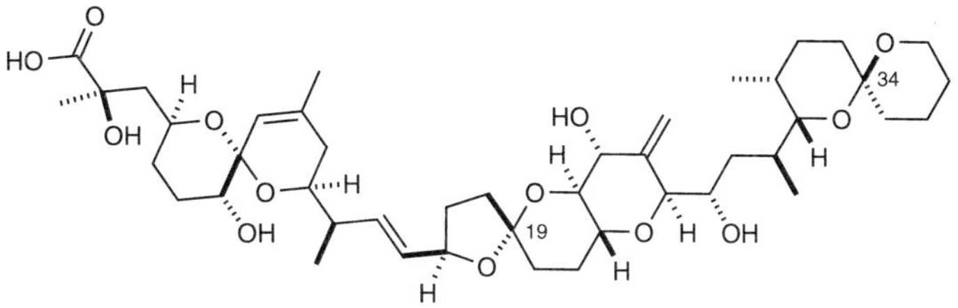

HO O OH H O O H OH HO H O 19 O H O H OH H O 34 O H

Other applications of hydroalkoxylation are involved in the synthesis of the spiroketal function of cephalosporolide H, which is isolated from the marine fungus *Penicillium* sp. and shows anti-inflammatory properties,[157] and the preparation of the spiroketal core of rubromycins (Scheme 105).[158]

Scheme 105 Spiroketal Core of Cephalosporide H and Rubromycins[157,158]

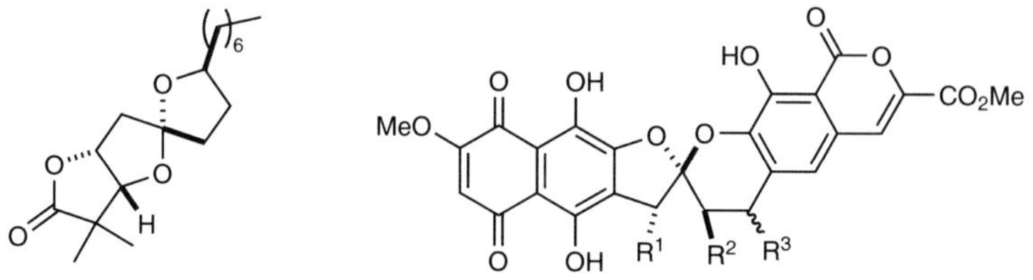

cephalosporolide H

γ-rubromycin (R^1 = R^2 = R^3 = H)
purpuromycin (R^1 = R^2 = H; R^3 = OH)
heliquinomycin (R^1 = *o*-cymarose; R^2 = OH; R^3 = H)

A double hydroalkoxylation on enynol **239** involving one hydroxy group and one methyl ether group as nucleophiles is reported during the preparation of spiroketal **240**, which contains the A–D rings of the marine toxin azaspiracid (Scheme 106).[159]

Scheme 106 Gold-Catalyzed Synthesis of an Intermediate of Azaspiracid[159]

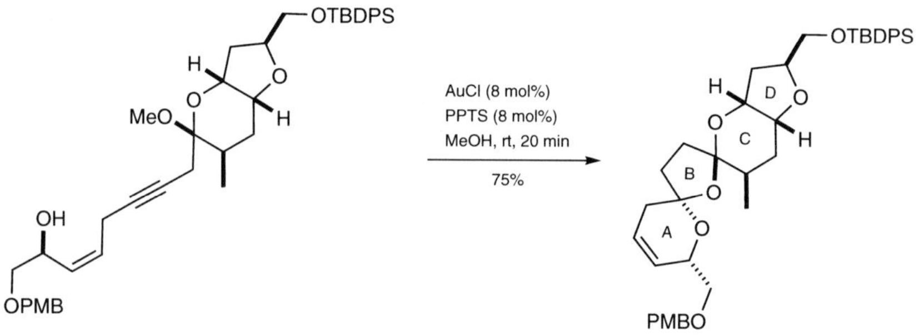

AuCl (8 mol%)
PPTS (8 mol%)
MeOH, rt, 20 min
75%

239 → **240**

An efficient application of a 5-*endo-dig* cyclization affords the spiroketal core of the eastern fragment of (+)-cephalostatin 1, a macrolide with interesting biological activity.[160] The efficacy of the gold-catalyzed oxacyclization of alkynols is also highlighted in the key step of the total synthesis of aurone skeletons, a class of natural flavonoids exhibiting many biological activities.[161]

Intramolecular formation of diaryl ethers starting from aryl bromides (e.g., **241**) as a result of an alkoxylation process is applied to the synthesis of aristoyagonine (**242**), an alkaloid with a benzoxepine skeleton (Scheme 107),[162] and other dibenzoxepine frameworks.[163] The *Z* configuration of the double bond does not hamper the formation of the dibenzoxepine ring system.

Scheme 107 Copper-Catalyzed Synthesis of Aristoyagonine[162]

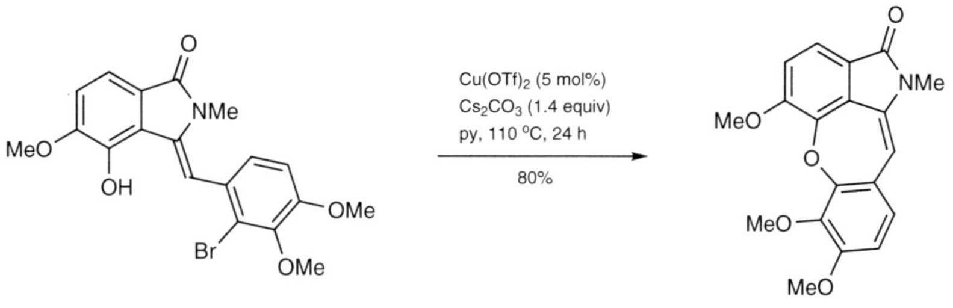

$Cu(OTf)_2$ (5 mol%)
Cs_2CO_3 (1.4 equiv)
py, 110 °C, 24 h
80%

241 **242**

Analogously, the arylation of phenols under copper catalysis is involved in the synthesis of macrocyclic ethers combrestatin D2[164] and piperazinomycin,[165] the cyclic hexapeptide bouvardin,[166] and the glycopeptide antibiotic vancomycin.[167]

Intramolecular tandem alkoxycarbonylation of alkenes is applied as a key step in the construction of the functionalized polycyclic portion of micrandilactone A (**246**), a triterpene isolated from the medicinal plant *Schisandra micrantha*. Use of the particular thiourea **244** as ligand ensures the stereoselective formation of the tricyclic lactone **245** from enediol **243** (Scheme 108).[168]

Scheme 108 Synthesis of an Intermediate of Micrandilactone A[168]

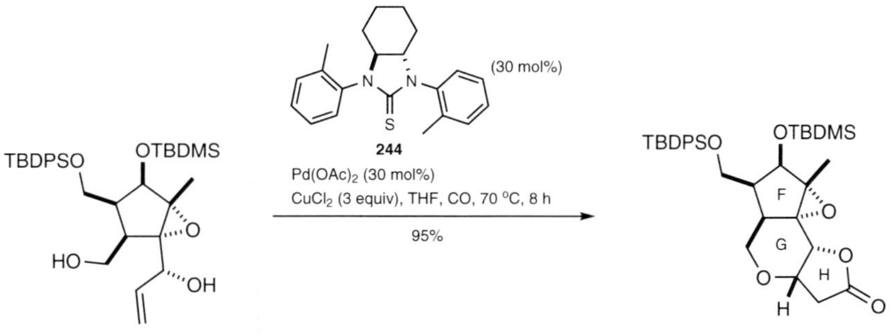

244 (30 mol%)
$Pd(OAc)_2$ (30 mol%)
$CuCl_2$ (3 equiv), THF, CO, 70 °C, 8 h
95%

243 **245**

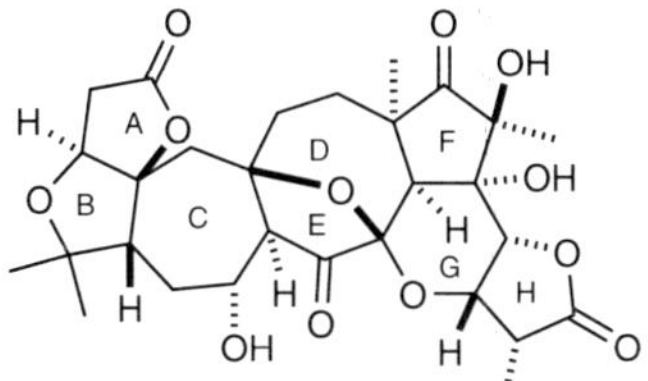

246

Trioxadispiroketal System 240; Typical Procedure:[159]
AuCl (0.9 mmol) and PPTS (0.9 mmol) were added to a stirred soln of enynol **239** (11 mmol) in MeOH (1 mL) under N_2. After the mixture had been stirred for 20 min, the solvent was removed and the residue was purified by flash chromatography (silica gel, hexane/EtOAc 3:1) to give the product as a colorless oil; yield: 75%.

for references see p 499

Aristoyagonine (242); Typical Procedure:[162]
A soln of aryl bromide **241** (0.6 mmol), Cs_2CO_3 (0.85 mmol), and $Cu(OTf)_2$ (0.03 mmol) in pyridine (10 mL) was heated at reflux for 24 h. The mixture was poured into 1 M aq HCl and extracted with EtOAc. The organic layer was stirred with 10% aq NaOH and then washed with H_2O and brine, dried ($MgSO_4$), filtered, and concentrated to give a yellow solid. Aristoyagonine was obtained by recrystallization (MeOH) as bright yellow crystals; yield: 80%.

(1*R*,2*S*,2a*R*,4a*S*,7a*R*,7b*S*,8a*S*)-1-(*tert*-Butyldimethylsiloxy)-2-[(*tert*-butyldiphenylsiloxy)-methyl]-8a-methylhexahydro-2*H*-furo[3,2-*b*]oxireno[2′,3′:1,5]cyclopenta[1,2-*d*]pyran-6(5*H*)-one (245); Typical Procedure:[168]

CAUTION: *Carbon monoxide is extremely flammable and toxic, and exposure to higher concentrations can quickly lead to a coma.*

To a soln of thiourea ligand **244** (0.103 mmol) in THF (15 mL) were added $Pd(OAc)_2$ (23.1 mg, 0.103 mmol) and $CuCl_2$ (1.03 mmol) and the mixture was stirred under N_2 at rt for 1 h. To this soln the diol **243** (0.34 mmol) in THF (3 mL) was added dropwise and the mixture was placed under CO (balloon) and stirred at 70 °C for 8 h. The solvent was removed under reduced pressure, and the residue was purified by flash chromatography (petroleum ether/EtOAc 6:1) to give the product as a white foam; yield: 95%.

1.8.4 Conclusions

This chapter demonstrates that the formation of C—N and C—O bonds through transition-metal-catalyzed processes constitutes an efficient strategy to synthesize polyheterocyclic systems. Among these processes, the addition of amines to aryl halides occurs with sufficient scope and efficiency to be a commonly used synthetic method in both academic and industrial settings. Nevertheless, the possibility to use unsaturated substrates without prefunctionalized C—X bonds (X = halogens, OTf, etc.) for the construction of complex structures in a few steps is of considerable importance. In particular, addition reactions to alkenes constitute some of the most desirable, and when successfully developed, most used catalytic reactions in the chemical industry, especially considering the favorable economical aspects. Such a goal may not be easily obtained by other methods. Likewise, the $C(sp^3)$ oxidative amination is one of the most recent enlightening results.

It is expected that the application of these methods in the context of fine chemical synthesis and medicinal chemistry, and in the total synthesis of natural products, will continue and at the same time will lead to further advances in the field. In fact, some challenges regarding the total regioselectivity and the use of new chiral ligands to improve asymmetric synthesis remain.

References

[1] Müller, T. E.; Beller, M., *Chem. Rev.*, (1998) **98**, 675.
[2] Kadota, I.; Shibuya, A.; Gyoung, Y. S.; Yamamoto, Y., *J. Am. Chem. Soc.*, (1998) **120**, 10262.
[3] Genin, E.; Antoniotti, S.; Michelet, V.; Genêt, J.-P., *Angew. Chem. Int. Ed.*, (2005) **44**, 4949.
[4] Guram, A. S.; Rennels, R. A.; Buchwald, S. L., *Angew. Chem. Int. Ed. Engl.*, (1995) **34**, 1348.
[5] Ley, S. V.; Thomas, A. W., *Angew. Chem. Int. Ed.*, (2003) **42**, 5400.
[6] Monnier, F.; Taillefer, M., *Angew. Chem. Int. Ed.*, (2009) **48**, 6954.
[7] Muzart, J., *Tetrahedron*, (2005) **61**, 9423.
[8] Yet, L., *Chem. Rev.*, (2000) **100**, 2963.
[9] Nakamura, I.; Yamamoto, Y., *Chem. Rev.*, (2004) **104**, 2127.
[10] D'Souza, D. M.; Müller, T. J. J., *Chem. Soc. Rev.*, (2007) **36**, 1095.
[11] Zhang, M., *Adv. Synth. Catal.*, (2009) **351**, 2243.
[12] Krause, N.; Aksin-Artok, Ö.; Breker, V.; Deutsch, C.; Gockel, B.; Poonoth, M.; Sawama, Y.; Sawama, Y.; Sun, T.; Winter, C., *Pure Appl. Chem.*, (2010) **82**, 1529.
[13] Magano, J.; Dunetz, J. R., *Chem. Rev.*, (2011) **111**, 2177.
[14] Müller, T. E., *Tetrahedron Lett.*, (1998) **39**, 5961.
[15] Arcadi, A.; Cacchi, S.; Marinelli, F., *Tetrahedron Lett.*, (1989) **30**, 2581.
[16] Müller, T. E.; Grosche, M.; Herdtweck, E.; Pleier, A.-K.; Walter, E.; Yan, Y.-K., *Organometallics*, (2000) **19**, 170.
[17] Müller, T. E.; Hultzsch, K. C.; Yus, M.; Foubelo, F.; Tada, M., *Chem. Rev.*, (2008) **108**, 3795.
[18] Hiroya, K.; Itoh, S.; Sakamoto, T., *J. Org. Chem.*, (2004) **69**, 1126.
[19] Kel'in, A. V.; Sromek, A. W.; Gevorgyan, V., *J. Am. Chem. Soc.*, (2001) **123**, 2074.
[20] Gabriele, B.; Salerno, G.; Fazio, A., *J. Org. Chem.*, (2003) **68**, 7853.
[21] Fukuda, Y.; Utimoto, K., *Synthesis*, (1991), 975.
[22] Villemin, D.; Goussu, D., *Heterocycles*, (1989) **29**, 1255.
[23] Arcadi, A.; Bianchi, G.; Marinelli, F., *Synthesis*, (2004), 610.
[24] Rudisill, D. E.; Stille, J. K., *J. Org. Chem.*, (1989) **54**, 5856.
[25] Mohr, F.; Binfield, S. A.; Fettinger, J. C.; Vedernikov, A. N., *J. Org. Chem.*, (2005) **70**, 4833.
[26] Patil, N. T.; Mpaka Lutete, L.; Wu, H.; Pahadi, N. K.; Gridnev, I. D.; Yamamoto, Y., *J. Org. Chem.*, (2006) **71**, 4270.
[27] Narsireddy, M.; Yamamoto, Y., *J. Org. Chem.*, (2008) **73**, 9698.
[28] Trost, B. M.; McClory, A., *Angew. Chem. Int. Ed.*, (2007) **46**, 2074.
[29] Kondo, T.; Okada, T.; Suzuki, T.; Mitsudo, T.-a., *J. Organomet. Chem.*, (2001) **622**, 149.
[30] Álvarez-Corral, M.; Muñoz-Dorado, M.; Rodríguez-García, I., *Chem. Rev.*, (2008) **108**, 3174.
[31] Carney, J. M.; Donoghue, P. J.; Wuest, W. M.; Wiest, O.; Helquist, P., *Org. Lett.*, (2008) **10**, 3903.
[32] Enomoto, T.; Girard, A.-L.; Yasui, Y.; Takemoto, Y., *J. Org. Chem.*, (2009) **74**, 9158.
[33] Zulys, A.; Dochnahl, M.; Hollmann, D.; Löhnwitz, K.; Herrmann, J.-S.; Roesky, P. W.; Blechert, S., *Angew. Chem. Int. Ed.*, (2005) **44**, 7794.
[34] Yin, Y.; Ma, W.; Chai, Z.; Zhao, G., *J. Org. Chem.*, (2007) **72**, 5731.
[35] Wang, H.; Li, Y.; Jiang, L.; Zhang, R.; Jin, K.; Zhao, D.; Duan, C., *Org. Biomol. Chem.*, (2011) **9**, 4983.
[36] Ohmiya, H.; Moriya, T.; Sawamura, M., *Org. Lett.*, (2009) **11**, 2145.
[37] Liu, X.-Y.; Li, C.-H.; Che, C.-M., *Org. Lett.*, (2006) **8**, 2707.
[38] Bender, C. F.; Widenhoefer, R. A., *Org. Lett.*, (2006) **8**, 5303.
[39] Hesp, K. D.; Stradiotto, M., *Org. Lett.*, (2009) **11**, 1449.
[40] Hesp, K. D.; Tobisch, S.; Stradiotto, M., *J. Am. Chem. Soc.*, (2010) **132**, 413.
[41] Komeyama, K.; Morimoto, T.; Takaki, K., *Angew. Chem. Int. Ed.*, (2006) **45**, 2938.
[42] Bender, C. F.; Widenhoefer, R. A., *J. Am. Chem. Soc.*, (2005) **127**, 1070.
[43] Bender, C. F.; Hudson, W. B.; Widenhoefer, R. A., *Organometallics*, (2008) **27**, 2356.
[44] Liu, Z.; Hartwig, J. F., *J. Am. Chem. Soc.*, (2008) **130**, 1570.
[45] Julian, L. D.; Hartwig, J. F., *J. Am. Chem. Soc.*, (2010) **132**, 13813.
[46] Shen, X.; Buchwald, S. L., *Angew. Chem. Int. Ed.*, (2010) **49**, 564.
[47] Hegedus, L. S.; Allen, G. F.; Bozell, J. J.; Waterman, E. L., *J. Am. Chem. Soc.*, (1978) **100**, 5800.
[48] Hegedus, L. S.; McKearin, J. M., *J. Am. Chem. Soc.*, (1982) **104**, 2444.
[49] Abbiati, G.; Beccalli, E. M.; Broggini, G.; Martinelli, M.; Paladino, G., *Synlett*, (2006), 73.
[50] Wu, L.; Qiu, S.; Liu, G., *Org. Lett.*, (2009) **11**, 2707.
[51] Broggini, G.; Barbera, V.; Beccalli, E. M.; Chiacchio, U.; Fasana, A.; Galli, S.; Gazzola, S., *Adv. Synth. Catal.*, (2013) **355**, 1640.

[52] Kondo, T.; Okada, T.; Mitsudo, T.-a., *J. Am. Chem. Soc.*, (2002) **124**, 186.
[53] Beccalli, E. M.; Broggini, G.; Martinelli, M.; Sottocornola, S., *Chem. Rev.*, (2007) **107**, 5318.
[54] Brasche, G.; Buchwald, S. L., *Angew. Chem. Int. Ed.*, (2008) **47**, 1932.
[55] Wang, H.; Wang, Y.; Peng, C.; Zhang, J.; Zhu, Q., *J. Am. Chem. Soc.*, (2010) **132**, 13217.
[56] Xiao, Q.; Wang, W.-H.; Liu, G.; Meng, F.-K.; Chen, J.-H.; Yang, Z.; Shi, Z.-J., *Chem.–Eur. J.*, (2009) **15**, 7292.
[57] Inamoto, K.; Saito, T.; Hiroya, K.; Doi, T., *Synlett*, (2008), 3157.
[58] Mei, T.-S.; Wang, X.; Yu, J.-Q., *J. Am. Chem. Soc.*, (2009) **131**, 10806.
[59] Miura, T.; Ito, Y.; Murakami, M., *Chem. Lett.*, (2009) **38**, 328.
[60] Wasa, M.; Yu, J.-Q., *J. Am. Chem. Soc.*, (2008) **130**, 14058.
[61] Tsang, W. C. P.; Zheng, N.; Buchwald, S. L., *J. Am. Chem. Soc.*, (2005) **127**, 14560.
[62] Tsang, W. C. P.; Munday, R. H.; Brasche, G.; Zheng, N.; Buchwald, S. L., *J. Org. Chem.*, (2008) **73**, 7603.
[63] Jordan-Hore, J. A.; Johansson, C. C. C.; Gulias, M.; Beck, E. M.; Gaunt, M. J., *J. Am. Chem. Soc.*, (2008) **130**, 16184.
[64] Inamoto, K.; Saito, T.; Katsuno, M.; Sakamoto, T.; Hiroya, K., *Org. Lett.*, (2007) **9**, 2931.
[65] Pearson, R.; Zhang, S.; He, G.; Edwards, N.; Chen, G., *Beilstein J. Org. Chem.*, (2013) **9**, 891.
[66] Inamoto, K.; Saito, T.; Hiroya, K.; Doi, T., *J. Org. Chem.*, (2010) **75**, 3900.
[67] Hsieh, T. H. H.; Dong, V. M., *Tetrahedron*, (2009) **65**, 3062.
[68] Yamada, K.; Kubo, T.; Tokuyama, H.; Fukuyama, T., *Synlett*, (2002), 231.
[69] Kwong, F. Y.; Buchwald, S. L., *Org. Lett.*, (2003) **5**, 793.
[70] Evindar, G.; Batey, R. A., *Org. Lett.*, (2003) **5**, 133.
[71] Liubchak, K.; Tolmachev, A.; Nazarenko, K., *J. Org. Chem.*, (2012) **77**, 3365.
[72] Cuny, G.; Bois-Choussy, M.; Zhu, J., *J. Am. Chem. Soc.*, (2004) **126**, 14475.
[73] Schlummer, B.; Scholz, U., *Adv. Synth. Catal.*, (2004) **346**, 1599.
[74] Yang, B. H.; Buchwald, S. L., *Org. Lett.*, (1999) **1**, 35.
[75] Brown, J. A., *Tetrahedron Lett.*, (2000) **41**, 1623.
[76] Abbiati, G.; Beccalli, E. M.; Broggini, G.; Paladino, G.; Rossi, E., *Synthesis*, (2005), 2881.
[77] He, G.; Zhao, Y.; Zhang, S.; Lu, C.; Chen, G., *J. Am. Chem. Soc.*, (2012) **134**, 3.
[78] Neumann, J. J.; Rakshit, S.; Dröge, T.; Glorius, F., *Angew. Chem. Int. Ed.*, (2009) **48**, 6892.
[79] Espino, C. G.; Du Bois, J., *Angew. Chem. Int. Ed.*, (2001) **40**, 598.
[80] Lwowski, W., In *Nitrenes*, Lwowski, W., Ed.; Interscience: New York, (1970); p 185.
[81] Shen, M.; Leslie, B. E.; Driver, T. G., *Angew. Chem. Int. Ed.*, (2008) **47**, 5056.
[82] Gorin, D. J.; Davis, N. R.; Toste, F. D., *J. Am. Chem. Soc.*, (2005) **127**, 11260.
[83] Stokes, B. J.; Dong, H.; Leslie, B. E.; Pumphrey, A. L.; Driver, T. G., *J. Am. Chem. Soc.*, (2007) **129**, 7500.
[84] Wrona, I. E.; Gozman, A.; Taldone, T.; Chiosis, G.; Panek, J. S., *J. Org. Chem.*, (2010) **75**, 2820.
[85] Wrona, I. E.; Gabarda, A. E.; Evano, G.; Panek, J. S., *J. Am. Chem. Soc.*, (2005) **127**, 15026.
[86] Stokes, B. J.; Jovanović, B.; Dong, H.; Richert, K. J.; Riell, R. D.; Driver, T. G., *J. Org. Chem.*, (2009) **74**, 3225.
[87] Shou, W. G.; Li, J.; Guo, T.; Lin, Z.; Jia, G., *Organometallics*, (2009) **28**, 6847.
[88] Dong, H.; Shen, M.; Redford, J. E.; Stokes, B. J.; Pumphrey, A. L.; Driver, T. G., *Org. Lett.*, (2007) **9**, 5191.
[89] Hiroya, K.; Matsumoto, S.; Ashikawa, M.; Ogiwara, K.; Sakamoto, T., *Org. Lett.*, (2006) **8**, 5349.
[90] Orito, K.; Horibata, A.; Nakamura, T.; Ushito, H.; Nagasaki, H.; Yuguchi, M.; Yamashita, S.; Tokuda, M., *J. Am. Chem. Soc.*, (2004) **126**, 14342.
[91] Inoue, S.; Shiota, H.; Fukumoto, Y.; Chatani, N., *J. Am. Chem. Soc.*, (2009) **131**, 6898.
[92] Cacchi, S.; Fabrizi, G.; Lamba, D.; Marinelli, F.; Parisi, L. M., *Synthesis*, (2003), 728.
[93] Dai, G.; Larock, R. C., *J. Org. Chem.*, (2003) **68**, 920.
[94] Arcadi, A.; Anacardio, R.; D'Anniballe, G.; Gentile, M., *Synlett*, (1997), 1315.
[95] Gabriele, B.; Salerno, G.; Fazio, A.; Campana, F. B., *Chem. Commun. (Cambridge)*, (2002), 1408.
[96] Wolfe, J. P., *Synlett*, (2008), 2913.
[97] Bertrand, M. B.; Neukom, J. D.; Wolfe, J. P., *J. Org. Chem.*, (2008) **73**, 8851.
[98] Nakhla, J. S.; Schultz, D. M.; Wolfe, J. P., *Tetrahedron*, (2009) **65**, 6549.
[99] Fritz, J. A.; Wolfe, J. P., *Tetrahedron*, (2008) **64**, 6838.
[100] Chiba, H.; Sakai, Y.; Ohara, A.; Oishi, S.; Fujii, N.; Ohno, H., *Chem.–Eur. J.*, (2013) **19**, 8875.
[101] Crawley, S. L.; Funk, R. L., *Org. Lett.*, (2006) **8**, 3995.
[102] Trost, B. M.; Dong, G., *J. Am. Chem. Soc.*, (2006) **128**, 6054.

[103] Wang, S.; Dong, Y.; Wang, X.; Hu, X.; Liu, J. O.; Hu, Y., *Org. Biomol. Chem.*, (2005) **3**, 911.
[104] Yamada, K.; Kurokawa, T.; Tokuyama, H.; Fukuyama, T., *J. Am. Chem. Soc.*, (2003) **125**, 6630.
[105] Okano, K.; Tokuyama, H.; Fukuyama, T., *J. Am. Chem. Soc.*, (2006) **128**, 7136.
[106] Ma, D.; Geng, Q.; Zhang, H.; Jiang, Y., *Angew. Chem. Int. Ed.*, (2010) **49**, 1291.
[107] Ma, D.; Xia, C., *Org. Lett.*, (2001) **3**, 2583.
[108] Toumi, M.; Couty, F.; Evano, G., *Angew. Chem. Int. Ed.*, (2007) **46**, 572.
[109] Beccalli, E. M.; Broggini, G.; Paladino, G.; Zoni, C., *Tetrahedron*, (2005) **61**, 61.
[110] Carril, M.; SanMartin, R.; Churruca, F.; Tellitu, I.; Domínguez, E., *Org. Lett.*, (2005) **7**, 4787.
[111] Takahashi, M.; Tanabe, H.; Nakamura, T.; Kuribara, D.; Yamazaki, T.; Kitagawa, O., *Tetrahedron*, (2010) **66**, 288.
[112] Bertrand, M. B.; Wolfe, J. P., *Org. Lett.*, (2006) **8**, 2353.
[113] Schwartz, R. E.; Liesch, J.; Hensens, O.; Zitano, L.; Honeycutt, S.; Garrity, G.; Fromtling, R. A.; Onishi, J.; Monaghan, R., *J. Antibiot.*, (1988) **41**, 1774.
[114] Kinzy, T. G.; Harger, J. W.; Carr-Schmid, A.; Kwon, J.; Shastry, M.; Justice, M.; Dinman, J. D., *Virology*, (2002) **300**, 60.
[115] Fukuda, Y.; Shiragami, H.; Utimoto, K.; Nozaki, H., *J. Org. Chem.*, (1991) **56**, 5816.
[116] Kel'in, A. V.; Gevorgyan, V., *J. Org. Chem.*, (2002) **67**, 95.
[117] Liu, Y.; Song, F.; Song, Z.; Liu, M.; Yan, B., *Org. Lett.*, (2005) **7**, 5409.
[118] Antoniotti, S.; Genin, E.; Michelet, V.; Genêt, J.-P., *J. Am. Chem. Soc.*, (2005) **127**, 9976.
[119] Hashmi, A. S. K.; Weyrauch, J. P.; Frey, W.; Bats, J. W., *Org. Lett.*, (2004) **6**, 4391.
[120] Li, X.; Chianese, A. R.; Vogel, T.; Crabtree, R. H., *Org. Lett.*, (2005) **7**, 5437.
[121] Li, X.; Chen, P.; Faller, J. W.; Crabtree, R. H., *Organometallics*, (2005) **24**, 4810.
[122] Gabriele, B.; Salerno, G.; Lauria, E., *J. Org. Chem.*, (1999) **64**, 7687.
[123] Beccalli, E. M.; Borsini, E.; Broggini, G.; Palmisano, G.; Sottocornola, S., *J. Org. Chem.*, (2008) **73**, 4746.
[124] Fürstner, A.; Davies, P. W., *J. Am. Chem. Soc.*, (2005) **127**, 15 024.
[125] Oh, C. H.; Reddy, V. R.; Kim, A.; Rhim, C. Y., *Tetrahedron Lett.*, (2006) **47**, 5307.
[126] Wolf, J.; Werner, H.; Serhadli, O.; Ziegler, M. L., *Angew. Chem. Int. Ed. Engl.*, (1983) **22**, 414.
[127] Garciá Alonso, F. J.; Höhn, A.; Wolf, J.; Otto, H.; Werner, H., *Angew. Chem. Int. Ed. Engl.*, (1985) **24**, 406.
[128] Trost, B. M.; Rhee, Y. H., *J. Am. Chem. Soc.*, (2003) **125**, 7482.
[129] Seiller, B.; Bruneau, C.; Dixneuf, P. H., *Tetrahedron*, (1995) **51**, 13 089.
[130] Trost, B. M.; Rhee, Y. H., *J. Am. Chem. Soc.*, (2002) **124**, 2528.
[131] Aschwanden, P.; Frantz, D. E.; Carreira, E. M.; *Org. Lett.*, (2000) **2**, 2331.
[132] Genin, E.; Toullec, P. Y.; Antoniotti, S.; Brancour, C.; Genêt, J.-P.; Michelet, V., *J. Am. Chem. Soc.*, (2006) **128**, 3112.
[133] Komeyama, K.; Morimoto, T.; Nakayama, Y.; Takaki, K., *Tetrahedron Lett.*, (2007) **48**, 3259.
[134] Qian, H.; Han, X.; Widenhoefer, R. A., *J. Am. Chem. Soc.*, (2004) **126**, 9536.
[135] Ohta, T.; Kataoka, Y.; Miyoshi, A.; Oe, Y.; Furukawa, I.; Ito, Y., *J. Organomet. Chem.*, (2007) **692**, 671.
[136] Yang, C.-G.; Reich, N. W.; Shi, Z.; He, C., *Org. Lett.*, (2005) **7**, 4553.
[137] Reiter, M.; Ropp, S.; Gouverneur, V., *Org. Lett.*, (2004) **6**, 91.
[138] Reiter, M.; Turner, H.; Mills-Webb, R.; Gouverneur, V., *J. Org. Chem.*, (2005) **70**, 8478.
[139] Smith, A. B., III; Guaciaro, M. A.; Schow, S. R.; Wovkulich, P. M.; Toder, B. H.; Hall, T. W., *J. Am. Chem. Soc.*, (1981) **103**, 219.
[140] Smith, A. B., III; Jerris, P. J., *Tetrahedron Lett.*, (1980) **21**, 711.
[141] Han, X.; Widenhoefer, R. A., *J. Org. Chem.*, (2004) **69**, 1738.
[142] Broggini, G.; Beccalli, E. M.; Borsini, E.; Fasana, A.; Zecchi, G., *Synlett*, (2011), 227.
[143] Larock, R. C.; Wei, L.; Hightower, T. R., *Synlett*, (1998), 522.
[144] Ueda, S.; Nagasawa, H., *Angew. Chem. Int. Ed.*, (2008) **47**, 6411.
[145] Fang, Y.; Li, C., *J. Am. Chem. Soc.*, (2007) **129**, 8092.
[146] Evindar, G.; Batey, R. A., *J. Org. Chem.*, (2006) **71**, 1802.
[147] Chen, C.-y.; Dormer, P. G., *J. Org. Chem.*, (2005) **70**, 6964.
[148] Fang, Y.; Li, C., *J. Org. Chem.*, (2006) **71**, 6427.
[149] Fang, Y.; Li, C., *Chem. Commun. (Cambridge)*, (2005), 3574.
[150] Kuwabe, S.-i.; Torraca, K. E.; Buchwald, S. L., *J. Am. Chem. Soc.*, (2001) **123**, 12 202.
[151] Nan, Y.; Miao, H.; Yang, Z., *Org. Lett.*, (2000) **2**, 297.
[152] Luo, F. T.; Schreuder, I.; Wang, R. T., *J. Org. Chem.*, (1992) **57**, 2213.
[153] Hu, Y.; Nawoschik, K. J.; Liao, Y.; Ma, J.; Fathi, R.; Yang, Z., *J. Org. Chem.*, (2004) **69**, 2235.

[154] Arcadi, A.; Cacchi, S.; Del Rosario, M.; Fabrizi, G.; Marinelli, F., *J. Org. Chem.*, (1996) **61**, 9280.
[155] Larock, R. C.; Yang, H.; Pace, P.; Narayanan, K.; Russell, C. E.; Cacchi, S.; Fabrizi, G., *Tetrahedron*, (1998) **54**, 7343.
[156] Fang, C.; Pang, Y.; Forsyth, C. J., *Org. Lett.*, (2010) **12**, 4528.
[157] Tlais, S. F.; Dudley, G. B., *Org. Lett.*, (2010) **12**, 4698.
[158] Zhang, Y.; Xue, J.; Xin, Z.; Xie, Z.; Li, Y., *Synlett*, (2008), 940.
[159] Li, Y.; Zhou, F.; Forsyth, C. J., *Angew. Chem. Int. Ed.*, (2007) **46**, 279.
[160] Fortner, K. C.; Kato, D.; Tanaka, Y.; Shair, M. D., *J. Am. Chem. Soc.*, (2010) **132**, 275.
[161] Harkat, H.; Blanc, A.; Weibel, J.-M.; Pale, P., *J. Org. Chem.*, (2008) **73**, 1620.
[162] Moreau, A.; Couture, A.; Deniau, E.; Grandclaudon, P., *J. Org. Chem.*, (2004) **69**, 4527.
[163] Olivera, R.; SanMartin, R.; Churruca, F.; Domínguez, E., *J. Org. Chem.*, (2002) **67**, 7215.
[164] Boger, D. L.; Sakya, S. M.; Yohannes, D., *J. Org. Chem.*, (1991) **56**, 4204.
[165] Boger, D. L.; Zhou, J., *J. Am. Chem. Soc.*, (1993) **115**, 11426.
[166] Boger, D. L.; Patane, M. A.; Zhou, J., *J. Am. Chem. Soc.*, (1994) **116**, 8544.
[167] Nicolaou, K. C.; Mitchell, H. J.; Jain, N. F.; Bando, T.; Hughes, R.; Winssinger, N.; Natarajan, S.; Koumbis, A. E., *Chem.–Eur. J.*, (1999) **5**, 2648.
[168] Tang, Y.; Zhang, Y.; Dai, M.; Luo, T.; Deng, L.; Chen, J.; Yang, Z., *Org. Lett.*, (2005) **7**, 885.

Keyword Index

A

Acetals, alkyne tethered, iron(III)-catalyzed Prins cyclization to give five-membered carbo- and heterocycles 268, 269
Acetamide enolates, stabilized, palladium-catalyzed intramolecular allylic alkylation to give 4-vinylpyrrolidin-2-ones 101
Acetates, allylic, palladium-catalyzed intramolecular allylic amination to give 2-vinylazepanes 126
Acetimidates, use in intramolecular allylic substitution 143, 144
Acrylamides, 2-halophenyl-, intramolecular Heck reaction/intermolecular reaction with hydrazones to give oxindoles 61, 62
Acrylamides, 2-halophenyl-, intramolecular Heck reaction/intramolecular C—H functionalization to give spiro-fused oxindoles 61
Acrylamides, 3,3-diaryl-, palladium/copper-catalyzed cyclization to give 4-arylquinolin-2-ones 423
Agelastatin A, precursor 456
Alcohols, aliphatic, use in intramolecular allylic substitution 139–141
Aldehydes, aliphatic, use in intramolecular allylic substitution 102–104
Aldehydes, alkenyl-, Schiff base/chromium(III)-catalyzed enantioselective intramolecular carbonyl-ene reactions to give vinylcarbo- and vinylheterocycles 271–273
Aldehydes, indium(III)-catalyzed Prins cyclization with phenol-tethered homoallyl alcohols to give spirotetrahydropyrans 263, 264
Aldehydes, indium(III)-catalyzed Prins cyclization/polyene cyclization with homoallyl alcohols to give 3-oxaterpenoids 264–266
Aldehydes, iron(III)-catalyzed Prins cyclization with homoallyl alcohols to give *cis*-2-alkyl-4-halotetrahydropyrans 261
Aldehydes, iron(III)-catalyzed Prins cyclization with homoallyl alcohols to give tetrahydro-2*H*-pyran-4-ols 266, 267
Aldehydes, nickel-catalyzed intramolecular cyclization with an alkenyl iodide to give 2-hydroxyoctahydro-2*H*-quinolizines 33
Aldehydes, nickel-catalyzed intramolecular cyclization with an alkynyl iodide to give hydroxycycloalkynes 35
Aldehydes, nickel-catalyzed intramolecular cyclization with an allyl bromide 36
Aldehydes, rhenium(VII)-catalyzed Prins cyclization with homoallyl alcohols to give tetrahydro-2*H*-pyran-4-ols 262, 263
Aldehydes, scandium(III)-catalyzed aza-Prins reaction with *N*,*N*′-(hex-3-ene-1,6-diyl)bis(4-toluenesulfonamide) to give octahydro-1*H*-pyrrolo[3,2-*c*]pyridines 276, 277
Alkaloid (+)-241D 146
Alkenes, heteroatom-substituted, transition-metal-catalyzed cyclization, mechanism 385–388
Alkenesulfonamides, gold-catalyzed cyclization to give 1,2-thiazinane 1,1-dioxides 403
Alkenoic acids, silver-catalyzed cyclization to give lactones 480, 481
Alkenols, silver-catalyzed cyclization to give cyclic ethers 478, 479
Alkenylamines, copper-catalyzed cyclization to give pyrrolidines and piperidines 401, 402
Alkenylamines, iridium-catalyzed cyclization to give pyrrolidines and piperidines 403, 404
Alkenylamines, rhodium-catalyzed cyclization to give pyrrolidines and piperidines 406–408
Alkenylamines, ruthenium-catalyzed cyclization to give 3,4-dihydro-2*H*-pyrroles and 2,3,4,5-tetrahydropyridines 413, 414
2-Alkenylaryl azides, rhodium-catalyzed cyclization to give indoles 439, 440
Alkenyl(benzyl)amines, platinum-catalyzed cyclization to give *N*-benzylpyrrolidines and -piperidines 405, 406
Alkenylchromium cyclization, nickel-catalyzed 32–34
Alkenyl halides, intramolecular Stille coupling with alkenylstannanes 14, 15
Alkenyl halides, palladium-catalyzed intramolecular Suzuki–Miyaura reaction to form five- and six-membered rings 3, 4
Alkenyl iodides, copper(I) chloride promoted Stille-type intramolecular cross coupling with alkenylstannanes to give 1,2-dialkylidenecyclopentanes 41, 42
Alkenyl iodides, nickel-catalyzed intramolecular cyclization with an aldehyde to give 2-hydroxyoctahydro-2*H*-quinolizines 33
Alkenyl iodides, palladium-catalyzed intramolecular Suzuki–Miyaura reaction to form medium and large rings 5
Alkenyl trifluoromethanesulfonates, intramolecular Stille coupling with alkenylstannanes 14, 15
Alkenylstannanes, copper(I) chloride promoted Stille-type intramolecular cross coupling with alkenyl iodides to give 1,2-dialkylidenecyclopentanes 41, 42

Alkenylstannanes, intramolecular Stille coupling with alkenyl halides and trifluoromethanesulfonates 14, 15
Alkenyl-*N*-tosylamines, iron-catalyzed cyclization to give *N*-tosylpyrrolidines 405
Alkenylureas, palladium-catalyzed cyclization to give fused piperazinones 412
Alkoxycarbonylation, of alkynes 490, 491
Alkoxylation, of aryl or vinyl halides 485–490
Alkyl bromides, intramolecular reductive coupling with aryl iodides to give benzo-fused rings 47, 48
Alkyl bromides, tandem cyclization–intermolecular cross coupling with aryl iodides to give 3-benzylhexahydrofuro[2,3-*b*]furans 48
Alkylidenecyclopropanes, homo-Nazarov-type cyclization 188, 189
Alkynes, heteroatom-substituted, transition-metal-catalyzed cyclization, mechanism 385–388
Alkynes, terminal, copper-catalyzed intramolecular coupling with aryl iodides to give fused cycloalkynes 39, 40
Alkynes, terminal, copper-catalyzed intramolecular coupling with vinyl iodides to give cyclic enynes 37, 38
Alkynes, terminal, intramolecular Sonogashira coupling with aryl bromides to give cyclic peptides 19, 20
Alkynoates, aryl, metal-catalyzed cyclization to give 2*H*-1-benzopyran-2-ones and quinolin-2(1*H*)-ones 300–302
Alkynoic acids, gold-catalyzed cyclization to give γ-lactones 475
Alkynols, rhodium-catalyzed cyclization to give dihydrofurans and dihydropyrans 470, 471
Alkynols, ruthenium-catalyzed cyclization to give dihydropyrans 471, 472
Alkynols, ruthenium-catalyzed cyclization to give δ-valerolactones 473
Alkynylamines, gold-catalyzed cyclization to give 3,4-dihydro-2*H*-pyrroles and 2,3,4,5-tetrahydropyridines 391, 392
Alkynylamines, palladium-catalyzed cyclization to give pyrrolidines and piperidines 394
Alkynylamines, ruthenium-catalyzed cyclization to give 3,4,5,6-tetrahydro-2*H*-azepines and 2-vinylidenepyrrolidines 396
Alkynylamines, silver-catalyzed cyclization to give 3,4-dihydro-2*H*-pyrroles and 3,4,5,6-tetrahydropyridines 397
Alkynylamines, zinc-catalyzed cyclization to give 3,4-dihydro-1,4-oxazines, 1,2,3,4-tetrahydropyrazines, and 1,2,3,4-tetrahydropyridines 399
Alkynylbenzenes, 2-substituted, iridium-catalyzed cyclization to give benzopyrans and benzofurans 465
(2-Alkynylbenzyl)carbamates and -sulfonamides, silver–gold catalyzed cyclization to give 1,2-dihydroisoquinolines 397, 398
Alkynylchromium cyclization, intramolecular, nickel catalyzed 35, 36
Alkynylcyclopropanes, silver-catalyzed ring expansion to give naphthalen-2(1*H*)-ones 223, 224
Alkynylcyclopropyl ketones, gold- or nickel-catalyzed cycloisomerization 194, 195
Alkynylcyclopropyl ketones, pentacarbonyltungsten-catalyzed cycloisomerization 197, 198
Alkynyl diols, gold-catalyzed cyclization to give bicyclic ketals 463
Alkynylimines, copper-catalyzed cyclization to give pyrroles 390
Alkynyl iodides, nickel-catalyzed intramolecular cyclization with an aldehyde to give hydroxycycloalkynes 35
Alkynyl ketones, copper-catalyzed cyclization to give furans 461, 462
2-Alkynylphenols, platinum-catalyzed cyclization to give benzo[*b*]furans 468
2-Alkynylphenyl ethers, platinum-catalyzed cyclization to give benzo[*b*]furans 468
Allenenes, rhodium-catalyzed cycloisomerization to give methylenecycloalkanes 372
Allenenes, rhodium-catalyzed cycloisomerization to give methylenecycloalkenes 373
Allenes, α-amino, gold-catalyzed cycloisomerization to give 2,5-dihydropyrroles 330
Allenes, aryl tethered, gold-catalyzed cyclohydroarylation to give vinyltetrahydronaphthalenes and an 8-vinyl-5,6,7,8-tetrahydroindolizine 337, 338
Allenes, dihydroxy, gold-catalyzed cycloisomerization to give 2,5-dihydrofurans 343
Allenes, α-hydroxy, gold-catalyzed dimeric coupling/cyclization to give 2,5-dihydrofurans 323, 324
Allenes, α-hydroxy, palladium-catalyzed heterodimeric coupling/cyclization to give 2,5-dihydrofurans 321, 322
Allenes, metal-catalyzed cyclization reactions 317–347
Allenes, sulfanyl substituted, gold-catalyzed cycloisomerization to give 2,5-dihydrothiophenes 342
Allenic diazo esters, copper-catalyzed cyclopropanation to give lactone-fused methylenecyclopropanes 334
Allenic diazo esters, intramolecular rhodium-catalyzed C—H insertion to give allenyl furan-2-ones 333
Allenoates, gold-catalyzed cyclization to give butenolides 322, 323

Allenyl aldehydes, palladium-catalyzed carbocyclization to give cyclopentanes or pyrrolidines 331, 332
Allenyl carbamates, silver-catalyzed cyclization to give 4-vinyloxazolidin-2-ones 328
Allenyl carbamates, silver-catalyzed dynamic kinetic enantioselective intramolecular hydroamination to give 2-vinylpyrrolidines 329
Allenyl sulfamates, rhodium-catalyzed cyclization to give nitrogen-containing stereotriads 324, 325
Allenyl sulfonamides, gold-catalyzed enantioselective intramolecular hydroamination to give 2-vinylpyrrolidines 327
Allenyl ureas, gold-catalyzed enantioselective intramolecular hydroamination to give 2-vinylpyrrolidines and -piperidenes 326
Allenynes, gold-catalyzed cycloisomerization to give methylenecycloalkenes 369, 370
Allenynes, rhodium-catalyzed cycloisomerization to give methylenecyclohexenes and heterocycles 368, 369
(+)-Allokainic acid 145
Allyl alcohols, aryl substituted, metal-catalyzed cyclization to give vinyltetrahydronaphthalenes and -carbazoles 290–292
Allyl alcohols, gold(I)-catalyzed enantioselective intramolecular amination to give 2-vinylpyrrolidines 133
Allyl alcohols, gold(I)-catalyzed intramolecular amination to give 2-vinylpiperidines and -pyrrolidines 132, 133
Allyl alcohols, hydroxyalkyl substituted, iron(III) chloride catalyzed intramolecular allylic etherification to give tetrahydropyrans 140, 141
Allyl alcohols, indol-2-yl substituted, gold-catalyzed intramolecular Friedel–Crafts-type alkylation to give tetrahydrocarbazoles 110
Allyl alcohols, indol-3-yl substituted, gold-catalyzed intramolecular Friedel–Crafts-type alkylation to give tetrahydrocarbazoles 109
Allyl alcohols, iron(III) chloride hexahydrate catalyzed intramolecular allylic amination to give quinolines and dihydroquinolines 135
Allyl alcohols, mercury(II) trifluoromethanesulfonate–BINAPHANE catalyzed cyclization to give 2-vinyl-2,3-dihydroindoles 134
Allyl alcohols, ruthenium-catalyzed intramolecular enantioselective allylic alkylation 98
Allyl bromides, nickel-catalyzed intramolecular cyclization with an aldehyde 36, 37
Allyl bromides, palladium-catalyzed coupling/cyclization with allenic β-keto esters to give 2,3-dihydrofurans 321
Allyl(2-bromoallyl)amines, intramolecular Heck cyclization/coupling with boronic acids to give pyrrolidines 62
Allyl carbonates, indol-2-yl substituted, iridium-catalyzed intramolecular Friedel–Crafts-type alkylation to give tetrahydrocarbolines 107, 108
Allyl carbonates, indol-2-yl substituted, palladium-catalyzed intramolecular Friedel–Crafts-type alkylation to give tetrahydrocarbolines 106–107
Allylchromium cyclization, intramolecular 36
N-Allylethane-1,2-diamines, palladium-catalyzed intramolecular carboamination to give piperazines 452
Allylic alkylation, intramolecular, enantioselective, of allylic alcohols, ruthenium catalyzed 98
Allylic dearomatization/migration reactions of indoles and pyrroles 115–117
Allylic substitution reactions, metal catalyzed, intramolecular 95–149
Allylic substitution reactions, metal catalyzed, intramolecular, of acetamides 101, 102
Allylic substitution reactions, metal catalyzed, intramolecular, of β-keto esters 99, 100
Allylic substitution reactions, metal catalyzed, intramolecular, of malonates 96–99
Allylic substitution reactions, metal catalyzed, intramolecular, with resonance-stabilized enolates 96–102
Allyl phenyl ethers, palladium(II)-catalyzed oxidative intramolecular coupling to give benzofurans 27–29
N-Allylureas, palladium-catalyzed intramolecular carboamination to give imidazolidinones 452, 453
Almorexant 130
Amidation, allylic, intramolecular, iridium catalyzed 129, 130
Amides, *N*-alkoxy-, palladium-catalyzed cyclization to give 1,3-dihydro-2*H*-indol-2-ones 420
Amides, *N*-phenyl-, allylic carbonate substituted, iridium-catalyzed intramolecular allylic substitution to give 4-vinyl-4*H*-benzo[*d*][1,3]-oxazines 143, 144
Amidines, aryl-, copper-catalyzed cyclization to give benzimidazoles 415, 416
Amidines, aryl-, palladium-catalyzed cyclization to give 2-arylbenzimidazoles 417, 418
Amidines, *N*′-(halopyrazolyl)-, copper-catalyzed cyclization to give imidazo[4,5-*c*]pyrazoles 429
Amination, allylic, intramolecular 126–137
Amination, allylic, intramolecular, asymmetric, ruthenium catalyzed 131, 132
Amines, protected, use in intramolecular allylic amination 126–135
Anilides, 2-alkyl-, palladium-catalyzed cyclization to give dihydroindoles 435, 436
Anilides, 2-halo-, copper-catalyzed cyclization to give benzoxazoles 487

Anilines, *N*-acetyl-2-alkynyl-, palladium-catalyzed cyclization to give indoles 393
Anilines, 2-alkynyl-, copper-catalyzed intramolecular hydroamination to give indoles 389
Anilines, 2-alkynyl-, gold-catalyzed cyclization to give indoles 392
Anilines, 2-alkynyl-, rhodium-catalyzed cyclization to give indoles 395
Anilines, 2-alkynyl-*N*-tosyl-, zinc-catalyzed cyclization to give *N*-tosylindoles 400
Anilines, 2-(arylalkynyl), zinc-catalyzed cyclization to give 2-arylindoles 400, 401
Anilines, *N*-allenyl-, gold-catalyzed cyclization to give 3,4-dihydroquinolines 336
Anilines, 2-allyl-, palladium-catalyzed cyclization to give indoles 409, 410
Apoptolidinone A 10, 11
Arenes, use in intramolecular allylic substitution 121–126
Aristoyagonine 497
Aryl boronates, intramolecular Suzuki–Miyaura cyclization to construct biaryl macrocycles 8
Aryl bromides, intramolecular Sonogashira coupling with terminal alkynes to give cyclic peptides 19, 20
Aryl halides, intramolecular Suzuki–Miyaura cyclization to construct biaryl macrocycles 8
Aryl halides, palladium-catalyzed intramolecular Suzuki–Miyaura reaction to form aryl-fused five- and six-membered rings 3, 4
Aryl iodides, copper-catalyzed intramolecular coupling with terminal alkynes to give fused cycloalkynes 39, 40
Aryl iodides, intramolecular reductive coupling with alkyl bromides to give benzo-fused rings 47, 48
Aryl iodides, palladium-catalyzed intramolecular Suzuki–Miyaura reaction to form aryl-fused medium and large rings 5
Aryl iodides, tandem cyclization–intermolecular cross coupling with alkyl bromides to give 3-benzylhexahydrofuro[2,3-*b*]furans 48
ω-Arylalkylamines, palladium-catalyzed cyclization to give benzolactams 445, 446
Asperolide C 145
(−)-Aurantioclavine 146
Autolytimycin, intermediate 457
3-Azaalk-5-enols, palladium-catalyzed cyclization to give dihydro-1,4-oxazines 483
3-Azabicyclo[4.1.0]heptanes, from alkene-tethered cyclopropenes by gold-catalyzed intramolecular cyclopropanation via in situ carbene formation 242
9-Azabicyclo[4.2.1]non-2-enes, from 6-amidocyclooct-2-en-1-yl carbonates by palladium-catalyzed intramolecular allylic amination 127
Aza-Prins reactions 274–277
Azaspiracid, intermediate 497
Azepanes, 2-vinyl-, from allylic acetates by palladium-catalyzed intramolecular allylic amination 126
Azepanes, from methylenecyclopropanes by intramolecular hydroamination 154, 155
Azepines, fused, by intramolecular Heck reaction 74
Azetidines, *N*-acyl-, from *N*-alkylcarboxamides by palladium-catalyzed cyclization 434

B

Benzaldimines, palladium-catalyzed carboamination to give 3,4-disubstituted isoquinolines 449
Benzamides, *N*-(amidomethyl)-2-iodo-, copper-catalyzed cyclization to give 1,4-benzodiazepine-2,5-diones 429, 430
Benzamides, *N*-aryl-, copper-catalyzed cyclization to give benzoxazoles 484, 485
Benzamides, ruthenium-catalyzed carbonylation to give phthalimides 446, 447
Benzazepines, by intramolecular Heck reaction 75
Benzenes, alk-3-ynyl-, metal-catalyzed cyclization to give 1,2-dihydronaphthalenes 293, 294
Benzenes, alkenyl tethered, ruthenium(III)-catalyzed cyclization to give arene-fused carbo- and heterocycles 289
Benzenes, alkynyl tethered, metal-catalyzed cyclization to give 2*H*-1-benzopyrans and dihydroquinolines 303–305
Benzenes, allylic alcohol substituted, silver(I)-catalyzed intramolecular allylic alkylation to give 1-vinyltetrahydronaphthalenes 122, 123
Benzenes, allylic carbonate substituted, molybdenum(II)-catalyzed intramolecular allylic alkylation to give 4-vinyl-3,4-dihydronaphthalene-2,2(1*H*)-dicarboxylates 122
Benzenes, electron deficient, π-activated allylic alcohol substituted, iron(III)-catalyzed intramolecular allylic alkylation to give 1-vinyltetrahydronaphthalenes 123
Benzimidazoles, 2-amino-, from arylguanidines by copper-catalyzed cyclization 428
Benzimidazoles, 2-aryl-, from arylamidines by palladium-catalyzed cyclization 417, 418
Benzimidazoles, from arylamidines by copper-catalyzed cyclization 415, 416
1,4-Benzodiazepine-2,5-diones, from *N*-(amidomethyl)-2-iodobenzamides by copper-catalyzed cyclization 429, 430
Benzodioxanes, from 2-(2-bromophenoxy)ethanols by palladium-catalyzed cyclization 489, 490
Benzo[*c*]fluorenes, from vinylidenecyclopropanes by Lewis acid catalyzed rearrangement 199, 200

Benzo[*b*]furan-3-carboxylates, methyl, from 2-alkynylphenols by palladium-catalyzed cyclization 490
Benzo[*b*]furan-7(3a*H*)-ones, from 3-cyclopropylideneprop-2-en-1-ones by copper-catalyzed tandem ring opening and cycloaddition 209, 210
Benzo[*b*]furans, 2,3-diaryl-, from 2-alkynylphenols by palladium-catalyzed cyclization 492
Benzofurans, from 2-alkynylphenols by iridium-catalyzed cyclization 465
Benzo[*b*]furans, from 2-alkynylphenols or 2-alkynylphenyl ethers by platinum-catalyzed cyclization 468
Benzofurans, from allyl phenyl ethers by palladium(II)-catalyzed oxidative intramolecular coupling 27–29
Benzo[*b*]furans, from 2-halobenzyl ketones by copper-catalyzed cyclization 487
Benzo[*b*]furans, 2-substituted, from 2-ethynylphenols by palladium-catalyzed cyclization 493, 494
Benzolactams, from ω-arylalkylamines by palladium-catalyzed cyclization 446
4*H*-Benzo[*d*][1,3]oxazines, 4-vinyl-, from allylic carbonate substituted *N*-phenylamides by iridium-catalyzed intramolecular allylic substitution 143, 144
2*H*-1-Benzopyran-2-ones, from aryl alkynoates by metal-catalyzed cyclization 300–302
Benzopyrans, from 2-alkynylbenzyl alcohols by iridium-catalyzed cyclization 465
2*H*-1-Benzopyrans, from alkynyl phenyl ethers by metal-catalyzed cyclization 303–305
Benzopyrans, from 2-(2-bromobenzyl)-β-keto esters by copper-catalyzed cyclization 488
2*H*-1-Benzopyrans, from homoallenyl phenyl ethers by gold-catalyzed cyclization 336, 337
Benzoxazoles, from *N*-arylbenzamides by copper-catalyzed cyclization 485
Benzoxazoles, from 2-haloanilides by copper-catalyzed cyclization 487
Benzyl ketones, 2-halo-, copper-catalyzed cyclization to give benzo[*b*]furans 486, 487
Biaryl azides, rhodium-catalyzed cyclization to give carbazoles 441
Biaryl azides, ruthenium-catalyzed cyclization to give carbazoles and pyrroles 442
Biaryl macrocycles, from aryl halides by intramolecular Suzuki–Miyaura cyclization 8
Bicyclo[3.1.0]hex-2-enes, 6-alkynyl-, gold-catalyzed rearrangement to give 6-alkynylcyclohexa-1,3-dienes and 4-methylenebicyclo[3.2.1]octa-2,6-dienes 211, 212
Bicyclo[3.2.1]octa-2,6-dienes, 4-methylene-, from 6-alkynylbicyclo[3.1.0]hex-2-enes by gold-catalyzed rearrangement 211, 212
BINAP, use in enantioselective intramolecular Heck reactions 78, 81–89
BINAP, use in intramolecular Heck reactions 59, 65
Biphenyl-2-amines, palladium-catalyzed cyclization to give carbazoles 420–422
Biphenyls, 2-alk-1-ynyl-, metal-catalyzed cyclization to give phenanthrenes and other polycyclic arenes 294–298
Bipinnatin J 36
Bis(1,3-dienes), palladium-catalyzed cycloisomerization to give 1,3-dienylcycloalkanes and -piperidines 379
(+)-Brasilenyne 24
Briarellin E 310
Briarellin F 310
3-Bromoalk-3-enols, copper-catalyzed cyclization to give 2-methyleneoxetanes 486
(2-Bromophenyl)alkanols, palladium-catalyzed cyclization to give dihydrobenzopyrans, benzodioxanes, or dihydrobenzoxazines 489, 490
Buchwald–Hartwig reaction 431–434
Butenolides, from allenoates by gold-catalyzed cyclization 322, 323

C

Carbamates, alkyl, rhodium-catalyzed cyclization to give oxazolidin-2-ones 436, 437
Carbamoylation of arenes 445–447
Carbazoles, from 2-alk-3-ynylindoles by gold-catalyzed cyclization 299, 300
Carbazoles, from biaryl azides by rhodium-catalyzed cyclization 441
Carbazoles, from biaryl azides by ruthenium-catalyzed cyclization 442
Carbazoles, from biphenyl-2-amines by palladium-catalyzed cyclization 420–422
Carbazoles, from indolyl–1-(alk-1-ynylcyclopropyl)alkanols by gold-catalyzed cycloisomerization 213–215
Carbenes, cyclopropyl, ring expansion to give cyclobutenes 224–226
Carboalkoxylation, of alkenes 494, 495
Carboalkoxylation, of alkynes 491–494
Carboamination, of alkenes 450–453
Carboamination, of alkynes 447–450
Carbonates, allylic, iridium-catalyzed intramolecular allylic amination to give 2-vinylpiperidines 128
Carboxamides, *N*-alkyl-, palladium-catalyzed cyclization to give *N*-acylazetidines and -pyrrolidines 433, 434
Carboxylates, unsaturated, gold(I)-catalyzed intramolecular allylic substitution to give five-membered lactones 142
(*R*)-(+)-Carnegine 127
(−)-Cedrelin A 145
Cephalosporide H 496

(−)-Chanoclavine I, synthesis by palladium-catalyzed intramolecular allylic alkylation of an aliphatic nitro compound 105
2-Chloroquinolin-3-yl homoallyl alcohols, intramolecular Heck cyclization to give 3-methylene-2,3-dihydro-1*H*-cyclopenta[*b*]quinolines 59
(+)-Clusifoliol 139, 147
Coeloginin 311
Communesin B, precursor 455
Complestatin 9
Conia-ene reactions, metal catalyzed 278
Corianin 381
Cycloalkadienes, from iodoalkenyl cycloalkenylsiloxanes by palladium-catalyzed intramolecular Hiyama coupling 23
Cycloalkanecarboxylates, 1-hydroxy-, from α-oxo esters by copper(II)-catalyzed asymmetric carbonyl-ene reactions 270
Cycloalkanes, 1,3-dienyl-, from bis(1,3-dienes) by palladium-catalyzed cycloisomerization 379
Cycloalkanes, from dihaloalkanes by nickel-catalyzed reductive cyclization 45, 46
Cycloalkanes, vinyl-, from (1,3-diene)enes by iron-catalyzed cycloisomerization 374–376
Cycloalkanes, vinyl-, from (1,3-diene)enes by rhodium-catalyzed enantioselective cycloisomerization 376–378
Cycloalkanes, vinyl-, from malonates by iridium-catalyzed intramolecular allylic alkylation 97
Cycloalkanols, from alkenyl aldehydes by Schiff base/chromium(III)-catalyzed enantioselective intramolecular carbonyl-ene reactions 271–273
Cycloalkenes, from yne– or ene–methylenecyclopropanes by intramolecular cycloaddition 155, 156
Cycloalkenes, nine- to twelve-membered, from allenes containing a nucleophilic functionality and organic halides by palladium-catalyzed coupling/cyclization 341, 342
Cycloalkynes, by copper-catalyzed intramolecular coupling of terminal alkynes with aryl and vinyl iodides 39, 40
Cycloalkynes, hydroxy-, by nickel-catalyzed intramolecular cyclization of alkynyl iodides with aldehydes 35
Cyclobutanols, 1-alkynyl-, from (1-alkynylcyclopropyl)methanols by iron-catalyzed selective ring expansion 222, 223
Cyclobutanones, by ruthenium-catalyzed ring expansion of 1-alkynylcyclopropanols 217
Cyclobutanones, from 1-(alk-1-ynyl)cyclopropanols by gold-catalyzed ring expansion 210, 211
Cyclobutenes, from cyclopropyl carbenes by ring expansion 224–226
Cyclobutenes, from methylenecyclopropanes by palladium-catalyzed ring expansion, mechanism 156, 157
Cyclobutenes, from [2-(methylene)cyclopropyl]carbinols by ring-opening cycloisomerization 157
Cyclobutenes, from methylenecyclopropanes by platinum-catalyzed ring expansion, mechanism 159
Cyclohex-2-enones, from silylated cyclopropenylcyclopropanols by zinc-catalyzed rearrangement 235, 236
Cyclohexa-1,3-dienes, 6-alkynyl-, from 6-alkynylbicyclo[3.1.0]hex-2-enes by gold-catalyzed rearrangement 211–213
Cyclohexadienones, spiro-fused, from phenols by iridium-catalyzed asymmetric allylic dearomatization 121
Cyclohexadienones, spiro-fused, from phenols by palladium-catalyzed allylic dearomatization 120
Cyclohexanones, 2-acyl-, from alkenyl-tethered 1,3-diketones by palladium-catalyzed 6-*endo-trig* cyclization 279
Cyclohexanones, vinyl-, from β-keto esters by palladium-catalyzed intramolecular allylic alkylation 99, 100
Cyclohexenols, from 1-acylcyclopropane-1-carboxylates by indium-catalyzed homo-Nazarov cyclization 187, 188
Cyclohexenones, from 1-acylcyclopropane-1-carboxylates by indium-catalyzed homo-Nazarov cyclization 187, 188
Cyclohexenones, from [1,1′-bi(cyclopropan)]-2′-en-1-ols by rhodium(I)-catalyzed rearrangement 232, 233
Cyclooct-2-en-1-yl carbonates, 6-amido-, palladium-catalyzed intramolecular allylic amination to give 9-azabicyclo[4.2.1]non-2-enes 127
Cyclopenta[*a*]indenes, from alkyne-substituted methylenecyclopropanes by nickel-catalyzed intramolecular cycloaddition 165, 166
Cyclopenta[*c*][1]benzopyrans, from alkyne-tethered cyclopropane-1,1-dicarboxylates by intramolecular parallel cycloaddition 187
1*H*-Cyclopenta[*c*]quinolines, from alkyne-tethered cyclopropane-1,1-dicarboxylates by intramolecular parallel cycloaddition 187
Cyclopentanecarbaldehydes, vinyl-, from allenyl aldehydes by palladium-catalyzed carbocyclization 331, 332
Cyclopentanes, vinyl-, by gold(I)/secondary amine cocatalyzed enantioselective intramolecular α-allylic alkylation of aldehydes 103
Cyclopentanes, vinyl-, from malonates by palladium-catalyzed intramolecular allylic alkylation 96, 97

Cyclopentenes, dimethyl-, from α,ω-dienes by palladium-catalyzed cycloisomerization 362, 363
Cyclopentenes, from alkenyl or aryl halides by palladium-catalyzed intramolecular Suzuki–Miyaura reaction 3, 4
Cyclopentenes, stannyl, from stannyl enol ethers by iron(III)-catalyzed stannyl Conia-ene cyclization 287, 288
Cyclopentenones, by ruthenium-catalyzed ring expansion of 1-alkynylcyclopropanols 217, 218
Cyclopentenones, from 1-bromopenta-1,4-dien-3-ols by intramolecular Heck reaction/oxidation 69, 70
Cycloprop-2-enecarboxylates, copper-catalyzed cycloisomerization to give 2-methoxyfurans 239, 240
Cycloprop-2-enecarboxylates, ruthenium-catalyzed cycloisomerization to give 2-methoxyfurans 234, 235
Cyclopropane aminals, palladium-catalyzed ring opening to give 3,4-dihydro-2(1*H*)-quinolinones 221, 222
Cyclopropane-1,1-dicarboxylates, alkenylamine tethered, cycloisomerization to give tetracyclic spiroindolines 196
Cyclopropane-1,1-dicarboxylates, alkenyl, intramolecular (3 + 2)-cross-cycloaddition reaction and ring opening to give bridged [n.2.1] carbocycles 182–186
Cyclopropane-1,1-dicarboxylates, alkyne tethered, intramolecular parallel cycloaddition to give cyclopenta[*c*][1]benzopyrans and 1*H*-cyclopenta[*c*]quinolines 187
Cyclopropane-1,1-dicarboxylates, allenyl, intramolecular cycloaddition to give [4.3.0]nonanes 185, 186
Cyclopropane-1,1-dicarboxylates, intramolecular (3 + 2)-cross-cycloadditions to give bridged [n.2.1] heterocycles 180–182
Cyclopropane-1,1-dicarboxylates, nitrone substituted, intramolecular cycloaddition to give tetrahydro-1,2-oxazines 177–179
Cyclopropane-1,1-dicarboxylates, oxime substituted, intramolecular cycloaddition to give pyrrolo[1,2-*b*]isoxazolidines 176, 177
Cyclopropane-1,1-dicarboxylates, parallel- and cross-type intramolecular (3 + 2) cycloadditions 179
Cyclopropane-1,1-dicarboxylates, 2-vinyl-, from allenic malonates and organic halides by palladium-catalyzed coupling/cyclization 338, 339
Cyclopropanecarbaldehydes, rearrangement to give [n,5]spiroketals 191
Cyclopropanecarboxylates, 1-acyl-, indium-catalyzed homo-Nazarov cyclizations to give cyclohexenols and cyclohexenones 187, 188
Cyclopropanecarboxylates, α-amido, tandem ring-opening/Friedel–Crafts alkylation to give indole-fused lactams 190
Cyclopropanes, acyl-, intramolecular (3 + 2) cycloaddition under visible-light photocatalysis 219, 220
Cyclopropanes, donor–acceptor substituted, copper-catalyzed cycloisomerization 196, 197
Cyclopropanes, donor–acceptor substituted, metal trifluoromethanesulfonate catalyzed cycloisomerization 176–193
Cyclopropanes, donor–acceptor substituted, tungsten-catalyzed hetero-cycloisomerization to give 4,5-dihydrobenzo[*b*]furans and 4,5-dihydroindoles 197, 198
Cyclopropanols, 1-(alk-1-ynyl)-, gold-catalyzed ring expansion to give cyclobutanones 210, 211
Cyclopropanols, 1-alkynyl-, ruthenium-catalyzed ring expansion to give methylenecyclobutanones and cyclopentenones 217, 218
Cyclopropanols, cyclopropene substituted, zinc-catalyzed cycloisomerization 235–237
Cyclopropanols, 1-cycloprop-2-enyl-, rhodium(I)-catalyzed rearrangement to give cyclohexenones 232, 233
Cyclopropanols, silyl protected, cyclopropene substituted, zinc-catalyzed rearrangement to give cyclohex-2-enones 235–237
Cyclopropenes, acyl-, rhodium-catalyzed cycloisomerization to give furans 227, 228
Cyclopropenes, alkene tethered, gold-catalyzed intramolecular cyclopropanation via in situ carbene formation to give 3-oxa- and 3-azabicyclo[4.1.0]heptanes 242
Cyclopropenes, alkyne-, alkene-, and allene-tethered, gold-catalyzed cycloisomerization to give indenes 248, 249
Cyclopropenes, copper-catalyzed cycloisomerization 237–240
Cyclopropenes, cyclopropanol substituted, zinc-catalyzed cycloisomerization 235–237
Cyclopropenes, gold-catalyzed cycloisomerization 241–249
Cyclopropenes, 3-(hydroxymethyl)-, O-protected, gold-catalyzed cycloisomerization to give 1-methylene-1*H*-indenes 243, 244
Cyclopropenes, phenyl-, gold-catalyzed rearrangement to give indene derivatives 241, 243
Cyclopropenes, 3-propargyl-, gold-catalyzed cycloisomerization to give phenols 244–247
Cyclopropenes, rhodium-catalyzed cycloisomerization 227–233
Cyclopropenes, ruthenium-catalyzed cycloisomerization 234, 235

Cyclopropenes, silylated cyclopropanol substituted, zinc-catalyzed rearrangement to give cyclohex-2-enones 235, 236
Cyclopropenes, 1-vinyl-, copper-catalyzed cycloisomerization to give indenes 237, 238
Cyclopropyl-fused tricycles, by 5-*exo-trig*/3-*exo-trig* Heck cyclization cascade 57
Cyclopropylmethanols, iron-catalyzed selective ring expansion to give 1-alkynylcyclobutanols 222, 223

D

Dearomatization reactions, allylic, of indoles and pyrroles 112–115
Dearomatization reactions, intramolecular, of phenols 119–121
Dehydrogenative arene–alkene cross coupling, intramolecular, palladium catalyzed 24–31
Dehydrogenative cross coupling, proposed mechanism 25
1,2-Dialkylidenecyclopentanes, by copper(I) chloride promoted Stille-type intramolecular cross coupling of alkenylstannanes and alkenyl iodides 41, 42
α-Diazo amides, *N*-propargyl-, rhodium-catalyzed cyclopropene formation and rearrangement 229
Diazo ketones, rhodium-catalyzed rearrangement via fused cyclopropenes to give indenones 228
Dibenzoazocines, by intramolecular Heck reaction 74
(−)-Dictyostatin 34
(1,3-Diene)enes, iron-catalyzed cycloisomerization to give vinyl-substituted cycloalkanes and heterocycles 374–376
(1,3-Diene)enes, rhodium-catalyzed enantioselective cycloisomerization to give vinyl-substituted cycloalkanes and heterocycles 376–378
1,6-Dienes, nickel-catalyzed enantioselective cycloisomerization to give methylenecyclopentanes 366, 367
α,ω-Dienes, palladium-catalyzed cycloisomerization to give 1,2-dimethylcyclopentenes 362
α,ω-Dienes, ruthenium-catalyzed cycloisomerization to give methylene-substituted cycloalkanes and heterocycles 365, 366
α,ω-Dienes, titanium-catalyzed cycloisomerization to give methylene-substituted cycloalkanes and heterocycles 363, 364
Dienones, from vinylidenecyclopropane-tethered alcohols by gold-catalyzed intramolecular rearrangement 202, 203
1,3-Dienyl azides, ruthenium-catalyzed cyclization to give pyrroles 442
1,3-Dienyl azides, zinc-catalyzed cyclization to give pyrroles 443
Dihaloalkanes, nickel-catalyzed reductive cyclization to give cycloalkanes 45, 46
5,12-Dihydrobenzo[4,5]cyclohepta[1,2-*b*]indoles, from 1*H*-indoles and nitroenynes by metal-catalyzed 7-*endo-dig* cyclization 306, 307
4,5-Dihydrobenzo[*b*]furans, by tungsten-catalyzed hetero-cycloisomerization of donor–acceptor cyclopropanes 197, 198
Dihydrobenzo[*b*]furans, from 2-alkenylphenols by palladium-catalyzed cyclization 494, 495
2,3-Dihydrobenzo[*b*]furans, from 2-allylphenols by ruthenium-catalyzed cyclization 478
Dihydrobenzofurans, hetarylmethyl substituted, by intramolecular Heck cyclization/direct arylation 62–64
2,3-Dihydrobenzo[*f*]indoles, from methylenecyclopropane–triazoles by copper-catalyzed intramolecular cycloisomerization 166, 167
Dihydrobenzopyrans, from 2-alkenylphenols by palladium-catalyzed cyclization 494, 495
Dihydrobenzopyrans, from 3-(2-bromophenyl)-propanols by palladium-catalyzed cyclization 489, 490
3,4-Dihydro-2*H*-1-benzopyrans, from 2-homoallylphenols by ruthenium-catalyzed cyclization 478
3,4-Dihydrobenzopyrans, 2-vinyl-, from allylic carbonate substituted phenols by palladium-catalyzed asymmetric allylic substitution 138
Dihydrobenzoxazines, from 3-(2-bromophenylamino)ethanols by palladium-catalyzed cyclization 489, 490
Dihydrocarbazoles, by enantioselective intramolecular Heck reaction 90
9,10-Dihydrocyclohepta[*b*]indol-8(5*H*)-ones, from indoles and enynones by gold-catalyzed cyclization 308
1,3-Dihydro-2*H*-fluorene-2,2-dicarboxylates, from 1,6-enynes by gold-catalyzed cycloisomerization 247, 248
Dihydrofurans, from but-3-yn-1-ols by rhodium-catalyzed cyclization 470, 471
2,3-Dihydrofurans, from allenic β-keto esters and allyl bromides by palladium-catalyzed coupling/cyclization 321
2,5-Dihydrofurans, from dihydroxyallenes by gold-catalyzed cycloisomerization 343
Dihydrofurans, from *Z*-enynols by gold-catalyzed cyclization 462, 463
2,5-Dihydrofurans, from α-hydroxyallenes by gold-catalyzed dimeric coupling/cyclization 323, 324
2,5-Dihydrofurans, from α-hydroxyallenes by palladium-catalyzed heterodimeric coupling/cyclization 321, 322
2,3-Dihydroinden-1-ones, 3-aryl substituted, by enantioselective intramolecular Heck reaction 79

2,3-Dihydroinden-1-ones, 2-methylene-, by enantioselective intramolecular Heck reaction 79
4,5-Dihydroindoles, by tungsten-catalyzed hetero-cycloisomerization of donor–acceptor cyclopropanes 197, 198
Dihydroindoles, from 2-alkylanilides by palladium-catalyzed cyclization 435, 436
Dihydroindoles, from bromoaryl-substituted secondary amides and carbamates by palladium-catalyzed cyclization 431, 432
Dihydroindoles, from phenethylamines by palladium-catalyzed cyclization 418, 419
2,3-Dihydroindoles, 2-vinyl-, from sulfonamide allyl alcohols by mercury(II) trifluoromethanesulfonate–BINAPHANE catalyzed cyclization 134
Dihydroindolizidinones, by enantioselective intramolecular Heck reaction 83
2,3-Dihydroindolizines, 3-vinyl-, from pyridines by iridium-catalyzed intramolecular asymmetric allylic dearomatization 136
1,3-Dihydro-2*H*-indol-2-ones, from *N*-alkoxy amides by palladium-catalyzed cyclization 419, 420
3,4-Dihydroisoquinolin-1-ones, 4-[(methoxycarbonyl)methyl]-, by carbonylative Heck cyclization 67
1,2-Dihydroisoquinolines, from (2-alkynylbenzyl)carbamates and -sulfonamides by silver–gold catalyzed cyclization 398
Dihydroisoxazoles, from *N*-propargylhydroxylamines by zinc-catalyzed cyclization 474
Dihydro-1,4-oxazines, from aminoalkynes by zinc-catalyzed cyclization 399
Dihydro-1,4-oxazines, from 3-azaalk-5-enols by palladium-catalyzed cyclization 483
1,2-Dihydronaphthalenes, from alk-3-ynylbenzenes by metal-catalyzed cyclization 293, 294
1,2-Dihydronaphthalenes, from vinylidenecyclopropanes by gold-catalyzed rearrangement 200, 201
2,3-Dihydro-4*H*-pyran-4-ones, from hydroxyenones by palladium/copper-catalyzed cyclizations 481, 482
Dihydropyrans, from pent-4-yn-1-ols by rhodium-catalyzed cyclization 470, 471
Dihydropyrans, from pent-4-yn-1-ols by ruthenium-catalyzed cyclization 472
1,2-Dihydropyridazines, cyclic, from methylenecyclopropane–hydrazones by Lewis acid catalyzed ring expansion 167, 168
3,4-Dihydro-2*H*-pyrroles, from alkenylamines by ruthenium-catalyzed cyclization 413, 414
3,4-Dihydro-2*H*-pyrroles, from alkynylamines by gold-catalyzed cyclization 391, 392
2,5-Dihydropyrroles, from α-aminoallenes by gold-catalyzed cycloisomerization 330
2,3-Dihydropyrroles, from *N*,*N*-diallyl amides by palladium-catalyzed cycloisomerization 361, 362
3,4-Dihydro-2*H*-pyrroles, from primary alkynylamines by silver-catalyzed cyclization 397
6,7-Dihydropyrrolo[1,2-*a*]pyrazines, 6-vinyl-, from pyrazines by iridium-catalyzed intramolecular asymmetric allylic dearomatization 136, 137
Dihydroquinolines, from alkynyl(phenyl)amines by metal-catalyzed cyclization 303–305
3,4-Dihydroquinolines, from *N*-allenylanilines by gold-catalyzed cyclization 337, 338
Dihydroquinolines, from allylic alcohols by iron(III) chloride hexahydrate catalyzed intramolecular allylic amination 135
3,4-Dihydro-2(1*H*)-quinolinones, from cyclopropane aminals by palladium-catalyzed ring opening 221, 222
2,5-Dihydrothiophenes from sulfanyl allenes by gold-catalyzed cycloisomerization 342
1,3-Diketones, alkenyl tethered, palladium-catalyzed 6-*endo-trig* cyclization to give 2-acylcyclohexanones 279
1,3-Diketones, alkyne tethered, zinc(II) chloride catalyzed Conia-ene reaction to give methylenecycloalkanes 286
1,3-Diketones, alkynyl tethered, gold(I)-catalyzed 5-*endo-dig* cyclization to give 3-acylcycloalkenes 285
β-Diketones, α-allyl, palladium-catalyzed alkoxylation to give furans 483
5,6-Dimethylenehexahydroazocines, from vinylidenecyclopropanes by rhodium-catalyzed intramolecular cycloaddition 208, 209
Diterpenoids, tetracyclic key intermediate, by enantioselective intramolecular Heck reaction 89
Dragmacidin F 30
Duocarmycin, natural product analogues 74

E

Elaiolide 43
Ene–yne coupling reactions, intramolecular, copper catalyzed 37–41
Ene–yne coupling reactions, intramolecular, copper catalyzed, mechanism 38, 39
Enolates, resonance stabilized, use in intramolecular allylic substitution 96–102
Enones, cyclic, from alkyne-tethered acetals by iron(III)-catalyzed Prins-type cyclization 269
Enones, hydroxy-, palladium/copper-catalyzed cyclizations to give 2,3-dihydro-4*H*-pyran-4-ones and furan-3(2*H*)-ones 482
Enynamines, copper-catalyzed cyclization to give pyrroles 390, 391

Enynes, cyclic, by copper-catalyzed intramolecular coupling of terminal alkynes with vinyl iodides 37, 38
1,6-Enynes, gold-catalyzed cycloisomerization to give 1,3-dihydro-2*H*-fluorene-2,2-dicarboxylates 247, 248
α,ω-Enynes, iron-catalyzed cycloisomerization to give methylenecycloalkanes 355–357
α,ω-Enynes, palladium-catalyzed cycloisomerization to give methylenecycloalkanes 351, 352
α,ω-Enynes, palladium-catalyzed enantioselective cycloisomerization to give 4-methylene-1,2,3,4-tetrahydroquinolines 359, 360
α,ω-Enynes, rhodium-catalyzed enantioselective cycloisomerization to give 3-methylenetetrahydrofurans and -pyrrolidin-2-ones 357, 358
α,ω-Enynes, ruthenium-catalyzed cycloisomerization to give methylene-substituted cycloalkanes or heterocycles 354, 355
α,ω-Enynes, titanium-catalyzed cycloisomerization to give methylenecycloalkanes 352–354
1,6-Enynes, transition-metal-catalyzed cycloisomerization, representative mechanisms 350
Z-Enynols, gold-catalyzed cyclization to give dihydrofurans or furans 462, 463
Z-Enynols, palladium-catalyzed cyclization to give furans 466
Enynones, gold-catalyzed annulation with indoles to give 9,10-dihydrocyclohepta[*b*]indol-8(5*H*)-ones 308
Enynones, platinum-catalyzed cyclization to give furans 469
(+)-Eptazocine 86
Ergot alkaloids, C/D ring system 345, 346
Esters, use in intramolecular allylic substitution 141–143
Etherification, allylic, intramolecular, asymmetric, ruthenium catalyzed 140

F

(+)-Fawcettimine 251, 252
Flinderoles B and C 344, 345
Friedel–Crafts-type reactions, intramolecular, of allylic indoles 106–112
Friedel–Crafts-type reactions, intramolecular, of phenols 118, 119
Furan-2-ones, allenyl-, from allenic diazo esters by intramolecular rhodium-catalyzed C—H insertion 333
Furan-3(2*H*)-ones, from hydroxyenones by palladium/copper-catalyzed cyclizations 482
Furans, from acylcyclopropenes by rhodium-catalyzed cycloisomerization 227, 228
Furans, from acyl-substituted methylenecyclopropanes by palladium(II)-catalyzed cycloisomerization 154
Furans, from alkynyl ketones by copper-catalyzed cyclization 462
Furans, from α-allyl β-diketones by palladium-catalyzed alkoxylation 483
Furans, from *Z*-enynols by gold-catalyzed cyclization 462, 463
Furans, from *Z*-enynols by palladium-catalyzed cyclization 466
Furans, from enynones by platinum-catalyzed cyclization 469
Furans, 2-methoxy-, from cyclopropenecarboxylates by ruthenium-catalyzed cycloisomerization 234, 235
Furans, 2-methoxy-, from methyl cycloprop-2-enecarboxylates by copper-catalyzed cycloisomerization 239, 240

G

Guanidines, aryl-, copper-catalyzed cyclization to give 2-aminobenzimidazoles 428

H

Heck reaction, intramolecular 55–91
Heck reaction, intramolecular, catalytic cycle 55
Heck reaction, intramolecular, enantioselective 77–91
Heck reaction, intramolecular, enantioselective formation of quaternary centers from acyclic alkenes 85–87
Heck reaction, intramolecular, enantioselective formation of quaternary centers from cyclic alkenes 88–91
Heck reaction, intramolecular, enantioselective formation of tertiary centers from acyclic alkenes 78–81
Heck reaction, intramolecular, enantioselective formation of tertiary centers from cyclic alkenes 81–85
Heck reaction, intramolecular, 5-*endo-trig* cyclizations 69–71
Heck reaction, intramolecular, 6-*endo-trig* cyclizations 72–73
Heck reaction, intramolecular, 7-*endo-trig* cyclizations 75
Heck reaction, intramolecular, 8-*endo-trig* cyclizations 76
Heck reaction, intramolecular, 3-*exo-trig* cyclizations 57
Heck reaction, intramolecular, 4-*exo-trig* cyclizations 58
Heck reaction, intramolecular, 5-*exo-trig* cyclizations 59–65
Heck reaction, intramolecular, 6-*exo-trig* cyclizations 65–69
Heck reaction, intramolecular, 7-*exo-trig* cyclizations 74
Heck reaction, intramolecular, 8-*exo-trig* cyclizations 74

Heck reaction, intramolecular, macrocycle synthesis 76
Helenaquinone 86
Hexa-1,5-dien-2-ol trifluoromethanesulfonates, intramolecular Heck reactions to give methylenecyclobutanes 58
Hexahydrofuro[2,3-*b*]furans, 3-benzyl-, by tandem cyclization–intermolecular cross coupling of alkyl bromides with aryl iodides 47, 48
4,6,6a,7,8,9-Hexahydroindolo[4,3-*fg*]quinolines, from allenic bromoindoles by palladium-catalyzed domino cyclization 345, 346
Hex-5-enamides, *N*-tosyl-, palladium-catalyzed cyclization to give lactams 410, 411
Hiyama reaction, intramolecular, palladium catalyzed 22–24
Hiyama reaction, mechanism 22
Homoallyl alcohol, iron(III)-catalyzed Prins cyclization with aldehydes to give *cis*-2-alkyl-4-halotetrahydropyrans 261
Homoallyl alcohols, indium(III)-catalyzed Prins cyclization/polyene cyclization with aldehydes to give 3-oxaterpenoids 264–266
Homoallyl alcohols, iron(III)-catalyzed Prins cyclization with aldehydes to give tetrahydro-2*H*-pyran-4-ols 266, 267
Homoallyl alcohols, phenol tethered, indium(III)-catalyzed Prins cyclization with aldehydes to give spirotetrahydropyrans 263, 264
Homoallyl alcohols, rhenium(VII)-catalyzed Prins cyclization with aldehydes to give tetrahydro-2*H*-pyran-4-ols 262, 263
Homo-Nazarov-type cyclizations of alkylidenecyclopropanes 187–189
Homopropargyl azides, platinum-catalyzed cyclization to give pyrroles 443, 444
Homopropargyl azides, silver/gold-catalyzed cyclization to give pyrroles 444, 445
Hydrazones, alkynyl-, palladium-catalyzed cyclization with aryl iodides to give 1,2,3,5-tetrasubstituted pyrroles 449, 450
Hydrazones, aryl-, palladium-catalyzed cyclization to give arylindazoles 424
Hydrazones, methylenecyclopropane substituted, Lewis acid catalyzed ring expansion to give cyclic dienamines 167, 168
Hydroacyloxylation, of alkenes 480, 481
Hydroacyloxylation, of alkynes 474, 475
Hydroalkoxylation, of alkenes 475–480
Hydroalkoxylation, of alkynes 461–474
Hydroamination, of alkenes 401–408
Hydroamination, of alkynes 389–401
γ-Hydroxyalkenes, iron-catalyzed cyclization to give tetrahydrofurans 476
Hydroxyalkenes, platinum-catalyzed cyclization to give cyclic ethers 476, 477
4-Hydroxydictyolactone 6
Hydroxylamines, allenic, silver-catalyzed cyclization to give isoxazolidines 344
Hydroxylamines, *O*-[(2-methylenecyclopropyl)-methyl]-, ring-opening hydroamination to give isoxazolidines 173, 174
Hydroxylamines, vinylidenecyclopropane-1,1-dicarboxylate tethered, gold(I)-catalyzed domino intramolecular hydroamination and ring-opening reaction 203, 204

I

epi-Illudol 35
Imidazolidinones, from *N*-allylureas by palladium-catalyzed intramolecular carboamination 452, 453
Imidazo[4,5-*c*]pyrazoles, from *N'*-(halopyrazolyl)amidines by copper-catalyzed cyclization 429
Indanones, 3-vinyl-, from ketones by palladium-catalyzed intramolecular asymmetric allylic alkylation 104
Indazoles, aryl-, from arylhydrazones by palladium-catalyzed cyclization 424
Indenes, from 1-vinylcyclopropenes by copper-catalyzed cycloisomerization 237, 238
Indenes, from alkyne-, alkene-, and allene-tethered cyclopropenes by gold-catalyzed cycloisomerization 248, 249
Indenes, from phenylcyclopropenes by gold-catalyzed rearrangement 241, 243
Indenones, from 1-(2-bromophenyl)prop-2-en-1-ols by intramolecular Heck reaction/oxidation 69
Indenones, from diazo ketones by rhodium-catalyzed rearrangement of fused cyclopropenes formed in situ 228
Indole-2-carboxamides, 1-allyl-, palladium-catalyzed cyclization to give pyrazino[1,2-*a*]indol-1(2*H*)-ones 410
Indole-2-carboxamides, *N*-(haloaryl)-, palladium-catalyzed cyclization to give indolo[1,2-*a*]-quinoxalines 432
Indole-2-carboxylates, from vinyl azides by rhodium-catalyzed cyclization 440
Indole-fused lactams, from cyclopropyl α-amido esters by tandem ring-opening/Friedel–Crafts alkylation 190
Indoles, 3-alkenyl-, palladium-catalyzed oxidative annulation to give fused indoles 26, 27
Indoles, 1-(alk-1-ynylcyclopropyl)alkanol substituted, gold-catalyzed cycloisomerization to give carbazoles 213–216
Indoles, 2-alk-3-ynyl-, gold-catalyzed cyclization to give carbazoles 299, 300
Indoles, allenic, gold-catalyzed hydroarylation to give a flinderole intermediate 344, 345
Indoles, 2-allenic, gold-catalyzed intramolecular hydroarylation to give fused indoles 335, 336

Indoles, allenic, palladium-catalyzed domino cyclization to give fused polycyclic indoles 345, 346
Indoles, 2-aryl-, from 2-alkynylanilines by zinc-catalyzed cyclization 400, 401
Indoles, 3-aryl-, from nitroalkenes by palladium-catalyzed cyclization 424, 425
Indoles, by intramolecular Heck reaction 55
Indoles, 2,3-disubstituted, from 2-iodoanilines and 2-diazobut-3-enoates by N—H bond insertion/intramolecular Heck reaction 60
Indoles, from *N*-acetyl-2-alkynylanilines by palladium-catalyzed cyclization 393
Indoles, from 2-alkenylaryl azides by rhodium-catalyzed cyclization 439, 440
Indoles, from 2-alkynylanilines by copper-catalyzed intramolecular hydroamination 389
Indoles, from 2-alkynylanilines by gold-catalyzed cyclization 392
Indoles, from 2-alkynylanilines by rhodium-catalyzed cyclization 395
Indoles, from 2-alkynylaryltrifluoroacetanilides by palladium-catalyzed carboamination 448
Indoles, from 2-allylanilines by palladium-catalyzed cyclization 409, 410
Indoles, from 2-bromoanilines and vinyl bromides by amination/intramolecular Heck reaction 70, 71
Indoles, fused, by iridium-catalyzed intramolecular Friedel–Crafts-type alkylation of indoles 111, 112
Indoles, fused, by palladium-catalyzed allylic dearomatization of indoles 115
Indoles, fused, by palladium-catalyzed intramolecular Friedel–Crafts-type alkylation of indoles 111
Indoles, fused, from 2-allenic indoles by gold-catalyzed intramolecular hydroarylation 334, 335
Indoles, fused, from 3-alkenylindoles by palladium-catalyzed oxidative annulation 26, 27
Indoles, gold-catalyzed annulation with enynones to give 9,10-dihydrocyclohepta[*b*]indol-8(5*H*)-ones 307, 308
Indoles, iridium-catalyzed enantioselective functionalization via a spiro intermediate to give vinyltetrahydrocarbolines 117
Indoles, iridium-catalyzed intramolecular asymmetric allylic dearomatization to give spiro-fused indoles 113
Indoles, iridium-catalyzed intramolecular asymmetric allylic dearomatization to give spiro-fused indoles and tetrahydrocarbazoles 116
Indoles, iridium-catalyzed intramolecular Friedel–Crafts-type alkylation 111, 112
Indoles, metal-catalyzed 7-*endo-dig* cyclization with nitroenynes to give 5,12-dihydrobenzo[4,5]cyclohepta[1,2-*b*]indoles 306, 307
Indoles, methylenecyclopropane substituted, rhodium(I)-catalyzed cycloisomerization to give pyrido[3,4-*b*]indoles 160
Indoles, palladium-catalyzed allylic dearomatization to give fused indoles 115
Indoles, palladium-catalyzed intramolecular Friedel–Crafts-type alkylation 111
Indoles, ruthenium-catalyzed intramolecular allylic dearomatization to give spiro-fused indoles 114
Indoles, spiro-fused, by iridium-catalyzed intramolecular asymmetric allylic dearomatization of indoles 113, 116
Indoles, spiro-fused, by ruthenium-catalyzed intramolecular allylic dearomatization of indoles 114
Indoles, *N*-tosyl-, from 2-alkynyl-*N*-tosylanilines by zinc-catalyzed cyclization 400
Indoles, use in intramolecular allylic substitution 106–117
Indolines, hetarylmethyl substituted, by intramolecular Heck cyclization/direct arylation 62–64
Indolizines, from 2-(cycloprop-2-enyl)pyridines by copper-catalyzed rearrangement 238, 239
Indolizines, from 2-(cycloprop-2-enyl)pyridines by rhodium(I)-catalyzed rearrangement 230, 231
2-Iodoanilines, N—H bond insertion with 2-diazobut-3-enoates/intramolecular Heck reaction to give 2,3-disubstituted indoles 60
2-Iodoanilines, N—H insertion with 4-aryl-2-diazobut-3-enoates/intramolecular Heck reaction to give quinolines 72, 73
Isoquinolines, by intramolecular Heck reaction of Ugi reaction products 66
Isoquinolines, 3,4-disubstituted, from benzaldimines by palladium-catalyzed carboamination 449
Isoretronecanol 101
Isoxazolidines, from allenic hydroxylamines by silver-catalyzed cyclization 344
Isoxazolidines, from *O*-[(2-methylenecyclopropyl)methyl]hydroxylamines by ring-opening hydroamination 173, 174

J

Josiphos, use in enantioselective intramolecular Heck reactions 81

K

K-13 18
(−)-α-Kainic acid 102, 250, 251
Kedarcidin chromophore 20, 21
Ketals, bicyclic, from alkynyl diols by gold-catalyzed cyclization 463
β-Keto amides, alkenyl tethered, gold(I)-catalyzed 5-*exo-trig* cyclization to give lactams 280, 281

β-Keto amides, alkyne tethered, zinc(II) chloride catalyzed Conia-ene reaction to give methylenecycloalkanes 286
β-Keto esters, alkenyl tethered, palladium-catalyzed 6-*endo-trig* cyclization to give 2-oxocyclohexanecarboxylates 279
β-Keto esters, alkyne tethered, zinc(II) chloride catalyzed Conia-ene reaction to give methylenecycloalkanes 286
β-Keto esters, alkynyl tethered, gold(I)-catalyzed 5-*endo-dig* cyclization to give cycloalk-2-enecarboxylates 285
β-Keto esters, alkynyl tethered, gold(I)-catalyzed Conia-ene reactions to give 2-methylenecycloalkanecarboxylates 284
β-Keto esters, allenic, palladium-catalyzed coupling/cyclization with allyl bromides to give 2,3-dihydrofurans 321
β-Keto esters, allenic, palladium-catalyzed coupling/cyclization with organic iodides to give lactones 340
β-Keto esters, 2-(2-bromobenzyl)-, copper-catalyzed cyclization to give benzopyrans 488
β-Keto esters, palladium-catalyzed cyclization 100
β-Keto esters, palladium-catalyzed intramolecular allylic alkylation to give vinylcyclohexanones 99, 100
Ketones, alkenyl, gold(I)-catalyzed *exo-trig* cyclization to give cycloalkyl ketones 281–283
Ketones, cyclic, unsaturated, chromium(III)-catalyzed enantioselective transannular ketone-ene reactions to give bicyclic alcohols 273, 274
Ketones, cycloalkyl, from alkenyl ketones by gold(I)-catalyzed *exo-trig* cyclization 281–283
Ketones, palladium-catalyzed intramolecular asymmetric allylic alkylation to give 3-vinylindanones 104
Ketones, use in intramolecular allylic substitution 104, 105

L

Lactams, from alkenyl β-oxo amides by gold(I)-catalyzed 5-*exo-trig* cyclization 280, 281
Lactams, from *N*-tosylhex-5-enamides by palladium-catalyzed cyclization 410, 411
Lactones, eight to ten membered, from allenic 3-oxoalkanoates and organic iodides by palladium-catalyzed coupling/cyclization 340
Lactones, five membered, from malonates by gold(I)-catalyzed intramolecular allylic alkylation 141, 142
Lactones, five membered, from unsaturated monoesters by gold(I)-catalyzed intramolecular allylic substitution 141, 142
Lactones, from alkenoic acids by silver-catalyzed cyclization 480, 481
γ-Lactones, from alkynoic acids by gold-catalyzed cyclization 475
Lavendamycin, analogues 312
(+)-Linalool oxide 343
Lotrafiban, intermediate 458
Lundurine A 313

M

Malonates, alkyne tethered, zinc(II) chloride catalyzed Conia-ene reaction to give methylenecycloalkanes 286
Malonates, allenic, palladium-catalyzed coupling/cyclization with organic iodides to give large rings 341
Malonates, gold(I)-catalyzed intramolecular allylic alkylation to give five-membered lactones 141, 142
Malonates, 2-homoallenyl-, palladium-catalyzed coupling/cyclization with organic halides to give 2-vinylcyclopropane-1,1-dicarboxylates 338, 339
Malonates, iridium-catalyzed intramolecular allylic alkylation to give vinylcycloalkanes 97
Malonates, palladium-catalyzed intramolecular allylic alkylation to give vinylcyclopentanes 96, 97
Methylenecycloalkanes, from allenenes by rhodium-catalyzed cycloisomerization 372
Methylenecycloalkanes, from α,ω-dienes by ruthenium-catalyzed cycloisomerization 365, 366
Methylenecycloalkanes, from α,ω-dienes by titanium-catalyzed cycloisomerization 363, 364
Methylenecycloalkanes, from α,ω-enynes by iron-catalyzed cycloisomerization 355–357
Methylenecycloalkanes, from α,ω-enynes by palladium-catalyzed cycloisomerization 351, 352
Methylenecycloalkanes, from α,ω-enynes by ruthenium-catalyzed cycloisomerization 354, 355
Methylenecycloalkanes, from α,ω-enynes by titanium-catalyzed cycloisomerization 352–354
Methylenecycloalkenes, from allenenes by rhodium-catalyzed cycloisomerization 371–373
Methylenecycloalkenes, from allenynes by gold-catalyzed cycloisomerization 369, 370
Methylenecyclobutanes, from hexa-1,5-dien-2-ol trifluoromethanesulfonates by intramolecular Heck reaction 58
Methylenecyclohexenes, from allenynes by rhodium-catalyzed cycloisomerization 368
Methylenecyclopentanes, from 1,6-dienes by nickel-catalyzed enantioselective cycloisomerization 366, 367

Methylenecyclopentanes, stannyl, from stannyl enol ethers by iron(III)-catalyzed stannyl Conia-ene cyclization 287, 288
Methylenecyclopropanes, acyl substituted, palladium(II)-catalyzed cycloisomerization, mechanism 153
Methylenecyclopropanes, acyl substituted, palladium(II)-catalyzed cycloisomerization to give furans 154
Methylenecyclopropanes, alkyne substituted, nickel-catalyzed intramolecular cycloaddition to give cyclopenta[*a*]indenes 165, 166
Methylenecyclopropanes, 1,5-dienyl, rearrangement 171, 172
Methylenecyclopropanes, 1,6-diyne substituted, cycloisomerization to give fused norbornanes 172, 173
Methylenecyclopropanes, ene, yne, or diene substituted, intramolecular cycloaddition to give cycloalkenes 155, 156
Methylenecyclopropanes, hydrazone substituted, Lewis acid catalyzed ring expansion to give cyclic dienamines 167, 168
Methylenecyclopropanes, hydroxyalkyl tethered, gold(I)-catalyzed cycloisomerization 174, 175
Methylenecyclopropanes, indole substituted, rhodium(I)-catalyzed cycloisomerization to give pyrido[3,4-*b*]indoles 160–162
Methylenecyclopropanes, indole substituted, rhodium(I)-catalyzed cycloisomerization, isotopic labeling experiments 161
Methylenecyclopropanes, indole substituted, rhodium(I)-catalyzed cycloisomerization, mechanism 162
Methylenecyclopropanes, intramolecular cycloisomerization reactions, palladium catalyzed 151–158
Methylenecyclopropanes, intramolecular cycloisomerization reactions, platinum catalyzed 159
Methylenecyclopropanes, intramolecular cycloisomerization reactions, rhodium catalyzed 160–165
Methylenecyclopropanes, intramolecular cycloisomerization reactions, nickel catalyzed 165, 166
Methylenecyclopropanes, intramolecular cycloisomerization reactions, copper catalyzed 166, 167
Methylenecyclopropanes, intramolecular cycloisomerization reactions, magnesium chloride catalyzed 167, 168
Methylenecyclopropanes, intramolecular cycloisomerization reactions, gold or silver catalyzed 168–176
Methylenecyclopropanes, intramolecular hydroamination to give tetrahydroquinolines and azepanes 154
Methylenecyclopropanes, lactone fused, from allenic diazo esters by copper-catalyzed cyclopropanation 334, 335
Methylenecyclopropanes, palladium(II)-catalyzed ring-opening cycloisomerization 151, 152
Methylenecyclopropanes, palladium-catalyzed ring expansion to give cyclobutenes, mechanism 156, 157
Methylenecyclopropanes, platinum-catalyzed ring expansion to give cyclobutenes, mechanism 159
Methylenecyclopropanes, propargylic alcohol tethered, gold(I)- and silver(I)-catalyzed cycloisomerization 170, 171
Methylenecyclopropanes, rhodium-catalyzed ene–cycloisomerization 250
Methylenecyclopropanes, rhodium-catalyzed ring opening through a C—H activation strategy 162, 163
Methylenecyclopropanes, triazole substituted, copper-catalyzed intramolecular cycloisomerization to give 2,3-dihydrobenzo[*f*]indoles 166, 167
Methylenecyclopropanes, triazole substituted, rhodium(II)-catalyzed intramolecular cycloisomerization 163–165
Methylenecyclopropanes, yne substituted, gold-catalyzed rearrangement 168, 169
[2-(Methylene)cyclopropyl]carbinols, ring-opening cycloisomerization to give cyclobutenes 157
3-Methylene-2,3-dihydro-1*H*-cyclopenta[*b*]quinolines, from 2-chloroquinolin-3-yl homoallyl alcohols by intramolecular Heck cyclization 59
1-Methylene-1*H*-indenes, from O-protected 3-(hydroxymethyl)cyclopropenes by gold-catalyzed cycloisomerization 243, 244
Methyleneoxetanes, from 3-bromoalk-3-enols by copper-catalyzed cyclization 486
Micrandilactone A 497
(+)-Minfiensine 90
Moluccanic acid methyl ester 311

N

Naphthalene-2,2(1*H*)-dicarboxylates, 4-vinyl-3,4-dihydro-, from arenes by molybdenum(II)-catalyzed intramolecular allylic alkylation 122
Naphthalenes, fused, from vinylidenecyclopropanes by titanium-catalyzed ring opening 206, 207
Naphthalen-2(1*H*)-ones, from alkynylcyclopropanes by silver-catalyzed ring expansion 223, 224
Negishi coupling, general mechanism 17
Negishi coupling, intramolecular, palladium catalyzed 16–18
Nigellamine A_2 34, 144

Nitidine, precursor 455
Nitro compounds, aliphatic, use in intramolecular allylic substitution 105
Nitroalkenes, 2,2-diaryl-, palladium-catalyzed cyclization to give 3-arylindoles 424, 425
Nitroenynes, metal-catalyzed 7-*endo-dig* cyclization with 1*H*-indoles to give 5,12-dihydrobenzo[4,5]cyclohepta[1,2-*b*]indoles 306, 307
Nitrones, cyclopropane-1,1-dicarboxylate substituted, intramolecular cycloaddition to give tetrahydro-1,2-oxazines 177–179
[4.3.0]Nonanes, from allenyl cyclopropane-1,1-dicarboxylates by intramolecular cycloaddition 185, 186
Norbornanes, fused, from methylenecyclopropane-bearing 1,6-diynes by cycloisomerization 172
Norepinephrine transporter (NET) inhibitor, intermediate 459
Norsesquiterpene 78
Nozaki–Hiyama–Kishi coupling, intramolecular 31–36

O

Octahydro-1*H*-pyrrolo[3,2-*c*]pyridines, from *N*,*N'*-(hex-3-ene-1,6-diyl)bis(4-toluenesulfonamide) and aldehydes by scandium(III)-catalyzed aza-Prins reaction 276, 277
Octahydro-2*H*-quinolizines, 2-hydroxy, by nickel-catalyzed intramolecular cyclization of an alkenyl iodide with an aldehyde 33
Okadaic acid 495
3-Oxabicyclo[4.1.0]heptanes, from alkene-tethered cyclopropenes by gold-catalyzed intramolecular cyclopropanation via in situ carbene formation 242
1-Oxa-2-silacycloalkenes, iodoalkenyl, palladium-catalyzed intramolecular Hiyama coupling to give cycloalkadienes 23
3-Oxaterpenoids, from homoallyl alcohols and aldehydes by indium(III)-catalyzed Prins cyclization/polyene cyclization 264–266
1,2,3-Oxathiazepane 2,2-dioxides, from allenyl sulfamates by rhodium-catalyzed cyclization 324, 325
Oxathiocins, by intramolecular Heck reaction 76
Oxazoles, 2,5-disubstituted, from *N*-propargylamides by gold-catalyzed cyclization 464
Oxazoles, from *N*-propargylamides by palladium/copper-catalyzed cyclization 466, 467
Oxazolidin-2-ones, from alkyl carbamates by rhodium-catalyzed cyclization 436, 437
Oxazolidin-2-ones, 4-vinyl-, from allenyl carbamates by silver-catalyzed cyclization 328, 329
Oxcarbazepine, intermediate 458
Oxidative alkoxylation, of alkenes 481–484
Oxidative alkoxylation, of arenes 484–485
Oxidative amination of alkenes, mechanism 409
Oxidative amination of arenes, mechanism 415
Oximes, cyclopropane-1,1-dicarboxylate substituted, intramolecular cycloaddition to give pyrrolo[1,2-*b*]isoxazolidines 176, 177
Oxindoles, by intramolecular Heck reaction of 2-halophenyl acrylamides/intermolecular reaction with hydrazones 61
Oxindoles, hetarylmethyl substituted, by intramolecular Heck cyclization/direct arylation 62–64
Oxindoles, 3-spiroannulated, by enantioselective intramolecular Heck reaction 88
Oxindoles, spiro-fused, by intramolecular Heck reaction of 2-halophenyl acrylamides/intramolecular C—H functionalization 61
Oxindoles, 3-vinyl-, by enantioselective intramolecular Heck reaction 85
Oxocins, by intramolecular Heck reaction 76
α-Oxo esters, copper(II)-catalyzed asymmetric carbonyl-ene reaction to give 1-hydroxycycloalkanecarboxylates 270

P

Penarolide sulfate A_1 20, 21
1,2,2,6,6-Pentamethylpiperidine, use in intramolecular Heck reactions 79, 85, 88–90
Pent-4-enamines, palladium-catalyzed intramolecular carboamination to give pyrrolidines 451
Pent-4-ynols, palladium-catalyzed carboalkoxylation with organic iodides to give 2-methylenetetrahydrofurans 492
Peptides, cyclic, from aryl bromides and terminal alkynes by intramolecular Sonogashira coupling 19, 20
Phenanthrenes, from 2-alk-1-ynylbiphenyls by metal-catalyzed cyclization 296–298
Phenanthridines, from *N*-(2-phenylbenzyl)-picolinamides by palladium-catalyzed cyclization 422, 423
Phenethylamines, palladium-catalyzed cyclization to give dihydroindoles 418, 419
Phenols, 2-alkenyl-, palladium-catalyzed cyclization to give dihydrobenzopyrans and dihydrobenzo[*b*]furans 494, 495
Phenols, 2-alkynyl-, palladium-catalyzed cyclization to give 2,3-diarylbenzo[*b*]furans 492, 493
Phenols, 2-alkynyl-, palladium-catalyzed cyclization to give methyl benzo[*b*]furan-3-carboxylates 490
Phenols, allylic carbonate substituted, palladium-catalyzed asymmetric allylic substitution to give 2-vinyl-3,4-dihydrobenzopyrans 138
Phenols, 2-allylic, ruthenium-catalyzed cyclization to give 2,3-dihydrobenzo[*b*]furans 478

Phenols, 2-ethynyl-, palladium-catalyzed cyclization to give 2-substituted benzo[*b*]furans 492, 493
Phenols, from yne–cyclopropenes by gold-catalyzed cycloisomerization 244–246
Phenols, iridium-catalyzed asymmetric allylic dearomatization to give spiro-fused cyclohexadienones 121
Phenols, iridium-catalyzed intramolecular Friedel–Crafts-type alkylation to give vinyltetrahydroisoquinolines and -naphthalenes 118
Phenols, palladium-catalyzed allylic dearomatization to give spiro-fused cyclohexadienones 120
Phenols, palladium-catalyzed asymmetric allylic dearomatization to give spiro-fused cyclohexadienones 120
Phenols, use in intramolecular allylic substitution, carbon nucleophiles 118–121
Phenols, use in intramolecular allylic substitution, oxygen nucleophiles 137–139
Phenyl ethers, allenic, gold-catalyzed cyclization to give 2*H*-benzopyrans 336, 337
Phenyl-1*H*-indenes, from vinylidenecyclopropanes by Lewis acid catalyzed rearrangement 199, 200
Phomactin A 35
Phthalimides, from benzamides by ruthenium-catalyzed carbonylation 446, 447
Picolinamides, *N*-(2-phenylbenzyl)-, palladium-catalyzed cyclization to give phenanthridines 422, 423
Picrotoxane sesquiterpenes 380, 381
Picrotoxinin 381
Piperazines, from *N*-allylethane-1,2-diamines by palladium-catalyzed intramolecular carboamination 451, 452
Piperazinones, fused, from alkenylureas by palladium-catalyzed cyclization 411, 412
Piperchabamide G 254, 255
Piperidines, *N*-benzyl-, from alkenyl(benzyl)-amines by platinum-catalyzed cyclization 405, 406
Piperidines, 1,3-dienyl-, from bis(1,3-dienes) by palladium-catalyzed cycloisomerization 379
Piperidines, enantioselective synthesis via gold(I)/secondary amine cocatalyzed intramolecular α-allylic alkylation of an aldehyde 103
Piperidines, from alkenylamines by copper-catalyzed cyclization 401, 402
Piperidines, from alkenylamines by iridium-catalyzed cyclization 403, 404
Piperidines, from alkenylamines by rhodium-catalyzed cyclization 406, 407
Piperidines, from alkynylamines by palladium-catalyzed cyclization 394
Piperidines, from *N*-homoallyl sulfonamides and aldehydes by iron(III)-catalyzed aza-Prins reaction 275
Piperidines, 2-vinyl-, from allenyl ureas by gold-catalyzed enantioselective intramolecular hydroamination 326
Piperidines, 2-vinyl-, from allylic alcohols by gold(I)-catalyzed intramolecular amination 132, 133
Piperidines, 2-vinyl-, from allylic carbonates by iridium-catalyzed intramolecular allylic amination 128
Platensimycin 253, 254, 381
Polyene cyclization, enantioselective, iridium catalyzed 124, 125
(+)-Preussin, intermediate 459
Prins cyclization, of homoallylic alcohols with aldehydes 260–268
Prins cyclization, of nucleophilic alkynes with aldehydes 268, 269
N-Propargylamides, gold-catalyzed cyclization to give 2,5-disubstituted oxazoles 464
N-Propargylamides, palladium/copper-catalyzed cyclization to give oxazoles 466, 467
N-Propargylhydroxylamines, *N*-benzyl-, zinc-catalyzed cyclization to give 2,3,5-trisubstituted dihydroisoxazoles 474
Pyrazines, iridium-catalyzed intramolecular asymmetric allylic dearomatization to give 6-vinyl-6,7-dihydropyrrolo[1,2-*a*]pyrazines 136, 137
Pyrazines, use in intramolecular allylic amination 136, 137
Pyrazino[1,2-*a*]indol-1(2*H*)-ones, from 1-allylindole-2-carboxamides by palladium-catalyzed cyclization 410
Pyridines, allylic carbonate substituted, iridium-catalyzed intramolecular asymmetric allylic dearomatization to give 3-vinyl-2,3-dihydroindolizines 136
Pyridines, 2-(arylamino)-, copper/iron-catalyzed cyclization to give pyrido[1,2-*a*]benzimidazoles 416, 417
Pyridines, 2-(cycloprop-2-enyl)-, copper-catalyzed rearrangement to give indolizines 238, 239
Pyridines, 2-(cycloprop-2-enyl)-, rhodium(I)-catalyzed rearrangement to give indolizines 230, 231
Pyridines, use in intramolecular allylic amination 136, 137
Pyrido[1,2-*a*]benzimidazoles, from 2-(arylamino)pyridines by copper/iron-catalyzed cyclization 416, 417
Pyrido[3,4-*b*]indoles, from nitrogen-tethered indolyl–methylenecyclopropanes by rhodium(I)-catalyzed cycloisomerization 160, 161

Pyrrole-2-carboxamides, *N*-(haloaryl)-, palladium-catalyzed cyclization to give pyrrolo[1,2-*a*]quinoxalines 432
Pyrroles, from alkynylimines by copper-catalyzed cyclization 390
Pyrroles, from cyclopropyl-tethered alkynyl imines by ruthenium-catalyzed cycloisomerization 218, 219
Pyrroles, from 1,3-dienyl azides by ruthenium-catalyzed cyclization 442
Pyrroles, from 1,3-dienyl azides by zinc-catalyzed cyclization 443
Pyrroles, from enynamines by copper-catalyzed cyclization 390, 391
Pyrroles, from homopropargyl azides by platinum-catalyzed cyclization 443, 444
Pyrroles, from homopropargyl azides by silver/gold-catalyzed cyclization 444, 445
Pyrroles, iridium-catalyzed enantioselective functionalization via a spiro intermediate to give 7-vinyl-4,5,6,7-tetrahydro-1*H*-pyrrolo[3,2-*c*]pyridines 117
Pyrroles, iridium-catalyzed intramolecular asymmetric allylic dearomatization to give spiro-fused pyrroles 114
Pyrroles, spiro-fused, by iridium-catalyzed intramolecular asymmetric allylic dearomatization of pyrroles 114
Pyrroles, 1,2,3,5-tetrasubstituted, from alkynyl hydrazones by palladium-catalyzed cyclization with aryl iodides 449, 450
Pyrroles, use in intramolecular allylic substitution 106–117
Pyrrolidines, *N*-acyl-, from *N*-alkylcarboxamides by palladium-catalyzed cyclization 433–435
Pyrrolidines, *N*-benzyl-, from alkenyl(benzyl)-amines by platinum-catalyzed cyclization 405, 406
Pyrrolidines, from alkenylamines by copper-catalyzed cyclization 401, 402
Pyrrolidines, from alkenylamines by iridium-catalyzed cyclization 403, 404
Pyrrolidines, from alkenylamines by rhodium-catalyzed cyclization 406–408
Pyrrolidines, from alkynylamines by palladium-catalyzed cyclization 394
Pyrrolidines, from allyl(2-bromoallyl)amines and boronic acids by intramolecular Heck cyclization/coupling 62
Pyrrolidines, from pent-4-enamines by palladium-catalyzed intramolecular carboamination 451
Pyrrolidines, 3-methylene-, from alkyne-tethered and acetyl halides by iron(III)-catalyzed Prins cyclization 268
Pyrrolidines, *N*-tosyl-, from alkenyl-*N*-tosylamines by iron-catalyzed cyclization 404, 405
Pyrrolidines, 3-vinyl-, by gold(I)/secondary amine cocatalyzed enantioselective intramolecular α-allylic alkylation of aldehydes 103
Pyrrolidines, 3-vinyl-, from allenyl aldehydes by palladium-catalyzed carbocyclization 331, 332
Pyrrolidines, 2-vinyl-, from allenyl carbamates by silver-catalyzed dynamic kinetic enantioselective hydroamination 329
Pyrrolidines, 2-vinyl-, from allenyl sulfonamides by gold-catalyzed enantioselective intramolecular hydroamination 327
Pyrrolidines, 2-vinyl-, from allenyl ureas by gold-catalyzed enantioselective intramolecular hydroamination 326
Pyrrolidines, 2-vinyl-, from allylic alcohols by gold(I)-catalyzed intramolecular amination 132–134
Pyrrolidines, 2-vinylidene-, from alkynylamines by ruthenium-catalyzed cyclization 396
Pyrrolidin-2-ones, 3-methylene-, from α,ω-enynes by rhodium-catalyzed enantioselective cycloisomerization 358
Pyrrolidin-2-ones, 4-vinyl-, from stabilized acetamide enolates by palladium-catalyzed intramolecular allylic alkylation 101
Pyrrolo[1,2-*b*]isoxazolidines, by intramolecular cycloaddition of oximes with cyclopropane 1,1-diesters 176, 177

Q

(−)-Quinocarcin, fragment 454
Quinolines, from 2-iodoanilines and 4-aryl-2-diazobut-3-enoates by N—H insertion/intramolecular Heck reaction 72
Quinolines, from allylic alcohols by iron(III) chloride hexahydrate catalyzed intramolecular allylic amination 135
Quinolinones, 4-aryl-, from 3,3-diarylacrylamides by palladium/copper-catalyzed cyclization 423
Quinolinones, from *N*-arylalkynamides by metal-catalyzed cyclization 300–302
Quinoxalines, heteroannulated, from *N*-(haloaryl)pyrrole-2-carboxamides and -indole-2-carboxamides by palladium-catalyzed cyclization 432

R

Reblastatin, intermediate 457
Reductive coupling of organo halides, intramolecular, nickel catalyzed 44–50
Reductive coupling of organo halides, intramolecular, nickel catalyzed, mechanism 45
Rhazinicine 30
Rhodium carbenes, generated in situ from cyclopropenylalkyl ethers, rearrangement to give 4-methylenetetrahydropyrans 231, 232
Rubromycins 496

S

Sanglifehrin A 15, 16
Sarain A 15, 16
(−)-Siccanin 148
Sonogashira coupling, intramolecular, palladium catalyzed 18–22
Sonogashira coupling, mechanism 19
Spiroindolines, tetracyclic, from alkenylamine-tethered cyclopropane-1,1-dicarboxylates by cycloisomerization 196
[n,5]Spiroketals, from cyclopropyl aldehydes by rearrangement 191
Spirotetrahydropyrans, from phenol-tethered homoallyl alcohols and aldehydes by indium(III)-catalyzed Prins cyclization 263, 264
Stannyl enol ethers, iron(III)-catalyzed stannyl Conia-ene cyclization to give stannylated cyclopentanes and cyclopentenes 287, 288
Stille coupling, general mechanism 12, 13
Stille coupling, intramolecular, palladium catalyzed 12–16
Stille-type cross coupling, copper catalyzed 41–43
Superstolide A 10, 11
Suzuki–Miyaura coupling, general mechanism 2
Suzuki–Miyaura coupling, intramolecular, palladium catalyzed 1–12
Suzuki–Miyaura reaction, intramolecular, with sp^2 organoboron reagents 7–12
Suzuki–Miyaura reaction, intramolecular, with sp^3 organoboron reagents 2–7

T

Taiwaniaquinone F 146
3,4,5,6-Tetrahydro-2*H*-azepines, from alkynylamines by ruthenium-catalyzed cyclization 396
Tetrahydrobenzo[*b*]azepines, from bromoaryl-substituted secondary amides and carbamates by palladium-catalyzed cyclization 431, 432
Tetrahydro-1*H*-benzo[*d*]azepines, vinyl-substituted, from iodoarene–alkenylsilanes by enantioselective intramolecular Heck reaction 78
Tetrahydrocarbazoles, from indoles by iridium-catalyzed intramolecular asymmetric allylic dearomatization 116
Tetrahydrocarbazoles, from indol-2-yl-substituted allyl alcohols by gold-catalyzed intramolecular Friedel–Crafts-type alkylation 110
Tetrahydrocarbazoles, from indol-3-yl-substituted allyl alcohols by gold-catalyzed intramolecular Friedel–Crafts-type alkylation 109
2,3,4,9-Tetrahydro-1*H*-carbazoles, vinyl-, from aryl allylic alcohols by metal-catalyzed cyclization 290–292
Tetrahydrocarbolines, from indol-2-yl-substituted allyl carbonates by iridium-catalyzed intramolecular Friedel–Crafts-type alkylation 107, 108
Tetrahydrocarbolines, from indol-2-yl-substituted allyl carbonates by palladium-catalyzed intramolecular Friedel–Crafts-type alkylation 106–107
Tetrahydrocarbolines, vinyl-, from indoles by iridium-catalyzed enantioselective functionalization via a spiro intermediate 117
Tetrahydrofuran-3-ols, from alkene-tethered aldehydes by Schiff base/chromium(III)-catalyzed enantioselective intramolecular carbonyl-ene reactions 270–273
Tetrahydrofurans, from alkenols by silver-catalyzed cyclization 479
Tetrahydrofurans, from γ-hydroxyalkenes by iron-catalyzed cyclization 476
Tetrahydrofurans, from γ-hydroxyalkenes by platinum-catalyzed cyclization 477
Tetrahydrofurans, 3-methylene-, from alkyne-tethered acetals and acetyl halides by iron(III)-catalyzed Prins cyclization 268
Tetrahydrofurans, 3-methylene-, from α,ω-enynes by rhodium-catalyzed enantioselective cycloisomerization 358
Tetrahydrofurans, 2-methylene-, from pent-4-ynols by palladium-catalyzed carboalkoxylation with organic iodides 491, 492
2,3,3a,7a-Tetrahydro-1*H*-indenes, by enantioselective intramolecular Heck reaction 83
5,6,7,8-Tetrahydroindolizines, 8-vinyl-, from allene-tethered pyrroles by gold-catalyzed cyclohydroarylation 338
Tetrahydroisoquinolines, vinyl-, from phenols by iridium-catalyzed intramolecular Friedel–Crafts-type alkylation 118
Tetrahydroisoquinolines, vinyl-substituted, from iodoarene–alkenylsilanes by enantioselective intramolecular Heck reaction 78
1,2,4a,8a-Tetrahydronaphthalenes, by enantioselective intramolecular Heck reaction 82
1,2,3,4-Tetrahydronaphthalenes, by enantioselective intramolecular Heck reaction 86, 87
Tetrahydronaphthalenes, from alkenylaryl trifluoromethanesulfonates by intramolecular Heck reaction 65, 66
1,2,3,4-Tetrahydronaphthalenes, from alk-3-enylbenzenes by ruthenium(III)-catalyzed cyclization 289
Tetrahydronaphthalenes, 1-vinyl-, from arenes by silver(I)-catalyzed intramolecular allylic alkylation of arenes with allylic alcohols 122
1,2,3,4-Tetrahydronaphthalenes, 1-vinyl-, from aryl allylic alcohols by metal-catalyzed cyclization 290–292

Tetrahydronaphthalenes, 1-vinyl-, from electron-deficient arenes with π-activated alcohols by iron(III)-catalyzed intramolecular allylic alkylation 123
Tetrahydronaphthalenes, vinyl-, from phenols by iridium-catalyzed intramolecular Friedel–Crafts-type alkylation 118
Tetrahydronaphthalenes, vinyl-, from phenyl-tethered allenes by gold-catalyzed cyclohydroarylation 338
Tetrahydronaphthalenes, vinyl-substituted, by enantioselective intramolecular Heck reaction of iodoarene–alkenylsilanes 78
Tetrahydro-1,2-oxazines, bridged, from nitrone-substituted cyclopropane 1,1-diesters by intramolecular cycloaddition 177–179
Tetrahydro-2*H*-pyran-4-ols, from homoallyl alcohols and aldehydes by iron(III)-catalyzed Prins cyclization 266, 267
Tetrahydro-2*H*-pyran-4-ols, from homoallyl alcohols and aldehydes by rhenium(VII)-catalyzed Prins cyclization 262, 263
Tetrahydropyrans, *cis*-2-alkyl-4-halo-, from homoallyl alcohol and aldehydes by iron(III)-catalyzed Prins cyclization 261
Tetrahydropyrans, from alkenols by silver-catalyzed cyclization 479
Tetrahydropyrans, from homoallylic alcohols and aldehydes by Prins cyclization, mechanism 260
Tetrahydropyrans, from δ-hydroxyalkenes by platinum-catalyzed cyclization 477
Tetrahydropyrans, from hydroxyalkyl allylic alcohols by iron(III) chloride catalyzed intramolecular allylic etherification 140
Tetrahydropyrans, from vinylidenecyclopropane-tethered alcohols by gold-catalyzed nucleophilic ring opening 202
Tetrahydropyrans, 4-methylene-, from cyclopropenylalkyl ethers via rhodium carbene intermediates 231, 232
1,2,3,4-Tetrahydropyrazines, from aminoalkynes by zinc-catalyzed cyclization 399
2,3,4,5-Tetrahydropyridines, from alkenylamines by ruthenium-catalyzed cyclization 414
2,3,4,5-Tetrahydropyridines, from alkynylamines by gold-catalyzed cyclization 391, 392
1,2,3,4-Tetrahydropyridines, from aminoalkynes by zinc-catalyzed cyclization 399
Tetrahydropyridines, from *N*-homopropargyl sulfonamides and aldehydes by iron(III)-catalyzed aza-Prins reaction 274, 275
3,4,5,6-Tetrahydropyridines, from primary alkynylamines by silver-catalyzed cyclization 397
4,5,6,7-Tetrahydro-1*H*-pyrrolo[3,2-*c*]pyridines, 7-vinyl-, from pyrroles by iridium-catalyzed enantioselective functionalization via a spiro intermediate 117
1,2,3,4-Tetrahydroquinolines, from alkyl bromides and aryl iodides by intramolecular reductive coupling 47, 48
Tetrahydroquinolines, from bromoaryl-substituted secondary amides and carbamates by palladium-catalyzed cyclization 431, 432
Tetrahydroquinolines, from methylenecyclopropanes by intramolecular hydroamination 154
1,2,3,4-Tetrahydroquinolines, 4-methylene-, from α,ω-enynes by palladium-catalyzed enantioselective cycloisomerization 359, 360
1,2-Thiazinane 1,1-dioxides, from alkenesulfonamides by gold-catalyzed cyclization 403
(−)-Trachelanthamidine 102
1,2,3-Triazoles, hetero- and carbocycle-fused, by intramolecular Heck reaction 68
Triazoles, methylenecyclopropane substituted, copper-catalyzed intramolecular cycloisomerization to give 2,3-dihydrobenzo[*f*]indoles 166, 167
Triazoles, methylenecyclopropane substituted, rhodium(II)-catalyzed intramolecular cycloisomerization 163–165
Trifluoroacetanilides, 2-alkynylaryl-, palladium-catalyzed carboamination to give indoles 448
Tsuji–Trost reaction, general mechanism 95

U

Ugi reaction, of allylamine, isonitriles, 2-halobenzaldehydes, and benzoic or alkanoic acids, intramolecular Heck reaction precursor synthesis 66
Ullmann condensation, intramolecular 427–431

V

δ-Valerolactones, from alkynols by ruthenium-catalyzed cyclization 473
Ventricosene 254
Vernolepin 82
(±)-Vincorine 252, 253
Vinyl azides, rhodium-catalyzed cyclization to give methyl indole-2-carboxylates 440
Vinyl iodides, copper-catalyzed intramolecular coupling with terminal alkynes to give cyclic enynes 37, 38, 40
Vinylidenecyclopropane-1,1-dicarboxylates, hydroxylamine tethered, gold(I)-catalyzed domino intramolecular hydroamination and ring-opening reaction 203, 204
Vinylidenecyclopropanes, acyl, copper-catalyzed tandem ring opening and cycloaddition to give benzo[*b*]furan-7(3a*H*)-ones 209, 210

Vinylidenecyclopropanes, alcohol tethered, gold-catalyzed intramolecular rearrangement to give dienones 202, 203
Vinylidenecyclopropanes, alcohol tethered, gold-catalyzed nucleophilic ring opening to give tetrahydropyrans 202
Vinylidenecyclopropanes, gold-catalyzed rearrangement to give 1,2-dihydronaphthalenes 200, 201
Vinylidenecyclopropanes, intramolecular cycloisomerization reactions, tin catalyzed 199, 200
Vinylidenecyclopropanes, intramolecular cycloisomerization reactions, gold catalyzed 200–206
Vinylidenecyclopropanes, intramolecular cycloisomerization reactions, titanium catalyzed 206, 207
Vinylidenecyclopropanes, intramolecular cycloisomerization reactions, rhodium catalyzed 208, 209
Vinylidenecyclopropanes, intramolecular cycloisomerization reactions, copper catalyzed 209, 210
Vinylidenecyclopropanes, Lewis acid catalyzed rearrangement to give benzo[*c*]fluorenes and phenyl-1*H*-indenes 199, 200
Vinylidenecyclopropanes, rhodium-catalyzed intramolecular cycloaddition to give 5,6-dimethylenehexahydroazocines 208
Vinylidenecyclopropanes, yne substituted, gold-catalyzed cycloisomerization 204, 205

X

(+)-Xestoquinone 86

Y

Yatakemycin A 456

Author Index

In this index the page number for that part of the text citing the reference number is given first. The number of the reference in the reference section is given in a superscript font following this.

A

Abbiati, G. 410[49], 413[49], 432[76], 433[76]
Adachi, M. 84[57]
Adams, C. S. 324[13], 325[13]
Adams, J. 66[24], 68[24]
Adduci, L. L. 172[18], 173[18], 176[18]
Aeluri, M. 77[37]
Aggen, J. B. 4[14], 9[14]
Agnusdei, M. 288[42]
Ahmed, G. 270[16]
Ahn, K. H. 62[20], 65[20]
Aillard, P. 300[67], 301[67], 302[67]
Aïssa, C. 159[8], 162[10], 163[10], 165[10]
Akashi, S. 126[36], 127[36]
Aksin-Artok, Ö. 388[12]
Alcarazo, M. 295[55], 296[55], 297[55], 311[55]
Allen, G. F. 409[47], 412[47]
Allerton, C. 43[96]
Allouchi, H. 75[34], 76[34]
Álvarez-Corral, M. 397[30]
Amijs, C. H. M. 305[75], 307[75]
Anacardio, R. 449[94], 450[94]
Anastasia, L. 17[43]
Andersen, N. G. 88[73], 89[73]
Anderson, L. G. 374[47], 374[48], 374[50], 375[47], 375[48], 375[50], 376[50], 378[48]
Andrews, B. I. 307[79], 308[79]
Antoniotti, S. 387[3], 463[118], 464[118], 475[132]
Aoyagi, S. 32[82], 33[82], 35[82]
Aponick, A. 290[47]
Aponte-Guzmán, J. A. 188[30], 189[30], 193[30]
Arcadi, A. 389[15], 392[23], 393[23], 449[94], 450[94], 492[154], 493[154], 494[154]
Archambeau, A. 231[66], 232[66], 233[66]
Ardizzoia, G. A. 67[26], 69[26]
Arisawa, M. 365[33]
Armstrong, R. W. 4[13], 9[13]
Arya, P. 77[37]
Asakawa, N. 100[9]
Aschwanden, P. 474[131]
Ashikawa, M. 443[89], 444[89]
Ashimori, A. 85[60], 85[62], 85[63], 86[60], 88[63], 88[76], 89[76]
Atkins, K. E. 95[2]
Aubert, C. 174[20], 175[20], 176[20], 368[35]
Aue, D. H. 303[74]
Aznar, F. 70[30], 71[30]

B

Baba, S. 16[36], 16[37]
Bachand, B. 85[60], 85[63], 86[60], 88[63]
Badorrek, J. 305[77], 306[77]
Bagheri, V. 227[62], 242[62]
Bai, Y. 194[34], 195[34], 196[34]
Balachandran, R. 34[84]
Baldwin, J. E. 56[9]
Ballesteros, A. 211[49], 212[49], 213[49], 216[49]
Balraju, V. 19[51], 21[51]
Ban, Y. 55[4], 56[4]
Bandini, M. 102[14], 103[14], 106[17], 107[17], 108[19], 109[19], 110[17], 110[19], 121[30], 122[30], 122[31], 122[32], 123[31], 123[32], 125[30], 125[31], 126[32], 141[48], 142[48], 143[48], 288[37], 288[42], 290[48], 291[48], 292[48]
Bando, T. 497[167]
Banwell, M. G. 300[66], 300[67], 300[68], 301[66], 301[67], 301[68], 302[67], 302[68], 303[66], 304[66]
Baran, P. S. 44[104]
Barber, D. M. 331[16], 332[16]
Barbera, V. 411[51], 412[51], 413[51]
Barluenga, J. 70[30], 71[30], 211[49], 212[49], 213[49], 216[49]
Barluenga, S. 15[33]
Bartley, G. S. 323[12]
Bates, R. W. 344[25], 344[26], 344[27]
Batey, R. A. 427[70], 428[70], 430[70], 486[146], 487[146], 488[146]
Bats, J. W. 464[119]
Battina, S. K. 65[22], 68[22]
Baudoin, O. 15[33]
Baum, J. 66[24], 68[24]
Beal, R. B. 179[24]
Beau, J.-M. 4[13], 9[13]
Beccalli, E. M. 24[68], 67[26], 69[26], 410[49], 411[51], 412[51], 413[49], 413[51], 415[53], 432[76], 433[76], 458[109], 466[123], 467[123], 483[142], 484[142]
Beck, E. M. 29[73], 30[73], 418[63], 421[63], 422[63], 426[63]
Becker, M. H. 15[35], 16[35]
Bédard, A.-C. 38[94], 39[94], 40[94], 41[94]
Beller, M. 385[1], 386[1], 389[1], 408[1], 409[1]
Belmont, P. 349[3], 382[3]
Bender, C. F. 317[5], 318[5], 329[5], 403[38], 405[42], 405[43], 406[43]
Benet-Buchholz, J. 300[65], 303[65]
Benito, D. 307[79], 308[79]
Bensoussan, C. 140[47], 141[47]
Bertrand, M. B. 451[97], 453[97], 459[112], 461[112]
Bethuel, Y. 167[14], 168[14]
Bhattasali, D. 20[52]
Bian, J. 34[83], 144[50]
Bianchi, G. 392[23], 393[23]
Biannic, B. 290[47]
Binfield, S. A. 394[25]
Bissember, A. C. 300[66], 301[66], 303[66], 304[66]
Blanc, A. 496[161]
Blechert, S. 398[33], 399[33], 401[33]
Blumenkopf, T. A. 251[78]
Boden, C. D. J. 82[48]
Boger, D. L. 74[32], 497[164], 497[165], 497[166]
Bohle, D. S. 24[67]
Böing, C. 366[34], 367[34]
Bois-Choussy, M. 9[18], 9[19], 429[72], 430[72]
Boito, S. C. 368[36], 381[65]
Boralsky, L. B. 324[13], 325[13]
Borkar, P. 276[21], 277[21]
Borsini, E. 67[26], 69[26], 466[123], 467[123], 483[142], 484[142]
Bozell, J. J. 409[47], 412[47]
Braga, A. A. C. 2[3]
Brahma, S. 69[29], 70[29], 71[29]
Brancour, C. 475[132]
Brand, C. 191[32], 194[32]
Brasche, G. 415[54], 416[54], 417[54], 418[62], 420[62]
Bräse, S. 81[45]
Bray, K. L. 361[27], 362[27], 363[27]
Brecker, L. 58[12], 59[12]
Breker, V. 388[12]
Brenna, S. 67[26], 69[26]
Brödner, K. 146[56], 147[56]
Broggini, G. 24[68], 67[26], 69[26], 101[12], 102[12], 410[49], 411[51], 412[51], 413[49], 413[51], 415[53], 432[76], 433[76], 458[109], 466[123], 467[123], 483[142], 484[142]
Brown, J. A. 432[75]
Brummond, K. M. 368[37], 368[38], 368[39], 368[40], 368[41], 369[38], 369[39], 369[40], 369[41], 370[37], 371[44], 373[44], 374[44]
Bruneau, C. 295[57], 295[58], 471[129]
Buchwald, S. L. 79[44], 80[44], 352[14], 353[14], 354[14], 387[4], 407[46], 408[46], 415[54], 416[54], 417[54], 418[61], 418[62], 420[61], 420[62], 421[61], 426[61], 427[69], 431[74], 432[74], 489[150], 490[150]
Bugni, T. S. 254[80], 255[80]
Busacca, C. A. 88[71], 88[75], 89[71]
Butera, J. 65[23], 66[23], 68[23]
Butler, E. M. 55[7]

C

Cacchi, S. 389[15], 448[92], 449[92], 450[92], 492[154], 493[154], 494[154], 494[155], 495[155]
Caddick, S. 60[14]
Caggiano, L. 17[44], 18[44]
Calter, M. A. 85[60], 86[60]
Camarco, D. P. 34[84]
Campagne, J.-M. 293[51], 294[51]
Campana, F. B. 449[95]
Campbell, S. J. 88[71], 89[71]
Cao, J. 209[45], 209[46], 210[45]
Cao, P. 357[19]
Carballo, R. M. 261[5], 268[5], 274[5], 274[20], 275[5], 275[20]
Carbery, D. R. 307[79], 308[79]
Carbonnelle, A. C. 7[16], 8[16]
Cárdenas, D. J. 302[70], 303[70]
Carney, J. M. 397[31]
Carpenter, N. E. 77[38], 78[38]
Carreira, E. M. 123[33], 124[33], 125[33], 126[33], 144[53], 145[53], 474[131]
Carreras, J. 295[55], 296[55], 297[55], 311[55]
Carril, M. 458[110], 460[110]
Carr-Schmid, A. 459[114]
Casarez, A. D. 371[44], 373[44], 374[44]
Caspi, D. D. 29[74], 30[74]
Castedo, L. 155[3], 155[4], 155[5], 156[4], 156[5], 158[3], 168[3]
Cavitt, M. A. 190[31], 193[31]
Cera, G. 102[14], 103[14], 141[48], 142[48], 143[48]
Cervi, A. 300[67], 301[67], 302[67]
Chai, C. L. L. 300[67], 301[67], 302[67]
Chai, Z. 399[34], 400[34], 401[34]
Chakravarthy, P. P. 276[21], 277[21]
Chamberlin, A. R. 4[14], 9[14]
Chan, C. 351[12], 351[13], 352[12], 352[13]
Chan, L. Y. 283[36], 287[36], 288[36]
Chandra, A. 59[13], 60[13], 64[13]
Chandramouli, S. V. 379[52], 380[52]
Chandrasekar, G. 77[37]
Chang, D.-J. 381[62]
Chang, H.-K. 295[59], 296[59], 298[59]
Chang, J. J. 251[77], 252[77]
Chang, M.-C. 143[49], 144[49]
Chatani, N. 349[8], 446[91], 447[91]
Chattopadhyay, B. 74[33], 75[33], 76[35], 76[36]
Che, C.-M. 278[28], 278[29], 280[29], 281[28], 281[29], 282[28], 283[28], 403[37]
Chemler, S. R. 2[5], 4[11], 5[11], 6[11], 7[11]
Chen, A. 213[51], 222[51], 223[51]
Chen, B. 69[28], 70[28], 71[28]
Chen, C.-y. 486[147], 487[147], 488[147]
Chen, D. 368[40], 369[40]
Chen, D. Y.-K. 380[60]
Chen, G. 218[53], 219[53], 220[53], 422[65], 423[65], 426[65], 433[77], 434[77], 435[77], 436[77]
Chen, H. 4[12], 9[12], 368[37], 370[37], 371[44], 373[44], 374[44]
Chen, J. 66[25], 67[25], 69[25], 234[68], 235[68], 239[68], 240[68], 497[168], 498[168]
Chen, J.-H. 417[56], 418[56], 425[56]
Chen, K. 163[11], 164[11], 165[11]
Chen, P. 465[121]
Chen, Y. 307[80]
Cheong, P. H.-Y. 369[42], 370[42]
Chiacchio, U. 228[63], 411[51], 412[51], 413[51]
Chianese, A. R. 465[120]
Chiang, P. K. 65[22], 68[22]
Chiarucci, M. 102[14], 103[14], 141[48], 142[48], 143[48]
Chiba, H. 454[100], 459[100]
Chiba, K. 55[4], 56[4]
Chinchilla, R. 18[47], 18[49]
Chiosis, G. 440[84], 456[84], 457[84], 460[84]
Chlenov, M. 65[23], 66[23], 68[23]
Cho, S. Y. 86[65]
Choi, D. S. 300[63]
Christ, W. J. 31[77]
Chua, P. 15[35], 16[35]
Chung, J. Y. L. 380[59]
Chuprakov, S. 230[65], 231[65], 233[65], 238[65], 239[65], 240[65]
Churruca, F. 458[110], 460[110], 496[163]
Ciesielski, J. 35[85], 36[85]
Clarke, M. L. 259[2], 270[2]
Clement, F. 379[52], 380[52]
Clément, M. 44[99]
Coeffard, V. 55[8]
Coleman, R. S. 38[93], 41[93]
Collins, S. K. 38[94], 39[94], 40[94], 41[94]
Conia, J. M. 278[22]
Connell, N. T. 333[17], 334[17]
Corey, E. J. 12[28]
Corminboeuf, O. 309[83], 310[83]
Cornella, J. 44[104]
Corte, J. R. 380[55]
Cossy, J. 140[47], 141[47], 231[66], 232[66], 233[66], 242[72], 242[73], 249[72]
Couture, A. 496[162], 497[162], 498[162]
Couty, F. 457[108]
Coyne, A. G. 88[70], 89[70]
Cozzi, P. G. 102[14], 103[14]
Crabtree, R. H. 465[120], 465[121]
Craig, D. 101[13], 102[13]
Crawley, S. L. 455[101], 460[101]
Creswell, M. W. 374[50], 375[50], 376[50]
Cui, J. 66[25], 67[25], 69[25]
Cuny, G. 429[72], 430[72]
Curran, D. P. 34[84]

D

Dai, G. 449[93], 450[93]
Dai, L.-X. 107[18], 108[18], 110[20], 111[20], 112[20], 115[20], 118[27]
Dai, M. 497[168], 498[168]
Dal Zotto, C. 293[51], 294[51]
Damrauer, R. 4[9]
Danishefsky, S. J. 2[5], 4[11], 5[11], 6[11], 7[11]
Dankwardt, J. W. 72[31], 73[31]
D'Anniballe, G. 449[94], 450[94]
Datta, S. 331[16], 332[16]
Davies, H. M. L. 227[60]
Davies, I. W. 61[18], 64[18]
Davies, P. W. 467[124], 468[124], 470[124]
Davis, N. R. 438[82], 439[82], 444[82], 445[82]
Day, B. W. 34[84]
Declerck, V. 75[34], 76[34]
Delgado, A. 155[3], 155[4], 156[4], 158[3], 168[3]
Del Rosario, M. 492[154], 493[154], 494[154]
de Mendoza, P. 288[38], 294[38]
Deng, C.-L. 283[34], 283[35], 286[34]
Deng, J. 144[54], 146[54]
Deng, K. 65[23], 66[23], 68[23]
Deng, L. 497[168], 498[168]
Deng, Y. 321[7], 322[7]
Deniau, E. 496[162], 497[162], 498[162]
Denmark, S. E. 22[61], 22[62], 23[61], 23[62], 23[63], 23[64], 24[61], 24[62], 24[63], 24[64]
Dennison, P. R. 85[64]
Desper, J. 65[22], 68[22]
Deutsch, C. 388[12]
Dewey, M. R. 344[26]
Dhimane, A.-L. 35[86]
Dias, D. A. 177[22], 178[22], 191[22]
di Lillo, M. 102[14], 103[14]
Dilmeghani, M. 66[24], 68[24]
Ding, C.-H. 104[15], 105[15]
Ding, D. 60[17], 61[17], 64[17], 72[17], 73[17]
Dinman, J. D. 459[114]
Dittrich, B. 191[32], 194[32]
Dixneuf, P. H. 295[57], 295[58], 471[129]
Dixon, D. J. 331[16], 332[16]
Dochnahl, M. 398[33], 399[33], 401[33]
Doi, T. 418[57], 423[66], 426[66]
Dolan, N. S. 333[17], 334[17]
Dolgii, I. E. 227[59]
Dombroski, M. A. 179[24]
Domínguez, E. 458[110], 460[110], 496[163]
Dong, G. 455[102], 456[102], 460[102]
Dong, H. 440[83], 441[83], 441[86], 442[86], 443[88]
Dong, J. 268[14], 269[14]
Dong, L. 137[46], 138[46], 139[46], 147[57], 380[61]
Dong, V. M. 24[69], 424[67], 425[67], 426[67]
Dong, Y. 88[71], 89[71], 456[103]
Donoghue, P. J. 397[31]
Doran, R. 55[7]

Dormer, P. G. 486[147], 487[147], 488[147]
Doucet, H. 18[48]
Douglas, C. J. 15[35], 16[35]
Dounay, A. B. 55[5], 81[5], 85[59], 86[59], 88[74], 89[74], 90[74]
Downham, R. 15[35], 16[35]
Doyle, M. P. 227[62], 242[62]
Driega, A. B. 176[21], 177[21], 179[21], 191[21]
Driver, T. G. 438[81], 439[81], 440[81], 440[83], 441[81], 441[83], 441[86], 442[86], 443[88]
Dröge, T. 435[78], 436[78]
D'Souza, D. M. 388[10]
Du, K. 173[19], 174[19], 176[19]
Du, X. 307[80]
Duan, C. 400[35], 401[35]
Dubé, P. 293[50]
Du Bois, J. 436[79], 437[79], 438[79]
Dudley, G. B. 496[157]
Dufour, J. 7[17], 8[17], 11[17]
Dunetz, J. R. 388[13]
Durán, J. 155[5], 156[5]
Durandetti, M. 44[98], 44[99]

E

Eastgate, M. D. 44[104]
Echavarren, A. M. 12[27], 288[38], 294[38], 300[65], 302[69], 302[70], 303[65], 303[70], 305[75], 305[76], 307[75], 307[76], 312[85], 313[85], 350[11], 382[11]
Edmonds, D. J. 381[63], 382[63]
Edwards, J. T. 44[104]
Edwards, N. 422[65], 423[65], 426[65]
Eichholzer, A. 108[19], 109[19], 110[19], 121[30], 122[30], 122[31], 123[31], 125[30], 125[31], 290[48], 291[48], 292[48]
Ellery, S. P. 381[63], 382[63]
Elliott, M. R. 35[86]
Emer, E. 288[37]
Enders, D. 305[77], 306[77]
England, D. B. 311[84], 312[84]
Enna, M. 38[91]
Enomoto, T. 398[32], 454[32], 455[32], 459[32]
Eom, D. 293[52], 294[52], 303[52], 305[52]
Erickson, L. W. 44[97]
Eriksson, M. C. 88[71], 89[71]
Eros, D. 349[7]
Escribano-Cuesta, A. 312[85], 313[85]
Espino, C. G. 436[79], 437[79], 438[79]
Evano, G. 440[85], 456[85], 457[108]
Evans, D. A. 288[41]
Evans, P. A. 250[76], 251[76]
Everson, D. A. 44[100]
Evindar, G. 427[70], 428[70], 430[70], 486[146], 487[146], 488[146]

F

Fabrizi, G. 448[92], 449[92], 450[92], 492[154], 493[154], 494[154], 494[155], 495[155]
Fagnou, K. 62[21], 63[21], 64[21], 65[21]
Fairlamb, I. J. S. 361[27], 362[27], 363[27], 365[32], 366[32]
Faller, J. W. 465[121]
Fañanás-Mastral, M. 128[38], 128[39], 129[38], 130[38], 130[39], 143[49], 144[49]
Fandrick, K. R. 288[41]
Fang, C. 495[156]
Fang, J. 182[26], 183[26], 184[26], 185[26], 192[26], 244[75], 245[75], 246[75], 247[75], 248[75], 249[75]
Fang, Y. 486[145], 486[148], 487[148], 488[145], 488[148], 488[149], 489[148]
Fasana, A. 411[51], 412[51], 413[51], 483[142], 484[142]
Fathi, R. 66[25], 67[25], 69[25], 492[153], 494[153]
Faul, M. M. 66[24], 68[24]
Fazio, A. 390[20], 391[20], 449[95]
Felix, R. J. 171[17], 172[17], 172[18], 173[18], 176[17], 176[18]
Feng, X. 266[11], 266[12], 267[12]
Feng, Z. 44[102]
Fensterbank, L. 174[20], 175[20], 176[20], 368[35]
Feringa, B. L. 128[38], 128[39], 129[38], 130[38], 130[39], 143[49], 144[49]
Fernández, M. A. 70[30], 71[30]
Fernández-Salas, J. A. 128[39], 130[39]
Ferreira, E. M. 25[70], 25[71], 26[70], 26[71], 27[70], 27[71], 27[72], 28[71], 28[72], 29[71], 30[70], 30[71], 290[44], 290[45]
Ferrer, C. 305[75], 305[76], 307[75], 307[76], 312[85], 313[85]
Fettinger, J. C. 394[25]
Findlay, A. D. 300[66], 301[66], 303[66], 304[66]
Fisher, E. L. 197[36], 198[36], 199[36]
Fitzpatrick, M. O. 88[70], 89[70]
Fleming, I. 18[46], 260[3]
Flippin, L. A. 72[31], 73[31]
Floreancig, P. E. 309[81]
Forsyth, C. J. 495[156], 496[159], 497[159]
Fortner, K. C. 496[160]
Foubelo, F. 389[17]
France, M. B. 259[2], 270[2]
France, S. 187[29], 188[29], 188[30], 189[30], 190[31], 193[29], 193[30], 193[31]
Franciò, G. 366[34], 367[34]
Frank, S. A. 4[12], 9[12]
Frantz, D. E. 474[131]
Frey, W. 464[119]
Friedman, A. A. 68[27], 69[27]
Friesen, R. W. 13[29], 13[31], 14[31]
Fritz, J. A. 451[99], 452[99], 453[99]
Fromtling, R. A. 459[113]
Frontier, A. J. 35[85], 36[85]
Fu, C. 299[60], 300[60], 341[22], 342[22]
Fugami, K. 317[4], 318[4], 328[4]
Fujii, N. 319[6], 320[6], 336[18], 337[18], 345[6], 346[6], 454[100], 459[100]
Fujiwara, Y. 24[65], 300[61], 300[62], 301[62], 302[62]
Fukuda, Y. 391[21], 392[21], 393[21], 461[115]
Fukumoto, Y. 446[91], 447[91]
Fukuyama, T. 427[68], 456[104], 456[105]
Fukuyama, Y. 380[56]
Funk, R. L. 455[101], 460[101]
Fürstner, A. 31[79], 31[80], 31[81], 159[8], 162[10], 163[10], 165[10], 294[53], 296[53], 297[53], 298[53], 355[17], 356[17], 356[18], 357[17], 357[18], 467[124], 468[124], 470[124]
Furukawa, I. 478[135]
Furuune, M. 38[91]
Furuyama, H. 60[16]

G

Gabarda, A. E. 440[85], 456[85]
Gabriele, B. 390[20], 391[20], 449[95], 466[122], 467[122]
Gaddam, J. 77[37]
Gagné, M. R. 171[17], 172[17], 172[18], 173[18], 176[17], 176[18], 337[19], 338[19]
Galli, S. 411[51], 412[51], 413[51]
Gandon, V. 35[85], 36[85], 174[20], 175[20], 176[20]
Gao, Q. 278[27], 283[33]
Gao, S. 18[50]
Gao, Z. 293[52], 294[52], 303[52], 305[52]
Garcia, P. 368[35]
Garciá Alonso, F. J. 470[127]
Garg, N. K. 15[34], 15[35], 16[35], 29[74], 30[74]
Garg, R. 38[93], 41[93]
Garreau, Y. 228[63]
Garrity, G. 459[113]
Gaunt, M. J. 29[73], 30[73], 418[63], 421[63], 422[63], 426[63]
Gazzola, S. 411[51], 412[51], 413[51]
Genet, J.-P. 105[16]
Genêt, J.-P. 349[9], 350[9], 387[3], 463[118], 464[118], 475[132]
Geng, Q. 456[106]
Genin, E. 387[3], 463[118], 464[118], 475[132]
Gentile, M. 449[94], 450[94]
Gevorgyan, V. 230[65], 231[65], 233[65], 238[65], 239[65], 240[65], 389[19], 390[19], 391[19], 461[116], 462[116]
Ghidu, V. P. 9[21], 10[21]

Giambastiani, G. 101[10], 101[11], 102[10]
Gianatassio, R. 44[104]
Gilbert, A. 65[23], 66[23], 68[23]
Gimeno, A. 295[55], 296[55], 297[55], 311[55]
Girard, A.-L. 398[32], 454[32], 455[32], 459[32]
Glorius, F. 435[78], 436[78]
Gnamm, C. 140[47], 141[47], 146[56], 147[56]
Gockel, B. 388[12]
Goddard, R. 356[18], 357[18]
Goldberg, I. 37[90]
Gong, H. 44[105], 44[106], 45[106], 46[106], 47[106], 50[106]
Gong, X. 379[52], 380[52]
González, J. M. 303[73], 304[73]
Gopakumar, G. 295[55], 296[55], 297[55], 311[55]
Gordon, M. S. 4[9]
Gorin, D. J. 293[50], 438[82], 439[82], 444[82], 445[82]
Goussu, D. 391[22]
Gouverneur, V. 481[137], 481[138], 482[137], 482[138], 484[137]
Govek, S. P. 85[60], 86[60]
Gozman, A. 440[84], 456[84], 457[84], 460[84]
Grachan, M. L. 270[18], 271[18], 272[18], 273[18]
Grandclaudon, P. 496[162], 497[162], 498[162]
Gray, D. L. F. 15[33]
Gree, R. 276[21], 277[21]
Gridnev, I. D. 394[26]
Grigg, R. 56[10], 57[11], 60[15], 360[25]
Grosche, M. 389[16], 391[16]
Grossbach, D. 88[71], 88[75], 89[71]
Gruden-Pavlovic, M. 144[51], 145[51]
Grzybowski, P. 190[31], 193[31]
Gu, L. 295[55], 296[55], 297[55], 311[55]
Gu, Z. 61[19], 62[19], 64[19]
Guaciaro, M. A. 481[139]
Gudiksen, M. S. 380[55]
Guérinot, A. 140[47], 141[47]
Guiry, P. J. 55[6], 55[7], 55[8], 82[46], 82[47], 88[70], 88[72], 89[70], 89[72], 90[72]
Gulias, M. 418[63], 421[63], 422[63], 426[63]
Gulías, M. 155[5], 156[5]
Guo, T. 442[87]
Guram, A. S. 387[4]
Gurjar, M. K. 20[52]
Gutierrez, O. 171[17], 172[17], 176[17]
Guzei, I. A. 324[13], 325[13]
Gyoung, Y. S. 386[2]

H

Haffner, C. D. 380[58], 381[58]
Hagihara, N. 18[45]
Hahne, J. 366[34], 367[34]
Hall, R. P. 349[6]
Hall, T. W. 481[139]
Hamada, Y. 119[28], 120[28], 121[28], 144[52], 145[52]
Han, X. 12[28], 278[26], 281[26], 309[81], 317[2], 334[2], 335[2], 476[134], 477[134], 478[134], 483[141], 484[141]
Harada, K. 380[56]
Hardou, L. 44[99]
Harger, J. W. 459[114]
Harkat, H. 496[161]
Harris, E. B. J. 300[68], 301[68], 302[68]
Harris, R. E. 88[71], 89[71]
Hartwig, J. F. 406[44], 406[45], 407[44], 408[44]
Hashmi, A. S. K. 464[119]
Hata, G. 95[3]
Hatanaka, K. 85[59], 86[59]
Hatanaka, Y. 22[56], 22[57], 22[58]
Hatano, M. 359[23], 360[23]
Hatley, R. 29[73], 30[73]
Hazeri, N. 300[67], 301[67], 302[67]
He, C. 300[64], 301[64], 478[136], 479[136], 480[136], 481[136]
He, G. 422[65], 423[65], 426[65], 433[77], 434[77], 435[77], 436[77]
He, H. 112[21], 113[21], 115[21]
He, M. 357[20], 357[21], 357[22], 358[20], 358[21], 358[22], 360[20]
Heathcock, C. H. 251[78]
Heck, R. F. 55[1], 55[2]
Heffernan, S. J. 307[79], 308[79]
Hegedus, L. S. 409[47], 409[48], 412[47]
Heijnen, D. 128[39], 130[39]
Helmchen, G. 97[6], 99[6], 128[37], 146[56], 147[56]
Helquist, P. 397[31]
Hennessy, A. J. 307[79], 308[79]
Hensens, O. 459[113]
Herdtweck, E. 389[16], 391[16]
Herrero-Gómez, E. 300[65], 303[65]
Herrmann, J.-S. 398[33], 399[33], 401[33]
Hesp, K. D. 403[39], 403[40], 404[40]
Hiebert, S. 15[34], 15[35], 16[35]
Hierso, J.-C. 18[48]
Hightower, T. R. 484[143]
Hirama, M. 20[53], 22[53], 60[16]
Hirano, S. 31[75]
Hiroya, K. 389[18], 391[18], 418[57], 418[64], 423[66], 424[64], 426[64], 426[66], 443[89], 444[89]
Hiyama, T. 22[56], 22[57], 22[58], 22[59], 22[60], 31[75], 31[76]
Ho, E. 134[43], 135[43]
Hoffmann-Röder, A. 323[11], 324[11], 343[24]
Höhn, A. 470[127]
Hollmann, D. 398[33], 399[33], 401[33]
Honda, T. 83[55], 84[55]
Honeycutt, S. 459[113]
Honzawa, S. 87[68], 87[69]
Horibata, A. 445[90], 446[90]
Hou, X.-L. 104[15], 105[15]
Houk, K. N. 196[35], 197[35], 369[42], 370[42]
Hsieh, T. H. H. 424[67], 425[67], 426[67]
Hu, C. 266[12], 267[12]
Hu, X. 456[103]
Hu, Y. 456[103], 492[153], 494[153]
Hua, D. H. 65[22], 68[22]
Huang, L. 66[24], 68[24]
Huang, X. 194[33], 206[43], 207[43], 209[45], 209[46], 210[45], 252[79], 253[79]
Huang, Y. 61[19], 62[19], 64[19]
Hudson, W. B. 405[43], 406[43]
Huffman, M. A. 61[18], 64[18]
Hughes, R. 497[167]
Hultzsch, K. C. 389[17]
Humphrey, J. M. 4[14], 9[14]
Humphreys, P. G. 88[74], 89[74], 90[74]
Hung, O. Y. 168[15], 169[15], 175[15], 254[15]
Huntsman, W. D. 349[6], 349[7]
Hyland, C. J. T. 101[13], 102[13]

I

Igawa, R. 295[56], 296[56], 298[56]
Ikeno, T. 210[47]
Imagawa, H. 134[43], 135[43], 290[46], 291[46], 292[46], 293[49]
Inamoto, K. 418[57], 418[64], 423[66], 424[64], 426[64], 426[66]
Inglesby, P. A. 250[76], 251[76]
Innitzer, A. 58[12], 59[12]
Inoue, M. 60[16]
Inoue, S. 446[91], 447[91]
Inuki, S. 319[6], 320[6], 345[6], 346[6]
Ioffe, A. I. 227[59]
Iqbal, J. 19[51], 21[51]
Ishige, Y. 119[28], 120[28], 121[28]
Ishikawa, M. 2[6], 2[7], 2[8], 3[7], 3[8], 4[7], 4[8], 6[7]
Ishiyama, T. 2[6], 2[7], 3[7], 4[7], 6[7]
Itano, W. 84[58]
Ito, H. 283[32], 284[32]
Ito, K. 126[36], 127[36]
Ito, Y. 418[59], 478[135]
Itoh, K. 365[30], 366[30], 371[43], 372[43], 373[43]
Itoh, S. 389[18], 391[18]
Iwamoto, M. 210[47]
Iwasawa, N. 210[47]
Iwata, A. 319[6], 320[6], 345[6], 346[6]

J

Jackson, R. F. W. 17[44], 18[44]
Jackson, S. K. 176[21], 177[21], 179[21], 191[21]
Jacobs, A. 9[21], 10[21]
Jacobsen, E. N. 270[18], 270[19], 271[18], 272[18], 273[18], 273[19], 274[19]
Jain, N. F. 497[167]
Jalal, S. 263[7], 264[7]
Jaroch, S. 15[35], 16[35]
Jarvo, E. R. 44[97]

Jebaratnam, D. J. 351[12], 352[12], 380[58], 381[58]
Jeker, O. F. 144[53], 145[53]
Jerris, P. J. 481[140]
Jia, C. 300[61], 300[62], 301[62], 302[62]
Jia, G. 340[21], 341[21], 442[87]
Jia, Y. 9[18], 9[19]
Jiang, H. 306[78], 307[78]
Jiang, L. 400[35], 401[35]
Jiang, X. 317[3], 318[3], 321[3], 341[22], 342[22]
Jiang, Y. 456[106]
Jiang, Z.-X. 44[102]
Jiao, N. 213[51], 222[51], 223[51], 338[20], 339[20]
Jimenez, A. L. 65[22], 68[22]
Jiménez, M. 34[84]
Jiménez-Núñez, E. 350[11], 382[11]
Jin, H. 31[77]
Jin, K. 400[35], 401[35]
Johansson, C. C. C. 418[63], 421[63], 422[63], 426[63]
Jones, P.-J. 88[71], 89[71]
Jordan-Hore, J. A. 418[63], 421[63], 422[63], 426[63]
Jovanović, B. 441[86], 442[86]
Julian, L. D. 406[45]
Jullien, H. 349[5], 382[5]
Jung, J.-W. 381[62]
Jung, M. E. 251[77], 252[77]
Jung, W.-H. 34[84]
Justice, M. 459[114]

K

Kablaoui, N. M. 352[14], 353[14], 354[14]
Kadota, I. 386[2]
Kadunce, N. T. 44[103]
Kagan, M. 65[23], 66[23], 68[23]
Kagechika, K. 84[56], 84[57], 85[56]
Kaiser, J.-P. 361[27], 362[27], 363[27]
Kajimoto, T. 89[77], 90[77]
Kakui, T. 22[55]
Kamada, M. 154[2], 158[2]
Kandur, W. V. 61[18], 64[18]
Kanematsu, M. 119[28], 120[28], 121[28]
Kang, E. J. 327[15], 328[15]
Kang, J.-E. 322[8], 323[8]
Karadeolian, A. 176[21], 177[21], 179[21], 191[21]
Kardos, N. 105[16]
Kassir, J. M. 227[61], 228[61], 228[63]
Kataoka, Y. 478[135]
Kato, D. 496[160]
Kato, H. 380[56]
Katsuki, T. 126[36], 127[36]
Katsuno, M. 418[64], 424[64], 426[64]
Kattnig, E. 356[18], 357[18]
Kavthe, R. D. 278[23]
Kawakami, S. 97[7], 98[7], 99[7], 139[7], 140[7]
Kawamura, S. 44[104]
Kawasaki, M. 221[55], 222[55]
Kawashima, K. 270[15]
Keay, B. A. 13[29], 88[73], 89[73]
Kel'in, A. V. 389[19], 390[19], 391[19], 461[116], 462[116]
Kennedy-Smith, J. J. 283[30], 283[31], 284[30], 285[31], 286[30]
Kennewell, P. 60[15]
Kerr, M. A. 176[21], 177[21], 177[22], 178[22], 179[21], 191[21], 191[22]
Khan, M. I. 20[52]
Kibayashi, C. 32[82], 33[82], 35[82]
Kiely, D. 82[46], 82[47], 88[72], 89[72], 90[72]
Kim, A. 468[125], 469[125], 470[125]
Kim, D.-D. 381[62]
Kim, J. H. 300[63]
Kim, J.-Y. 88[71], 89[71]
Kim, K. S. 62[20], 65[20]
Kim, M. M. 61[18], 64[18]
Kim, N.-J. 381[62]
Kim, S. 283[36], 287[36], 288[36], 293[52], 294[52], 303[52], 305[52]
Kimura, M. 317[4], 318[4], 328[4]
Kinder, R. E. 317[2], 334[2], 335[2]
King, A. O. 16[38], 16[39], 16[40]
Kingsbury, C. L. 381[66]
Kinzy, T. G. 459[114]
Kisanga, P. 361[26], 362[26], 363[26]
Kishi, Y. 4[13], 9[13], 31[77]
Kitagawa, O. 458[111], 459[111], 460[111]
Kitambi, S. S. 77[37]
Kitamura, M. 97[7], 98[7], 99[7], 131[40], 132[40], 139[7], 140[7]
Kitamura, T. 288[39], 292[39], 300[61], 300[62], 301[62], 302[62]
Knochel, P. 55[7]
Koch, G. 96[5]
Koch, O. 128[37]
Kočovský, P. 270[16]
Kodanko, J. J. 85[59], 85[64], 86[59]
Kofie, W. 60[14]
Kohno, Y. 119[28], 120[28], 121[28]
Kojima, A. 82[48], 86[66]
Komeyama, K. 295[56], 296[56], 298[56], 405[41], 476[133]
Kondo, K. 82[50], 82[53], 82[54], 84[50]
Kondo, T. 396[29], 413[52], 414[52]
Kong, W. 299[60], 300[60]
Kotrusz, P. 121[30], 122[30], 122[31], 123[31], 125[30], 125[31]
Koumbis, A. E. 497[167]
Koyama, Y. 20[53], 22[53]
Kozmin, S. A. 350[10], 382[10]
Krause, H. 356[18], 357[18]
Krause, N. 322[10], 323[11], 324[11], 330[10], 342[23], 343[23], 343[24], 388[12]
Krauter, C. M. 146[56], 147[56]
Krautwald, S. 123[33], 124[33], 125[33], 126[33]
Kravina, A. G. 144[53], 145[53]
Krische, M. J. 126[34], 349[2], 380[57], 380[58], 381[58], 382[2]
Krumpe, K. E. 228[63]
Kubo, T. 427[68]
Kucera, D. J. 77[38], 78[38]
Kumada, M. 22[54], 22[55]
Kumar, R. 59[13], 60[13], 64[13]
Kumar Singarapu, K. 263[7], 264[7]
Kunz, R. K. 4[12], 9[12]
Kuribara, D. 458[111], 459[111], 460[111]
Kuroda, T. 31[78]
Kurokawa, T. 456[104]
Kutsuna, H. 97[7], 98[7], 99[7], 139[7], 140[7]
Kuwabe, S.-i. 489[150], 490[150]
Kwochka, W. R. 4[9], 4[10]
Kwon, J. 459[114]
Kwong, F. Y. 427[69]

L

Lai, Y.-C. 263[9], 264[9], 265[9], 266[9], 310[9], 311[9]
Lalli, C. 263[8]
LaLonde, R. L. 327[15], 328[15]
Lamaty, F. 75[34], 76[34]
Lamba, D. 448[92], 449[92], 450[92]
Lapointe, D. 62[21], 63[21], 64[21], 65[21]
Larhed, M. 55[8]
Larock, R. C. 449[93], 450[93], 484[143], 494[155], 495[155]
Larsen, R. 66[24], 68[24]
Lauria, E. 466[122], 467[122]
Lautens, M. 68[27], 69[27], 167[14], 168[14], 351[12], 351[13], 352[12], 352[13]
Lear, M. J. 20[53], 22[53]
Lee, C.-W. 62[20], 65[20]
Lee, E.-S. 322[8], 323[8]
Lee, J. Y. 300[63]
Lee, P. H. 283[36], 287[36], 288[36], 293[52], 294[52], 303[52], 305[52]
Lee, S. D. 326[14], 327[14]
Lee, S. I. 349[8]
Lehmann, C. W. 356[18], 357[18]
Lei, A. 357[20], 357[21], 357[22], 358[20], 358[21], 358[22], 360[20]
Leitner, W. 366[34], 367[34]
Lemaire, S. 101[11]
Le Perchec, P. 278[22]
Leslie, B. E. 438[81], 439[81], 440[81], 440[83], 441[81], 441[83]
Ley, S. V. 387[5], 427[5]
Li, A. 144[54], 146[54], 381[63], 382[63]
Li, B. 263[9], 264[9], 265[9], 266[9], 310[9], 311[9]
Li, C. 232[67], 233[67], 235[67], 236[67], 237[67], 242[74], 243[74], 244[74], 244[75], 245[75], 246[75], 247[75], 248[75], 249[74], 249[75], 486[145], 486[148], 487[148], 488[145], 488[148], 488[149], 489[148]
Li, C.-H. 403[37]
Li, C.-J. 24[66], 24[67], 260[4], 261[4]
Li, D. 268[14], 269[14]
Li, G. 194[33], 202[39], 205[39]
Li, H. 326[14], 327[14]

Li, J. 60[17], 61[17], 64[17], 72[17], 73[17], 144[54], 146[54], 321[7], 322[7], 442[87]
Li, J.-H. 278[27], 283[33], 283[34], 283[35], 286[34]
Li, M. 331[16], 332[16]
Li, Q. 376[51], 377[51], 378[51]
Li, R. 144[54], 146[54]
Li, S. 135[44], 136[44], 166[13], 167[13]
Li, W. 202[39], 205[39]
Li, X. 44[102], 465[120], 465[121]
Li, X.-H. 104[15], 105[15]
Li, Y. 165[12], 166[12], 180[25], 268[14], 269[14], 380[54], 400[35], 401[35], 496[158], 496[159], 497[159]
Li, Z. 24[67], 180[25]
Liang, Y. 196[35], 197[35]
Liang, Y.-M. 187[28], 193[28]
Liang, Z. 44[106], 45[106], 46[106], 47[106], 50[106], 165[12], 166[12]
Liao, Y. 492[153], 494[153]
Liesch, J. 459[113]
Lim, W. 295[54], 296[54], 297[54]
Lin, L. 266[12], 267[12]
Lin, R. 213[51], 222[51], 223[51]
Lin, Z. 442[87]
Linowski, P. 295[55], 296[55], 297[55], 311[55]
Lipowsky, G. 97[6], 99[6], 128[37]
Liu, C. 179[23], 180[23], 180[25], 181[23], 192[23], 253[23], 254[23], 317[2], 334[2], 335[2]
Liu, F. 38[92]
Liu, G. 60[17], 61[17], 64[17], 72[17], 73[17], 410[50], 411[50], 413[50], 417[56], 418[56], 425[56]
Liu, H. 306[78], 307[78]
Liu, J. O. 456[103]
Liu, L. 223[56], 224[56], 226[56]
Liu, L.-P. 156[6], 158[6], 199[37], 200[37]
Liu, M. 462[117], 463[117], 464[117]
Liu, P. 66[24], 68[24]
Liu, Q. 206[43], 207[43], 213[51], 222[51], 223[51]
Liu, Q. S. 9[20], 9[21], 10[20], 10[21], 11[20]
Liu, R. 224[58], 225[58], 226[58], 227[58], 254[80], 255[80]
Liu, R.-S. 295[59], 296[59], 298[59]
Liu, W.-B. 112[21], 112[22], 112[23], 113[21], 114[22], 114[23], 115[21], 119[29], 121[29]
Liu, X. 61[19], 62[19], 64[19], 266[11], 266[12], 267[12]
Liu, X.-Y. 187[28], 193[28], 278[28], 281[28], 282[28], 283[28], 403[37]
Liu, Y. 182[26], 183[26], 184[26], 185[26], 192[26], 307[80], 462[117], 463[117], 464[117]
Liu, Y.-L. 283[34], 286[34]
Liu, Z. 406[44], 407[44], 408[44]
Liubchak, K. 429[71], 430[71]
Livinghouse, T. 363[28], 363[29], 364[28], 364[29], 365[28], 365[29]
Lloyd-Jones, G. C. 361[27], 362[27], 363[27]
Locritani, M. 141[48], 142[48], 143[48]
Loh, C. C. J. 305[77], 306[77]
Loh, T.-P. 263[9], 264[9], 265[9], 266[9], 310[9], 311[9]
Löhnwitz, K. 398[33], 399[33], 401[33]
Lombart, H.-G. 43[96]
López, F. 155[4], 155[5], 156[4], 156[5]
López, S. 300[65], 303[65]
Lorenz, J. C. 88[71], 89[71]
Lormann, M. E. P. 81[45]
Lou, K. 65[22], 68[22]
Lu, B.-L. 202[40], 203[40], 205[40], 208[44], 209[44], 371[45], 373[45]
Lu, C. 433[77], 434[77], 435[77], 436[77]
Lu, J. 69[28], 70[28], 71[28]
Lu, J.-M. 199[37], 200[37], 201[38], 205[38]
Lu, K. 66[25], 67[25], 69[25]
Lu, L. 151[1], 152[1], 153[1], 154[1], 158[1]
Lu, P. 371[46], 373[46]
Lu, W. 300[61]
Lu, Y. 66[24], 68[24], 344[25]
Lu, Y.-F. 13[30], 14[30]
Lu, Z. 219[54], 220[54], 221[54]
Luo, F. T. 491[152], 492[152], 493[152]
Luo, L. 44[109], 47[109]
Luo, T. 66[25], 67[25], 69[25], 497[168], 498[168]
Luo, Y. 166[13], 167[13]
Luzung, M. R. 369[42], 370[42]
Lwowski, W. 436[80]

M

Ma, D. 38[92], 456[106], 457[107], 458[107], 460[107]
Ma, J. 492[153], 494[153]
Ma, S. 151[1], 152[1], 153[1], 154[1], 158[1], 234[68], 234[69], 235[68], 239[68], 240[68], 299[60], 300[60], 317[1], 317[3], 318[3], 321[3], 321[7], 322[7], 338[20], 339[20], 340[21], 341[21], 341[22], 342[22], 371[46], 373[46]
Ma, W. 399[34], 400[34], 401[34]
Ma, X. 61[19], 62[19], 64[19], 317[3], 318[3], 321[3]
McCartney, D. 55[6]
McClory, A. 395[28], 396[28]
McDonald, R. 88[73], 89[73]
McGlacken, G. P. 365[32], 366[32]
McKearin, J. M. 409[48]
McKellop, K. B. 88[71], 89[71]
MacPherson, D. T. 351[13], 352[13]
Maddaluno, J. 44[99]
Madec, D. 101[12], 102[12]
Magano, J. 388[13]
Mahon, M. F. 307[79], 308[79]
Majima, K. 355[17], 356[17], 356[18], 357[17], 357[18]
Majumdar, K. C. 74[33], 75[33], 76[35], 76[36]
Makida, Y. 283[32], 284[32]
Makino, T. 371[43], 372[43], 373[43]
Malacria, M. 35[86], 174[20], 175[20], 176[20], 368[35]
Malkov, A. V. 270[16]
Malone, J. F. 360[25]
Mamane, V. 294[53], 296[53], 297[53], 298[53]
Manyik, R. M. 95[2]
Mao, S. 368[40], 369[40]
Marinelli, F. 389[15], 392[23], 393[23], 448[92], 449[92], 450[92], 492[154], 493[154], 494[154]
Marinetti, A. 349[5], 382[5]
Markham, J. P. 210[48], 211[48], 216[48]
Marnett, L. J. 9[21], 10[21]
Marshall, J. A. 322[9], 323[12]
Martin, R. 355[17], 356[17], 356[18], 357[17], 357[18]
Martín, V. S. 261[5], 268[5], 268[13], 274[5], 274[20], 275[5], 275[20]
Martinelli, M. 24[68], 410[49], 413[49], 415[53]
Martinez, J. 75[34], 76[34]
Martín-Matute, B. 302[70], 303[70]
Mascareñas, J. L. 155[3], 155[4], 155[5], 156[4], 156[5], 158[3], 168[3]
Mase, T. 221[55], 222[55]
Maseras, F. 2[3]
Mashimo, T. 20[53], 22[53]
Matovic, R. 144[51], 145[51]
Matsumoto, S. 443[89], 444[89]
Matsuo, N. 144[52], 145[52]
Matsuo, T. 210[47]
Matsuoka, R. T. 15[35], 16[35]
Matsuura, T. 85[62]
Mattes, J. 65[23], 66[23], 68[23]
Maulide, N. 217[52], 218[52], 220[52]
Mehrman, S. J. 381[66]
Mei, T.-S. 418[58], 419[58], 425[58]
Meige, F. 231[66], 232[66], 233[66]
Melloni, A. 106[17], 107[17], 110[17], 288[42]
Meng, F.-K. 417[56], 418[56], 425[56]
Menon, R. S. 300[66], 301[66], 303[66], 304[66]
Meyer, C. 231[66], 232[66], 233[66], 242[72], 242[73], 249[72]
Miao, H. 490[151], 491[151]
Miao, M. 209[45], 209[46], 210[45]
Michaut, M. 365[31], 366[31]
Michelet, V. 349[9], 350[9], 387[3], 463[118], 464[118], 475[132]
Middleton, J. A. 15[35], 16[35]
Miege, F. 242[72], 242[73], 249[72]
Mikami, K. 359[23], 360[23]
Miller, N. 97[6], 99[6]
Miller, N. A. 6[15]
Mills-Webb, R. 481[138], 482[138]
Milstein, D. 12[24], 12[25]
Minatti, A. 79[44], 80[44]
Miranda, P. O. 261[5], 268[5], 268[13], 274[5], 275[5]
Mitasev, B. 368[38], 368[39], 369[38], 369[39], 371[44], 373[44], 374[44]

Mitchell, H. J. 497[167]
Mitchell, T. R. B. 360[25]
Mitsudo, T.-a. 396[29], 413[52], 414[52]
Miura, M. 38[91]
Miura, T. 418[59]
Miyake, A. 95[3]
Miyata, K. 97[7], 98[7], 99[7], 139[7], 140[7]
Miyaura, N. 1[1], 2[4], 2[6], 2[7], 2[8], 3[7], 3[8], 4[7], 4[8], 6[7]
Miyoshi, A. 478[135]
Mizoroki, T. 55[3]
Mizutani, T. 87[68], 87[69]
Mo, J. 293[52], 294[52], 303[52], 305[52]
Mohapatra, D. K. 20[52]
Mohr, F. 394[25]
Molander, G. A. 55[7]
Monaghan, R. 459[113]
Monnier, F. 387[6], 427[6]
Moon, H. 381[62]
Morán-Poladura, P. 303[73], 304[73]
Moreau, A. 496[162], 497[162], 498[162]
Morganelli, P. 369[42], 370[42]
Morgon, N. H. 2[3]
Mori, D. 89[77], 90[77]
Mori, K. 55[3], 290[46], 291[46], 292[46]
Mori, M. 55[4], 56[4], 82[53], 82[54]
Morikawa, M. 95[1]
Morimoto, T. 405[41], 476[133]
Morita, N. 322[10], 330[10], 342[23], 343[23]
Moritani, I. 24[65]
Moriya, T. 401[36], 402[36]
Mpaka Lutete, L. 394[26]
Mueller, T. 351[12], 352[12]
Mukherjee, P. 132[41], 133[41], 133[42], 134[41]
Müller, P. 227[62], 242[62]
Müller, T. E. 385[1], 386[1], 389[1], 389[14], 389[16], 389[17], 391[16], 408[1], 409[1]
Müller, T. J. J. 388[10]
Mulzer, J. 58[12], 59[12]
Munday, R. H. 418[62], 420[62]
Muñoz-Dorado, M. 397[30]
Murakami, M. 418[59]
Murphy, F. 15[33]
Muzart, J. 387[7], 388[7]
Myoung, Y. C. 381[64]
Myoung, Y.-C. 368[36], 374[48], 375[48], 378[48]

N

Nagasaki, H. 445[90], 446[90]
Nagasawa, H. 484[144], 485[144]
Nájera, C. 18[47], 18[49]
Nakagai, Y. 365[30], 366[30]
Nakagawa, M. 365[33]
Nakamura, I. 154[2], 158[2], 388[9]
Nakamura, M. 263[10]
Nakamura, T. 445[90], 446[90], 458[111], 459[111], 460[111]
Nakano, M. 89[77], 90[77]
Nakao, Y. 22[59], 22[60]
Nakaoka, M. 22[59]
Nakatani, Y. 270[15]
Nakayama, Y. 476[133]
Nakhla, J. S. 451[98], 452[98], 453[98]
Namba, K. 134[43], 135[43], 290[46], 291[46], 292[46]
Nan, Y. 490[151], 491[151]
Narayanan, K. 494[155], 495[155]
Narsireddy, M. 394[27]
Natarajan, S. 497[167]
Nawoschik, K. J. 492[153], 494[153]
Nazarenko, K. 429[71], 430[71]
Nédélec, J.-Y. 44[98]
Nédellec, Y. 75[34], 76[34]
Nefedov, O. M. 227[59]
Negishi, E. 16[36], 16[37], 16[38], 16[39], 16[40], 16[41], 16[42], 17[43]
Nemeth, J. 344[27]
Nemoto, T. 119[28], 120[28], 121[28], 144[52], 145[52]
Neukom, J. D. 451[97], 453[97]
Neumann, J. J. 435[78], 436[78]
Neuville, L. 7[17], 8[17], 11[17]
Nevado, C. 302[69], 302[70], 303[70]
Ng, F. W. 15[35], 16[35]
Nicolaou, K. C. 15[33], 381[63], 382[63], 497[167]
Nieger, M. 81[45]
Nieto-Oberhuber, C. 300[65], 303[65]
Niiyama, K. 263[10]
Nishida, A. 365[33]
Nishiguchi, H. 146[55]
Nishizawa, M. 134[43], 135[43], 290[46], 291[46], 292[46], 293[49]
Node, M. 89[77], 90[77]
Nolasco, L. 17[44], 18[44]
Nolley, J. P. 55[2]
Nomura, M. 38[91]
Nozaki, H. 31[75], 31[76], 31[78], 461[115]
Nukui, S. 82[49], 83[49], 85[49]

O

O'Brien, E. M. 88[71], 88[75], 89[71]
Ochida, A. 283[32], 284[32]
Oe, Y. 478[135]
Oestreich, M. 85[59], 85[64], 86[59]
Ogiwara, K. 443[89], 444[89]
Oh, C. H. 468[125], 469[125], 470[125]
Oh, K. S. 62[20], 65[20]
Ohara, A. 454[100], 459[100]
Ohashi, I. 20[53], 22[53]
Ohkoshi, N. 365[30], 366[30]
Ohmiya, H. 283[32], 284[32], 401[36], 402[36]
Ohnishi, Y. 22[59]
Ohno, H. 319[6], 320[6], 336[18], 337[18], 345[6], 346[6], 454[100], 459[100]
Ohrai, K. 82[50], 84[50]
Ohshima, T. 15[33], 84[57], 84[58]
Ohta, T. 478[135]
Oishi, S. 319[6], 320[6], 336[18], 337[18], 345[6], 346[6], 454[100], 459[100]
Okada, T. 396[29], 413[52], 414[52]
Okamoto, S. 363[28], 363[29], 364[28], 364[29], 365[28], 365[29]
Okano, K. 456[105]
Okude, Y. 31[75], 31[76]
Okukado, N. 16[38], 16[39]
Okuro, K. 38[91]
Olivera, R. 496[163]
Onishi, J. 459[113]
Orito, K. 445[90], 446[90]
Oshima, K. 31[78]
Oslob, J. D. 126[35], 127[35]
Otten, E. 143[49], 144[49]
Otto, H. 470[127]
Overman, L. E. 15[34], 15[35], 16[35], 55[5], 77[38], 78[38], 81[5], 85[59], 85[60], 85[61], 85[62], 85[63], 85[64], 86[59], 86[60], 86[61], 88[63], 88[74], 88[76], 89[74], 89[76], 90[74], 309[82], 309[83], 310[83]
Oyamada, J. 300[61]
Ozaki, A. 55[3]
Ozeki, M. 89[77], 90[77]

P

Pace, P. 494[155], 495[155]
Pacini, B. 101[10], 102[10]
Padrón, J. I. 261[5], 268[5], 268[13], 274[5], 274[20], 275[5], 275[20]
Padwa, A. 227[61], 228[61], 228[63], 229[64], 233[64], 311[84], 312[84]
Pahadi, N. K. 394[26]
Painter, T. O. 368[39], 368[40], 369[39], 369[40]
Paladino, G. 410[49], 413[49], 432[76], 433[76], 458[109]
Pale, P. 496[161]
Palmisano, G. 466[123], 467[123]
Pan, C.-M. 44[104]
Pan, W. 179[23], 180[23], 181[23], 192[23], 253[23], 254[23]
Pan, X. 69[28], 70[28], 71[28]
Panek, J. S. 440[84], 440[85], 456[84], 456[85], 457[84], 460[84]
Pang, Y. 495[156]
Panteleev, J. 68[27], 69[27]
Parisi, L. M. 448[92], 449[92], 450[92]
Park, S.-I. 322[8], 323[8]
Park, T. 381[62]
Park, Y. 283[36], 287[36], 288[36]
Parker, E. 349[3], 382[3]
Parrain, J.-L. 365[31], 366[31]
Pastine, S. J. 288[43], 289[43], 290[43], 302[71], 303[71], 303[72], 304[71], 304[72], 305[71]
Pastor, I. M. 259[1], 260[1]
Patane, M. A. 497[166]
Paterson, I. 43[96]
Patil, D. V. 187[29], 188[29], 190[31], 193[29], 193[31]
Patil, N. T. 278[23], 394[26]
Paul, S. 69[29], 70[29], 71[29]
Pautex, N. 227[62], 242[62]
Pearson, R. 422[65], 423[65], 426[65]
Peh, G. 309[81]

Pei, T. 278[24], 278[25], 278[26], 279[24], 279[25], 281[26]
Peixoto, P. A. 380[60]
Peng, C. 416[55], 417[55]
Peng, Y. 44[107], 44[108], 44[109], 47[107], 47[108], 47[109], 48[107], 49[107], 50[107]
Pennington, L. D. 309[82], 309[83], 310[83]
Perchellet, E. M. 65[22], 68[22]
Perchellet, J.-P. H. 65[22], 68[22]
Perez Gonzalez, M. 17[44], 18[44]
Periasamy, M. 19[51], 21[51]
Périchon, J. 44[98]
Petit, L. 300[67], 301[67], 302[67]
Petuskova, J. 295[55], 296[55], 297[55], 311[55]
Pfaltz, A. 96[5]
Pfeifer, L. A. 85[59], 86[59]
Phan, L. T. 380[53]
Phelps, A. M. 333[17], 334[17]
Phun, L. H. 187[29], 188[29], 193[29]
Piao, D. 300[61], 300[62], 301[62], 302[62]
Piccinelli, F. 106[17], 107[17], 110[17]
Piedrafita, M. 303[73], 304[73]
Piers, E. 13[29], 13[30], 13[31], 14[30], 14[31], 41[95], 42[95], 43[95], 44[95]
Pindi, S. 202[39], 205[39]
Planas, L. 89[77], 90[77]
Pleier, A.-K. 389[16], 391[16]
Poli, G. 101[10], 101[11], 101[12], 102[10], 102[12]
Poon, D. J. 85[60], 85[61], 85[62], 85[63], 86[60], 86[61], 88[63]
Poonoth, M. 388[12]
Porcelloni, M. 101[10], 102[10]
Prestat, G. 101[11], 101[12], 102[12]
Probst, D. A. 368[39], 369[39]
Protopopova, M. N. 227[59]
Pu, J. 65[23], 66[23], 68[23]
Pumphrey, A. L. 440[83], 441[83], 443[88]
Purino, M. 274[20], 275[20]

Q

Qi, M.-H. 237[70], 238[70], 240[70]
Qian, H. 317[2], 334[2], 335[2], 476[134], 477[134], 478[134]
Qian, Q. 44[106], 45[106], 46[106], 47[106], 50[106]
Qin, S. 266[12], 267[12]
Qin, T. 44[104]
Qin, Y. 252[79], 253[79]
Qiu, F. 88[71], 89[71]
Qiu, S. 410[50], 411[50], 413[50]
Qiu, Y. 299[60], 300[60]
Queru, M. E. 307[79], 308[79]

R

Raabe, G. 305[77], 306[77]
Radinov, R. 126[34]
Rajapaksa, N. S. 270[19], 273[19], 274[19]
Rakshit, S. 435[78], 436[78]
Ramasubbu, A. 360[25]
Ramírez, M. A. 268[13]
Raschke, T. 78[40], 78[41], 79[40], 79[41]
Rauch, G. 191[32], 194[32]
Ray, D. 69[29], 70[29], 71[29]
Ray, J. K. 69[29], 70[29], 71[29]
Ready, J. M. 34[83], 144[50]
Reddy, B. V. S. 263[7], 264[7], 276[21], 277[21]
Reddy, R. S. 19[51], 21[51]
Reddy, V. R. 468[125], 469[125], 470[125]
Redford, J. E. 443[88]
Reich, N. W. 478[136], 479[136], 480[136], 481[136]
Reisman, S. E. 44[103]
Reiter, M. 481[137], 481[138], 482[137], 482[138], 484[137]
Ren, J. 179[23], 180[23], 180[25], 181[23], 185[27], 186[27], 187[27], 192[23], 193[27], 194[34], 195[34], 196[34], 253[23], 254[23]
René, O. 62[21], 63[21], 64[21], 65[21]
Rennels, R. A. 387[4]
Resnick, L. 65[23], 66[23], 68[23]
Reymond, S. 140[47], 141[47]
Rhee, Y. H. 295[54], 296[54], 297[54], 470[128], 471[128], 471[130], 472[130], 473[130]
Rhim, C. Y. 468[125], 469[125], 470[125]
Ribière, P. 75[34], 76[34]
Richard, J.-A. 380[60]
Richert, K. J. 441[86], 442[86]
Riell, R. D. 441[86], 442[86]
Rigamonti, M. 67[26], 69[26]
Rodrígues, J. R. 155[3], 155[4], 156[4], 158[3], 168[3]
Rodríguez-García, I. 397[30]
Roesky, P. W. 398[33], 399[33], 401[33]
Roethle, P. A. 36[87], 37[87]
Romaniello, A. 102[14], 103[14]
Romines, K. R. 227[60]
Rong, Z.-Q. 119[29], 121[29]
Ropp, S. 481[137], 482[137], 484[137]
Rossi, E. 432[76], 433[76]
Roush, W. R. 4[12], 9[12], 9[22], 9[23], 10[22], 10[23]
Rubio, E. 303[73], 304[73]
Ruck, R. T. 61[18], 64[18]
Rudisill, D. E. 393[24], 394[24]
Russell, C. E. 494[155], 495[155]
Rychnovsky, S. D. 261[6], 262[6], 263[6]

S

Sacchi, K. L. 100[9]
Saicic, R. N. 144[51], 145[51]
Saito, B. 126[36], 127[36]
Saito, T. 418[57], 418[64], 423[66], 424[64], 426[64], 426[66]
Sakai, Y. 454[100], 459[100]
Sakaki, S. 22[59]
Sakamoto, T. 389[18], 391[18], 418[64], 424[64], 426[64], 443[89], 444[89]
Sakazaki, H. 60[16]
Sakee, U. 57[11]
Sakya, S. M. 497[164]
Salerno, G. 390[20], 391[20], 449[95], 466[122], 467[122]
Samanta, S. 74[33], 75[33]
Sames, D. 288[43], 289[43], 290[43], 302[71], 303[71], 303[72], 304[71], 304[72], 305[71]
SanMartin, R. 458[110], 460[110], 496[163]
Sannicolo, F. 78[43], 79[43]
Santandrea, J. 38[94], 39[94], 40[94], 41[94]
Santelli, M. 365[31], 366[31]
Sarlah, D. 123[33], 124[33], 125[33], 126[33]
Sarpong, R. 197[36], 198[36], 199[36]
Sasai, H. 86[67], 87[67]
Sasaki, H. 2[7], 3[7], 4[7], 6[7]
Sasaki, I. 290[46], 291[46], 292[46]
Sato, H. 22[59]
Sato, Y. 77[39], 78[39], 81[39], 82[49], 82[51], 82[52], 83[49], 83[55], 84[55], 85[49]
Satoh, M. 2[7], 3[7], 4[7], 6[7]
Sawama, Y. 388[12]
Sawamura, M. 283[32], 284[32], 401[36], 402[36]
Schafroth, M. A. 123[33], 124[33], 125[33], 126[33]
Schelwies, M. 97[6], 99[6]
Schimpf, R. 78[42], 78[43], 79[42], 79[43], 80[42]
Schlummer, B. 431[73]
Schmidt, M. 44[104]
Schmidt, M. W. 4[9]
Schnaderbeck, M. J. 4[12], 9[12]
Schoffstall, A. M. 228[63]
Scholz, U. 431[73]
Schomaker, J. M. 324[13], 325[13], 333[17], 334[17]
Schow, S. R. 481[139]
Schreuder, I. 491[152], 492[152], 493[152]
Schroeder, G. M. 100[9], 380[61]
Schulman, J. M. 68[27], 69[27]
Schultz, D. M. 451[98], 452[98], 453[98]
Schwartz, R. E. 459[113]
Scott, M. E. 167[14], 168[14]
Scott, R. M. 360[25]
Seifried, D. D. 368[40], 369[40]
Seiller, B. 471[129]
Seki, T. 131[40], 132[40]
Serhadli, O. 470[126]
Serra-Muns, A. 140[47], 141[47]
Sethofer, S. G. 168[15], 169[15], 175[15], 254[15]
Severin, R. 380[60]
Shair, M. D. 496[160]
Shao, L.-X. 237[70], 238[70], 240[70]
Shapiro, E. A. 227[59]
Sharma, N. 59[13], 60[13], 64[13]
Shashidhara, K. S. 20[52]
Shastry, M. 459[114]

She, X. 69[28], 70[28], 71[28]
Shen, H. C. 137[46], 138[46], 139[46], 147[57], 147[58], 148[58]
Shen, H.-C. 295[59], 296[59], 298[59]
Shen, L. 252[79], 253[79]
Shen, M. 219[54], 220[54], 221[54], 438[81], 439[81], 440[81], 441[81], 443[88]
Shen, X. 407[46], 408[46]
Shen, Z.-L. 263[9], 264[9], 265[9], 266[9], 310[9], 311[9]
Sherry, B. D. 327[15], 328[15]
Shevlin, M. 61[18], 64[18]
Shi, J. 266[11]
Shi, M. 156[6], 157[7], 158[6], 158[7], 160[9], 161[9], 162[9], 163[11], 164[11], 165[9], 165[11], 170[16], 171[16], 173[19], 174[19], 175[16], 176[16], 176[19], 199[37], 200[37], 201[38], 202[39], 202[40], 203[40], 203[41], 204[41], 204[42], 205[38], 205[39], 205[40], 205[42], 206[41], 206[42], 208[44], 209[44], 213[50], 214[50], 215[50], 216[50], 218[53], 219[53], 220[53], 237[70], 238[70], 240[70], 241[71], 249[71], 371[45], 373[45]
Shi, N. 31[79], 31[80]
Shi, X. 66[25], 67[25], 69[25]
Shi, Z. 300[64], 301[64], 478[136], 479[136], 480[136], 481[136]
Shi, Z.-J. 417[56], 418[56], 425[56]
Shibasaki, M. 77[39], 78[39], 81[39], 82[48], 82[49], 82[50], 82[51], 82[52], 82[53], 82[54], 83[49], 83[55], 84[50], 84[55], 84[56], 84[57], 84[58], 85[49], 85[56], 86[65], 86[66], 86[67], 87[67], 87[68], 87[69]
Shibuya, A. 386[2]
Shin, S. 322[8], 323[8]
Shin, U. S. 300[63]
Shinde, V. S. 278[23]
Shiota, H. 446[91], 447[91]
Shiragami, H. 461[115]
Shou, W. G. 442[87]
Shrestha, R. 44[100]
Shu, D. 224[57], 225[57]
Sill, P. 368[37], 370[37]
Silvanus, A. C. 307[79], 308[79]
Silveira, A., Jr. 16[40]
Simonneau, A. 368[35]
Simpson, R. D. 88[71], 89[71]
Singh, B. 59[13], 60[13], 64[13]
Singh, R. M. 59[13], 60[13], 64[13]
Sinha, B. 76[35], 76[36]
Sinisi, R. 106[17], 107[17], 110[17]
Siriwardana, A. I. 154[2], 158[2]
Slatford, P. A. 361[27], 362[27], 363[27]
Smith, A. B., III 481[139], 481[140]
Smith, B. B. 4[10]
Smith, K. M. 251[78]
Smith, L. 88[71], 89[71]
Smyth, J. R. 4[10]
Snell, R. H. 344[27]
Snider, B. B. 179[24], 260[3]
So, R. C. 88[71], 88[75], 89[71]
Sodeoka, M. 77[39], 78[39], 81[39], 82[49], 82[50], 82[52], 82[53], 82[54], 83[49], 84[50], 84[57], 85[49], 86[66], 86[67], 87[67]
Solomon, V. C. 349[7]
Song, C. E. 300[63]
Song, F. 462[117], 463[117], 464[117]
Song, H.-J. 288[41]
Song, R.-J. 283[34], 283[35], 286[34]
Song, X.-R. 187[28], 193[28]
Song, Y. 206[43], 207[43]
Song, Z. 462[117], 463[117], 464[117]
Sonogashira, K. 18[45], 18[46]
Sottocornola, S. 24[68], 101[12], 102[12], 415[53], 466[123], 467[123]
Spinelli, E. M. 88[71], 88[75], 89[71]
Sridharan, V. 57[11]
Šrogl, J. 270[16]
Sromek, A. W. 389[19], 390[19], 391[19]
Staben, S. T. 168[15], 169[15], 175[15], 210[48], 211[48], 216[48], 254[15], 283[30], 283[31], 284[30], 285[31], 286[30]
Steele, J. 270[16]
Stevenson, P. 56[10]
Stille, J. K. 12[24], 12[25], 12[26], 12[27], 13[32], 14[32], 15[32], 393[24], 394[24]
Stokes, B. J. 440[83], 441[83], 441[86], 442[86], 443[88]
Stoltz, B. M. 12[28], 25[70], 25[71], 26[70], 26[71], 27[70], 27[71], 27[72], 28[71], 28[72], 29[71], 29[74], 30[70], 30[71], 30[74], 290[44], 290[45]
Stradiotto, M. 403[39], 403[40], 404[40]
Straub, C. S. 229[64], 233[64]
Streiff, S. 97[6], 99[6]
Sturla, S. J. 352[14], 353[14], 354[14]
Su, C. 206[43], 207[43]
Suárez-Pantiga, S. 303[73], 304[73]
Suetsugu, S. 146[55]
Sugihara, T. 293[49]
Sugiyama, A. 22[59]
Suh, Y.-G. 381[62]
Sukirthalingam, S. 57[11]
Sulikowski, G. A. 9[20], 9[21], 10[20], 10[21], 11[20]
Sun, J. 60[17], 61[17], 64[17], 72[17], 73[17], 350[10], 382[10]
Sun, T. 388[12]
Sun, X. 135[44], 136[44]
Surivet, J.-P. 137[46], 138[46], 139[46], 147[57], 147[58], 148[58]
Suzuki, A. 1[1], 1[2], 2[6], 2[7], 2[8], 3[7], 3[8], 4[7], 4[8], 6[7]
Suzuki, T. 396[29]
Suzuki, Y. 144[52], 145[52]
Swope, R. J. 4[10]
Sylvain, C. 147[57]

T

Tada, M. 389[17]
Tadpetch, K. 261[6], 262[6], 263[6]
Tagashira, M. 31[78]
Taillefer, M. 387[6], 427[6]
Takacs, B. E. 374[49], 374[50], 375[49], 375[50], 376[49], 376[50]
Takacs, J. M. 368[36], 374[47], 374[48], 374[49], 374[50], 375[47], 375[48], 375[49], 375[50], 376[49], 376[50], 378[48], 379[52], 380[52], 381[64], 381[65], 381[66]
Takahashi, H. 95[1]
Takahashi, K. 95[3], 365[33]
Takahashi, M. 22[54], 458[111], 459[111], 460[111]
Takai, K. 31[78]
Takaki, K. 295[56], 296[56], 298[56], 405[41], 476[133]
Takao, H. 293[49]
Takemoto, T. 86[66], 86[67], 87[67]
Takemoto, Y. 146[55], 398[32], 454[32], 455[32], 459[32]
Takita, H. 89[77], 90[77]
Taldone, T. 440[84], 456[84], 457[84], 460[84]
Tamao, K. 22[54], 22[55]
Tamaru, Y. 317[4], 318[4], 328[4]
Tamatani, S. 89[77], 90[77]
Tanabe, H. 458[111], 459[111], 460[111]
Tanaka, M. 13[32], 14[32], 15[32]
Tanaka, S. 131[40], 132[40], 317[4], 318[4], 328[4]
Tanaka, Y. 496[160]
Tang, J. 156[6], 158[6]
Tang, J.-M. 295[59], 296[59], 298[59]
Tang, S. 69[28], 70[28], 71[28]
Tang, W. 224[57], 224[58], 225[57], 225[58], 226[58], 227[58], 254[80], 255[80]
Tang, X.-Y. 160[9], 161[9], 162[9], 163[11], 164[11], 165[9], 165[11], 204[42], 205[42], 206[42], 213[50], 214[50], 215[50], 216[50], 218[53], 219[53], 220[53]
Tang, Y. 196[35], 197[35], 497[168], 498[168]
Tanoury, G. J. 351[13], 352[13]
Tantillo, D. J. 171[17], 172[17], 176[17]
Tao, W. 194[34], 195[34], 196[34]
Tarselli, M. A. 337[19], 338[19]
Tasker, A. 66[24], 68[24]
Taylor, J. E. 188[30], 189[30], 193[30]
Teasdale, A. J. 60[15]
Teichert, J. F. 128[38], 128[39], 129[38], 130[38], 130[39]
Tellam, J. P. 307[79], 308[79]
Tellitu, I. 458[110], 460[110]
Terada, Y. 365[33]
Thangavelauthum, R. 57[11]
Thede, K. 78[43], 79[43]
Thiel, W. 295[55], 296[55], 297[55], 311[55]

Thomas, A. P. 380[58], 381[58]
Thomas, A. W. 387[5], 427[5]
Thuong, M. B. T. 101[12], 102[12]
Tian, G.-Q. 157[7], 158[7]
Tietze, L. F. 78[40], 78[41], 78[42], 78[43], 79[40], 79[41], 79[42], 79[43], 80[42]
Tillman, J. E. 188[30], 189[30], 193[30]
Tlais, S. F. 496[157]
Tobisch, S. 403[40], 404[40]
Toder, B. H. 481[139]
Tohda, Y. 18[45]
Tokuda, M. 445[90], 446[90]
Tokuyama, H. 427[68], 456[104], 456[105]
Tollefson, E. J. 44[97]
Tolmachev, A. 429[71], 430[71]
Tomás, M. 211[49], 212[49], 213[49], 216[49]
Tomilov, Yu. V. 227[59]
Tommasi, S. 106[17], 107[17], 110[17], 288[37]
Tonoi, Y. 380[56]
Torraca, K. E. 489[150], 490[150]
Tortosa, M. 9[22], 9[23], 10[22], 10[23]
Tosaki, S.-y. 87[69]
Toste, F. D. 168[15], 169[15], 175[15], 210[48], 211[48], 216[48], 254[15], 283[30], 283[31], 284[30], 285[31], 286[30], 293[50], 327[15], 328[15], 344[28], 345[28], 349[4], 354[15], 354[16], 355[15], 355[16], 369[42], 370[42], 382[4], 438[82], 439[82], 444[82], 445[82]
Toullec, P. Y. 349[9], 350[9], 475[132]
Toumi, M. 457[108]
Tragni, M. 122[31], 122[32], 123[31], 123[32], 125[31], 126[32]
Trauner, D. 2[5], 36[87], 37[87]
Tria, G. S. 381[63], 382[63]
Trinath, D. V. K. S. 77[37]
Troisi, S. 122[31], 123[31], 125[31]
Trost, B. 380[57]
Trost, B. M. 18[46], 95[4], 100[9], 126[34], 126[35], 127[35], 137[46], 138[46], 139[46], 147[57], 147[58], 148[58], 217[52], 218[52], 220[52], 260[3], 349[1], 349[2], 351[12], 351[13], 352[12], 352[13], 354[15], 354[16], 355[15], 355[16], 380[53], 380[54], 380[55], 380[58], 380[59], 380[61], 381[58], 382[2], 395[28], 396[28], 455[102], 456[102], 460[102], 470[128], 471[128], 471[130], 472[130], 473[130]
Tsang, W. C. P. 418[61], 418[62], 420[61], 420[62], 421[61], 426[61]
Tsuji, J. 95[1], 99[8]
Tsukano, C. 146[55]
Tsuritani, T. 221[55], 222[55]
Tudela, E. 211[49], 212[49], 213[49], 216[49]
Tudge, M. T. 270[18], 271[18], 272[18], 273[18]
Turnbull, P. 74[32]
Turner, H. 481[138], 482[138]

U

Uchida, T. 22[55]
Ueda, S. 484[144], 485[144]
Uenishi, J.-i. 4[13], 9[13], 31[77]
Ujaque, G. 2[3]
Ullmann, F. 37[88], 37[89]
Umani-Ronchi, A. 106[17], 107[17], 110[17], 121[30], 122[30], 122[31], 122[32], 123[31], 123[32], 125[30], 125[31], 126[32], 288[37], 288[42]
Ushito, H. 445[90], 446[90]
Utimoto, K. 31[78], 391[21], 392[21], 393[21], 461[115]

V

Valdés, C. 70[30], 71[30]
Valdomir, G. 274[20], 275[20]
van de Weghe, P. 263[8]
Van Horn, D. E. 16[41]
Vayalakkada, S. 381[66]
Vedernikov, A. N. 394[25]
Vicente, R. 211[49], 212[49], 213[49], 216[49]
Villani, F. J., Jr. 16[40]
Villemin, D. 391[22]
Virieux, D. 293[51], 294[51]
Viswanathan, G. S. 260[4], 261[4]
Vitous, J. 88[71], 89[71]
Vogel, T. 465[120]
Vogt, A. 34[84]
Voituriez, A. 349[5], 382[5]
Volz, F. 343[24]
Vulovic, B. 144[51], 145[51]

W

Wadman, S. H. 343[24]
Waldkirch, J. P. 357[21], 358[21]
Walker, W. E. 95[2]
Wallace, E. M. 322[9]
Walsh, M. J. 6[15]
Walter, E. 389[16], 391[16]
Wan, B. 340[21], 341[21]
Wang, B. 232[67], 233[67], 235[67], 236[67], 237[67], 357[19]
Wang, D. 18[50]
Wang, G. 60[17], 61[17], 64[17], 72[17], 73[17]
Wang, H. 400[35], 401[35], 416[55], 417[55]
Wang, J. 44[104], 232[67], 233[67], 235[67], 236[67], 237[67], 242[74], 243[74], 244[74], 244[75], 245[75], 246[75], 247[75], 248[75], 249[74], 249[75]
Wang, J. Q. 9[21], 10[21]
Wang, L. 196[35], 197[35]
Wang, Q. 69[28], 70[28], 71[28]
Wang, R. T. 491[152], 492[152], 493[152]
Wang, S. 44[105], 456[103]
Wang, T.-C. 32[82], 33[82], 35[82]
Wang, W.-H. 417[56], 418[56], 425[56]
Wang, X. 44[105], 278[26], 281[26], 418[58], 419[58], 425[58], 456[103]
Wang, Y. 135[44], 136[44], 303[74], 416[55], 417[55]
Wang, Y.-W. 44[107], 44[108], 44[109], 47[107], 47[108], 47[109], 48[107], 49[107], 50[107]
Wang, Z. 9[18], 135[44], 136[44], 179[23], 180[23], 180[25], 181[23], 182[26], 183[26], 184[26], 185[26], 185[27], 186[27], 187[27], 192[23], 192[26], 193[27], 194[34], 195[34], 196[34], 253[23], 254[23]
Wang, Z.-Q. 283[35]
Ward, S. E. 101[13], 102[13]
Wasa, M. 418[60], 419[60], 420[60], 425[60]
Watanabe, S. 82[51]
Watanabe, T. 336[18], 337[18]
Waterman, E. L. 409[47], 412[47]
Watson, I. D. G. 349[4], 382[4]
Wehbe, J. 293[51], 294[51]
Wei, H.-X. 15[33]
Wei, L. 484[143]
Wei, Y. 160[9], 161[9], 162[9], 165[9], 170[16], 171[16], 175[16], 176[16], 202[40], 203[40], 204[42], 205[40], 205[42], 206[42], 218[53], 219[53], 220[53]
Weibel, J.-M. 496[161]
Weiss, M. M. 85[59], 86[59]
Weissberger, F. 365[32], 366[32]
Weix, D. J. 44[100], 44[101]
Welter, C. 97[6], 99[6], 128[37]
Werner, H. 470[126], 470[127]
Werness, J. B. 224[57], 225[57]
Werz, D. B. 191[32], 194[32]
Weyrauch, J. P. 464[119]
Widenhoefer, R. A. 132[41], 133[41], 133[42], 134[41], 278[24], 278[25], 278[26], 279[24], 279[25], 281[26], 317[2], 317[5], 318[5], 326[14], 327[14], 329[5], 334[2], 335[2], 361[26], 362[26], 363[26], 403[38], 405[42], 405[43], 406[43], 476[134], 477[134], 478[134], 483[141], 484[141]
Wiest, O. 397[31]
Wilkerson-Hill, S. M. 197[36], 198[36], 199[36]
Williams, D. R. 6[15]
Willis, A. C. 300[67], 300[68], 301[67], 301[68], 302[67], 302[68]
Wilson, C. M. 55[7]
Wingerden, M. V. 34[83], 144[50]
Winssinger, N. 497[167]
Winston-McPherson, G. 224[58], 225[58], 226[58], 227[58]
Winston-McPherson, G. N. 254[80], 255[80]
Winter, C. 388[12]
Wolf, J. 470[126], 470[127]
Wolf, M. A. 322[9]
Wolfe, J. P. 451[96], 451[97], 451[98], 451[99], 452[98], 452[99], 453[97], 453[98], 453[99], 459[112], 461[112]
Wong, T. 41[95], 42[95], 43[95], 44[95]
Wong, Y.-H. 263[9], 264[9], 265[9], 266[9], 310[9], 311[9]
Worakun, T. 56[10]

Wovkulich, P. M. 481[139]
Wrobleski, A. D. 88[74], 89[74], 90[74]
Wrona, I. E. 440[84], 440[85], 456[84], 456[85], 457[84], 460[84]
Wu, B. 9[20], 9[21], 10[20], 10[21], 11[20]
Wu, H. 135[44], 136[44], 394[26]
Wu, J. 166[13], 167[13]
Wu, L. 201[38], 203[41], 204[41], 205[38], 206[41], 209[45], 209[46], 210[45], 410[50], 411[50], 413[50]
Wu, Q.-F. 112[21], 112[22], 112[23], 113[21], 114[22], 114[23], 115[21], 115[24], 116[24], 116[25], 117[25], 119[29], 121[29], 136[45], 137[45]
Wu, S. 357[20], 358[20], 360[20]
Wuest, W. M. 397[31]
Wyche, T. P. 254[80], 255[80]

X

Xia, C. 457[107], 458[107], 460[107]
Xia, X.-F. 187[28], 193[28]
Xiang, T. 66[24], 68[24]
Xiang, Z. 66[25], 67[25], 69[25]
Xiao, J. 44[108], 47[108]
Xiao, Q. 417[56], 418[56], 425[56]
Xiao, Y.-P. 278[28], 281[28], 282[28], 283[28]
Xie, G. 232[67], 233[67], 235[67], 236[67], 237[67]
Xie, J. 217[52], 218[52], 220[52]
Xie, M. 266[12], 267[12]
Xie, X. 69[28], 70[28], 71[28], 307[80]
Xie, Z. 496[158]
Xin, Z. 496[158]
Xing, S. 179[23], 180[23], 180[25], 181[23], 192[23], 253[23], 254[23]
Xu, G. 60[17], 61[17], 64[17], 72[17], 73[17]
Xu, G.-C. 199[37], 200[37]
Xu, H. 44[106], 45[106], 46[106], 47[106], 50[106], 224[57], 225[57]
Xu, J. 306[78], 307[78]
Xu, J. Y. 15[33]
Xu, Q. 213[50], 214[50], 215[50], 216[50]
Xu, Q.-L. 107[18], 108[18], 110[20], 111[20], 112[20], 115[20], 116[25], 117[25], 118[27]
Xu, S. 306[78], 307[78]
Xu, S. L. 227[61], 228[61]
Xu, T. 268[14], 269[14]
Xu, X.-B. 44[107], 44[108], 47[107], 47[108], 48[107], 49[107], 50[107]
Xue, J. 496[158]
Xue, W. 44[105], 44[106], 45[106], 46[106], 47[106], 50[106]

Y

Yadav, J. S. 276[21], 277[21]
Yadav, V. K. 293[49]
Yakelis, N. A. 9[22], 9[23], 10[22], 10[23]
Yamada, K. 427[68], 456[104]
Yamakawa, T. 263[10]
Yamamoto, H. 22[55], 134[43], 135[43], 290[46], 291[46], 292[46]
Yamamoto, K. 99[8]
Yamamoto, Y. 154[2], 158[2], 221[55], 222[55], 288[40], 360[24], 361[24], 365[30], 366[30], 386[2], 388[9], 394[26], 394[27]
Yamashita, S. 445[90], 446[90]
Yamazaki, T. 458[111], 459[111], 460[111]
Yan, B. 368[41], 369[41], 462[117], 463[117], 464[117]
Yan, C.-S. 44[107], 44[109], 47[107], 47[109], 48[107], 49[107], 50[107]
Yan, Y.-K. 389[16], 391[16]
Yan, Y.-L. 278[27]
Yanagi, T. 1[1]
Yang, B. H. 431[74], 432[74]
Yang, C.-G. 478[136], 479[136], 480[136], 481[136]
Yang, C.-W. 295[59], 296[59], 298[59]
Yang, D. 270[17], 271[17], 278[27], 283[33]
Yang, H. 494[155], 495[155]
Yang, M. 270[17], 271[17]
Yang, Q. 268[14], 269[14], 338[20], 339[20], 341[22], 342[22]
Yang, S.-M. 22[61], 22[62], 23[61], 23[62], 23[63], 23[64], 24[61], 24[62], 24[63], 24[64]
Yang, Z. 66[25], 67[25], 69[25], 417[56], 418[56], 425[56], 490[151], 491[151], 492[153], 494[153], 497[168], 498[168]
Yang, Z.-P. 116[26], 136[45], 137[45]
Yao, B. 165[12], 166[12]
Yao, L.-F. 170[16], 171[16], 175[16], 176[16]
Yasui, Y. 398[32], 454[32], 455[32], 459[32]
Ye, K.-Y. 119[29], 121[29]
Ye, L. 303[74]
Yet, L. 388[8]
Yeung, C. S. 24[69]
Yin, Y. 399[34], 400[34], 401[34]
Yohannes, D. 497[164]
Yoon, M. Y. 300[63]
Yoon, T. P. 219[54], 220[54], 221[54]
Yoshida, J.-i. 22[54], 22[55]
Yoshida, M. 119[28], 120[28], 121[28]
Yoshimura, F. 20[53], 22[53]
You, L. 368[37], 370[37]
You, S.-L. 107[18], 108[18], 110[20], 111[20], 112[20], 112[21], 112[22], 112[23], 113[21], 114[22], 114[23], 115[20], 115[21], 115[24], 116[24], 116[25], 116[26], 117[25], 118[27], 119[29], 121[29], 136[45], 137[45]
Youn, S. W. 288[43], 289[43], 290[43], 302[71], 303[71], 304[71], 305[71]
Yu, B. 135[44], 136[44]
Yu, J.-Q. 418[58], 418[60], 419[58], 419[60], 420[60], 425[58], 425[60]
Yu, S. 38[92], 317[1]
Yu, Y. 341[22], 342[22]
Yu, Z. 268[14], 269[14]
Yu, Z.-X. 376[51], 377[51], 378[51]
Yuan, W. 202[39], 204[42], 205[39], 205[42], 206[42]
Yuan, Z.-L. 157[7], 158[7]
Yuguchi, M. 445[90], 446[90]
Yun, H. 381[62]
Yus, M. 259[1], 260[1], 389[17]

Z

Zanoni, G. 126[34]
Zanoni, M. 191[32], 194[32]
Zavattaro, C. 88[71], 89[71]
Zecchi, G. 483[142], 484[142]
Zeldin, R. M. 344[28], 345[28]
Zeng, Y. 242[74], 243[74], 244[74], 244[75], 245[75], 246[75], 247[75], 248[75], 249[74], 249[75]
Zhang, D.-H. 160[9], 161[9], 162[9], 165[9], 173[19], 174[19], 176[19]
Zhang, G. 194[33]
Zhang, H. 25[71], 26[71], 27[71], 27[72], 28[71], 28[72], 29[71], 30[71], 232[67], 233[67], 235[67], 236[67], 237[67], 244[75], 245[75], 246[75], 247[75], 248[75], 249[75], 290[44], 456[106]
Zhang, J. 69[28], 70[28], 71[28], 151[1], 152[1], 153[1], 154[1], 158[1], 209[45], 209[46], 210[45], 223[56], 224[56], 226[56], 234[69], 416[55], 417[55]
Zhang, J.-J. 44[109], 47[109]
Zhang, L. 194[33], 303[74], 350[10], 382[10]
Zhang, M. 224[58], 225[58], 226[58], 227[58], 252[79], 253[79], 254[80], 255[80], 388[11]
Zhang, R. 400[35], 401[35]
Zhang, S. 422[65], 423[65], 426[65], 433[77], 434[77], 435[77], 436[77]
Zhang, W. 144[54], 146[54], 224[57], 225[57]
Zhang, W.-C. 260[4], 261[4]
Zhang, X. 44[102], 112[22], 114[22], 357[19], 357[20], 357[21], 357[22], 358[20], 358[21], 358[22], 360[20]
Zhang, X.-N. 218[53], 219[53], 220[53]
Zhang, Y. 165[12], 166[12], 232[67], 233[67], 235[67], 236[67], 237[67], 244[75], 245[75], 246[75], 247[75], 248[75], 249[75], 496[158], 497[168], 498[168]
Zhang, Y.-P. 237[70], 238[70], 240[70]
Zhang, Z. 213[50], 214[50], 215[50], 216[50], 317[2], 317[5], 318[5], 329[5], 334[2], 335[2]
Zhao, D. 143[49], 144[49], 400[35], 401[35]
Zhao, G. 399[34], 400[34], 401[34]
Zhao, H. 65[22], 68[22]
Zhao, Q. 116[25], 117[25]
Zhao, Y. 263[9], 264[9], 265[9], 266[9], 310[9], 311[9], 433[77], 434[77], 435[77], 436[77]
Zheng, B.-F. 283[33]
Zheng, B.-H. 104[15], 105[15]
Zheng, C. 115[24], 116[24]
Zheng, H. 172[18], 173[18], 176[18]
Zheng, K. 266[11], 266[12], 267[12]

Zheng, N. 418[61], 418[62], 420[61], 420[62], 421[61], 426[61]
Zheng, X. 79[44], 80[44]
Zheng, Z. 317[3], 318[3], 321[3], 338[20], 339[20]
Zheng, Z.-B. 196[35], 197[35]
Zhou, C.-Y. 278[29], 280[29], 281[29]
Zhou, F. 496[159], 497[159]
Zhou, J. 497[165], 497[166]
Zhou, S. 144[54], 146[54]
Zhou, Y. 306[78], 307[78]
Zhu, J. 7[16], 7[17], 8[16], 8[17], 9[18], 9[19], 11[17], 196[35], 197[35], 379[52], 380[52], 429[72], 430[72]
Zhu, N.-Y. 270[17], 271[17]
Zhu, Q. 416[55], 417[55]
Zhu, W. 34[84], 182[26], 183[26], 184[26], 185[26], 192[26]
Zhu, Z.-B. 157[7], 158[7], 241[71], 249[71]
Zhu, Z.-Z. 163[11], 164[11], 165[11]
Zhuo, C.-X. 107[18], 108[18], 112[23], 114[23], 116[25], 116[26], 117[25], 119[29], 121[29]
Ziegler, M. L. 470[126]
Zitano, L. 459[113]
Zoni, C. 458[109]
Zou, T. 283[35]
Zriba, R. 174[20], 175[20], 176[20]
Zulys, A. 398[33], 399[33], 401[33]

Abbreviations

Chemical

Name Used in Text	Abbreviation Used in Tables and on Arrow in Schemes	Abbreviation Used in Experimental Procedures
(*R*)-1-amino-2-(methoxymethyl)pyrrolidine	RAMP	RAMP
(*S*)-1-amino-2-(methoxymethyl)pyrrolidine	SAMP	SAMP
ammonium cerium(IV) nitrate	CAN	CAN
2,2′-azobisisobutyronitrile	AIBN	AIBN
barbituric acid	BBA	BBA
benzyltriethylammonium bromide	TEBAB	TEBAB
benzyltriethylammonium chloride	TEBAC	TEBAC
N,*O*-bis(trimethylsilyl)acetamide	BSA	BSA
9-borabicyclo[3.3.1]nonane	9-BBNH	9-BBNH
borane–methyl sulfide complex	BMS	BMS
N-bromosuccinimide	NBS	NBS
tert-butyldimethylsilyl chloride	TBDMSCl	TBDMSCl
tert-butyl peroxybenzoate	TBPB	*tert*-butyl peroxybenzoate
10-camphorsulfonic acid	CSA	CSA
chlorosulfonyl isocyanate	CSI	chlorosulfonyl isocyanate
3-chloroperoxybenzoic acid	MCPBA	MCPBA
N-chlorosuccinimide	NCS	NCS
chlorotrimethylsilane	TMSCl	TMSCl
1,4-diazabicyclo[2.2.2]octane	DABCO	DABCO
1,5-diazabicyclo[4.3.0]non-5-ene	DBN	DBN
1,8-diazabicyclo[5.4.0]undec-7-ene	DBU	DBU
dibenzoyl peroxide	DBPO	dibenzoyl peroxide
dibenzylideneacetone	dba	dba
di-*tert*-butyl azodicarboxylate	DBAD	di-*tert*-butyl azodicarboxylate
di-*tert*-butyl peroxide	DTBP	DTBP
2,3-dichloro-5,6-dicyanobenzo-1,4-quinone	DDQ	DDQ
dichloromethyl methyl ether	DCME	DCME
dicyclohexylcarbodiimide	DCC	DCC
N,*N*-diethylaminosulfur trifluoride	DAST	DAST
diethyl azodicarboxylate	DEAD	DEAD
diethyl tartrate	DET	DET
2,2′-dihydroxy-1,1′-binaphthyllithium aluminum hydride	BINAL-H	BINAL-H
diisobutylaluminum hydride	DIBAL-H	DIBAL-H
diisopropyl tartrate	DIPT	DIPT

Chemical (cont.)

Name Used in Text	Abbreviation Used in Tables and on Arrow in Schemes	Abbreviation Used in Experimental Procedures
1,2-dimethoxyethane	DME	DME
dimethylacetamide	DMA	DMA
dimethyl acetylenedicarboxylate	DMAD	DMAD
2-(dimethylamino)ethanol	$Me_2N(CH_2)_2OH$	2-(dimethylamino)ethanol
4-(dimethylamino)pyridine	DMAP	DMAP
dimethylformamide	DMF	DMF
dimethyl sulfide	DMS	DMS
dimethyl sulfoxide	DMSO	DMSO
1,3-dimethyl-3,4,5,6-tetrahydropyrimidin-2(1*H*)-one	DMPU	DMPU
ethyl diazoacetate	EDA	EDA
ethylenediaminetetraacetic acid	edta	edta
hexamethylphosphoric triamide	HMPA	HMPA
hexamethylphosphorous triamide	HMPT	HMPT
iodomethane	MeI	MeI
N-iodosuccinimide	NIS	NIS
lithium diisopropylamide	LDA	LDA
lithium hexamethyldisilazanide	LiHMDS	LiHMDS
lithium isopropylcyclohexylamide	LICA	LICA
lithium 2,2,6,6-tetramethylpiperidide	LTMP	LTMP
lutidine	lut	lut
methylaluminum bis(2,6-di-*tert*-butyl-4-methylphenoxide)	MAD	MAD
methyl ethyl ketone	MEK	methyl ethyl ketone
methylmaleimide	NMM	NMM
4-methylmorpholine *N*-oxide	NMO	NMO
1-methylpyrrolidin-2-one	NMP	NMP
methyl vinyl ketone	MVK	methyl vinyl ketone
petroleum ether	PE[a]	petroleum ether
N-phenylmaleimide	NPM	NPM
polyphosphoric acid	PPA	PPA
polyphosphate ester	PPE	polyphosphate ester
potassium hexamethyldisilazanide	KHMDS	KHMDS
pyridine	pyridine[b]	pyridine
pyridinium chlorochromate	PCC	PCC
pyridinium dichromate	PDC	PDC
pyridinium 4-toluenesulfonate	PPTS	PPTS
sodium bis(2-methoxyethoxy)aluminum hydride	Red-Al	Red-Al
tetrabutylammonium bromide	TBAB	TBAB

[a] Used to save space; abbreviation must be defined in a footnote.
[b] py used on arrow in schemes.

Chemical (cont.)

Name Used in Text	Abbreviation Used in Tables and on Arrow in Schemes	Abbreviation Used in Experimental Procedures
tetrabutylammonium chloride	TBACl	TBACl
tetrabutylammonium fluoride	TBAF	TBAF
tetrabutylammonium iodide	TBAI	TBAI
tetracyanoethene	TCNE	tetracyanoethene
tetrahydrofuran	THF	THF
tetrahydropyran	THP	THP
2,2,6,6-tetramethylpiperidine	TMP	TMP
trimethylamine *N*-oxide	TMANO	trimethylamine *N*-oxide
N,*N*,*N*′,*N*′-tetramethylethylenediamine	TMEDA	TMEDA
tosylmethyl isocyanide	TosMIC	TosMIC
trifluoroacetic acid	TFA	TFA
trifluoroacetic anhydride	TFAA	TFAA
trimethylsilyl cyanide	TMSCN	TMSCN

Ligands

acetylacetonato	acac
2,2′-bipyridyl	bipy
1,2-bis(dimethylphosphino)ethane	DMPE
2,3-bis(diphenylphosphino)bicyclo[2.2.1]hept-5-ene	NORPHOS
2,2′-bis(diphenylphosphino)-1,1′-binaphthyl	BINAP
1,2-bis(diphenylphosphino)ethane	dppe (not diphos)
1,1′-bis(diphenylphosphino)ferrocene	dppf
bis(diphenylphosphino)methane	dppm
1,3-bis(diphenylphosphino)propane	dppp
1,4-bis(diphenylphosphino)butane	dppb
2,3-bis(diphenylphosphino)butane	Chiraphos
bis(salicylidene)ethylenediamine	salen
cyclooctadiene	cod
cyclooctatetraene	cot
cyclooctatriene	cte
η^5-cyclopentadienyl	Cp
dibenzylideneacetone	dba
6,6-dimethylcyclohexadienyl	dmch
2,4-dimethylpentadienyl	dmpd
ethylenediaminetetraacetic acid	edta
isopinocampheyl	Ipc
2,3-*O*-isopropylidene-2,3-dihydroxy-1,4-bis(diphenylphosphino)butane	Diop
norbornadiene (bicyclo[2.2.1]hepta-2,5-diene)	nbd
η^5-pentamethylcyclopentadienyl	Cp*

Radicals

acetyl	Ac
aryl	Ar
benzotriazol-1-yl	Bt
benzoyl	Bz
benzyl	Bn
benzyloxycarbonyl	Cbz
benzyloxymethyl	BOM
9-borabicyclo[3.3.1]nonyl	9-BBN
tert-butoxycarbonyl	Boc
butyl	Bu
sec-butyl	*s*-Bu
tert-butyl	*t*-Bu
tert-butyldimethylsilyl	TBDMS
tert-butyldiphenylsilyl	TBDPS
cyclohexyl	Cy
3,4-dimethoxybenzyl	DMB
ethyl	Et
ferrocenyl	Fc
9-fluorenylmethoxycarbonyl	Fmoc
isobutyl	iBu
mesityl	Mes
mesyl	Ms
4-methoxybenzyl	PMB
(2-methoxyethoxy)methyl	MEM
methoxymethyl	MOM
methyl	Me
4-nitrobenzyl	PNB
phenyl	Ph
phthaloyl	Phth
phthalimido	NPhth
propyl	Pr
isopropyl	iPr
tetrahydropyranyl	THP
tolyl	Tol
tosyl	Ts
triethylsilyl	TES
triflyl, trifluoromethanesulfonyl	Tf
triisopropylsilyl	TIPS
trimethylsilyl	TMS
2-(trimethylsilyl)ethoxymethyl	SEM
trityl [triphenylmethyl]	Tr

General

absolute	abs
anhydrous	anhyd
aqueous	aq
boiling point	bp
catalyst	no abbreviation
catalytic	cat.
chemical shift	δ
circular dichroism	CD
column chromatography	no abbreviation
concentrated	concd
configuration (in tables)	Config
coupling constant	*J*
day	d
density	*d*
decomposed	dec
degrees Celsius	°C
diastereomeric ratio	dr
dilute	dil
electron-donating group	EDG
electron-withdrawing group	EWG
electrophile	E^+
enantiomeric excess	ee
enantiomeric ratio	er
equation	eq
equivalent(s)	equiv
flash-vacuum pyrolysis	FVP
gas chromatography	GC
gas chromatography-mass spectrometry	GC/MS
gas–liquid chromatography	GLC
gram	g
highest occupied molecular orbital	HOMO
high-performance liquid chromatography	HPLC
hour(s)	h
infrared	IR
in situ	in situ
in vacuo	in vacuo
lethal dosage, e.g. to 50% of animals tested	LD_{50}
liquid	liq
liter	L
lowest unoccupied molecular orbital	LUMO
mass spectrometry	MS
medium-pressure liquid chromatography	MPLC
melting point	mp
milliliter	mL
millimole(s)	mmol
millimoles per liter	mM
minute(s)	min
mole(s)	mol
nuclear magnetic resonance	NMR
nucleophile	Nu^-
optical purity	op
phase-transfer catalysis	PTC
proton NMR	^1H NMR

General (cont.)

quantitative	quant
reference (in tables)	Ref
retention factor (for TLC)	R_f
retention time (chromatography)	t_R
room temperature	rt
saturated	sat.
solution	soln
temperature (in tables)	Temp (°C)
thin layer chromatography	TLC
ultraviolet	UV
volume (literature)	Vol.
via	via
vide infra	*vide infra*
vide supra	*vide supra*
yield (in tables)	Yield (%)

List of All Volumes

Science of Synthesis, Houben–Weyl Methods of Molecular Transformations

Category 1: Organometallics

1 Compounds with Transition Metal—Carbon π-Bonds and Compounds of Groups 10–8 (Ni, Pd, Pt, Co, Rh, Ir, Fe, Ru, Os)

2 Compounds of Groups 7–3 (Mn···, Cr···, V···, Ti···, Sc···, La···, Ac···)

3 Compounds of Groups 12 and 11 (Zn, Cd, Hg, Cu, Ag, Au)

4 Compounds of Group 15 (As, Sb, Bi) and Silicon Compounds

5 Compounds of Group 14 (Ge, Sn, Pb)

6 Boron Compounds

7 Compounds of Groups 13 and 2 (Al, Ga, In, Tl, Be ··· Ba)

8a Compounds of Group 1 (Li ··· Cs)

8b Compounds of Group 1 (Li ··· Cs)

Category 2: Hetarenes and Related Ring Systems

9 Fully Unsaturated Small-Ring Heterocycles and Monocyclic Five-Membered Hetarenes with One Heteroatom

10 Fused Five-Membered Hetarenes with One Heteroatom

11 Five-Membered Hetarenes with One Chalcogen and One Additional Heteroatom

12 Five-Membered Hetarenes with Two Nitrogen or Phosphorus Atoms

13 Five-Membered Hetarenes with Three or More Heteroatoms

14 Six-Membered Hetarenes with One Chalcogen

15 Six-Membered Hetarenes with One Nitrogen or Phosphorus Atom

16 Six-Membered Hetarenes with Two Identical Heteroatoms

17 Six-Membered Hetarenes with Two Unlike or More than Two Heteroatoms and Fully Unsaturated Larger-Ring Heterocycles

Category 3: Compounds with Four and Three Carbon—Heteroatom Bonds

18 Four Carbon—Heteroatom Bonds: X—C≡X, X=C=X, X_2C=X, CX_4

19 Three Carbon—Heteroatom Bonds: Nitriles, Isocyanides, and Derivatives

20a Three Carbon—Heteroatom Bonds: Acid Halides; Carboxylic Acids and Acid Salts

20b Three Carbon—Heteroatom Bonds: Esters and Lactones; Peroxy Acids and R(CO)OX Compounds; R(CO)X, X = S, Se, Te

21 Three Carbon—Heteroatom Bonds: Amides and Derivatives; Peptides; Lactams

22 Three Carbon—Heteroatom Bonds: Thio-, Seleno-, and Tellurocarboxylic Acids and Derivatives; Imidic Acids and Derivatives; Ortho Acid Derivatives

23 Three Carbon—Heteroatom Bonds: Ketenes and Derivatives

24 Three Carbon—Heteroatom Bonds: Ketene Acetals and Yne—X Compounds